Small-signal stability, control and dynamic performance of power systems

If a reader wishes to cite this publication the authors would be grateful if the following format were used:

M.J. Gibbard, P. Pourbeik and D.J. Vowles, *Small-signal stability, control and dynamic performance of power systems*, University of Adelaide Press, Adelaide, 2015.

Small-signal stability, control and dynamic performance of power systems

by

M.J. Gibbard
The University of Adelaide

P. Pourbeik

Electric Power Research Institute, USA

D.J. Vowles
The University of Adelaide

THE UNIVERSITY
of ADELAIDE

UNIVERSITY OF
ADELAIDE PRESS

Published in Adelaide by

University of Adelaide Press
The University of Adelaide
Level 14, 115 Grenfell Street
South Australia 5005
press@adelaide.edu.au
www.adelaide.edu.au/press

The University of Adelaide Press publishes externally refereed scholarly books by staff of the University of Adelaide. It aims to maximise access to the University's best research by publishing works through the internet as free downloads and for sale as high quality printed volumes.

For the full Cataloguing-in-Publication data please contact the National Library of Australia: cip@nla.gov.au

ISBN (paperback) 978-1-925261-02-8
ISBN (ebook) 978-1-925261-03-5

Book design: M.J. Gibbard and D.J.Vowles
Cover design: Emma Spoehr
Cover image: M.J. Gibbard

Contents

Preface .... **xvii**

List of Symbols, Acronyms and Abbreviations .... **xxi**

1 Introduction 1

1.1 Why analyse the small-signal dynamic performance of power systems? 1
1.2 The purpose and features of the book 2
1.3 Synchronizing and damping torques 5
1.4 Definitions of power system stability 7
1.5 Types of modes. 10
1.6 Synchronous generator and transmission system controls 12
1.7 Power system and controls performance criteria and measures. 13
1.7.1 Power system damping performance criteria 13
1.7.2 Control system performance measures 14
1.8 Validation of power system models 15
1.9 Robust controllers 15
1.10 How small is 'small' in small-signal analysis? 16
1.11 Units of Modal Frequency 18
1.12 Advanced control methods 18
1.13 References 19

2 Control systems techniques for small-signal dynamic performance analysis 23

2.1 Introduction 23
2.1.1 Purpose and aims of the chapter 23
2.2 Mathematical model of a dynamic plant or system 24
2.3 The Laplace Transform 27
2.4 The poles and zeros of a transfer function. 30
2.5 The Partial Fraction Expansion and Residues 31
2.5.1 Calculation of Residues 31
2.5.2 A simple check on values of the residues 32
2.6 Modes of Response 32
2.7 The block diagram representation of transfer functions 37
2.8 Characteristics of first- and second-order systems 39
2.8.1 First-order system 39
2.8.2 The second-order system 40
2.9 The stability of linear systems 44
2.10 Steady-state alignment and following errors 45
2.10.1 Steady-state alignment error. 46
2.10.2 The steady-state following error 48
2.11 Frequency response methods 49
2.12 The frequency response diagram and the Bode Plot 51
2.12.1 Plotting the frequency response of the open-loop transfer function 51
2.12.2 Stability Analysis of the closed-loop system from the Bode Plot 59
2.13 The Q-filter, a passband filter 61
2.14 References 62

3 State equations, eigen-analysis and applications 63

3.1 Introduction 63
3.1.1 Example 3.1. 63
3.1.2 Example 3.2 65
3.2 The concept of state and the state equations 67
3.3 The linearized model of the non-linear dynamic system 68
3.3.1 Linearization procedure 69
3.4 Solution of the State Equations 71
3.4.1 The Natural Response 71
3.4.2 Example 3.3 72
3.4.3 Example 3.4: Natural response 73
3.4.4 The Forced Response 74
3.4.5 Example 3.4 (continued). 74

3.5 Eigen-analysis 74
3.5.1 The eigenvalues of the state matrix, *A* 74
3.5.2 A note on eigenvalues, modes and stability 76

3.6 Decoupling the state equations 76

3.7 Determination of residues from the state equations 77

3.8 Determination of zeros of a SISO sub-system 78

3.9 Mode shapes 80
3.9.1 Example 3.5: Mode shapes and modal responses 82

3.10 Participation Factors 83
3.10.1 The relative participation of a mode in a selected state 83
3.10.2 The relative participation of a state in a selected mode 84
3.10.3 Example 3.6: Participation factors 85

3.11 Eigenvalue sensitivities 85

3.12 References 87

4 Small-signal models of synchronous generators, FACTS devices and the power system 89

4.1 Introduction 89

4.2 Small-signal models of synchronous generators 90
4.2.1 Structure of the per-unit linearized synchronous generator models 90
4.2.2 Generator modelling assumptions 93
4.2.3 Electromagnetic model in terms of the per-unit coupled-circuit parameters 94
4.2.4 Alternative d- and q-axis rotor structures 111
4.2.5 Per-unit electromagnetic torque and electrical power output 113
4.2.6 Per-unit rotor equations of motion 114
4.2.7 Non-reciprocal definition of the per-unit field voltage and current ...115
4.2.8 Modelling generator saturation 118
4.2.9 Balanced steady-state operating conditions of the coupled-circuit model 125
4.2.10 Interface between the generator Park/Blondel reference frame and the synchronous network reference frame 130
4.2.11 Linearized coupled-circuit formulation of the generator model equations 133
4.2.12 Transfer-function representation of the electromagnetic equations ...138
4.2.13 Electromagnetic model in terms of classically-defined standard parameters 145
4.2.14 Generator parameter conversions 156

4.3 Small-signal models of FACTS Devices 157
4.3.1 Linearized equations of voltage, current and power at the AC terminals of FACTS Devices: general results 159
4.3.2 Model of a Static VAR Compensator (SVC) 163
4.3.3 Model of a Voltage Sourced Converter (VSC) 165

4.3.4 Simplified STATCOM model 172
4.3.5 Modelling of HVDC Transmission Systems 177
4.3.6 Model of a distributed-parameter HVDC transmission line or cable 178
4.3.7 Model of HVDC transmission with Voltage Sourced Converters (VSCX) 180
4.3.8 Model of HVDC transmission with Voltage Commutated Converters 181
4.3.9 Thyristor Controlled Series Capacitor (TCSC) 192

4.4 Linearized power system model 195
4.4.1 General form of the linearized DAEs for a device and its controls ...196
4.4.2 General form of the network nodal current equations 197
4.4.3 General form of the linearized DAEs of the interconnected power system 198
4.4.4 Example demonstrating the structure of the linearized DAEs 199

4.5 Load models 201
4.5.1 Types of load models 201
4.5.2 Linearized load models 202

4.6 References 205

App. 4–I Linearization of the classical parameter model of the generator. 209

App. 4–II Forms of the equations of motion of the rotors of a generating unit 216

4–II.1 Introduction 216

4–II.2 Shaft equations expressed in terms of per-unit angular speed and torques 218

4–II.3 Per-unit shaft acceleration equation in terms of rotor-speed and power 219

4–II.4 Shaft acceleration equation neglecting speed perturbations in the torque/power relationship 220

4–II.5 A common misunderstanding in calculating the accelerating torque and power 220

5 Concepts in the tuning of power system stabilizers for a single machine system 223

5.1 Introduction 223

5.2 Heffron and Phillips' Model of single machine - infinite bus system 226

5.3 Synchronizing and damping torques acting on the rotor of a synchronous generator 227

5.4 The role of the Power System Stabilizer - some simple concepts 230

5.5 The inherent synchronizing and damping torques in a SMIB system 232
5.5.1 Example 5.1 233

5.6 Effect of the excitation system gain on stability 235

5.7 Effect of an idealized PSS on stability 236

5.8 Tuning concepts for a speed-PSS for a SMIB system 237

5.8.1 Determination of compensating transfer function 238
5.8.2 The nature of the P-Vr characteristic 242
5.8.3 Example 5.2: Evaluate the P-Vr characteristics of the generator and determine the PSS compensating transfer function. 243
5.8.4 Determination of the damping gain *k* of the PSS 245
5.8.5 Example 5.3. Calculation of the damping gain setting for the PSS 245
5.8.6 Washout and low-pass filters 246

5.9 Implementation of the PSS in a SMIB System 248
5.9.1 The transfer function of the PSS 248
5.9.2 Example 5.4. The dynamic performance of the speed-PSS 249
5.9.3 Analysis of the variation in the mode shifts over the range of operating conditions 251

5.10 Tuning of a PSS for a higher-order generator model in a SMIB system 257
5.10.1 The power system model 258
5.10.2 Calculation of the synchronizing and damping torque coefficients 259
5.10.3 Calculation of the P-Vr characteristics for a SMIB system with high-order generator models 260
5.10.4 Example 5.5: Tuning and analysis of the performance of the PSS for the higher-order generator model 261
5.10.5 The P-Vr characteristics for a SMIB system with a 6th order generator model 264
5.10.6 Tuning a speed-PSS for a SMIB system with a 6th order generator model 265

5.11 Performance of the PSS for a higher-order generator model 273

5.12 Alternative form of PSS compensation transfer function 278

5.13 Tuning an electric power-PSS based on the P-Vr approach 279
5.13.1 Example 5.6: Tuning of a power-based PSS 281

5.14 Summary: P-Vr approach to the tuning of a fixed-parameter PSS 284

5.15 References 286

App. 5–I **288**

5–I.1 K-coefficients, Heffron and Phillips Model of SMIB System 288

5–I.2 Transfer function of the SMIB system with closed-loop control of terminal voltage 289

5–I.3 Model of the 6th-order generator and excitation system 289
5–I.3.1 State-space model 289
5–I.3.2 Calculation of the inherent torque coefficients. 291

6 Tuning of PSSs using methods based on Residues and the GEP transfer function 293

6.1 Introduction 293

6.2 Method of Residues 294

6.2.1 Theoretical basis for the Method 294

6.3 Tuning a speed-PSS using the Method of Residues 297
6.3.1 Calculation of the compensation transfer function of the PSS 297
6.3.2 Design Case C. Performance of the PSS with increasing PSS gain 299
6.3.3 Significance of the PSS gains 300

6.4 Conclusions, Method of Residues 300

6.5 The GEP Method 302

6.6 Tuning a speed-PSS using the GEP Method 303
6.6.1 Example 2. Performance of the PSS based on Design Case C 303

6.7 Conclusions, GEP method 306

6.8 References 307

App. 6–I **309**

6–I.1 Algorithm for the calculation of stabilizer parameters 309

6–I.2 Calculation of the nominal upper limit of the range of stabilizer gains 311

7 Introduction to the Tuning of Automatic Voltage Regulators 313

7.1 Introduction 313
7.1.1 Purposes 313
7.1.2 Coverage of the topic 314

7.2 The excitation control system of a synchronous generator 314

7.3 Types of compensation and methods of analysis 316

7.4 Steady-state and dynamic performance requirements on the generator and excitation system 316

7.5 A single-machine infinite-bus test system 319

7.6 Transient Gain Reduction (TGR) Compensation 320
7.6.1 Introduction 320
7.6.2 The performance of the generator and compensated excitation system on-line 321
7.6.3 The performance of the generator and compensated excitation system off-line 324
7.6.4 Comparison of performance of the excitation control system on- and off-line 326

7.7 PID compensation 327
7.7.1 PID Compensation: Theoretical Background 329
7.7.2 Tuning methodology for PID Compensation Types 1 and 2A 335

7.8 Type 2B PID Compensation: Theory and Application to AVR tuning 347
7.8.1 Tuning of Type 2B PID compensation 348
7.8.2 Example: Evaluation of Type 2B PID parameters. 349

7.9 Proportional plus Integral Compensation 351

7.9.1 Simple PI Compensation ...351
7.9.2 Conversion to a PID Compensator with an additional lead-lag block ...352

7.10 Rate feedback compensation ...354
7.10.1 Method of analysis ...354
7.10.2 Tuning of the Excitation System (ES) ...354
7.10.3 Rate feedback compensation using Frequency Response Methods. ...356
7.10.4 Rate feedback compensation using the Root Locus Method ...368

7.11 Tuning of AVRs with Type 2B PID compensation in a three-generator system ...371
7.11.1 The three-generator, 132 kV power system ...371
7.11.2 The frequency response characteristics of the brushless exciter and generator ...373

7.12 Summary, Chapter 7 ...383

7.13 References ...384

App. 7–I ...**386**

7–I.1 Generator and exciter parameters ...386
7–I.1.1 Parameters for the 6th order generator and a simple exciter ...386
7–I.1.2 Parameters for the 5th order salient-pole generator and a brushless AC exciter ...386

7–I.2 Models of the brushless AC exciter ...386

7–I.3 PI Compensation using positive feedback ...388

7–I.4 Integrator Wind-up Limiting ...391

7–I.5 A 'phase-matching' method for constant phase margin over an appropriate frequency range ...392

8 Types of Power System Stabilizers 397

8.1 Introduction ...397

8.2 Dynamic characteristics of washout filters ...399
8.2.1 Time-domain responses ...399
8.2.2 Frequency-domain responses ...401
8.2.3 Comparison of dynamic performance between a single and two washout filters. ...403

8.3 Performance of a PSS with electric power as the stabilizing signal. ...404
8.3.1 Transfer function and parameters of the electric power pre-filter. ...404
8.3.2 Dynamic performance of a speed-PSS with an electric power pre-filter. ...406

8.4 Performance of a PSS with bus-frequency as the stabilizing signal. ...407
8.4.1 Dynamic performance of a speed-PSS with a bus-frequency pre-filter ...408
8.4.2 Degradation in damping with the bus-frequency pre-filter ...410

8.5 Performance of the "Integral-of-accelerating-power" PSS 413
8.5.1 Introduction 413
8.5.2 Torsional modes introduced by the speed stabilizing signal 414
8.5.3 The electric power signal supplied to the pre-filter 414
8.5.4 The Ramp Tacking Filter (RTF) 415

8.6 Conceptual explanation of the action of the pre-filter in the IAP PSS 416
8.6.1 Action of the pre-filter, no washout filters 416
8.6.2 Effect of the washout filters and integrators on the performance of the pre-filter 419
8.6.3 Dynamic performance of the complete pre-filter 424
8.6.4 Potential causes of degradation in performance of the pre-filter of the IAP PSS 429

8.7 The Multi-Band Power System Stabilizer 433

8.8 Concluding remarks 436

8.9 References 438

App. 8–I 441

8–I.1 Action of the Ramp Tracking Filter (RTF) 441

8–I.2 Steady-state conditions at the input and output of the RTF and associated tracking errors for mechanical power input 442
8–I.2.1 With and without an Ideal Integrator 442
8–I.2.2 With a Pseudo Integrator 444

8–I.3 Multi-Band PSS transfer function 444

9 Basic Concepts in the Tuning of PSSs in Multi-Machine Applications 447

9.1 Introduction 447
9.1.1 Eigenvalues and Modes of the system 448

9.2 Mode Shape Analysis 449
9.2.1 Example 1: Two-mass spring system 450
9.2.2 Example 2: Four-mass spring system 456

9.3 Participation Factors 459
9.3.1 Example 4.3 460

9.4 Determination of the PSS parameters based on the P-Vr approach with speed perturbations as the stabilizing signal 462
9.4.1 The P-Vr transfer function in the multi-machine environment 462
9.4.2 Transfer function of the PSS of generator i in a multi-machine system 466

9.5 Synchronising and damping torque coefficients induced by PSS i on generator i 469

9.6 References 471

10 Application of the PSS Tuning Concepts to a Multi-Machine Power System 475

10.1 Introduction ...475
10.2 A fourteen-generator model of a longitudinal power system ...477
10.2.1 Power flow analysis ...479
10.2.2 Dynamic performance criterion ...479
10.3 Eigen-analysis, mode shapes and participation factors of the 14-generator system, no PSSs in service ...481
10.3.1 Eigenvalues of the system with no PSSs in service ...481
10.3.2 Application of Participation Factor and Mode Shape Analyses to Case 1 ...482
10.4 The P-Vr characteristics of the generators and the associated synthesized characteristics ...487
10.5 The synthesized P-Vr and PSS transfer functions ...493
10.6 Synchronising and damping torque coefficients induced by PSS i on generator i ...496
10.7 Dynamic performance of the system with PSSs in service ...500
10.7.1 Assessment of dynamic performance based on eigen-analysis ...500
10.7.2 Assessment of dynamic performance based on participation and mode-shape analysis ...503
10.7.3 Assessment of dynamic performance based on time responses ...505
10.8 Intra-station modes of rotor oscillation ...507
10.9 Correlation between small-signal dynamic performance and that following a major disturbance ...509
10.9.1 A transient stability study based on the fourteen-generator system ...509
10.9.2 The analysis of modal interactions ...513
10.10 Summary: Tuning of PSSs based on the P-Vr approach ...516
10.11 References ...518
App. 10–I ...**520**
10–I.1 Modes of rotor oscillation for Cases 2, 3, 5 and 6 ...520
10–I.2 Data for steady-state power flow analysis ...522
10–I.3 Data for dynamic performance analysis ...527
10–I.3.1 Excitation System Parameters ...528

11 Tuning of FACTS Device Stabilizers 531

11.1 Introduction ...531
11.2 A 'simplistic' tuning procedure for a SVC ...533
11.3 Theoretical basis for the tuning of FACTS Device Stabilizers ...536
11.4 Tuning SVC stabilizers using bus frequency as a stabilizing signal ...538

	11.4.1 Use of bus frequency as a stabilizing signal for the SVC, BSVC_4	539
11.5	Use of line real-power flow as a stabilizing signal for a SVC	545
11.6	Use of bus frequency as a stabilizing signal for the SVC, PSVC_5	548
11.7	Tuning a FDS for a TCSC using a power flow stabilizing signal	551
	11.7.1 Gain range for the stability of TCSC with the FDS in service	555
	11.7.2 Inter-area mode trajectories with increasing stabilizer gain	556
11.8	Concluding comments	557
	11.8.1 Improving the damping of inter-area modes using FACTS devices	557
	11.8.2 Robustness of FDSs	558
	11.8.3 Estimated versus calculated mode shifts	559
	11.8.4 The notion of a 'nominal upper gain' for FDSs	559
11.9	References	559

12 The Concept, Theory, and Calculation of Modal Induced Torque Coefficients 563

12.1	Introduction	563
12.2	The Concept of Modal Induced Torque Coefficients (MITCs)	565
	12.2.1 Conventional frequency response techniques versus modal analysis	565
	12.2.2 Modal torque coefficients induced by the action of a power system stabilizer	565
	12.2.3 Modal torque coefficients induced by the action of a FACTS device stabilizer	567
	12.2.4 Modal torque coefficients induced by centralized stabilizers	569
12.3	Transfer function matrix representation of a linearized multi-machine power system and its controllers	569
12.4	Modal torque coefficients induced by a centralized speed PSS	574
12.5	Modal torque coefficients induced by a centralized FDS	577
12.6	General expressions for the torque coefficients induced by conventional, decentralized PSSs & FDSs	580
	12.6.1 The total modal induced torque coefficients for systems with both PSSs and FDSs	582
	12.6.2 A relationship between modal induced torque coefficients and incremental stabilizer gains	582
12.7	References	583
App. 12–I		**585**
12–I.1	Appendix: System response at a single modal frequency	585
12–I.2	Reducing the TFM model in Figure 12.2 to those in 12.3 and 12.4	586
12–I.3	Elements of the output matrix, *C*	587

13 Interactions between, and effectiveness of, PSSs and FDSs in a multi-machine power system 589

13.1 Introduction 589

13.2 Relationship between rotor mode shifts and stabilizer gain increments 590
13.2.1 Relationship between residues and MITCs in calculation of mode shifts 593
13.2.2 Concept of 'interactions' 594
13.2.3 Relationships between mode shifts, MITCs, participation factors and stabilizer gains 595

13.3 Case Study: Contributions to MITCs/Mode Shifts by PSSs and generators 596
13.3.1 Contributions to the MITC of each generator, local mode B 599
13.3.2 Contributions of the mode shifts of each generator to mode B damping 602
13.3.3 Contributions to the MITCs of each generator, inter-area mode M ..604
13.3.4 Contributions of the mode shifts of each generator to the Mode M damping 606

13.4 Stabilizer damping contribution diagrams 609

13.5 Comparison of the estimated and actual mode shifts for increments in stabilizer gain settings 614

13.6 Summary 616
13.6.1 Interactions 616
13.6.2 Relative Effectiveness of Stabilizers 617

13.7 References 618

14 Coordination of PSSs and FDSs using Heuristic and Linear Programming Approaches 621

14.1 Introduction 621

14.2 The 14-generator power system 623
14.2.1 Evaluation of the transfer function for the SVC at bus 212. 624

14.3 A Heuristic Coordination Approach 626
14.3.1 Coordination of stabilizers for damping the inter-area modes 626
14.3.2 Coordination of local-area modes. 633

14.4 Simultaneous Coordination of PSSs and FDSs using Linear Programming 634
14.4.1 Introduction 634
14.4.2 Comment on the LP solution: optimality versus uniqueness 635
14.4.3 Coordination of PSSs and FDSs 636

14.5 Case study: Simultaneous coordination in a multi-machine power system of PSSs and FDSs using linear programming 639
14.5.1 Scenario 1: Inter-area modes. Maximum PSS & FDS gain 40 pu. 642
14.5.2 Scenario 2: Inter-area modes. Limits PSS gains 40 pu, FDSs 20 pu ...644
14.5.3 Scenario 3: Inter-area modes. Limits: all stabilizer gains 20 pu 645

14.5.4 Scenario 4: Inter-area modes. Various limits 645
14.5.5 Scenario 5: Inter- and local-area modes 645
14.6 Concluding remarks 647
14.7 References 650

Preface

We have written this book in the hope that the following engineers, or potential engineers, will benefit from it:

- Recent graduates in electrical engineering who need to understand the tools and techniques currently available in the analysis of small-signal dynamic performance and design.
- Practicing electrical engineers who need to understand the significance of more recent developments and techniques in the field of small-signal dynamic performance.
- Postgraduate students in electrical engineering who need to understand current developments in the field and the need to orient their research to achieve practical, useful outcomes.
- Undergraduate electrical engineering students in courses oriented towards electric power engineering in which there is an introductory subject in power system dynamics (for access to basic material).
- Managerial staff with responsibilities in power system planning, and system stability and control.

An aim of the book is to provide a bridge between the mathematical/theoretical and physical/practical significance to the topic. Some of the fundamental background relevant to the main topics of the book is presented in the early chapters so that the necessary material is readily available to the reader in the one book.

- Because the emphasis is on controllers for generators, for FACTS and other devices, the pertinent topics in classical control and eigenanalysis techniques are provided in Chapters 2 and 3.
- The authors have covered in Chapter 4 a wide range of small-signal generator models, equations, and associated material. Third- to eighth-order generator models in their coupled-circuit and operational parameter versions are described. The following features are also included in the generator models: (i) the 'classical' and 'exact' definitions

of the operational parameters; (ii) the various approaches to the modelling of saturation; (iii) the formulation of the differential-algebraic generator equations to exploit sparsity. These models and features are employed in the Mudpack software package. Small-signal equations and models of FACTS devices employed in the software are also described. Devices covered include SVCs, STATCOMS, Thyristor Controlled Series Compensators, HVDC links with Voltage Source Converters or with line-commutated converters.

- In Chapters 5, 9 and 10 there is an emphasis on practical robust techniques, based on the P-Vr method, for the design of robust stabilizers for generators in multi-machine systems.

- Two other techniques for the tuning of stabilizers, the GEP and the Method of Residues, are examined in Chapter 6. The benefits and limitations of the two as well as those of the P-Vr method are reviewed.

- An introduction to the tuning of Automatic Voltage Regulators is provided in Chapter 7. The authors have attempted to outline in some detail potential design approaches for (i) TGR (transient gain reduction), (ii) PI and various types of PID control; (iii) brushless, static and conventional excitation systems.

- A detailed analysis of the 'Integral of accelerating power' stabilizer for generators, not previously published, is provided in Chapter 8, together with other practical PSS structures.

- In Chapter 11 the tuning of stabilizers for various FACTS devices in a large power system for operation over a range of conditions is described and illustrated, together with the merits and limitations of the design.

- The concept, theory, and calculation of Modal Induced Torque Coefficients are outlined in Chapter 12,. This is a new method of analysis, developed by one of the authors and forms the basis for Chapter 13. The synchronizing and damping torques induced on generator shafts at the modal frequencies by both PSSs and FACTS device stabilizers are derived.

- The interactions between, and effectiveness of, PSSs and FACTS device stabilizers in a multi-machine power system are analysed in Chapter 13. A new and potentially valuable tool, which is based on the Stabilizer Damping Contribution Diagrams (SDCD) and developed by the authors, enables the engineer to assess the effectiveness of stabilizers installed on generators or FACTS devices in enhancing the damping and stability of the power system.

- In Chapter 14 the coordination of PSSs with PSSs, or PSSs with FACTS device stabilizers is achieved by either heuristic or an optimization techniques. In either case it is based on the newly developed tool, the SDCD, of Chapter 13.

A number of chapters in the book are based on a PhD Thesis by Pouyan Pourbeik, 'Design and Coordination of Stabilisers for Generators and FACTS devices in Multi-machine Power Systems', The University of Adelaide, Australia, 1997. The comprehensive, small-signal modelling of devices is a major contribution by David Vowles to the development of the Mudpack software, as is the architecture of - and graphics in - the package. Many of these developments were based on the earliest versions of the software written by Rainer Korte.

The authors are indebted to the following organizations for their support through R & D grants for the development of Mudpack software and research in the field of power system dynamics: Australian Energy Market Operator, Powerlink Queensland, TransGrid (New South Wales), Transend Networks (Tasmania) and ElectraNet (South Australia).

Finally, we wish to thank our families for their support and patience over the long period of the gestation of the book.

Michael Gibbard Pouyan Pourbeik David Vowles

List of Symbols, Acronyms and Abbreviations

Mathematical symbols

a_{ij}, $\boldsymbol{a}_{i*}$, $\boldsymbol{a}_{*i}$, $\boldsymbol{A}$	$(i,j)^{th}$ element, the ith row and the ith column vector of the matrix $\boldsymbol{A}$, respectively. All elements may have real of complex values.
$\boldsymbol{A}^{-1}, \boldsymbol{A}^{T}, \boldsymbol{A}^{*}$	Inverse, the transpose complex conjugate transpose of the matrix $\boldsymbol{A}$.
$\Re\{\ \}, \Im\{\ \}$	Real and imaginary parts of a complex number.
$\mathbf{I}$, $\mathbf{0}$, $diag\{a_i\}$	Identity, null and diagonal matrices.
j; s	$\sqrt{-1}$; the Laplace operator.
$\dot{x}$, x_0, Δx	Time derivative, initial condition, and perturbation of variable x
$X_3 = \mathfrak{D}(x1, x2, x3)$	Diagonal matrix X_3, elements $x1, x2, x3$

Variables and parameters

$\boldsymbol{A}$, $\boldsymbol{B}$	System state and input matrices.
$\boldsymbol{C}$, $\boldsymbol{D}$	System output and direct-transmission matrices.
$\hat{\boldsymbol{B}}, \hat{\boldsymbol{C}}$	Mode controllability and observability matrices.
D, K_d	Damping coefficients
b	Susceptance
E_{fd}', E_q'	Field voltage, and voltage proportional to d-axis flux linkages.

f	Frequency of a sinusoidal signal
f_0	System frequency in Hertz (50 Hz unless otherwise stated)
$G(s)$	A transfer function
H	Inertia constant of the rotating generating or motoring unit
$\boldsymbol{H}_i(s)$	Transfer Function Matrix (see Section 12.3 for definitions of various TFMs)
$H_{ij}(s)$	Transfer function from input j to output i
I_{dc}	DC current through HVDC link.
k	PSS damping gain.
K_d , D	Damping coefficients
M	Twice the inertia constant, $M = 2H$
p	differential operator $d/(dt)$
P_e, P_m	Torque of electromagnetic origin; prime-mover torque (power, if in per unit)
Q	Reactive power output or flow
r	General symbol for various resistances
r_h	Residue of the mode λ_h
s	Laplace operator
t	Time (s)
T	General symbol for various time constants (s)
T_{ij}^h	Complex modal induced torque coefficient
$\boldsymbol{V}$, $\boldsymbol{W}$	Right and left modal matrices of the state matrix A.
U_r, V_r	Reference signals at the controller summing junction of an AVR of a FACTS device; of a generator
V_t	Terminal voltage
x	General symbol for various reactances
α	Real part of a complex quantity, e.g. of an eigenvalue in Np/s
β	Imaginary part of a complex quantity, e.g. of an eigenvalue in rad/s
Γ	Total modal induced torque coefficient $(\sum_j T_j)$
δ	Generator rotor angle (degrees)
Δ	Δx implies perturbation in variable x ; also Δk implies perturbation in parameter k
ζ	Damping ratio

θ	Angle (normally in degrees)
λ	Real or complex eigenvalue or mode
Λ	Matrix of diagonal eigenvalues
ξ	Damping ratio
ρ	Participation factor
σ	Real part of an eigenvalue or mode
τ	Speed-torque coefficient transfer function; also a time constant
φ	Flux linkages
ψ	Stabilizing signal for FACTS device
ω_b	Base value of speed
ω	Rotational speed in rad/s, or imaginary part of a complex quantity, e.g. of an eigenvalue
ω_f	Frequency of an exciting sinusoidal signal in rad/s
ω_n	Undamped natural frequency in rad/s
ω_0	Synchronous speed ($=2\pi f_0$) in rad/s
Ω	ohm
$\Re$	Symbol for real part of a complex number
$\Im$	Symbol for imaginary part of a complex number

List of Acronyms and Abbreviations

AVR	Automatic Voltage Regulator
cc	Couples-circuit
dB	decibel
DAE	Differential-algebraic equation
DC	Direct current (or may imply zero frequency)
ECS	Excitation control system
em	Electromechanical
ES	Excitation system
FACTS	Flexible AC transmission systems
FDS	FACTS device stabilizer
FVT	Final Value Theorem
GEP(s)	Generator, excitation system and power system transfer function
GEPSDD	GEP with shaft dynamics disabled

HVDC	High voltage direct current
IAP	Integral of accelerating power (PSS)
IPE	Integral of electric power signal, $\frac{1}{2H}\int \Delta P_e dt$
IPM	Integral of mechanical power signal, $\frac{1}{2H}\int \Delta P_m dt$
LPF	Low pass (filter)
MB-PSS	Multi-band PSS
MIMO	Multi-input multi-output
MITC	Modal induced torque coefficient
MMPS	Multi-machine power system
MS	Mode shape
o.c	open circuit
O.C.C.	Open-circuit characteristic
*op	n^{th} order operation impedance (model of a generator)
PF	Participation factor
PI	Proportional plus integral (compensation)
PID	Proportional plus integral plus derivative (compensation)
POD	Power Oscillation Damper (referred to as a FACTS device stabilizer, FDS, in the following chapters)
PSS	Power system stabilizer
PSS®E	Power System Simulator (Software package from PTI)
PTI	Power Technologies International.
pu	per-unit
P-Vr TF	P-Vr transfer function
PWM	Pulse Width Modulation
RMS	root-mean-square
RI	A synchronously rotating network reference frame
RTF	Ramp tracking filter
SDCD	Stabilizer damping contribution diagram
SI	International System of Units
SISO	Single-input single-output
SMIB	Single machine infinite bus (system)
STATCOM	Static compensator
SVC	Static var compensator
TCCX	Voltage-commutated thyristor-controlled converters with a HVDC link
TCR	Thyristor controlled reactor

TCSC	Thyristor controlled series compensator
TF	Transfer function
TFM	Transfer function matrix
TGR	Transient gain reduction
TSP	Transmission Service Provider
VSC	Voltage sourced converter
VSCX	Voltage source converters with HVDC transmission
XPVR	Extended P-Vr
WO	Washout (filter)

Chapter 1

Introduction

1.1 Why analyse the small-signal dynamic performance of power systems?

We shall be concerned mainly with the analysis of the dynamic performance and control of large, interconnected electric power systems in the following chapters. The differential-algebraic equations which describe the behaviour of a power system are inherently non-linear. Among the non-linearities are functional types (e.g. $\sin\delta$), product types (e.g. voltage × current), limits on controller action, saturation in magnetic circuits, etc. The general method of assessing the performance of the system, with all its non-linearities, is through a time-domain simulation which reveals the response of the system to a specific disturbance, e.g. a fault, the loss of a generating unit, line switching. Typically, it may be necessary to conduct many such studies with disturbances applied in various locations in the system to ascertain its stability and dynamic characteristics. Even with many such studies, many of the characteristics of the dynamic behaviour may be missed and insights into system performance lost. In small-signal analysis of dynamic performance of multi-machine systems the stability and characteristics of the system are readily derived from eigenanalysis and other tools. Furthermore, in such linear analysis the design of controllers and their integration into the dynamics of plant are facilitated.

Modern linear control system theory contains many powerful techniques, not only for determining the stability and dynamic characteristics of large linear systems, but also for tuning controllers that satisfy steady-state and dynamic performance specifications. Fortunately and importantly, Henri Poincaré [1] showed that if the linearized form of the non-linear system is stable, so is the non-linear system stable at the steady-state operating condition at which the system is linearized. Moreover, the dynamic characteristics of the system at the selected operating condition can be established from linear control system theory and, as long as the perturbations are small, the time-domain responses can be calculated. With such information the design of linear controllers may be undertaken and the resulting controls embedded in the non-linear system. In practice, if the modelling of the devices is adequate, small-signal tests involving generator controls, for example, have revealed close agreement between simulation and test results. Continuously-acting controllers of interest for synchronous generating units are Automatic Voltage Regulators (AVRs), Power System Stabilizers (PSSs) and speed governors. In a later chapter an analysis of the controls and stabilizers for FACTS-based devices is conducted; such stabilizers are commonly called a Power Oscillations Dampers (PODs). Many of the techniques and control concepts are also applicable to the small-signal analysis of dynamic performance of wind turbine-generators and other technologies.

1.2 The purpose and features of the book

The main purpose of this book is to introduce the graduate engineer to the concepts and applications of small-signal analysis and controller design for the enhancement of the dynamic performance of multi-machine power systems. To this end, the analyses of the control and dynamic performance are illustrated by examples based on an interconnected high-voltage system comprising fourteen generating stations and various types of FACTS devices. An emphasis in the book is on more recent theoretical developments and application to practical issues which are amenable to small-signal analysis using a comprehensive software package. In addition, the tools in - and features of - such a software package for analysis and controller design are illustrated.

The aim and features of the book are illustrated in the following summary.

1. In the following chapters it is assumed that the reader has already been introduced to the basics of: (i) the steady-state and dynamic performance of power systems [2], [3], [4], [5], [6], [7], [8] and (ii) control system theory [9], [10]. However, because controller design and tuning are described in later chapters of this book, Chapters 2 and 3 are devoted particularly to those aspects of basic control theory and the associated analysis and design techniques which are employed in later. *Practical insights and limitations in control theory and analysis are emphasized in order to isolate that material which is important for application in later chapters.*

2. For the practical design of robust power system stabilizers (PSSs), a tuning approach based on the generator P-Vr characteristics, which are a development of the GEP Method for single generator applications, is applied to multi-machine systems [11], [12]. The uses of the P-Vr characteristics are (a) analysed in detail in Chapter 5 for a single generator power system in order to explain the features of the P-Vr Method

and its limitations; (b) applied in Chapters 9 and 10 to the multi-machine system over a range of operating conditions.

3. It is shown in Chapters 5 and 10 that the rationale in the tuning of PSSs based on the P-Vr method is that:

   (i) there are *two important components in the PSS transfer function $kG(s)$ which are essentially decoupled for practical purposes* [1]:

      (a) the rotor modes are more-or-less directly left-shifted [2] by the PSS compensating transfer function $G(s)$ with increase in the PSS damping gain, k [3];

      (b) the extent of the left-shift of the rotor modes is determined by the damping gain, k;

      (c) the incremental left-shifts of the rotor modes are linearly related to increments in damping gain for changes about the nominal values (typically ± 5 to ± 10 pu on generator MVA rating) [11];

   (ii) the tuning of the PSS is based on a more extreme set of *encompassing* operating conditions;

   (iii) the PSS damping gain has special significance: it is also the *damping torque coefficient induced by the PSS on the shaft of the generator*. It forms the basis for the theoretical developments in Chapters 5, 10, 12 to 14.

   (iv) the PSS damping gain can be adjusted to 'swamp out' any inherent negative damping torques;

   (v) as a result of (i), (ii) and (iv) above, the PSS transfer function $kG(s)$ is *robust over the encompassing range of normal and outage operating conditions* [12];

   (vi) the PSS damping gain, when expressed in per unit on generator MVA rating, is a meaningful quantity, unlike the term "PSS gain" currently used. PSS damping gains less than 10 pu are low, are normal between 10 and 30 pu, and greater than 30 pu tend to be high.

   As opposed to the application of advanced control techniques, the significance of the above rationale is that the *natural characteristics* of the generator and the system are employed and thus meaningful insight and explanations for the dynamic behaviour of the system can be established.

---

1. The features described in items (i)(a) and (i)(b) are illustrated for six operating conditions in Figure 10.26
2. By 'direct left-shift' is implied that the mode shift is $-\alpha \pm j0$, $\alpha \geq 0$. As explained in Chapter 13, deviations from the 'direct left-shift' of modes are mainly due to interactions between multi-machine PSSs and non-real generator participation factors.
3. The 'damping' gain of the PSS is defined in Section 5.4.

4. Methods other than the P-Vr method for the design of PSSs, namely the commonly-used Method of Residues [13] and the GEP Method [14], are described in Chapter 6. By means of an example the merits, deficiencies and limitations of the latter Methods and the P-Vr approach are examined [15]. (See item 4 in Section 6.7.)

5. Various concepts and methods for the tuning of automatic voltage regulators (AVRs) are *introduced* and examined in Chapter 7. Some simplifications in the approaches to the commonly-used techniques are suggested.

6. A more fundamental and detailed examination is undertaken - than previously conducted - to explain, and understand more fully, not only the performance of certain devices but also the theory behind certain tools. Examples are: (a) the performance of the 'Integral of accelerating power PSS' in Chapter 8; (b) the characteristics of two tools, Mode Shapes and Participation Factors in Chapter 9 (these are used in the analysis of the performance of multi-machine systems).

7. The tuning of power oscillation dampers for FACTS devices (PODs, also referred to as FACTS Device Stabilizers (FDSs)) is described in Chapter 11. Some of the problems encountered in the design are revealed in the case of a multi-machine system for a wide range of operating conditions [16], [17].

   Due to the short-comings of existing techniques for the tuning of FACTS Device Stabilizers, their robustness is more difficult to achieve compared the tuning of PSSs for robustness (see item 3 above).

8. The concept, theory, and calculation of Modal Induced Torque Coefficients (MITCs) are described in Chapter 12. In this chapter the synchronizing and damping torques induced on a generator by both PSSs and FDSs at the modal frequencies are explained; this chapter provides the theoretical basis for Chapter 13 [16], [17].

9. The interactions between, and effectiveness of, PSSs and FDSs in a multi-machine power system are analysed in Chapter 13. A valuable aid in establishing the relative effectiveness of stabilizers are the Stabilizer Damping Contribution Diagrams (SDCDs) [18]. Extending the concepts introduced in Chapter 13 the SDCDs form a basis for the heuristic coordination of power system stabilizers and FACTs device stabilizers. Both the latter approach as well as an optimization approach based on linear programming are illustrated in Chapter 14 and [20].

10. A comprehensive set of the various small-signal models of synchronous generators and FACTS devices are provided in Chapter 4. These models are intended to be embedded in a set of differential and algebraic equations (DAEs) which are employed to take advantage of sparsity in the system equations [21], [22].

11. The practical theme throughout this book is based on consulting projects for industry and queries raised by industry on practical problems that they have encountered. Many of the queries relate to some lack of understanding of the theoretical or practical backgrounds to the issues raised.

An aim of the design of controllers is to enhance the damping of the rotor modes of oscillation, either to stabilize unstable oscillations or to ensure that the damping criteria for the power system are satisfied. Therefore the concepts of synchronizing and damping torques [23] - which operate on the shafts of generating units - are introduced in Section 1.3, followed by the concepts and definitions of stability in Section 1.4.

1.3 Synchronizing and damping torques

It is assumed that the reader is generally familiar with the basic concepts of power system stability, such as the equal area criterion; these concepts are covered in many texts [2] - [7]. However, let us consider a somewhat simplistic scenario which may reveal not only the nature of dynamic interactions between a generator rotor and a power system following a disturbance but also the actions that occur within the rotating system of the generator itself.

Assume a synchronous generator is connected through its transformer by two parallel transmission lines to a receiving-end transformer and a large system. This scenario is summarized in Figure 1.1(a).

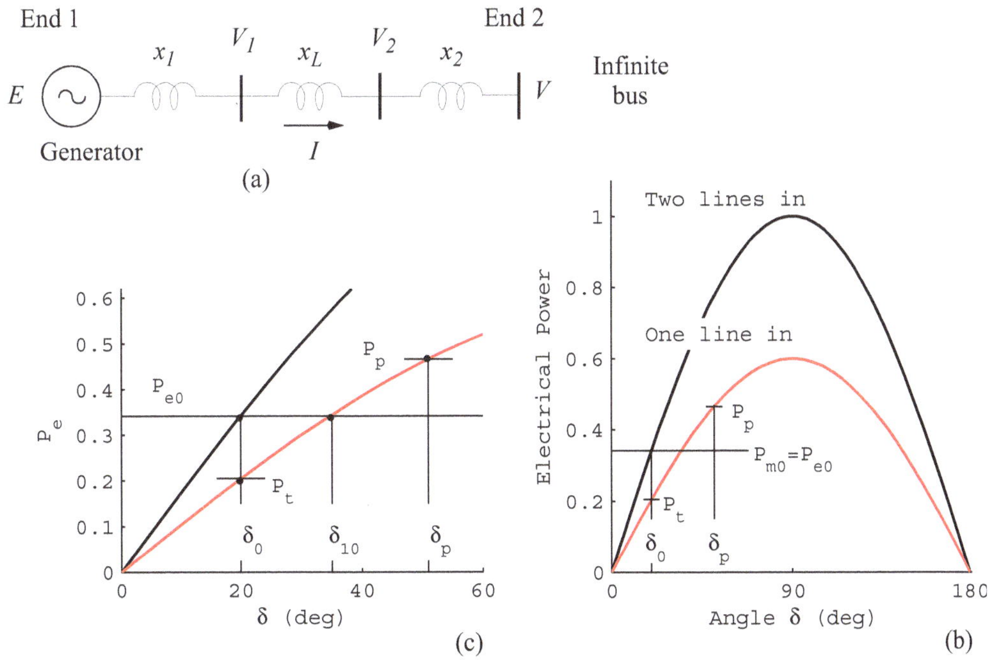

Figure 1.1 (a) Single-machine infinite-bus power system. (b) Power-angle characteristics. (c) Expanded view about δ_{10}, P_{e0}.

The per unit voltage behind transient reactance of the generator is E, and the large system is represented by an infinite bus, voltage V per unit. The reactance x_1 comprises the transient and transformer reactances of the generating unit; x_2 represents the reactances of the receiving-end transformer plus the Thévenin equivalent of the large system. The effective reactance of the parallel lines is x_L; their series resistance and shunt capacitance are negligible. All reactances are in per-unit (pu) on the generator MVA rating. The direction of current flow I is consistent with the generator producing power. The system is assumed to be lossless.

Because the power flow is from the generator to the large system, the rotor of the generator leads that of the infinite bus by an angle δ (rad). The power output of the generator is

$$P_e = \frac{E \cdot V}{x_1 + x_L + x_2} \cdot \sin\delta \quad \text{(per unit).} \tag{1.1}$$

The associated power-angle curve is shown in Figure 1.1(b). The power output of the generator increases from zero at zero rotor angle and reaches a maximum value of

$$P_{em} = \frac{E \cdot V}{x_1 + x_L + x_2} \quad \text{at } \delta = \pi/2 \text{ rad., or } 90° \text{ (pu).}$$

Let us assume that (i) when two lines are in service an equilibrium or steady-state condition exists in which the power output of the prime-mover P_{m0} is equal to the electrical power output of the generator, P_{e0}, at synchronous speed and the rotor angle is δ_0; (ii) the power output of the prime-mover remains constant during a disturbance on the electrical system; (iii) at time zero, one of the two lines is opened. Because the effective reactance of the lines is now $2x_L$, it follows from (1.1) that immediately after the disturbance the electric power output of the generator falls to P_t at δ_0 on the one-line-in power-angle characteristic shown in Figure 1.1(c). The net torque acting on the shaft of the generator will cause it to accelerate with respect to the system. The rotor angle of the generator, δ, immediately starts to increase from δ_0 on the latter characteristic thereby increasing the electrical power flow from the generator. Once the electrical power output exceeds the prime-mover power output P_{m0} at δ_{10} the generator decelerates but, due to the inertia of the rotor, the rotor angle continues to increase until the speed falls to synchronous. At this time the electric power output and the rotor angle are at their peak values, P_p and δ_p. However, the net decelerating torque continues acting on the shaft to reduce both the electrical power flow and the rotor angle along the lower characteristic until zero net accelerating torque once more arises at δ_{10} in Figure 1.1(c). Due to inertia, the electric power output and rotor angle continue to decrease and reach their minimum values at P_t and δ_0 at synchronous speed. Thereafter the process repeats itself with the electric power output and rotor angle oscillating about P_{m0}, δ_{10} between peak and trough values P_p, δ_p and P_t, δ_0, respectively.

In the absence of damping, these oscillations will continue indefinitely. Synchronism in this scenario is maintained by the electrical power flow, given by (1.1), between the generator and the system, resulting in a *synchronizing torque* $P_e(t)$) being produced on the shaft of the generator. For the oscillations to decay away a *damping torque* must also be established on the rotor of the generator, typically by means of a PSS. The *inherent damping torques* acting on the shaft of the generator are typically associated with eddy currents flowing in the rotor iron and/or the damper (amortisseur) windings installed on the rotor, together with windage, friction and other losses.

Throughout the analysis which follows in the later chapters it will be shown that the production of sufficient positive synchronizing and damping torques on the shaft of a generator is a continuing requirement for stable dynamic performance. Following a disturbance, the perturbation in the electromagnetic torque of a synchronous machine, either a generator or a motor, can be resolved into the two components defined as follows:

1. synchronizing torque component: a torque in phase with rotor angle perturbations, and
2. damping torque component: a torque in phase with rotor speed perturbations.

Those familiar with the equal area criterion will recognize that for the scenario shown in Figure 1.1(b) the power system is stable if positive damping torques are present. What are implied by power system stability and some associated concepts are reviewed in the following section.

1.4 Definitions of power system stability

> "**Power system stability** is the ability of an electric power system, for a given initial operating condition, to regain a state of operating equilibrium after being subjected to a physical disturbance, with most system variables bounded so that practically the entire system remains intact." [24].

This definition is intended to apply to an interconnected system in its entirety, however, it also includes the instability and timely disconnection of an element such as a generator without the system itself becoming unstable.

There are three main categories of power system stability: rotor-angle stability, voltage stability and frequency stability. Ensuring stability in all its categories is the primary focus of the design of power system controllers. Such controllers are designed to ensure, through the specifications on the dynamic performance of the system, that adequate margins of stable operation are attained over a range of normal operating conditions and contingencies.

In the following chapters we are concerned with rotor-angle stability which is defined as follows [24]:

> "**Rotor angle stability** refers to the ability of synchronous machines of an interconnected power system to remain in synchronism after being subjected to a disturbance. It depends on the ability to maintain/restore equilibrium between electromagnetic torque and mechanical torque of each synchronous machine in the system. Instability that may result occurs in the form of increasing angular swings of some generators leading to their loss of synchronism with other generators."

Rotor-angle stability, and oscillations of the rotors of synchronous generators, are essentially governed by the equations of motion of the unit; the relevant versions of the equations are derived in Chapter 4. In terms of the per unit rotor speed ω, synchronous speed ω_0 and the per unit prime-mover torque and the torque of electromagnetic origin, T_m and T_g, the equations of motion are given by (4.58) and (4.59) which are repeated below:

$$p\delta = \omega_b(\omega - \omega_0) \text{ and}$$

$$p\omega = \frac{1}{2H}(T_m - T_g - D(\omega - \omega_0)).$$

.H is the inertia constant (MWs/MVA) of the generating unit and D (pu torque/pu speed) is the damping torque coefficient. From the above equation it is apparent that a steady-state condition exists when the torques are in balance and thus there is no change in rotor angle or in speed about synchronous speed. However, a disturbance on an element of the electrical system will result in an imbalance in the torques and cause the rotor to accelerate or decelerate, in turn causing the rotor angle to increase or decrease. The shaft equation is linear so it is applicable to large and small disturbances. For large disturbances the term 'transient stability' is defined as follows [24]:

> **Large-disturbance rotor angle stability or transient stability**, as it is commonly referred to, is concerned with the ability of the power system to maintain synchronism when subjected to a severe disturbance, such as a short circuit on a transmission line.

On the other hand, rotor-angle stability for small disturbances is defined as [24]:

> **Small-disturbance (or small-signal) rotor-angle stability** is concerned with the ability of the power system to maintain synchronism under small disturbances. The disturbances are considered to be sufficiently small that linearization of system equations is permissible for purposes of analysis.

Associated with **transient stability** are severe events or disturbances such as system faults, the opening of a faulted line - or a heavily loaded circuit, the tripping of a large generator, the loss of a large load. As indicated in Section 1.1, for transient stability analysis the dynamic behaviour of certain devices are modelled by their non-linear differential and algebraic equations. The presence of the various types of non-linearities in the equations results in transient stability analysis, in practice, being conducted by simulation studies in the time domain. The basis of such analyses is a power flow study, an equilibrium or steady-state operating condition to which the relevant disturbance is applied.

Stable, large-disturbance performance of a multi-machine power system depends on adequate synchronizing power flows being established between synchronous generators to prevent loss of synchronism of any generator on the system. High gain excitation systems are employed to increase synchronizing power flows and torques. The decay of oscillations, not only following the initial transient (usually the first swing) but also following cessation of limiting action by controllers, is dependent on the development of damping torques of an electro-magnetic origin acting on the generator rotors. Damping torques may be degraded significantly by high gain excitation systems such that, if the net damping torque is negative, instability occurs. To counter this type of instability, positive damping torques can be induced on generators by installing continuously-acting controllers known as stabilizers.

Small-disturbance or **small-signal rotor-angle stability** is associated with disturbances such as the more-or-less continuous switching on and off of relatively small loads. The analysis of small-signal rotor-angle stability is conducted for a selected steady-state operating condition about which the non-linear differential and algebraic equations and other non-linearities are linearized. This process produces a set of equations in a new set of variables, the perturbed variables. Important features of small-signal analysis are: (i) as shown by Poincaré, information on the stability of the non-linear model at the selected operating condition, based on the stability of the linearized system, is exact; and (ii) all the powerful tools and techniques in linear control system analysis are available for the design and analysis of dynamic performance. The design of power system stabilizers for inducing damping torques under normal and post-contingency conditions is conducted using such facilities.

Two forms of spontaneous small-signal instability may be: (i) a steady increase in rotor angle due to inadequate synchronizing torque, or (ii) rotor oscillations of increasing amplitude due to insufficient damping torque. Most generally in practice, however, the latter is of concern in small-signal rotor-angle stability analysis.

While we are mainly interested in small-signal rotor-angle stability in the following chapters, the definitions of voltage and frequency stability are quoted from [24] for information and for the sake of completeness.

> **Voltage stability** refers to the ability of a power system to maintain steady voltages at all buses in the system after being subjected to a dis-

> turbance from a given initial operating condition. It depends on the ability to maintain/restore equilibrium between load demand and load supply from the power system. Instability that may result occurs in the form of a progressive fall or rise of voltages of some buses.

(In system planning studies it is occasionally found that analyses which suggested rotor angle instability are, in fact, associated with voltage instability; at times, it may be difficult to discriminate between them.)

> **Frequency stability** refers to the ability of a power system to maintain steady frequency following a severe system upset resulting in a significant imbalance between generation and load. It depends on the ability to maintain/restore equilibrium between system generation and load, with minimum unintentional loss of load. Instability that may result occurs in the form of sustained frequency swings leading to tripping of generating units and/or loads.

1.5 Types of modes.

The term 'mode' is used to refer to the natural or characteristic response to a disturbance of the small-signal dynamics of the power system. Such modes may be oscillatory or monotonic and in the time domain are of the forms:

$$y_1(t) = A_1 e^{\alpha_1 t} \sin(\omega_1 t + \phi_1) \quad \text{or} \quad y_2(t) = A_2 e^{\alpha_2 t}, \text{ respectively.}$$

A mode, and its typical frequency range, is usually identified with a phenomenon of one of the following types.

- *Global Mode.* This is a low-frequency mode of 0.05 - 0.2 Hz (approx. 0.3 - 1.2 rad/s) in which all generating units move in unison. Such a phenomenon is observed, for example, in isolated systems connected to an AC system through a HVDC link [25].

- *Low-frequency mode.* This is a localized oscillatory mode of frequency 0.01 - 0.05 Hz (approx. 0.05 - 0.3 rad/s). For example, they have been associated with interactions between the water column and the governors on hydro-generators [26], [27].

- *Local-area mode.* This is an oscillatory electro-mechanical mode and is usually associated with the rotors of synchronous generating units [1] in a station swinging against the rest of the power system, or against electrically-close generating station(s). Its frequency range is normally in the range 6 - 12 rad/s (1 - 2 Hz).

---

1. By a 'synchronous generating unit' is implied the synchronous generator, its prime mover and their controls.

- *Intra-station or intra-plant mode.* This oscillatory electro-mechanical mode is associated with units within a generating station swinging against each other. The range of modal frequencies is normally 10 - 15 rad/s (1.5 - 2.5 Hz).

- *Inter-area mode.* This is an oscillatory electro-mechanical mode and is associated with a group of generating stations in one area of the system swinging against a group of stations in one or more other areas of the system. Inter-area modes are usually associated with (possibly) weak interconnecting ties between geographically separated areas of the system. The range of modal frequencies is typically 1.5 - 6 rad/s (0.25 - 1.0 Hz).

- *Torsional modes.* These modes are normally associated with oscillations between the rotating masses on the prime-mover - generator shaft. The frequencies of these modes are normally greater than 50 rad/s (8 Hz) for nuclear units, and greater than 95 rad/s (15 Hz) for other small and large generating units.

- *Control modes.* These modes may be oscillatory or monotonic and may be identified with the controls of generating units or FACTS devices.

A number of the above modes are referred to as the 'electro-mechanical' modes. As stated earlier the associated oscillations are the characteristic or natural modes of the system, the frequency and damping of which generally change with the operating conditions, i.e. changes in the system configuration and the loading conditions. Such changes in operating condition may cause the system to drift - or be forced [1]- towards a small-signal stability limit. Instability of torsional modes caused by interactions with other controllers or other devices has been observed [2].

In recent times it has been necessary to operate power systems closer to stability limits because of environmental and economic considerations. Furthermore, lightly damped inter-area modes are becoming more common since interconnections between power systems are also increasing. This is because interconnections allow adjacent systems (i) to share spinning reserve, (ii) to reduce costs by better utilisation of the more efficient generating stations, and (iii) to reduce the environmental impact by using the most efficient units, thereby facilitating the postponement of investment in new generation. Methods are continually being sought for increasing the power transfer over existing (possibly weak) interconnections thereby reducing the damping of the already lightly-damped modes. Consequently, ensuring that the damping of modes of rotor oscillation in power systems provides adequate margins of stability has been - and still is - of concern to system planners and operators.

A most important feature of small-signal analysis is that it provides an understanding of the underlying modal structure of a power system and gives insights into a system's dynamic characteristics that cannot easily be derived from time-domain simulations. For example, in the time-domain response following a major fault shown in Figure 10.32 only three of the thirteen modes appear to be excited; the nature and location of the fault does not significantly excite the local-area modes outside the faulted area at all. Understanding the nature

1. In the aftermath of a major disturbance on the system.

of the modal behaviour as revealed from small-signal analysis yields a synoptic view of the system characteristics which would require many large-signal studies of faults and other disturbances in different locations to gain similar, but not exact, information.

1.6 Synchronous generator and transmission system controls

In Figure 1.2 are shown the basic controllers and control signals for a prime mover, synchronous generator and static VAR compensator (SVC). These are the basic devices with which we will be concerned in later chapters in the context of the design of controllers, however, other Flexible Alternating Current Transmission System (FACTS) devices such as Thyristor Controlled Series Capacitors (TCSCs), HVDC transmission links, and their controls will also be considered [28], [29].

The control objectives are to ensure that system voltages and frequency lie within specified ranges during normal and abnormal operating conditions. For system voltage control this is achieved by adjusting the voltage references (or set points) to appropriate levels on generators and SVCs, by on-load tap-changing transformers, or by injecting or absorbing reactive power by switching of capacitor banks or reactors. System frequency control may be implemented in several ways (e.g. through generator dispatch by adjustment of the speed or load set-points on governors). For the purpose of small-signal stability analysis our focus is on the dynamic behaviour of continuously-acting control systems. Discontinuous controls such as transformer tap-changers, or switching operations of capacitor banks or reactors, typically incorporate dead-bands, hysteresis, and time delays. For small-signal stability analysis at a given operating condition, the outputs of discontinuous controllers are usually assumed to be fixed at their initial steady-state values. This is legitimate because for small-disturbances the changes in the discontinuous controller inputs are assumed to be negligibly small and insufficient to trigger changes in the controller output.

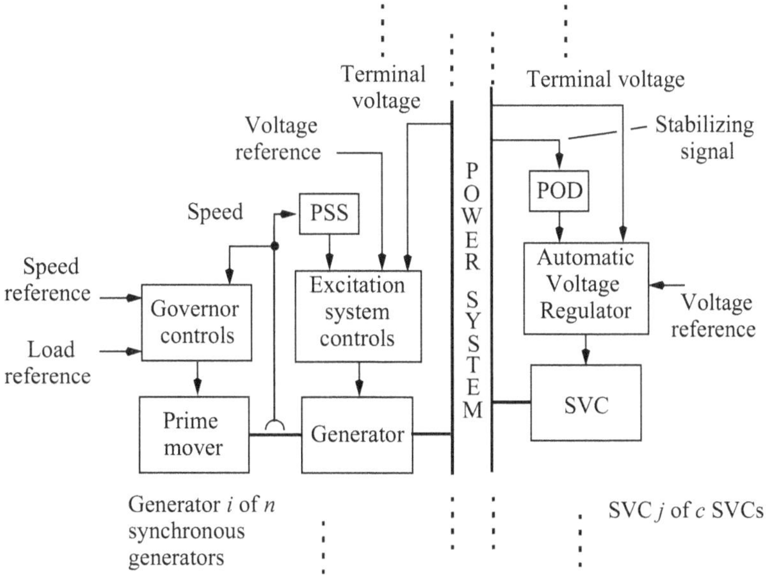

Figure 1.2 Basic controls for a synchronous generating unit and a SVC. Other reactive and voltage controls include other types of FACTs devices, reactors and capacitor banks.

When investigating the small-signal performance of a system, not only is its behaviour under normal conditions of interest but also is its performance in the immediate post-fault condition before tap-changer and reactive switching operations have had time to occur and also when all such operations have been completed following the disturbance. Any investigations would include establishing if the margin of stability for such conditions is adequate.

1.7 Power system and controls performance criteria and measures.

1.7.1 Power system damping performance criteria

The criteria for small-signal damping performance of the power system are the more relevant in the analyses in the following chapters. These measures take different forms [1], namely:

- The damping ratios of the dominant local or inter-area modes of rotor oscillation should exceed a specified value, e.g. 0.03 or 3%.
- The time constant of the dominant mode should less than a nominated value. (The time constant of the mode is time taken for the mode to decay to 37% of its initial value).

---

1. Values for these measures used by a number of organizations are given in a 1996 report [30].

- The settling time of the dominant mode should be shorter than a specified value. For example, depending on the definition, the mode will decay to within 10%, 5% or 2% of its initial value within 2.3, 3 or 4 times the time constant of the mode.

As an example, a slightly different form of the criterion used in South-Eastern Australia is, as stated in detail in the National Electricity Rules [31]:

> "Damping of power system oscillations must be assessed for planning purposes according to the design criteria which states that power system damping is considered adequate if after the most critical credible contingency event, simulations calibrated against past performance indicate that the halving time of the least damped electromechanical mode of oscillation is not more than five seconds.
>
> "To assess the damping of power system oscillations during operation, or when analysing results of tests ..... , the Network Service Provider must take into account statistical effects. Therefore, the power system damping operational performance criterion is that at a given operating point, real-time monitoring or available test results show that there is less than a 10 percent probability that the halving time of the least damped mode of oscillation will exceed ten seconds, and that the average halving time of the least damped mode of oscillation is not more than five seconds."

The above criteria imply that (i) the damping constant of the mode should be less than 0.139 Np/s for a halving time of 5 s, and (ii) for an inter-area mode of 2 rad/s (0.32 Hz), say, the damping ratio should be greater than 0.07.

1.7.2 Control system performance measures

The measures commonly quoted to characterize the performance of a control system are (i) in the time domain: rise time, percentage overshoot, settling time, steady-state error; (ii) in the *s*-domain: damping ratio, damping constant; (iii) in the frequency domain: phase and gain margins, gain-crossover frequency, bandwidth. However, these measures are determined not only by the controller(s) but also by the device under control; such measures are considered in more detail in Chapter 2.

For specific control systems such as synchronous generator excitation systems, governing systems, etc., detailed performance criteria are the subject of various technical standards. For example, in the case of excitation systems, two sources of relevant information are (i) the IEEE Std. 421.2-1990 [32] which is a guide that presents dynamic performance criteria, definitions, and test objectives for excitation control systems as applied in power systems; (ii) Clause S5.2.5.13 "Voltage and reactive power control" of the Australian National Electricity

Rules [31] which specify dynamic performance criteria for excitation systems of generation connected to the Australian grid.

1.8 Validation of power system models

Power system simulations for the purpose of detailed dynamic performance analysis are required for a variety of purposes including real-time operational control (e.g. dynamic security assessment), operational and long-term planning, controller design and tuning. If the results of such dynamic simulations are to be trusted then the models of the generating plants and their controls, the system loads and the interconnecting network, on which on which the simulations are based must be accurate. In a number of jurisdictions, to achieve the required level of modelling accuracy it is required that field tests be performed to establish the parameters of generating plants and their controls, and to verify that the model accurately represents the dynamic performance of the plant at its point of connection to the network. For example, in Australia the Generating System Model Guidelines [33] - which are enforceable under the National Electricity Rules [31] - stipulate requirements for steady-state, short-circuit and dynamic models of generating plant that are connected to the grid. In 2007 the Western Electricity Coordinating Council (WECC) in the United States similarly developed and maintained guidelines for generating system technical data, testing and model validation [34]. An IEEE Task Force published guidelines for validation testing of generator models for rotor-angle stability analysis [35]. Models for rotor-angle stability analysis are typically required to be accurate in the frequency range from DC to at least 5 Hz (31.4 rad/s). In some jurisdictions the minimum accuracy requirements are expressed quantitatively; in others engineering judgment is applied on a case-by-case basis to assess the accuracy of models.

In addition to validating the models of individual generating plant and control systems, system wide tests are typically undertaken from time to time to establish the validity of the integrated system models. Furthermore, the results of such tests may be used to calibrate the system models. Such system tests may be staged with the intent of exciting key system dynamic responses, for example, by switching transmission circuits, energizing braking resistors. System disturbances may be unstaged with important system dynamic responses being excited, say, by a system fault. Recently, on-line modal estimation schemes have also been employed to assist in validating and/or calibrating small-signal models of power systems [36].

1.9 Robust controllers

In what follows it is assumed - based on some design procedure or technique - that the configuration and parameter values of a robust controller are fixed, i.e. gain scheduling, conditional path switching or some such form of variation is not employed. For the purposes of the analyses in later chapters a fixed-parameter controller is said to be robust if over the defined range of modal frequencies, and over a prudently-chosen, encompassing range of *N* and *N*-1 operating conditions, the stabilizers induce adequate positive levels of damping and synchronizing torques on the shafts of the generators such that the power system is stable subject to the relevant stability margins. (See Section 1.2, item 3) (Note: as is shown in Chapter 13, for some inter-area modes the stabilizer may degrade the damping torques on other generators.)

1.10 How small is 'small' in small-signal analysis?

For the purpose of small-signal analysis it is often convenient or necessary to express the shaft equation for the synchronous machine in terms of the perturbations in the per-unit mechanical and airgap torques ΔT_m and ΔT_g, acting on the shaft. The equation for the small-signal motion of the shaft, evaluated at a steady-steady operating condition, is derived in (4.59), and is expressed in a commonly-used per-unit form as:

$$2H \cdot \frac{d}{dt}\Delta\omega + D \cdot \Delta\omega = \Delta T_m - \Delta T_g ,$$

where $\Delta\omega$ is the per-unit perturbation in the angular speed of the shaft, H is the inertia constant of the prime mover, shaft and generator (MWs/MVA), and D is a damping coefficient (pu torque/pu speed perturbation). It should be noted that for small perturbations in speed: (i) $\Delta P_m = \Delta T_m$ and $\Delta P_g = \Delta T_g$ (per-unit); (ii) the perturbation in the electrical power output of the generator (ΔP_e) is related to the air-gap power (ΔP_g) by $\Delta P_e = \Delta P_g - 2r_a I_0 \Delta I$ where r_a is the stator resistance (per-unit), I_0 and ΔI are the steady-state and the perturbations in stator current (per-unit), respectively. Typically, r_a is very small and consequently $\Delta P_e = \Delta P_g$.

In using the linearized equations for the generator and other elements of the power system, or when conducting field tests on devices, it is necessary to decide for what size of disturbance the system response can deemed to be linear. For example, for what peak-peak swing in electrical power output is the generator response essentially small signal? For the particular application an analysis may be required to determine for what size perturbations the performance of the system can be considered 'small signal'.

In the context of the 'small-signal analysis' of the dynamic performance of power systems the question often arises in practice 'how small is *small*'. As mentioned earlier, we know from Poincaré that the non-linear model of the power system at a selected operating condition is stable if its linearized system at that operating condition is stable. However, information on the perturbations in the variables in the linearized system becomes exact only as their magnitudes tend to zero. In practical applications such as staged tests, this implies that the magnitude of any perturbations must be kept 'small', e.g. for step-changes in the reference input signal for the testing of a closed-loop control system whose design is based on linear control theory.

For illustrative purposes consider the classical power-angle characteristics, treated earlier, of a simple transmission line, reactance X in per-unit. The non-linear power-angle function is

$$P = (V_s V_r / X)\sin\delta = P_{max}\sin\delta\,, \tag{1.2}$$

in which P is the power (per-unit) transmitted from the sending to receiving end, δ is the difference in voltage angle (rad) between the sending- and receiving-end buses; V_s and V_r are the respective voltage magnitudes (per-unit). For convenience we define $P_{max} = V_s V_r / X$.

For small-disturbances the non-linear power-angle function is linearized about an initial steady-state operating point (P_0, δ_0) employing, for example, the method used in Section 3.3. The resulting small-signal characteristic is described by:

$$\Delta P = P_{max}\cos\delta_0 \Delta\delta \quad \text{per unit at} \quad (P_0, \delta_0). \tag{1.3}$$

Let $P_{max} = 1000$ MW $\equiv 1$ pu. In Figure 1.3 two regions of linear operation on the non-linear characteristic are shown. For these regions the steady-state operating conditions are respectively $P_0 = 500$ MW, $\delta_0 = 30°$ and $P_0 = 940$ MW, $\delta_0 = 70°$. The criterion employed for a linear range is, say, that the maximum power deviation between the linear and non-linear characteristics is less than 8.5 MW.

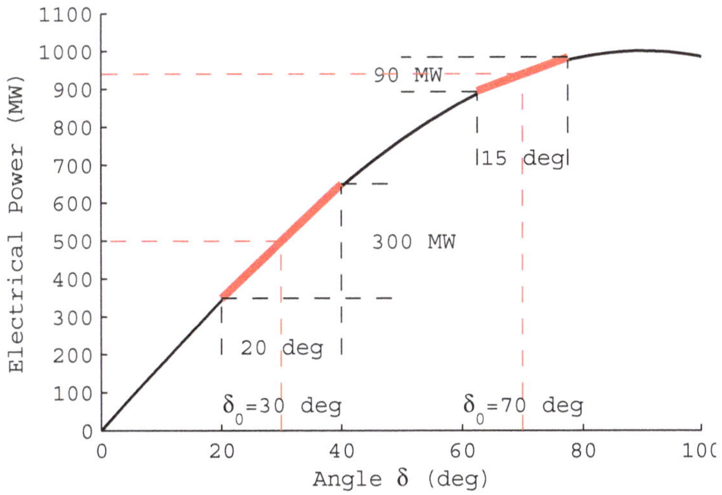

Figure 1.3 Regions of the power-angle characteristic about 30° and 70° for which peak to peak oscillatory swings in power and rotor angle can be considered linear. Over these ranges the maximum deviations between the linear and non-linear power characteristics are less than 8.5 MW (i.e. 0.0085 pu)

Based on our criterion, for a continuous oscillatory angular swing of 20° peak-peak about the steady-state angle of 30°; the power swing of 300 MW peak-peak is 'linearly related' to the angular swing. Similarly, at the steady-state angle of 70° the maximum power swing is restricted to 90 MW peak-peak and is 'linearly related' to a smaller angular swing of 15° peak-peak. Clearly, under more stressed conditions the range of perturbations over which the system performance can be considered more-or-less linear is much smaller. Therefore, depending on the application and the type of disturbance, engineering judgement - and analysis - is required to establish 'how small is small'.

1.11 Units of Modal Frequency

Throughout this book the preferred unit of frequency ω of the mode $\alpha \pm j\omega$ is in radian/second (rad/s) and the damping constant α is in Neper/second (Np/s). The damping ratio is defined in Section 2.8.2.1 as:

$$\xi = -\alpha/\sqrt{\alpha^2+\omega^2} \approx -\alpha/\omega \text{ if } 0 < -\alpha < 0.3\omega \text{, say.}$$

Knowledge of the value of α is important, in particular the system is stable if $\alpha < 0$ and unstable if $\alpha > 0$. Moreover, the value of α not only provides a measure of the margin of stability but yields the settling time of the modes, e.g. the time for the envelope of an oscillatory mode to decay to a value of 5% of the final value is three time constants, $-3/\alpha$.

For lightly damped modes it is useful to remember that if $0 < -\alpha < 0.3\omega$, and if ω is in rad/s, then $\xi \approx -\alpha/\omega$.

In reports and papers of some organizations, it is common to characterize a mode by its values of ξ and ω (in Hz) rather than $-\alpha$ (in Np/s) and ω (in rad/s). Therefore, in order to estimate (mentally) the value of α for a lightly damped stable mode it is necessary multiply ω (in Hz) by $2\pi\omega\xi$. For example, if $\xi = 0.03$ and $\omega = 0.2$ Hz, then $\omega = 1.26$ rad/s and $\alpha = -0.038$ Np/s; it may then be considered that the associated margin of stability of 0.038 Np/s is too low.

If the frequency is specified in rad/s an approximate value is sometimes given in Hz, and vice-versa; for example $\omega = 6.3$ rad/s or 1 Hz.

1.12 Advanced control methods

It should be emphasized that, in this book, techniques for the analysis and design of controllers employ the 'inherent' characteristics of the components of the power system, for example the 'P-Vr' characteristic of the generator for the tuning of its power system stabilizer. These techniques are mainly based on the so-called 'classical' control theory. Other techniques, sometimes called 'Advanced Control Methods' [37], tend not to utilize the 'inherent' system characteristics and in the case of large, multi-machine systems advanced methods of control have had limited application in practice to date. Nevertheless, combining the ideas in advanced control methods, and utilizing the 'inherent' characteristics of the system and its devices, may not only be a fruitful line of research but may also lead to practical outcomes.

1.13 References

[1] Jules Henri Poincaré, b. 1853, d. 1912. See web references to 'Poincaré maps and stability'.

[2] P. Kundur, *Power system stability and control.* New York: McGraw-Hill, 1994.

[3] J. J. Grainger, and W. Stevenson, *Power system analysis*, McGraw-Hill Education - Europe, 1994.

[4] A. R. Bergen and V. Vittal, *Power system analysis.* Second Edition, Prentice Hall, 1999.

[5] H. Saadat, *Power system analysis.* Third Edition. PSA Publishing, 2010.

[6] J. D. Glover, M. S. Sarmi, and T. Overbye, *Power system analysis and design.* Prentice-Hall, 1998.

[7] P. W. Sauer and M. A. Pai, *Power System Dynamics and Stability*, Prentice Hall, New Jersey, 1998.

[8] L. L. Grigsby (Ed.), *Power System Stability and Control*, Electric Power Engineering Handbook, 3rd Edition, CRC Press, Taylor & Francis Group, Boca Raton, 2012.

[9] G. F. Franklin, J. D. Powell and Abbas Emami-Naeini, *Feedback Control of Dynamic Systems*, Prentice Hall, 6th Edition, Oct. 2009.

[10] R. C. Dorf and R. H. Bishop, *Modern Control Systems*, Prentice Hall, 12th Edition, July 2010.

[11] M. J. Gibbard, "Coordinated design of multimachine power system stabilisers based on damping torque concepts," *IEE Proceedings C Generation, Transmission and Distribution*, vol. 135, pp. 276-284, 1988.

[12] M. J. Gibbard, "Robust design of fixed-parameter power system stabilizers over a wide range of operating conditions," *Power Systems, IEEE Transactions on*, vol. 6, pp. 794-800, 1991.

[13] F. L. Pagola, I. J. Perez-Arriaga, and G. C. Verghese, "On sensitivities, residues and participations: applications to oscillatory stability analysis and control," *Power Systems, IEEE Transactions on*, vol. 4, pp. 278-285, 1989.

[14] E. V. Larsen and D. A. Swann, "Applying power system stabilizers: Part I – III," *Power Apparatus and Systems, IEEE Transactions on*, vol. PAS-100, pp. 3017–3046, June 1981.

[15] M. J. Gibbard and D. J. Vowles, "Reconciliation of methods of compensation for PSSs in multimachine systems," *Power Systems, IEEE Transactions on*, vol. 19, pp. 463-472, 2004.

[16] P. Pourbeik and M. J. Gibbard, "Damping and synchronizing torques induced on generators by FACTS stabilizers in multimachine power systems," *Power Systems, IEEE Transactions on*, vol. 11, pp. 1920-1925, 1996.

[17] P. Pourbeik, M. J. Gibbard, and D. J. Vowles, "Proof of the Equivalence of Residues and Induced Torque Coefficients for Use in the Calculation of Eigenvalue Shifts," *Power Systems, IEEE Transactions on*, IEEE, vol. 22, pp. 58-60, 2002.

[18] M. J. Gibbard, D. J. Vowles, and P. Pourbeik, "Interactions between, and effectiveness of, power system stabilizers and FACTS device stabilizers in multimachine systems," *Power Systems, IEEE Transactions on*, vol. 15, pp. 748-755, 2000.

[19] M. J. Gibbard and D. J. Vowles, "Reconciliation of methods of compensation for PSSs in multimachine systems," *Power Systems, IEEE Transactions on*, vol. 19, pp. 463-472, 2004.

[20] P. Pourbeik and M. J. Gibbard, "Simultaneous coordination of power system stabilizers and FACTS device stabilizers in a multimachine power system for enhancing dynamic performance," *Power Systems, IEEE Transactions on*, vol. 13, pp. 473-479, 1998.

[21] K. E. Brennan, S. L. Campbell and L. R. Petzold, *Numerical Solution of Initial-Value Problems in Differential-Algebraic Equations*, Society for Industrial and Applied Mathematics, Philadelphia, 1996.

[22] I. S. Duff, A. M. Erisman and J. K. Reid, *Direct Methods for Sparse Matrices*, Oxford University Press, Oxford, 2003.

[23] F. P. de Mello and C. Concordia, "Concepts of Synchronous Machine Stability as Affected by Excitation Control," *Power Apparatus and Systems, IEEE Transactions on*, vol. PAS-88, pp. 316-329, 1969.

[24] IEEE/CIGRE Joint Task Force on Stability Terms and Definitions, "Definition and Classification of Power System Stability", *Power Systems, IEEE Transactions on*, vol. 19, no. 2, May 2004, pp. 1387 - 1401.

[25] R. Grondin, I. Kamwa, G. Trudel, L. Gerin-Lajoie, and J. Taborda, "Modeling and closed-loop validation of a new PSS concept, the multi-band PSS," in *Power Engineering Society General Meeting*, 2003, IEEE, 2003, p. 1809 Vol. 3.

[26] F. R. Schleif, G. E. Martin, and R. R. Angell, "Damping of System Oscillations with a Hydrogenerating Unit," *Power Apparatus and Systems, IEEE Transactions on*, vol. PAS-86, pp. 438-442, 1967.

[27] H. V. Pico, J. D. McCalley, A. Angel, R. Leon, and N. J. Castrillon, "Analysis of Very Low Frequency Oscillations in Hydro-Dominant Power Systems Using Multi-Unit Modeling," *Power Systems, IEEE Transactions on*, vol. 27, pp. 1906-1915, 2012.

[28] Narian G. Hingorani and Laszlo Gyugyi, *Understanding FACTS: concepts and technology of flexible AC transmission systems*, IEEE Press, Piscataway, NJ, 2000.

[29] Xiao-Ping Zhang, C. Rehtanz and B. Pal, *Flexible AC Transmission Systems: Modelling and Control*, Springer-Verlag Berlin Heidelberg, 2012.

[30] CIGRE Task Force 07.01.38, *Analysis and Control of Power System Oscillations*, CIGRE Technical Brochure No. 111, Dec. 1996.

[31] Australian Energy Market Commission, *National Electricity Rules*, Vers. 61, March 2014.

[32] "*IEEE Guide for Identification, Testing, and Evaluation of the Dynamic Performance of Excitation Control Systems,*" IEEE Std. 421.2-1990, 1990.

[33] Australian Energy Market Operator, *Generating System Model Guidelines*, Planning Department, Doc. No. 118-0009,Vers. No. 1.0, Feb. 2008.

[34] "*WECC Guideline: Generating Facility Data, Testing and Model Validation Requirements,*" Western Electricity Coordinating Council, 13 July 2013. Available: http://www.wecc.biz/library/WECC%20Documents/Documents%20for%20Generators/Generator%20Testing%20Program/WECC%20Gen%20Fac%20Testing%20and%20Model%20Validation%20Rqmts%20v%207-13-2012.pdf.

[35] L. Hajagos, J. Barton, R. Berube, M. Coultes, et al., "Guidelines for Generator Stability Model Validation Testing," in *Power Engineering Society General Meeting,* 2007. IEEE, pp. 1-16.

[36] Juan Sanchez-Gasca (Task Force Chairman), "*Identification of Electromechanical Modes in Power Systems*", Special Publication, TP462, Task Force on Identification of Electromechanical Modes. Power System Stability Subcommittee of the Power System Dynamic Performance Committee of the IEEE PES, July 2012, 282 pages.

[37] B. Pal and B. Chaudhuri, *Robust Control in Power Systems*, Springer Science+Business Media, Inc., New York, 2005.

Chapter 2

Control systems techniques for small-signal dynamic performance analysis

2.1 Introduction

2.1.1 Purpose and aims of the chapter

As emphasized in the Section 1.1 the equations describing an electric power system and its components are inherently non-linear. The equations contain non-linearities such as the product of voltage and current, functional non-linearities such as sine and cosine, and non-linear characteristics such as magnetic saturation in machines. The analysis of dynamic systems with non-linearities is complex, particularly for power systems which are large and have a variety of non-linear elements. On the other hand, in the case of linear control systems, there is a comprehensive body of theory and a wide range of techniques and tools for assessing both the performance and stability of dynamic systems.

For small-signal analysis of power systems, the non-linear differential and algebraic equations are linearized about a selected steady-state operating condition. A set of linear equations in a *new* set of variables, the *perturbed* variables, result. For example, on linearization, the non-linear equation $y = f(x_1, x_2, \ldots, x_n) = f(\boldsymbol{x})$ becomes a linear equation in the perturbed variables, $\Delta y = k_1 \Delta x_1 + k_2 \Delta x_2 + \ldots + k_n \Delta x_n$, at the initial steady-state operating condition $Y_0, X_{10}, X_{20}, \ldots, X_{n0}$. The constant coefficients k_i depend on the initial condition. The question now is: how does the assessment of stability and dynamic performance based on the analysis of the linearized system relate to those aspects of the non-linear system? As also mentioned earlier, a theorem by Poincaré states that information on the stability of the non-linear system, based on a stability analysis of the linearized equations, is exact at the steady-state operating condition selected. However, information on the variable $x_i = \Delta x_i + X_{i0}$ becomes exact only as $\Delta x_i \to 0$. That is, for practical purposes, the perturbations must be small - typically a few percent of the steady-state value.

Small-signal analysis of power systems, based on the linearized dynamic equations, provides a means not only of assessing the stability and the damping performance of the system (through eigenanalysis and other techniques), but also for designing controllers and determining their effectiveness. The various applications of small-signal analysis in the field of power systems dynamics and control are the subjects of later chapters. The purpose of this chapter is to introduce and extend some of the concepts in linear control theory, analysis and design which are particularly relevant to understanding of later material.

2.2 Mathematical model of a dynamic plant or system

Why model? Maybe a reason is that we wish to describe the behaviour of the plant when subjected to some disturbance or to the action of a control signal. One means of characterizing its behaviour is by determining its time response to a test signal such as a step input. However, in order to calculate the response we require a mathematical model of the plant. Such a model can be derived from tests but often is most simply described by a set of differential equations which are derived from first principles. Let us consider two simple examples the results of which will be of interest in later cases. In these examples let p represent the differential operator $d/(dt)$, i.e. $p = d/(dt)$. Note that we can manipulate expressions in p as we would any algebraic variable.

Example 1.
A simple resistive-inductive circuit is shown in Figure 2.1. Write down the equations which describe the behaviour of the current $i(t)$ when an arbitrary voltage $v_s(t)$ is applied to the circuit. The circuit resistance is r (ohm) and its inductance is L (Henry); the inductor is air-cored.

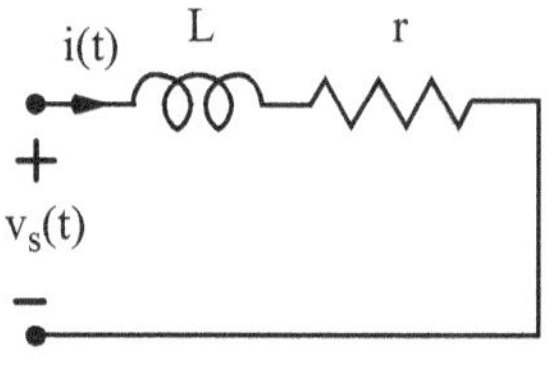

Figure 2.1 Simple resistive-inductive circuit

Because there is no iron in the magnetic circuit the voltage drop across the inductance is linearly related to the current through it. Thus, from Kirchoff's voltage law, the voltage-current relationship at time *t* is

$$v_s(t) = i(t)r + L\frac{di(t)}{dt},$$

$$\text{or} \quad \frac{di(t)}{dt} + i(t)\frac{r}{L} = \frac{v_s(t)}{L}.$$

The latter equation can be expressed in a more convenient form involving the operator $p = d/(dt)$, i.e.

$$(p + r/L)i = v_s/L. \tag{2.1}$$

Note that in (2.1) we recognise that the variables are instantaneous quantities in the time domain and therefore we have dropped the dependency on time, (*t*).

Example 2

A load is driven by a d.c. motor at an angular speed of $\omega(t)$ (rad/s) as illustrated in Figure 2.2. The speed of the motor and load is controlled by varying the DC supply voltage $v_s(t)$ (volts), the field current being held constant. The back-emf developed by the motor is $v_b(t)$; the torque of electromagnetic origin developed by the motor is $T_e(t)$ (Nm) and the opposing load torque, $T_L(t)$ (Nm), is proportional to shaft speed. The combined polar moment of inertia of the rotors of the load and motor is *J* (kg-m^2). The resistance and inductance of the armature winding are *r* (ohm) and *L* (Henry), respectively. The effect of armature reaction on the field flux is negligible.

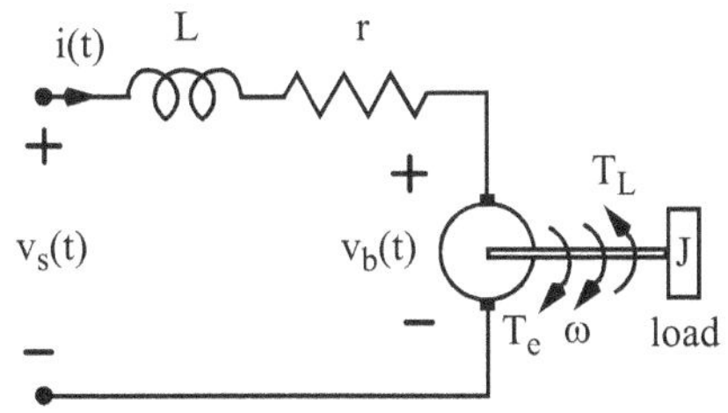

Figure 2.2 DC motor and load

(i) Write down the equation that describes the behaviour of the load current *i*(*t*) when an arbitrary voltage $v_s(t)$ is applied to the circuit.

Based on Kirchoff's voltage law, the voltage-current relationship at time *t* is:

$$v_s = v_b + ir + L\frac{di}{dt}, \quad \text{or}$$

$$i = \frac{1/L}{p + r/L}(v_s - v_b). \tag{2.2}$$

(ii) Derive an expression which describes the behaviour of the shaft speed when the motor torque is varied.

Based on Newton's Second Law of Motion, the accelerating torque on the shaft is

$$T_e - T_m = J\frac{d\omega}{dt}. \tag{2.3}$$

Let the load torque-speed characteristic be defined by $T_m = k_L\omega$. Following substitution of the latter expression for T_m in (2.3), we find:

$$\omega = \frac{1/J}{p + k_L/J}T_e. \tag{2.4}$$

(iii) Derive the differential equation which describes the variation of motor speed with changes in supply voltage.

Because the field flux is independent of the load current the electrical torque is proportional to load current, $T_e = k_i i$. The back e.m.f. is proportional to speed, $v_b = k_\omega\omega$.
Substitute for T_e in (2.4), with the result that:

$$\omega = \frac{1/J}{p + k_L/J}k_i i. \tag{2.5}$$

Substitution of (2.2) in (2.5) yields

$$\omega = k_i\left(\frac{1/L}{p + r/L}\right)\left(\frac{1/J}{p + k_L/J}\right)(v_s - k_\omega\omega). \tag{2.6}$$

Speed ω, the dependent variable, can then be expressed in terms of the supply voltage, the independent variable v_s, in the following form:

$$\left[p^2 + p\left(\frac{r}{L} + \frac{k_L}{J}\right) + \frac{k_L r + k_\omega k_i}{LJ}\right]\omega(t) = \frac{k_i}{LJ}v_s(t). \tag{2.7}$$

The plant equations such as (2.1) and (2.7) are simple; more complex cases will be considered later. The significance of these equations is that they represent simple forms of the following general form of the differential equation:

$$(p^n + a_{n-1}p^{n-1} + \ldots + a_1 p + a_0)y(t) = (b_m p^m + b_{m-1}p^{m-1} + \ldots + b_0)u(t), \quad m \le n. \tag{2.8}$$

Notice all the terms in the output or dependent variable $y(t)$ are collected on the left-hand side of the equation, those in the input variable $u(t)$ on the right-hand side. Importantly, the resulting equation is a single, n^{th} order differential equation in the dependent variable $y(t)$, i.e. the nature of the time-domain response depends on the form of the input variable $u(t)$.

For example, the input signal may be a step function or a sinusoidal function of time, both of which are commonly-used test signals.

In the general case $m \le n$. The system of order n is said to be "proper" when $m = n$, or "strictly proper" when $m < n$.

Example 3
For the plants described in (2.1) and (2.7), $n = 1$, $m = 0$ and $n = 2$ and $m = 0$, respectively; both systems are therefore strictly proper.

2.3 The Laplace Transform

The theory and application of Laplace Transforms are covered in detail in the literature [1] [2], hence only features of significance to the understanding of the material in this and the following chapters are reviewed.

A valuable application of Laplace Transforms is the solution of linear differential equations of the form of (2.8).

Example 4
Form the Laplace Transform of the following second-order differential equation which describes the dynamics of a plant:

$$p^2y + a_1py + a_0y = b_1pu + b_0u. \tag{2.9}$$

Let $F(s)$ be the Laplace Transform of a function $f(t)$. The following results are derived from a table of Laplace Transforms:

$L\,[af(t)] = aF(s)$, a is a constant coefficient;

$L\,[pf(t)] = sF(s) - f(0)$, $f(0)$ is the value of $f(t)$ at time zero;

$L\,[p^2f(t)] = s^2F(s) - sf(0) - pf(0)$, $pf(0)$ is the value of the derivative at time zero.

A convention is adopted that the input $u(t)$ commences at time $t(0^+)$, hence at time zero $u(t)$ and all its derivatives are zero, i.e. $u(0) = pu(0) = \ldots = 0$. Initial conditions on the dependent variable are specified at $t = 0$.

Substituting the expressions for the Laplace Transforms in (2.9), and accounting for initial conditions on $u(t)$, we find

$$(s^2 + a_1 s + a_0)Y(s) - (s + a_1)y(0) - py(0)$$
$$= (b_1 s + b_0)U(s) - b_1 u(0)$$

Thus the output response $Y(s)$ can be expressed in terms of both the forcing function $U(s)$ and the initial conditions on the output as:

$$Y(s) = \underbrace{\frac{b_1 s + b_0}{s^2 + a_1 s + a_0} U(s)}_{\text{forced response}} + \underbrace{\frac{(s + a_1)y(0) + py(0)}{s^2 + a_1 s + a_0}}_{\text{natural response}} \quad \text{and} \quad u(0) = 0. \tag{2.10}$$

Note that we can manipulate the Laplace operator s as any other algebraic variable.

There are two terms in the output response $Y(s)$ in (2.10). The first term, called the forced response, is determined by the nature of the forcing function, the input $U(s)$. The second term, called the natural response, is determined only by initial conditions on the dependent variable; in this case the output signal and its derivatives at time zero. If all initial conditions are zero, only the forced response is present in the output. Similarly, only the natural response exists in the output in the absence of an input signal ($U(s) = 0$).

If all initial conditions are zero, equations of the form of (2.10) can be written as:

$$\frac{Y(s)}{U(s)} = \frac{b_1 s + b_0}{s^2 + a_1 s + a_0} = G(s).$$

$G(s)$ is called the Transfer Function (TF) between the input $U(s)$ and output $Y(s)$.

In summary, therefore, if it is assumed that:

- the initial conditions on the dependent variable and all its derivatives are zero, i.e. $y(t_0) = py(t_0) = p^2 y(t_0) = \ldots = 0$, and

- the input signal is applied at time $t(0^+)$ (so that $u(t_0)$ and all its derivatives are zero),

then the *Laplace Transform of* (2.8) and (2.9) *can be simply formed from the differential equation by replacing the differential operator p by the complex Laplace operator, s*. The time-domain variables $y(t)$ and $u(t)$ become the Laplace variables $Y(s)$ and $U(s)$, respectively. Applying these results directly to (2.8), we find the general form of the differential equation describing the plant becomes

$$(s^n + a_{n-1}s^{n-1} + \ldots + a_1 s + a_0)Y(s) = (b_m s^m + b_{m-1}s^{m-1} + \ldots + b_0)U(s), \quad m \le n. \tag{2.11}$$

The transfer function, which can be expressed a ratio of two polynomials in s, is thus

$$G(s) = \frac{Y(s)}{U(s)} = \frac{b_m s^m + b_{m-1} s^{m-1} + \dots + b_0}{s^n + a_{n-1} s^{n-1} + \dots + a_1 s + a_0}, \quad \text{for } (m < n). \tag{2.12}$$

Notice that a significant short-cut for writing the transfer function directly from the differential equation is demonstrated.

Note that if $m = n$ then the transfer function can be written in the form:

$$b_n + \frac{(b_{n-1} - b_n a_{n-1}) s^{n-1} + \dots + (b_1 - b_n a_1) s + (b_0 - b_n a_0)}{s^n + a_{n-1} s^{n-1} + \dots + a_1 s + a_0}.$$

The above transfer function consists of two paths, one being the direct path between input and output through the gain b_n; this path is simple to accommodate in any analysis - for example in that of lead and lag transfer functions. Note that in the following analysis and chapters it will be assumed that $m \le n$, i.e. all transfer functions are proper or strictly proper. It will be found in the analyses that follow in this chapter that the transfer function (2.12) is a more useful and practical form of the plant or system model than the form described by the nth order differential equation, (2.8).

An example illustrating an application of the above results is discussed below.

Example 5
Find the time response $y(t)$ of a plant described by (2.12) given an input signal $u(t)$.

If the input function $U(s)$ is known, the solution for the response $Y(s)$ is

$$Y(s) = [(b_m s^m + b_{m-1} s^{m-1} + \dots + b_0)/(s^n + a_{n-1} s^{n-1} + \dots + a_1 s + a_0)]U(s).$$

The Laplace Transforms for a range of input signals are given in tables of Laplace Transforms; for example:

step input: $u(t) = R_0$ for all $t > 0$, $u(t) = 0$ for $t < 0$: $U(s) = R_0/s$;

sinusoidal input: $u(t) = A \sin \omega t$: $U(s) = (A\omega)/(s^2 + \omega^2)$.

As an illustration, let us determine the response of the current through the inductor in Example 1 to a step increase in the input voltage of V_0 volts. Assuming zero initial conditions, replacing p by s in (2.1) and setting $U(s) = V_0/s$, we find

$$I(s) = \frac{1/L}{s + r/L} \cdot \frac{V_o}{s} = \frac{V_o}{r}\left\{\frac{1}{s} - \frac{1}{s + r/L}\right\}.$$

From a table of inverse Laplace Transforms [1] [2], the time responses for the terms $1/s$ and $1/(s + r/L)$ are found to be 1 and $\exp[-(r/L)t]$, respectively. Hence the time response of the inductor current is:

$$i(t) = (V_o/r)\{1 - e^{-(r/L)t}\}.$$

The concepts and results discussed above are of particular value in analyzing and designing power system controllers. The so-called "classical" methods for the analysis and design of linear control systems are based on the single-input, single-output transfer-function model of a plant or system. The examination of the properties of the transfer function, and of the information on system dynamic performance derived from those properties, is the subject of the following sections.

2.4 The poles and zeros of a transfer function.

Considerable information on the stability and dynamic performance of a plant is derived from the knowledge of the location of the poles and zeros of its transfer function. Importantly, such knowledge is obtained *without* having to solve the differential equation of the form of (2.8) for the plant. Not only may a solution of the differential equation be time-consuming to conduct, but the information derived from the associated time response is not as extensive as that extracted from an analysis of its associated transfer function (2.12).

The transfer function $G(s)$ of (2.12) can be expressed in the following pole-zero form, namely,

$$G(s) = \frac{P(s)}{Q(s)} = \frac{A(s - z_1)(s - z_2)\dots(s - z_m)}{(s - p_1)(s - p_2)\dots(s - p_n)}, \quad m < n,$$

when $P(s)$ and $Q(s)$ are factorized into factors with real roots or complex-conjugate pairs of roots. The denominator polynomial $Q(s)$, when set equal to zero, is known as the *characteristic equation*:

$$Q(s) = a_n s^n + a_{n-1} s^{n-1} + \dots + a_0 = 0. \qquad (2.13)$$

The roots of the characteristic equation $p_1, \dots, p_n$ are known as the *zeros* of $\boldsymbol{Q(s)}$ or the *poles* of $\boldsymbol{G(s)}$; for example the poles lie at $s = p_1, \dots, s = p_n$ in the complex s-plane. Likewise, $z_1, \dots, z_m$ are the *zeros* of both $\boldsymbol{P(s)}$ and $\boldsymbol{G(s)}$ and lie at $s = z_1, \dots, s = z_m$ in the complex s-plane.

Note that the poles and zeros of $G(s)$ may be real or complex. For example the poles of the transfer function,

$$G(s) = \frac{13(s+5)}{s^2+4s+13},$$

are a complex-conjugate pair and lie at $s_1 = -2+j3$, $s_2 = -2-j3$ in the complex s-plane; the zero lies at $s = -5$.

The significance of the poles and zeros of $G(s)$ will be examined in more detail in the following sections.

2.5 The Partial Fraction Expansion and Residues

2.5.1 Calculation of Residues

If all the poles are distinct, i.e. there is not more than one pole at any location, the transfer function given by (2.12) can be written as

$$G(s) = \frac{P(s)}{Q(s)} = \frac{K_1}{s-p_1} + \frac{K_2}{s-p_2} + \ldots + \frac{K_i}{s-p_i} + \ldots + \frac{K_n}{s-p_n}, \quad m < n. \tag{2.14}$$

If p_1 and p_2, say, are a complex-conjugate pair, K_1 and K_2 are also a complex-conjugate pair.

The coefficient K_i is calculated by multiplying both sides of (2.14) by $(s-p_i)$ (this isolates K_i in (2.14)), and then setting s equal to p_i (this sets all other terms, except the ith, to zero). Thus, for example, to isolate and calculate K_1 we evaluate the expression,

$$K_1 = \left[(s-p_1)\frac{P(s)}{Q(s)}\right]\Bigg|_{s=p_1} = \frac{A(p_1-z_1)(p_1-z_2)\ldots(p_1-z_m)}{(p_1-p_2)(p_1-p_3)\ldots(p_1-p_n)}, \quad m < n, \tag{2.15}$$

using the procedure illustrated in the following example.

Example 6

Evaluate the partial fraction expansion for the following transfer function.

$$G(s) = \frac{A(s+1)}{(s+2)(s+3)} = \frac{K_1}{(s+2)} + \frac{K_2}{(s+3)}, \quad \text{hence } p_1 = -2,\ p_2 = -3.$$

Multiply terms two and three by $(s+2)$ to isolate K_1,

$$\frac{A(s+1)}{(s+3)} = K_1 + \frac{K_2(s+2)}{(s+3)}. \quad \text{Then set } s = -2 \text{, thus}$$

$$\frac{A(-2+1)}{(-2+3)} = K_1 + K_2\frac{(-2+2)}{(-2+3)} = K_1, \quad \text{hence} \quad K_1 = -A.$$

Following the same procedure for the pole at $s = -3$, we find $K_2 = -2A$.

When the transfer function $G(s)$ is expressed as a summation of first-order transfer functions, as in (2.14), the constant K_i is also known as the *residue*, r_i, of the pole at $(s = p_i)$. For a pair of complex poles, the residue K_i and its complex conjugate exists.

The case when there is a zero in close proximity to a pole is of particular interest. Say, there is a zero at $s = z_1$, close to the pole at p_1. The factor in $(p_1 - z_1)$ in (2.15), will be small or negligible and *so too* will be the residue r_1. This useful result will be employed in different contexts later. (To confirm this observation, try evaluating the residues when the zero at $s = -1$ in the transfer function $G(s)$ of Example 6 is replaced by one at $s = -2.01$.)

The special case of multiple poles at any location is not considered here; see [1] or [2].

2.5.2 A simple check on values of the residues

The following result assists in the calculation - or in checking the calculation - of residues.

Let us multiply out the right-hand side of (2.14), i.e.

$$G(s) = \frac{r_1(s-p_2)(s-p_3)\dots(s-p_n) + r_2(s-p_1)(s-p_3)\dots(s-p_n) + \dots}{(s-p_1)(s-p_2)\dots(s-p_n)}, \tag{2.16}$$

where $r_i = K_i$. The numerator term of (2.16) is thus

$$(r_1 + r_2 + \dots + r_n)\, s^{n-1} + [\text{lower order terms in } s]. \tag{2.17}$$

The order of the denominator is n. Hence, if the order of the numerator is:

* one less than the denominator, i.e. $m = n - 1$, the coefficient of s^{n-1} must be

$$\sum_{i=1}^{n} r_i = A,$$

* two or more less than the denominator, the coefficient of s^{n-1} vanishes, i.e

$$\sum_{i=1}^{n} r_i = 0.$$

What can you say when $m = n$?

2.6 Modes of Response

Let us examine the response of a plant to a step input. If $U(s)$ is a unit step input, i.e. $U(s) = 1/s$, then the response $Y(s)$ is

$$Y(s) = G(s)\frac{1}{s}. \tag{2.18}$$

This can be expressed in partial fraction form as:

$$Y(s) = \frac{K_0}{s} + \frac{K_1}{s-p_1} + \frac{K_2}{s-p_2} + \dots + \frac{K_n}{s-p_n}. \tag{2.19}$$

Using a table of inverse Laplace transforms yields the time response to the step input,

$$y(t) = K_0 + K_1 e^{p_1 t} + K_2 e^{p_2 t} + \dots\dots + K_n e^{p_n t}. \tag{2.20}$$

Note that

- the response $y(t)$ comprises a steady-state component K_0 and transient terms; K_0 is the response after all the transients components have decayed away;
- K_i, $i > 0$, is the amplitude at time zero of the transient terms, $K_i e^{p_i t}$;
- if the input step-size were increased by a factor A, all the terms in (2.19) and (2.20) are multiplied by the same factor;
- the form of the transient response $y(t)$ is determined by the n roots of the characteristic polynomial, $Q(s)$, or by the *n poles* of $G(s)$.
- if p_1 and p_2 are a complex-conjugate pole pair, i.e. $p_2 = p_1^*$, then the associated residues are also a complex-conjugate pair, $K_2 = K_1^*$.

For the present time, assume all roots of $Q(s)$ have negative real parts. In general, there will be real and complex roots of $Q(s)$ of the form,

$$p_i = \alpha_i \text{ for the } i^{\text{th}} \text{ real root,}$$
$$\text{and } p_k = \alpha_k - j\omega_k \text{ and } p_{k+1} = \alpha_k + j\omega_k \text{ for the complex pair } k,\ k+1.$$

Associated with each real root $p_i = \alpha_i$, there is a term in the partial fraction expansion

$$K_i/(s-\alpha_i).$$

Taking the inverse transform of the latter term, as in (2.19) - (2.20), results in a term in the time-domain response $K_i e^{\alpha_i t}$. If α_i is negative, *this response is a monotonically, exponentially-decaying **mode*** [1].

Likewise, for the complex pair, $p_k = \alpha_k - j\omega_k$ and, $p_{k+1} = \alpha_k + j\omega_k$ there are terms in the partial fraction expansion of the form

---

1. See a note on the term 'mode' in Section 3.5.2.

$$\frac{K_k}{s-(\alpha_k - j\omega_k)} + \frac{K_k^*}{s-(\alpha_k + j\omega_k)},$$ where K_k and K_k^* are complex conjugates.

Correspondingly, there will be a term in the time-domain response of the form $Ae^{\alpha_k t}\sin(\omega_k t + \phi_k)$. If α_k is negative, *this response is an oscillatory, exponentially-decaying, sinusoidal* ***mode.*** In power systems analysis the term "mode" usually refers to a broader set of properties that characterize the physical behaviour of the natural system responses. Other modal characterisations include whether it is an electro-mechanical mode, for example, or a controller mode, etc.

The *significance* of the above analysis is that it reveals, by a simple examination of the poles of the transfer function, the nature of the transient response in the time domain. The real part of the pole, α_k, measured in Neper/s (Np/s), indicates how rapidly the modes in the response decay away. The imaginary part, ω_k, of a complex pole pair is the frequency in rad/s of the damped sinusoidal oscillation. These results provide another valuable short-cut in linear analysis: *there is no need to solve the differential equations to determine the nature of the transients in the response.*

Let us illustrate some of the important concepts outlined above by means of examples. They are intended to provide some useful insights into the dynamic behaviour of systems.

Example 7

In the following cases, find the time-response of the plant to a step input of magnitude A units.

Case 1. The plant is described by the second-order differential equation

$$p^2y(t) + 5py(t) + 6y(t) = 6pu(t) + 6u(t),$$

where $u(t)$ and $y(t)$ are the input and output signals, respectively.

Assuming initial conditions are all zero, the plant transfer function is found by replacing the differential operator p by the Laplace operator s, i.e.

$$\frac{Y(s)}{U(s)} = G(s) = \frac{6(s+1)}{s^2+5s+6} = \frac{6(s+1)}{(s+2)(s+3)}.$$

For a step input, $U(s) = A/s$. The response can be found using a partial fraction expansion, i.e.

$$Y(s) = G(s)U(s) = \frac{6A(s+1)}{s(s+2)(s+3)} = \frac{A}{s} + \frac{3A}{(s+2)} - \frac{4A}{(s+3)}.$$

Note that the sum of the residues in this case is zero since $(n-m \geq 2)$. Using the inverse Laplace transform tables, the response in the time domain is found to be

$$y(t) = A + 3Ae^{-2t} - 4Ae^{-3t}.$$

Note that the coefficients of the modes e^{-2t} and e^{-3t}, 3A and 4A respectively, are the initial amplitudes of the transient response and are of comparable magnitude.

Case 2. Assume that the zero in the previous plant transfer function lies at $s = -2$ Np/s instead of at $s = -1$ Np/s. The coefficient of the mode e^{-2t} is then zero. (Note that the pole at $s = -2$ Np/s still exists in the plant; this aspect is considered later.) This case shows that pole-zero cancellation causes the amplitude of mode to become small or negligible. Pole-zero cancellation, or close cancellation, is sometimes used in control system design. However, it should be used with caution because the mode e^{-2t} is then only partially observable - or even unobservable - in the output.

Case 3. Let the plant transfer function be:

$$\frac{Y(s)}{U(s)} = G(s) = \frac{20}{s^2 + 12s + 20} = \frac{20}{(s+2)(s+10)}.$$

The time response to the step input $U(s) = A/s$ is

$$y(t) = A - 1.25Ae^{-2t} + 0.25Ae^{-10t}.$$

In this case the pole at $s = -2$ Np/s is much closer to the origin of the complex s-plane than that at $s = -10$ Np/s. The response of the *fast* mode e^{-10t} at a time equal to its time constant, i.e. $t = T = 1/10$ s, is about 37% of its initial amplitude (see Section 2.8.1). The response of the *slow* mode e^{-2t} at $t = 1/10$ s is about 82% of its initial amplitude. In this case the contribution to the overall response of the fast mode rapidly diminishes with time. Thus the response of the slow mode with a time constant of 1/2 s, dominates until it itself decays after a time equal to four time constants (2 s).

Case 4. Furthermore, it can be shown that the initial amplitude of the response of the fast mode becomes smaller as the pole at $s = -10$ is moved further into the left-half of the s-plane relative to the location of the pole of the slow mode. For example, for the transfer function:

$$\frac{Y(s)}{U(s)} = \frac{200}{(s+2)(s+100)},$$

the step response is $y(t) = A - 1.0204Ae^{-2t} + 0.0204Ae^{-100t}$.

Note that the initial amplitude $(0.0204A)$ – and therefore the time response – of the fast mode is almost negligible in comparison to the slower mode.

Case 5. Let the plant transfer function be:

$$\frac{Y(s)}{U(s)} = G(s) = \frac{20(s+1)}{s^2+12s+20} = \frac{20(s+1)}{(s+2)(s+10)}.$$

The time response to the step input is $y(t) = A + 1.25Ae^{-2t} - 2.25Ae^{-10t}$.

Although the poles in this case are the same as those in Case 3, the initial amplitude of the fast mode e^{-10t} is relatively much larger. Due to this mode the overall response of the plant is much faster, although it settles in a time determined by the slow mode.

The significance of the results illustrated in Cases 3 and 5 is that the slower modes tend to dominate the response. Case 5 reveals that the placement of a zero at an appropriate position can speed up the response of a sluggish system. In this case the location of the zero at $s = -1$ Np/s relatively close to the pole $s = -2$ Np/s diminishes the effect of the slow mode. This concept is commonly used in classical control system design for speeding up the response of a sluggish system.

Case 6. The plant transfer function of Case 1 is modified so that the damping ratio of its second-order poles is less than one, say,

$$\frac{Y(s)}{U(s)} = \frac{6(s+1)}{s^2+2s+6} = 6\frac{(s+1)}{(s+1)^2+5} = \frac{6(s+1)}{(s+1+j\sqrt{5})(s+1-j\sqrt{5})}.$$

There are a pair of complex poles at $s_{1,2} = -1 \pm j\sqrt{5}$. The response to a step of magnitude A results in the following partial fraction expansion

$$Y(s) = A\left[\frac{1}{s} - \frac{(1+j\sqrt{5})}{2((s+1)+j\sqrt{5})} - \frac{(1-j\sqrt{5})}{2((s+1)-j\sqrt{5})}\right].$$

Replacing each term by its inverse Laplace transform, we find the closed-form expression for the oscillatory response is:

$$y(t) = A\left[1 - \frac{(1+j\sqrt{5})e^{-(1+j\sqrt{5})t}}{2} - \frac{(1-j\sqrt{5})e^{-(1-j\sqrt{5})t}}{2}\right]$$

$$= A + Ae^{-t}\left[\sqrt{5}\left(\frac{e^{j\sqrt{5}t} - e^{-j\sqrt{5}t}}{j2}\right) - \left(\frac{e^{j\sqrt{5}t} + e^{-j\sqrt{5}t}}{2}\right)\right]$$

$$= A + Ae^{-t}[\sqrt{5}\sin(\sqrt{5}t) - \cos(\sqrt{5}t)]$$

$$= A[1 + \sqrt{6}e^{-t}\sin(\sqrt{5}t - 24.1°)]$$

The denominator of the transfer function, $s^2 + 2s + 6$, is of the form $s^2 + 2\xi\omega_n s + \omega_n^2$. (The following results are described in Section 2.8.2.1). The undamped natural frequency is

$\omega_n = \sqrt{6}$ rad/s. The oscillatory mode thus has a damping ratio $\xi = 2/(2\sqrt{6}) = 0.41$; the damping constant is -1 Np/s. If this were a rotor mode having a frequency of oscillation of $\sqrt{5}$ rad/s, it would be considered well damped.

As has been pointed out earlier, the form of the mode $e^{-t}\sin(\sqrt{5}t-\varphi)$ can be derived by *inspection of the poles of the transfer function* without having to solve for the time response to a step input of the system described by the second-order differential equation,

$$(p^2 + 2p + 6)y(t) = 6(p + 1)u(t)\,.$$

Let us emphasize some important results:

- The response to an input of a linear system consists of a steady-state response and a transient response.
- The *steady-state* term bears a direct relationship to the input function (e.g. doubling the amplitude of the input signal doubles the amplitude of the response).
- The *transient* terms are determined by the *initial magnitude of the input function* (at time $t(0^+)$. However the transient response has a form which is characteristic of the system, and may by identified with the *position of the poles* of the transfer function; these poles are. the zeros of the characteristic equation.
- The concept not only of modes, poles and zeros, together with information provided through the partial fraction expansion, provide important engineering short-cuts for predicting the characteristic response of the plant in the time domain. By *inspection* of the factorised denominator of the plant transfer function - thus revealing the pole positions - we can ascertain: whether the plant responses will contain monotonic or oscillatory (i.e. sinusoidal) components, how rapidly transients decay away, and the frequency of any oscillation.

2.7 The block diagram representation of transfer functions

Block diagrams are a very convenient way of communicating knowledge about the structure of the plant and its control system. Typically, a block diagram contains a number of blocks that represent the transfer functions of elements or components in the system. Moreover, such blocks can be combined or eliminated in a series of operations that reduce or modify the system to a form that is amenable to analysis. Both the basic transfer function blocks and the operations on the blocks are outlined below.

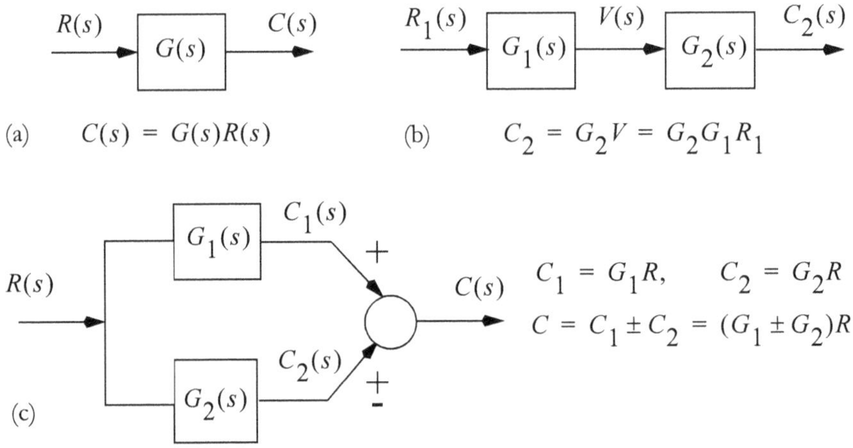

(a) $C(s) = G(s)R(s)$ (b) $C_2 = G_2V = G_2G_1R_1$

(c) $C_1 = G_1R, \quad C_2 = G_2R$

$C = C_1 \pm C_2 = (G_1 \pm G_2)R$

Figure 2.3 (a) The basic transfer function block. (b) Combining blocks in series. (c) Combining blocks in parallel.

Assume the input to a block such as that in Figure 2.3(a) represents the transfer function of system elements that have high input impedances and negligible output impedances (i.e. there is no loading by the elements). Figure 2.3(b) shows that the cascading of two or more blocks can be represented by a single block whose transfer function is the product of the individual transfer functions, $G_2(s)G_1(s)$. Likewise, in Figure 2.3(c) it is demonstrated that blocks in parallel can be represented by the sum or difference of the individual transfer functions, $G_1(s) \pm G_2(s)$. These building blocks form the basis for the analyzing, manipulating or reducing block diagrams representing more complex systems.

A transfer function of particular interest is that representing a closed-loop control system with negative feedback as shown in Figure 2.4. A purpose of the automatic control of stable closed-loop systems is to minimise the error $E(s)$, i.e. the difference between the input and feedback signals, so that the output signal $C(s)$ aligns closely with the reference input $R(s)$. In the following the transfer functions $G(s)$, $H(s)$, $G(s)H(s)$ and $W(s)$ are known as the forward-loop, feedback-loop, open-loop and closed-loop transfer functions, respectively.

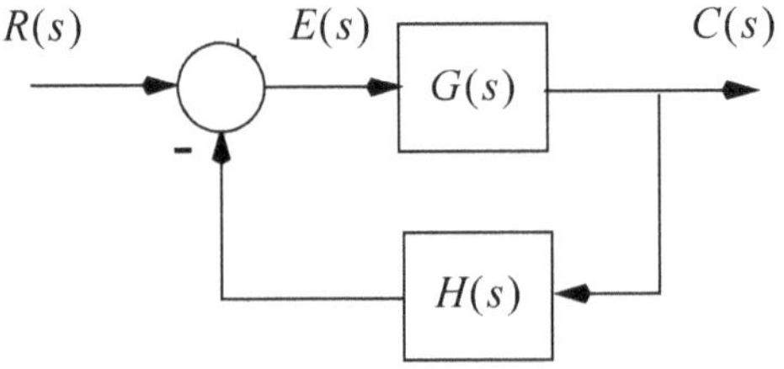

Figure 2.4 The elements of the basic closed-loop control system, $W(s)$.

The relation between the error signal and the output and reference signals are:

$$C(s) = G(s)E(s), \qquad E(s) = R(s) - H(s)C(s).$$

Eliminating the error signal and rearranging the terms, the closed-loop transfer function is found, i.e.

$$W(s) = C(s)/R(s) = G(s)/[1 + G(s)H(s)]. \tag{2.21}$$

This result is basic to the analysis of closed-loop control systems and is referred to frequently in the following sections and chapters.

2.8 Characteristics of first- and second-order systems

In an earlier section we noticed that the poles of the system determined its modes - and thus the form of the transient response. We will now examine how the pole locations determine the characteristics of that response. It is important to note that, by understanding the nature of the transient responses for simple first- and second-order systems, it is possible to predict the characteristics of the dynamic behaviour of higher-order systems.

2.8.1 First-order system

The transfer function of the first-order system has the forms:

$$\frac{Y(s)}{U(s)} = \frac{1}{1+sT} = \frac{a}{s+a} \quad \text{where} \quad a = \frac{1}{T}; \tag{2.22}$$

T is called the time constant of the system, $U(s)$ and $Y(s)$ are its input and output signals, respectively.

The response to a step input, $U(s) = A/s$, is $Y(s) = \frac{a}{s+a} \cdot \frac{A}{s} = \left(\frac{1}{s} - \frac{1}{s+a}\right)A$,

$$\text{or} \qquad y(t) = A(1 - e^{-at}). \tag{2.23}$$

The time-domain response $y(t)$ is shown if Figure 2.5. Note the following important properties of the first-order system:

- At a time t equal to the *time constant,* $t = T = 1/a$, the term Ae^{-at} in the above response is $Ae^{-at} = Ae^{-a/a} = Ae^{-1} = 0.368A$, i.e. this term has decayed to 36.8% of its initial value, $Ae^{-at}\big|_{t=0} = A$. The value of the response, however, is $y(t = T) = 0.632A$, i.e. 63.2% of its final value.

- After a time equal to four time constants, the response $y(t)$ lies within 2% [1] of the final value, A units, i.e. in effect, the transient response Ae^{-at} has completely decayed away. This time is known as the "2% settling time", $t_s = 4T$. Similarly, a "5% settling time" is often quoted for which $t_s = 3T$.

---

1. Actual value at four time constants is 1.83%; for three time constants it is 4.98%.

From the denominator of (2.22) we note there is a real pole at $s = -a$ Np/s. The associated term in the transient response is Ke^{-at}. Thus a *real pole* is associated with an *exponentially decaying mode* in the response if $-a$ is negative.

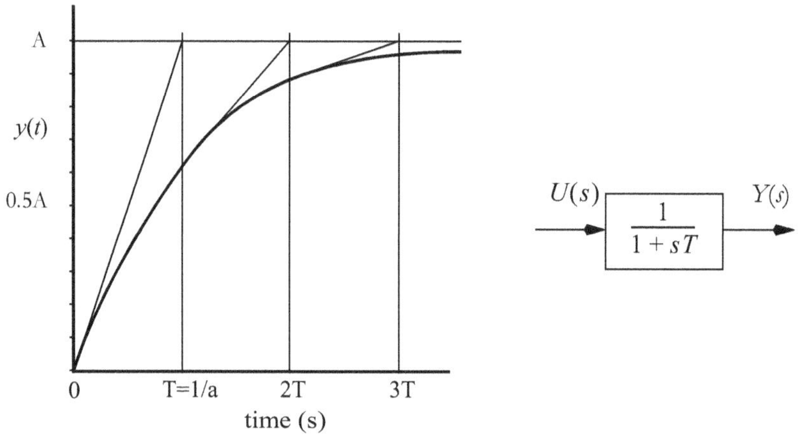

Figure 2.5 Time response of a first-order system to a step input of magnitude *A* units.

2.8.2 The second-order system

2.8.2.1 The characteristics of the second-order system

The typical form of the differential equation of a second-order system is:

$$(p^2 + a_1 p + a_0)y(t) = b_0 u(t).$$

Having taken the Laplace Transform and assuming zero initial conditions, we can express the transfer function in the following two forms:

$$\frac{Y(s)}{U(s)} = \frac{b_0}{s^2 + a_1 s + a_0} = \frac{\omega_n^2}{s^2 + 2\xi\omega_n s + \omega_n^2} \quad \text{if} \quad a_0 = b_0. \tag{2.24}$$

The second form is the ideal or classical form of the second-order transfer function which has complex poles. The parameters ξ and ω_n in (2.24), and associated quantities marked as important (*), are defined below:

- * ω_n is the *undamped natural frequency* (*rad/s*), $\omega_n = \sqrt{a_0}$;

- * ξ is the *damping ratio*, $0 \le \xi < 1$; $\xi = a_1/(2\omega_n)$.

The *poles* of the above transfer function are complex when $0 \le \xi < 1$ and are of the form $s_{1,2} = \alpha \pm j\omega_d$ where

- * ω_d (*rad/s*) is the *frequency of the damped oscillations* in the transient response,
- * α is the *damping constant* (*Neper/s*), and
- * the relation between radian frequency and frequency f in Hertz (Hz) is $\omega = 2\pi f$.

The *characteristic equation* for the system of (2.24) is

$$s^2 + 2\xi\omega_n s + \omega_n^2 = 0\text{, with roots } s_{1,2} = -\xi\omega_n \pm \sqrt{(\xi\omega_n)^2 - \omega_n^2}.$$

Since the damping ratio lies in the range $0 \le \xi < 1$ the complex poles of the second-order transfer function are:

$$s_{1,2} = \alpha \pm j\omega_d = -\xi\omega_n \pm j\omega_n\sqrt{1-\xi^2}. \tag{2.25}$$

Based on the above definitions the *frequency of the damped oscillations* and the *damping constant* are given by $\omega_d = \omega_n\sqrt{1-\xi^2}$ rad/s and $\alpha = -\xi\omega_n$ Np/s, respectively. Solving for ξ from the latter two relations it is found that the damping ratio is:

$$\xi = -\alpha/(\sqrt{\alpha^2 + \omega_d^2}) \approx -\alpha/\omega_d \quad \text{if} \quad \alpha < 0.3\omega_d.$$

The time-domain response of the second-order transfer function (2.24) to a step input of magnitude A units is:

$$Y(s) = \frac{\omega_n^2}{s^2 + 2\xi\omega_n s + \omega_n^2} \cdot \frac{A}{s}.$$

The associated time response consists of two terms, the steady-state term of value A and a transient component (an exponentially decaying sinusoid); it is:

$$y(t) = A - \frac{A}{\sqrt{1-\xi^2}} \cdot e^{-\xi\omega_n t} \cdot \sin(\omega_d t + \phi), \quad 0 \le \xi < 1, \tag{2.26}$$

where $\cos\phi = \xi$ and $\sin\phi = \sqrt{1-\xi^2}$.

It is often useful to refer to the characteristics of the step response of the ideal second-order system with a complex pole pair. The damped oscillatory response is shown in Figure 2.6 in which some meaningful measures that characterize the response are defined.

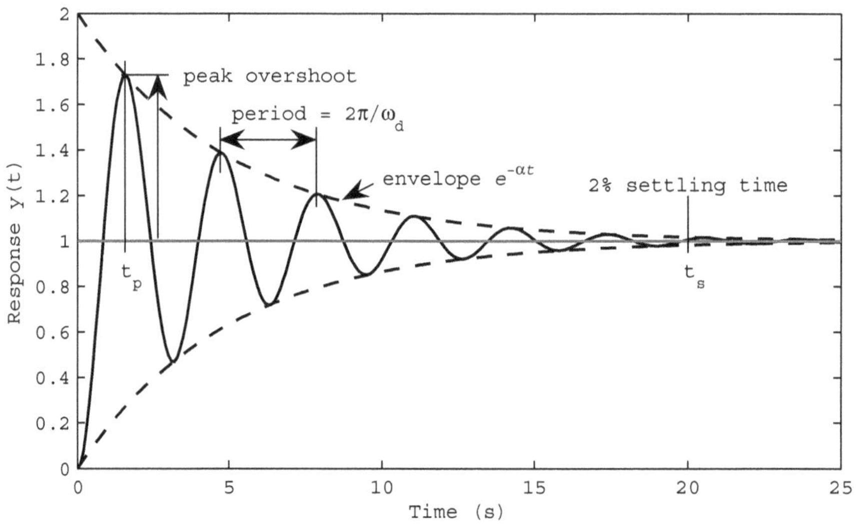

Figure 2.6 Characteristics of the response to a unit step input of the ideal second-order system with a complex pole-pair.

The *frequency in Hertz* of the damped oscillation is: $f_d = \omega_d/(2\pi)$ Hz.

The *period* of the oscillation is: $t_f = 1/f_d = 2\pi/\omega_d$ s.

The *time to the first peak* is 1/2 of the period: $t_p = \pi/\omega_d = \pi/(\omega_n\sqrt{1-\xi^2})$ s.

For a *unit* step input the *peak overshoot* occurs at $t = t_p$ and is:

$$M_0 = y(t_p) - 1 = e^{\alpha t_p} = e^{-\xi\pi/\sqrt{1-\xi^2}}.$$

As already stated, the *settling time* t_s is the time for envelope to decay to a value of 2% (actually 1.8%) of the final value of $y(t)$ and is equal to four time constants:

$$\text{i.e.} \quad t_s = -4/\alpha = 4/(\xi\omega_n) \text{ s.}$$ [1]

The useful reference family of normalised-time responses for a step input is shown in Figure 2.7 for values of ξ between 0.1 and 1. The figure can be interpreted as follows. If, say, $\omega_n = 1$ rad/s and $\xi = 0.1$ then the first peak in the transient occurs at 3.2 s, however, if $\omega_n = 2$ the first peak is reached at 1.6 s, etc.

1. A 5% *settling time* is equal to three time constants

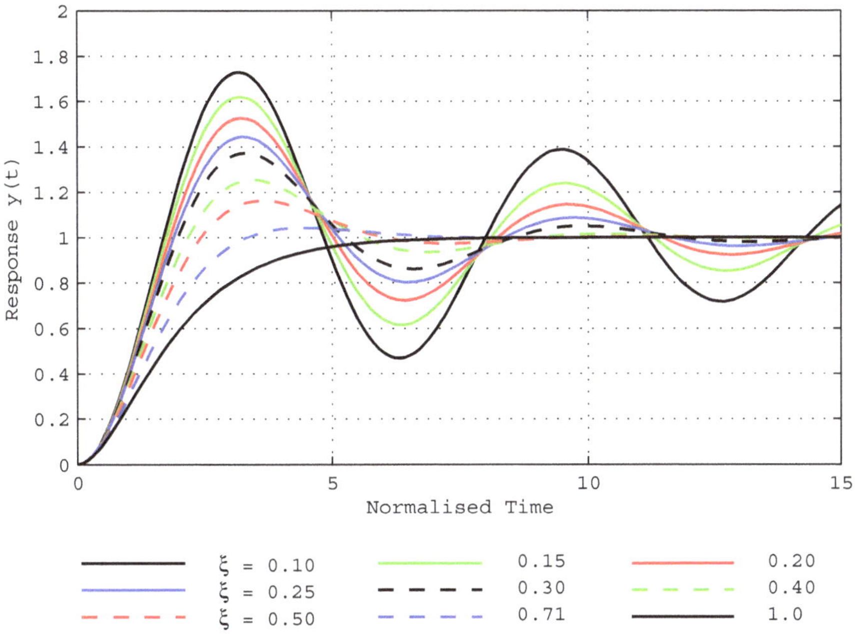

Figure 2.7 Response of the second-order transfer function to a unit step input as a function of normalised time, $\omega_n t$, for damping ratios between 0.1 and 1.

2.8.2.2 Implications for the dynamic performance of power systems

Certain electro-mechanical modes of oscillation, associated with the rotors of generators, are typically complex and lightly damped and of the form $\alpha \pm j\omega_d$. The important features of these modes to which frequent reference will be made are listed in Section 2.8.2.1.

In the analysis of power system dynamic performance the damping ratio, ξ, is used in several contexts, e.g. a criterion for the dynamic performance of the system is that the damping ratio for all rotor modes should be better than, say, 0.05 (or 5%). For the second-order transfer function given in (2.24), the complex poles $\alpha \pm j\omega_d$ vary as shown in Figure 2.8 for $\xi \geq 0$. A line of constant damping ratio makes an angle ϕ with the negative real axis such that $\xi = \cos\phi$. Note that the loci of the poles in the complex *s*-plane is along a semi-circle of radius ω_n. The nature of the time-domain response for a mode located on the semi-circle can be ascertained directly by inspection.

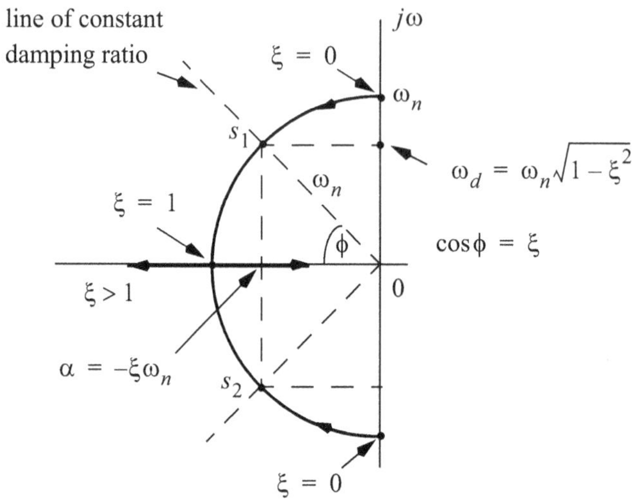

Figure 2.8 Trajectory of the complex poles of the second-order transfer function, (2.24), for the damping ratio $\xi \geq 0$.

2.9 The stability of linear systems

A plant or a system is described by a mathematical model such as a nth order differential equation or transfer function. If the system is passive there is no internal source of energy. A bounded input signal is the only available source of the energy to provide the energy necessary in the output signal; moreover, some of the input energy may be dissipated internally within the system. Such a system is inherently stable since the output must also be bounded. A system is defined as stable if a bounded input always produces a bounded output.

In an active system there is an internal energy source from which the energy contained in the output signal is derived. In this case the possibility of a bounded input producing an unbounded output exists. Such a system may therefore be unstable.

We saw earlier, for linear systems, that the response consists of a steady-state and a transient component. If the system is to be stable all transient terms must decay to zero. For this to occur, we noted in the examination of the modes for first and second order system that the real part α of the mode must be negative. Thus for a system to be stable, the poles of the transfer function - or the roots of the characteristic equation (2.13) - *must all have negative real parts.* An alternative way of stating the same result is: for stability, *all the poles of the transfer function must lie in the left-half of the complex s-plane.*

If, in the characteristic equation, the real part of a complex pair of poles, $s_k = \alpha_k - j\omega_k$ and $s_{k+1} = \alpha_k + j\omega_k$ is positive ($\alpha_k > 0$), the pair lies in the right-half of the s-plane. The associated instability is manifested in a term in the time response which is an *exponentially in-*

creasing sinusoid. A similar result applies to a first order pole, i.e. an *exponentially increasing response.*

If any poles lie on the imaginary axis, all others being in the left-half of the s-plane, the system is said to be marginally stable. However, in practical linear control systems it is not possible to locate and maintain poles exactly on the imaginary and therefore marginal stability is of academic interest only.

2.10 Steady-state alignment and following errors

One reason for the use of closed-loop control systems is to automatically control the output of a system to align with a reference input or a set-point as closely as possible. If the reference is fixed, i.e. set to a constant value, the difference between the set-point and the controlled output in the steady-state is called the *alignment error*. However, the reference input may be time varying; in this case it is necessary for the controlled output to track or follow the reference as closely as possible. In order to assess how well a closed-loop control system aligns with - or follows - a reference input a set of test reference signals is devised that provides a measure of the quality of system performance. These tests signals are analysed in the following sections.

The steady-state value of a time-varying signal $x(t)$, i.e. its value after all oscillations associated with any transients have died away, x_{ss}, can be derived from the final-value theorem (FVT), i.e.

$$x_{ss} = \lim_{t \to \infty} x(t) = \lim_{s \to 0} sX(s). \tag{2.27}$$

From this result the steady-state value of the output of a system, the alignment and following errors for a given test reference-input can be determined.

The closed-loop control system under study is shown in Figure 2.9, where $C(s)$ is the controlled output signal, $R(s)$ is the reference input signal and the error signal is $E(s) = R(s) - H(s)C(s)$. With forward-loop and feedback-path transfer functions $G(s)$ and $H(s)$, respectively, the transfer function of the closed-loop system of Figure 2.9 has been shown in (2.21) to be

$$W(s) = \frac{C(s)}{R(s)} = \frac{G(s)}{1 + G(s)H(s)}.$$

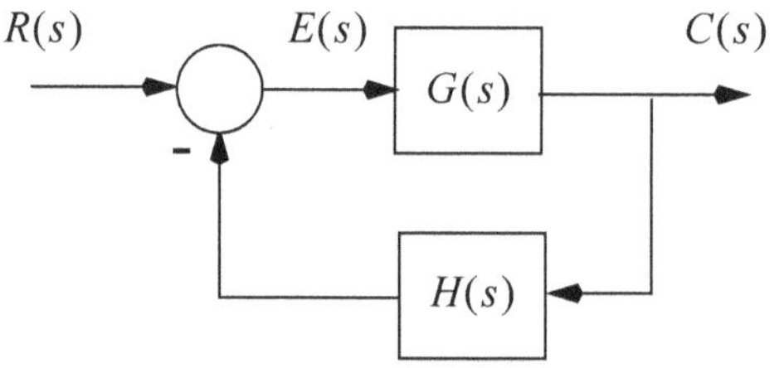

Figure 2.9 Block diagram of the classical closed-loop system

2.10.1 Steady-state alignment error.

The response of the closed-loop system to a step change in the reference input is shown in Figure 2.10. Under steady-state conditions following the change in the reference input $r(t)$, the steady-state output c_{ss} may not be equal to the constant value of the reference, r_{ss}. Based on the final value theorem the steady-state alignment error e_{ass} is defined as:

$$e_{ass} = r_{ss} - c_{ss} = \lim_{s \to 0} s\left[\frac{R_0}{s} - W(s)\frac{R_0}{s}\right], \text{ where } C(s) = W(s)R(s). \qquad (2.28)$$

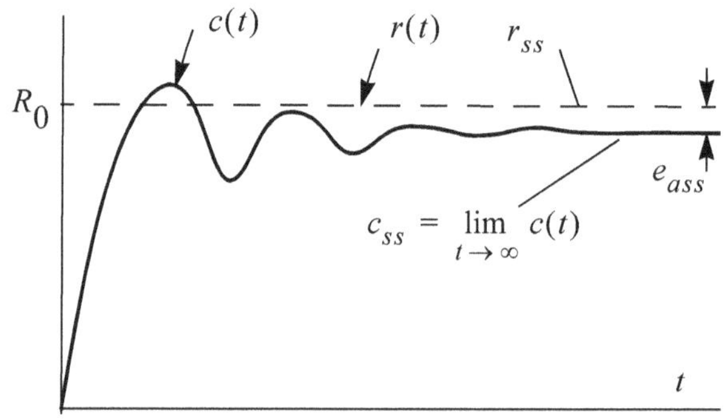

Figure 2.10 The steady-state alignment error e_{ass} following a step input.

Thus from (2.28) a general result for the alignment error follows, i.e.

$$e_{ass} = \lim_{s \to 0} R_0[1 - W(s)]. \qquad (2.29)$$

In the following we will derive the alignment errors for a unity-feedback closed-loop system, i.e. when $H(s) = 1$. In the case of a closed-loop system in which the feedback back is *not* unity gain, the alignment errors can be derived by using the general result (2.29) or by direct application of the final value theorem. Let us consider two special cases for the unity-feedback system, namely without and with integration in the forward path.

2.10.1.1 (a) No integration in the forward path, H(s)=1.

Let us assume $G(s)$ takes the form $G(s) = \dfrac{K(1+sT_{b1})(1+sT_{b2})\ldots}{(1+sT_{a1})(1+C_{a1}s+C_{a2}s^2)\ldots}$.

In the steady-state following a transient, $G(s) \to K$ as $s \to 0$. Note the form of the factors $(1+sT)$ - not $(s+a)$ - in the transfer function is important; with no integration in the forward path the gain K is often called the *static gain.* According to (2.29) the associated alignment error is

$$e_{ass} = \lim_{s \to 0} R_0(1-W(s)) = \lim_{s \to 0} R_0\left[1 - \frac{G(s)}{1+G(s)\cdot 1}\right],$$

$$\text{i.e.} \quad e_{ass} = \lim_{s \to 0} R_0\left[\frac{1}{1+G(s)}\right], \tag{2.30}$$

$$e_{ass} = R_0\frac{1}{1+K}, \tag{2.31}$$

since $G(s) \to K$ as $s \to 0$.

The result in (2.31) provides a very useful insight, namely, the higher the static gain K the smaller is the alignment error. However, there is a downside to high gain settings without suitable compensation, i.e. the dynamic performance of the closed-loop system may become more oscillatory and even unstable. The latter effect will revealed through the analysis of the stability and performance of the closed-loop system using the Bode plot in Section 2.12.2.

2.10.1.2 (b) Single integration in the forward path, H(s)=1.

In this case $G(s)$ takes the form $G(s) = \dfrac{K}{s}\cdot\left(\dfrac{(1+sT_{b1})(1+sT_{b2})\ldots}{(1+sT_{a1})(1+sC_{a1}+C_{a2}s^2)\ldots}\right)$. Let $s \to 0$, then $G(s) \to K/s$. Substitute this limit for $G(s)$ in (2.30) above; the steady-state error is thus

$$e_{ass} = \lim_{s \to 0} R_0\frac{1}{1+\frac{K}{s}} = \lim_{s \to 0} R_0\frac{s}{s+K} = 0.$$

This is a very useful result; it shows that a single integration in the forward path "integrates out" to zero any error $e(t)$ that develops between the reference input and the controlled output. If there is no integration in the forward path the introduction of proportional plus integral (PI) compensation [1], [2] into that path ensures zero alignment error in the steady state. This is highly desirable in some types of closed-loop control systems, however, like the case above, there is a disadvantage. For example, introduction of pure integration $1/s$ also introduces a phase-lag of $90°$ in the open-loop transfer function. In turn, the Phase Margin is reduced, and consequently the closed-loop system may become unstable. This is discussed in Section 2.12.2.

2.10.2 The steady-state following error

A test reference input of a suitable form for defining the following error is the ramp signal, $r(t) = R_0 \cdot t$. Based on the final-value theorem the steady-state following error e_{fss} is given by:

$$e_{fss} = \lim_{t \to \infty} [r(t) - c(t)] = \lim_{s \to 0} s\left[\frac{R_0}{s^2} - W(s)\frac{R_0}{s^2}\right], \text{ or}$$

$$e_{fss} = \lim_{s \to 0} \frac{R_0}{s}[1 - W(s)]. \tag{2.32}$$

This is a general result for the following error. However, let us again consider a *unity*-feedback system and two special cases, without and with integration in the forward path.

2.10.2.1 (a) No integration in the forward path, H(s)=1.

Assume $G(s)$ again takes the form $G(s) = \dfrac{K(1 + sT_{b1})(1 + sT_{b2})\ldots}{(1 + sT_{a1})(1 + sC_{a1} + C_{a2}s^2)\ldots}$. In the steady-state following the decay of the transient we find $G(s) \to K$ as $s \to 0$. By substitution of this limit in (2.32), the following error becomes

$$e_{fss} = \lim_{s \to 0} \frac{R_0}{s}\left[1 - \frac{G(s)}{1 + G(s) \cdot 1}\right] = \lim_{s \to 0} \frac{R_0}{s}\left[\frac{1}{1 + G(s)}\right], \tag{2.33}$$

$$e_{fss} = \lim_{s \to 0} \frac{R_0}{s + sK}, \text{ i.e. } e_{fss} \to \infty. \tag{2.34}$$

The result reveals that the following error becomes increasingly large with time; the latter is illustrated in Fig. 2.11. The controlled output thus cannot follow a reference input that changes linearly with time. Such a closed-loop system cannot track, for example, a satellite passing overhead.

2.10.2.2 (b) Single integration in the forward path, H(s)=1.

As shown in Section 2.10.1.2 above, $G(s) \to \dfrac{K}{s}$ as $s \to 0$. Substitution for $G(s)$ in (2.33) above yields a following error given by:

$$e_{fss} = \lim_{s \to 0} \frac{R_0}{s}\left[\frac{1}{1 + \frac{K}{s}}\right] = \lim_{s \to 0} \frac{R_0}{s + K} = \frac{R_0}{K}. \tag{2.35}$$

Thus after any oscillations have died away following the application of the ramp, there is a constant difference, or error, between the ramp input and the controlled output given by (2.35), as illustrated in Figure 2.11.

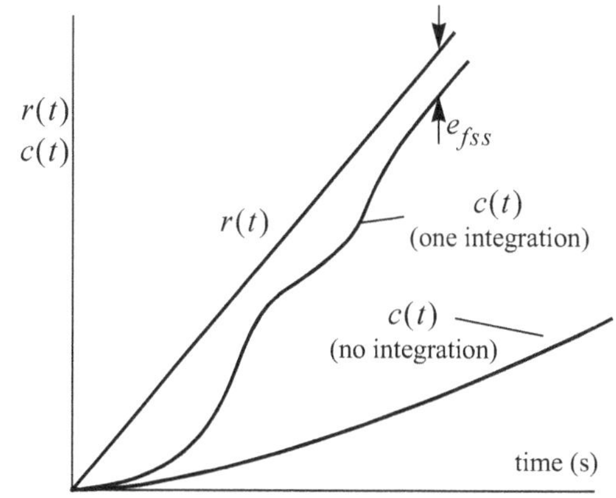

Figure 2.11 The following error e_{fss} resulting from a ramp input.

Exercise. Show that if the forward-loop transfer function $G(s)$ in a unity feedback system contains two integrations, the following error for a ramp reference input is zero, but is finite for a parabolic input $r(t) = R_0 t^2/2$.

Example 8
A unity feedback system has a forward-loop transfer function:

$$G(s) = K'(s+2)/[s(s+5)(s+10)].$$

Find the value of the gain K' such that the steady-state following error $e_{fss} < 0.01$ units for a ramp input of 2 unit/second.

Change the factors in $G(s)$ into the form $(1 + sT)$, i.e

$$G(s) = K\frac{1+0.5s}{s(1+0.2s)(1+0.1s)}, \quad \text{where the 'static' gain is } K = \frac{2K'}{50}.$$

In order to satisfy the specification, the error $e_{fss} = \frac{R_0}{K} = \frac{2}{K} < 0.01$, i.e. $K > 200$, or $K' > 5000$.

2.11 Frequency response methods

The purpose of this section is to outline briefly frequency response methods of analysis and to provide insight and understanding of those features of the analysis that are relevant to power system dynamics and control.

In frequency response analysis the injected frequency will be represented by ω_f (rad/s) rather than by ω - which will refer to rotor speed in later chapters.

Frequency response methods assume that a sinusoidal signal of constant amplitude is applied to system. That is, when a signal $r(t) = A\sin\omega_f t$ of frequency ω_f (rad/s) is injected into the system $G(s)$, a steady-state sinusoidal signal $y(t)$ of the same frequency appears at the output after the transient terms have died away. In the Laplace domain the response of the system is

$$Y(s) = G(s)U(s),$$

where, for a sinusoidal signal, $U(s) = A\omega_f/(s^2 + \omega_f^2)$. The response is thus

$$Y(s) = G(s)\,A\frac{\omega_f}{s^2 + \omega_f^2}.$$

Forming the partial fraction expansion as described in Section 2.5, we find

$$Y(s) = \frac{AG(j\omega_f)}{2j(s - j\omega_f)} + \frac{AG(-j\omega_f)}{(-2j)(s + j\omega_f)} + \begin{bmatrix}\text{terms of the form}\\ K_i/(s - p_i)\end{bmatrix},$$

where p_i are the poles of $G(s)$ and K_i are their residues.

Taking the inverse Laplace transforms, the time-domain response is

$$y(t) = \frac{AG(j\omega_f)}{2j}e^{j\omega_f t} + \frac{AG(-j\omega_f)}{-2j}e^{-j\omega_f t} + \begin{bmatrix}\text{transient terms which}\\ \text{tend to zero as t} \to \infty\end{bmatrix}. \quad (2.36)$$

Let $G(j\omega_f) = |G(j\omega_f)|e^{j\phi(\omega_f)}$ and $G(-j\omega_f) = |G(-j\omega)|e^{-j\phi(\omega_f)}$, where the phase angle ϕ of $G(j\omega_f)$ is a function of frequency. Following substitution for these terms in (2.36), the time-domain response tends to

$$y(t) = A|G(j\omega_f)|\left[\frac{e^{j(\omega_f t + \phi(\omega_f))} - e^{-j(\omega_f t + \phi(\omega_f))}}{2j}\right] = A\,|G(j\omega_f)|\sin(\omega_f t + \phi(\omega_f)),$$

after all transient terms have decayed away.

Hence $y(t) = A|G(j\omega_f)|\sin[\omega_f t + \phi(j\omega_f)]$ is the steady-state response. The steady-state output $y(t)$ is a sinusoid of amplitude $A|G(j\omega_f)|$ with phase shift $\phi(j\omega_f)$ with respect to the input signal. The 'frequency response' of $G(s)$ is a plot of $|G(j\omega_f)|$ and $\phi(j\omega_f)$ as ω_f (rad/s) is varied over a range of the frequencies. Typically the frequency response is plotted in two forms, the polar plot and the Bode plot; we shall employ the Bode-type responses in later chapters.

It is interesting to note that the analysis so far has allowed us to take several short-cuts. Firstly, $G(s)$ was formed by replacing, for zero initial conditions, the differential operator *p by s* in the differential equations for the system. Secondly, the frequency response is obtained simply by replacing s by $j\omega_f$ in $G(s)$.

2.12 The frequency response diagram and the Bode Plot

The following treatment of the Bode Plot contains two main features, firstly, the graphical plotting of the frequency response and, secondly, its application to the determination of the stability of closed-loop control systems. Although the frequency response plot of a given transfer function is readily obtained using the appropriate software, the ability to visualise the frequency response plot is a very useful skill, or conversely, to deduce the form of a transfer function when its frequency response plot is presented. By understanding the basis of the frequency response plot it becomes easier to carry out the required visualisation or interpretation expeditiously.

For the analysis of stability of closed-loop control systems the frequency response plot of interest is that of the open-loop transfer function, $G(j\omega_f)H(j\omega_f)$, where $G(s)$ and $H(s)$ are the forward-loop and feedback-path transfer functions, respectively, shown in Figure 2.9. The associated log-magnitude and phase responses of $G(j\omega_f)H(j\omega_f)$ are known as the Bode Plot from which information on stability can be deduced - subject to certain conditions. Consequently, the following is a brief description of the basis of techniques not only for drawing the Bode Plot, but also for interpreting the Plot to assess both the margin of stability of the closed-loop control system and its dynamic performance.

The system is excited by a sinusoidal signal $r(t)$ of unity amplitude and frequency ω_f (rad/s). The two plots of the Bode diagram are both plotted as a logarithmic function of frequency, $\log\omega_f$. The first plot is of the log (on base 10) of the magnitude (LM) of the open-loop transfer function, i.e. $20\log|G(j\omega_f)H(j\omega_f)|$ in dB; the second plot is of the argument (*Arg*) - or phase - $\phi(j\omega_f) = \angle G(j\omega_f)H(j\omega_f)$ of the open-loop transfer function in degrees.

It should be noted that for any fairly simple transfer function the frequency response can be drawn manually using the graphical technique described in the following section.

2.12.1 Plotting the frequency response of the open-loop transfer function

Any transfer function can be divided into a number of basic first- or second-order factors which form the numerators or denominators of the element. The magnitude and phase response of the basic factors are simple to derive and recall. By combining the responses of the factors, the overall frequency response of the transfer function is generated.

Any transfer function (with $s = j\omega_f$) contains factors, $F(j\omega_f)$, of the following types in its numerator or denominator.

$F(j\omega_f) =$			$F(s) =$
(i)	K	from	K, a scalar gain
(ii)	$(j\omega_f)^{\pm n}$	from	$s^{\pm n}$
(iii)	$(1 + j\omega_f T)^{\pm m}$	from	$(1 + sT)^{\pm m}$
(iv)	$\left\{1 - \left(\frac{\omega_f}{\omega_n}\right)^2 + j\,2\delta\frac{\omega_f}{\omega_n}\right\}^{\pm p}$	from	$\left\{1 + 2\delta\frac{s}{\omega_n} + \frac{s^2}{\omega_n^2}\right\}^{\pm p}$

Note the forms of the type (iii) *and* (iv) *factors, i.e. the polynomial form* $(1 + sa_1 + \ldots)$, *rather than pole-zero form* $(s + a)$.

Example 9

Find log-magnitude and phase of the open-loop transfer function:

$$G(j\omega)H(j\omega) = \frac{K}{j\omega_f(1 + j\omega_f T)}.$$

The log-magnitude and phase responses are:

$$LM = 20\log|GH| = 20\{\log K - \log|j\omega_f| - \log|1 + j\omega_f T|\};$$
$$Arg(GH) = Arg(K) - Arg(j\omega_f) - Arg(1 + j\omega_f T).$$

This example demonstrates that multiplication or division of the four types of factors become *addition or subtraction* of their log-magnitudes and of their phase contributions. This simplifies analysis because it involves simple addition or subtraction of the component terms; such operations are the basis for plotting manually the frequency response of transfer functions.

Let us consider the log-magnitude and phase plots of each of the factors $F(j\omega_f)$ as a function of $\log\omega_f$.

2.12.1.1 (i) A factor in the transfer function is a constant gain K.

The transfer function is $F(j\omega_f) = K$, and the associated log-magnitude and phase are: $LM = 20\log|K|$ and $\phi(j\omega_f) = 0$ if $K > 0$, (or $\phi(j\omega_f) = 180°$ if $K < 0$) respectively. The magnitude response is shown in Figure 2.12.

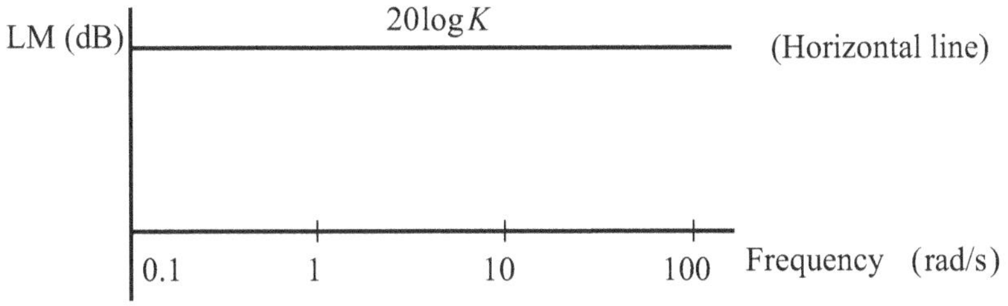

Figure 2.12 Transfer function of a constant gain K

Increasing or decreasing the gain K results in the plot of $20\log K$ moving vertically up or down.

2.12.1.2 (ii) A factor in the transfer function contains pure differentiation or integration of multiplicity n

The transfer function is of the form: $F(j\omega_f) = (j\omega_f)^{\pm n}$, $n = 1, 2, \ldots$. Differentiation is associated with $+n$ and integration with $-n$. The log magnitude of the transfer function is:

$$LM = 20\ \log\left|(j\omega_f)^{\pm n}\right| = \pm 20 n \log\omega_f \ \text{dB}. \tag{2.37}$$

This asymptote, when plotted against $\log\omega_f$, is a straight-line having a slope $\pm 20n$ db/decade. According to (2.37) it intersects the frequency axis when LM = 0 dB at $\log\omega_f = 0$, i.e. at $\omega_f = 1$ rad/s. The phase response is $Arg[(j\omega_f)^{\pm n}] = \pm n90^\circ$, i.e. it is a constant for a given n over the entire frequency range. The log-magnitude and phase responses are plotted in Figure 2.13.

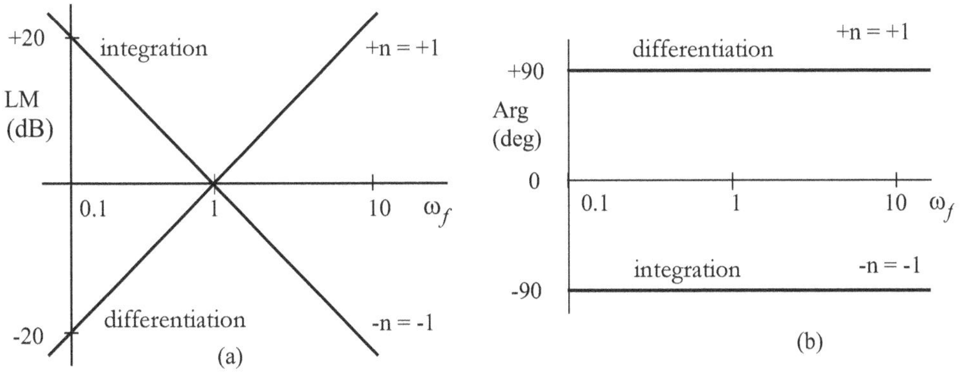

Figure 2.13 (a) Magnitude and (b) phase responses of $(j\omega)^{\pm n}$ when n=1

2.12.1.3 (iii) A factor in the transfer function contains a real pole / zero of multiplicity n:

The transfer function is of the form: $F(j\omega_f) = (1 + j\omega_f T)^{\pm n}$, $n = 1, 2, \ldots$. Its log magnitude is:

$$LM = 20\log|F| = \pm 20n\log(1+\omega^2T^2)^{1/2} \text{ dB.} \tag{2.38}$$

Consider the asymptotes for (a) a low frequency case when $\omega_f T \ll 1$, (b) a high-frequency case when $\omega_f T \gg 1$.

(a) $\omega_f T \ll 1$, $LM = \pm 20n\log 1 = 0$ dB. This low-frequency asymptote is a horizontal line when plotted against $\log\omega_f$.

(b) $\omega_f T \gg 1$, $LM = \pm 20n\log\omega_f T = \pm 20n(\log\omega_f + \log T)$ dB.

This log-magnitude asymptote, when plotted against $\log\omega_f$, is a straight-line. The slope of the line is $\pm 20n$ dB for each decade of frequency (i.e. for each $\log 10 = 1$ unit). The low-frequency and high-frequency straight-line asymptotes of the log-magnitude plots are shown in Figure 2.14 for zeros (+*n*) or poles (-*n*). The actual plot is also drawn for a transfer function $1/(1+j\omega_f T)^n$ over a frequency range about the so-called *corner frequency* ω_c where the high- and low-frequency asymptotes intersect; $\omega_c = 1/T$ rad/s. The differences between the actual plot and the straight-line asymptotes are easily remembered. In the case of multiple poles the actual log-magnitude plot is 3*n* dB down at the corner frequency and *n* dB down at an octave above and below the corner.

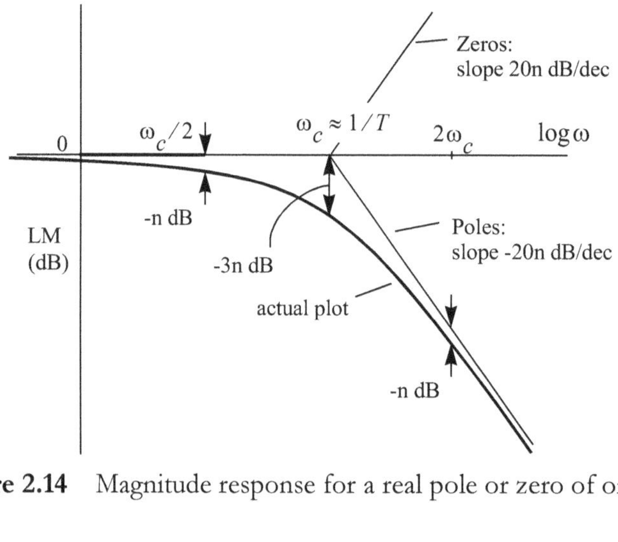

Figure 2.14 Magnitude response for a real pole or zero of order *n*.

Consideration of (2.38) reveals that for multiple zeros $(1+j\omega_f T)^n$ the actual log-magnitude plot and the straight-line asymptotes are the mirror image about the frequency axis of those for multiple poles as shown in Figure 2.14.

Consider now the plot of the phase shift $\phi(j\omega_f)$ for the transfer function $(1+j\omega_f T)^{\pm n}$.

(a) for $\omega_f T \ll 1$: $\phi = \pm n \cdot \operatorname{atan}(\omega_f T \to 0)$, hence $\phi \to 0^\circ$;

(b) for $\omega_f T \gg 1$: $\phi = \pm n \cdot \operatorname{atan}(\omega_f T \to \infty)$, hence $\phi \to \pm n 90^\circ$;

(c) for $\omega_f T = 1$ (corner): $\phi = \pm n \cdot \operatorname{atan} 1$, hence $\phi = \pm n 45^\circ$.

The phase response for real poles of order n is shown in Figure 2.15. A straight-line approximation of the phase response is employed from a decade below the corner, phase 0°, to a decade above the corner at $-n90^\circ$. The straight-line approximation to the response, which passes through the corner frequency at $-n45^\circ$, differs at most from the actual by 5 to 6° (for $n = 1$) over a decade in frequency on either side of the corner. The phase response for multiple zeros is the mirror image about the frequency axis of those for the multiple poles shown in Figure 2.15.

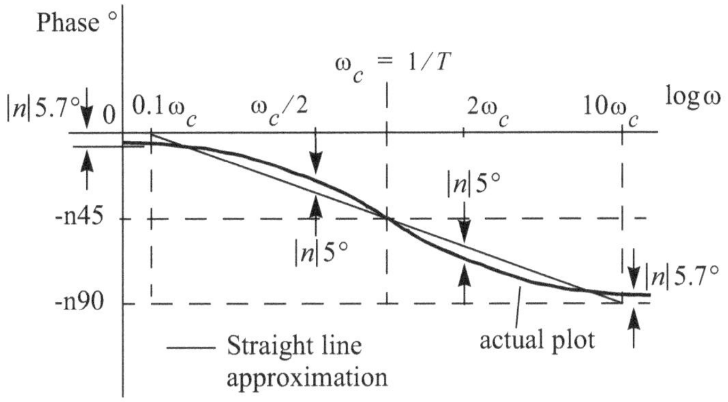

Figure 2.15 Phase response for real poles of order n.

2.12.1.4 (iv) A factor in the transfer function contains a complex pair of poles or zeros of multiplicity n:

The transfer function is of the form: $F(j\omega) = \left\{1 - \left(\frac{\omega_f}{\omega_n}\right)^2 + j\,2\xi\frac{\omega_f}{\omega_n}\right\}^{\pm n}$.

The frequency responses of this factor for a single pair of complex poles ($n = -1$) are shown in Figure 2.16 for damping ratios $0.1 \le \xi \le 1$. The responses are given for a normalised frequency ω_f/ω_n, where ω_n is the undamped natural frequency. The straight-line approximations which can be employed are crude and therefore the more accurate plots shown in the figure are used as templates when sketching the frequency responses of complex poles or zeros. Note that for a single pair of complex zeros the magnitude and phase plots are those shown in Figure 2.16 rotated a half-turn about their respective frequency axes; note that for $n = 1$ the associated phase varies between zero and $180°$.

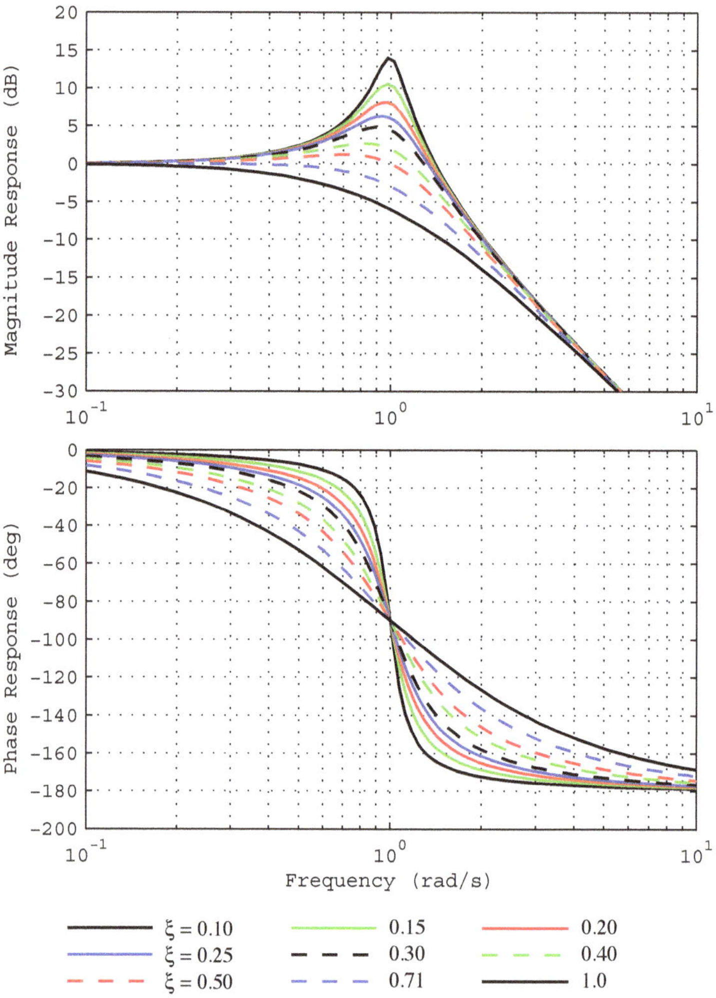

Figure 2.16 Magnitude and phase responses as a function of the normalized frequency ω_f/ω_n for a pair of complex poles.

Example 10. Lead compensation

Lead compensation is often employed as a more practical form of derivative compensation over a range of frequencies. In its application in power system dynamics and control it is used to provide phase lead over a desired range of frequencies; the design of cascade phase-lead compensation for conventional closed-loop control systems is covered in [1].

The form of the lead transfer function is $A \cdot G_{LD}(s) = A\dfrac{1+sT}{1+s(\alpha T)}$, where A is an adjustable gain. The range of values for α is typically $0.1 < \alpha < 1$ when the transfer function is implemented using analog devices. For $\alpha = 0.1$ the phase lead is $56°$. For values of $\alpha < 0.1$

the additional phase lead provided is small. For example, cascading two identical lead networks with $\alpha = 0.25$ produces the same maximum phase shift as a single lead network with $\alpha = 0.025$.

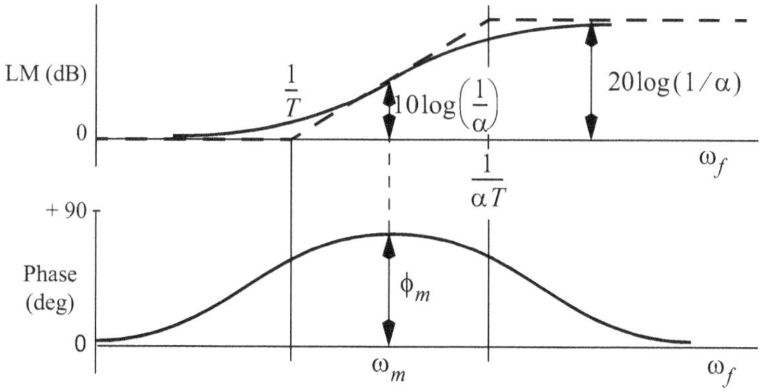

Figure 2.17 Frequency response for a lead transfer function.

The form of the frequency response of the lead compensator is shown in Figure 2.17; the important feature of this response is the phase lead introduced by the compensator. The maximum phase lead, ϕ_m, occurs at the geometric mean of its corners, ω_m, where $\omega_m = \left(\frac{1}{\alpha T} \cdot \frac{1}{T}\right)^{1/2} = \frac{1}{\sqrt{\alpha}T}$. The maximum phase lead can be shown to be $\phi_m = \operatorname{asin}\left(\frac{1-\alpha}{1+\alpha}\right)$. Note that at ω_m the log-magnitude is $LM = 10\log\left(\frac{1}{\alpha}\right)$.

2.12.1.5 Exercise.

Show that the magnitude and phase plots of the lag block,

$$A \cdot G_{LG}(s) = A\frac{1+sT}{1+s(\alpha T)}, \quad (\alpha > 1),$$

are those shown in Figure 2.17 reflected in their respective frequency axes. It is a practical form of integral compensation over a range of frequencies; identify that range.

Example 11. Plotting the frequency response

It is often useful to visualize or sketch the frequency response of a given transfer function based on the straight-line approximations to the frequency responses of the component factors.

Draw the straight-line approximations to the frequency response of the following open-loop transfer function of a unity-feedback control system. Show both the straight-line asymptotes and the actual plot.

$$B(s) = \frac{10^4}{s(s+10)(s+100)}. \tag{2.39}$$

Note the linear factors are in the form $(s+a)$ rather than $(1+sT)$ form required for plotting the asymptotes. By dividing the denominator factors by 10 and 100 the form of (2.39) is changed to $B(s) = \frac{10}{s(1+s0.1)(1+s0.01)}$; note the gain is 10, or 20 dB. Set $s = j\omega_f$.

The corner frequencies of the transfer function are $\omega_c = 1/T$, i.e $1/0.1 = 10$ and $1/0.01 = 100$ rad/s for the two factors $(1+s0.1)^{-1}$ and $(1+s0.01)^{-1}$, respectively. The straight-line representations for the four factors are shown in Figure 2.18. These are combined, making allowance for the deviations of the actual responses from the straight-line approximations as shown in Figure 2.14 and Figure 2.15, to form the response of the transfer function.

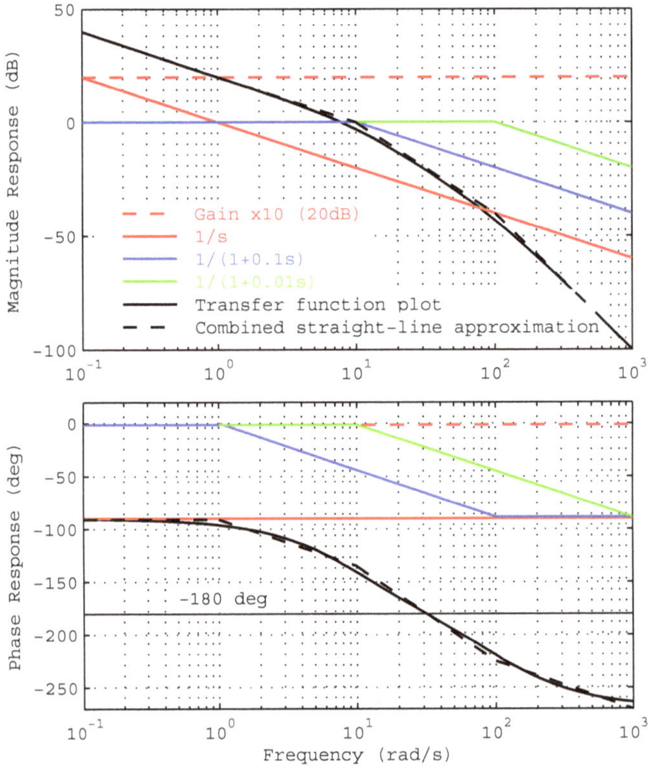

Figure 2.18 Frequency response plots of the transfer function $B(s)$ using straight-line approximations.

2.12.2 Stability Analysis of the closed-loop system from the Bode Plot

The transfer function $W(s)$ of the closed-loop system shown Figure 2.9 was derived in Section 2.7; the transfer function is $W(s) = \dfrac{G(s)}{1+G(s)H(s)}$.

As mentioned earlier, the important feature of the Bode Plot is that the stability of the *closed-loop* system can be derived from the plot of the *open-loop* transfer function $G(s)H(s)$. The theoretical basis for this result requires that no poles or zeros of $G(s)H(s)$ lie in the right-half of the complex *s*-plane, i.e. it is open-loop stable and 'minimum phase' [1]. (If zeros of $G(s)H(s)$ lie in the right-half of the *s*-plane the latter transfer function is called 'non-minimum phase'.) The criterion for the stability of closed-loop systems based on the Bode Plot of open-loop stable transfer functions follows from the more generally applicable Nyquist criterion that covers both open-loop unstable and non-minimum phase systems [1].

Assume the Bode Plot in Figure 2.19 is drawn in the vicinity of the gain cross-over frequency, ω_{co}, for the open-loop transfer function $G(j\omega_f)H(j\omega_f)$. (This analysis may be carried out for the transfer function $B(s)$ in Figure 2.18; $B(s)$ satisfies the condition that it is open-loop stable and minimum phase.)

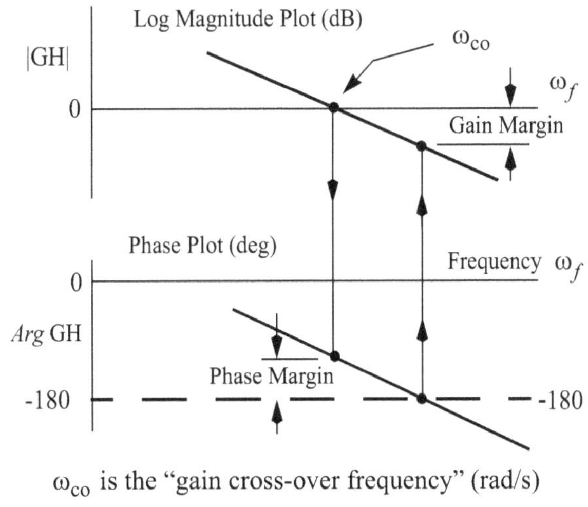

Figure 2.19 Gain and Phase Margins defined on the Bode Plot of the $G(s)H(s)$.

It can then be shown that when the phase shift $\phi(\omega_f) = -180^\circ$, the corresponding value of the LM must be negative for stability.The amount by which the gain can be increased before instability results, is called the 'Gain Margin'.

The term, 'Phase Margin' is defined as the difference between the (-180°) line and the phase plot when the Log Magnitude Plot crosses the zero dB axis, i.e. when $LM = 0$. The Phase Margin for a minimum-phase system must be positive for stability. The Phase Margin can also be interpreted as the amount of phase lag that can be introduced at unity loop-gain before instability of the closed-loop system results.

Example 12. Derive the information on stability and dynamic performance of the closed-loop system from the Bode Plot

Assume that the transfer function $B(s)$ in the Example 11 represents the open-loop transfer function of a unity gain feedback system, i.e. $B(s) = G(s)H(s) = G(s)$. The Gain and Phase Margins for this system are illustrated in Figure 2.20 for the Bode Plot for $G(s)H(s)$. The following information can be extracted from the Plot.

- Because the Gain and Phase Margins are positive the closed-loop system is stable.

- If the gain in the forward-loop is increased by 21 dB the closed-loop system becomes marginally stable and, ideally, it would oscillate with a constant amplitude at a frequency of 32 rad/s. Note that the gain-crossover frequency shifts to the right - increasing in frequency from 8 to 32 rad/s as the magnitude plot shifts vertically upwards.

- Further indicative data on the dynamic performance of the closed-loop system is revealed by the Phase Margin (PM°). A rule of thumb is, if the closed-loop system has a pair of dominant complex poles, the damping ratio of the closed-loop poles is approximately $PM^{\circ}/100$ [1].

For a good servo-system transient response, the Phase Margin should be about 70° for a closed-loop system that has a dominant pair of complex poles. For such a Phase Margin it would be necessary to reduce the gain by 9.3 dB in the case of Figure 2.20. Note the gain crossover frequency is reduced to 3.3 rad/s; this implies that the frequency of the damped sinusoidal response to a step change at the input of the closed-loop system would also be reduced, possibly to 4 - 5 rad/s.

The significance of this example is that it illustrates how a variety of useful information for the *design of the performance of the closed-loop system can be derived from the Bode Plot of the open-loop system.*

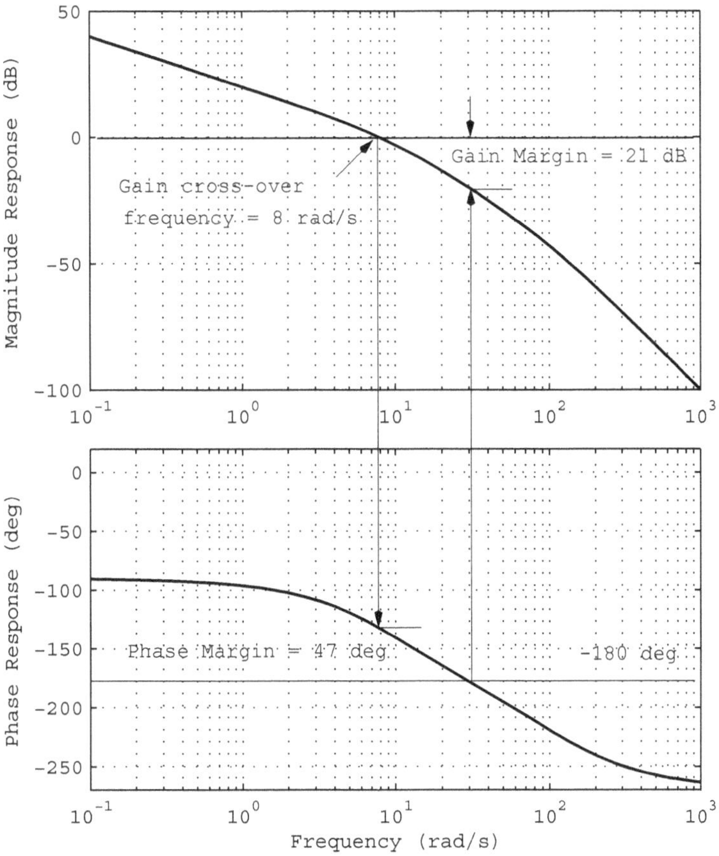

Figure 2.20 Bode Plot of the open-loop transfer function showing the gains and phase margins.

As stated, in applications to power system analysis one should be aware of open-loop systems that are non-minimum phase when using the Bode Plot for stability analysis. An example of such a case is the model of a Francis turbine in a hydro-electric plant. This model contains a right-half plane zero (which causes the turbine power output to rise initially as the wicket gates are closed).

2.13 The Q-filter, a passband filter

The Q-filter is a bandpass filter which passes a selected band of frequencies and attenuates those which lie outside the bandwidth of the filter. The transfer function of the filter is

$$Q(s) = \frac{(2\xi/\omega_m)s}{1 + (2\xi/\omega_m)s + (s/\omega_m)^2}, \tag{2.40}$$

where ω_m (rad/s) is the centre or resonant frequency, i.e. the frequency at which the magnitude of $Q(s)$ is a maximum; ξ is the damping ratio. The 3 dB bandwidth is $\omega_2 - \omega_1 = 2\omega_m\xi$ where $\omega_1\omega_2 = \omega_m^2$; the quality factor of the filter is defined as $Q = 1/(2\xi)$. The frequency response characteristics of the filter are shown in Figure 2.21 as a function of the damping ratio; note that the phase responses pass through zero degrees at resonance. These characteristics are relevant to a type of stabilizer in section Appendix 8–I.3.

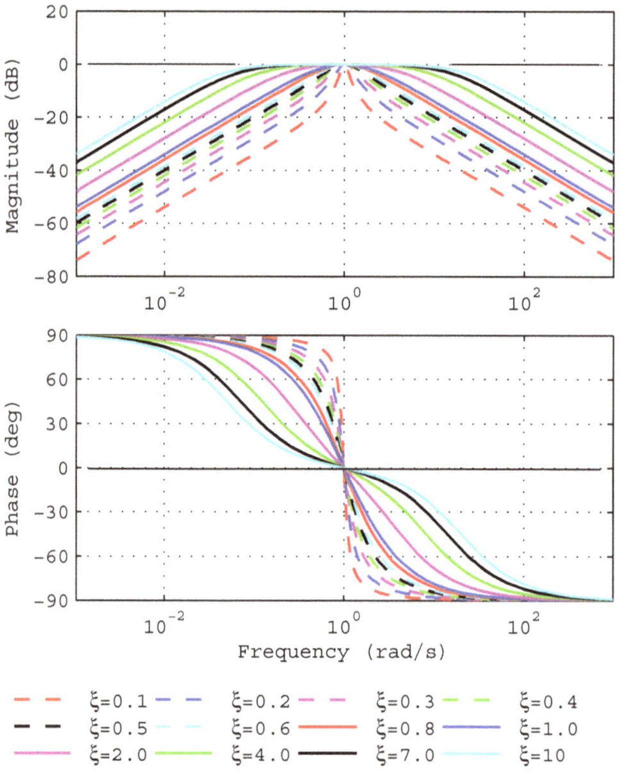

Figure 2.21 Frequency response of the Q-Filter for damping ratios ξ from 0.1 to 10.

2.14 References

[1] G. Franklin, J.D. Powell and A. Emami-Naeini, *Feedback Control of Dynamic Systems*, 6th edition, Prentice Hall, 2009.

[2] R.C. Dorf and R.H. Bishop, *Modern Control Systems,* 12th edition, Prentice Hall, 2010.

[3] P. Kundur, *Power System Stability and Control,* McGraw-Hill, Inc., 1994.

Chapter 3

State equations, eigen-analysis and applications

3.1 Introduction

The description of the dynamics of large systems, such as power systems, by their transfer functions is unsatisfactory for a number of reasons. For example, for a system of order n, say 100, the characteristic polynomial has degree 100 and 101 coefficients of s. Moreover, such systems typically have more than one output variable and more than one input signal. modelling based on the multi-input multi-output state equations of the system is simpler and problems of loss of accuracy are reduced. Moreover, such modelling has a number of advantages and features some of which are described in the following sections. To illustrate the formation of the state equations of a plant or an electro-mechanical system, let us consider two examples.

Much of the material on linear systems analysis in the later sections is covered by [1].

3.1.1 Example 3.1.

Find a set of state and output equations for the simple RLC circuit shown in Figure 3.1. The voltage supplied by an ideal source is $v_s(t)$, and the required outputs are the capacitor voltage are $v_C(t)$ and inductor current $i_L(t)$.

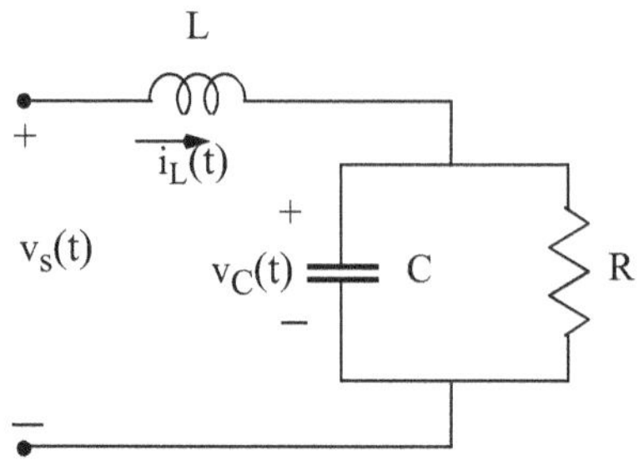

Figure 3.1 RLC circuit

$$\text{Loop voltages:}\quad v_s = v_L + v_C = Lpi_L + v_C \quad\text{or}\quad pi_L = -\frac{v_C}{L} + \frac{v_s}{L}, \tag{3.1}$$

where p is the differential operator d/dt.

$$\text{Current flow:}\quad i_L = \frac{v_C}{R} + Cpv_C \quad\text{or}\quad pv_C = -\frac{v_C}{RC} + \frac{i_L}{C}. \tag{3.2}$$

Note that each of the right-hand equations is a *first-order* differential equation with the *derivative specifically sited on the left-hand side* of the equation.

There are two independent energy storage elements, C and L. Because the instantaneous energy stored in C and L is $\frac{1}{2}Cv_C^2$ and $\frac{1}{2}Li_L^2$, respectively, the variables $x_1 = v_C$ and $x_2 = i_L$ are 'natural' selections for states. Hence, the *state equations* are formed as follows:

From (3.1), $pi_L = -\frac{v_C}{L} + \frac{v_s}{L}$, which becomes $\dot{x}_2 = -\frac{1}{L}x_1 + \frac{1}{L}v_s$,

and, from (3.2), $pv_C = -\frac{v_C}{RC} + \frac{i_L}{C}$, we have $\dot{x}_1 = -\frac{1}{RC}x_1 + \frac{1}{C}x_2$.

The two output equations required are $y_1 = v_C = x_1$ and $y_2 = i_L = x_2$.

The state and output equations can thus be written in matrix form as follows:

$$\begin{bmatrix}\dot{x}_1\\ \dot{x}_2\end{bmatrix} = \begin{bmatrix}-\frac{1}{RC} & \frac{1}{C}\\ -\frac{1}{L} & 0\end{bmatrix}\cdot\begin{bmatrix}x_1\\ x_2\end{bmatrix} + \begin{bmatrix}0\\ \frac{1}{L}\end{bmatrix}\cdot v_s \quad\text{or}\quad \begin{bmatrix}\dot{x}_1\\ \dot{x}_2\end{bmatrix} = \boldsymbol{A}\cdot\begin{bmatrix}x_1\\ x_2\end{bmatrix} + \boldsymbol{b}\cdot v_s,\ \text{and}$$

$$\begin{bmatrix}y_1\\ y_2\end{bmatrix} = \begin{bmatrix}1 & 0\\ 0 & 1\end{bmatrix}\cdot\begin{bmatrix}x_1\\ x_2\end{bmatrix} + \begin{bmatrix}0\\ 0\end{bmatrix}\cdot v_s \quad\text{or}\quad \begin{bmatrix}y_1\\ y_2\end{bmatrix} = \boldsymbol{C}\cdot\begin{bmatrix}x_1\\ x_2\end{bmatrix} + \boldsymbol{d}\cdot v_s,$$

where $A = \begin{bmatrix} -\frac{1}{RC} & \frac{1}{C} \\ -\frac{1}{L} & 0 \end{bmatrix}$, $b = \begin{bmatrix} 0 \\ \frac{1}{L} \end{bmatrix}$, $C = \begin{bmatrix} 1 & 0 \\ 0 & 1 \end{bmatrix}$, and $d = \begin{bmatrix} 0 \\ 0 \end{bmatrix}$.

3.1.2 Example 3.2

A drive system shown in Figure 3.2 consists of a DC motor driving an inertial load through a speed-reducing gearbox. The controlled DC supply voltage to the armature is supplied by a power amplifier. The motor field current is maintained constant (i.e. the flux/pole is constant). Write down the equations of motion for this system.

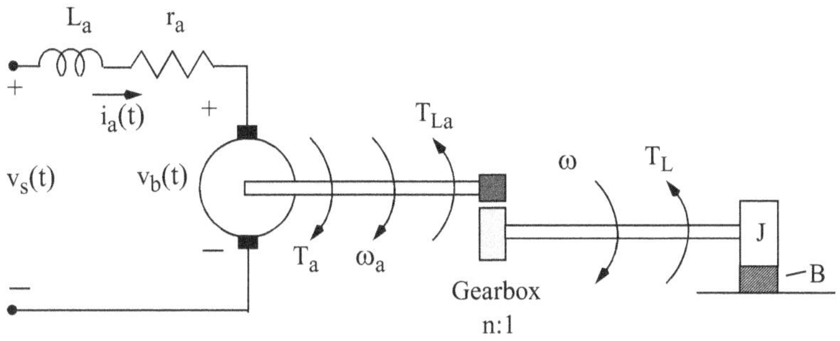

Figure 3.2 Load driven by a DC motor through a gearbox

The equations describing the dynamic behaviour of the system are developed below. The variables, parameters and their units are:

$v_s(t)$ and $v_b(t)$ are respectively the supply voltage and the back emf of the motor (V);

$i_a(t)$ is the armature current (A);

$T_a(t)$ is the motor (armature) torque (Nm);

$T_L(t)$ and $T_{La}(t)$ are load torques referred to the load and motor shafts, respectively (Nm);

$\omega_a(t)$ and $\omega(t)$ are the motor and load speeds (rad/s), n:1 is the gearbox ratio;

$r_a(t)$ and L_a are the resistance (ohm) and self-inductance (Henry) of the armature circuit;

J is the inertia constant of the rotating system (kg-m^2);

B is the coefficient of viscous friction (Nm/rad/s);

k_ω, k_t are constants of proportionality.

Armature circuit equation:	$v_s = r_a i_a + L_a p i_a + v_b$	
Back emf of motor:	$v_b = k_\omega \omega_a$	($v_b \propto$ flux/pole $\times \omega_a$)
Torque developed by motor:	$T_a = k_t i_a$	($T_a \propto$ flux/pole $\times i_a$)

Opposing load torque: $T_L = Jp\omega + B\omega$

Speed of motor shaft: $\omega_a = n\omega$

Power transfer across gearbox: $T_{La}\omega_a = T_L\omega; \quad T_{La} = T_L/n$

Eliminating $T_a(t)$, $T_L(t)$, $T_{La}(t)$, $\omega_a(t)$ and $v_b(t)$ from the above equations, and inserting the derivative terms on the left-hand side of the relevant equations, we find

$$pi_a = -\frac{r_a}{L_a}i_a - \frac{nk_\omega}{L_a}\omega + \frac{v_s}{L_a}, \tag{3.3}$$

$$p\omega = -\frac{B}{J}\omega + \frac{nk_t}{J}i_a. \tag{3.4}$$

Assume the outputs of interest are: i_a, ω, ω_a, T_a.

Let the states be $x_1 = i_a$ and $x_2 = \omega$, then substituting these quantities in (3.3) and (3.4) we find:

$$\begin{bmatrix} \dot{x}_1 \\ \dot{x}_2 \end{bmatrix} = \begin{bmatrix} -\frac{r_a}{L_a} & -\frac{nk_\omega}{L_a} \\ \frac{nk_t}{J} & -\frac{B}{J} \end{bmatrix} \cdot \begin{bmatrix} x_1 \\ x_2 \end{bmatrix} + \begin{bmatrix} \frac{1}{L_a} \\ 0 \end{bmatrix} \cdot v_s(t) \quad \text{or} \quad \begin{bmatrix} \dot{x}_1 \\ \dot{x}_2 \end{bmatrix} = \boldsymbol{A} \cdot \begin{bmatrix} x_1 \\ x_2 \end{bmatrix} + \boldsymbol{b} \cdot v_s(t). \tag{3.5}$$

The outputs of interest can be expressed as:

$$y_1 = i_a, \quad y_2 = \omega, \quad y_3 = \omega_a = n\omega, \quad y_4 = T_a = k_t i_a, \quad \text{then}$$

$$\begin{bmatrix} y_1 \\ y_2 \\ y_3 \\ y_4 \end{bmatrix} = \begin{bmatrix} 1 & 0 \\ 0 & 1 \\ 0 & n \\ k_t & 0 \end{bmatrix} \cdot \begin{bmatrix} x_1 \\ x_2 \end{bmatrix} + \begin{bmatrix} 0 \\ 0 \\ 0 \\ 0 \end{bmatrix} \cdot v_s \quad \text{or} \quad \begin{bmatrix} y_1 \\ y_2 \\ y_3 \\ y_4 \end{bmatrix} = \boldsymbol{C} \cdot \begin{bmatrix} x_1 \\ x_2 \end{bmatrix} + \boldsymbol{d} \cdot v_s, \quad \boldsymbol{d} = \underline{0}. \tag{3.6}$$

Note there are 2 state equations, 1 input, 4 outputs. Thus A is dimension 2×2, b is 2×1, C is 4×2, d is 4×1 and is a zero vector.

Equations (3.5) are called the *state equations* for this second-order system, (3.6) are its *output equations*. There are two independent energy storage elements, the field-circuit inductance and the inertia of the rotating system, L and J. As stated, the energy stored in these elements is $Li_a^2/2$ and $J\omega^2/2$, respectively at any time *t*, and are uniquely determined by the instantaneous values of $i_a(t)$ and $\omega(t)$. Thus $x_1 = i_a$ and $x_2 = \omega$ are thus 'natural' selections for the state variables. In general there are as many states in the system as there are independent energy-storage elements - with the addition of those states and state equations representing pure integration (e.g. the relation $pd = v$ between speed v and distance d).

3.2 The concept of state and the state equations [1]

Definition: The state of a system at any time $t = t_i$ is a minimum set of numbers which, together with the input function $\boldsymbol{u}(t)$ for $t \geq t_i$ and the equations describing the dynamics, are sufficient to determine the future behaviour of the states and the output of the system.

The state of the system at time t is described by a vector of n state variables $\left[x_1(t)\ x_2(t)\ \ldots\ x_n(t)\right]^T$. A knowledge of the initial values of the state variables $\left[x_1(t_0)\ x_2(t_0)\ \ldots\ x_n(t_0)\right]^T$ at time t_0 and the m input signals $\left[u_1(t)\ u_2(t)\ \ldots\ u_m(t)\right]^T$ for $t \geq t_0$, is sufficient to determine the future values of the state variables and the output.

These concepts lead to the *state-space model* of a system in terms of a set of *n first-order differential equations* (in contrast to a *single* n^{th} - order differential equation from which the conventional transfer function is derived). Recall that in each first-order equation the derivative of the state variable is placed on the left-hand side of the equation and all terms in the state variables and inputs on the right-hand side. A general form of the n state equations, with m input signals, is thus

$$\begin{aligned}
\dot{x}_1 &= a_{11}x_1 + a_{12}x_2 + \ldots + a_{1n}x_n + b_{11}u_1 + b_{12}u_2 + \ldots + b_{1m}u_m \\
\dot{x}_2 &= a_{21}x_1 + a_{22}x_2 + \ldots + a_{2n}x_n + b_{21}u_1 + b_{22}u_2 + \ldots + b_{2m}u_m \\
&\vdots \qquad\qquad\qquad\qquad\qquad \vdots \\
\dot{x}_n &= a_{n1}x_1 + a_{n2}x_2 + \ldots + a_{nn}x_n + b_{n1}u_1 + b_{n2}u_2 + \ldots + b_{nm}u_m
\end{aligned} \tag{3.7}$$

The p output equations are:

$$\begin{aligned}
y_1 &= c_{11}x_1 + c_{12}x_2 + \ldots + c_{1n}x_n + d_{11}u_1 + d_{12}u_2 + \ldots + d_{1m}u_m \\
&\vdots \qquad\qquad\qquad\qquad\qquad \vdots \\
y_p &= c_{p1}x_1 + c_{p2}x_2 + \ldots + c_{pn}x_n + d_{p1}u_1 + d_{p2}u_2 + \ldots + d_{pm}u_m
\end{aligned} \tag{3.8}$$

A more compact arrangement of the above equations is the matrix form:

$$\begin{aligned}
\dot{\boldsymbol{x}}(t) &= \boldsymbol{A}\boldsymbol{x}(t) + \boldsymbol{B}\boldsymbol{u}(t) \\
\boldsymbol{y}(t) &= \boldsymbol{C}\boldsymbol{x}(t) + \boldsymbol{D}\boldsymbol{u}(t)
\end{aligned}, \tag{3.9}$$

$$\text{where} \quad \boldsymbol{A} = \begin{bmatrix} a_{11} & a_{12} & \cdots & a_{1n} \\ a_{21} & a_{22} & \cdots & a_{2n} \\ \cdot & \cdot & \cdot & \cdot \\ a_{n1} & a_{n2} & \cdots & a_{nn} \end{bmatrix}, \quad \boldsymbol{B} = \begin{bmatrix} b_{11} & b_{12} & \cdots & b_{1m} \\ b_{21} & b_{22} & \cdots & b_{2m} \\ \cdot & \cdot & \cdot & \cdot \\ b_{n1} & b_{n2} & \cdots & b_{nm} \end{bmatrix}, \tag{3.10}$$

of dimension $n \times n$ and $n \times m$, respectively, and

$$\text{where} \quad \boldsymbol{C} = \begin{bmatrix} c_{11} & c_{12} & \cdots & c_{1n} \\ c_{21} & c_{22} & \cdots & c_{2n} \\ \cdot & \cdot & \cdot & \cdot \\ c_{p1} & c_{p2} & \cdots & c_{pn} \end{bmatrix}, \quad \boldsymbol{D} = \begin{bmatrix} d_{11} & d_{12} & \cdots & d_{1m} \\ d_{21} & d_{22} & \cdots & d_{2m} \\ \cdot & \cdot & \cdot & \cdot \\ d_{p1} & d_{p2} & \cdots & d_{nm} \end{bmatrix}, \tag{3.11}$$

of dimension $p \times n$ and $p \times m$.

The $n \times 1$ state vector is $\quad \boldsymbol{x}(t) = \left[x_1(t)\ x_2(t)\ \ldots\ x_n(t)\right]^T,$

the $m \times 1$ input vector is $\quad \boldsymbol{u}(t) = \left[u_1(t)\ u_2(t)\ \ldots\ u_m(t)\right]^T$, and

the $p \times 1$ output vector is $\quad \boldsymbol{y}(t) = \left[y_1(t)\ y_2(t)\ \ldots\ y_p(t)\right]^T.$

Examples of state and output equations have been given in Examples 1 and 2. (Also see [1], [2] and [3].)

3.3 The linearized model of the non-linear dynamic system

The equations describing the power system and its dynamics are non-linear. For analyzing the dynamic performance of the non-linear plant and the system, typically following a large disturbance such as a fault, a step-by-step integration of the non-linear equations is carried out to calculate the time-domain responses of the system variables. Such variables are generator speeds and rotor angles, bus voltages, controller outputs, etc. Because the dynamic behaviour of the system depends very much on the location and the severity of the disturbance, as well as the operating conditions, it is necessary to conduct a large number of so-called transient stability studies to characterise the dynamics of the system [4].

Linearizing the set of non-linear equations for a selected operating condition results in a *new* set of equations in a *new* set of variables. These variables are the perturbations $\Delta\boldsymbol{x}$ about the steady-state quantities $\boldsymbol{x}_0$. The variable $\boldsymbol{x}$ in the non-linear equations is related to the former pair by $\boldsymbol{x} = \Delta\boldsymbol{x} + \boldsymbol{x}_0$. The advantages of forming the linearized equations of a system are:

- All the powerful analytical methods developed in linear control theory are available for the analysis of the linearized dynamic system.

- If the linearized system is stable at the selected steady-state operating point then, according to a theorem by Poincaré [5], the non-linear system is also stable at that operating point.

- The dynamic performance of the linearized system can be characterized by the location of its poles in the complex s-plane. Based on results in the theory covered in Sections 2.8 and 2.9, the real parts of these poles (or damping constants) provide the information on stability, how well damped the modes [1] are, the nature of the transient response, etc. Such information cannot be gleaned directly from the results of time-domain analysis of the non-linear system.

If the transient responses of the linearized system to a disturbance are calculated, a question is 'how accurate are the responses'. Poincaré proved that the response of the linearized system to a disturbance is exactly the same as that of the original non-linear system if the disturbance is vanishingly small. As explained in Section 1.1, the responses predicted by the linearized model are often sufficiently accurate for practical analysis and design purposes. However, care must be exercised to take into account the nature of the non-linearities, the operating point and the size of the perturbation when deciding if the linearized model is practically applicable to the analysis being performed.

3.3.1 Linearization procedure

In [6] the set of nonlinear differential-algebraic equations (DAEs) describing the dynamic behaviour of the integrated power system are derived and are shown to be of the form:

$$\dot{x} = f(x, \varepsilon, u), \; 0 = g(x, \varepsilon, u), \; y = h(x, \varepsilon),$$

where the vector x represents the n states of the system, ε the r algebraic variables, u the m system input variables, and y the p output variables.

At the steady-state operating condition, which is the equilibrium point about which the system is to be linearized, implies by definition that all rates of change are zero, $\dot{x} = 0$, thus

$$f(x_0,\varepsilon_0,u_0) = 0, \quad 0 = g(x_0,\varepsilon_0,u_0), \quad y_0 = h(x_0,\varepsilon_0). \tag{3.12}$$

Assume the system is subjected to a small perturbation from the steady state such that

$$x = x_0 + \Delta x, \quad \varepsilon = \varepsilon_0 + \Delta\varepsilon, \quad u = u_0 + \Delta u, \quad y = y_0 + \Delta y. \tag{3.13}$$

The perturbed the variables must satisfy (3.12). For example, in the case of the output y in (3.13)

$$y = y_0 + \Delta y = h(x_0 + \Delta x, \varepsilon_0 + \Delta\varepsilon).$$

Because the perturbations are small, the nonlinear function $y = h(x,\varepsilon)$ can be expressed as a first-order Taylor's series expansion. Consider the ith output, y_i, $i = 1,\ldots,p$:

---

1. See a note on eigenvalues, modes and stability in Section 3.5.2.

$$y_i = y_{i0} + \Delta y_i = h_i(\boldsymbol{x}_0, \boldsymbol{\varepsilon}_0) + \frac{\partial h_i}{\partial x_1}\Delta x_1 + \ldots + \frac{\partial h_i}{\partial x_n}\Delta x_n + \frac{\partial h_i}{\partial \varepsilon_1}\Delta \varepsilon_1 + \ldots + \frac{\partial h_i}{\partial \varepsilon_r}\Delta \varepsilon_r$$

where the partial derivatives $\frac{\partial h_i}{\partial x_a}$, $a = 1,\ldots,n$; and $\frac{\partial h_i}{\partial \varepsilon_b}$, $b = 1,\ldots,r$ are evaluated at the initial steady-state operating point $(\boldsymbol{x}_0,\boldsymbol{\varepsilon}_0,\boldsymbol{u}_0)$.

Because $y_{i0} = h_i(\boldsymbol{x}_0, \boldsymbol{\varepsilon}_0)$ in the above equation, it reduces to an equation in the perturbed variables:

$$\Delta y_i = \frac{\partial h_i}{\partial x_1}\Delta x_1 + \ldots + \frac{\partial h_i}{\partial x_n}\Delta x_n + \frac{\partial h_i}{\partial \varepsilon_1}\Delta \varepsilon_1 + \ldots + \frac{\partial h_i}{\partial \varepsilon_r}\Delta \varepsilon_r . \qquad (3.14)$$

Similar expressions can be derived for the two remaining functions in (3.12), for $j=1, \ldots, n$:

$$\Delta \dot{x}_j = \sum_{a=1}^{n} \frac{\partial f_j}{\partial x_a}\Delta x_a + \sum_{b=1}^{r} \frac{\partial f_j}{\partial \varepsilon_b}\Delta \varepsilon_b + \sum_{c=1}^{m} \frac{\partial f_j}{\partial u_c}\Delta u_c , \qquad (3.15)$$

$$0 = \sum_{a=1}^{n} \frac{\partial g_k}{\partial x_a}\Delta x_a + \sum_{b=1}^{r} \frac{\partial g_k}{\partial \varepsilon_b}\Delta \varepsilon_b + \sum_{c=1}^{m} \frac{\partial g_k}{\partial u_c}\Delta u_c . \qquad (3.16)$$

The sets of linearized equations, (3.14) to (3.16) are more conveniently expressed in matrix form,

$$\begin{aligned} \Delta \dot{\boldsymbol{x}} &= \boldsymbol{J}_{fx}\Delta \boldsymbol{x} + \boldsymbol{J}_{f\varepsilon}\Delta \boldsymbol{\varepsilon} + \boldsymbol{J}_{fu}\Delta \boldsymbol{u} \\ \boldsymbol{0} &= \boldsymbol{J}_{gx}\Delta \boldsymbol{x} + \boldsymbol{J}_{g\varepsilon}\Delta \boldsymbol{\varepsilon} + \boldsymbol{J}_{gu}\Delta \boldsymbol{u} \\ \Delta \boldsymbol{y} &= \boldsymbol{J}_{hx}\Delta \boldsymbol{x} + \boldsymbol{J}_{h\varepsilon}\Delta \boldsymbol{\varepsilon} + \boldsymbol{J}_{hu}\Delta \boldsymbol{u} \end{aligned} \qquad (3.17)$$

or more compactly:

$$\begin{bmatrix} \Delta \dot{\boldsymbol{x}} \\ \boldsymbol{0} \\ \Delta \boldsymbol{y} \end{bmatrix} = \begin{bmatrix} \boldsymbol{J}_{fx} & \boldsymbol{J}_{f\varepsilon} & \boldsymbol{J}_{fu} \\ \boldsymbol{J}_{gx} & \boldsymbol{J}_{g\varepsilon} & \boldsymbol{J}_{gu} \\ \boldsymbol{J}_{hx} & \boldsymbol{J}_{h\varepsilon} & \boldsymbol{J}_{hu} \end{bmatrix} \begin{bmatrix} \Delta \boldsymbol{x} \\ \Delta \boldsymbol{\varepsilon} \\ \Delta \boldsymbol{u} \end{bmatrix} \qquad (3.18)$$

where the $(i, a)^{th}$ element of the $\boldsymbol{J}_{fx}$ sub-matrix of the system Jacobian matrix is $\frac{\partial f_j}{\partial x_a}$ evaluated at the initial steady-state operating point. The elements of the other sub-matrices are similarly defined.

The formulation of the linearized equations of the system as a set of DAEs possesses a number of significant advantages, including:.

- The 'natural' formulation of the equations for devices and their controllers is exploited when building the set of system equations;

- The equations for large systems are inherently highly modular and extremely sparse (i.e. the Jacobian matrix contains very large numbers of zeros). Highly efficient computational algorithms for processing modular and sparse matrices are exploited when computing frequency responses, transfer-function residues - as well as computing a subset of the system eigenvalues within a selected region of the complex s-plane.

- The modularity and sparsity of the system equations can be exploited when computing eigenvalue sensitivities.

Elimination of the algebraic variables from the DAEs yields the conventional 'ABCD' form of the state equations, i.e.

$$\Delta\dot{\boldsymbol{x}}(t) = \boldsymbol{A}\Delta\boldsymbol{x}(t) + \boldsymbol{B}\Delta\boldsymbol{u}(t), \quad \Delta\boldsymbol{y}(t) = \boldsymbol{C}\Delta\boldsymbol{x}(t) + \boldsymbol{D}\Delta\boldsymbol{u}(t), \tag{3.19}$$

$$\text{,where} \quad \begin{array}{ll} \boldsymbol{A} = \boldsymbol{J}_{fx} - \boldsymbol{J}_{f\varepsilon}\boldsymbol{J}_{g\varepsilon}^{-1}\boldsymbol{J}_{gx} & \boldsymbol{B} = \boldsymbol{J}_{fu} - \boldsymbol{J}_{f\varepsilon}\boldsymbol{J}_{g\varepsilon}^{-1}\boldsymbol{J}_{gu} \\ \boldsymbol{C} = \boldsymbol{J}_{hx} - \boldsymbol{J}_{h\varepsilon}\boldsymbol{J}_{g\varepsilon}^{-1}\boldsymbol{J}_{gx} & \boldsymbol{D} = \boldsymbol{J}_{hu} - \boldsymbol{J}_{h\varepsilon}\boldsymbol{J}_{g\varepsilon}^{-1}\boldsymbol{J}_{gu} \end{array}. \tag{3.20}$$

Equations (3.19) are the state and output equations and are a form commonly used in the literature on linear control theory to describe a system.

3.4 Solution of the State Equations

The solution of the state and output equations (3.19) in matrix form can be obtained by taking the Laplace Transform of each of its first-order differential equations, i.e.

$$s\boldsymbol{X}(s) - \boldsymbol{x}(0) = \boldsymbol{A}\boldsymbol{X}(s) + \boldsymbol{B}\boldsymbol{U}(s), \quad \text{or}$$

$$(s\boldsymbol{I} - \boldsymbol{A})\boldsymbol{X}(s) = \boldsymbol{x}(0) + \boldsymbol{B}\boldsymbol{U}(s), \tag{3.21}$$

where $\boldsymbol{I}$ is the nth order identity matrix. Pre-multiplying both sides by $(s\boldsymbol{I} - \boldsymbol{A})^{-1}$, we obtain:

$$\boldsymbol{X}(s) = (s\boldsymbol{I} - \boldsymbol{A})^{-1}\boldsymbol{x}(0) + (s\boldsymbol{I} - \boldsymbol{A})^{-1}\boldsymbol{B}\boldsymbol{U}(s). \tag{3.22}$$

The first term on the right of (3.22) is the *natural response* and the second is the *forced response.* The time-domain solution is of the form:

$$\boldsymbol{x}(t) = e^{\boldsymbol{A}t}\boldsymbol{x}(0) + \int_0^t e^{\boldsymbol{A}(t-\tau)}\boldsymbol{B}\boldsymbol{u}(\tau)d\tau \qquad \boldsymbol{y}(t) = \boldsymbol{C}\boldsymbol{x}(t) + \boldsymbol{D}\boldsymbol{u}(t).$$

The solution involves the matrix-exponential $e^{\boldsymbol{A}t}$, whose numerical computation for high order systems is very challenging [7]. The following analysis reveals important theoretical aspects of the time response. However, it is emphasized that these methods are not employed in practice to numerically calculate time-responses.

3.4.1 The Natural Response

The Natural Response is derived from (3.22):

$$\begin{aligned} X(s) &= (sI-A)^{-1}x(0) \quad \text{or} \quad x(t) = e^{At}x(0). \\ Y(s) &= CX(s) \quad \text{or} \quad y(t) = Cx(t) \end{aligned} \tag{3.23}$$

The matrix $(sI-A)^{-1}$ is known as the *resolvent matrix*. The solution for $X(s)$ in (3.23) can be expressed in the form:

$$X(s) = (sI-A)^{-1}x(0) = \frac{Adjoint(sI-A)}{Determinant(sI-A)}x(0). \tag{3.24}$$

Let $B = Adjoint(A)$. The $(i,j)^{th}$ element of B, b_{ij} is defined as $b_{ij} = (-1)^{(i+j)}det(M_{ji})$ where M_{ji} is the $(j, i)^{th}$ minor of A; M is the $(n-1) \times (n-1)$ matrix obtained by deleting the jth row and i^{th} column of A [1].

3.4.2 Example 3.3

Let $A = \begin{bmatrix} 1 & -3 & 4 \\ -2 & 1 & 3 \\ 4 & 2 & 3 \end{bmatrix}$; compute $B = Adjoint(A)$.

$$b_{11} = (-1)^{(1+1)}det\left(\begin{bmatrix} 1 & 3 \\ 2 & 3 \end{bmatrix}\right) = (1)(1 \times 3 - 3 \times 2) = -3,$$

$$b_{12} = (-1)^{(1+2)}det\left(\begin{bmatrix} -3 & 4 \\ 2 & 3 \end{bmatrix}\right) = (-1)(-3 \times 3 - 4 \times 2) = 17, \text{ and so on, to yield}$$

$$B = \begin{bmatrix} -3 & 17 & -13 \\ 18 & -13 & -11 \\ -8 & -14 & -5 \end{bmatrix}.$$

For a solution which is non-trivial it is necessary that:

$$Determinant(sI-A) = 0. \tag{3.25}$$

This is also the *characteristic equation* of the system; the zeros of (3.25) are the poles of the system defined by the state matrix A. This is an important result because it reveals that the characteristic dynamic behaviour of system is encapsulated in the state matrix. Unfortunately, the solution for $X(s)$ based on (3.24) becomes unmanageable for fourth- and higher-order systems.

We can write the natural response, $x(t) = e^{At}x(0)$ in the form $x(t) = \Phi(t-t_0)x(0)$ where $\Phi(t-t_0)$ is known as the $n \times n$ *state transition matrix*. It describes the transition of the n states over the interval $t_0 \to t$ and is equal to e^{At}. If $t_0 = 0$, $\Phi(t-t_0)$ is replaced by $\Phi(t)$,

$$x(t) = \Phi(t)x(0). \tag{3.26}$$

We note, by examination of (3.24) and (3.26), that the Laplace Transform of $\Phi(t)$ is $\phi(s) = (sI - A)^{-1}$. Hence $\Phi(t)$ can be determined by evaluating the resolvent matrix, $(sI - A)^{-1}$, and taking the inverse Laplace Transform.

3.4.3 Example 3.4: Natural response

The state, input, output and direct-transmission matrices in this example are:

$$A = \begin{bmatrix} 0 & -2 \\ 1 & -3 \end{bmatrix}, \; B = \begin{bmatrix} 0 \\ 1 \end{bmatrix}, \; C = \begin{bmatrix} 2 & 0 \\ 0 & 1 \end{bmatrix}, \; D = 0 \text{ respectively.}$$

(a) Calculate the state transition matrix $\Phi(t)$ for the state-matrix.

Firstly, we evaluate the resolvent matrix $(sI - A)^{-1}$ using the result for the inverse given in (3.24). The Adjoint (Adj) matrix is the transpose of the matrix formed by replacing the elements a_{ij} by their cofactors (signed minors). Thus,

$$(sI - A) = \begin{bmatrix} s & 2 \\ -1 & (s+3) \end{bmatrix}, \quad \text{and} \quad \det(sI - A) = s^2 + 3s + 2 = (s+1)(s+2).$$

Hence
$$\phi(s) = (sI - A)^{-1} = \begin{bmatrix} \dfrac{s+3}{(s+1)(s+2)} & \dfrac{-2}{(s+1)(s+2)} \\ \dfrac{1}{(s+1)(s+2)} & \dfrac{s}{(s+1)(s+2)} \end{bmatrix}.$$

Note the poles of the system, $s_{1,2} = -1, -2$, are associated with each element of the matrix.

Taking the inverse Laplace Transform, we find: $\Phi(t) = e^{At} = \begin{bmatrix} 2e^{-t} - e^{-2t} & -2e^{-t} + 2e^{-2t} \\ e^{-t} - e^{-2t} & -e^{-t} + 2e^{-2t} \end{bmatrix}$.

(b) Find the responses of the outputs $y = \begin{bmatrix} y_1(t) \\ y_2(t) \end{bmatrix} = Cx = \begin{bmatrix} 2 & 0 \\ 0 & 1 \end{bmatrix} \cdot \begin{bmatrix} x_1(t) \\ x_2(t) \end{bmatrix}$ when $x_1(0) = 1$ and $x_2(0) = -1$.

The responses of the two states are $\begin{bmatrix} x_1(t) \\ x_2(t) \end{bmatrix} = \begin{bmatrix} \Phi_{11} & \Phi_{12} \\ \Phi_{21} & \Phi_{22} \end{bmatrix} \cdot \begin{bmatrix} x_1(0) \\ x_2(0) \end{bmatrix}$, i.e.

$$x_1(t) = (2e^{-t} - e^{-2t})x_1(0) + (-2e^{-t} + 2e^{-2t})x_2(0) = 4e^{-t} - 3e^{-2t}$$

$$x_2(t) = (e^{-t} - e^{-2t})x_1(0) + (-e^{-t} + 2e^{-2t})x_2(0) = 2e^{-t} - 3e^{-2t}$$

The output responses are thus: $$\begin{aligned} y_1(t) &= 2x_1(t) = 8e^{-t} - 6e^{-2t} \\ y_2(t) &= x_2(t) = 2e^{-t} - 3e^{-2t} \end{aligned}.$$

Note for this second-order system there are two eigenvalues in the responses, e^{-t} and e^{-2t}.

3.4.4 The Forced Response

Assuming all initial conditions are zero in (3.22), the state and output responses are:

$X(s) = (sI - A)^{-1}BU(s)$, $Y(s) = CX(s) + DU(s)$. Eliminating X(s), we find:

$$Y(s) = G(s)U(s) \quad \text{or} \quad Y(s) = [C(sI - A)^{-1}B + D]U(s). \tag{3.27}$$

From (3.27) the multi-input, multi-output (MIMO) transfer function is defined as:

$$G(s) = C(sI - A)^{-1}B + D = \begin{bmatrix} G_{11}(s) & \dots & G_{1m}(s) \\ \dots & \dots & \dots \\ G_{p1}(s) & \dots & G_{pm}(s) \end{bmatrix}.$$

Clearly, if there are no *pure gain* paths directly between input and output, then $D = 0$. The MIMO transfer function is then:

$$G(s) = C(sI - A)^{-1}B. \tag{3.28}$$

3.4.5 Example 3.4 (continued).

(c) Find G(s) given: $(sI - A)^{-1} = \dfrac{1}{(s+1)(s+2)}\begin{bmatrix} s+3 & -2 \\ 1 & s \end{bmatrix}$, $B = \begin{bmatrix} 0 \\ 1 \end{bmatrix}$.

The elements of C are given in part (b) of Example 3.4. Hence, by substitution in (3.28),

$$G(s) = \frac{1}{(s+1)(s+2)}\begin{bmatrix} 2 & 0 \\ 0 & 1 \end{bmatrix} \cdot \begin{bmatrix} s+3 & -2 \\ 1 & s \end{bmatrix} \cdot \begin{bmatrix} 0 \\ 1 \end{bmatrix} = \begin{bmatrix} \dfrac{-4}{(s+1)(s+2)} \\ \dfrac{s}{(s+1)(s+2)} \end{bmatrix}.$$

3.5 Eigen-analysis

3.5.1 The eigenvalues of the state matrix, A

The hth eigenvalue of a real, $n \times n$ matrix A is the real or complex scalar quantity, λ_h, which is the non-trivial solution of the equation

$$Av_h = \lambda_h v_h. \tag{3.29}$$

The n-element column vector, $v_h = \begin{bmatrix} v_{1h} & v_{2h} & \dots & v_{nh} \end{bmatrix}^T$, is called the *right* eigenvector of the matrix A corresponding to the eigenvalue λ_h.

To calculate the eigenvalue, let us rewrite (3.29) in the form

$$(\lambda_h I - A) v_h = 0.$$

For a solution which is non-trivial, $Determinant(\lambda_h I - A) = 0$.

This is also the *characteristic equation* of the system and is the same as (3.25) with s replaced by λ_h. *The eigenvalues of the state matrix A are thus the same as the poles of the transfer function,* and are independent of which variables in the system are selected to be inputs or outputs. For the nth-order system there are n eigenvalues which are real or exist in complex-conjugate pairs.

Likewise, there exists a n-element row vector, $w_h = \begin{bmatrix} w_{h1} & w_{h2} & \dots & w_{hn} \end{bmatrix}$, which satisfies the equation

$$w_h A = w_h \lambda_h; \tag{3.30}$$

w_h is called the *left* eigenvector of the matrix A corresponding to the eigenvalue λ_h.

Let us form the $n \times n$ matrices of right and left eigenvectors V and W, respectively, corresponding to the eigenvalues $\lambda_1, \lambda_2, \dots, \lambda_h, \dots, \lambda_n$, i.e.

$$V = \begin{bmatrix} v_1 & v_2 & \dots & v_n \end{bmatrix}, \text{ and } W = \begin{bmatrix} w_1^T & w_2^T & \dots & w_n^T \end{bmatrix}^T. \tag{3.31}$$

Also let us form the diagonal matrix Λ of eigenvalues, $diag\{\lambda_h\}$, $h = 1 \dots n$.

Equations (3.29) and (3.30) can then be expressed respectively as

$$AV = V\underline{\Lambda}, \text{ and } \quad WA = \underline{\Lambda}W.$$

If V is non-singular, which is usually the case for realistic power system models, then:

$$V^{-1}AV = \underline{\Lambda} \text{ and } WA = V^{-1}AVW,$$

from which we may conclude that $W = V^{-1}$ or $WV = I$.

The latter equation reveals a useful orthogonality property of eigenvectors corresponding to a selected eigenvalue, λ_h, namely

$$w_h v_h = 1, \tag{3.32}$$

whereas, for eigenvectors corresponding to different eigenvalues, λ_g and λ_h, $w_g v_h = 0$.

V and W are respectively known as the right and left modal matrices of the state matrix A.

3.5.2 A note on eigenvalues, modes and stability

As described in Section 3.5.1, an eigenvalue λ_i of the $n \times n$ system matrix A is a real or complex number. The system matrix has *n* eigenvalues, some real and some in complex-conjugate pairs. Each real eigenvalue (σ_1) is associated with a monotonic mode which in the time-domain has the form $y_1(t) = A_1 e^{\sigma_1 t}$. Each complex-conjugate pair of eigenvalues ($\sigma_2 \pm j\omega$) is associated with an oscillatory mode which in the time-domain has the form $y_2(t) = A_2 e^{\sigma_2 t} \sin(\omega t + \phi_2)$. The unforced response of a linear time-invariant system is a (weighted) superposition of the response of each of the system modes. In power systems analysis the term *mode* usually refers to a broader set of properties than just the damping and frequency of oscillation in order to characterise more completely the physical behaviour of the natural modes of system response *in the time domain.* Other modal characterisations include, for example, whether it is an electro-mechanical mode, or a controller mode, etc. In the case of electro-mechanical modes we refer to sub-classifications such as inter-area modes, local modes, etc. (see Section 1.5). In conjunction with the engineer's detailed knowledge of the system structure, the eigen-decomposition of the system, including eigenvalues, eigenvectors and participation factors, is a tool employed to characterize the system modes. In the following text the term *mode* is, at times, used instead of *eigenvalue.* Note that there are occasions when the symbol λ^h refers to either the hth mode or the hth eigenvalue; the application depends on the context. See Section 9.1.1 for further details.

It has been noted in Section 2.9 that the system is unstable if a pole or a pair of complex poles lies in the right-half of the *s*-plane. Correspondingly, instability arises if the real part of *any* eigenvalue is positive; the associated mode therefore increases exponentially with time.

3.6 Decoupling the state equations

Let us consider the response of the states in the state equations of (3.9) when the system responds to a set of initial conditions with zero input. Substituting $x = Vz$ and $\dot{x} = V\dot{z}$ in the resulting state equation,

$$\dot{x}(t) = Ax(0)\text{, we find} \tag{3.33}$$

$$\dot{z} = V^{-1}AVz(0) = \underline{\Lambda}z(0). \tag{3.34}$$

Cross-coupling between states exists in the state equations (3.33) which represent the physical system. However, because the matrix Λ is a diagonal matrix of eigenvalues, no cross-coupling terms exist in (3.34); the latter equation is a said to be a *decoupled* form of the state equations, with pseudo-states $z(t)$.

In the above analysis we have invoked two transformations,

$$x = Vz \quad \text{and} \quad z(0) = V^{-1}x(0) = Wx(0).$$

3.7 Determination of residues from the state equations

In the following sections much of the material relating to the applications to power system dynamics is covered in [8] to [12].

The full set of state equations (3.9) can be expressed in a decoupled form by letting $x = Vz$. The resulting state equations in the new state vector z are:

$$\dot{z} = \underline{\Lambda} \cdot z + \hat{B} \cdot U, \qquad y = \hat{C} \cdot z + \hat{D} \cdot U, \tag{3.35}$$

where $\underline{\Lambda} = WAV$ is a diagonal matrix of the system eigenvalues,

$\hat{B} = WB$ is the mode-controllability matrix,

$\hat{C} = CV$ is the mode-observability matrix, and

$\hat{D} = D$ (usually a null matrix in power system applications).

A mode λ_h is observable in an output signal y_m if and only if $c_{mh} \neq 0$. Consequently, $\hat{c}_{mh} = c_{m\circ} v_{\circ h}$ is a measure of the *observability* of the mode in the output y_m. Similarly, $\hat{b}_{hq} = w_{h\circ} \hat{b}_{\circ q}$ is a measure of the *controllability* of the mode from an input u_q.

Assume that there are n distinct eigenvalues. The transfer function between the qth input and the mth output is

$$\frac{y_m(s)}{u_q(s)} = \hat{c}_{m\circ}(sI - \underline{\Lambda})^{-1}\hat{b}_{\circ q}$$

$$= \hat{c}_{m\circ}\, dag\left\{\frac{1}{s-\lambda_1}, \frac{1}{s-\lambda_2}, \ldots, \frac{1}{s-\lambda_n}\right\}\hat{b}_{\circ q}$$

$$= \sum_{h=1}^{n} \frac{\hat{c}_{mh} \cdot \hat{b}_{hq}}{s - \lambda_h}$$

The form of the transfer function displayed in the last equation is the same as that in (2.14) in which K_h is the residue associated with the eigenvalue at $s = \lambda_h$. Here, the residue for the eigenvalue $s = \lambda_h$ is defined as

$$r_{mq}^{h} = \hat{c}_{mh} \cdot \hat{b}_{hq}, \tag{3.36}$$

which, according to the above definitions, is a combined measure of observability and controllability.

Based on the concepts of controllability the mode of concern is 'highly' controllable from the qth input of the state-space model of the system if the magnitude of $\hat{b}_{hq}$ is large relative to that evaluated at all other inputs - subject to a note of caution [1]. Likewise, the signal at m^{th} output of the model is 'highly' observable - and is suitable as a feedback or stabilizing signal for the selected mode - if the magnitude of $\hat{c}_{mh}$ is large relative to the values found for other possible feedback or stabilizing signals. Hence, for a candidate controller or a stabilizer in the path from the mth output to the qth input to be effective, *the magnitude of the residue* r_{mq}^{h} *must be relatively large* at the modal frequency λ_h.

3.8 Determination of zeros of a SISO sub-system

Although power systems are MIMO systems there is often interest in the analysis of a single-input, single-output (SISO) subsystem. Consider the SISO from the ith input to the jth output of the MIMO system described by the general set of state-equations in (3.9). Let $\boldsymbol{b}_i$ be the ith column of the input matrix $\boldsymbol{B}$, $\boldsymbol{c}_j$ the jth row of the output matrix $\boldsymbol{C}$ and d_{ij} be the ij^{th} element of the direct-transmission matrix. The resulting SISO sub-system is described by:

$$\begin{aligned} \dot{\boldsymbol{x}}(t) &= \boldsymbol{A}\boldsymbol{x}(t) + \boldsymbol{b}_i u_i(t) \\ y_j(t) &= \boldsymbol{c}_j \boldsymbol{x}(t) + d_{ij} u_i(t) \end{aligned} \tag{3.37}$$

in which $u_i(t)$ and $y_j(t)$ are respectively the ith and j^{th} input and output.

The objective is to find the set of zeros of the SISO subsystem: that is, complex frequencies ψ_h, $h = 1,\ldots,m \le n$ for which the forced response to the non-zero input $u_{ih}(t) = u_{ih0}e^{\psi_h t}$ results in the output $y_{jh}(t) \equiv 0$ for all time. It is also assumed that none of the zeros are equal to any of the system poles, i.e. there are no pole-zero cancellations. The forced response of the state-variables is $\boldsymbol{x}_h(t) = \boldsymbol{x}_{h0}e^{\psi_h t}$ and their rate of change is $\dot{\boldsymbol{x}}_h(t) = \boldsymbol{x}_{h0}\psi_h e^{\psi_h t}$. Substituting for the driving input and associated forced responses in (3.37) results in:

---

1. A note of caution: because the various output or input signals will be different (e.g. output signals may be voltage, power, speed, etc.), care should be taken in assessing the effects of scaling.

$$\begin{bmatrix} (\boldsymbol{A} - \psi_h \boldsymbol{I}) & \boldsymbol{b}_i \\ \boldsymbol{c}_j & d_{ij} \end{bmatrix} \begin{bmatrix} \boldsymbol{x}_{h0} \\ u_{ih0} \end{bmatrix} = \begin{bmatrix} \boldsymbol{0} \\ 0 \end{bmatrix}. \tag{3.38}$$

The non-trivial solution of (3.38) requires:

$$\text{Det}\left(\begin{bmatrix} (\boldsymbol{A} - \psi_h \boldsymbol{I}) & \boldsymbol{b}_i \\ \boldsymbol{c}_j & d_{ij} \end{bmatrix} \right) = 0. \tag{3.39}$$

Expansion of the above determinant results in a polynomial in ψ_h and the associated roots correspond to the zeros of the SISO sub-system.

Equation (3.39) does not provide a tractable means for computing the zeros of systems with more than a few state variables. Thus, (3.38) is rewritten in the form of a generalized eigenvalue problem as follows:

$$\boldsymbol{H}\boldsymbol{v}_h = \psi_h \boldsymbol{M}\boldsymbol{v}_h, \tag{3.40}$$

$$\text{where } \boldsymbol{H} = \begin{bmatrix} \boldsymbol{A} & \boldsymbol{b}_i \\ \boldsymbol{c}_j & d_{ij} \end{bmatrix}, \ \boldsymbol{M} = \begin{bmatrix} \boldsymbol{I} & \boldsymbol{0} \\ \boldsymbol{0} & 0 \end{bmatrix} \text{ and } \boldsymbol{v}_h = \begin{bmatrix} \boldsymbol{x}_{h0} \\ u_{ih0} \end{bmatrix}; \tag{3.41}$$

the objective is to compute all finite values of ψ_h for which there exist non-trivial solutions of (3.40).

The well-known QZ algorithm developed by Moler and Stewart [13] is a numerically robust procedure for computing ψ_h. The Fortran LAPACK library [14] provides a suite of subroutines in the public domain that implement the QZ algorithm. The Matlab function `qz` provides an interface to the appropriate LAPACK library routines.

At the heart of the QZ algorithm is the determination of unitary matrices $\boldsymbol{Q}$ and $\boldsymbol{Z}$ such that both $\boldsymbol{S} = \boldsymbol{QHZ}$ and $\boldsymbol{T} = \boldsymbol{QMZ}$ are upper diagonal. As stated in [13] the two eigenvalue problems $\boldsymbol{S}\boldsymbol{y}_h = \psi_h \boldsymbol{T}\boldsymbol{y}_h$ and $\boldsymbol{H}\boldsymbol{v}_h = \psi_h \boldsymbol{M}\boldsymbol{v}_h$ are unitarily equivalent: they both have the same eigenvalues ψ_h and their eigenvectors are related by $\boldsymbol{v}_h = \boldsymbol{Z}\boldsymbol{y}_h$. Suppose the hth diagonal entries of $\boldsymbol{S}$ and $\boldsymbol{T}$ are respectively s_{hh} and t_{hh} then $\psi_h = s_{hh}/t_{hh}$. If t_{hh} is zero, or computationally very close to zero, then $\psi_h = \infty$. If both s_{hh} and t_{hh} are zero, or computationally very close to zero, then the system is said to be degenerate.

Extensions of the above approach to the determination of the properties and computation of the various kinds of zeros of MIMO systems have been devised by a number of investigators [15, 16, 17]. Methods for computing dominant zeros in large systems have been developed and implemented by Martins, *et al.* [18, 19, 20].

3.9 Mode shapes

Let us assume that we able both to excite a particular mode and to evaluate the time responses of the states of the system. In (3.22) it was noted that the response can be separated into its natural and forced components. Assuming (i) non-zero initial conditions on the states, and (ii) no forcing signals applied at the inputs to the system, we can write the equation for the natural response of the states in the form $\boldsymbol{X}(s) = (s\boldsymbol{I}-\boldsymbol{A})^{-1}\boldsymbol{x}(0)$.

In the previous section a decoupled form of the state equations is derived (3.35),

$$\dot{z} = \Lambda \cdot z + \hat{\boldsymbol{B}} \cdot \boldsymbol{U},$$

assuming the pseudo-states and the original states, z and $\boldsymbol{x}$ respectively, are related by $\boldsymbol{x} = Vz$. With no external excitation at the inputs and initial conditions $z(0)$, the natural response is

$$z(t) = \boldsymbol{L}^{-1}\left\{(s\boldsymbol{I}-\underline{\Lambda})^{-1}\right\}\cdot z(0) = \boldsymbol{L}^{-1}\left[diag\left\{\frac{1}{s-\lambda_1}, \frac{1}{s-\lambda_2}, \ldots, \frac{1}{s-\lambda_n}\right\}\right]\cdot z(0),$$

where $\boldsymbol{L}\{f(t)\}$ is the Laplace transform of $f(t)$;

$$\text{i.e.} \quad z(t)= \begin{bmatrix} e^{\lambda_1 t} & 0 & \ldots & 0 \\ 0 & e^{\lambda_2 t} & \ldots & 0 \\ \ldots & \ldots & \ldots & \ldots \\ 0 & 0 & \ldots & e^{\lambda_n t} \end{bmatrix} \cdot z(0) = diag\left\{e^{\lambda_1 t}, e^{\lambda_2 t}, \ldots, e^{\lambda_n t}\right\}\cdot z(0). \tag{3.42}$$

On expressing the latter equation in terms of the original state variables, but retaining the decoupled modes, the response becomes

$$\boldsymbol{x}(t) = V\, diag\left\{e^{\lambda_1 t}, e^{\lambda_2 t}, \ldots, e^{\lambda_n t}\right\} \boldsymbol{W}\cdot \boldsymbol{x}(0),$$

the right and left modal matrices ($\boldsymbol{V}$ and $\boldsymbol{W}$) being defined in (3.31). An alternative form of (3.42) is

$$\boldsymbol{x}(t)= \sum_{h=1}^{n} \boldsymbol{v}_h\, e^{\lambda_h t}\, \boldsymbol{w}_h\, \boldsymbol{x}(0). \tag{3.43}$$

If we account for the fact that the inner product of two vectors is a scalar we can rewrite (3.43) as

$$\boldsymbol{x}(t)= \sum_{h=1}^{n} \boldsymbol{v}_h\, [\boldsymbol{w}_h\, \boldsymbol{x}(0)]e^{\lambda_h t}.$$

Let us assume that initial conditions on the states are set equal to the right eigenvector of the ith eigenvalue, i.e. $\boldsymbol{x}(0) = \boldsymbol{v}_i$. The latter equation becomes

$$x(t) = \sum_{h=1}^{n} v_h\, [w_h\, v_i] e^{\lambda_h t} . \tag{3.44}$$

From (3.32), $w_h\ v_i\ =\ 1$ when $h = i$ and is zero when $h \neq i$. Hence the above equation reduces to

$$x(t) = v_i\, e^{\lambda_i t} . \tag{3.45}$$

Thus *only mode i* is excited. Moreover,

$$x_1(t) = v_{1i}\, e^{\lambda_i t}, \ldots, \quad x_k(t) = v_{ki}\, e^{\lambda_i t}, \ldots, \quad x_n(t) = v_{ni}\, e^{\lambda_i t} . \tag{3.46}$$

Note that each of the modal responses, $x_1(t)$ to $x_n(t)$, has an identical form but their shapes are determined by the initial amplitude v_{ki} in each response. Thus for a given mode, the relative amplitudes or *shapes* of the responses are determined by the associated *right eigenvector*. Consequently, we can plot the *mode shapes* for selected modes, these shapes revealing not only the relative amplitude of the states in the mode, but also the relative phase between the responses of the states. From (3.46) the relative amplitudes of states at time *t* are

$$x_1 : x_k,\ x_2 : x_k, \ldots, 1, \ldots\ x_n : x_k, \text{ or } |v_{1i}| : |v_{ki}|,\ |v_{2i}| : |v_{ki}|, \ldots, 1, \ldots\ |v_{ni}| : |v_{ki}|, \tag{3.47}$$

where v_{ki} is the element with the largest magnitude among the selected states whose mode shapes are to be displayed; this result will be employed later in Chapters 9 and 10.

A note of caution. Prior to the development of participation factors (see Section 3.10), the element $|v_{ki}|$ of the right eigenvector was employed to determine the 'involvement' of the state variable x_k in mode *i*. A large relative value of $|v_{ki}|$ was assessed as representing a significant involvement of x_k in the ith mode. However, this is misleading as the numerical values of the elements $|v_{ki}|$ depend on the units selected (e.g. speed in pu, angle in rad.) for the associated state variables, i.e. they are not dimensionless, they are scaling dependent. It should be noted, therefore, that the relative amplitude of the component $|v_{jh}|/|v_{mh}|$ revealed in (3.47) should not be interpreted as implying the relative participations of states *j* and *m* in mode *h*. The concept of 'participation' is considered in Section 3.10.

In a practical application, the elements in the right eigenvector corresponding to the speed states of all generators are selected to reveal the speed mode-shape, say, for an inter-area mode. For such a mode the relative phase between the speed states reveal, for example, that machines in areas A and B swing against generators in area C. The plots of mode shapes and their significance in the analysis of dynamic behaviour of power systems will be discussed in more detail in Chapters 9 and 10.

The significance of modal response and the mode shapes are most simply illustrated by means of a numerical example.

3.9.1 Example 3.5: Mode shapes and modal responses

The state matrix of a system is given by $\boldsymbol{A} = \begin{bmatrix} 0 & 2 \\ -4 & -6 \end{bmatrix}$.

Determine its modal responses and mode shapes.

The eigenvalues of $\boldsymbol{A}$ are evaluated from the characteristic equation, det $(\lambda_h \boldsymbol{I} - \boldsymbol{A}) = 0$; there are two eigenvalues, a slower at $\lambda_1 = -2$ and a faster at $\lambda_2 = -4$.

Let the right modal matrix be $\boldsymbol{V} = \begin{bmatrix} \alpha_1 & \beta_1 \\ \alpha_2 & \beta_2 \end{bmatrix}$. The right eigenvector associated with the eigenvalue $\lambda_1 = -2$ is found from (3.29),

$$\boldsymbol{A}\boldsymbol{v}_h = \lambda_h \boldsymbol{v}_h, \quad \text{i.e.} \quad \begin{bmatrix} 0 & 2 \\ -4 & -6 \end{bmatrix} \cdot \begin{bmatrix} \alpha_1 \\ \alpha_2 \end{bmatrix} = -2 \begin{bmatrix} \alpha_1 \\ a_2 \end{bmatrix} \quad \text{or} \quad \alpha_2 = -\alpha_1 .$$

Likewise, for $\lambda_2 = -4$,

$$\begin{bmatrix} 0 & 2 \\ -4 & -6 \end{bmatrix} \cdot \begin{bmatrix} \beta_1 \\ \beta_2 \end{bmatrix} = -4 \begin{bmatrix} \beta_1 \\ \beta_2 \end{bmatrix} \quad \text{or} \quad \beta_2 = -2\beta_1 .$$

Let $\alpha_1 = \beta_1 = 1$, then $\boldsymbol{V} = \begin{bmatrix} 1 & 1 \\ -1 & -2 \end{bmatrix} = \begin{bmatrix} \boldsymbol{v}_1 & \boldsymbol{v}_2 \end{bmatrix}$, and $\boldsymbol{W} = \boldsymbol{V}^{-1} = \begin{bmatrix} 2 & 1 \\ -1 & -1 \end{bmatrix} = \begin{bmatrix} \boldsymbol{w}_1 \\ \boldsymbol{w}_2 \end{bmatrix}$.

The time response, as given by (3.43), is

$$\boldsymbol{x}(t) = \begin{bmatrix} 1 & 1 \\ -1 & -2 \end{bmatrix} \cdot \begin{bmatrix} e^{-2t} & 0 \\ 0 & e^{-4t} \end{bmatrix} \cdot \begin{bmatrix} 2 & 1 \\ -1 & -1 \end{bmatrix} \cdot \begin{bmatrix} x_1(0) \\ x_2(0) \end{bmatrix}, \quad \text{or}$$

$$\begin{bmatrix} x_1(t) \\ x_2(t) \end{bmatrix} = \begin{bmatrix} 1 \\ -1 \end{bmatrix} [2x_1(0) + x_2(0)] e^{-2t} + \begin{bmatrix} 1 \\ -2 \end{bmatrix} [-x_1(0) - x_2(0)] e^{-4t} .$$

The form of this response illustrates a number of significant points.

As stated earlier, if the initial value of the states is equal to either of the right eigenvectors, the associated mode is the only mode present in the response, e.g. for eigenvalue 1, $\lambda_1 = -2$,

$$\begin{bmatrix} x_1(0) \\ x_2(0) \end{bmatrix} = \begin{bmatrix} \alpha_1 \\ \alpha_2 \end{bmatrix} = \begin{bmatrix} 1 \\ -1 \end{bmatrix}, \quad \begin{bmatrix} x_1(t) \\ x_2(t) \end{bmatrix} = \begin{bmatrix} 1 \\ -1 \end{bmatrix} [2x_1(0) + x_2(0)] e^{-2t} = \begin{bmatrix} 1 \\ -1 \end{bmatrix} e^{-2t} .$$

Correspondingly, for eigenvalue 2, $\lambda_2 = -4$:

$$\begin{bmatrix} x_1(0) \\ x_2(0) \end{bmatrix} = \begin{bmatrix} \beta_1 \\ \beta_2 \end{bmatrix} = \begin{bmatrix} 1 \\ -2 \end{bmatrix}, \quad \begin{bmatrix} x_1(t) \\ x_2(t) \end{bmatrix} = \begin{bmatrix} 1 \\ -2 \end{bmatrix} e^{-4t}.$$

Note that in each of the two modal responses, the responses $x_1(t)$ and $x_2(t)$ are related by a *constant* factor for all t, i.e. $x_1(t) = -x_2(t)$ for 1, and $x_1(t) = -(1/2)x_2(t)$ for 2. We observe that

(i) for both modes, the responses of the states $x_1(t)$ and $x_2(t)$ vary in anti-phase;

(ii) for the slow mode $\lambda_1 = -2$ the relative amplitudes of the two states are same, but for the fast mode $\lambda_2 = -4$ they differ by a factor of two.

Thus, for a *given* mode, the *mode shape* reveals not only relative phase between the time responses of the states but also the *relative amplitudes* of the states in the modal responses; furthermore, the mode shape is determined by the *right eigenvector* of the associated eigenvalue.

Further insight into the physical significance of mode shapes is provided in Section 9.2.

3.10 Participation Factors

We will make fairly extensive use of participation factors later, mainly to determine the degree to which certain states of certain generators or other devices participate in a selected mode. For example, by examining the speed states of generators, the generators which are involved in a selected mode of rotor oscillation can be found.

3.10.1 The relative participation of a mode in a selected state

We will analyse participation factors in two stages. In the first it is assumed that the initial conditions on the states are such that only the kth state is excited, i.e. $\boldsymbol{x}(0) = \boldsymbol{e}_k$, the unit vector. Then (3.43) becomes:

$$\begin{bmatrix} x_1(t) \\ \cdots \\ x_k(t) \\ \cdots \\ x_n(t) \end{bmatrix} = \begin{bmatrix} v_{11} \\ \cdots \\ v_{k1} \\ \cdots \\ v_{n1} \end{bmatrix} (w_{1k}\, e^{\lambda_1 t}) + \ldots + \begin{bmatrix} v_{1h} \\ \cdots \\ v_{kh} \\ \cdots \\ v_{nh} \end{bmatrix} (w_{hk}\, e^{\lambda_h t}) + \ldots + \begin{bmatrix} v_{1n} \\ \cdots \\ v_{kn} \\ \cdots \\ v_{nn} \end{bmatrix} (w_{nk}\, e^{\lambda_n t}).$$

Note that, although only the kth state is excited, all eigenvalues are excited by the unit vector.

Let us consider the state $x_k(t)$, i.e

$$
\begin{aligned}
x_k(t) &= v_{k1}\, w_{1k}\, e^{\lambda_1 t} + \ldots + v_{kh}\, w_{hk}\, e^{\lambda_h t} + \ldots v_{kn}\, w_{nk}\, e^{\lambda_n t} \\
&= w_{1k}\, v_{k1}\, e^{\lambda_1 t} + \ldots + w_{hk}\, v_{kh}\, e^{\lambda_h t} + \ldots w_{nk}\, v_{kn}\, e^{\lambda_n t}\,.
\end{aligned} \tag{3.48}
$$

The participation of hth eigenvalue in state k is defined as

$$p_{hk} = w_{hk}\, v_{kh} = v_{kh}\, w_{hk}\,. \tag{3.49}$$

Thus, (3.48) becomes $x_k(t) = p_{1k}\, e^{\lambda_1 t} + \ldots + p_{hk}\, e^{\lambda_h t} + \ldots p_{nk}\, e^{\lambda_n t} = \sum_{h=1}^{n} p_{jk} e^{\lambda_h t}$.

However, from (3.32) we know

$$\sum_{k=1}^{n} w_{hk}\, v_{kh} = \sum_{k=1}^{n} p_{hk} = 1\,. \tag{3.50}$$

The inner product $w_h v_h$ (= 1) is dimensionless, therefore the numbers $w_{hk}\, v_{kh}$ are also dimensionless and are invariant under changes in the units of the state variables of the system. *Hence* p_{hk} *in* (3.49) *provides a measure of the relative extent to which* h^{th} *eigenvalue participates in state* k *at time* $t = 0$; p_{hk} is therefore known as the *participation factor of the* h^{th} *eigenvalue in the* k^{th} *state.* (Note that since each mode in the system decays at a different rate the relative amplitude of each mode in the response (3.48) does change with time.)

3.10.2 The relative participation of a state in a selected mode

For the second stage it is assumed that the initial conditions on the states are such that each state is excited *in turn* by the unit vector, i.e. $x(0) = e_j$, $j = 1, \ldots, n$. Using (3.43) and the principle of superposition, the form of (3.48) is modified to the following,

$$
\begin{array}{lll}
\text{for } j = 1: & x_1(t) = w_{11}\, v_{11}\, e^{\lambda_1 t} + \ldots + w_{h1}\, v_{1h}\, e^{\lambda_h t} + \ldots + w_{n1}\, v_{1n}\, e^{\lambda_n t} \\
& \ldots \quad \ldots \qquad \ldots \qquad \ldots \\
\text{for } j = k: & x_k(t) = w_{1k}\, v_{k1}\, e^{\lambda_1 t} + \ldots + w_{hk}\, v_{kh}\, e^{\lambda_h t} + \ldots + w_{nk}\, v_{kn}\, e^{\lambda_n t} \\
& \ldots \quad \ldots \qquad \ldots \qquad \ldots \\
\text{for } j = n: & x_n(t) = w_{1n}\, v_{n1}\, e^{\lambda_1 t} + \ldots + w_{hn}\, v_{nh}\, e^{\lambda_h t} + \ldots + w_{nn}\, v_{nn}\, e^{\lambda_n t}\,.
\end{array}
$$

If we replace $w_{hk}\, v_{kh}$ by the participation factors, p_{hk}, the coefficients of exponential terms in the above set of equations can be formed into the following column arrays,

$$e^{\lambda_1 t}: \begin{bmatrix} p_{11} \\ \dots \\ p_{1k} \\ \dots \\ p_{1n} \end{bmatrix}, \quad \dots, e^{\lambda_h t}: \begin{bmatrix} p_{h1} \\ \dots \\ p_{hk} \\ \dots \\ p_{hn} \end{bmatrix}, \quad \dots, e^{\lambda_n t}: \begin{bmatrix} p_{n1} \\ \dots \\ p_{nk} \\ \dots \\ p_{nn} \end{bmatrix}.$$

Consider the hth eigenvalue. According to (3.50), the sum of the participation factors in the column array p_{hk} is unity. Hence, when each state is excited in turn by the unit vector, *the participation factor* p_{hk} *also provides a measure of the relative extent to which each of the n states participates in the* h^{th} *eigenvalue at time* $t = 0$.

3.10.3 Example 3.6: Participation factors

In Example 3.4 the right and left modal matrices are

$$V = \begin{bmatrix} 1 & 1 \\ -1 & -2 \end{bmatrix} = \begin{bmatrix} v_1 & v_2 \end{bmatrix}, \quad \text{and} \quad W = \begin{bmatrix} 2 & 1 \\ -1 & -1 \end{bmatrix} = \begin{bmatrix} w_1 \\ w_2 \end{bmatrix};$$

these are associated with the eigenvalues $\lambda_1 = -2$ and $\lambda_2 = -4$.

The participation factors, $p_{hk} = w_{hk}\, v_{kh}$, for states x_1 and x_2 in each of the two eigenvalues are:

$$p_{11} = 2,\ p_{12} = -1,\ \text{for}\ \lambda_1 = -2,\ \text{and}\quad p_{21} = -1,\ p_{22} = 2,\ \text{for}\ \lambda_2 = -4.$$

Likewise, the participations of the eigenvalues λ_1 and λ_2 in each of the two states are:

$$p_{11} = 2,\ p_{21} = -1,\ \text{for}\ x_1,\ \text{and}\ p_{12} = -1,\ p_{22} = 2,\ \text{for}\ x_2.$$

Note for both eigenvalues the participation factors sum to one.

Further insight into the significance of participation factors is described in Section 9.3.

3.11 Eigenvalue sensitivities

In the analysis of power system dynamics, it is of interest to assess the effect of the change of a system parameter, or some element of the state matrix. Later we will need to examine the effect on certain modes of a change in an element a_{ij} of the state matrix, $\boldsymbol{A}$. Earlier we defined the relationship between the state matrix, the eigenvalues and right eigenvectors, i.e.

$$\boldsymbol{A}\, v_h = \lambda_h\, v_h.$$

Differentiation of this expression with respect to a_{ij} yields,

$$\frac{\partial \boldsymbol{A}}{\partial a_{ij}} \boldsymbol{v}_h + \boldsymbol{A} \frac{\partial \boldsymbol{v}_h}{\partial a_{ij}} = \frac{\partial \lambda_h}{\partial a_{ij}} \boldsymbol{v}_h + \lambda_h \frac{\partial \boldsymbol{v}_h}{\partial a_{ij}} .$$

Collecting terms and pre-multiplying by the left eigenvector, we find

$$\boldsymbol{w}_h (\boldsymbol{A} - \lambda_h \boldsymbol{I}) \frac{\partial \boldsymbol{v}_h}{\partial a_{ij}} = \boldsymbol{w}_h \left(\frac{\partial \lambda_h}{\partial a_{ij}} - \frac{\partial \boldsymbol{A}}{\partial a_{ij}} \right) \boldsymbol{v}_h .$$

Since, by definition $\boldsymbol{w}_h \boldsymbol{A} = \boldsymbol{w}_h \lambda_h$, the left-hand side vanishes. Furthermore $\boldsymbol{w}_h \boldsymbol{v}_h = 1$, and all elements except the ij^{th} of $\partial \boldsymbol{A} / \partial a_{ij}$ are zero. The sensitivity of the eigenvalue λ_h to a change in the element a_{ij} is thus

$$\frac{\partial \lambda_h}{\partial a_{ij}} = w_{hi} \frac{\partial \boldsymbol{A}}{\partial a_{ij}} v_{jh} . \tag{3.51}$$

In its simplest form $\partial A / \partial a_{ij} = 1$, and the sensitivity is the product of the elements w_{hi} and v_{jh} of the left and right eigenvectors,

$$\frac{\partial \lambda_h}{\partial a_{ij}} = w_{hi} v_{jh} . \tag{3.52}$$

In the case $i = j$, and $a_{ij} = a_{ii}$, the eigenvalue sensitivity is $\dfrac{\partial \lambda_h}{\partial a_{ij}} = w_{hi} v_{ih} = p_{hk}$, i.e. the participation factor of the hth eigenvalue in state i.

Note, however, that the element a_{ij} in (3.52) may be a function of a device parameter, q, i.e. $a_{ij} = f(q)$. Moreover, several elements of $\boldsymbol{A}$ may be functions of the same parameter. The above analysis can then be extended to evaluate the sensitivity of λ_h to changes in the parameter q.

3.12 References

[1] T. Kailath. *Linear Systems,* Prentice Hall, 1980.

[2] R.C. Dorf and R.H. Bishop, *Control Systems,* 12th edition, Prentice Hall, 2010.

[3] J. D. Aplevich, *Linear State Space Systems,* John Wiley & Sons Inc., 2000.

[4] P. Kundur, *Power system stability and control.* New York: McGraw-Hill, 1994.

[5] Jules Henri Poincaré, b. 1853, d. 1912. See web references to 'Poincaré maps and stability'.

[6] B. Stott, "Power system dynamic response calculations," *Proceedings of the IEEE,* vol. 67, pp. 219-241, 1979.

[7] C. Moler and C. V. Loan, "Nineteen Dubious Ways to Compute the Exponential of a Matrix, Twenty-Five Years Later," *SIAM Review,* vol. 45, pp. 3-49, 2003.

[8] I. J. Perez-Arriaga, F. L. Pagola, G. C. Verghese, and F. C. Schweppe, "Selective Modal Analysis in Power Systems," in *American Control Conference,* 1983, 1983, pp. 650-655.

[9] I. J. Perez-Arriaga, L. Rouco, F. L. Pagola, and J. L. Sancha, "The role of participation factors in reduced order eigenanalysis of large power systems," *in Circuits and Systems, 1988., IEEE International Symposium on, 1988,* pp. 923-927 vol.1.

[10] F. L. Pagola, I. J. Perez-Arriaga, and G. C. Verghese, "On sensitivities, residues and participations: applications to oscillatory stability analysis and control," *Power Systems, IEEE Transactions on,* vol. 4, pp. 278-285, 1989.

[11] G. C. Verghese, I. J. Perez-Arriaga, and F. C. Schweppe, "Selective Modal Analysis With Applications to Electric Power Systems, Part II: The Dynamic Stability Problem," *Power Apparatus and Systems, IEEE Transactions on,* vol. PAS-101, pp. 3126-3134, 1982.

[12] L. Rouco and F. L. Pagola, "Eigenvalue sensitivities for design of power system damping controllers," in *Decision and Control, 2001. Proceedings of the 40th IEEE Conference on,* 2001, pp. 3051-3055 vol.4.

[13] C. B. Moler and G. W. Stewart, "An Algorithm for Generalized Matrix Eigenvalue Problems," *SIAM Journal on Numerical Analysis,* vol. 10, pp. 241-256, 1973.

[14] E. Anderson, Z. Bai, C. Bischof, S. Blackford, et al., *LAPACK Users' Guide,* Third ed.: Society for Industrial and Applied Mathematics, 1999.

[15] H.H. Rosenbrock, *State-Space and Multivariable Theory,* Nelson, 1970.

[16] E. J. Davison and S. H. Wang, "Properties and calculation of transmission zeros of linear multivariable systems," *Automatica,* vol. 10, pp. 643-658, 1974.

[17] A. J. Laub and B. C. Moore, "Calculation of transmission zeros using QZ techniques," *Automatica*, vol. 14, pp. 557-566, 1978.

[18] N. Martins, H. J. C. P. Pinto and L. T. G. Lima, "Efficient methods for finding transfer function zeros of power systems," *Power Systems, IEEE Transactions on*, vol. 7, pp. 1350-1361, 1992.

[19] N. Martins, P. C. Pellanda and J. Rommes, "Computation of Transfer Function Dominant Zeros With Applications to Oscillation Damping Control of Large Power Systems," *Power Systems, IEEE Transactions on*, vol. 22, pp. 1657-1664, 2007.

[20] J. Rommes, *Methods for eigenvalue problems with applications in model order reduction*, PhD Thesis, University of Utrecht, 2007.

Chapter 4

Small-signal models of synchronous generators, FACTS devices and the power system

4.1 Introduction

In this chapter various models of synchronous generators, FACTS devices and of the power system are developed in forms which are employed in software for the analysis of the small-signal dynamic performance of multi-machine systems. Small-signal models for the synchronous generator are formulated in Section 4.2. An essential feature of this analysis is that the higher order coupled-circuit representation of the generator electromagnetic dynamic behaviour is formulated in Section 4.2.3. This is treated as the fundamental model from which the following two alternative but equivalent formulations of the electromagnetic model are derived. The first is the Operational Parameter formulation described in Section 4.2.12. The second, described in Section 4.2.13, is referred to as the Classical Parameter formulation and is expressed directly in terms of the classically-defined standard parameters of the generator. The Classical Parameter formulation is presented because it is employed in widely used power system simulation software packages such as Siemens PTI PSS®E [1] and GE PSLF™ [2]. The parameters for the fundamental coupled-circuit formulation are the resistances and inductances of the *d*- and *q*-axis circuits. The parameters for the Operational Parameter representation of the electromagnetic equations are the gains and time constants of the transfer-function representations of the respective axes and are collectively referred to as the 'exactly-defined standard parameters'. The Classical Parameter formulation requires the classically-defined standard parameters. The relationship and conversion between the three parameter sets are outlined in Section 4.2.14.

Small-signal models of a range of FACTS devices are formulated in Section 4.3 and include those of the Static VAR Compensator (SVC), Voltage Sourced Converter (VSC), Static Synchronous Compensator (STATCOM), and HVDC transmission links. The general purpose VSC model formulated in Section 4.3.3 is used as a component in the simplified STATCOM model in Section 4.3.4 as well as for the rectifier and inverter in the model of the VSC HVDC transmission link in Section 4.3.7. A general model for a voltage-commutated thyristor-controlled AC/DC converter is formulated Section 4.3.8; this model is then used in a modular fashion to represent the rectifier and inverter of a line-commutated HVDC transmission link. A methodology to formulate the small-signal equations of the power system is described in Section 4.4. Finally, in Section 4.5 a general purpose small-signal representation of a static load model is described.

4.2 Small-signal models of synchronous generators

4.2.1 Structure of the per-unit linearized synchronous generator models

Before developing the details of the per-unit linearized model of the synchronous generator the overall mathematical structure of the model is described by means of the block diagram in Figure 4.1. The linearized model is formulated in the rotating direct- and quadrature-axis *(dq)* coordinate system in which the *d*-axis is aligned with the magnetic north pole of the field winding and the *q*-axis leads the *d*-axis by 90 deg. (electrical). The model is linearized about an initial steady-state operating point that is defined by the initial stator terminal quantities which are typically obtained from a power flow solution. The calculation of the generator initial conditions from the specified terminal quantities is described in Section 4.2.9. The two principal components of the linearized generator model in Figure 4.1 are the electromagnetic (em) equations and the shaft equations of motion. The em equations comprise a set of differential equations that describe the dynamic characteristics of the *d*- and *q*-axis rotor-winding flux linkages together with algebraic equations for the stator voltage components.

As described in Section 4.2.10 the connection of the generator stator to the network requires the transformation of the perturbations in the stator voltage and current components in the generator *dq* coordinate system to the corresponding components in the synchronously rotating *RI*-network coordinate system[1].

There are two control inputs to the generator. The first is the perturbation in the field voltage (ΔE_{fd}) developed by the excitation system. As explained in Section 4.2.7 it is important to note that ΔE_{fd} is expressed in the non-reciprocal per-unit system of the generator field. The field voltage input in this per-unit system must be converted to the reciprocal per-unit system which is employed in the formulation of the generator electromagnetic equations. Similarly, the field current in the reciprocal per-unit system must be converted to the non-

1. *RI* refers to the Real and Imaginary components in the network coordinate system.

reciprocal per-unit system for use in the excitation system model. The second input is the perturbation in the mechanical torque (ΔT_m) developed by the turbine / governor system which is expressed in per-unit on the generator base value of mechanical torque. Models for the excitation and turbine / governor systems are not included in this chapter.

The generator models are designated by a code of the form *ndmq-c{0,1}* in which *n* and *m* are the number of *d*- and *q*-axis rotor-windings respectively; *c1* and *c0* are used to indicate, respectively, that unequal mutual coupling between the *d*-axis rotor windings is represented or neglected. The *1d0q-c0* model comprises three state-variables: the rotor-angle, rotor-speed and *d*-axis field flux-linkages. This is the basis for the Heffron-Phillips model [3, 4] that is frequently used for developing concepts for generator controls. It is, however, not recommended for use in power system analysis. The *3d3q-c1* model, the most complex model considered in this work, comprises a field winding, and two damper windings in the *d*-axis and three *q*-axis damper windings; unequal mutual coupling between the *d*-axis rotor windings is represented. This model – with eight state-variables comprising the six rotor-winding flux-linkage variables and rotor angle and rotor speed – is the most complex model encountered in small-signal analysis of large power systems. The most commonly employed models in large scale small-signal stability studies are the fifth and sixth order models *2d1q-c0* and *2d2q-c0* in which unequal mutual coupling effects are neglected.

The formulation of the em equations for the *3d3q-c1* model described in Section 4.2.3 is based on the ideal coupled-circuit representation of the synchronous machine for which the model parameters are the resistances, mutual and leakage inductances of the windings. As explained in Section 4.2.4 the em equations developed for this model are readily modified to represent machine models with fewer damper windings in the respective axes. In particular, the structure of the em equations and their interface with other components in the overall model of the generator are unaffected by changes in the number of damper windings.

The linearized coupled-circuit formulation of the state- and algebraic equations of the complete generator model are given in matrix form in equation (4.117) on page 133 followed in Table 4.9 by a step-by-step procedure for calculating the associated coefficient matrices.

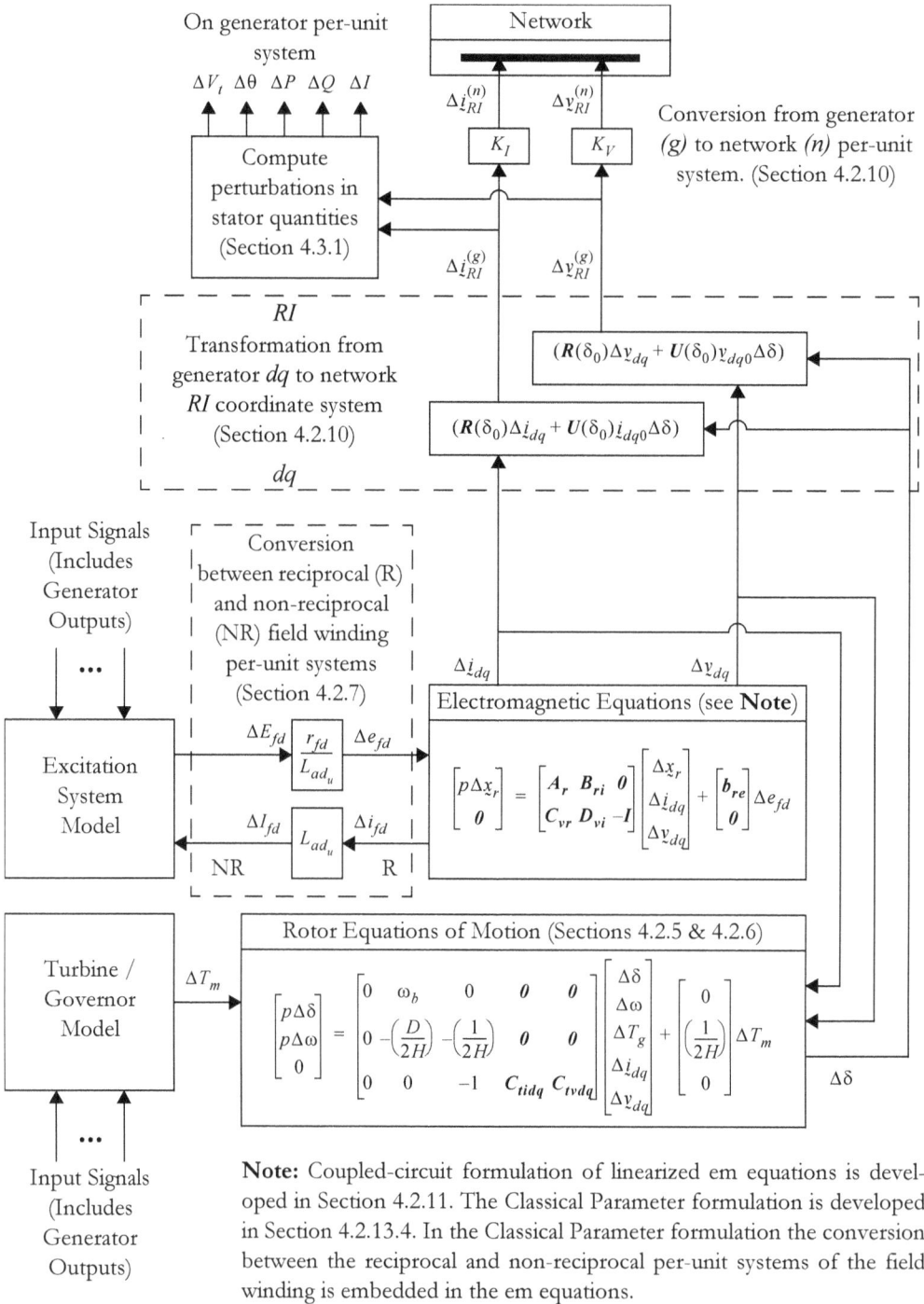

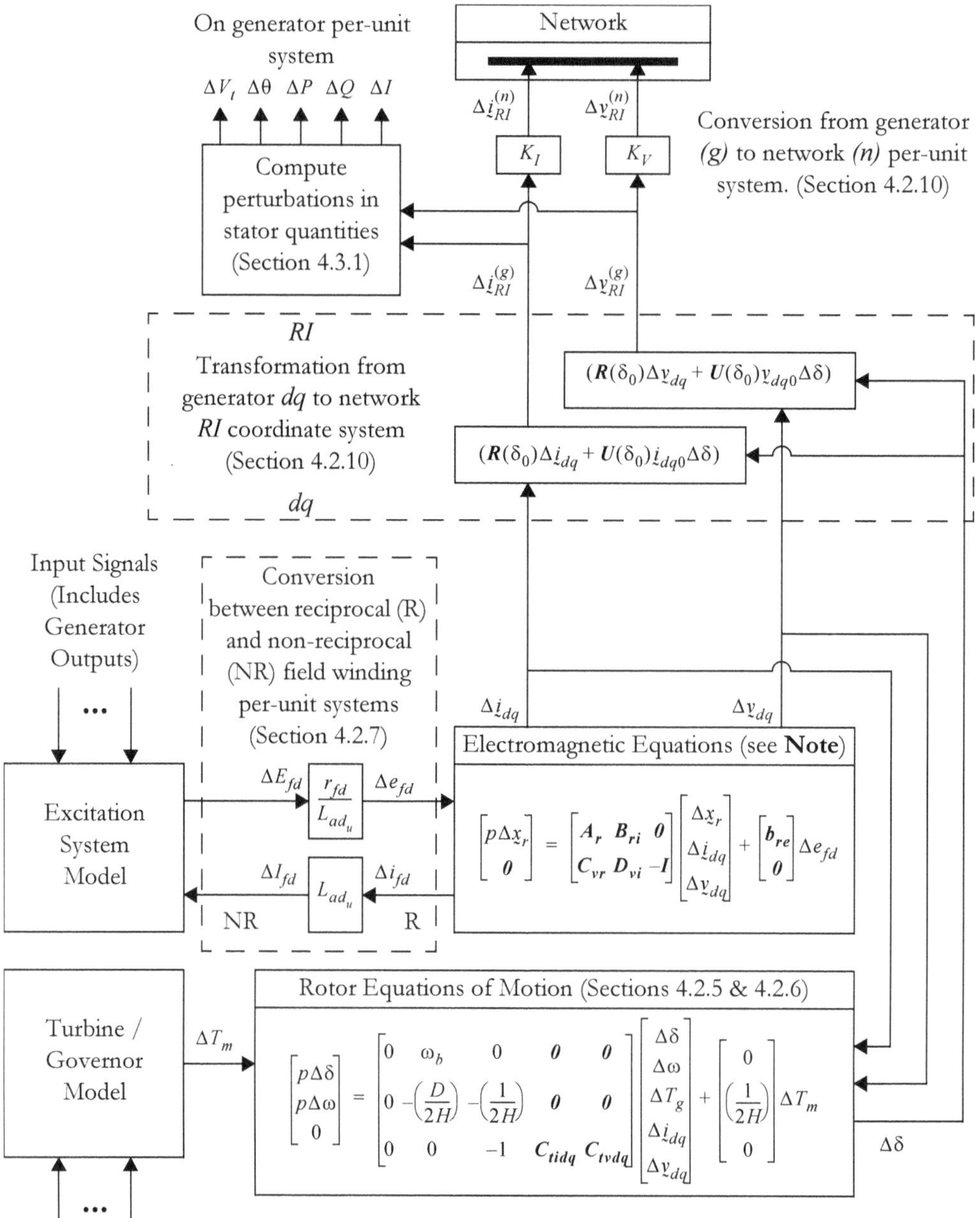

$$\begin{bmatrix} p\Delta x_r \\ 0 \end{bmatrix} = \begin{bmatrix} A_r & B_{ri} & 0 \\ C_{vr} & D_{vi} & -I \end{bmatrix} \begin{bmatrix} \Delta x_r \\ \Delta i_{dq} \\ \Delta v_{dq} \end{bmatrix} + \begin{bmatrix} b_{re} \\ 0 \end{bmatrix} \Delta e_{fd}$$

$$\begin{bmatrix} p\Delta\delta \\ p\Delta\omega \\ 0 \end{bmatrix} = \begin{bmatrix} 0 & \omega_b & 0 & 0 & 0 \\ 0 & -\left(\frac{D}{2H}\right) & -\left(\frac{1}{2H}\right) & 0 & 0 \\ 0 & 0 & -1 & C_{tidq} & C_{tvdq} \end{bmatrix} \begin{bmatrix} \Delta\delta \\ \Delta\omega \\ \Delta T_g \\ \Delta i_{dq} \\ \Delta v_{dq} \end{bmatrix} + \begin{bmatrix} 0 \\ \left(\frac{1}{2H}\right) \\ 0 \end{bmatrix} \Delta T_m$$

Note: Coupled-circuit formulation of linearized em equations is developed in Section 4.2.11. The Classical Parameter formulation is developed in Section 4.2.13.4. In the Classical Parameter formulation the conversion between the reciprocal and non-reciprocal per-unit systems of the field winding is embedded in the em equations.

Figure 4.1 Structure of the per-unit linearized model of the synchronous generator. (Refer to Tables 4.3 and 4.4 for descriptions of the parameter and variable symbols in this figure).

Test procedures that are used to identify synchronous generator models for dynamic analysis commonly employ the Operational Parameter representation of the generator. As explained in Section 4.2.12 this representation comprises three *d*-axis transfer-functions and one *q*-axis transfer-function to completely characterise the machine. The test procedures identify the gains and time constants of these transfer-functions. These transfer-function constants are referred to as the "standard parameters" such as T_{d0}', T_{d0}'', L_d', L_d'', etc. In order to employ the coupled-circuit formulation of the em equations when only the standard parameters are provided it is necessary to transform the standard parameters to the coupled-circuit parameters as outlined in Section 4.2.14. A troublesome aspect of using the standard parameters is that over the years two alternative and inconsistent definitions of the parameters have evolved. The 'Exact' definitions correspond to the exact roots of the above transfer-functions. In the 'Classical' definitions the *d*-axis standard parameters are related to the parameters of the equivalent circuit of the machine by the classical relationships which are based on the assumptions that (i) during the transient period the damper winding resistances are infinite; (ii) during the subtransient period the resistance of the field winding is zero and the resistances of the second damper winding is infinite; (iii) finally, during the sub-subtransient period the resistances of the field and first damper winding are assumed to be zero. In the *q*-axis, analogous assumptions are made to arrive at the classical definitions of the *q*-axis standard parameters in terms of the coupled-circuit parameters. It is important to know if the generator standard parameters that are provided conform to the 'Exact' or 'Classical' definitions and if necessary to transform them appropriately to suit the requirements of the simulation model in use. This is particularly important for the *q*-axis parameters.

The em equations in some widely-used simulation packages are formulated directly in terms of the classically-defined standard parameters. This is referred to as the Classical Parameter formulation in this book. *It is emphasised that the Classical Parameter formulation is exactly equivalent to the coupled-circuit formulation provided: (i) that the unequal coupling between the d-axis rotor windings is neglected, and (ii) that the same method for representing magnetic saturation is employed in the two models.*

4.2.2 Generator modelling assumptions

For the purpose of rotor-angle small-signal analysis the following generator modelling assumptions are made:

- The *d*-axis is aligned with magnetic axis of the field winding and the *q*-axis leads the *d*-axis by 90 degrees (electrical).

- The following non-reciprocal Park/Blondel transform [5, 6] is used to transform variables in the stationary *abc* coordinate system to the rotating *dq* coordinate system:

$$\boldsymbol{T}(\theta) = \frac{2}{3}\begin{bmatrix} \cos(\theta) & \cos(\theta-\alpha) & \cos(\theta+\alpha) \\ -\sin(\theta) & -\sin(\theta-\alpha) & -\sin(\theta+\alpha) \\ \frac{1}{2} & \frac{1}{2} & \frac{1}{2} \end{bmatrix}, \quad \alpha = \frac{2\pi}{3}. \tag{4.1}$$

- In the stator voltage equations the transformer voltage terms, $(p\varphi_d)$ and $(p\varphi_q)$, are omitted because, in the bandwidth of interest, they are negligible in comparison with the speed voltage terms, $(\omega\varphi_d)$ and $(\omega\varphi_q)$ [7].

- For the purpose of calculating the *d*- and *q*-axis stator voltages the perturbation of the per-unit rotor-speed ω from the per-unit synchronous speed ω_0 is assumed to be negligible and thus, for this purpose, $\omega = \omega_0$. Rotor angle and speed perturbations are, necessarily, represented in the rotor equations of motion.

- A consequence of the above assumption is that the per-unit power transferred across the airgap is independent of perturbations in the rotor speed. Thus, it is also necessary to assume that the mechanical power developed by the turbine is independent of perturbations in rotor speed.

- The generator equations are expressed in per-unit form in which the base quantities are summarized in Section 4.2.3.2. In particular, the L_{ad}-base reciprocal per-unit system is chosen for the rotor windings [8, 9, 10].

- As recommended in IEEE Std. 421.5 [11] a non-reciprocal per-unit system is assumed to be employed for the representation of generator excitation systems. As explained in Section 4.2.7 it is therefore necessary to appropriately adjust the scaling of the field current and voltage at the interface between the generator and excitation system models.

4.2.3 Electromagnetic model in terms of the per-unit coupled-circuit parameters

The development of the per-unit generator equations in the rotating *dq* coordinate system from the ideal coupled-circuit representation of the synchronous machine in the stationary *abc* coordinate system is provided in reference [12]. In this section the per-unit coupled-circuit formulation of the electromagnetic equations of the machine are listed. These equations are then linearized about the initial steady-state operating point of the machine in Section 4.2.11. The Operational Parameter and Classical Parameter formulations for the em equations are summarized in Sections 4.2.12 and 4.2.13 respectively.

The third to eighth-order coupled-circuit machine models require data in the form of the *d*- and *q*- axis equivalent circuit winding resistances and leakage and mutual inductances.

Though not commonly used the *3d3q-c1* model has been included because there is evidence in the literature that this number of rotor circuits may be required to adequately represent some machines. Test methods have already been developed to identify machine models with this number of rotor windings, for example [13, 14, 15, .16] Figure 4.2 shows the stator and rotor winding flux linkages in terms of the winding mutual and leakage inductances and the winding currents. The unequal mutual coupling between the *d*-axis rotor windings is represented by the inductances L_{c1} and L_{c2} which are referred to as the Canay inductances. The shielding of the field winding by the damper windings has been identified in the literature as important in correctly predicting the field voltage and current [13, 17, 18]. Unequal mutual coupling between the q-axis rotor windings is not represented since these windings are not directly observable. The d- and q-axis equivalent circuits for the *3d3q-c1* model are shown in Figure 4.3.

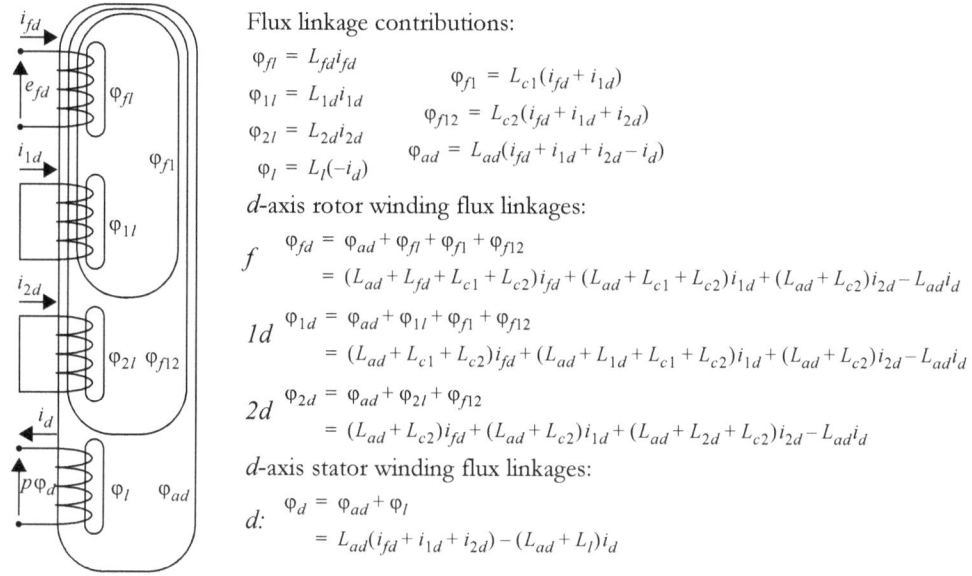

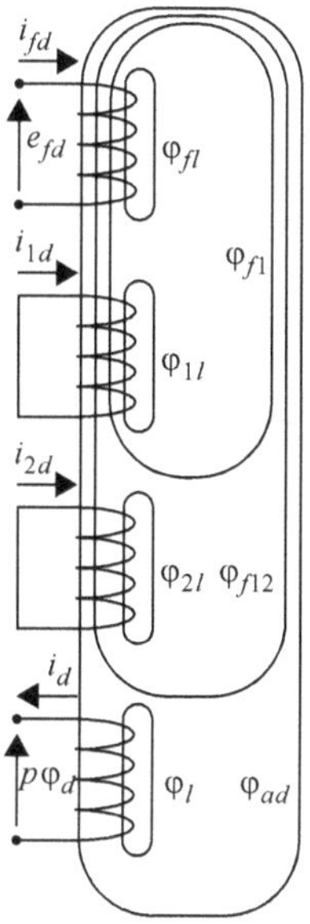

Flux linkage contributions:

$$\varphi_{fl} = L_{fd} i_{fd} \qquad \varphi_{1l} = L_{1d} i_{1d} \qquad \varphi_{2l} = L_{2d} i_{2d} \qquad \varphi_l = L_l(-i_d)$$

$$\varphi_{f1} = L_{c1}(i_{fd} + i_{1d})$$

$$\varphi_{f12} = L_{c2}(i_{fd} + i_{1d} + i_{2d})$$

$$\varphi_{ad} = L_{ad}(i_{fd} + i_{1d} + i_{2d} - i_d)$$

d-axis rotor winding flux linkages:

$$f \quad \begin{aligned}\varphi_{fd} &= \varphi_{ad} + \varphi_{fl} + \varphi_{f1} + \varphi_{f12} \\ &= (L_{ad} + L_{fd} + L_{c1} + L_{c2}) i_{fd} + (L_{ad} + L_{c1} + L_{c2}) i_{1d} + (L_{ad} + L_{c2}) i_{2d} - L_{ad} i_d\end{aligned}$$

$$1d \quad \begin{aligned}\varphi_{1d} &= \varphi_{ad} + \varphi_{1l} + \varphi_{f1} + \varphi_{f12} \\ &= (L_{ad} + L_{c1} + L_{c2}) i_{fd} + (L_{ad} + L_{1d} + L_{c1} + L_{c2}) i_{1d} + (L_{ad} + L_{c2}) i_{2d} - L_{ad} i_d\end{aligned}$$

$$2d \quad \begin{aligned}\varphi_{2d} &= \varphi_{ad} + \varphi_{2l} + \varphi_{f12} \\ &= (L_{ad} + L_{c2}) i_{fd} + (L_{ad} + L_{c2}) i_{1d} + (L_{ad} + L_{2d} + L_{c2}) i_{2d} - L_{ad} i_d\end{aligned}$$

d-axis stator winding flux linkages:

$$d: \quad \begin{aligned}\varphi_d &= \varphi_{ad} + \varphi_l \\ &= L_{ad}(i_{fd} + i_{1d} + i_{2d}) - (L_{ad} + L_l) i_d\end{aligned}$$

Figure 4.2 Per-unit d-axis flux linkage distribution showing the unequal mutual coupling between the rotor windings as described by Canay [17, 18].

Importantly, in the following analysis, *once the equations for the eighth-order model are defined, the lower-order coupled-circuit models are readily derived.* All lower-order coupled-circuit models are formed by deleting the equations and variables associated with those damper windings to be omitted in the formulation of the simpler model.

Two equivalent approaches to the representation of magnetic saturation are accommodated in the formulation of the model. The specific details of the non-linear saturation functions and their linearization are provided in Section 4.2.8.

4.2.3.1 Notes on vector and matrix nomenclature

The nomenclature in the following table applies to matrices and vectors that are extensively employed in the formulation of the models.

Table 4.1 Vector and matrix nomenclature

Symbol	Meaning
$\underset{\sim}{x}$	Denotes a column vector
$\boldsymbol{X}$	Denotes a matrix or, depending on the context, a vector.
$\underset{\sim}{x}^T$, $\boldsymbol{X}^T$	Denotes vector and matrix transposition respectively
$\boldsymbol{\emptyset}$	Null matrix or vector. The null matrix has zero rows and/ or zero columns. The possibility of a non-zero number of rows or columns ensures dimensional consistency of matrix equations. The dimension is to be inferred from the context.
$\boldsymbol{I}$	The identity matrix. The dimension is to be inferred from the context and in some situations may be the scalar identity (i.e. 1) or the null or empty matrix.
$\boldsymbol{0}$	The zero matrix or vector. The dimension is to be inferred from the context and in some situations may be the scalar zero (i.e. 0) or the null matrix.
$\boldsymbol{u} = \begin{bmatrix} 1 & 1 & \dots & 1 \end{bmatrix}^T$	Denotes a column vector whose entries are all ones. The dimension is to be inferred from the context. • If $\boldsymbol{u}$ is a n x 1 vector then $\boldsymbol{u}^T\boldsymbol{u}$ is a n x n matrix of all ones. • $\boldsymbol{u}^T\underset{\sim}{x}$ is the sum of all the elements in $\underset{\sim}{x}$
$\mathfrak{D}(X, Y) = \begin{bmatrix} X & 0 \\ 0 & Y \end{bmatrix}$	Denotes block diagonalization which is the result of appending the m_x x n_x matrix $\boldsymbol{X}$ and the m_y x n_y matrix $\boldsymbol{Y}$ to create the $(m_x + m_y)$ x $(n_x + n_y)$ matrix as shown. If $\boldsymbol{X}$ and $\boldsymbol{Y}$ are scalars or diagonal matrices the result is a diagonal matrix.

4.2.3.2 Summary of the generator per-unit system

The principal base quantities for the machine are normally V_{usb} (kV), S_{usb} (MVA) and f_{usb} (Hz) which are respectively values for the stator RMS line-to-line voltage, the stator three-phase apparent power and the stator frequency. Usually, but not necessarily, the generator rated values of these quantities are chosen. Additionally, the base value of time t_b is chosen to be one-second. Finally, the relationship between mechanical and electrical angles requires knowledge of the number of rotor pole pairs, n_{pp}.

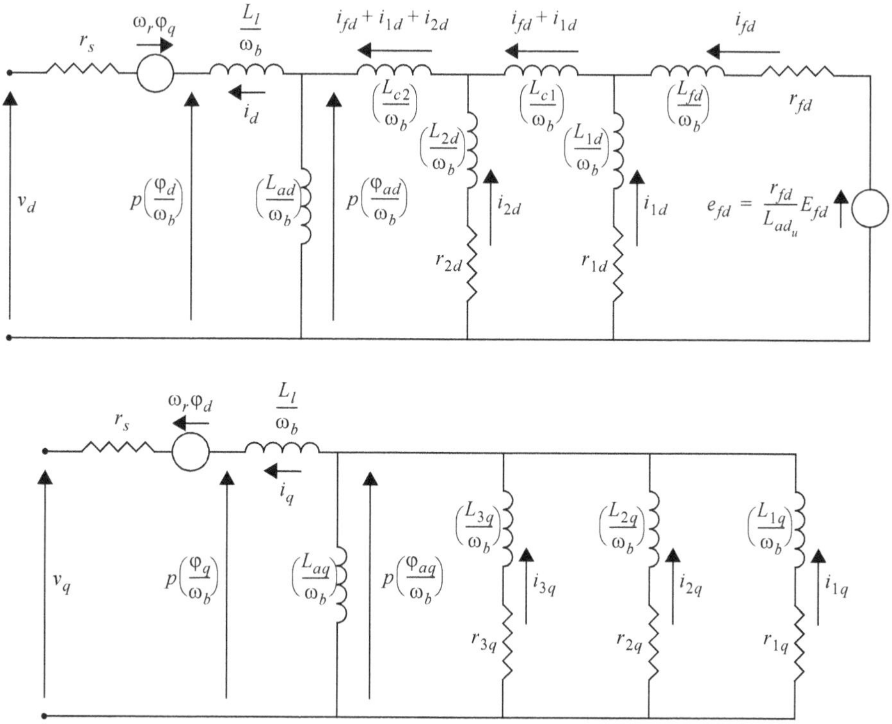

Figure 4.3 The *d*- (top) and *q*-axis (bottom) equivalent circuits for the *3d3q-c1* generator model represented by three rotor windings in each axis and unequal mutual coupling between the *d*-axis rotor windings. (Note: inductance & flux-linkage values are scaled by $1/\omega_b$ because the base value of time is one second).

The base values of the rotor winding currents and flux-linkages are determined such that (i) the per-unit mutual inductances between all pairs of windings are reciprocal; (ii) the mutual inductances between all *d*-axis rotor windings and the stator are equal to the per-unit *d*-axis mutual inductance L_{ad}; (iii) the mutual inductances between all *q*-axis rotor windings and the stator are equal to L_{aq}. Furthermore, the base values of rotor-winding voltages are chosen such that the form of the rotor winding voltage equations in SI units and in per-unit are identical. This choice of base values for the rotor quantities is equivalent to that recommended by Rankin in 1945 [8, 9] and is referred to as the "L_{ad}-base reciprocal per-unit system" [10]. It has gained very wide, if not universal, acceptance in the power system analysis field.

On the above basis for the generator per-unit system are derived the base values for the mechanical, stator winding and rotor winding quantities in Table 4.2.

Table 4.2 Base values for generator quantities (Note: a bar above a quantity (e.g. $\bar{x}$) means the SI value and the subscript 'b' denotes the base value of a quantity.).

Base Quantity	SI Units	Description
Principal base quantities from which all other base quantities are derived		
V_{usb}	kV (rms, ph-ph)	Arbitrary choice, but usually rated RMS phase-to-phase stator voltage (sometimes referred to as VBASE).
S_{usb}	MVA	Arbitrary choice, but usually three-phase MVA rating of the machine (sometimes referred to as MBASE).
f_{usb}	Hz	Arbitrary choice, but usually rated generator frequency. (This is not necessarily the same as nominal frequency of the system to which the generator is connected. For example, when a generator rated at 60 Hz is connected to a 50 Hz system or vice-versa).
n_{pp}		Number of pole pairs.
t_b	s	Base value of time is chosen to be 1 second.
Derived base quantities		
p_b	s^{-1}	Base value of the time differential operator: $\bar{p} = \frac{d}{d\bar{t}} = \frac{d}{d(t_b t)} = \frac{1}{t_b}\frac{d}{dt} = p_b p$ where $p_b = \frac{1}{t_b} = 1\ s^{-1}$.
ω_b	(elec) rad/s	Base electrical frequency: $\omega_b = 2\pi f_{usb}$.
v_{sb}	V(peak, ph-n)	Stator base voltage: peak value of phase to neutral voltage $v_{sb} = \sqrt{(2/3)}V_{usb} \times 10^3$.
S_b	VA	Machine three-phase VA (apparent power) base: $S_b = S_{usb} \times 10^6$.
U_b	Joules	Base energy: $U_b = S_b \times t_b = S_b \times 1$.
i_{sb}	A (peak, line)	Stator base current: peak value of line current $i_{sb} = (2/3)(S_b/v_{sb})$.
Z_{sb}	Ω	Stator base resistance / impedance: $Z_{sb} = v_{sb}/i_{sb}$.
L_{sb}	H	Stator base inductance: $L_{sb} = Z_{sb}/\omega_b = \varphi_{sb}/i_{sb}$.
φ_{sb}	Wb-turns	Stator base flux-linkages: $\varphi_{sb} = L_{sb} \times i_{sb} = v_{sb}/\omega_b$.
ω_{mb}	(mech) rad/s	Base mechanical rotor speed: $\omega_{mb} = \omega_b/n_{pp}$.
T_{mb}	Nm	Base mechanical torque: $T_{mb} = S_b/\omega_{mb}$.
T_b	Nm	Base electrical torque: $T_m = T_{mb}/T_{mb}$.

Base Quantity	SI Units	Description
d-axis rotor quantities (per-unit *unsaturated* d-axis mutual inductance $L_{ad} = \bar{L}_{ad}/L_{sb}$ is used in the following.)		
i_{fdb}	A	Base field current in the reciprocal per-unit system of units: $i_{fdb} = (L_{ad}\varphi_{sb})/\bar{L}_{afd} = (\bar{L}_{ad}/\bar{L}_{afd})i_{sb}$.
e_{fdb}	V	Base field voltage in the reciprocal per-unit system: $e_{fdb} = S_b/i_{fdb}$.
I_{fdb}	A	Base field current in the non-reciprocal per-unit system: $I_{fdb} = i_{fdb}/L_{ad}$.
E_{fdb}	V	Base field voltage in the non-reciprocal per-unit system: $E_{fdb} = \bar{r}_{fd} \times I_{fdb}$, $\bar{r}_{fd}$ is the field resistance in Ω at the specified temperature.
i_{kdb}	A	Base current of the d-axis damper windings: $i_{kdb} = (L_{ad}\varphi_{sb})/\bar{L}_{akd} = (\bar{L}_{ad}/\bar{L}_{akd})i_{sb}$, $k = 1, 2$.
v_{kdb}	V	Base voltage of the d-axis damper windings: $v_{kdb} = S_b/i_{kdb}$, $k = 1, 2$.
φ_{fdb}, φ_{1db}, φ_{2db}	Wb-turns	Base flux-linkages of d-axis rotor windings: $\varphi_{fdb} = v_{fdb}/\omega_b$, $\varphi_{1db} = v_{1db}/\omega_b$ & $\varphi_{2db} = v_{2db}/\omega_b$.
r_{fdb}, r_{1db}, r_{2db}	Ω	Base resistance of the d-axis rotor windings: $r_{fdb} = v_{fdb}/i_{fdb}$, $r_{1db} = v_{1db}/i_{1db}$ & $r_{2db} = v_{2db}/i_{2db}$.
q-axis rotor quantities (per-unit *unsaturated* q-axis mutual inductance $L_{aq} = \bar{L}_{aq}/L_{sb}$ is used in the following.)		
i_{kqb}	A	Base current of the q-axis damper windings: $i_{kqb} = (L_{aq}\varphi_{sb})/\bar{L}_{akq} = (\bar{L}_{ad}/\bar{L}_{akq})i_{sb}$, $k = 1, \ldots, 3$.
v_{kqb}	V	Base voltage of the q-axis damper windings: $v_{kqb} = S_b/i_{kqb}$, $k = 1, \ldots, 3$.
φ_{kqb}	Wb-turns	Base flux-linkages of the q-axis damper windings: $\varphi_{kqb} = v_{kqb}/\omega_b$, $k = 1, \ldots, 3$.
r_{kqb}	Ω	Base resistance of the q-axis damper windings: $r_{kqb} = v_{kqb}/i_{kqb}$, $k = 1, \ldots, 3$.

4.2.3.3 Parameter and variable definitions

The parameters and variables used in the formulation of the model are listed in Tables 4.3 and 4.4 respectively together with their base values as defined in Table 4.2.

Table 4.3 Summary of the parameters in the per-unit coupled-circuit representation of the *3d3q-c1* synchronous machine model.

Per-unit Parameter	Base Value (see Tab. 4.2)	Description
ω_b	n/a	The base frequency (elec. rad/s) which appears explicitly in the per-unit equations due to the choice of one second as the base value of time.
H	$(2U_b)/\omega_b^2$	Aggregate inertia constant of the generating unit. Refer to Appendix 4–II.2 for derivation.
D	T_{mb}/ω_{mb}, T_b/ω_b	Aggregate incremental mechanical damping torque coefficient of the generating unit. Refer to Appendix 4–II.2 for derivation.
r_s	Z_{sb}	Stator resistance, assumed identical in the *d*- and *q*-axes.
L_l	L_{sb}	Stator leakage inductance, assumed identical in the *d*- and *q*- axes.
L_{ad_u}, L_{aq_u}	L_{sb}	Respectively the *d*- and *q*-axis *unsaturated* airgap mutual inductance between the corresponding stator and rotor windings.
$\underset{\sim}{L}_{adq} = [L_{ad}\ L_{aq}]^T$	L_{sb}	The operating point dependent values of the *d*- and *q*-axis mutual inductances. (Note: $\underset{\sim}{L}_{adq}$ may be a variable depending on the method used to represent magnetic saturation).
r_{fd}, r_{1d}, r_{2d}	r_{fdb}, r_{1db}, r_{2db}	Resistances of the field winding and the first and second *d*-axis damper windings respectively.
L_{fd}, L_{1d}, L_{2d}		Leakage inductances of the field winding and the first and second *d*-axis damper windings respectively. These inductances represent flux that links only their respective windings.
L_{c1}	See Note (1)	Mutual inductance between the field and first damper winding which represents flux linkages between these windings but which do not link the stator or the second damper winding. To neglect unequal coupling between the *d*-axis rotor windings $L_{c1} = L_{c2} = 0$.
L_{c2}		Mutual inductance between the three *d*-axis rotor windings which represents flux that links all three *d*-axis rotor windings but not the stator.
r_{1q}, r_{2q}, r_{3q}	r_{1qb}, r_{1qb}, r_{2qb}	Resistances of the three *q*-axis damper windings.

Per-unit Parameter	Base Value (see Tab. 4.2)	Description
L_{1q}, L_{2q}, L_{3q}	See Note (2)	Leakage inductances of the three q-axis damper windings. Note that unequal coupling between the q-axis rotor windings is not represented since the q-axis is observable only from the stator.

(1) The d-axis rotor-winding per-unit inductance matrix $\boldsymbol{L}_{rd} = L_{ad}\boldsymbol{u}\boldsymbol{u}^T + \boldsymbol{L}_{lrd} = \left(\frac{\omega_b}{S_b}\right)\boldsymbol{i}_{rdb}\bar{\boldsymbol{L}}_{rd}\boldsymbol{i}_{rdb}$ is defined in terms of the corresponding matrix $\bar{\boldsymbol{L}}_{rd}$ in terms of SI units where the d-axis rotor-winding leakage inductance matrix $\boldsymbol{L}_{lrd}$ is defined in (4.19), $\bar{\boldsymbol{L}}_{rd} = \begin{bmatrix} \bar{L}_{ffd} & \bar{L}_{f1d} & \bar{L}_{f2d} \\ \bar{L}_{f1d} & \bar{L}_{11d} & \bar{L}_{12d} \\ \bar{L}_{f2d} & \bar{L}_{12d} & \bar{L}_{22d} \end{bmatrix}$ (H) and $\boldsymbol{i}_{rdb} = \mathfrak{D}(i_{fdb}, i_{1db}, i_{2db})$.

(2) The q-axis rotor-winding per-unit inductance matrix $\boldsymbol{L}_{rq} = L_{aq}\boldsymbol{u}\boldsymbol{u}^T + \boldsymbol{L}_{lrq} = \left(\frac{\omega_b}{S_b}\right)\boldsymbol{i}_{rdq}\bar{\boldsymbol{L}}_{rq}\boldsymbol{i}_{rqb}$ is defined in terms of the corresponding matrix $\bar{\boldsymbol{L}}_{rq}$ in terms of SI units where the q-axis rotor-winding leakage inductance matrix $\boldsymbol{L}_{lrq}$ is defined in (4.20), $\bar{\boldsymbol{L}}_{rq} = \begin{bmatrix} \bar{L}_{11q} & \bar{L}_{12q} & \bar{L}_{13q} \\ \bar{L}_{12q} & \bar{L}_{22q} & \bar{L}_{23q} \\ \bar{L}_{13q} & \bar{L}_{23q} & \bar{L}_{33q} \end{bmatrix}$ (H) and $\boldsymbol{i}_{rqb} = \mathfrak{D}(i_{1qb}, i_{2qb}, i_{3qb})$.

In the literature on models of generators reference is often made to per-unit machine reactances (e.g. X_d, X_{ad}, etc.) rather than per-unit machine inductances (e.g. L_d, L_{ad}). In this book we adopt per-unit machine inductances. It should be noted that in the per-unit system used the values of per-unit reactances and inductances can be used interchangeably.

Table 4.4 Summary of variables in the per-unit coupled-circuit representation of the *3d3q-c1* synchronous machine.

Variable (in per-unit)	Base Value (see sec. Tab. 4.2)	Description
$\underline{v}_{dq} = \begin{bmatrix} v_d & v_q \end{bmatrix}^T$	v_{sb}	*d*- and *q*-axis stator terminal voltage respectively.
$\underline{i}_{dq} = \begin{bmatrix} i_d & i_q \end{bmatrix}^T$	i_{sb}	*d*- and *q*-axis stator winding current respectively. Direction of positive stator current is from the generator into the network.
$\underline{\varphi}_{adq} = \begin{bmatrix} \varphi_{ad} & \varphi_{aq} \end{bmatrix}^T$	φ_{sb}	*d*- and *q*-axis airgap flux linkages respectively.
$\underline{\varphi}_{dq} = \begin{bmatrix} \varphi_d & \varphi_q \end{bmatrix}^T$	φ_{sb}	*d*- and *q*-axis stator flux linkages respectively.
$\underline{\varphi}_{rd} = \begin{bmatrix} \varphi_{fd} & \varphi_{1d} & \varphi_{2d} \end{bmatrix}^T$	$\begin{bmatrix} \varphi_{fdb} & \varphi_{1db} & \varphi_{2db} \end{bmatrix}^T$	*d*-axis rotor flux linkages in which subscripts 'fd' refers to the field winding and '1d' and '2d' refer respectively to the first and second *d*-axis damper windings.
$\underline{\varphi}_{rq} = \begin{bmatrix} \varphi_{1q} & \varphi_{2q} & \varphi_{3q} \end{bmatrix}^T$	$\begin{bmatrix} \varphi_{1qb} & \varphi_{2qb} & \varphi_{3qb} \end{bmatrix}^T$	*q*-axis rotor flux linkages in which the subscripts '1q', '2q' and '3q' refer respectively to the three *q*-axis damper windings.
$\underline{v}_{rd} = \begin{bmatrix} e_{fd} & 0 & 0 \end{bmatrix}^T$	$\begin{bmatrix} e_{fdb} & v_{1db} & v_{2db} \end{bmatrix}^T$	*d*-axis rotor-winding voltages in which e_{fd} is the per-unit field winding voltage in the reciprocal base system. The damper windings are short-circuit so their voltages are zero.
$\underline{v}_{rq} = \begin{bmatrix} 0 & 0 & 0 \end{bmatrix}^T$	$\begin{bmatrix} v_{1qb} & v_{2qb} & v_{3qb} \end{bmatrix}^T$	*q*-axis rotor winding voltages are identically zero since the damper windings are short-circuit.
$\underline{i}_{rd} = \begin{bmatrix} i_{fd} & i_{1d} & i_{2d} \end{bmatrix}^T$	$\begin{bmatrix} i_{fdb} & i_{1db} & i_{2db} \end{bmatrix}^T$	*d*-axis rotor winding currents. i_{fd} is the per-unit field winding current in the reciprocal base system.
$\underline{i}_{rq} = \begin{bmatrix} i_{1q} & i_{2q} & i_{3q} \end{bmatrix}^T$	$\begin{bmatrix} i_{1qb} & i_{2qb} & i_{3qb} \end{bmatrix}^T$	*q*-axis rotor winding currents.
Note: For compactness the *d*- and *q*-axis rotor winding variables are aggregated as follows: $\underline{\varphi}_{rdq} = \begin{bmatrix} \underline{\varphi}_{rd} \\ \underline{\varphi}_{rq} \end{bmatrix}, \ \underline{v}_{rdq} = \begin{bmatrix} \underline{v}_{rd} \\ \underline{v}_{rq} \end{bmatrix}, \ \underline{i}_{rdq} = \begin{bmatrix} \underline{i}_{rd} \\ \underline{i}_{rq} \end{bmatrix}.$		
I_{fd}, E_{fd}	I_{fdb}, E_{fdb}	Per-unit field current and voltage respectively in the non-reciprocal base system.

Variable (in per-unit)	Base Value (see sec. Tab. 4.2)	Description
$\underset{\sim}{i}_{sdq} = \begin{bmatrix} i_{sd} & i_{sq} \end{bmatrix}^T$;	$\begin{bmatrix} i_{fdb} & i_{1qb} \end{bmatrix}^T$	The demagnetizing components of the *d*- and *q*-axis excitation current that is required to account for the effects of magnetic saturation in the respective axes. This saturation excitation current is incorporated in the model only if the second method of saturation modelling in Section 4.2.8.2 is employed.
$\underset{\sim}{I}_{sdq} = \begin{bmatrix} I_{sd} & I_{sq} \end{bmatrix}^T$ where $I_{sd} = L_{ad_u} i_{sd}$ $I_{sq} = L_{aq_u} i_{sq}$	$\begin{bmatrix} i_{fdb}/L_{ad_u} \\ i_{1qb}/L_{aq_u} \end{bmatrix}$	As above, but the non-reciprocal per-unit system is employed. This representation of the demagnetizing effects of magnetic saturation is employed in the Classical Parameter formulation of the em equations in Section 4.2.13.
δ	(elec. rad)	Relative rotor angle being the angular position of the *d*-axis with respect to the synchronously rotating network reference (in elec. rad).
θ	(elec. rad)	Stationary rotor angle being the angular position of the *d*-axis with respect to a stationary reference (in elec. rad).
ω	ω_b	Rotor-speed.
ω_0	ω_b	Synchronous speed. Note, if the nominal system frequency is equal to the generator base frequency then $\omega_0 = 1$.
P_e	S_b	Electrical power output.
T_g	T_{mb}	Electromagnetic (or airgap) torque.

4.2.3.4 Summary of the coupled-circuit formulation of the generator electromagnetic equations

Summarized below are the per-unit coupled-circuit equations describing the electromagnetic behaviour of the generator in the rotating *dq* coordinate system. These equations are developed from first principles in [12].

The *d*- and *q*-axis rotor-winding voltage equations are respectively:

$$\frac{1}{\omega_b} p \begin{bmatrix} \varphi_{fd} \\ \varphi_{1d} \\ \varphi_{2d} \end{bmatrix} = \begin{bmatrix} 1 \\ 0 \\ 0 \end{bmatrix} e_{fd} - \begin{bmatrix} r_{fd} & 0 & 0 \\ 0 & r_{1d} & 0 \\ 0 & 0 & r_{2d} \end{bmatrix} \begin{bmatrix} i_{fd} \\ i_{1d} \\ i_{2d} \end{bmatrix} \text{ and } \tag{4.2}$$

$$\frac{1}{\omega_b}p\begin{bmatrix}\varphi_{1q}\\ \varphi_{2q}\\ \varphi_{3q}\end{bmatrix} = -\begin{bmatrix}r_{1q} & 0 & 0\\ 0 & r_{2q} & 0\\ 0 & 0 & r_{3q}\end{bmatrix}\begin{bmatrix}i_{1q}\\ i_{2q}\\ i_{3q}\end{bmatrix}, \tag{4.3}$$

These equations are conveniently expressed in the following compact forms:

$$p\underset{\sim}{\varphi}_{rd} = \boldsymbol{b}_{red}e_{fd} - \omega_b \boldsymbol{r}_{rd}\underset{\sim}{i}_{rd} \text{ in which } \boldsymbol{r}_{rd} = \mathfrak{D}(r_{fd},\ r_{1d},\ r_{2d});\ \boldsymbol{b}_{red} = \begin{bmatrix}\omega_b & 0 & 0\end{bmatrix}^T \tag{4.4}$$

$$\text{and } p\underset{\sim}{\varphi}_{rq} = -\omega_b \boldsymbol{r}_{rq}\underset{\sim}{i}_{rq} \text{ in which } \boldsymbol{r}_{rq} = \mathfrak{D}(r_{1q},\ r_{2q},\ r_{3q}). \tag{4.5}$$

The above equations are combined as follows.

Rotor Voltage Equations

$$p\underset{\sim}{\varphi}_{rdq} = -(\omega_b \boldsymbol{r}_{rdq})\underset{\sim}{i}_{rdq} + \boldsymbol{b}_{re}e_{fd}, \tag{4.6}$$

$$\text{in which } \boldsymbol{r}_{rdq} = \mathfrak{D}(\boldsymbol{r}_{rd}, \boldsymbol{r}_{rq}) \text{ and} \tag{4.7}$$

$$\boldsymbol{b}_{re} = \begin{bmatrix}\boldsymbol{b}_{red}\\ \boldsymbol{0}\end{bmatrix} \text{ where } \boldsymbol{b}_{red} = \begin{bmatrix}\omega_b & 0 & 0\end{bmatrix}^T. \tag{4.8}$$

The per-unit flux linkage equations for the d-axis are presented in terms of the winding mutual and leakage inductances, and the winding currents based on Figure 4.2 on page 95. As mentioned earlier, the flux linkages of the q-axis windings neglect unequal coupling between the q-axis rotor windings. This is valid because the q-axis is observable only from the stator. Consequently an equivalent circuit that assumes equal coupling between the q-axis rotor windings can be identified that represents the observable q-axis behaviour.

Referring to Figure 4.2 the d-axis mutual (or airgap) flux linkages are:

$$\varphi_{ad} = L_{ad}(i_{fd} + i_{1d} + i_{2d} - i_d) = L_{ad}(\boldsymbol{u}^T\underset{\sim}{i}_{rd} - i_d), \text{ where } \boldsymbol{u} = \begin{bmatrix}1 & 1 & 1\end{bmatrix}^T \tag{4.9}$$

and analogously the q-axis mutual flux linkages are:

$$\varphi_{aq} = L_{aq}(i_{1q} + i_{2q} + i_{3q} - i_q) = L_{aq}(\boldsymbol{u}^T\underset{\sim}{i}_{rq} - i_q). \tag{4.10}$$

The values of L_{ad} and L_{aq} depend on the method used to represent magnetic saturation. Two mathematically equivalent methods for representing magnetic saturation are considered. In the following, the method is denoted by the parameter $s_m = 1$ if the first method is being used, or $s_m = 2$ for the second method. If magnetic saturation is to be neglected then $s_m = 0$.

The first method for representing magnetic saturation ($s_m = 1$) is described in detail in Section 4.2.8.1. In this method it is assumed that the airgap mutual inductances are non-linear functions of the airgap flux linkages whereas the leakage inductances in both axes and the unequal mutual inductances between the d-axis rotor windings are assumed to independent of the fluxes linking them and are therefore constant parameters. Thus, in this method:

$$L_{ad} = L_{ad}(\varphi_{ad}, \varphi_{aq}), \quad L_{aq} = L_{aq}(\varphi_{ad}, \varphi_{aq}). \tag{4.11}$$

In the second method ($s_m = 2$), for which the details are provided in Section 4.2.8.2, the unsaturated values of the airgap mutual inductances are retained; instead the components of the excitation current, i_{sd} and i_{sq}, necessary to represent the demagnetizing effect of saturation in the respective axes, are deducted from the excitation current in equations (4.9) and (4.10) respectively, i.e.

$$\begin{bmatrix} \varphi_{ad} \\ \varphi_{aq} \end{bmatrix} = \begin{bmatrix} L_{ad} & 0 \\ 0 & L_{aq} \end{bmatrix} \left(\begin{bmatrix} \boldsymbol{u}^T & \boldsymbol{0} \\ \boldsymbol{0} & \boldsymbol{u}^T \end{bmatrix} \begin{bmatrix} \underset{\sim}{i}_{rd} \\ \underset{\sim}{i}_{rq} \end{bmatrix} - \begin{bmatrix} i_d \\ i_q \end{bmatrix} - s_2 \begin{bmatrix} i_{sd} \\ i_{sq} \end{bmatrix} \right), \tag{4.12}$$

$$\text{where } s_2 = \begin{cases} 1 & \text{if } s_m = 2 \\ 0 & \text{otherwise} \end{cases}. \tag{4.13}$$

If method 2 is being used to represent magnetic saturation then the unsaturated values of the airgap mutual inductances are used in the equations (i.e. $L_{ad} = L_{ad_u}$ *and* $L_{aq} = L_{aq_u}$*).*

As explained in Section 4.2.8.2 the demagnetizing currents which account for the effect of saturation are non-linear functions of the airgap flux linkages:

$$i_{sd} = i_{sd}(\varphi_{ad}, \varphi_{aq}), \quad i_{sq} = i_{sq}(\varphi_{ad}, \varphi_{aq}). \tag{4.14}$$

Defining $\boldsymbol{u}_2 = \mathfrak{D}(\boldsymbol{u}, \boldsymbol{u})$ results in the following compact matrix equation for the d- and q-axis airgap flux linkages:

<u>Airgap Mutual Flux Linkage Equations</u>

$$\underset{\sim}{\varphi}_{adq} = \boldsymbol{L}_{\boldsymbol{adq}}(\boldsymbol{u}_2^T \underset{\sim}{i}_{rdq} - \underset{\sim}{i}_{dq} - s_2 \underset{\sim}{i}_{sdq}), \quad \text{in which} \tag{4.15}$$

$$\boldsymbol{L}_{\boldsymbol{adq}} = \begin{cases} \mathfrak{D}(L_{ad}(\varphi_{ad}, \varphi_{aq}), L_{aq}(\varphi_{ad}, \varphi_{aq})) & \text{if } s_m = 1 \\ \mathfrak{D}(L_{ad_u}, L_{aq_u}) = \boldsymbol{L}_{\boldsymbol{adq_u}} & \text{otherwise} \end{cases} \tag{4.16}$$

$$\text{and } s_2 = \begin{cases} 1 & \text{if } s_m = 2 \\ 0 & \text{otherwise} \end{cases}. \tag{4.17}$$

The d- and q-axis rotor winding flux-linkages are expressed in terms of the mutual and leakage flux linkages as follows:

$$\underset{\sim}{\varphi}_{rd} = \varphi_{ad}\boldsymbol{u} + \boldsymbol{L}_{lrd}\underset{\sim}{i}_{rd} \text{ and } \underset{\sim}{\varphi}_{rq} = \varphi_{aq}\boldsymbol{u} + \boldsymbol{L}_{lrq}\underset{\sim}{i}_{rq}. \tag{4.18}$$

in which the d- and q-axis rotor leakage inductance matrices are respectively:

$$\boldsymbol{L}_{lrd} = \begin{bmatrix} (L_{fd}+L_{c1}+L_{c2}) & (L_{c1}+L_{c2}) & L_{c2} \\ (L_{c1}+L_{c2}) & (L_{1d}+L_{c1}+L_{c2}) & L_{c2} \\ L_{c2} & L_{c2} & (L_{2d}+L_{c2}) \end{bmatrix} \text{ and} \tag{4.19}$$

$$\boldsymbol{L}_{lrq} = \mathfrak{D}(L_{1q}, \; L_{2q}, \; L_{3q}). \tag{4.20}$$

From (4.18) the rotor winding flux linkage equations are written in the following compact form:

Rotor winding flux-linkage equations

$$\underset{\sim}{\varphi}_{rdq} = \boldsymbol{u}_2\underset{\sim}{\varphi}_{adq} + \boldsymbol{L}_{lrdq}\underset{\sim}{i}_{rdq} \text{ in which } \boldsymbol{L}_{lrdq} = \mathfrak{D}(\boldsymbol{L}_{lrd}, \boldsymbol{L}_{lrq}). \tag{4.21}$$

The stator winding flux linkage equations are:

$$\begin{bmatrix} \varphi_d \\ \varphi_q \end{bmatrix} = \begin{bmatrix} \varphi_{ad} \\ \varphi_{aq} \end{bmatrix} - \begin{bmatrix} L_l & 0 \\ 0 & L_l \end{bmatrix}\begin{bmatrix} i_d \\ i_q \end{bmatrix}, \tag{4.22}$$

which have the following compact form:

Stator winding flux-linkage equations

$$\underset{\sim}{\varphi}_{dq} = \underset{\sim}{\varphi}_{adq} - L_l\underset{\sim}{i}_{dq} \tag{4.23}$$

The generator d- and q-axis stator voltage equations in which, consistently with the modelling assumptions in Section 4.2.2, both the transformer voltages and the rotor-speed perturbations are neglected are:

$$\begin{bmatrix} v_d \\ v_q \end{bmatrix} = -\begin{bmatrix} r_s & 0 \\ 0 & r_s \end{bmatrix}\begin{bmatrix} i_d \\ i_q \end{bmatrix} + \omega_0\begin{bmatrix} 0 & -1 \\ 1 & 0 \end{bmatrix}\begin{bmatrix} \varphi_d \\ \varphi_q \end{bmatrix}, \tag{4.24}$$

which are expressed in the following compact form:

Stator winding voltage equations

$$\underset{\sim}{v}_{dq} = -r_s \underset{\sim}{i}_{dq} + \omega_0 \boldsymbol{W}_{dq} \underset{\sim}{\varphi}_{dq} \quad \text{where } \boldsymbol{W}_{dq} = \begin{bmatrix} 0 & -1 \\ 1 & 0 \end{bmatrix}. \tag{4.25}$$

The electromagnetic behaviour of the machine is characterised by the differential equations for the rotor winding voltages (4.6) and the algebraic equations for the airgap flux linkages (4.15), the rotor winding flux linkages (4.21), the stator winding flux linkages (4.23) and the stator voltages (4.25).

4.2.3.5 Linearization of the coupled-circuit formulation of the electromagnetic equations

The coupled-circuit formulation of the generator electromagnetic equations in the previous section are now linearized about the initial steady-state operating point of the machine. The procedure for determining the initial steady-state values of the generator quantities from the power flow solution of the generator stator terminal quantities is given in Section 4.2.9. All of the electromagnetic equations are linear, apart from the airgap flux-linkage equations. The linear equations are linearized trivially by replacing the variables with their perturbed values (i.e. replace x with Δx).

Equation (4.15) for the airgap mutual flux-linkages is firstly linearized for the case in which saturation method 1 (i.e. $s_m = 1$) is used. The initial steady-state values of the saturated air-gap mutual inductances are (L_{ad_0}, L_{aq_0}) and the corresponding flux linkages are $(\varphi_{ad_0}, \varphi_{aq_0})$. The perturbations in the airgap flux linkages are:

$$\Delta\underset{\sim}{\varphi}_{adq} = \boldsymbol{L}_{\boldsymbol{adq_0}}(\boldsymbol{u}_2^T \Delta\underset{\sim}{i}_{rdq} - \Delta\underset{\sim}{i}_{dq}) + s_1 \boldsymbol{C}_{\boldsymbol{aldq_0}} \Delta\underset{\sim}{L}_{adq}, \tag{4.26}$$

$$\text{in which } \boldsymbol{L}_{\boldsymbol{adq_0}} = \mathfrak{D}(L_{ad_0}, L_{aq_0}), \quad \boldsymbol{C}_{\boldsymbol{aldq_0}} = \mathfrak{D}\left(\frac{\varphi_{ad_0}}{L_{ad_0}}, \frac{\varphi_{aq_0}}{L_{aq_0}}\right) \text{ and} \tag{4.27}$$

$$s_1 = \begin{cases} 1 & \text{if } s_m = 1 \\ 0 & \text{otherwise} \end{cases}. \tag{4.28}$$

If saturation method 2 (i.e. $s_m = 2$) is used then the linearized airgap flux linkage equations are:

$$\Delta\underset{\sim}{\varphi}_{adq} = \boldsymbol{L}_{\boldsymbol{adq_u}}(\boldsymbol{u}_2^T \Delta\underset{\sim}{i}_{rdq} - \Delta\underset{\sim}{i}_{dq} - s_2 \Delta\underset{\sim}{i}_{sdq}). \tag{4.29}$$

It is noted that the two methods of representing magnetic saturation are mutually exclusive. Thus, it is convenient to define the "saturation variable" $\underset{\sim}{z}_s$ depending on the method used to represent magnetic saturation:

$$\Delta \underset{\sim}{z}_s = \begin{cases} \Delta \underset{\sim}{L}_{adq} & \text{if } s_m = 1 \\ \Delta \underset{\sim}{i}_{sdq} & \text{if } s_m = 2 \\ \emptyset & \text{if } s_m = 0 \end{cases} \tag{4.30}$$

As mentioned earlier the definition of the airgap mutual inductance matrix also depends on which saturation method is employed, as formalized by the following definition:

$$\boldsymbol{L}_{adq} = \begin{cases} \boldsymbol{L}_{adq_0} = \mathfrak{D}(L_{ad_0}, L_{aq_0}) & \text{if } s_m = 1 \\ \boldsymbol{L}_{adq_u} = \mathfrak{D}(L_{ad_u}, L_{aq_u}) & \text{otherwise} \end{cases} \tag{4.31}$$

From equations (4.26) and (4.29) the saturation coefficient matrix $\boldsymbol{C}_{asdq_0}$ is defined as:

$$\boldsymbol{C}_{asdq_0} = \begin{cases} \boldsymbol{C}_{aldq_0} & \text{if } s_m = 1 \\ -\boldsymbol{L}_{adq} & \text{if } s_m = 2 \\ \boldsymbol{\emptyset} & \text{if } s_m = 0 \end{cases} \tag{4.32}$$

Thus the linearized matrix equation for the airgap mutual flux linkages, which is applicable to either method of representing magnetic saturation, is:

$$\Delta \underset{\sim}{\varphi}_{adq} = \boldsymbol{L}_{adq}(\boldsymbol{u}_2^T \Delta \underset{\sim}{i}_{rdq} - \Delta \underset{\sim}{i}_{dq}) + \boldsymbol{C}_{asdq_0} \Delta \underset{\sim}{z}_s . \tag{4.33}$$

It will be shown in Section 4.2.8 that the perturbation in the saturation variable is related to perturbations in the airgap mutual flux linkages by the operating point dependent matrix $\boldsymbol{C}_{sadq_0}$ as follows:

$$\Delta \underset{\sim}{z}_s = \boldsymbol{C}_{sadq_0} \Delta \underset{\sim}{\varphi}_{adq} , \tag{4.34}$$

$$\text{in which } \boldsymbol{C}_{sadq_0} = \begin{cases} \boldsymbol{C}_{ladq_0} & \text{if } s_m = 1 \\ \boldsymbol{C}_{madq_0} & \text{if } s_m = 2 \\ \boldsymbol{\emptyset} & \text{if } s_m = 0 \end{cases} \tag{4.35}$$

where $\boldsymbol{C}_{ladq_0}$ is defined in (4.73) on page 123 and $\boldsymbol{C}_{madq_0}$ is defined in (4.80) on page 124.

Thus, from equations (4.6) on page 104, (4.21), (4.23), (4.25), (4.33) and (4.34) the differential and algebraic equations describing the electromagnetic behaviour of the machine are linearized about the initial steady-state operating point of the machine to yield:

$$
\begin{aligned}
p\Delta\underset{\sim}{\varphi}_{rdq} &= -(\omega_b \boldsymbol{r}_{\boldsymbol{rdq}})\Delta\underset{\sim}{i}_{rdq} + \boldsymbol{b}_{\boldsymbol{re}}\Delta e_{fd} \\
\boldsymbol{0} &= \boldsymbol{u}_2\Delta\underset{\sim}{\varphi}_{adq} + \boldsymbol{L}_{\boldsymbol{lrdq}}\Delta\underset{\sim}{i}_{rdq} - \Delta\underset{\sim}{\varphi}_{rdq} \\
\boldsymbol{0} &= \boldsymbol{L}_{\boldsymbol{adq}}(\boldsymbol{u}_2^T\Delta\underset{\sim}{i}_{rdq} - \Delta\underset{\sim}{i}_{dq}) + \boldsymbol{C}_{\boldsymbol{asdq_0}}\Delta\underset{\sim}{z}_s - \Delta\underset{\sim}{\varphi}_{adq} \\
\boldsymbol{0} &= \Delta\underset{\sim}{\varphi}_{adq} - L_l\Delta\underset{\sim}{i}_{dq} - \Delta\underset{\sim}{\varphi}_{dq} \\
\boldsymbol{0} &= \boldsymbol{C}_{\boldsymbol{sadq_0}}\Delta\underset{\sim}{\varphi}_{adq} - \Delta\underset{\sim}{z}_s \\
\boldsymbol{0} &= -r_s\Delta\underset{\sim}{i}_{dq} + \omega_0\boldsymbol{W}_{\boldsymbol{dq}}\Delta\underset{\sim}{\varphi}_{dq} - \Delta\underset{\sim}{v}_{dq}
\end{aligned}
\tag{4.36}
$$

The above equations are rewritten in the following matrix form:

$$
\begin{bmatrix} p\Delta\underset{\sim}{\varphi}_{rdq} \\ \boldsymbol{0} \\ \boldsymbol{0} \\ \boldsymbol{0} \\ \boldsymbol{0} \\ \boldsymbol{0} \end{bmatrix}
=
\begin{bmatrix}
\boldsymbol{0} & -(\omega_b\boldsymbol{r}_{\boldsymbol{rdq}}) & \boldsymbol{0} & \boldsymbol{0} & \boldsymbol{0} & \boldsymbol{0} & \boldsymbol{0} \\
-\boldsymbol{I} & \boldsymbol{L}_{\boldsymbol{lrdq}} & \boldsymbol{u}_2 & \boldsymbol{0} & \boldsymbol{0} & \boldsymbol{0} & \boldsymbol{0} \\
\boldsymbol{0} & \boldsymbol{L}_{\boldsymbol{adq}}\boldsymbol{u}_2^T & -\boldsymbol{I} & \boldsymbol{0} & \boldsymbol{C}_{\boldsymbol{asdq_0}} & -\boldsymbol{L}_{\boldsymbol{adq}} & \boldsymbol{0} \\
\boldsymbol{0} & \boldsymbol{0} & \boldsymbol{I} & -\boldsymbol{I} & \boldsymbol{0} & -L_l\boldsymbol{I} & \boldsymbol{0} \\
\boldsymbol{0} & \boldsymbol{0} & \boldsymbol{C}_{\boldsymbol{sadq_0}} & \boldsymbol{0} & -\boldsymbol{I} & \boldsymbol{0} & \boldsymbol{0} \\
\boldsymbol{0} & \boldsymbol{0} & \boldsymbol{0} & \omega_0\boldsymbol{W}_{\boldsymbol{dq}} & \boldsymbol{0} & -r_s\boldsymbol{I} & -\boldsymbol{I}
\end{bmatrix}
\begin{bmatrix} \Delta\underset{\sim}{\varphi}_{rdq} \\ \Delta\underset{\sim}{i}_{rdq} \\ \Delta\underset{\sim}{\varphi}_{adq} \\ \Delta\underset{\sim}{\varphi}_{dq} \\ \Delta\underset{\sim}{z}_s \\ \Delta\underset{\sim}{i}_{dq} \\ \Delta\underset{\sim}{v}_{dq} \end{bmatrix}
+
\begin{bmatrix} \boldsymbol{b}_{\boldsymbol{re}} \\ \boldsymbol{0} \\ \boldsymbol{0} \\ \boldsymbol{0} \\ \boldsymbol{0} \\ \boldsymbol{0} \end{bmatrix}
\Delta e_{fd}
\tag{4.37}
$$

In order to consolidate the structure of the above equations it is convenient to define the following consolidated vector of n_z algebraic variables:

$$
\Delta\underset{\sim}{z}_e = \begin{bmatrix} \Delta\underset{\sim}{i}_{rdq} \\ \Delta\underset{\sim}{\varphi}_{adq} \\ \Delta\underset{\sim}{\varphi}_{dq} \\ \Delta\underset{\sim}{z}_s \end{bmatrix}
\tag{4.38}
$$

and the associated consolidated matrix coefficients:

$$C_{re} = \begin{bmatrix} -(\omega_b r_{rdq}) & 0 & 0 & 0 \end{bmatrix},\; C_{er} = \begin{bmatrix} -I \\ 0 \\ 0 \\ 0 \end{bmatrix},\; C_{ee} = \begin{bmatrix} L_{lrdq} & u_2 & 0 & 0 \\ L_{adq}u_2^T & -I & 0 & C_{asdq_0} \\ 0 & I & -I & 0 \\ 0 & C_{sadq_0} & 0 & -I \end{bmatrix},$$

$$C_{ei} = \begin{bmatrix} 0 \\ -L_{adq} \\ L_l I \\ 0 \end{bmatrix},\; C_{ve} = \begin{bmatrix} 0 & 0 & \omega_0 W_{dq} & 0 \end{bmatrix} \tag{4.39}$$

Substitution of the quantities in (4.38) and (4.39) into (4.37) results in the following compact form of the linearized electromagnetic equations:

$$\begin{bmatrix} p\Delta\varphi_{rdq} \\ 0 \\ 0 \end{bmatrix} = \begin{bmatrix} 0 & C_{re} & 0 & 0 \\ C_{er} & C_{ee} & C_{ei} & 0 \\ 0 & C_{ve} & -r_s I & -I \end{bmatrix} \begin{bmatrix} \Delta\varphi_{rdq} \\ \Delta z_e \\ \Delta i_{dq} \\ \Delta v_{dq} \end{bmatrix} + \begin{bmatrix} b_{re} \\ 0 \\ 0 \end{bmatrix} \Delta e_{fd} \tag{4.40}$$

These equations preserve the important structural characteristics of the underlying coupled-circuit model and the elements in the coefficient matrices have a simple form. For these and other reasons it may be desirable to implement the generator model in the form of (4.40) without eliminating the internal algebraic variables Δz_e. However, it is also straight forward to eliminate Δz_e from (4.40) as follows:

$$\begin{bmatrix} p\Delta\varphi_{rdq} \\ 0 \end{bmatrix} = \begin{bmatrix} A_r & B_{ri} & 0 \\ C_{vr} & D_{vi} & -I \end{bmatrix} \begin{bmatrix} \Delta\varphi_{rdq} \\ \Delta i_{dq} \\ \Delta v_{dq} \end{bmatrix} + \begin{bmatrix} b_{re} \\ 0 \end{bmatrix} \Delta e_{fd} \tag{4.41}$$

in which $A_r = C_{re}J_{er}$, $B_{ri} = C_{re}J_{ei}$, $C_{vr} = C_{ve}J_{er}$ and $D_{vi} = C_{ve}J_{ei} - r_s I$, (4.42)

$$\text{where } J_{er} = -C_{ee}^{-1}C_{er} \text{ and } J_{ei} = -C_{ee}^{-1}C_{ei}. \tag{4.43}$$

The perturbations in the algebraic variables are given by:

$$\Delta z_e = J_{er}\Delta\varphi_{rdq} + J_{ei}\Delta i_{dq}. \tag{4.44}$$

Suppose a subset of the generator algebraic-variables in (4.38) is required as output variables Δy_g from the generator. Let k_i, $i = 1, \ldots, n_y$ be the index in Δz_e of the ith output variable. Then,

$$\Delta y_g = S_{gz}\Delta z_e, \tag{4.45}$$

in which $\boldsymbol{S}_{gz} \in \Re^{n_y \times n_z}$, $\boldsymbol{S}_{gz}(i, k_i) = 1$, $i = 1, \ldots, n_y$ and all other elements of $\boldsymbol{S}_{gz}$ are zero. Substituting for $\Delta \underset{\sim}{z}_e$ from (4.44) into (4.45) yields the following output equation in terms of the generator state-variables and stator current components:

$$\Delta \underset{\sim}{y}_g = (\boldsymbol{S}_{gz}\boldsymbol{J}_{er})\Delta \underset{\sim}{\varphi}_{rdq} + (\boldsymbol{S}_{gz}\boldsymbol{J}_{ei})\Delta \underset{\sim}{i}_{dq} . \tag{4.46}$$

For example, in order to monitor (i) the generator field-current set $k_i = 1$, or (ii) the *d*-axis airgap flux-linkages set $k_i = (n_d + n_q) + 1$.

The equations for perturbations in AC terminal quantities such as the voltage magnitude and angle, real and reactive power output, current magnitude, are formulated for FACTS Devices in Section 4.3.1. These equations, which are calculated in terms of the perturbations in the voltage and current components in the *RI* network frame of reference are also applicable to the generator stator terminal.

4.2.4 Alternative *d*- and *q*-axis rotor structures

The model development so far has been based on a representation with three rotor windings in each axis and in which the unequal mutual coupling between the *d*-axis rotor windings is represented by means of the Canay inductances (i.e. model *3d3q-c1*). A suite of simpler model structures that are commonly employed in practice is readily extracted from the *3d3q-c1* model. The modifications to the rotor winding variables and parameters required to represent a range of alternative rotor structures are summarized in the following. Once these modifications are made the formulation of the linearized equations of the generator in matrix form proceeds independently of the number of rotor windings.

Models of *d*-axis rotor structures with a field winding and respectively $\mathrm{n_{kd}} = 0$, 1 and 2 damper windings for a total of $\mathrm{n_d} = 1$, 2 and 3 rotor windings are developed. Cases with two or three *d*-axis rotor windings represented are provided which either neglect (c = 0) or include (c = 1) unequal coupling between the rotor windings. Definitions are given in Table 4.5 for the *d*-axis rotor-winding variables $\underset{\sim}{\varphi}_{rd}$ and $\underset{\sim}{i}_{rd}$, the associated resistance and leakage inductance parameters $\boldsymbol{r}_{rd}$ and $\boldsymbol{L}_{lrd}$, and the input matrix $\boldsymbol{b}_{red}$ for the resulting five alternative structures.

Table 4.5 Summary of *d*-axis rotor variables and parameters with one, two or three rotor windings and with the inclusion (c = 1) or exclusion (c = 0) of unequal coupling between the rotor windings.

	n_d	$\underline{\varphi}_{rd}$	$\underline{i}_{rd}$	$\boldsymbol{b}_{red}$	$\boldsymbol{r}_{rd}$	c	$\boldsymbol{L}_{lrd}$
1	1	φ_{fd}	i_{fd}	ω_b	r_{fd}	0	L_{fd}
2	2	$\begin{bmatrix}\varphi_{fd}\\ \varphi_{1d}\end{bmatrix}$	$\begin{bmatrix}i_{fd}\\ i_{1d}\end{bmatrix}$	$\begin{bmatrix}\omega_b\\ 0\end{bmatrix}$	$\mathfrak{D}(r_{fd}, r_{1d})$	0	$\mathfrak{D}(L_{fd}, L_{1d})$
3						1	$\begin{bmatrix}(L_{fd}+L_{c1}) & L_{c1}\\ L_{c1} & (L_{1d}+L_{c1})\end{bmatrix}$
4	3	$\begin{bmatrix}\varphi_{fd}\\ \varphi_{1d}\\ \varphi_{2d}\end{bmatrix}$	$\begin{bmatrix}i_{fd}\\ i_{1d}\\ i_{2d}\end{bmatrix}$	$\begin{bmatrix}\omega_b\\ 0\\ 0\end{bmatrix}$	$\mathfrak{D}(r_{fd}, r_{1d}, r_{2d})$	0	$\mathfrak{D}(L_{fd}, L_{1d}, L_{2d})$
5						1	$\begin{bmatrix}(L_{fd}+L_{c1}+L_{c2}) & (L_{c1}+L_{c2}) & L_{c2}\\ (L_{c1}+L_{c2}) & (L_{1d}+L_{c1}+L_{c2}) & L_{c2}\\ L_{c2} & L_{c2} & (L_{2d}+L_{c2})\end{bmatrix}$

Note that the leakage inductance matrices for rotor structures #2 and #4 which neglect unequal coupling between the rotor windings (i.e. c = 0) can be obtained from the corresponding matrices for rotor structures #3 and #5 (i.e. c = 1) by setting, in the latter structures, the rotor mutual inductances L_{c1} and L_{c2} to zero.

Table 4.6 presents definitions of the *q*-axis rotor-variables $\underline{\varphi}_{rq}$, $\underline{i}_{rq}$ and the associated resistance and inductance parameters $\boldsymbol{r}_{rq}$ and $\boldsymbol{L}_{lrq}$ for four *q*-axis rotor structures with n_q = 0, 1, 2 or 3 damper windings.

Table 4.6 Summary of *q*-axis rotor variables and parameters with zero to three rotor windings.

#	n_q	$\underline{\varphi}_{rq}$	$\underline{i}_{rq}$	$\boldsymbol{r}_{rq}$	$\boldsymbol{L}_{lrq}$
1	0	Ø	Ø	Ø	Ø
2	1	φ_{1q}	i_{1q}	r_{1q}	L_{1q}
3	2	$\begin{bmatrix}\varphi_{1q} & \varphi_{2q}\end{bmatrix}^T$	$\begin{bmatrix}i_{1q} & i_{2q}\end{bmatrix}^T$	$\mathfrak{D}(r_{1q}, r_{2q})$	$\mathfrak{D}(L_{1q}, L_{2q})$
4	3	$\begin{bmatrix}\varphi_{1q} & \varphi_{2q} & \varphi_{3q}\end{bmatrix}^T$	$\begin{bmatrix}i_{1q} & i_{2q} & i_{3q}\end{bmatrix}^T$	$\mathfrak{D}(r_{1q}, r_{2q}, r_{3q})$	$\mathfrak{D}(L_{1q}, L_{2q}, L_{3q})$

4.2.5 Per-unit electromagnetic torque and electrical power output

The per-unit equations for the electromagnetic torque and electrical power output of the machine presented below are developed from first principles in [12].

Under balanced conditions the non-linear per-unit equations for the electrical power output and electromagnetic torque are respectively:

$$P_e = i_d v_d + i_q v_q \text{ and } T_g = \varphi_d i_q - \varphi_q i_d. \tag{4.47}$$

It is instructive to express the per-unit generator power output in terms of the stator currents and flux linkages by eliminating from equation (4.47) the stator voltages by substitution from equation (4.24) on page 106 to yield:

$$P_e = \omega_0(\varphi_d i_q - \varphi_q i_d) - r_s(i_d^2 + i_q^2). \tag{4.48}$$

It is important to recall at this point that (4.24) neglects the transformer voltage terms and rotor-speed perturbations in the speed voltage terms. Therefore the electrical power equation in (4.48) also neglects these effects.

Then, substituting for $(\varphi_d i_q - \varphi_q i_d)$ in the preceding equation from (4.47) gives the following relationship between the per-unit electrical power and electromagnetic torque:

$$T_g = (P_e + r_s(i_d^2 + i_q^2))/\omega_0. \tag{4.49}$$

Substituting for P_e from (4.47) into (4.49) gives the electromagnetic torque in terms of the stator voltage and current components:

$$T_g = (i_d v_d + i_q v_q + r_s(i_d^2 + i_q^2))/\omega_0. \tag{4.50}$$

The airgap power is the electrical power output of the generator inclusive of the stator resistive losses:

$$P_g = P_e + r_s(i_d^2 + i_q^2) \text{, and thus} \tag{4.51}$$

$$P_g = \omega_0 T_g. \tag{4.52}$$

This confirms that the relationship between the airgap power and torque neglects the perturbations in rotor speed *which occurs as a consequence of neglecting perturbations in rotor speed in the calculation of the stator voltage.*

The electromagnetic torque equation (4.50) is linearized about the initial steady-state operating point $(v_{d_0}, v_{q_0}, i_{d_0}, i_{q_0})$ to yield:

$$\Delta T_g = \boldsymbol{C}_{\boldsymbol{tvdq}} \Delta \underset{\sim}{v}_{dq} + \boldsymbol{C}_{\boldsymbol{tidq}} \Delta \underset{\sim}{i}_{dq}, \tag{4.53}$$

in which $C_{tvdq} = \frac{1}{\omega_0}\begin{bmatrix} i_{d_0} & i_{q_0} \end{bmatrix}$ and $C_{tidq} = \frac{1}{\omega_0}\begin{bmatrix} (v_{d_0} + 2r_s i_{d_0}) & (v_{q_0} + 2r_s i_{q_0}) \end{bmatrix}$. (4.54)

Note that if, as normally is the case, the per-unit synchronous speed $\omega_0 = 1$ then from (4.49) the per-unit electrical power output of the generator is equal to the electromagnetic torque less the resistive losses in the stator winding.

An essential point, that is overlooked in some commercial software packages, is that when calculating the mechanical torque developed by the turbine from the mechanical power it necessary to neglect perturbations in the rotor speed. That is:

$$P_m = \omega_0 T_m . \tag{4.55}$$

This is to be consistent with neglecting the rotor speed perturbations in the relationship between airgap power and torque revealed in equation (4.52). It is shown in Appendix 4–II.5 that if the relationship between mechanical power and torque does include rotor speed perturbations then the effect is to erroneously increase the generator damping constant D by P_{m_0}/ω_0^2 per-unit.

4.2.6 Per-unit rotor equations of motion

The per-unit rotor equations of motion in which the rotor-position, δ, is measured with respect to the synchronously rotating network reference frame (see Section 4.2.10 and Appendix 4–II) are:

$$p\delta = \omega_b(\omega - \omega_0) \text{ and} \tag{4.56}$$

$$p\omega = \frac{1}{2H}(T_m - T_g - D(\omega - \omega_0)) . \tag{4.57}$$

in which T_g is defined in (4.47) and equivalently in (4.50). The mechanical torque T_m developed by the turbine is treated as a generator model input.

The equations of motion are linearized about the steady-state operating point, in which it is assumed that the machine is rotating synchronously with the network reference frame at a speed $\omega = \omega_0$ per-unit to yield:

$$p\Delta\delta = \omega_b(\Delta\omega) \text{ and} \tag{4.58}$$

$$p\Delta\omega = \frac{1}{2H}(\Delta T_m - \Delta T_g - D\Delta\omega). \tag{4.59}$$

Substituting the expression for the perturbation in the electromagnetic torque ΔT_g from (4.53) into the preceding equations yields the following formulation of the shaft acceleration equation in terms of perturbations in the rotor-speed, the stator voltage and current components and mechanical torque:

$$p\Delta\omega = -\left(\frac{D}{2H}\right)\Delta\omega - \left(\frac{1}{2H}\right)\boldsymbol{C}_{tvdq}\Delta\underset{\sim}{v}_{dq} - \left(\frac{1}{2H}\right)\boldsymbol{C}_{tidq}\Delta\underset{\sim}{i}_{dq} + \left(\frac{1}{2H}\right)\Delta T_m . \tag{4.60}$$

Importantly, the above formulation is independent of the model employed to represent the rotor-winding structure.

To facilitate analysis in later chapters an alternative formulation of the linearized acceleration equation is now developed in terms of the perturbation of acceleration power. Multiplying the acceleration equation (4.57) by the per-unit rotor-speed yields:

$$\omega(p\omega) = \frac{1}{2H}(P_m - P_g - D\omega(\omega - \omega_0)) . \tag{4.61}$$

Substituting for P_g from (4.51) in the preceding equation gives:

$$\omega(p\omega) = \frac{1}{2H}(P_m - P_e - D\omega(\omega - \omega_0) - r_s(i_d^2 + i_q^2)) . \tag{4.62}$$

Linearizing the preceding equation yields:

$$p\Delta\omega = \frac{1}{2H\omega_0}(\Delta P_m - \Delta P_e - D\omega_0\Delta\omega - 2r_s(i_{d_0}\Delta i_d + i_{q_0}\Delta i_q)) . \tag{4.63}$$

If, as is normally the case $\omega_0 = 1$ and if $r_s \approx 0$, it follows that:

$$p\Delta\omega = \frac{1}{2H}(\Delta P_m - \Delta P_e - D\Delta\omega) . \tag{4.64}$$

In the preceding equation $2H$ is sometimes replaced by $M = 2H$.

4.2.7 Non-reciprocal definition of the per-unit field voltage and current

Although the L_{ad}-base reciprocal per-unit system has a number of advantages from the perspective of representing the generator it is usually the case that a different per-unit system, referred to as the "non-reciprocal" or "unity-slope" per-unit system, is used when representing the excitation system of the generator. Thus, it is necessary to establish the relationship between these two per-unit systems for the purpose of interfacing between the field winding of the generator and excitation system models.

To begin, consider the generator represented by the per-unit equations in the reciprocal per-unit system and neglecting magnetic saturation. Suppose that the generator is on open-circuit and rotating steadily at one per-unit speed. Under this steady-state condition $i_d = i_q = 0$, $\omega = \omega_0 = 1$ and the rates of change of all variables in the *dq* coordinate system are zero. From the *d*-axis rotor voltage equations (4.2) on page 103 it is deduced that $e_{fd} = r_{fd}i_{fd}$ and the *d*-axis damper winding currents are zero (i.e. $i_{1d} = i_{2d} = 0$). From the corresponding *q*-axis equations (4.3) the *q*-axis damper winding currents are also found to be zero (i.e. $\underset{\sim}{i}_{rq} = 0$). Given these initial values of the winding currents it is deduced from the *d*- and *q*-axis airgap flux-linkage equations (4.15) on page 105 and the *d*- and *q*-axis flux-linkage equations (4.23) on page 106 that $\varphi_d = L_{ad_u} i_{fd}$ and $\varphi_q = 0$. From the *d*- and *q*-axis stator voltage equations (4.24) it follows that $v_d = \varphi_q = 0$ and $v_q = \varphi_d = L_{ad_u} i_{fd}$. If the field current in the reciprocal per-unit system is one per-unit then the stator voltage is L_{ad_u} per-unit (neglecting saturation) and the field voltage is $e_{fd} = r_{fd}$ per-unit.

Thus, an equivalent way of defining the base field current in the L_{ad}-base reciprocal per-unit system is:

> The base field current i_{fdb} in the reciprocal per-unit system is that field current, in Amperes, which is required to generate L_{ad_u} per-unit stator voltage on the airgap line when the machine is open-circuit and rotating steadily at one per-unit speed. The base field voltage e_{fdb} is the corresponding field voltage in Volts divided by the per-unit field winding resistance [1] at the specified field winding temperature.

The above reciprocal definition of the base field current and voltage is *not* consistent with the non-reciprocal definition of the base values of the field quantities which is recommended in Annex B of IEEE Std. 421.5 [11] for the modelling of excitation systems. The following definition of the non-reciprocal per-unit system for the field current and voltage is consistent with that given in IEEE Std. 421.5. Note that in the reciprocal per-unit system quantities related to the field current and voltage are denoted by lower case '*i*' and '*e*' respectively whereas the corresponding quantities in the non-reciprocal per-unit system are denoted by upper-case '*I*' and '*E*'.

> The base field current I_{fdb} in the non-reciprocal per-unit system is that field current, in Amperes, which is required to generate 1.0 per-unit stator voltage on the airgap line when the machine is open-circuit and rotating steadily at one per-unit speed. The base field voltage E_{fdb} in this per-unit system is the field voltage in Volts, corrected to the specified field winding temperature, required to generate the base field current I_{fdb}.

---

1. From Table 4.2, the per-unit field resistance is $r_{fd} = (\bar{r}_{fd})/r_{fdb}$ where $\bar{r}_{fd}$ is the field resistance in ohms and $r_{fdb} = S_b/i_{fdb}^2$ (ohm) is the base value of field resistance in the reciprocal per-unit system.

The above definitions lead to the following mathematical conversions between the per-unit field current I_{fd} (and voltage E_{fd}) in the non-reciprocal per-unit system and the corresponding value of i_{fd} (e_{fd}) in the reciprocal per-unit system:

$$I_{fd} = L_{ad_u} i_{fd} \text{ and } E_{fd} = \frac{L_{ad_u}}{r_{fd}} e_{fd}. \tag{4.65}$$

The conversion between the reciprocal and non-reciprocal definitions of the field current is shown graphically in the generator open-circuit characteristic in Figure 4.4 in which three field current scales are shown: (i) Amperes, (ii) per-unit on the reciprocal base system; and (iii) per-unit on the non-reciprocal base system.

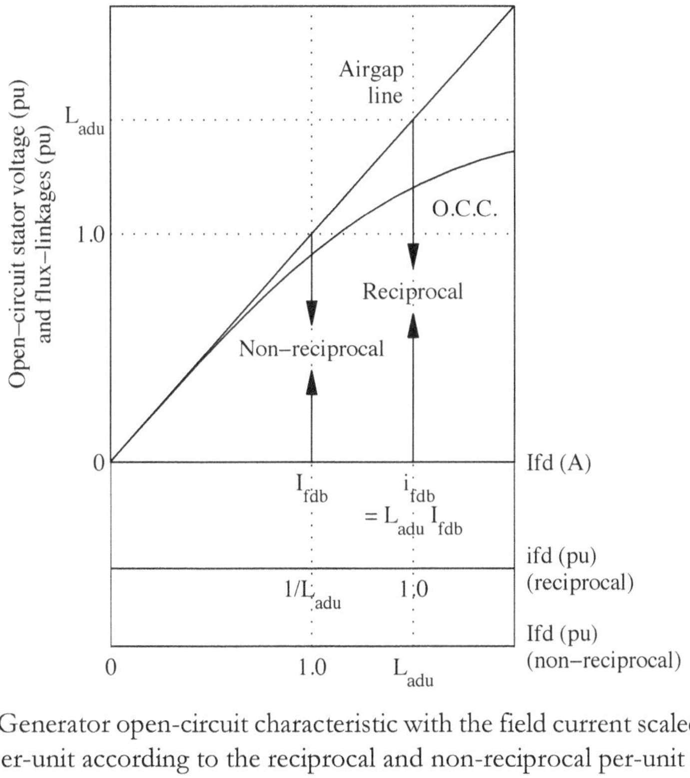

Figure 4.4 Generator open-circuit characteristic with the field current scaled in Amperes, and in per-unit according to the reciprocal and non-reciprocal per-unit systems.

The scaling required at the interface between the model of the exciter and the generator field winding is depicted in Figure 4.5. It is assumed that the output from the exciter is the field voltage in per-unit in the non-reciprocal per-unit system and the input to the generator is the per-unit field voltage in the reciprocal system. It is assumed that the generator per-unit field current in the reciprocal system is, from model signal flow perspective, an output signal from the generator which is input to the model of the exciter in per-unit in the non-reciprocal per-unit system.

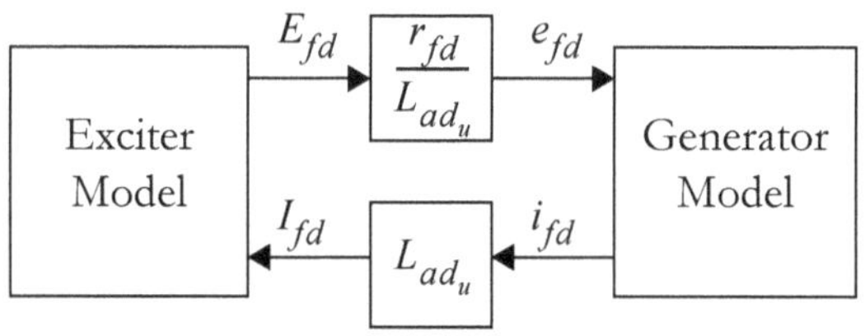

Figure 4.5 Interface between the generator and exciter model taking account of the conversion between the reciprocal and non-reciprocal per-unit systems in the respective models.

In the reciprocal per-unit system it follows from (4.2) on page 103 that under steady-state condition $e_{fd_0} = r_{fd} i_{fd_0}$. By applying the conversion in (4.65) to this relationship it follows that in the non-reciprocal per-unit system the steady-state value of the field voltage and current are equal (i.e. $E_{fd_0} = I_{fd_0}$).

Some of the reasons why the non-reciprocal per-unit system is preferred [10] are:

- The measured generator open-circuit characteristic (O.C.C.) rarely extends beyond a stator voltage of 1.1 per-unit and never to L_{ad_u}. Thus, direct graphical determination of the base field current I_{fdb} from the measured O.C.C. is straight-forward in the non-reciprocal system whereas supplementary calculation is required to determine the base field current i_{fdb} in the reciprocal system.

- The numerical value of per-unit field-voltage e_{fd} in the reciprocal per-unit system is very small whereas, under steady-state conditions, $E_{fd} = I_{fd}$ in the non-reciprocal system.

Although IEEE Std. 421.5 recommends the use of the non-reciprocal per-unit system for modelling of the excitation system, it is sometimes the case that vendors or testing contractors provide excitation system model parameters on a different per-unit system. For example, sometimes the base value of field current is defined as that field current, in Amperes, that is required to produce rated stator voltage when the generator is operating at *rated output* and frequency. Therefore, it is essential that those who are entering data into simulation programs understand the basis on which model parameters are supplied and, if necessary, adjust parameter values to comply with the per-unit system assumed by the simulation program being used.

4.2.8 Modelling generator saturation

Before proceeding further the two methods for representing the effects of magnetic saturation introduced in Section 4.2.3.4 are described in some detail. *It is emphasised that these two methods are strictly equivalent and yield identical results.* Both methods are employed in different simulation packages. In the first method the airgap mutual inductances in the respective axes are assumed to be subject to magnetic saturation. The objective in this case is to show that the perturbations in these inductances, ΔL_{ad} and ΔL_{aq}, about their steady-state saturated values L_{ad_0} and L_{aq_0} can be expressed in terms of the perturbations in the airgap flux linkage components in the respective axes. In the second method, the component of excitation current necessary to account for the demagnetizing effect of magnetic saturation is deducted from the respective axes. The *d*-and *q*-axis components of the "saturation demagnetizing current" are referred to as i_{sd} and i_{sq} respectively. The objective in the following is show that the perturbations in the saturation demagnetizing currents Δi_{sd} and Δi_{sq} about their steady-state values of i_{sd_0} and i_{sq_0} can also be expressed in terms of the perturbations in the airgap mutual flux-linkages. Provision is made for these two representations when formulating the generator equations in Section 4.2.3.4 by including the perturbations in either the mutual inductances or saturation demagnetizing currents depending on the method employed.

In both methods, the saturation level is determined from the user-supplied, open-circuit saturation characteristic(s). The user may choose to supply only the *d*-axis characteristic and select one of several functions for determining the *q*-axis characteristic from the *d*-axis characteristic. Alternatively, the manufacturer or testing contractor may supply a separate characteristic for each axis.

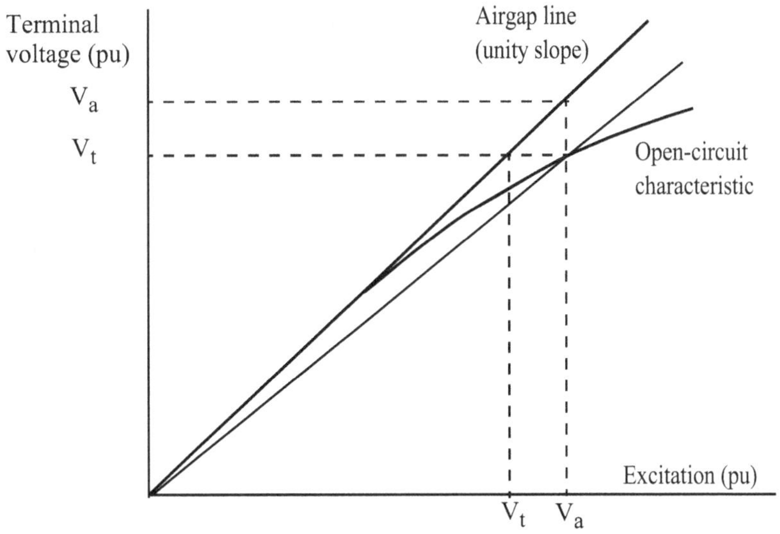

Figure 4.6 Open-circuit characteristic of the generator in the non-reciprocal per-unit system. (Note that the per-unit values of terminal voltage and airgap flux linkages are equal when the machine is open-circuit and rotating steadily at base rotor speed.)

An open-circuit saturation characteristic is shown in Figure 4.6 in the non-reciprocal or unity-slope per-unit system. If V_t is the terminal voltage and V_a is the corresponding voltage on the airgap line, then the saturation function $S(V_t)$ is defined as

$$S(V_t) = (V_a - V_t)/V_t. \tag{4.66}$$

For the given open-circuit characteristic of the generator, saturation is characterized by the values of $S(\varphi_1)$ and $S(\varphi_2)$ such that $\varphi_2 > \varphi_1 > 0$ together with the selection of a function to interpolate between the latter two points on the saturation characteristic. Typically, $\varphi_1 = 1.0$ pu and $\varphi_2 = 1.2$ pu. Several commonly employed interpolation functions are detailed in Table 4.7 although other functions may be used.

Table 4.7 Interpolation functions for saturation characteristics.

Type	$S(\varphi) = f(A, B)$	A, B	$\partial S/\partial\varphi$
Exponential	$A\varphi^B/\varphi$	$B = \ln\left(\frac{\varphi_2 S(\varphi_2)}{\varphi_1 S(\varphi_1)}\right)/\left(\ln\left(\frac{\varphi_2}{\varphi_1}\right)\right)$ $A = (\varphi_1 S(\varphi_1))/(\varphi_1^B)$	$A(B-1)\varphi^{(B-2)}$
Quadratic	$B(\varphi - A)^2/\varphi$ if $\varphi > A$ 0 otherwise	$A = (\varphi_2 - a\varphi_1)/(1-a)$ $B = \varphi_1 S(\varphi_1)/(\varphi_1 - A)^2$, where $a = \sqrt{(\varphi_2 S(\varphi_2))/(\varphi_1 S(\varphi_1))}$	$B\left(1.0 - \left(\frac{A}{\varphi}\right)^2\right)$
Ontario Hydro [10]	$Ae^{B(\varphi - \varphi_I)}/\varphi$ if $\varphi > \varphi_I$ 0 otherwise; $\varphi_I < \varphi_1$	$B = \ln\left(\frac{\varphi_2 S(\varphi_2)}{\varphi_1 S(\varphi_1)}\right)/(\varphi_2 - \varphi_1)$ $A = \varphi_1 S(\varphi_1)e^{-B(\varphi_1 - \varphi_I)}$	$S(\varphi)\left[B - \frac{1}{\varphi}\right]$
Linear	$\frac{B(\varphi - A)}{\varphi}$	$A = (\varphi_2 - a\varphi_1)/(1-a)$, $B = (\varphi_1 S(\varphi_1))/(\varphi_1 - A)$, where $a = (\varphi_2 S(\varphi_2))/(\varphi_1 S(\varphi_1))$	$\frac{AB}{\varphi^2}$

So far only the open-circuit characteristic has been considered. However, when the generator is loaded the open-circuit characteristics no longer apply. A common approximation is that the resultant airgap flux φ_{ag} is indicative of the level of saturation when the generator is on-load. This is based on the fact that when the generator is on open-circuit and rotating at one per-unit speed the terminal voltage and airgap flux are equal in the per-unit system used. Other approximations for the level of saturation which are employed in widely-used software packages are described in Section 4.2.13.2.

Note that in Table 4.7 the value of φ to be used depends on the context. When determining the parameters A and B of the interpolation function the values of φ are the o.c. flux-linkages (equivalently o.c. voltages) obtained from the open-circuit-characteristic. When evaluating the saturation function when the machine is loaded then φ is the value of the selected saturation level indicator such as the resultant airgap flux (φ_{ag}), the resultant k^{th}-transient flux-linkages (φ^{k}), etc.

The resultant airgap flux linkages are defined as:

$$\varphi_{ag} = \sqrt{\varphi_{ad}^2 + \varphi_{aq}^2}. \tag{4.67}$$

To determine the steady-state operating value of the airgap flux, it is noted that φ_{ag} is also equal to the voltage behind the stator resistance and leakage inductance in the per-unit system used, taking into account any difference between the synchronous speed and base frequency of the generator (i.e. to account for the situation when $\omega_0 \neq 1$).

$$\varphi_{ag_0} = \frac{v_{ag_0}}{\omega_0} = \left(\frac{1}{\omega_0}\right)\sqrt{\left(V_{t_0} + \left(\frac{r_s P_0 + \omega_0 L_l Q_0}{V_{t_0}}\right)\right)^2 + \left(\frac{\omega_0 L_l P_0 - r_s Q_0}{V_{t_0}}\right)^2}, \tag{4.68}$$

in which V_{t_0}, P_0 and Q_0 are respectively the initial steady-state values of the generator stator terminal voltage, and real and reactive power output.

The perturbations in the resultant airgap flux linkages about the operating point φ_{ag_0} defined in (4.68) and in which the corresponding steady-state values of the *d*- and *q*-axis flux linkages are φ_{ad_0} and φ_{aq_0} respectively are obtained by linearizing equation (4.67) to give:

$$\Delta\varphi_{ag} = \left(\frac{\varphi_{ad_0}}{\varphi_{ag_0}}\right)\Delta\varphi_{ad} + \left(\frac{\varphi_{aq_0}}{\varphi_{ag_0}}\right)\Delta\varphi_{aq} = \left(\frac{1}{\varphi_{ag_0}}\right)\begin{bmatrix}\varphi_{ad_0} & \varphi_{aq_0}\end{bmatrix}\Delta\underset{\sim}{\varphi}_{adq}. \tag{4.69}$$

It should be noted that the saturation characteristic interpolation functions listed in Table 4.7 are intended to be used when the machine is operating within its normal range of steady-state operating conditions, i.e. φ_{ag_0} is expected to range at most between about 0.8 and 1.3 pu. The interpolation functions may require modification at higher flux levels that may occur under some large disturbance conditions.

The *d*- and *q*-axis saturation characteristics are denoted by S_d and S_q respectively. In the situation where a *q*-axis saturation characteristic is not provided one of the rules in Table 4.8 can be used to derive the *q*-axis characteristic from the *d*-axis characteristic provided.

Table 4.8 q-Axis saturation characteristics as a function of those for the d-axis

	$S_q(S_d(\varphi))$	$\partial S_q / \partial S_d$
A	$S_d(\varphi)$	1
	Note A: $L_{aq}/L_{ad} = L_{aqu}/L_{adu}$	
B	N/A	N/A
	Note B: The points $S_q(1.0)$ and $S_q(1.2)$ on the q-axis open-circuit saturation characteristic are specified. The same interpolation function used for the d-axis characteristic is employed.	
C	$\dfrac{S_d}{S_d[(L_{aqu}/L_{adu})-1]+(L_{aqu}/L_{adu})}$	$\dfrac{S_d}{S_q}\left[1 - S_d S_q\left(\dfrac{L_{aqu}}{L_{adu}} - 1\right)\right]$
	Note C: $L_{ad} - L_{aq} = L_{adu} - L_{aqu} = L_{sal}$. This characteristic is based on empirical results reported by Shackshaft [19]; the variation of L_{ad} and L_{aq} with rotor position is neglected.	
D	0	0
	Note D: The q-axis is unsaturated. Useful in modelling salient pole machines.	
E	$(L_{du}/L_{qu})S_d$	L_{du}/L_{qu}
	Note E: This option is employed in the saturation model of some software.	
F	$(L_{du}/L_{qu})^Z S_d(\varphi)$, where $Z = \{\ln[S_d(1.2)/S_d(1.0)]\}/\ln(1.2)$	$(L_{du}/L_{qu})^Z$
	Note F: This option is employed in the saturation model of some software.	

4.2.8.1 Method 1: Non-linear airgap mutual inductances

As mentioned earlier the first method for representing generator magnetic saturation is to treat the d- and q-axis airgap mutual inductances as non-linear parameters that depend on the resultant airgap flux linkages. It is assumed that the leakage inductances are not subject to magnetic saturation and are thus assumed to be constant parameters.

The values of the d- and q-axis airgap mutual inductances are expressed in terms of their respective saturation characteristics by:

$$L_{ad} = \frac{L_{ad_u}}{1+S_d(\varphi_{ag})} \text{ and } L_{aq} = \frac{L_{aq_u}}{1+S_q(\varphi_{ag})}. \tag{4.70}$$

The steady-state saturated values of the airgap mutual inductances L_{ad_0} and L_{aq_0} are obtained by substituting $\varphi_{ag} = \varphi_{ag_0}$ in (4.70).

The perturbations in the airgap mutual inductances about the operating point φ_{ag_0} defined in (4.68) are obtained by linearization of the equations for the non-linear airgap mutual inductances in (4.70) to yield:

$$\begin{bmatrix} \Delta L_{ad} \\ \Delta L_{aq} \end{bmatrix} = -\begin{bmatrix} \left(\frac{L_{ad_0}^2}{L_{ad_u}}\right) \frac{\partial S_d(\varphi_{ag})}{\partial \varphi_{ag}}\bigg|_0 \\ \left(\frac{L_{aq_0}^2}{L_{aq_u}}\right) \frac{\partial S_q(\varphi_{ag})}{\partial \varphi_{ag}}\bigg|_0 \end{bmatrix} \Delta\varphi_{ag} \tag{4.71}$$

Substituting for the perturbations in the resultant airgap flux linkages from equation (4.69) into the preceding equation results in the following expression for the perturbations in the airgap mutual inductances in terms of the perturbations in the airgap flux-linkages:

Saturation Method 1:
Perturbations in airgap mutual inductances

$$\Delta \underset{\sim}{L}_{adq} = \boldsymbol{C}_{\boldsymbol{ladq_0}} \Delta \underset{\sim}{\varphi}_{adq}, \tag{4.72}$$

in which

$$\boldsymbol{C}_{\boldsymbol{ladq_0}} = -\left(\frac{1}{\varphi_{ag_0}}\right) \begin{bmatrix} \left(\frac{L_{ad_0}^2}{L_{ad_u}}\right) \frac{\partial S_d(\varphi_{ag})}{\partial \varphi_{ag}}\bigg|_0 \\ \left(\frac{L_{aq_0}^2}{L_{aq_u}}\right) \frac{\partial S_q(\varphi_{ag})}{\partial \varphi_{ag}}\bigg|_0 \end{bmatrix} \begin{bmatrix} \varphi_{ad_0} & \varphi_{aq_0} \end{bmatrix} \tag{4.73}$$

4.2.8.2 *Method 2: Saturation demagnetization current*

The second method for representing the effects of magnetic saturation involves deducting non-linear components of *d*- and *q*-axis saturation demagnetization current i_{sd} and i_{sq} from the excitation of the *d*- and *q*-axis windings respectively. In this formulation the model utilizes the fixed *unsaturated* airgap mutual inductances. This is especially advantageous when representing saturation in the Classical Parameter Formulation of the generator model because it is unnecessary to adjust the classically-defined standard parameters to account for the effects of saturation.

To determine the expression for i_{sd} it can be deduced from (4.12) on page 105 that:

$$i_{sd} = (\boldsymbol{u}^T \underset{\sim}{i}_{rd} - i_d) - \frac{\varphi_{ad}}{L_{ad_u}}. \tag{4.74}$$

Alternatively, if the effects of magnetic saturation are represented by non-linear mutual air-gap inductances according to (4.70) (i.e. by Method 1) then from (4.12) with $s_2 = 0$ and with L_{ad} defined according to (4.70) it follows that:

$$(\boldsymbol{u}^T \underset{\sim}{i}_{rd} - i_d) = \frac{(1 + S_d(\varphi_{ag}))\varphi_{ad}}{L_{ad_u}}. \tag{4.75}$$

Substituting from the preceding equation for $(\boldsymbol{u}^T \underset{\sim}{i}_{rd} - i_d)$ into (4.74) yields:

$$i_{sd} = S_d(\varphi_{ag})\frac{\varphi_{ad}}{L_{ad_u}}. \tag{4.76}$$

The q-axis saturation demagnetization current component is similarly derived:

$$i_{sq} = S_q(\varphi_{ag})\frac{\varphi_{aq}}{L_{aq_u}}. \tag{4.77}$$

Equations (4.76) and (4.77) are combined to yield:

$$\underset{\sim}{i}_{sdq} = \boldsymbol{S}_{dq}(\varphi_{ag})\underset{\sim}{\varphi}_{adq} \text{ in which } \boldsymbol{S}_{dq}(\varphi_{ag}) = \mathfrak{D}\left(\frac{S_d(\varphi_{ag})}{L_{ad_u}}, \frac{S_q(\varphi_{ag})}{L_{aq_u}}\right). \tag{4.78}$$

Linearizing the preceding equation about the operating point $(\varphi_{ag_0}, \varphi_{ad_0}, \varphi_{aq_0})$ yields the following expression for the perturbations in the saturation demagnetizing current components in terms of the perturbations in the airgap flux-linkages.

<u>Saturation Method 2:</u>

<u>Perturbations in saturation demagnetizing currents</u>

$$\Delta\underset{\sim}{i}_{sdq} = \boldsymbol{C}_{madq_0}\Delta\underset{\sim}{\varphi}_{adq}, \tag{4.79}$$

where

$$\boldsymbol{C}_{madq_0} = \boldsymbol{S}_{dq_0} + \begin{bmatrix} \frac{\varphi_{ad_0}}{L_{ad_u}}\left(\frac{\partial S_d(\varphi_{ag})}{\partial \varphi_{ag}}\right)_0 \\ \frac{\varphi_{aq_0}}{L_{aq_u}}\left(\frac{\partial S_q(\varphi_{ag})}{\partial \varphi_{ag}}\right)_0 \end{bmatrix} \begin{bmatrix} \left(\frac{\varphi_{ad_0}}{\varphi_{ag_0}}\right) & \left(\frac{\varphi_{aq_0}}{\varphi_{ag_0}}\right) \end{bmatrix} \tag{4.80}$$

$$\text{and } \boldsymbol{S}_{dq_0} = \mathfrak{D}\left(\frac{S_d(\varphi_{ag_0})}{L_{ad_u}}, \frac{S_q(\varphi_{ag_0})}{L_{aq_u}}\right). \tag{4.81}$$

4.2.9 Balanced steady-state operating conditions of the coupled-circuit model

The initial steady-state values of the coupled-circuit generator model variables when operating under balanced conditions are now calculated. In the following analysis the subscript '0' denotes the steady-state value of the variable.

It is assumed that the steady-state generator stator voltage magnitude, V_{t0}, and the real and reactive power *output* $P_0 = P_{e_0}$ and Q_0 of the generator are given, in per-unit on the generator base quantities. These initial values are usually obtained from the power flow solution on which the dynamic analysis is to be based.

Under steady-state conditions $p\delta = 0$ so from (4.56) on page 114 it follows that $\omega = \omega_0$ per-unit; and $p\omega = 0$ so from (4.57) on page 114 it follows that $T_m = T_g$. Note, that normally the generator rated frequency is the same as the system nominal frequency and so normally $\omega_0 = 1$. However, if, for example, a generator rated for 60 Hz is connected to a 50 Hz system and ω_b is chosen to be $2\pi(60)$, then $\omega_0 = 5/6$ per-unit.

Under balanced steady-state operating conditions the stator voltage is represented as a phasor $\hat{V}_t$ in the complex plane in which the *d*-axis corresponds to the real axis of the complex plane and the *q*-axis to that of the imaginary axis so that:

$$\hat{V}_t = v_{d0} + jv_{q0} = V_{t0}e^{j\gamma_0} = V_{t0}\cos\gamma_0 + jV_{t0}\sin\gamma_0 . \tag{4.82}$$

where γ_0 is the angle by which the voltage phasor leads the *d*-axis. (Note that γ_0 is defined differently than the "load angle" which is the angle by which the voltage phasor lags the *q*-axis. The use of γ_0 is convenient analytically and the results are consistent.).

The phasor representing the generator current output is:

$$\hat{I} = i_{d0} + ji_{q0} = \frac{P_0 - jQ_0}{V_{t0}e^{-j\gamma_0}} = I_0 e^{j(\gamma_0 + \beta_0)}, \tag{4.83}$$

where

$$I_0 = \frac{\sqrt{P_0^2 + Q_0^2}}{V_{t0}} \tag{4.84}$$

is the magnitude of the current; and

$$\beta_0 = \text{atan2}(-Q_0, P_0) \; ^1 \tag{4.85}$$

is the angle by which the current phasor *leads* that of the voltage.

If the effects of magnetic saturation are being represented then, for the purpose of calculating the initial steady-state operating conditions, the following steady-state saturated values of the *d*- and *q*-axis airgap mutual inductances obtained from (4.70) are used. This applies to both of the methods of representing the effects of magnetic saturation in the dynamic model of the machine.

$$L_{ad_0} = L_{ad_u}/(1 + S_d(\varphi_{ag_0})) \text{ and } L_{aq_0} = L_{aq_u}/(1 + S_q(\varphi_{ag_0})), \tag{4.86}$$

in which the resultant airgap flux-linkages, φ_{ag_0}, are obtained from (4.68) on page 121.

The saturated values of the *d*- and *q*-axis synchronous inductances at the steady-state operating point are:

$$L_{d_0} = L_{ad_0} + L_l \text{ and } L_{q_0} = L_{aq_0} + L_l. \tag{4.87}$$

Since, under steady-state conditions, $p\underset{\sim}{\varphi}_{rq} = 0$ and since $\underset{\sim}{v}_{rq} = \underset{\sim}{0}$ it is deduced from (4.5) on page 104 that $\underset{\sim}{i}_{rq_0} = \underset{\sim}{0}$. Consequently, from equation (4.12) on page 105,

$$\varphi_{aq_0} = -L_{aq_0} i_{q_0} \tag{4.88}$$

and then from (4.22) on page 106 it follows that:

$$\varphi_{q_0} = \varphi_{aq_0} - L_l i_{q_0} = -(L_{aq_0} + L_l) i_{q_0} = -L_{q_0} i_{q_0}, \tag{4.89}$$

a result that is independent of the number of rotor windings.

From (4.24) and (4.89), under steady-state conditions,

$$v_{d_0} = -r_s i_{d_0} - \omega_0 \varphi_{q_0} = -r_s i_{d_0} + \omega_0 L_{q_0} i_{q_0}; \tag{4.90}$$

which again is independent of the number of rotor windings.

The *d*-axis damper winding currents are zero in the steady-state so from (4.9) on page 104, (4.22) on page 106 and (4.65) on page 117 it follows that:

$$\varphi_{ad_0} = L_{ad_0}(i_{fd_0} - i_{d_0}) = (L_{ad_0}/L_{ad_u}) I_{fd_0} - L_{ad_0} i_{d_0} \text{ and} \tag{4.91}$$

---

1. Definition of $\theta = \text{atan2}(y, x)$:
   If $x = y = 0$ then arbitrarily define $\theta = 0$;
   else if $x = 0$ then if $y > 0$, $\theta = \pi/2$; else $\theta = -\pi/2$,
   otherwise let $z = |y/x|$ and define $\phi = \text{atan}(z)$ then
   First quadrant: $x > 0$ and $y \geq 0$ then $\theta = \phi$;
   Second quadrant: $x < 0$ and $y \geq 0$ then $\theta = \pi - \phi$;
   Third quadrant: $x < 0$ and $y < 0$ then $\theta = -\pi + \phi$; and
   Fourth quadrant: $x > 0$ and $y < 0$ then $\theta = -\phi$.

$$\varphi_{d_0} = \varphi_{ad_0} - L_l i_{d_0} = -L_{d_0} i_{d_0} + L_{ad_0} i_{fd_0} = -L_{d_0} i_{d_0} + \left(\frac{L_{ad_0}}{L_{ad_u}}\right) I_{fd_0}. \tag{4.92}$$

From (4.24) on page 106 and (4.92) the following expression for the steady-state q-axis voltage is obtained:

$$v_{q_0} = -r_s i_{q_0} + \omega_0 \varphi_{d_0} = -r_s i_{q_0} - \omega_0 L_{d_0} i_{d_0} + \omega_0 \left(\frac{L_{ad_0}}{L_{ad_u}}\right) I_{fd_0}. \tag{4.93}$$

The stator voltage phasor is obtained by combining (4.90) and (4.93) to yield:

$$\hat{V}_t = v_{d_0} + j v_{q_0} = -(r_s + j\omega_0 L_{q_0})\hat{I} + j\omega_0 \left\{ \left(\frac{L_{ad_0}}{L_{ad_u}}\right) I_{fd_0} - (L_{d_0} - L_{q_0}) i_{d_0} \right\} \tag{4.94}$$

$$\text{in which } \hat{I} = (i_{d_0} + j i_{q_0}) \tag{4.95}$$

Rearranging (4.94) yields:

$$\hat{E}_q = \hat{V}_t + (r_s + j\omega_0 L_{q_0})\hat{I} = j\omega_0 \left(\left(\frac{L_{ad_0}}{L_{ad_u}}\right) I_{fd_0} - (L_{d_0} - L_{q_0}) i_{d_0} \right) = jE_q. \tag{4.96}$$

The artificial voltage phasor $\hat{E}_q = jE_q$ is aligned with q-axis and corresponds to the voltage behind the impedance $Z = (r_s + j\omega_0 L_{q_0})$.

Now, substituting for $\hat{V}_t$ and $\hat{I}$ from equations (4.82) and (4.83) into the preceding equation yields:

$$\begin{aligned} E_q e^{j(\pi/2)} &= \left(V_{t_0} + (r_s + j\omega_0 L_{q_0}) I_0 e^{j\beta_0}\right) e^{j\gamma_0} \\ \therefore E_q e^{j(\pi/2 - \gamma_0)} &= (V_{t_0} + r_s I_0 \cos\beta_0 - \omega_0 L_{q_0} I_0 \sin\beta_0) + j(r_s I_0 \sin\beta_0 + \omega_0 L_{q_0} I_0 \cos\beta_0) \end{aligned} \tag{4.97}$$

By equating the arguments of both sides of (4.97) yields γ_0, the angle by which voltage phasor leads the d-axis:

$$\gamma_0 = \frac{\pi}{2} - \operatorname{atan2}(I_0(r_s \sin\beta_0 + \omega_0 L_{q_0} \cos\beta_0), (V_{t_0} + I_0(r_s \cos\beta_0 - \omega_0 L_{q_0} \sin\beta_0)))$$

in which, from equations (4.84) & (4.85) respectively:

$$I_0 = \frac{\sqrt{P_0^2 + Q_0^2}}{V_{t0}} \text{ and } \beta_0 = \operatorname{atan2}(-Q_0, P_0). \quad (4.98)$$

This result is independent of the number of *d*- or *q*-axis rotor-windings or of the representation of coupling between the *d*-axis rotor windings. Figure 4.7 is a phasor diagram showing the computation of $\hat{E}_q$, γ_0 and the associated location of the *d*- and *q*-axes with respect to the voltage phasor. The *d*- and *q*-axis components of the voltage and current phasors are also shown in this diagram.

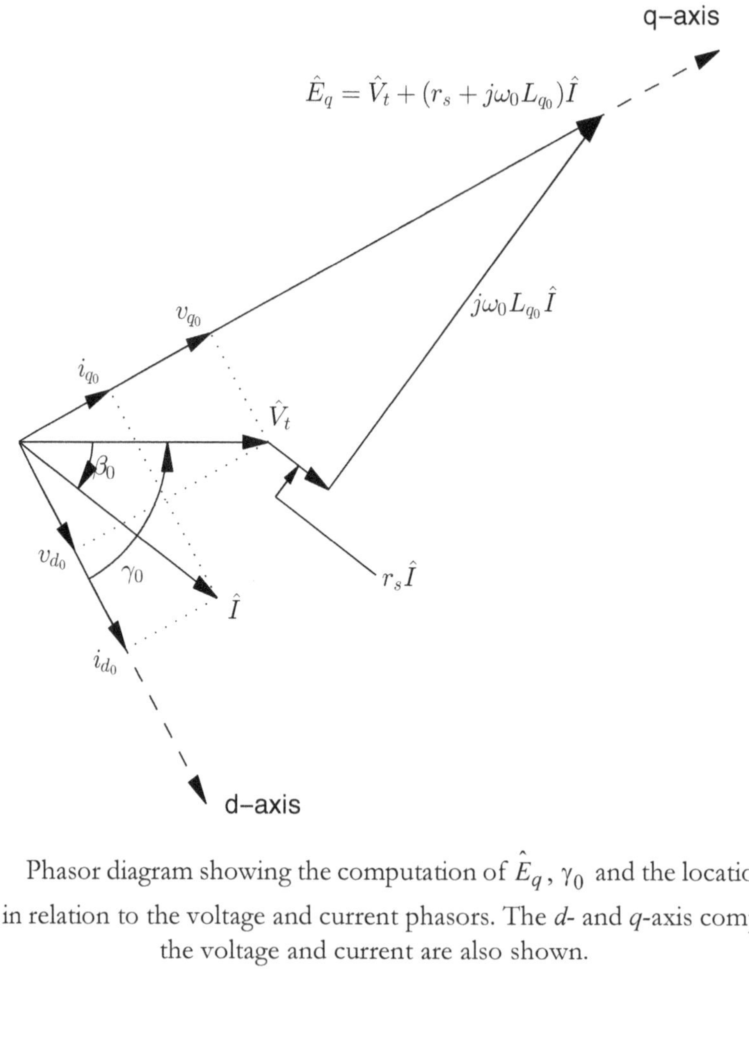

Figure 4.7 Phasor diagram showing the computation of $\hat{E}_q$, γ_0 and the location of the *d*- and *q*-axes in relation to the voltage and current phasors. The *d*- and *q*-axis components of the voltage and current are also shown.

Having calculated the steady-state saturated values of the airgap mutual inductances and synchronous inductances according to (4.86) and (4.87) and the values of γ_0, I_0 and β_0 in (4.98) the initial steady-state values of those generator variables that are independent of the rotor winding structure are readily found to be:

$$
\begin{array}{llll}
v_{d_0} = V_{t_0}\cos\gamma_0 & & \\
v_{q_0} = V_{t_0}\sin\gamma_0 & i_{fd_0} = (\varphi_{d_0} + L_{d_0} i_{d_0})/L_{ad_0} & \\
i_{d_0} = I_0\cos(\gamma_0 + \beta_0) & I_{fd_0} = L_{ad_u} i_{fd_0} & \\
i_{q_0} = I_0\sin(\gamma_0 + \beta_0) & e_{fd_0} = r_{fd} i_{fd_0} & \\
\varphi_{d_0} = (v_{q_0} + r_s i_{q_0})/\omega_0 & E_{fd_0} = I_{fd_0} & \\
\varphi_{ad_0} = \varphi_{d_0} + L_l i_{d_0} & T_{g_0} = \varphi_{d_0} i_{q_0} - \varphi_{q_0} i_{d_0} = (P_0 + r_s I_0^2)/\omega_0 & \\
\varphi_{q_0} = -(v_{d_0} + r_s i_{d_0})/\omega_0 & T_{m_0} = T_{g_0} = P_{m_0}/\omega_0 & \\
\varphi_{aq_0} = \varphi_{q_0} + L_l i_{q_0} & &
\end{array}
\qquad (4.99)
$$

The calculation of δ_0, the initial steady-state value of the angle by which the *d*-axis leads the *R*-axis of the synchronously rotating network reference frame is deferred until Section 4.2.10.

For the *3d3q-c1* model the steady-state values of the following *d*- and *q*-axis rotor winding current and flux-linkage and variables are:

$$
\begin{array}{ll}
i_{1d_0} = i_{2d_0} = 0 & \\
\varphi_{fd_0} = \varphi_{ad_0} + (L_{fd} + L_{c1} + L_{c2}) i_{fd_0} & i_{1q_0} = i_{2q_0} = i_{3q_0} = 0 \\
\varphi_{1d_0} = \varphi_{ad_0} + (L_{c1} + L_{c2}) i_{fd_0} & \varphi_{1q_0} = \varphi_{2q_0} = \varphi_{3q_0} = \varphi_{aq_0} \\
\varphi_{2d_0} = \varphi_{ad_0} + L_{c2} i_{fd_0} &
\end{array}
\qquad (4.100)
$$

For generator models which neglect unequal coupling between the *d*-axis rotor windings $L_{c1} = L_{c2} = 0$. For models with only one *d*-axis damper winding the variables i_{2d} and φ_{2d} do not exist and the non-existence of L_{c2} is represented by setting its value to zero in the above equations. Similar trivial modifications are made to (4.100) so they can be applied to coupled-circuit models with other rotor structures.

4.2.10 Interface between the generator Park/Blondel reference frame and the synchronous network reference frame

The generator equations are developed in the Park/Blondel co-ordinate system in which, as mentioned earlier, the *d*-axis is aligned with the magnetic axis of the rotor field winding and the *q*-axis leads the *d*-axis by 90 electrical degrees. The *dq* reference frame rotates in an anti-clockwise direction at the speed of the generator rotor ω per-unit. As explained in the development of the generator equations of motion in [12] the generator rotor angle $\delta(t)$ (elec. rad) is measured relative to a synchronously rotating reference. In the analysis of multi-machine systems the *R*-axis of the synchronously rotating network *RI* reference frame is chosen as the reference for the rotor angle of each generator.

To facilitate the analysis of multi-machine systems it is necessary to transform the stator current and voltage at the machine terminals between the generator *dq* reference frame and network *RI* reference frame.

In Figure 4.8 the stator current phasor, $\hat{I}^{(g)}$, is shown at the instant *t*. The superscript *(g)* denotes that the current phasor is in per-unit of the generator base quantities. At this instant the *d*-axis leads the *R*-axis by $\delta(t)$ (rad) and the current phasor leads the *d*-axis by $\alpha(t)$ (rad). In the generator *dq* reference frame the current phasor $\hat{I}^{(g)}$ is expressed as:

$$\hat{I}^{(g)} = i_d + ji_q = \left|\hat{I}^{(g)}\right| e^{j\alpha(t)},$$

and in the *RI* reference frame it is:

$$\hat{I}^{(g)} = i_R^{(g)} + ji_I^{(g)} = \left|\hat{I}^{(g)}\right| e^{j(\alpha(t)+\delta(t))} = e^{j\delta(t)}\left|\hat{I}^{(g)}\right| e^{j\alpha(t)} = e^{j\delta(t)}(i_d + ji_q).$$

Expanding the preceding equation yields the following relationship between the current phasor components in the *dq* and *RI* reference frames.

$$\begin{aligned} i_R^{(g)} + ji_I^{(g)} &= (i_d + ji_q)e^{j\delta} = (i_d + ji_q)(\cos(\delta) + j\sin(\delta)) \\ &= \{\cos(\delta)i_d - \sin(\delta)i_q\} + j\{\sin(\delta)i_d + \cos(\delta)i_q\} \end{aligned}$$

Equating respectively the real and imaginary components in the above equation yields the following matrix relationship between the current components in the respective reference frames:

$$\begin{bmatrix} i_R^{(g)} \\ i_I^{(g)} \end{bmatrix} = \begin{bmatrix} \cos(\delta) & -\sin(\delta) \\ \sin(\delta) & \cos(\delta) \end{bmatrix} \begin{bmatrix} i_d \\ i_q \end{bmatrix}, \tag{4.101}$$

In compact matrix form this equation is denoted as:

$$\underset{\sim}{i}_{RI}^{(g)} = \boldsymbol{R}(\delta)\underset{\sim}{i}_{dq}, \tag{4.102}$$

in which $\underset{\sim}{i}_{RI}^{(g)} = \begin{bmatrix} i_R^{(g)} & i_I^{(g)} \end{bmatrix}^T$, $\underset{\sim}{i}_{dq} = \begin{bmatrix} i_d & i_q \end{bmatrix}^T$ and $\boldsymbol{R}(\delta) = \begin{bmatrix} \cos(\delta) & -\sin(\delta) \\ \sin(\delta) & \cos(\delta) \end{bmatrix}$. (4.103)

The inverse relationship is:

$$\underset{\sim}{i}_{dq} = \boldsymbol{R}(\delta)^{-1} \underset{\sim}{i}_{RI}^{(g)}. \tag{4.104}$$

The unitary matrix $\boldsymbol{R}(\delta)$ is referred to as the rotation matrix and its inverse is equal to its transpose:

$$\boldsymbol{R}(\delta)^{-1} = \boldsymbol{R}(\delta)^T = \boldsymbol{R}(-\delta) = \begin{bmatrix} \cos(\delta) & \sin(\delta) \\ -\sin(\delta) & \cos(\delta) \end{bmatrix}. \tag{4.105}$$

When linearizing the model the partial derivative of the rotation matrix is required:

$$\boldsymbol{U}(\delta) = \frac{\partial \boldsymbol{R}(\delta)}{\partial \delta} = \begin{bmatrix} -\sin(\delta) & -\cos(\delta) \\ \cos(\delta) & -\sin(\delta) \end{bmatrix} \quad \text{and} \quad \frac{\partial \boldsymbol{R}(\delta)^{-1}}{\partial \delta} = \boldsymbol{U}(\delta)^T. \tag{4.106}$$

The transformation of the stator terminal voltage between the respective reference frames is similar to that for the stator current:

$$\underset{\sim}{v}_{RI}^{(g)} = \boldsymbol{R}(\delta)\underset{\sim}{v}_{dq} \quad \text{and} \quad \underset{\sim}{v}_{dq} = \boldsymbol{R}(\delta)^{-1} \underset{\sim}{v}_{RI}^{(n)} \tag{4.107}$$

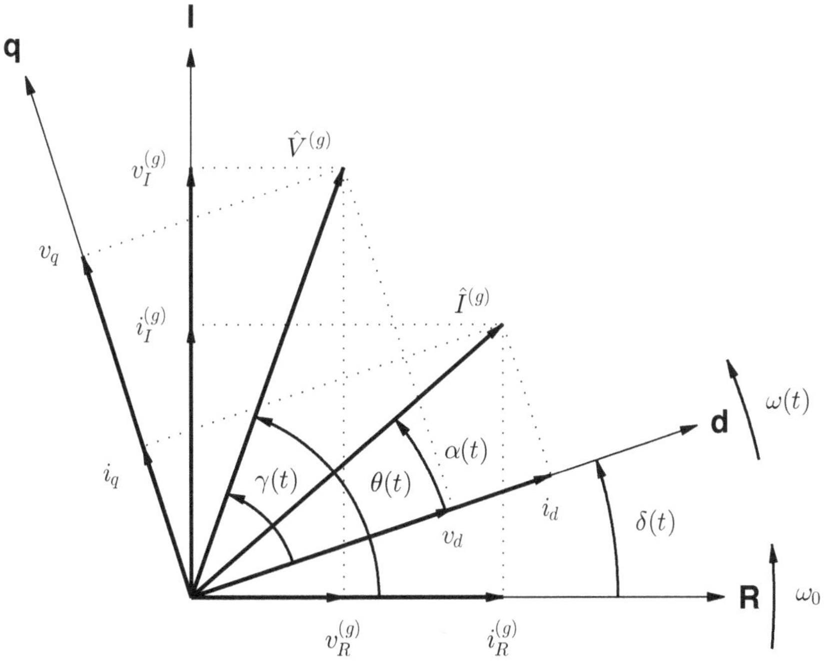

Figure 4.8 Relationship between the Park/Blondel *(dq)* and Network *(RI)* reference frames.

Again referring to Figure 4.8 the angle by which the voltage phasor leads the R-axis is $\theta(t)$. The initial steady-state value of this angle, θ_0, is typically obtained from the power flow solution. Given this angle, it follows that the initial steady-state value of the rotor angle δ_0 is:

$$\delta_0 = \theta_0 - \gamma_0, \tag{4.108}$$

where γ_0 is given by (4.98). From the above value of δ_0 and the initial steady-state values $\underset{\sim}{i}_{dq_0}$ and $\underset{\sim}{v}_{dq_0}$ from (4.99), the initial steady-state values $\underset{\sim}{i}^{(g)}_{RI_0}$ and $\underset{\sim}{v}^{(g)}_{RI_0}$ are deduced from equations (4.102) and (4.107) respectively.

$$\underset{\sim}{i}^{(g)}_{RI_0} = \boldsymbol{R}(\delta_0)\underset{\sim}{i}_{dq_0} \quad \text{and} \quad \underset{\sim}{v}^{(g)}_{RI_0} = \boldsymbol{R}(\delta_0)\underset{\sim}{v}_{dq_0} \tag{4.109}$$

Linearizing the current transformation equations (4.102) and (4.104) about the steady-state operating point $(\delta_0, \underset{\sim}{i}_{dq_0}, \underset{\sim}{i}^{(g)}_{RI_0})$ yields:

$$\Delta\underset{\sim}{i}^{(g)}_{RI} = \boldsymbol{R}(\delta_0)\Delta\underset{\sim}{i}_{dq} + \boldsymbol{U}(\delta_0)\underset{\sim}{i}_{dq_0}\Delta\delta \tag{4.110}$$

$$\text{and } \Delta\underset{\sim}{i}_{dq} = \boldsymbol{R}^T(\delta_0)\Delta\underset{\sim}{i}^{(g)}_{RI} + \boldsymbol{U}^T(\delta_0)\underset{\sim}{i}^{(g)}_{RI_0}\Delta\delta. \tag{4.111}$$

The voltage transformation equations (4.107) are similarly linearized about the steady-state operating point $(\delta_0, \underset{\sim}{v}_{dq_0}, \underset{\sim}{v}^{(g)}_{RI_0})$ to give:

$$\Delta\underset{\sim}{v}^{(g)}_{RI} = \boldsymbol{R}(\delta_0)\Delta\underset{\sim}{v}_{dq} + \boldsymbol{U}(\delta_0)\underset{\sim}{v}_{dq_0}\Delta\delta \tag{4.112}$$

$$\text{and } \Delta\underset{\sim}{v}_{dq} = \boldsymbol{R}^T(\delta_0)\Delta\underset{\sim}{v}^{(g)}_{RI} + \boldsymbol{U}^T(\delta_0)\underset{\sim}{v}^{(g)}_{RI_0}\Delta\delta. \tag{4.113}$$

There is normally a change in the apparent power base between the generator model and that of the network. For generality it will also be assumed that there is a change in base voltage between the generator and that of the network bus to which the generator is connected. However, most simulation programs assume that the respective base voltages are identical.

Let $S^{(g)}_{usb}$ MVA and $V^{(g)}_{usb}$ kV be respectively the generator three-phase MVA base and line-to-line RMS voltage base values of the generator (see Table 4.2 on page 98). Let $S^{(n)}_{usb}$ and $V^{(n)}_{usb}$ be the corresponding quantities for the network bus to which the generator is connected. The base stator currents in the respective per-unit systems are then deduced as:

$$I_b^{(g)} = \frac{S_{usb}^{(g)} \times 10^3}{\sqrt{3}\,V_{usb}^{(g)}} \text{ (A) and } I_b^{(n)} = \frac{S_{usb}^{(n)} \times 10^3}{\sqrt{3}\,V_{sub}^{(n)}} \text{ (A).} \tag{4.114}$$

Let $\underset{\sim}{v}_{RI}^{(g)}$ and $\underset{\sim}{v}_{RI}^{(n)}$ be the stator voltage components in the network *RI* reference frame in the per-unit systems of the generator and network, respectively. It follows that the two are related by:

$$\underset{\sim}{v}_{RI}^{(n)} = K_V \underset{\sim}{v}_{RI}^{(g)} \text{ in which } K_V = V_{usb}^{(g)}/V_{usb}^{(n)}, \tag{4.115}$$

and similarly for the current components in the respective per-unit systems:

$$\underset{\sim}{i}_{RI}^{(n)} = K_I \underset{\sim}{i}_{RI}^{(g)} \text{ in which } K_I = I_b^{(g)}/I_b^{(n)} = K_S/K_V \text{ and } K_S = S_{usb}^{(g)}/S_{usb}^{(n)}. \tag{4.116}$$

4.2.11 Linearized coupled-circuit formulation of the generator model equations

The linearized state- and algebraic-equations of the coupled-circuit generator model incorporating generator saturation and the transformation between the *dq* generator and network *RI* reference frames that were developed in the previous sections are now combined in matrix form. A systematic procedure for formulating the coefficient matrices is also provided.

The generator equations are linearized about the initial steady-state operating condition calculated in Sections 4.2.9 and 4.2.10.

The linearized generator equations comprising (i) the electromagnetic state and algebraic equations (4.41) on page 110; (ii) the rotor equations of motion in (4.58) and (4.60) on page 115; (iii) the transformation of the stator current and voltage components from the *dq* to *RI* coordinate systems in equations (4.110) and (4.112) on page 132; and (iv) the conversion from the generator to network per-unit systems for the voltage and current components in equations (4.115) and (4.116) are combined to form the following matrix equation,

$$\begin{bmatrix} p\Delta\underset{\sim}{x}_m \\ 0 \\ 0 \\ 0 \\ 0 \\ 0 \end{bmatrix} = \begin{bmatrix} A_m & B_{mi} & B_{mv} & 0 & 0 \\ C_{vgx} & D_{vgi} & -I & 0 & 0 \\ C_{inx} & D_{ing} & 0 & -I & 0 \\ C_{vnx} & 0 & D_{vng} & 0 & -I \\ 0 & 0 & 0 & K_I I & 0 \\ 0 & 0 & 0 & 0 & K_V I \end{bmatrix} \begin{bmatrix} \Delta\underset{\sim}{x}_m \\ \Delta\underset{\sim}{i}_{dq} \\ \Delta\underset{\sim}{v}_{dq} \\ \Delta\underset{\sim}{i}_{RI}^{(g)} \\ \Delta\underset{\sim}{v}_{RI}^{(g)} \end{bmatrix} + \begin{bmatrix} 0 & 0 \\ 0 & 0 \\ 0 & 0 \\ 0 & 0 \\ -I & 0 \\ 0 & -I \end{bmatrix} \begin{bmatrix} \Delta\underset{\sim}{i}_{RI}^{(n)} \\ \Delta\underset{\sim}{v}_{RI}^{(n)} \end{bmatrix} + \begin{bmatrix} B_m \\ 0 \\ 0 \\ 0 \\ 0 \\ 0 \end{bmatrix} \Delta\underset{\sim}{u}_m. \tag{4.117}$$

In the preceding equation the generator state- and input-variable vectors are respectively:

$$\Delta\underset{\sim}{x}_m = \begin{bmatrix} \Delta\delta \\ \Delta\omega \\ \Delta\underset{\sim}{x}_r \end{bmatrix} \text{ and } \Delta\underset{\sim}{u}_m = \begin{bmatrix} \Delta e_{fd} \\ \Delta T_m \end{bmatrix} \tag{4.118}$$

and the vector of electromagnetic state-variables are:

$$\Delta \underset{\sim}{x}_r = \Delta \underset{\sim}{\varphi}_{rdq} = \begin{bmatrix} \Delta \underset{\sim}{\varphi}_{rd} \\ \Delta \underset{\sim}{\varphi}_{rq} \end{bmatrix}. \tag{4.119}$$

The step-by-step procedure to formulate the coefficient matrices in (4.117) is given in the following table.

Table 4.9 Step-by-step procedure to compute the coefficient matrices in the linearized coupled-circuit equations of the generator.

Step	Operation	Source
1	Obtain the model parameters according to the number of *d*- and *q*-axis rotor windings n_d and n_q respectively.	
2	If magnetic saturation is neglected set $s_m = 0$. Otherwise obtain the saturation model parameters and set (i) $s_m = 1$ if the incremental inductance; or (ii) set $s_m = 2$ if the demagnetizing current representation of saturation is employed.	
3	Compute the generator steady-state initial conditions in the *dq* reference frame.	Section 4.2.9
4	Compute the initial values of the generator rotor-angle (δ_0) and stator voltage and current in the synchronously rotating *RI* reference frame.	Section 4.2.10
5	If $s_m = 1$ then in the following $L_{ad} = L_{ad_0}$ and $L_{aq} = L_{aq_0}$ otherwise $L_{ad} = L_{ad_u}$ and $L_{aq} = L_{aq_u}$	(4.31) pg. 108
6	Define $\Delta \underset{\sim}{\varphi}_{rd}$, $\Delta \underset{\sim}{i}_{rd}$, $\boldsymbol{r_{rd}}$, $\boldsymbol{L_{lrd}}$ and $\boldsymbol{b_{red}}$ depending on n_d the number of *d*-axis rotor windings and whether or not unequal mutual coupling between the *d*-axis rotor windings is to be represented.	Tab. 4.5 pg. 112
7	Define $\Delta \underset{\sim}{\varphi}_{rq}$, $\Delta \underset{\sim}{i}_{rq}$, $\boldsymbol{r_{rq}}$ and $\boldsymbol{L_{lrq}}$ depending on the number of *q*-axis damper windings.	Tab. 4.6 pg. 112
8	Construct $\boldsymbol{r_{rdq}} = \mathfrak{D}(\boldsymbol{r_{rd}}, \boldsymbol{r_{rq}})$ and $\boldsymbol{L_{lrdq}} = \mathfrak{D}(\boldsymbol{L_{lrd}}, \boldsymbol{L_{lrq}})$	(4.7) pg. 104 & (4.21) pg. 106
9	Construct $\boldsymbol{b_{re}} = \begin{bmatrix} \boldsymbol{b_{red}} \\ \boldsymbol{0} \end{bmatrix}$ in which $\boldsymbol{b_{red}} = \begin{bmatrix} \omega_b & 0 & 0 \end{bmatrix}^T$	(4.8) pg. 104
10	Construct the airgap mutual inductance matrix $\boldsymbol{L_{adq}} = \mathfrak{D}(L_{ad}, L_{aq})$	(4.31) pg. 108

Step	Operation	Source
11	Construct the coefficient matrices in the stator voltage equations: $$\boldsymbol{W}_{dq} = \begin{bmatrix} 0 & -1 \\ 1 & 0 \end{bmatrix}$$	(4.25) pg. 107
12	Calculate the coefficient of the saturation variable in the airgap mutual flux linkage equation: $$\boldsymbol{C}_{asdq_0} = \begin{cases} \boldsymbol{C}_{aldq_0} = \mathfrak{D}\left(\dfrac{\varphi_{ad_0}}{L_{ad_0}}, \dfrac{\varphi_{aq_0}}{L_{aq_0}}\right) & \text{if } s_m = 1 \\ -\boldsymbol{L}_{adq} & \text{if } s_m = 2 \\ \boldsymbol{\emptyset} & \text{if } s_m = 0 \end{cases}$$	(4.32) pg. 108
13	Calculate the coefficient matrix $\boldsymbol{C}_{sadq_0}$ of the saturation variable equation (4.34) on page 108 depending on the method used to represent saturation. $$\boldsymbol{C}_{sadq_0} = \begin{cases} \boldsymbol{C}_{ladq_0} & \text{if } s_m = 1 \\ \boldsymbol{C}_{madq_0} & \text{if } s_m = 2 \\ \boldsymbol{\emptyset} & \text{if } s_m = 0 \end{cases}, \text{ where}$$	(4.35) pg. 108
13(a)	$$\boldsymbol{C}_{ladq_0} = -\left(\frac{1}{\varphi_{ag_0}}\right)\begin{bmatrix} \left(\dfrac{L_{ad_0}^2}{L_{ad_u}}\right)\left(\dfrac{\partial S_d(\varphi_{ag})}{\partial \varphi_{ag}}\right)_0 \\ \left(\dfrac{L_{aq_0}^2}{L_{aq_u}}\right)\left(\dfrac{\partial S_q(\varphi_{ag})}{\partial \varphi_{ag}}\right)_0 \end{bmatrix}\begin{bmatrix} \varphi_{ad_0} & \varphi_{aq_0} \end{bmatrix} \text{ and}$$	(4.73) pg. 123
13(b)	$$\boldsymbol{C}_{madq_0} = \boldsymbol{S}_{dq_0} + \begin{bmatrix} \dfrac{\varphi_{ad_0}}{L_{ad_u}}\left(\dfrac{\partial S_d(\varphi_{ag})}{\partial \varphi_{ag}}\right)_0 \\ \dfrac{\varphi_{aq_0}}{L_{aq_u}}\left(\dfrac{\partial S_q(\varphi_{ag})}{\partial \varphi_{ag}}\right)_0 \end{bmatrix}\begin{bmatrix} \left(\dfrac{\varphi_{ad_0}}{\varphi_{ag_0}}\right) & \left(\dfrac{\varphi_{aq_0}}{\varphi_{ag_0}}\right) \end{bmatrix} \text{ in which}$$ $$\boldsymbol{S}_{dq_0} = \mathfrak{D}\left(\frac{S_d(\varphi_{ag_0})}{L_{ad_u}}, \frac{S_q(\varphi_{ag_0})}{L_{aq_u}}\right).$$ ... Continued on following page ...	(4.80) pg. 124 & (4.81) pg. 124

Step	Operation	Source
	The d- and q-axis saturation functions $S_d(\varphi_{ag})$, $S_q(\varphi_{ag})$ and their derivatives $\frac{\partial S_d}{\partial \varphi_{ag}}$ and $\frac{\partial S_q}{\partial \varphi_{ag}}$ depend on the selected interpolation function in Table 4.7 on page 120. If parameters for the q-axis saturation characteristic are not specified then one of the relationships in Table 4.8 may be used to express S_q as a function of S_d.	
14	Compute the coefficient sub-matrices in the electromagnetic state- and algebraic-equations of the machine (4.41) on page 110. $\boldsymbol{C}_{re} = \begin{bmatrix} -(\omega_b \boldsymbol{r}_{rdq}) & \boldsymbol{0} & \boldsymbol{0} & \boldsymbol{0} \end{bmatrix}$, $\boldsymbol{C}_{er} = \begin{bmatrix} -\boldsymbol{I} \\ \boldsymbol{0} \\ \boldsymbol{0} \\ \boldsymbol{0} \end{bmatrix}$, $\boldsymbol{C}_{ei} = \begin{bmatrix} \boldsymbol{0} \\ -\boldsymbol{L}_{adq} \\ L_l \boldsymbol{I} \\ \boldsymbol{0} \end{bmatrix}$ $\boldsymbol{C}_{ee} = \begin{bmatrix} \boldsymbol{L}_{lrdq} & \boldsymbol{u}_2 & \boldsymbol{0} & \boldsymbol{0} \\ \boldsymbol{L}_{adq} \boldsymbol{u}_2^T & -\boldsymbol{I} & \boldsymbol{0} & \boldsymbol{C}_{asdq_0} \\ \boldsymbol{0} & \boldsymbol{I} & -\boldsymbol{I} & \boldsymbol{0} \\ \boldsymbol{0} & \boldsymbol{C}_{sadq_0} & \boldsymbol{0} & -\boldsymbol{I} \end{bmatrix}$, $\boldsymbol{C}_{ve} = \begin{bmatrix} \boldsymbol{0} & \boldsymbol{0} & \omega_0 \boldsymbol{W}_{dq} & \boldsymbol{0} \end{bmatrix}$ $\boldsymbol{J}_{er} = -\boldsymbol{C}_{ee}^{-1} \boldsymbol{C}_{er}$ and $\boldsymbol{J}_{ei} = -\boldsymbol{C}_{ee}^{-1} \boldsymbol{C}_{ei}$ $\boldsymbol{A}_r = \boldsymbol{C}_{re} \boldsymbol{J}_{er}$, $\boldsymbol{B}_{ri} = \boldsymbol{C}_{re} \boldsymbol{J}_{ei}$, $\boldsymbol{C}_{vr} = \boldsymbol{C}_{ve} \boldsymbol{J}_{er}$ & $\boldsymbol{D}_{vi} = \boldsymbol{C}_{ve} \boldsymbol{J}_{ei} - r_s \boldsymbol{I}$	(4.39) pg. 110 (4.43) pg. 110 (4.42) pg. 110
15	Compute the coefficient matrices $\boldsymbol{C}_{tidq}$ and $\boldsymbol{C}_{tvdq}$ in the shaft acceleration equation (4.60) on page 115 $\boldsymbol{C}_{tidq} = \frac{1}{\omega_0} \begin{bmatrix} (v_{d_0} + 2r_s i_{d_0}) & (v_{q_0} + 2r_s i_{q_0}) \end{bmatrix}$ and $\boldsymbol{C}_{tvdq} = \frac{1}{\omega_0} \begin{bmatrix} i_{d_0} & i_{q_0} \end{bmatrix}$	(4.54) pg. 114
16	Compute the transformation matrices between the generator dq and network synchronously rotating RI coordinate systems. $\boldsymbol{R}(\delta_0) = \begin{bmatrix} \cos(\delta_0) & -\sin(\delta_0) \\ \sin(\delta_0) & \cos(\delta_0) \end{bmatrix}$, $\boldsymbol{U}(\delta_0) = \left(\frac{\partial \boldsymbol{R}(\delta)}{\partial \delta}\right)_0 = \begin{bmatrix} -\sin(\delta_0) & -\cos(\delta_0) \\ \cos(\delta_0) & -\sin(\delta_0) \end{bmatrix}$.	(4.103) pg. 131, (4.106) pg. 131

Step	Operation	Source
17	Construct the coefficient matrices in the generator state- and algebraic-equations in (4.117): $\boldsymbol{A}_m = \begin{bmatrix} 0 & \omega_b & \boldsymbol{0} \\ 0 & -\left(\frac{D}{2H}\right) & \boldsymbol{0} \\ \boldsymbol{0} & \boldsymbol{0} & \boldsymbol{A}_r \end{bmatrix}$, $\boldsymbol{B}_{mi} = \begin{bmatrix} \boldsymbol{0} \\ -\left(\frac{1}{2H}\right)\boldsymbol{C}_{tidq} \\ \boldsymbol{B}_{ri} \end{bmatrix}$, $\boldsymbol{B}_{mv} = \begin{bmatrix} \boldsymbol{0} \\ -\left(\frac{1}{2H}\right)\boldsymbol{C}_{tvdq} \\ \boldsymbol{0} \end{bmatrix}$, $\boldsymbol{B}_m = \begin{bmatrix} 0 & 0 \\ 0 & \left(\frac{1}{2H}\right) \\ \boldsymbol{b}_{re} & \boldsymbol{0} \end{bmatrix}$, $\boldsymbol{C}_{vgx} = \begin{bmatrix} \boldsymbol{0} & \boldsymbol{0} & \boldsymbol{C}_{vr} \end{bmatrix}$, $\boldsymbol{D}_{vgi} = \boldsymbol{D}_{vi}$, $\boldsymbol{C}_{inx} = \begin{bmatrix} \boldsymbol{U}(\delta_0)\underset{\sim}{i}_{dq_0} & \boldsymbol{0} & \boldsymbol{0} \end{bmatrix}$, $\boldsymbol{D}_{ing} = \boldsymbol{R}(\delta_0)$, $\boldsymbol{C}_{vnx} = \begin{bmatrix} \boldsymbol{U}(\delta_0)\underset{\sim}{v}_{dq_0} & \boldsymbol{0} & \boldsymbol{0} \end{bmatrix}$ and $\boldsymbol{D}_{vng} = \boldsymbol{R}(\delta_0)$	(4.41) pg. 110, (4.58) pg. 114 & (4.60) pg. 115
18	Compute the generator to network voltage and current scaling factors: $K_V = V_{ub}^{(g)}/V_{ub}^{(n)}$, $K_S = S_{ub}^{(g)}/S_{ub}^{(n)}$, $K_I = I_b^{(g)}/I_b^{(n)} = K_S/K_V$,	(4.115) pg. 133, (4.116) pg. 133

4.2.12 Transfer-function representation of the electromagnetic equations

In the previous sections fundamental electromagnetic circuit concepts were employed to provide a sound physical basis for modelling the dynamic performance characteristics of synchronous generators in terms of coupled circuits. However, tests to directly identify the coupled-circuit parameters (resistances and inductances) of these models are not routinely performed in practice. Rather, parameter identification is commonly based on the so-called "operational" or, in modern terminology, "transfer-function" representation of the machine depicted in Figure 4.9. In the transfer-function representation the *d*-axis of the machine is represented by a two-port network comprising the stator and field terminals and the *q*-axis by a single port network.

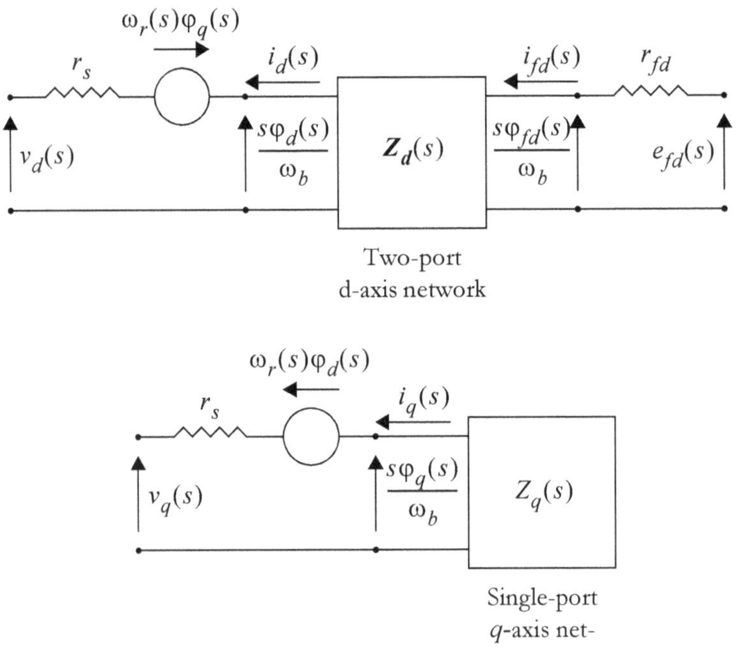

Figure 4.9 General transfer-function representations of the *d*- and *q*-axis of the synchronous generator.

The transfer impedance representation of the incremental form of the *d*-axis two port network can be shown to be [22]

$$\begin{bmatrix} \Delta\varphi_d(s) \\ \Delta\varphi_{fd}(s) \end{bmatrix} = \frac{\omega_b}{s} \begin{bmatrix} Z_{11}(s) & Z_{12}(s) \\ Z_{12}(s) & Z_{22}(s) \end{bmatrix} \begin{bmatrix} -\Delta i_d(s) \\ \Delta i_{fd}(s) \end{bmatrix}. \tag{4.120}$$

Note that the impedance matrix is symmetrical since the network is passive and linear. The scaling of the impedance matrix by the base electrical frequency occurs because one second is the base value of time. The following hybrid formulation of the transfer characteristic is useful because the input quantities are the *d*-axis stator current and field voltage and is a formulation frequently encountered in the literature (e.g. [6, 15, 20, 21, 22, 23]):

$$\begin{bmatrix}\Delta\varphi_d(s)\\ \Delta i_{fd}(s)\end{bmatrix} = \begin{bmatrix} L_d(s) & G(s)\\ -\frac{s}{\omega_b}G(s) & Y_{fd}(s)\end{bmatrix}\begin{bmatrix}-\Delta i_d(s)\\ \Delta e_{fd}(s)\end{bmatrix}. \quad (4.121)$$

The three *d*-axis operational parameters are:

$$\begin{aligned} L_d(s) &= \frac{\omega_b}{s}(Z_{11}(s) - Z_{12}(s)Y_{fd}(s)Z_{12}(s))\\ G(s) &= \frac{\omega_b}{s}Z_{12}(s)Y_{fd}(s)\\ Y_{fd}(s) &= (r_{fd} + Z_{22}(s))^{-1}\end{aligned} \quad . \quad (4.122)$$

in which $L_d(s)$ is referred to as the *"d-axis operational inductance"* of the stator with constant field voltage; $G(s)$ is the *"stator-to-field transfer-function"* and $Y_{fd}(s)$ is the *"operational admittance of the field with the stator open-circuit"*.

The single port representation of the *q*-axis is:

$$\Delta\varphi_q(s) = -\frac{\omega_b}{s}Z_q(s)\Delta i_q(s) = -L_q(s)\Delta i_q(s). \quad (4.123)$$

$L_q(s)$ is referred to as the *"q-axis operational inductance"*.

It is important to note that no assumptions have been made at this stage about the internal electromagnetic structure of the machine.

4.2.12.1 Exact relationship between coupled-circuit and transfer-function representations of the electromagnetic equations

In [12] the exact relationships between the coupled-circuit formulation of the electromagnetic equations in Section 4.2.3.4 and the transfer-function representations are established. It is found that for a machine with n_d *d*-axis rotor windings and n_q *q*-axis rotor-windings the transfer-functions have the following forms:

$$L_d(s) = L_d\frac{N_{Ld}(s)}{D_d(s)},\ G(s) = G_0\frac{N_G(s)}{D_d(s)},\ Y_{fd}(s) = Y_0\frac{N_Y(s)}{D_d(s)} \text{ and } L_q(s) = L_q\frac{N_{Lq}(s)}{D_q(s)} \quad (4.124)$$

in which the *d*-axis numerator polynomials are:

$$N_{Ld}(s) = 1 + \sum_{k=1}^{n_d}(a_{kd}s^k),\ N_G(s) = 1 + \sum_{k=1}^{n_d-1}(b_{kd}s^k),\ N_Y(s) = 1 + \sum_{k=1}^{n_d-1}(c_{kd}s^k). \quad (4.125)$$

The *d*-axis denominator polynomial is:

$$D_d(s) = 1 + \sum_{k=1}^{n_d} (d_{kd}s^k). \tag{4.126}$$

The q-axis numerator and denominator polynomials are:

$$N_{Lq}(s) = 1 + \sum_{k=1}^{n_q} (a_{kq}s^k) \text{ and } D_q(s) = 1 + \sum_{k=1}^{n_q} (d_{kq}s^k). \tag{4.127}$$

The coefficients in the above polynomials (i.e. a_{kd}, b_{kd}, etc.) are functions of the coupled-circuit parameters.

It is normal practice to factorize the d-axis numerator and denominator polynomials such that:

$$N_{Ld}(s) = \prod_{k=1}^{n_d} (1 + sT_{dk}),\quad N_G(s) = \prod_{k=1}^{n_d-1} (1 + sT_{Gk}),\quad N_Y(s) = \prod_{k=1}^{n_d-1} (1 + sT_{Yk}), \tag{4.128}$$

$$D_d(s) = \prod_{k=1}^{n_d} (1 + sT_{d0k}) \tag{4.129}$$

where $T_{dk} > T_{d(k+1)}$, $T_{d0k} > T_{d0(k+1)}$, $T_{Gk} > T_{G(k+1)}$ and $T_{Yk} > T_{Y(k+1)}$.

Similarly, for the q-axis:

$$N_{Lq}(s) = \prod_{k=1}^{n_q} (1 + sT_{qk}) \text{ and } D_q(s) = \prod_{k=1}^{n_q} (1 + sT_{q0k}) \tag{4.130}$$

such that $T_{qk} > T_{q(k+1)}$, $T_{q0k} > T_{q0(k+1)}$

The time constants T_{dk}, $k = 1, \ldots, n_d$ are referred to as the "d-axis principal short-circuit time constants" and are denoted as the d-axis transient (T_d'), subtransient (T_d'') and sub-subtransient (T_d''') short-circuit time constants. The denominator time constants T_{d0k}, $k = 1, \ldots, n_d$ are referred to as the "d-axis principal open-circuit time constants" and are denoted as the d-axis transient (T_{d0}'), subtransient (T_{d0}'') and sub-subtransient (T_{d0}''') open-circuit time constants. Analogous definitions apply to the q-axis principal short- and open-circuit time constants.

As shown in [12] the d- and q-axis operational-inductances can also be expressed as sums of partial fractions as follows:

$$\frac{1}{L_d(s)} = \frac{1}{L_d} + \sum_{k=1}^{n_d} \left(\frac{1}{L_{dk}} - \frac{1}{L_{d(k-1)}}\right)\left(\frac{sT_{dk}}{1+sT_{dk}}\right), \text{ where } L_{d0} = L_d \text{ and} \quad (4.131)$$

$$\frac{1}{L_q(s)} = \frac{1}{L_q} + \sum_{k=1}^{n_q} \left(\frac{1}{L_{qk}} - \frac{1}{L_{q(k-1)}}\right)\left(\frac{sT_{qk}}{1+sT_{qk}}\right), \text{ where } L_{q0} = L_q. \quad (4.132)$$

The inductances L_{dk}, $k = 1, \ldots, n_d$ in (4.131) are referred to as the "*d*-axis principal dynamic inductances" and are denoted as the *d*-axis transient (L_d'), subtransient (L_d'') and sub-subtransient (L_d''') inductances. Analogous definitions apply to the *q*-axis principal dynamic inductances.

The exact mathematical relationships between (i) the coupled-circuit parameters and (ii) the time constants and principal dynamic inductances in the above factorizations of the operational parameters are very complex when there is more than one rotor-winding in an axis. Exact bi-directional transformations have been derived between the coupled-circuit parameters of the machines and the transfer-function parameters in [12]. Software for performing these exact transformations are available on the website of this eBook [24].

4.2.12.2 Classical definitions of the standard parameters in terms of the coupled-circuit parameters

The opportunity for considerable simplification in the relationships between the coupled-circuit and standard parameters of the *d*-axis has long been recognized (e.g. Concordia [20]). If the *d*-axis is assumed to be represented by a field winding and one damper winding the simplification is based on the observation that the *d*-axis damper winding resistance r_{1d} is usually much larger than the field winding resistance r_{fd}. Consequently, following a disturbance the *d*-axis damper winding flux-linkages tend to decay much more rapidly than those of the field. Thus, the relatively rapid rate of decay of the damper winding flux-linkages in the so-called *subtransient period* immediately following a disturbance can be *approximated* by assuming that $r_{fd} = 0$ and that the field flux linkages remain constant during this period. The behaviour of the generator in the *transient period* which follows the subtransient period is approximated by neglecting the damper winding (i.e. $r_{1d} = L_{1d} = \infty$) on the assumption that the damper winding current has decayed to zero. Finally, the behaviour in the *steady-state period,* following the decay of the transient response, is approximated by neglecting both the damper and field windings. By applying these approximations, the *"classical definitions"* of the standard-parameters in terms of the coupled-circuit parameters result. By applying analogous approximations to the *q*-axis and extending the approximations to axes with three rotor-windings, classical definitions of the standard parameters in terms of the coupled-circuit parameters are obtained in [12]. In the case of a *d*-axis represented with two rotor-windings it is usually found that there is little difference between the standard parameters which are specified in accordance with the exact definitions and those which are specified in accordance with the classical definitions. However, if the q-axis is represented with two or more windings, or if the d-axis is represented by three rotor windings, the accuracy of the classical approximations of the standard parameters may be considerably diminished. This is because there is greater overlap between the transient and subtransient periods of the q-axis than occurs in the d-axis. Similarly, when three rotor windings are represented there is a tendency, particularly in the q-axis, for there to be overlap between the transient, sub-transient and sub-subtransient periods with consequential inaccuracy of the classical approximations of the standard parameters.

Table 4.10 gives the transformation from given coupled-circuit parameters to the classically-defined principal open- and short-circuit time constants and the principal dynamic inductances for the d- or q-axis represented by one, two or three rotor-windings respectively. Table 4.11 gives the inverse transformation (i.e. classically-defined standard parameters to coupled-circuit parameters). If the transformations are being performed for the d-axis the subscript 'a' in these tables is substituted with 'd'; '1' with 'f'; '2' with '1' and '3' with '2'. Thus, L_a is replaced by L_d; L_{1a} by L_{fd}; r_{2a} by r_{1d}; T_a' by T_d', etc. If the q-axis parameters are being converted then the subscript 'a' is substituted with 'q'. An exception to the above rules is that the airgap mutual inductance L_{aa} is substituted with L_{ad} or L_{aq} depending on which axis is being converted. These transformations assume that unequal coupling between the d-axis rotor windings is neglected (i.e. $L_{c1} = L_{c2} = 0$).

When the q-axis is represented with a single rotor winding it is conventional practice to label the standard parameters as being the subtransient parameters (i.e. (i.e. T_{q0}'', T_q'' and L_q'') instead of the transient parameters T_{q0}', T_q' and L_q' . This is because the principal q-axis time constants for a machine whose q-axis is adequately represented by a single rotor winding tend to be short. Thus, in this scenario, in order to apply the conversions in Tables 4.10 and 4.11 it is necessary to substitute the transient parameters in the tables with the given sub-transient q-axis parameters.

The electromagnetic equations in some widely-used simulation packages are formulated directly in terms of the classically-defined standard parameters. This classical parameter formulation of the model is summarized in Section 4.2.13. *It is emphasised that the classical parameter formulation is exactly equivalent to the coupled-circuit formulation provided (i) that the unequal coupling between the d-axis rotor windings is neglected, and (ii) that the same method for representing magnetic saturation is employed in the two models.*

Table 4.10 Classical definitions of the principal short- and open-circuit time constants and dynamic inductances in terms of the coupled-circuit parameters for the *d*- or *q*-axis represented with $n_a = 0, 1, 2$ or 3 rotor windings. [CC2CS]

Inputs: Coupled-circuit parameters	Outputs: Classically-defined standard parameters
	Synchronous parameters ($n_a \geq 0$)
L_{aa}, L_l	$L_a = L_{aa} + L_l$
	Transient parameters ($n_a > 0$)
r_{1a}, L_{1a}	$T_a' = \frac{1}{\omega_b r_{1a}}\left(L_{1a} + \frac{L_{aa}L_l}{L_a}\right)$
	$T_{a0}' = \frac{L_{aa} + L_{1a}}{\omega_b r_{1a}}$
	$L_a' = \frac{1}{1/L_{aa} + 1/L_{1a}} + L_l = \left(\frac{T_a'}{T_{a0}'}\right)L_a$
	Subtransient parameters ($n_a > 1$)
r_{2a}, L_{2a}	$T_a'' = \frac{1}{\omega_b r_{2a}}\left(L_{2a} + \frac{L_{aa}L_lL_{1a}}{L_{aa}L_l + L_aL_{1a}}\right)$
	$T_{a0}'' = T_{a011} = \frac{1}{\omega_b r_{2a}}\left(L_{2a} + \frac{L_{aa}L_{1a}}{L_{aa} + L_{1a}}\right)$
	$L_a'' = \frac{1}{\frac{1}{L_{aa}} + \frac{1}{L_{1a}} + \frac{1}{L_{2a}}} + L_l = \left(\frac{T_a''}{T_{a0}''}\right)L_a'$
	Sub-subtransient parameters ($n_a > 2$)
r_{3a}, L_{3a}	$T_a''' = \frac{1}{\omega_b r_{3a}}\left(L_{3a} + \frac{1}{\frac{1}{L_l} + \frac{1}{L_{aa}} + \frac{1}{L_{1a}} + \frac{1}{L_{2a}}}\right)$
	$T_{a0}''' = \frac{1}{\omega_b r_{3a}}\left(L_{3a} + \frac{1}{\frac{1}{L_{aa}} + \frac{1}{L_{1a}} + \frac{1}{L_{2a}}}\right)$
	$L_a''' = \frac{1}{\frac{1}{L_{aa}} + \frac{1}{L_{1a}} + \frac{1}{L_{2a}} + \frac{1}{L_{3a}}} + L_l = \left(\frac{T_a'''}{T_{a0}'''}\right)L_a''$

Table 4.11 Derivation of coupled-circuit parameters from the classically-defined open-circuit time constants and dynamic inductances for the *d*- or *q*-axis represented with $n_a = 0, 1, 2$ or 3 rotor windings. [CS2CC]

Inputs: Classically-defined standard parameters	**Outputs: Coupled-circuit parameters**
	$\mathbf{n_a \geq 0}$
L_a, L_l	$L_{aa} = L_a - L_l$
	$\mathbf{n_a > 0}$
T_{a0}', L_a'	$L_{1a} = \dfrac{(L_a - L_l)(L_a' - L_l)}{(L_a - L_a')}$
	$r_{1a} = \dfrac{1}{\omega_b T_{a0}'} \dfrac{(L_a - L_l)^2}{(L_a - L_a')}$
	$\mathbf{n_a > 1}$
T_{a0}'', L_a''	$L_{2a} = \dfrac{(L_a' - L_l)(L_a'' - L_l)}{(L_a' - L_a'')}$
	$r_{2a} = \dfrac{1}{\omega_b T_{a0}''} \dfrac{(L_a' - L_l)^2}{(L_a' - L_a'')}$
	$\mathbf{n_a > 2}$
T_{a0}''', L_a'''	$L_{3a} = \dfrac{(L_a'' - L_l)(L_a''' - L_l)}{(L_a'' - L_a''')}$
	$r_{3a} = \dfrac{1}{\omega_b T_{a0}'''} \dfrac{(L_a'' - L_l)^2}{(L_a'' - L_a''')}$
Use $L_a' = (T_a'/T_{a0}')L_a$, $L_a'' = (T_a''/T_{a0}'')L_a'$ and $L_a''' = (T_a'''/T_{a0}''')L_a''$ to derive the principal dynamic inductances if the principal short- and open-circuit time constants are given; or use $T_{a0}' = (L_a/L_a')T_a'$, $T_{a0}'' = (L_a'/L_a'')T_a''$ and $T_{a0}''' = (L_a''/L_a''')T_a'''$ if the principal dynamic inductances and short-circuit time constants are given.	

4.2.13 Electromagnetic model in terms of classically-defined standard parameters

Some widely used rotor angle stability analysis software packages employ generator models which are formulated directly in terms of the classically-defined standard parameters. *It is shown in [12] that the "Classical Parameter Model" is exactly equivalent to the coupled-circuit formulation of the electromagnetic equations provided (i) unequal coupling between the d-axis rotor windings is neglected; and (ii) the same representation of magnetic saturation is employed.*

The transfer-function block-diagram representation of the classical parameter model shown in Figure 4.10 is often presented in the literature [34, 13 (vol 3)]. The associated state- and algebraic-equations for the model, which are rigorously derived from the underlying coupled-circuit formulation of the model in [12], are summarized in Section 4.2.13.1. The relationships between the classically-defined standard parameters and the underlying coupled-circuit parameters are given in Tables 4.10 and 4.11 of Section 4.2.12.2.

Please turn over to Figure 4.10.

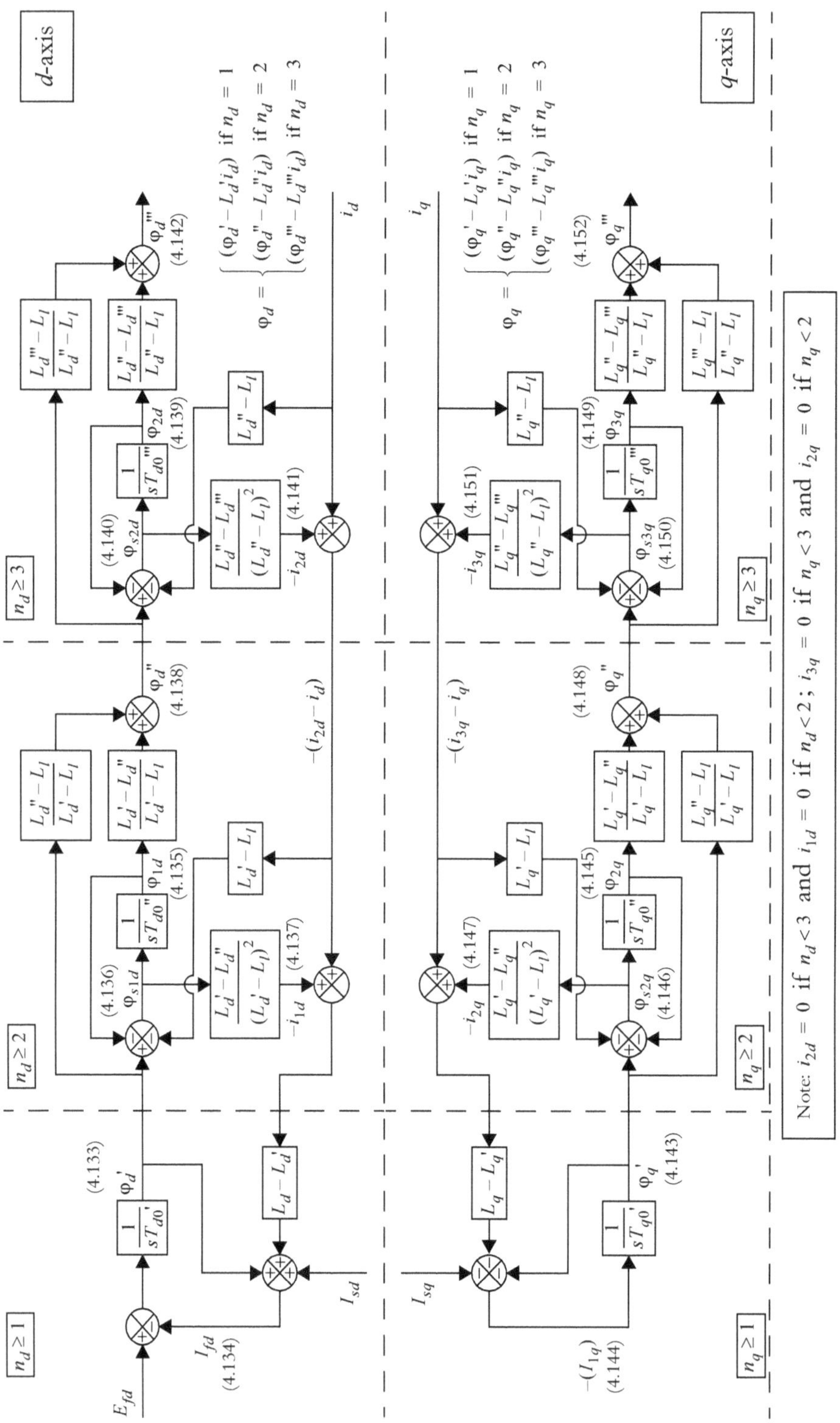

Figure 4.10 Transfer-function block diagram representation of the *d*- and *q*-axis rotor-winding equations in terms of the classically-defined standard parameters. The effects of magnetic saturation is represented by the demagnetizing I_{sd} and I_{sq} in the respective axes.

4.2.13.1 Summary of classical parameter electromagnetic model equations

The *d*-axis rotor winding equations in terms of the classically-defined standard parameters are summarized below. The equations are applicable to models with up to three rotor windings (i.e. $n_d = 3$). Note that the field voltage and current are expressed in the non-reciprocal per-unit system as is the demagnetizing current, I_{sd}, representing the effects of magnetic saturation in the *d*-axis.

Field Winding Equations

Note: $i_{2d} = 0$ if $n_d < 3$ and $i_{1d} = 0$ if $n_d < 2$

$$\frac{d\varphi_d'}{dt} = \frac{1}{T_{d0}'}(E_{fd} - I_{fd}) \tag{4.133}$$

$$I_{fd} = \varphi_d' + (L_d - L_d')(i_d - (i_{1d} + i_{2d})) + I_{sd} \tag{4.134}$$

First *d*-axis Damper Winding Equations (include if nd > 1)

$$\frac{d\varphi_{1d}}{dt} = \frac{1}{T_{d0}''}\varphi_{s1d}, \tag{4.135}$$

$$\varphi_{s1d} = \varphi_d' - \varphi_{1d} + (L_d' - L_l)(i_{2d} - i_d), \tag{4.136}$$

$$i_{1d} = -\left(\frac{L_d' - L_d''}{(L_d' - L_l)^2}\right)\varphi_{s1d}, \tag{4.137}$$

$$\varphi_d'' = \left(\frac{L_d'' - L_l}{L_d' - L_l}\right)\varphi_d' + \left(\frac{L_d' - L_d''}{L_d' - L_l}\right)\varphi_{1d}. \tag{4.138}$$

Second *d*-axis Damper Winding Equations (include if nd > 2)

$$\frac{d\varphi_{2d}}{dt} = \frac{1}{T_{d0}'''}\varphi_{s2d}, \tag{4.139}$$

$$\varphi_{s2d} = \varphi_d'' - \varphi_{2d} - (L_d'' - L_l)i_d, \tag{4.140}$$

$$i_{2d} = -\left(\frac{L_d'' - L_d'''}{(L_d'' - L_l)^2}\right)\varphi_{s2d}, \tag{4.141}$$

$$\varphi_d''' = \left(\frac{L_d''' - L_l}{L_d'' - L_l}\right)\varphi_d'' + \left(\frac{L_d'' - L_d'''}{L_d'' - L_l}\right)\varphi_{2d}. \tag{4.142}$$

The *q*-axis rotor winding equations in terms of the classically-defined standard parameters are summarized below. The equations are applicable to models with up to three *q*-axis damper windings (i.e. $n_q = 3$). By analogy with the field winding, the current in the first *q*-axis damper winding, I_{1q}, is expressed in the non-reciprocal per-unit system such that

$I_{1q} = L_{aq_u} i_{1q}$. Similarly, the q-axis saturation demagnetizing excitation current, I_{sq}, is in per-unit in the non-reciprocal base system (i.e. $I_{sq} = L_{aq_u} i_{sq}$).

First q-axis Damper Winding Equations

Note: $i_{3d} = 0$ if $n_q < 3$ and $i_{2q} = 0$ if $n_q < 2$.

$$\frac{d\varphi_q'}{dt} = \frac{1}{T_{q0}'}(-I_{1q}), \tag{4.143}$$

$$(-I_{1q}) = -\varphi_q' + (L_q - L_q')((i_{2q} + i_{3q}) - i_q) - I_{sq}. \tag{4.144}$$

Second q-axis Damper Winding Equations

$$\frac{d\varphi_{2q}}{dt} = \frac{1}{T_{q0}''}\varphi_{s2q}, \tag{4.145}$$

$$\varphi_{s2q} = \varphi_q' - \varphi_{2q} + (L_q' - L_l)(i_{3q} - i_q), \tag{4.146}$$

$$i_{2q} = -\left(\frac{L_q' - L_q''}{(L_q' - L_l)^2}\right)\varphi_{s2q}, \tag{4.147}$$

$$\varphi_q'' = \left(\frac{L_q'' - L_l}{L_q' - L_l}\right)\varphi_q' + \left(\frac{L_q' - L_q''}{L_q' - L_l}\right)\varphi_{2q}. \tag{4.148}$$

Third q-axis Damper Winding Equations

$$\frac{d\varphi_{3q}}{dt} = \frac{1}{T_{q0}'''}\varphi_{s3q}, \tag{4.149}$$

$$\varphi_{s3q} = \varphi_q'' - \varphi_{3q} - (L_q'' - L_l)i_q, \tag{4.150}$$

$$i_{3q} = -\left(\frac{L_q'' - L_q'''}{(L_q'' - L_l)^2}\right)\varphi_{s3q}, \tag{4.151}$$

$$\varphi_q''' = \left(\frac{L_q''' - L_l}{L_q'' - L_l}\right)\varphi_q'' + \left(\frac{L_q'' - L_q'''}{L_q'' - L_l}\right)\varphi_{3q}. \tag{4.152}$$

It is noted that the d-axis transient flux linkages φ_d' are related to the field winding flux linkages φ_{fd} in the underlying coupled circuit model as follows:

$$\varphi_d' = \left(\frac{L_d - L_d'}{L_d - L_l}\right)\varphi_{fd}. \tag{4.153}$$

Furthermore, it is common practice to define the transient q-axis voltage E_q' which is in quadrature with φ_d' such that $E_q' = \omega\varphi_d'$. Since, for the purpose of small-signal modelling, perturbations in the rotor-speed are neglected in the formulation of the electromagnetic equations it follows that $E_q' = \omega_0 \varphi_d'$ and, in the usual case when $\omega_0 = 1$ then $E_q' = \varphi_d'$.

By analogy, in the q-axis $\varphi_q' = \left(\frac{L_q - L_q'}{L_q - L_l}\right)\varphi_{1q}$ and $E_d' = \omega\varphi_q'$.

To facilitate the formulation of the stator-winding d-axis flux linkage equations the following d-axis "k^{th}-transient flux-linkage" ($\overset{k}{\varphi_d}$) and "k^{th}-transient inductance" ($\overset{k}{L_d}$) quantities are defined depending on the number of d-axis rotor-windings:

$$\overset{k}{\varphi_d} = \begin{cases} \varphi_d' & \text{if } n_d = 1 \\ \varphi_d'' & \text{if } n_d = 2 \\ \varphi_d''' & \text{if } n_d = 3 \end{cases} \text{ and } \overset{k}{L_d} = \begin{cases} L_d' & \text{if } n_d = 1 \\ L_d'' & \text{if } n_d = 2 \\ L_d''' & \text{if } n_d = 3 \end{cases} \tag{4.154}$$

- and similarly for the q-axis.

The d- and q-axis stator winding flux linkage equations are respectively:

$$\varphi_d = \overset{k}{\varphi_d} - \overset{k}{L_d} i_d \text{ and } \varphi_q = \overset{k}{\varphi_q} - \overset{k}{L_q} i_q. \tag{4.155}$$

The following expressions for the d- and q-axis airgap mutual flux linkages are derived in terms of the kth-transient flux-linkages and inductances from (i) the coupled-circuit equations for the airgap flux linkages in (4.12) on page 105 with $s_m = 2$ (i.e. Method 2 for representing magnetic saturation); and (ii) from the equations for the d- and q-axis flux linkages in (4.23) on page 106 and from (4.155).

$$\varphi_{ad} = I_{fd} + L_{ad_u}(i_{1d} + i_{2d} - i_d) - I_{sd} = \varphi_d + L_l i_d = \overset{k}{\varphi_d} - (\overset{k}{L_d} - L_l) i_d \text{ and} \tag{4.156}$$

$$\varphi_{aq} = I_{1q} + L_{aq_u}(i_{2q} + i_{3q} - i_q) - I_{sq} = \varphi_q + L_l i_q = \overset{k}{\varphi_q} - (\overset{k}{L_q} - L_l) i_q. \tag{4.157}$$

From the coupled-circuit equations for the stator-voltage components in (4.25) on page 107 and the preceding equations for the airgap flux linkages the following equations for the d- and q-axis stator voltage equations are derived.

$$\begin{aligned} v_d &= -r_s i_d - \omega_0 \varphi_q = -r_s i_d + \omega_0 \overset{k}{L_q} i_q - \omega_0 \overset{k}{\varphi_q} \\ &= -r_s i_d - \omega_0 (I_{1q} + L_{aq_u}(i_{2q} + i_{3q}) - ((L_{aq_u} + L_l) i_q + I_{sq}))) \end{aligned} \text{ and} \tag{4.158}$$

$$\begin{aligned} v_q &= -r_s i_q + \omega_0 \varphi_d = -r_s i_q - \omega_0 \overset{k}{L_d} i_d + \omega_0 \overset{k}{\varphi_d} \\ &= -r_s i_q + \omega_0 (I_{fd} + L_{ad_u}(i_{1d} + i_{2d}) - ((L_{ad_u} + L_l) i_d + I_{sd}))) \end{aligned}. \tag{4.159}$$

4.2.13.2 Magnetic saturation in the classical parameter model

Saturation Method 2 in Section 4.2.8.2, in which the demagnetizing effect of magnetic saturation is represented by deducting a component of excitation current from each axis, is used in the classical parameter model. In that section the resultant airgap flux linkages were chosen to represent the level of flux-linkages on the saturation characteristic. This choice can also be applied to the classical parameter model. However, some widely used commercial software packages choose other flux quantities to represent the level of saturation. These alternative choices change both the initial steady-state operating condition of the machine as well as the dynamic response. Two such alternative choices are considered below.

4.2.13.2.1 Resultant k^{th}-transient flux-linkages as the saturation level indicator

When saturation of both the *d*- and *q*-axis is to be represented the resultant k^{th}-transient flux-linkages φ^k is sometimes used as the saturation level indicator, particularly when representing round-rotor machines.

$$\varphi^k = \sqrt{(\varphi_d^k)^2 + (\varphi_q^k)^2}, \tag{4.160}$$

and the demagnetizing currents in the respective axes are

$$I_{sd} = L_{ad_u} i_{sd} = \varphi_d^k S_d(\varphi^k) \text{ and } I_{sq} = L_{aq_u} i_{sq} = \varphi_q^k S_q(\varphi^k). \tag{4.161}$$

The initial steady-state values of the saturation level indicator and the saturated *d*- and *q*-axis airgap inductances are now derived. Once these values are determined the procedure outlined in Section 4.2.13.3 can be used to calculate the initial steady-state values of the other generator variables.

The following expression for the initial steady-state value of the *d*-axis saturation demagnetizing current is obtained by substituting for the yet unknown value of $\varphi_{d_0}^k$ from(4.161) into (4.156) and recognizing that under steady-state conditions the damper winding currents are zero.

$$I_{sd_0} = \left(\frac{S_d(\varphi_0^k)}{1 + S_d(\varphi_0^k)}\right)(I_{fd_0} - (L_{ad_u} + L_l - L_d^k)i_{d_0}). \tag{4.162}$$

Similarly, the steady-state *q*-axis saturation demagnetizing current is given by:

$$I_{sq_0} = -\left(\frac{S_q(\varphi_0^k)}{1 + S_q(\varphi_0^k)}\right)(L_{aq_u} + L_l - L_q^k)i_{q_0}. \tag{4.163}$$

The following expressions for the steady-state values of the *d*- and *q*-axis stator voltages are obtained by substituting the saturation demagnetizing current components from equations (4.162) and (4.163) into equations (4.158) and (4.159) to yield:

$$v_{d_0} = -r_s i_{d_0} + \omega_0 L_{q_0} i_{q_0} \text{ and } v_{q_0} = -r_s i_{q_0} - \omega_0 L_{d_0} i_{d_0} + \omega_0 \left(\frac{I_{fd_0}}{1 + S_d(\varphi_0^k)} \right), \quad (4.164)$$

in which

$$L_{d_0} = \left(\frac{L_{ad_u} + (L_d^k - L_l) S_d(\varphi_0^k)}{1 + S_d(\varphi_0^k)} \right) + L_l \text{ and } L_{q_0} = \left(\frac{L_{aq_u} + (L_q^k - L_l) S_q(\varphi_0^k)}{1 + S_q(\varphi_0^k)} \right) + L_l \quad (4.165)$$

From equations (4.82) on page 125 and (4.164) the initial steady-state voltage phasor is:

$$\hat{V}_t = V_{t0} e^{j\gamma_0} = v_{d0} + jv_{q0} = -(r_s + j\omega_0 L_{q_0})\hat{I} + jE_{q_0}, \quad (4.166)$$

in which $\hat{I} = I_0 e^{j(\gamma_0 + \beta_0)} = i_{d0} + ji_{q0}$ is obtained from (4.83) in terms of the initial steady-state real and reactive power output and stator voltage magnitude and

$$E_{q_0} = \omega_0 \left(\frac{I_{fd_0}}{1 + S_d(\varphi_0^k)} - (L_{d_0} - L_{q_0}) i_{d_0} \right). \quad (4.167)$$

Equation (4.166) is rearranged into the same form as equation (4.97) on page 127

$$\left(V_{t_0} + (r_s + j\omega_0 L_{q_0}) I_0 e^{j\beta_0} \right) e^{j\gamma_0} = E_{q_0} e^{j(\pi/2)}, \quad (4.168)$$

from which the initial steady-state value of the angle by which the voltage phasor leads the *d*-axis, is given by:

$$\gamma_0 = \frac{\pi}{2} - \operatorname{atan2}(I_0(r_s \sin\beta_0 + \omega_0 L_{q_0} \cos\beta_0), (V_{t_0} + I_0(r_s \cos\beta_0 - \omega_0 L_{q_0} \sin\beta_0))). \quad (4.169)$$

This equation is identical in form to the calculation of γ_0 in equation (4.98) on page 128 for the coupled-circuit formulation of the model. However, the value of L_{q_0} is different between the coupled-circuit and classical parameter formulations because different representations of saturation are used in the respective formulations.

In order to calculate the initial steady-state value of the *q*-axis synchronous reactance L_{q_0} in equation (4.165) it is necessary determine the initial value of the resultant k^{th}-transient flux-linkages φ_0^k and thence the value of the *q*-axis saturation characteristic $S_q(\varphi_0^k)$. In the case where $L_q^k = L_d^k$ it can be shown that:

$$\varphi_0^k = \left(\frac{1}{\omega_0} \right) \sqrt{\left(V_{t_0} + \left(\frac{r_s P_0 + \omega_0 L_d^k Q_0}{V_{t_0}} \right) \right)^2 + \left(\frac{\omega_0 L_d^k P_0 - r_s Q_0}{V_{t_0}} \right)^2}. \quad (4.170)$$

However, when the kth-transient inductances in the respective axes are unequal a closed form solution for φ_0^k is not possible. In this case the following set of three simultaneous non-linear equations must be solved iteratively to determine the initial steady-state values of L_{q_0}, γ_0 and φ_0^k.

$$L_{q_0} = \left(\frac{L_{aq_u} + (L_q^k - L_l)S_q(\varphi_0^k)}{1 + S_q(\varphi_0^k)}\right) + L_l$$

$$\gamma_0 = \frac{\pi}{2} - \text{atan2}(I_0(r_s \sin\beta_0 + \omega_0 L_{q_0} \cos\beta_0), (V_{t_0} + I_0(r_s \cos\beta_0 - \omega_0 L_{q_0} \sin\beta_0))) \tag{4.171}$$

$$\varphi_0^k = \left|\left(\frac{1}{\omega_0}\right)\left\{V_{t_0} e^{j\gamma_0} + (r_s + j\omega_0 L_q^k) I_0 e^{j(\gamma_0 + \beta_0)} + j\omega_0 (L_d^k - L_q^k) I_0 \cos(\gamma_0 + \beta_0)\right\}\right|$$

Having solved for the initial value of the saturation level indicator φ_0^k from equation (4.170) or (4.171) the initial values L_{d_0} and L_{q_0} are obtained from (4.165).

4.2.13.2.2 Transient *d*-axis flux-linkages as saturation level indicator

When saturation only of the *d*-axis is to be represented the transient flux-linkages φ_d' are sometimes used as the saturation level indicator. This approach is sometimes used to represent saturation in salient pole machines. In this case the *d*- and *q*-axis saturation demagnetizing currents are respectively:

$$I_{sd} = \varphi_d' S_d(\varphi_d') \text{ and } I_{sq} = 0. \tag{4.172}$$

Since it is assumed that the *q*-axis is unsaturated it follows that:

$$L_{q_0} = L_{aq_u} + L_l, \quad S_q(\varphi_d') = 0 \tag{4.173}$$

and the initial steady-state value of γ_0 is calculated according to (4.98) on page 128; the initial steady-state values $\underset{\sim}{v}_{dq_0}$ and $\underset{\sim}{i}_{dq_0}$ are calculated from (4.99).

The objective of the following steps is to calculate the initial steady-state value φ_{d_0}' taking account of saturation. The following expression for the initial steady-state value of the *d*-axis saturation demagnetizing current is obtained by substituting for φ_{d_0}' from (4.134) into (4.172) and recognizing that under steady-state conditions the damper winding currents are zero.

$$I_{sd_0} = \left(\frac{S_d(\varphi_{d_0}')}{1 + S_d(\varphi_{d_0}')}\right)(I_{fd_0} - (L_{ad_u} + L_l - L_d')i_{d_0}). \tag{4.174}$$

The following expressions for the steady-state values of the d- and q-axis stator voltages are obtained by substituting for the above expression of the d-axis saturation demagnetizing current component, together with $I_{sq_0} = 0$, into equations (4.158) and (4.159) to yield:

$$v_{d_0} = -r_s i_{d_0} + \omega_0 (L_{aq_u} + L_l) i_{q_0}, \quad v_{q_0} = -r_s i_{q_0} - \omega_0 L_{d_0} i_{d_0} + \omega_0 \left(\frac{I_{fd_0}}{1 + S_d(\varphi_{d_0}')} \right) \tag{4.175}$$

$$\text{and } L_{d_0} = \left(\frac{L_{ad_u} + (L_d' - L_l) S_d(\varphi_{d_0}')}{1 + S_d(\varphi_{d_0}')} \right) + L_l . \tag{4.176}$$

Following a procedure similar to that in equations (4.166) to (4.168) the following relationship is obtained

$$(v_{d_0} + j v_{q_0}) + (r_s + j\omega_0 L_{q_0})(i_{d_0} + j i_{q_0}) = j\omega_0 \left(\frac{I_{fd_0}}{1 + S_d(\varphi_{d_0}')} - (L_{d_0} - L_{q_0}) i_{d_0} \right) . \tag{4.177}$$

By equating the imaginary components of the above expression the following equation for the field current is obtained:

$$I_{fd_0} = \left(\frac{1 + S_d(\varphi_{d_0}')}{\omega_0} \right) (v_{q_0} + r_s i_{q_0} + \omega_0 L_{d_0} i_{d_0}) . \tag{4.178}$$

By substituting for I_{fd_0} and I_{sd_0} from equations (4.178) and (4.172) into equation (4.134) on page 147 the following expression for φ_{d_0}' is derived:

$$\varphi_{d_0}' = \left(\frac{v_{q_0} + r_s i_{q_0} + L_d' i_{d_0}}{\omega_0} \right) . \tag{4.179}$$

The value of φ_{d_0}' is now back-substituted into equation (4.176) to obtain the initial saturated value of the d-axis synchronous inductance L_{d_0}. This value of L_{d_0} together with the value of L_{q_0} in (4.173) are used to calculate the initial values in (4.99) on page 129.

4.2.13.2.3 <u>Resultant airgap flux-linkages expressed in terms of kth-transient flux-linkages and stator-current components</u>

As explained in Section 4.2.8.2, when the resultant airgap flux-linkages are chosen to represent the saturation level, the saturation demagnetizing current components are given by:

$$\underset{\sim}{I}_{sdq} = \boldsymbol{S_{dq}}(\varphi_{ag}) \underset{\sim}{\varphi}_{adq} , \tag{4.180}$$

$$\text{in which } \boldsymbol{S_{dq}}(\varphi_{ag}) = \mathfrak{D}(S_d(\varphi_{ag}), S_q(\varphi_{ag})) \text{ and } \varphi_{ag} = \sqrt{(\underset{\sim}{\varphi}_{adq})^T (\underset{\sim}{\varphi}_{adq})} . \tag{4.181}$$

In the present context of directly representing the machine in terms of the classically-defined standard parameters it is convenient to express the airgap flux-linkages in terms of the kth-transient flux linkages and stator winding current components. Thus, from equations (4.156) and (4.157) on page 149 it follows that:

$$\underset{\sim}{\varphi}_{adq} = \underset{\sim}{\varphi}^{k}_{dq} - \boldsymbol{L}_{aidq}\underset{\sim}{i}_{dq} \text{ in which } \boldsymbol{L}_{aidq} = \mathfrak{D}((L^{k}_{d} - L_{l}), (L^{k}_{q} - L_{l})) . \tag{4.182}$$

Substituting for $\underset{\sim}{\varphi}_{adq}$ from the preceding equation in (4.180) yields the following expression for the saturation demagnetizing current components:

$$\underset{\sim}{I}_{sdq} = \boldsymbol{S}_{dq}(\varphi_{ag})\underset{\sim}{\varphi}^{k}_{dq} - \boldsymbol{S}_{dq}(\varphi_{ag})\boldsymbol{L}_{aidq}\underset{\sim}{i}_{dq} . \tag{4.183}$$

4.2.13.3 Balanced steady-state operating-conditions of the classical parameter model

As in the case of the coupled-circuit representation of the generator model it is assumed that the steady-state generator stator voltage magnitude, V_{t0}, and the real and reactive power *output* $P_0 = P_{e_0}$ and Q_0 of the generator are given in per-unit of the generator base quantities.

The procedure for calculating the generator steady-state initial conditions for the classical parameter model is the same as that described for the coupled-circuit formulation of the generator model in Section 4.2.9 except that the initial saturated values of the *d*- and *q*-axis synchronous inductances L_{ad_0} and L_{aq_0} in equation (4.86) on page 126 are modified depending on the method used to represent the level of magnetic saturation as summarized in Table 4.12.

Table 4.12 Initial values of the saturation level and the corresponding saturated values of the *d*- and *q*-axis synchronous inductances depending on the method of representing the level of magnetic saturation.

#	φ_{m_0}	**Calculation of** φ_{m_0}	**Calculation of** L_{ad_0} **and** L_{aq_0}
1	φ_{ag_0}	Calculated according to (4.68) pg. 121	Calculated according to (4.86) pg. 126
2	φ^{k}_{0}	If $L^{k}_{q} = L^{k}_{d}$ then φ^{k}_{0} is calculated according to (4.170); otherwise φ^{k}_{0} is obtained from the iterative solution of (4.171).	Calculated according to (4.165)
3	φ_{d_0}'	Calculated according to (4.179)	Calculated according to equations (4.176) and (4.173) respectively.

The initial values of the saturation demagnetizing currents in the non-reciprocal per-unit system are

$$I_{sd_0} = S_d(\varphi_{m_0})\varphi_{m_0} \quad \text{and} \quad I_{sq_0} = S_q(\varphi_{m_0})\varphi_{m_0}, \tag{4.184}$$

where φ_{m_0} is the initial value of the saturation level indicator from Table 4.12. The initial values of the following d- and q-axis rotor winding variables are deduced from equations (4.133) on page 147 to (4.152) on page 148.

$$\begin{aligned}
\varphi_{d_0}' &= (I_{fd_0} - I_{sd_0}) - (L_d - L_d')i_{d_0} & I_{1q_0} &= 0 \\
\varphi_{1d_0} &= (I_{fd_0} - I_{sd_0}) - L_{ad_u}i_{d_0} & \varphi_{q_0}' &= -I_{sq_0} - (L_q - L_q')i_{q_0} \\
\varphi_{s1d_0} &= 0 & \varphi_{2q_0} &= -I_{sq_0} - L_{aq_u}i_{q_0} \\
i_{1d_0} &= 0 & \varphi_{s2q_0} &= 0 \\
\varphi_{d_0}'' &= (I_{fd_0} - I_{sd_0}) - (L_d - L_d'')i_{d_0} & i_{2q_0} &= 0 \\
\varphi_{2d_0} &= \varphi_{1d_0} & \varphi_{q_0}'' &= -I_{sq_0} - (L_q - L_q'')i_{q_0} \\
\varphi_{s2d_0} &= 0 & \varphi_{3q_0} &= \varphi_{2q_0} \\
i_{2d_0} &= 0 & i_{3q_0} &= 0 \\
\varphi_{d_0}''' &= (I_{fd_0} - I_{sd_0}) - (L_d - L_d''')i_{d_0} & \varphi_{q_0}''' &= -I_{sq_0} - (L_q - L_q''')i_{q_0}
\end{aligned} \tag{4.185}$$

4.2.13.4 Linearization of the classical parameter formulation of the electromagnetic equations
In this section the generator d- and q-axis rotor-winding equations (4.133) on page 147 to (4.152) on page 148, the generator stator voltage equations (4.158) and (4.159) on page 149 and equations in Section 4.2.13.2 representing the effects of magnetic saturation are linearized about the initial steady-state operating point determined in Section 4.2.13.3. The objective is to reduce the linearized electromagnetic equations of the machine, expressed in terms of the classically-defined standard parameters, to the following matrix form. This formulation is structurally similar to that for the coupled-circuit model equations in (4.41) on page 110.

$$\begin{bmatrix} p\Delta\underset{\sim}{\varphi}_{dqc} \\ \boldsymbol{0} \end{bmatrix} = \begin{bmatrix} \boldsymbol{A}_r & \boldsymbol{B}_{ri} & \boldsymbol{0} \\ \boldsymbol{C}_{vr} & \boldsymbol{D}_{vi} & -\boldsymbol{I} \end{bmatrix} \begin{bmatrix} \Delta\underset{\sim}{\varphi}_{dqc} \\ \Delta\underset{\sim}{i}_{dq} \\ \Delta\underset{\sim}{v}_{dq} \end{bmatrix} + \begin{bmatrix} \boldsymbol{b}_{re} \\ \boldsymbol{0} \end{bmatrix} \Delta E_{fd}. \tag{4.186}$$

The differences between the above classical parameter formulation and that of the coupled-circuit parameter formulation in (4.41) are (i) the vector of rotor-winding state-variables $\underset{\sim}{\varphi}_{dqc}$ in the classical parameter formulation is different to that in the coupled-circuit formulation $\underset{\sim}{\varphi}_{rdq}$; (ii) the field-voltage input in the classical parameter formulation is in per-unit in the non-reciprocal base system (i.e. E_{fd}) whereas in the coupled-circuit formulation it is in per-unit in the reciprocal base system (i.e. e_{fd}); (iii) although the coefficient sub-matrices in the two formulations have the same names their values are different. Since the two alternative formulations of the generator electromagnetic equations are structurally the same they can be used interchangeably within the complete model of the generator which is depicted in Figure 4.1 on page 92. The linearized rotor-equations of motion developed in Sections 4.2.5 and 4.2.6 are applicable to the classical model as are the equations in Section 4.2.10 which provide the interface between the *dq* reference frame and the synchronously rotating network *RI* reference frame.

The rotor *d*- and *q*-axis state-variables in the classical formulation are respectively:

$$\underset{\sim}{\varphi}_{cd} = \begin{bmatrix}\varphi_d' & \varphi_{1d} & \varphi_{2d}\end{bmatrix}^T \text{ and } \underset{\sim}{\varphi}_{cq} = \begin{bmatrix}\varphi_q' & \varphi_{1q} & \varphi_{2q}\end{bmatrix}^T. \quad (4.187)$$

The detailed derivation of the coefficient matrices in the linearized electromagnetic equations of the generator in (4.186) is given in Appendix 4–I. Once these equations are formed the linearized state- and algebraic equations of the generator including the rotor equations of motion and interface with the network are formulated as in (4.117) on page 133 but with $\Delta\underset{\sim}{x}_r$ and $\Delta\underset{\sim}{u}_m$ redefined as follows:

$$\Delta\underset{\sim}{x}_r = \begin{bmatrix}\Delta\underset{\sim}{\varphi}_{cd} \\ \Delta\underset{\sim}{\varphi}_{cq}\end{bmatrix} \text{ and } \Delta\underset{\sim}{u}_m = \begin{bmatrix}\Delta E_{fd} \\ \Delta T_m\end{bmatrix}. \quad (4.188)$$

4.2.14 Generator parameter conversions

Conventionally synchronous generator parameters are provided in the form of standard parameters. These are based on manufacturers' design calculations or the results of actual transient response tests, such as sudden short-circuit or load-rejection tests.

As explained in Section 4.2.12 the standard parameters are ambiguous if they are specified without also specifying whether they conform to their 'Classical' or 'Exact' definitions. Furthermore, it is important to know if the standard parameters are 'saturated' or 'unsaturated'. If saturation is being modelled explicitly then the 'unsaturated' machine parameters should be used. Conversely, if saturation is not being modelled explicitly, the 'saturated' machine parameters should be used, though this approach to the representation of saturation in small-signal stability analysis is not recommended.

It is frequently the case that data supplied by manufacturers is based on the 'Classical' definitions, although this is not always so. While the difference may be insignificant in the case of salient pole machines, it may not be so when round rotor machines are modelled with two *q*-axis damper windings. In the latter case it is possible that significant errors in predicted system responses will occur if the software assumes the parameters conform to the 'Classical' definition when the standard parameters input to the program are actually specified according to the 'Exact' definitions or vice-versa.

Some software packages represent the machine internally using the coupled-circuit formulation although the user is permitted to input standard parameters. The software package then internally converts the user-supplied standard parameters to coupled-circuit parameters. Again, to avoid ambiguity, it is important that the user is aware whether the software requires that the standard parameters conform to the exact or classical definitions. Some software packages internally formulate machine electromagnetic equations directly in terms of the classically-defined standard parameters as summarized in Section 4.2.13. In such a case, if the user has standard parameters for the machine which conform to the exact definitions then they should first be converted to conform with the classical definitions.

Figure 4.11 shows a parameter conversion roadmap. It is important to note that direct conversion between the classical and exact definitions of the standard parameters is unnecessary. Rather, if for example, conversion from exact to classical parameters is required then the ES2CC transformation is applied to the exact standard parameters to yield the coupled-circuit parameters. These are then converted to the classical parameters using the CC2CS transformation.

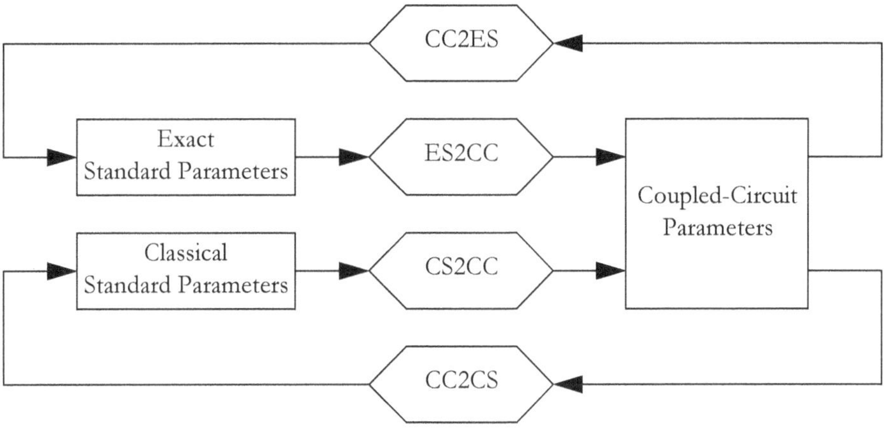

Figure 4.11 Generator parameter conversion roadmap.
(Note: The conversions CS2CC and CC2CS are given in tables 4.11 and 4.10 respectively. The conversions ES2CC and CC2ES are developed in [12] and software implementations are provided in [24]).

4.3 Small-signal models of FACTS Devices

Equipment and plant which incorporate power electronics are referred to as components in *Flexible AC Transmission Systems* or as FACTS devices. The linearized equations will be formulated for the small-signal models of five commonly-used FACTS devices, namely, the Static VAR Compensator (SVC), the Static Synchronous Compensator (STATCOM), a High Voltage Direct Current link with Voltage Source Converters (VSCX), a HVDC link with line-commutated thyristor controlled converters (TCCX), and Thyristor-Controlled

Series Capacitors (TCSC). The core of many modern FACTS devices is the Voltage Sourced Converter (VSC) for which the linearized equations are also summarized. The HVDC link may comprise overhead lines, or underground or submarine cables. Details of these devices and their operation, items which not directly relevant to the following sections, are described in [10, 25, 26, 27, 29, 30, 31].

The following assumptions are made which are consistent with the requirements for rotor-angle small-signal analysis. (i) The FACTS device is not operating at a limiting condition in the steady state. (ii) The losses in power electronic converters are constant for small perturbations about the steady-state operating condition. (iii) Only the fundamental-frequency components of the AC voltages and currents are relevant to the analysis. (iv) DC currents are free of ripple. (v) Depending on the device, the DC voltage may contain some ripple - the average value of the ripple voltage is then used. (vi) When analysing any device, all AC related parameters are in per-unit on the specified base quantities of the device; at the AC interface between the device and the network, conversions between the device and network base quantities may be required. (vii) All DC related quantities are in SI units. As a consequence of these assumptions, (i) the very fast switching processes at the heart of power electronic devices are approximated by algebraic equations; and (ii) many of the sophisticated limiting controls which are required under large-disturbance conditions are omitted from the linearized model of the device.

The linearized DAEs for the devices are summarized in this section. However, to integrate the DAEs of these device models with those of the AC network and the control systems to which they are connected they must first be rewritten in the form described in Section 4.4. The details of how such reformulation can be performed are provided in [12].

As shown in Figure 4.12 the general form of a linearized representation of a FACTS device comprises n_t AC terminals that are connected to AC network buses. For generality it is assumed that the base values of the interface quantities between the network buses and the corresponding device terminals differ, as will be discussed in Section 4.3.1. For the kth terminal the perturbations in the real and imaginary components of the network bus voltage $\Delta \underset{\sim}{v}_{RI}^{(n,\,k)}$ are applied to the device terminal through the base conversion factor $1/K_V^{(k)}$; the current outputs $\Delta \underset{\sim}{i}_{RI}^{(d,\,k)}$ from the device are injected into the network, again through a base conversion factor of $K_I^{(k)}$. The steady-state operating condition about which the model is linearized is determined from the power flow solution for the buses to which the device terminals are connected.

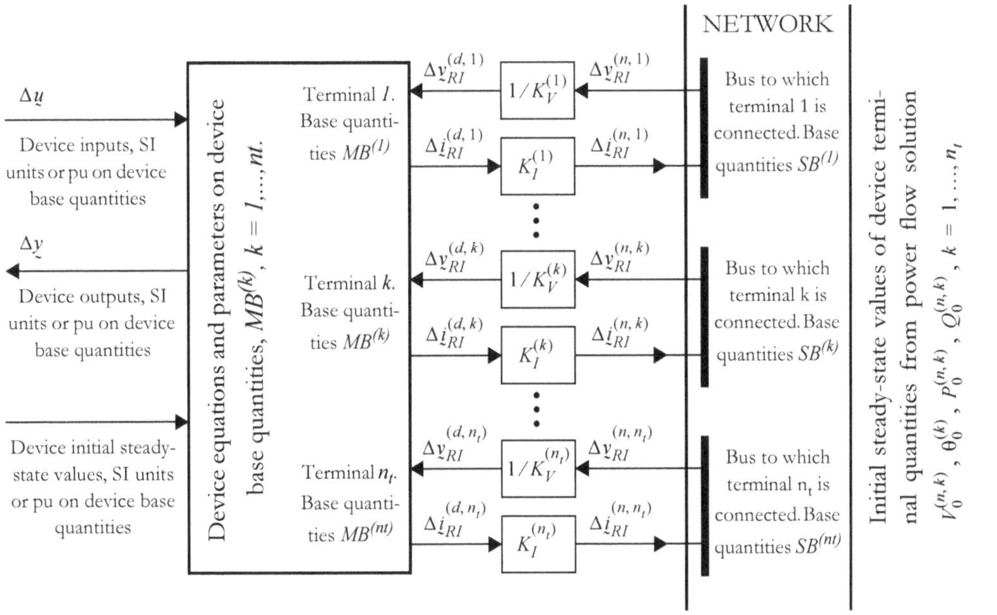

Figure 4.12 Interface between the network and a general FACTS device with n_t AC terminals.

Prior to considering a particular device, some results generally applicable to the AC terminals of FACTS devices are first derived.

4.3.1 Linearized equations of voltage, current and power at the AC terminals of FACTS Devices: general results

As shown in Figure 4.12, a general FACTS device has multiple points of connection to the AC network. In the following the connection of just one terminal, *k*, of the FACTS device to its network bus is considered. The results are applicable to the connection of the other terminals of the device to their respective network buses. As noted in Section 4.2.3.5 the results in this section are also applicable to calculating the quantities at the stator terminals of generator models.

Normally the principal base quantities for the kth terminal of the device and the network bus to which the terminal is connected are the RMS line-to-line voltage, the three-phase apparent power and the fundamental frequency as listed in Table 4.13. These base quantities are specified by the user (as indicated by the subscript *'usb'*). The base value of time, for both the device and network, t_b, is chosen to be one-second.

Table 4.13 Principal base quantities for the kth device terminal and the network bus to which it is connected.

Base Quantity		Units	Description
Device Terminal	Network Bus		
$V_{usb}^{(d,\,k)}$	$V_{usb}^{(n,\,k)}$	kV (RMS, line-to-line)	The base value of the fundamental-frequency positive-phase sequence component of the RMS line-to-line voltage of the kth terminal of the device. Normally $V_{usb}^{(d,\,k)} = V_{usb}^{(n,\,k)}$.
$S_{usb}^{(d,\,k)}$	$S_{usb}^{(n,\,k)}$	MVA	The base value of the three-phase apparent power of the kth terminal of the device. Normally, $S_{usb}^{(d,\,k)}$ is related to the device rating. The value of $S_{usb}^{(n,\,k)}$ is normally the apparent power base of the power flow analysis.
$f_{usb}^{(d,\,k)}$	$f_{usb}^{(n,\,k)}$	Hz	The base value of fundamental frequency of the kth terminal of the device. Normally $f_{usb}^{(d,\,k)} = f_{usb}^{(n,\,k)}$. It is assumed that $f_{usb}^{(n,\,k)}$ is the nominal operating frequency of the network.

The set of base quantities for the kth terminal of the device is denoted as $MB^{(k)}$ and the corresponding set of base quantities for the network bus to which the kth terminal is connected is denoted as $SB^{(k)}$.

The per-unit values of voltage, real and reactive power and current in the base system of the of the kth terminal of the device (denoted by the superscript *(d,k)*) are related to the corresponding quantities in the network base system (denoted by the superscript *(n,k)*) as follows:

$$\hat{V}^{(n,\,k)} = K_V^{(k)}\hat{V}^{(d,\,k)} \text{ in which } K_V^{(k)} = V_{usb}^{(d,\,k)}/V_{usb}^{(n,\,k)}, \tag{4.189}$$

$$P^{(n,\,k)} + jQ^{(n,\,k)} = K_S^{(k)}(P^{(d,\,k)} + jQ^{(d,\,k)}) \text{ in which } K_S^{(k)} = S_{usb}^{(d,\,k)}/S_{usb}^{(n,\,k)}, \tag{4.190}$$

$$\hat{I}^{(n,\,k)} = K_I\hat{I}^{(d,\,k)} \text{ in which } K_I^{(k)} = K_S^{(k)}/K_V^{(k)}. \tag{4.191}$$

It is assumed that the initial steady-state values $V_0^{(n,\,k)}$, $\theta^{(k)}$, $P_0^{(n,\,k)}$ and $Q_0^{(n,\,k)}$ at the bus to which the kth terminal of the device is connected are obtained from the power flow solution.

For the purposes of analysis the voltages and currents are represented by the real and imaginary components of their phasor quantities in the synchronously rotating network frame of reference. Thus, in per unit on the base quantities of the kth terminal of the device,

$$\hat{V}^{(d,k)} = v_R^{(d,k)} + jv_I^{(d,k)} = V^{(d,k)} e^{j\theta^{(k)}} = V^{(d,k)}(\cos(\theta^{(k)}) + j\sin(\theta^{(k)})), \tag{4.192}$$

in which

$$V^{(d,k)} = \sqrt{(v_R^{(d,k)})^2 + (v_I^{(d,k)})^2}, \quad \theta^{(k)} = \operatorname{atan2}(v_I^{(d,k)}, v_R^{(d,k)}), \text{ and} \tag{4.193}$$

$$v_R^{(d,k)} = V^{(d,k)}\cos(\theta^{(k)}), \quad v_I^{(d,k)} = V^{(d,k)}\sin(\theta^{(k)}) \text{ and } \underset{\sim}{v}_{RI}^{(d,k)} = \begin{bmatrix} v_R^{(d,k)} & v_I^{(d,k)} \end{bmatrix}^T. \tag{4.194}$$

The initial steady-state value of the voltage magnitude is $V_0^{(d,k)} = V_0^{(n,k)}/K_V^{(k)}$ per-unit on $V_{usb}^{(d,k)}$; the initial value $\underset{\sim}{v}_{RI_0}^{(d,k)}$ is then obtained by substitution of $V^{(d,k)} = V_0^{(d,k)}$ and $\theta^{(k)} = \theta_0^{(k)}$ in (4.194). By substituting superscript *d* with *n* in equations (4.192) to (4.194) the voltage magnitude and *RI* components are obtained in per-unit of the base voltage of the network bus to which the kth terminal is connected.

The linearized forms of the voltage magnitude $\Delta V^{(d,k)}$ and angle $\Delta\theta^{(k)}$ of the kth terminal of the device are written, respectively, as

$$\Delta V = (v_{R_0}/V_0)\Delta v_R + (v_{I_0}/V_0)\Delta v_I = (1/V_0)\underset{\sim}{v}_{RI_0}^T \Delta\underset{\sim}{v}_{RI} \quad \text{(pu), and} \tag{4.195}$$

$$\Delta\theta = -(v_{I_0}/V_0^2)\Delta v_R + (v_{R_0}/V_0^2)\Delta v_I = (1/V_0^2)\begin{bmatrix} -v_{I_0} & v_{R_0} \end{bmatrix}\Delta\underset{\sim}{v}_{RI}, \quad \text{(rad)} \tag{4.196}$$

in which the superscript *(d,k)* is omitted from the voltage quantities.

The terminal current phasor is expressed in terms of the terminal voltage and the real and reactive power injected by the device from its kth terminal into the network as follows. All quantities are in per-unit on the base quantities of either the kth terminal of the device *(d,k)* or those of the network bus to which the kth terminal is connected *(n,k)*.

$$\begin{aligned} \hat{I} = (i_R + ji_I) &= \left(\frac{P+jQ}{Ve^{j\theta}}\right)^* \\ &= \left(\frac{1}{V}\right)((P\cos\theta + Q\sin\theta) + j(P\sin\theta - Q\cos\theta)) \quad \text{(pu)} \\ &= \left(\frac{1}{V^2}\right)((Pv_R + Qv_I) + j(Pv_I - Qv_R)) \end{aligned} \tag{4.197}$$

Equating the real and imaginary components of the preceding equation results in the following matrix relationship:

$$\begin{bmatrix} i_R \\ i_I \end{bmatrix} = \left(\frac{1}{V}\right)\begin{bmatrix} \cos\theta & \sin\theta \\ \sin\theta & -\cos\theta \end{bmatrix}\begin{bmatrix} P \\ Q \end{bmatrix} = \left(\frac{1}{V^2}\right)\begin{bmatrix} v_R & v_I \\ v_I & -v_R \end{bmatrix}\begin{bmatrix} P \\ Q \end{bmatrix}. \tag{4.198}$$

The initial values of the current components $\underset{\sim}{i}_{RI_0}^{(n,k)}$, in the network per-unit system, are found by substituting the initial values $V^{(n,k)}$, $\theta^{(k)}$, $P_0^{(n,k)}$ and $Q_0^{(n,k)}$ obtained from the power flow solution into equation (4.198) and then in the per-unit system of the device terminal by setting $\underset{\sim}{i}_{RI_0}^{(d,k)} = \underset{\sim}{i}_{RI_0}^{(n,k)}/K_I^{(k)}$.

The perturbations in the current components about their initial values are obtained by linearizing equation (4.198) and by eliminating perturbations in the voltage magnitude using (4.195) to give:

$$\Delta\underset{\sim}{i}_{RI} = \boldsymbol{J}_{is}\Delta\underset{\sim}{S} + \boldsymbol{J}_{iv}\Delta\underset{\sim}{v}_{RI} \quad \text{(pu), in which } \underset{\sim}{S} = \begin{bmatrix} P & Q \end{bmatrix}^T, \tag{4.199}$$

$$\boldsymbol{J}_{is} = \left(\frac{1}{V_0^2}\right)\begin{bmatrix} v_{R_0} & v_{I_0} \\ v_{I_0} & -v_{R_0} \end{bmatrix} \text{ and } \boldsymbol{J}_{iv} = \left(\frac{1}{V_0^2}\right)\begin{bmatrix} (P_0 - 2i_{R_0}v_{R_0}) & (Q_0 - 2i_{R_0}v_{I_0}) \\ -(Q_0 - 2i_{I_0}v_{R_0}) & (P_0 - 2i_{I_0}v_{I_0}) \end{bmatrix}. \tag{4.200}$$

The current magnitude:

$$I = \sqrt{i_R^2 + i_I^2} \tag{4.201}$$

is linearized about the initial steady-state operating point to yield:

$$\Delta I = (i_{R_0}/I_0)\Delta i_R + (i_{I_0}/I_0)\Delta i_I = (1/I_0)\underset{\sim}{i}_{RI_0}^T\Delta\underset{\sim}{i}_{RI} \quad \text{(pu)}; \tag{4.202}$$

I_0 is obtained from (4.201) by substitution of the initial values of the current components $\underset{\sim}{i}_{RI_0}$.

The apparent power is $P + jQ = \hat{V}\hat{I}^*$, where all quantities are in per-unit on the bases of either (i) the kth terminal of the device *(d,k)*; or (ii) the network bus to which the terminal is connected *(n,k)*. From this relationship it follows that:

$$P = v_R i_R + v_I i_I \text{ (pu) and } Q = v_I i_R - v_R i_I \text{ (pu)}. \tag{4.203}$$

The perturbations in the real and reactive power are thus given by:

$$\begin{bmatrix} \Delta P \\ \Delta Q \end{bmatrix} = \begin{bmatrix} v_{R_0} & v_{I_0} \\ v_{I_0} & -v_{R_0} \end{bmatrix} \begin{bmatrix} \Delta i_R \\ \Delta i_I \end{bmatrix} + \begin{bmatrix} i_{R_0} & i_{I_0} \\ -i_{I_0} & i_{R_0} \end{bmatrix} \begin{bmatrix} \Delta v_R \\ \Delta v_I \end{bmatrix} \text{ (pu).} \tag{4.204}$$

Consider the general case of an AC current flow $\hat{I}$ from terminal 1 to terminal 2 through a reactance X connected between the terminals, the terminal voltage phasors being $\hat{V}_1$ and $\hat{V}_2$; all quantities are in pu on the appropriate bases. Thus, $\hat{V}_1 = \hat{V}_2 + jX\hat{I}$ from which the components of the terminal 1 voltage are:

$$v_{1R} = v_{2R} - Xi_I, \text{ and } v_{1I} = v_{2I} + Xi_R. \tag{4.205}$$

Because the equations (4.205) are linear, the linearized equations are formed by replacing the variables by their perturbed quantities.

4.3.2 Model of a Static VAR Compensator (SVC)

The purpose of the SVC, through its control system, is to maintain the voltage of a busbar constant by acting as a source or sink of reactive power. From small-signal considerations, the SVC model comprising a controllable susceptance B, as shown in Figure 4.13, is the simplest to construct. Since a wide variety of SVC control systems exist in practice the control system model is omitted from the following.

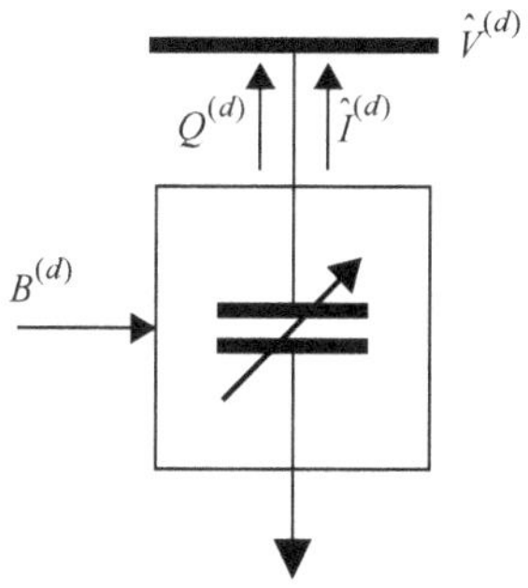

Figure 4.13 Model of a SVC
(All quantities are shown in per-unit of the SVC base values)

Since the SVC is a single-terminal device the terminal descriptor k is redundant and is therefore omitted from the following equations. The SVC equations are $\hat{I}^{(d)} = -jB^{(d)} \cdot \hat{V}^{(d)}$, $P^{(d)} = 0$ and $Q^{(d)} = B^{(d)}(V^{(d)})^2$ each in per-unit of the SVC base quantities. The positive direction of current and reactive power flow is from the SVC and into the network. The base value of admittance on the SVC base quantities is $Y_b^{(d)} = S_{usb}^{(d)}/(V_{usb}^{(d)})^2$ and on the network bus base quantities is $Y_b^{(n)} = S_{usb}^{(n)}/(V_{usb}^{(n)})^2$. Thus, the SVC susceptance in per-unit of the network bus base admittance is related to the per-unit value on the SVC base admittance by $B^{(n)} = K_Y B^{(d)}$ in which $K_Y = K_S/K_V^2$ and where K_V and K_S are defined in (4.189) and (4.190) respectively. The initial value of the SVC susceptance is $B_0^{(d)} = (Q_0^{(n)}/(V_0^{(n)})^2)/K_Y$ where $Q_0^{(n)}$ and $V_0^{(n)}$ are the initial steady-state values obtained from the power-flow solution.

The real and imaginary components of the SVC current are respectively $i_R^{(d)} = B^{(d)}v_I^{(d)}$ and $i_I^{(d)} = -B^{(d)}v_R^{(d)}$ which upon linearization yield the following perturbations of the SVC current components:

$$\begin{bmatrix} \Delta i_R^{(d)} \\ \Delta i_I^{(d)} \end{bmatrix} = \begin{bmatrix} 0 & B_0^{(d)} \\ -B_0^{(d)} & 0 \end{bmatrix} \begin{bmatrix} \Delta v_R^{(d)} \\ \Delta v_I^{(d)} \end{bmatrix} + \begin{bmatrix} v_{I_0}^{(d)} \\ -v_{R_0}^{(d)} \end{bmatrix} \Delta B^{(d)} . \qquad (4.206)$$

The perturbed output equations for the voltage magnitude and angle are given by equations (4.195) and (4.196) respectively; the current magnitude is given by (4.202) and the reactive power by (4.204).

The quantity (Q/V) is the signed current generated by the SVC and is commonly used as an input signal to the SVC control system for representing the current droop feedback. Noting that $Q/V = BV$ it follows that the perturbed variable is:

$$\Delta(Q/V) = B_0\Delta V + V_0\Delta B = (B_0 v_{R_0}/V_0)\Delta v_R + (B_0 v_{I_0}/V_0)\Delta v_I + V_0\Delta B \qquad (4.207)$$

in which all quantities are in per-unit on the SVC base values.

The linearized equations of the SVC are written in the following matrix form in preparation for interconnection with the transmission network, as described in Section 4.4. Note that the model does not have any state-variables.

$$\begin{bmatrix} \mathbf{0} \\ \mathbf{0} \end{bmatrix} = \begin{bmatrix} \mathbf{J}_{gz} & \mathbf{J}_{gi} & \mathbf{J}_{gv} \\ \mathbf{J}_{iz} & \mathbf{J}_{ii} & \mathbf{J}_{iv} \end{bmatrix} \begin{bmatrix} \Delta \underset{\sim}{z} \\ \Delta \underset{\sim}{i}_{RI}^{(n)} \\ \Delta \underset{\sim}{v}_{RI}^{(n)} \end{bmatrix} + \begin{bmatrix} \mathbf{J}_{gu} \\ \mathbf{J}_{iu} \end{bmatrix} \Delta \underset{\sim}{u} \qquad (4.208)$$

in which

$$\Delta \underset{\sim}{z} = \left[\Delta i_R^{(d)}\ \Delta i_I^{(d)}\ \Delta v_R^{(d)}\ \Delta v_I^{(d)}\ \Delta V^{(d)}\ \Delta\theta\ \Delta Q^{(d)}\ \Delta(Q/V)^{(d)}\ \Delta I^{(d)}\right]^T, \quad \Delta \underset{\sim}{u} = \Delta B^{(d)}. \quad (4.209)$$

$$\boldsymbol{J}_{gz} = \begin{bmatrix} -1 & 0 & 0 & B_0^{(d)} & 0 & 0 & 0 & 0 & 0 \\ 0 & -1 & -B_0^{(d)} & 0 & 0 & 0 & 0 & 0 & 0 \\ 0 & 0 & K_V & 0 & 0 & 0 & 0 & 0 & 0 \\ 0 & 0 & 0 & K_V & 0 & 0 & 0 & 0 & 0 \\ 0 & 0 & \left\{v_{R_0}^{(d)}/V_0^{(d)}\right\} & \left\{v_{I_0}^{(d)}/V_0^{(d)}\right\} & -1 & 0 & 0 & 0 & 0 \\ 0 & 0 & \left\{-v_{I_0}^{(d)}/(V_0^{(d)})^2\right\} & \left\{v_{R_0}^{(d)}/(V_0^{(d)})^2\right\} & 0 & -1 & 0 & 0 & 0 \\ v_{I_0}^{(d)} & -v_{R_0}^{(d)} & -i_{I_0}^{(d)} & i_{R_0}^{(d)} & 0 & 0 & -1 & 0 & 0 \\ 0 & 0 & 0 & 0 & B_0^{(d)} & 0 & 0 & -1 & 0 \\ \left\{i_{R_0}^{(d)}/I_0^{(d)}\right\} & \left\{i_{I_0}^{(d)}/I_0^{(d)}\right\} & 0 & 0 & 0 & 0 & 0 & 0 & -1 \end{bmatrix}$$

$$\boldsymbol{J}_{gi} = \boldsymbol{0}, \quad \boldsymbol{J}_{gv} = \begin{bmatrix} 0 & 0 & -1 & 0 & 0 & 0 & 0 & 0 & 0 \\ 0 & 0 & 0 & -1 & 0 & 0 & 0 & 0 & 0 \end{bmatrix}^T, \quad \boldsymbol{J}_{iz} = \begin{bmatrix} K_I & 0 & 0 & 0 & 0 & 0 & 0 & 0 & 0 \\ 0 & K_I & 0 & 0 & 0 & 0 & 0 & 0 & 0 \end{bmatrix}, \quad \boldsymbol{J}_{ii} = \begin{bmatrix} -1 & 0 \\ 0 & -1 \end{bmatrix},$$

$$\boldsymbol{J}_{iv} = \boldsymbol{0}, \quad \boldsymbol{J}_{gu} = \begin{bmatrix} v_{I_0}^{(d)} & -v_{R_0}^{(d)} & 0 & 0 & 0 & 0 & 0 & V_0^{(d)} & 0 \end{bmatrix}^T, \text{ and } \boldsymbol{J}_{iu} = \boldsymbol{0}.$$

4.3.3 Model of a Voltage Sourced Converter (VSC)

The Voltage Sourced Converter (VSC) is a core component of a number of FACTS devices including the Static Synchronous Compensator (STATCOM), VSC based HVDC transmission links, etc. [25, 26, 27, 28]. A generic VSC model, as depicted in Figure 4.14, is developed based on concepts in [28]. While this model is suitable for small-signal rotor-angle stability analysis it is not applicable when the dynamic behaviour of a fast acting Pulse Width Modulation (PWM) scheme, switching controls, and such-like is required.

The VSC acts as an AC voltage source where both magnitude and phase angle of the source are controllable. As shown in Figure 4.14 the VSC model has two terminals, *c1* and *c2* on the AC side. This allows the model to be used as a series connected element as well as a shunt connected element. In the latter case the voltage $\hat{V}^{(d,\,c2)}$ of terminal *c2* is constrained to zero.

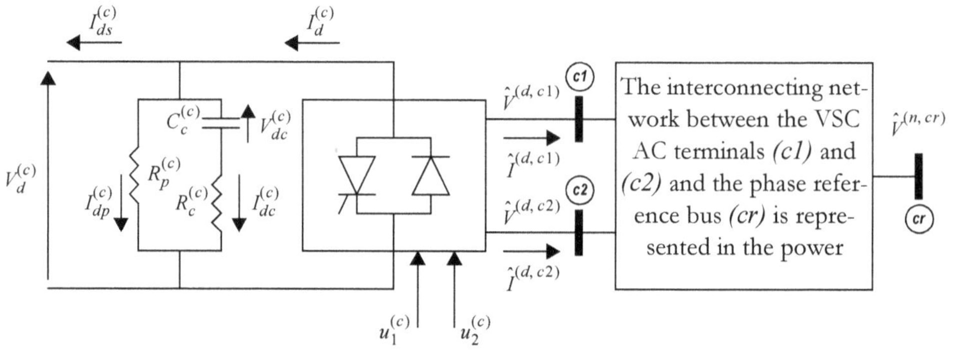

Figure 4.14 Generic VSC model.

The VSC base voltages and apparent power for terminals *c1* and *c2* are assumed to be,

$$V_{usb}^{(d,c1)} = V_{usb}^{(d,c2)} = V_{usb}^{(d,c)} \text{ (kV)}, S_{usb}^{(d,c1)} = S_{usb}^{(d,c2)} = S_{usb}^{(d,c)} \text{ (MVA)}; \tag{4.210}$$

the corresponding network bus base values are $V_{usb}^{(n,c1)}$ and $S_{usb}^{(n,c1)}$ for terminal *c1*, and $V_{usb}^{(n,c2)}$, $S_{usb}^{(n,c2)}$ for terminal *c2*.

From equations (4.189) and (4.190) on page 160 the conversion factors between the network and VSC base values for the *c1* terminal voltage and apparent power are

$$K_V^{(c1)} = V_{usb}^{(d,c)}/V_{usb}^{(n,c1)} \text{ and } K_S^{(c1)} = S_{usb}^{(d,c)}/S_{usb}^{(n,c1)}, \tag{4.211}$$

- and similarly for terminal *c2*:

$$K_V^{(c2)} = V_{usb}^{(d,c)}/V_{usb}^{(n,c2)} \text{ and } K_S^{(c2)} = S_{usb}^{(d,c)}/S_{usb}^{(n,c2)} \tag{4.212}$$

The voltage magnitude and phase angle at terminals *c1* and *c2* are found from (4.193) on page 161 in terms of the real and imaginary voltage components of terminals *c1* and *c2*, $\underset{\sim}{v}_{RI}^{(d,c1)}$ and $\underset{\sim}{v}_{RI}^{(d,c2)}$.

The source voltage magnitude and phase are controlled by the modulation ratio $m^{(c)}$ and phase angle control $\alpha^{(c)}$ both of which are inputs to the VSC. The magnitude of the source voltage, which is the magnitude of the voltage difference between the two AC terminals *(c1)* and *(c2),* is related to the modulation ratio as follows:

$$V_s^{(c)} = (k_c^{(c)} m^{(c)} V_d^{(c)})/V_{usb}^{(c)}. \tag{4.213}$$

In the latter equation $V_d^{(c)}$ is the DC voltage across the converter in kV and $k_c^{(c)}$ is the AC to DC transformation ratio, which for a linear PWM scheme is $(\sqrt{6})/4$; $V_{usb}^{(c)}$ in kV is the internal AC base voltage of the voltage source [28].

The voltage angle at bus *cr*, $\theta^{(cr)}$ rad, is the reference angle for the VSC phase-angle control signal input $\alpha^{(c)}$ rad. Thus, the phase-angle of the voltage source is:

$$\theta_s^{(c)} = \theta^{(cr)} + \alpha^{(c)} \quad \text{(rad)}. \tag{4.214}$$

The internal voltage base $V_{usb}^{(c)}$ is usually chosen to equal the base voltage of the VSC terminals, $V_{usb}^{(d,c)}$, when the VSC is shunt connected. However, for a series connected VSC the internal base voltage may be selected in relation to the series voltage rating of the device. The VSC internal and terminal voltage bases are thus related by the following,

$$K_C^{(c)} = V_{usb}^{(d,c)} / V_{usb}^{(c)}. \tag{4.215}$$

The internal VSC apparent power base, $S_b^{(c)}$, is defined to be equal to the VSC terminal VA base, $S_{usb}^{(d,c)}$ MVA. Thus, the VSC internal base value AC current, $I_b^{(c)}$ A, is defined as,

$$I_b^{(c)} = \frac{S_{usb}^{(d,c)} \times 10^3}{\sqrt{3} \times V_{usb}^{(c)}} = \left(\frac{S_{usb}^{(d,c)} \times 10^3}{\sqrt{3} \times V_{usb}^{(d,c)}}\right) \times \left(\frac{V_{usb}^{(d,c)}}{V_{usb}^{(c)}}\right) = K_C^{(c)} I_b^{(d,c)} \quad \text{(A)}, \tag{4.216}$$

where $I_b^{(d,c)}$ (A) is the VSC base current at its respective terminals.

The voltage source phasor is:

$$\hat{V}_s^{(c)} = V_s^{(c)} e^{j\theta_s^{(c)}} = v_{sR}^{(c)} + j v_{sI}^{(c)} \text{ pu on } V_{usb}^{(c)}, \tag{4.217}$$

where
$$\underset{\sim}{v}_{sRI}^{(c)} = \left[v_{sR}^{(c)} \ v_{sI}^{(c)}\right]^T = V_s^{(c)}\left[\cos(\theta_s^{(c)}) \ \sin(\theta_s^{(c)})\right]^T = K_C^{(c)}(\underset{\sim}{v}_{RI}^{(d,c1)} - \underset{\sim}{v}_{RI}^{(d,c2)}); \tag{4.218}$$

The initial steady-state values $\underset{\sim}{v}_{RI_0}^{(d,c1)}$ and $\underset{\sim}{v}_{RI_0}^{(d,c2)}$ are calculated from the power flow solution as described in Section 4.3.1. These values are then substituted in (4.218) to calculate the initial values of the voltage source components, $\underset{\sim}{v}_{sRI_0}^{(c)}$.

The voltage source magnitude $V_s^{(c)}$ and phase $\theta_s^{(c)}$, in terms of the real and imaginary components of the source voltage are:

$$V_s^{(c)} = \sqrt{(v_{sR}^{(c)})^2 + (v_{sI}^{(c)})^2} = \sqrt{(\underset{\sim}{v}_{sRI}^{(c)})^T \underset{\sim}{v}_{sRI}^{(c)}}, \quad \theta_s^{(c)} = \operatorname{atan2}(v_{sI}^{(c)}, v_{sR}^{(c)}), \tag{4.219}$$

which are linearized to yield the following expressions for these quantities about their initial steady-state values:

$$\Delta V_s^{(c)} = \left(\frac{1}{V_{s_0}^{(c)}}\right)(\underset{\sim}{v}_{sRI_0}^{(c)})^T \Delta \underset{\sim}{v}_{sRI}^{(c)} \text{ (pu)}, \quad \Delta\theta^{(c)} = \left(\frac{1}{(V_{s_0}^{(c)})^2}\right)\begin{bmatrix} -v_{sI_0}^{(c)} & v_{sR_0}^{(c)} \end{bmatrix} \Delta \underset{\sim}{v}_{sRI}^{(c)} \text{ (rad)}, \tag{4.220}$$

where $V_{s_0}^{(c)}$ is obtained by substituting $\underset{\sim}{v}_{sRI_0}^{(c)}$ in (4.219).

It is assumed that the initial steady-state value of the DC voltage, $V_{d_0}^{(c)}$, is a specified quantity. Thus from (4.213) it follows that the initial value of the modulation ratio is,

$$m_0^{(c)} = (V_{s_0}^{(c)} V_{usb}^{(c)})/(k_c^{(c)} V_{d_0}^{(c)}). \tag{4.221}$$

The perturbations in the VSC source voltage magnitude are obtained by linearizing equation (4.213) to give:

$$\Delta V_s^{(c)} = \left(\frac{V_{s_0}^{(c)}}{V_{d_0}^{(c)}}\right)\Delta V_d^{(c)} + \left(\frac{V_{s_0}^{(c)}}{m_0^{(c)}}\right)\Delta m^{(c)} \quad \text{pu on } V_{usb}^{(c)}. \tag{4.222}$$

The VSC source current components $\underset{\sim}{i}_{sRI}^{(c)} = \begin{bmatrix} i_{sR}^{(c)} & i_{sI}^{(c)} \end{bmatrix}^T$, in per-unit on the internal VSC value of base current $I_b^{(c)}$ A, are related to VSC terminal currents on the terminal base current $I_b^{(d,c)}$, by the following relationships,

$$\underset{\sim}{i}_{sRI}^{(c)} = \underset{\sim}{i}_{RI}^{(d,c1)}/K_C^{(c)} = -\underset{\sim}{i}_{RI}^{(d,c2)}/K_C^{(c)}. \tag{4.223}$$

Furthermore, in accordance with (4.191) on page 160 the terminal *c1* and *c2* current components in the respective network bus per-unit systems are related to the corresponding values on the VSC terminal base current as follows:

$$\underset{\sim}{i}_{RI}^{(n,c1)} = K_I^{(c1)} \underset{\sim}{i}_{RI}^{(d,c1)} \text{ in which } K_I^{(c1)} = K_S^{(c1)}/K_V^{(c1)} \text{ and} \tag{4.224}$$

$$\underset{\sim}{i}_{RI}^{(n,c2)} = K_I^{(c2)} \underset{\sim}{i}_{RI}^{(d,c2)} \text{ in which } K_I^{(c2)} = K_S^{(c2)}/K_V^{(c2)}. \tag{4.225}$$

The net real and reactive power, $P_s^{(c)}$ and $Q_s^{(c)}$, generated by the VSC are given by:

$$P_s^{(c)} + jQ_s^{(c)} = \hat{V}_s^{(c)}(\hat{I}_s^{(c)})^* \quad \text{pu on } S_b^{(c)} = S_{usb}^{(d,c)} \text{ MVA} \tag{4.226}$$

which are calculated in terms of the source voltage and current components in accordance with equation (4.203) on page 162 and the associated perturbations in $P_s^{(c)}$ and $Q_s^{(c)}$ are derived from (4.204).

The often substantial switching losses in a VSC are represented by a user supplied fixed quantity $P_{sl}^{(c)}$ pu on $S_{usb}^{(d,\,c)}$. Thus, the DC power output from the VSC is,

$$P_d^{(c)} = -(P_s^{(c)} + P_{sl}^{(c)})S_{usb}^{(d,\,c)} = V_d^{(c)} I_d^{(c)} \times 10^3 \ \text{MW}, \tag{4.227}$$

where $I_d^{(c)}$ is in A and, as mentioned earlier, $V_d^{(c)}$ is in kV.

The initial DC power and current are respectively:

$$P_{d_0}^{(c)} = -(P_{s_0}^{(c)} + P_{sl}^{(c)})S_{usb}^{(d,\,c)} \ \text{MW and} \ I_{d_0}^{(c)} = P_{d_0}^{(c)}/V_{d_0}^{(c)} \times 10^{-3} \ \text{A}. \tag{4.228}$$

Linearizing equation (4.227) gives the perturbation in the DC power output from the VSC:

$$\Delta P_d^{(c)} = -S_{usb}^{(d,\,c)} \Delta P_s^{(c)} = (P_{d_0}^{(c)}/V_{d_0}^{(c)})\Delta V_d^{(c)} + (P_{d_0}^{(c)}/I_{d_0}^{(c)})\Delta I_d^{(c)} \ \text{MW}. \tag{4.229}$$

As shown in Figure 4.14 there are two shunt elements across the DC terminals (i) a resistance $R_p^{(c)}$ Ω across the terminals, (ii) a capacitance $C_c^{(c)}$ F and a resistance $R_c^{(c)}$ Ω in series across the terminals. In the most general case the relevant equations are:

capacitor state equation:
$$pV_{dc}^{(c)} = I_{dc}^{(c)}/C_c^{(c)} \times 10^{-3} \ \text{kV/s}, \tag{4.230}$$

voltage drop across shunt resistor:
$$V_d^{(c)} = R_p^{(c)} I_{dp}^{(c)} \times 10^{-3} \ \text{kV}, \tag{4.231}$$

voltage drop across RC element:
$$V_d^{(c)} = V_{dc}^{(c)} + R_c^{(c)} I_{dc}^{(c)} \times 10^{-3} \ \text{kV}, \tag{4.232}$$

Kirchoff's Current Law:
$$I_d^{(c)} = I_{dp}^{(c)} + I_{dc}^{(c)} + I_{ds}^{(c)} \ \text{A}. \tag{4.233}$$

Because (4.230) to (4.233) are linear equations the variables can be replaced by their perturbed values. There are some special cases for which the above equations can be modified, namely,

(a) no shunt elements are modelled;

(b) only the resistance across the terminals is connected;

(c) only the resistance and series capacitance combination is present.

It is convenient to define a variable $U_d^{(c)}$ representing the external input to the DC terminal of the VSC which is normally defined as:

$$U_d^{(c)} = I_{ds}^{(c)} \text{ (A) or } U_d^{(c)} = V_d^{(c)} \text{ (kV)} \tag{4.234}$$

depending on whether the source on the DC side of the VSC is represented as a current or voltage source.

This VSC model has two control input signals denoted by $u_1^{(c)}$ and $u_2^{(c)}$. From a detailed modelling perspective these two variables are typically (i) the VSC phase-angle measured with respect to the phase reference bus *cr*, $\alpha^{(c)}$, and (ii) the VSC voltage modulation ratio $m^{(c)}$ resulting in the following constraints:

$$u_1^{(c)} = \alpha^{(c)} \text{ (rad) and } u_2^{(c)} = m^{(c)}. \tag{4.235}$$

It is often desirable to provide a simplified functional representation of the VSC control systems, for example, (i) in scoping studies before the details of the VSC control systems are known, or (ii) in studies where detailed representation of the VSC controls have an insignificant effect on the dynamic performance of the system. For example, if the VSC is equipped with a fast acting control system whose objective is to maintain constant DC power flow then control input $u_1^{(c)}$ can be set to $P_d^{(c)}$ rather than $\alpha^{(c)}$. By doing so it is unnecessary to represent the details of the DC power control loop of the VSC.

4.3.3.1 Summary of linearized VSC equations
Listed below is a consolidated list of the linearized DAEs of the VSC.

DC side equations

Note that the DC side equations can be modified depending on the representation of the DC shunt network. In all such alternative representations the variables $\Delta P_d^{(c)}$, $\Delta V_d^{(c)}$, $\Delta I_d^{(c)}$ and $\Delta I_{ds}^{(c)}$ are retained.

$$p\Delta V_{dc}^{(c)} = \frac{1}{(C_c^{(c)} \times 10^3)} \Delta I_{dc}^{(c)} \text{ (kV/s) from (4.230),} \tag{4.236}$$

$$0 = \Delta P_d^{(c)} + S_{usb}^{(d,c)} \Delta P_s^{(c)} \text{ (MW), power conservation from (4.229),} \tag{4.237}$$

$$0 = (P_{d_0}^{(c)} / V_{d_0}^{(c)}) \Delta V_d^{(c)} + (P_{d_0}^{(c)} / I_{d_0}^{(c)}) \Delta I_d^{(c)} - \Delta P_d^{(c)} \text{ (MW) from (4.229),} \tag{4.238}$$

$$0 = \Delta V_{dc}^{(c)} - \Delta V_d^{(c)} + \left\{ R_c^{(c)} \times 10^{-3} \right\} \Delta I_{dc}^{(c)} \text{ (kV) from (4.232),} \tag{4.239}$$

$$0 = -\Delta V_d^{(c)} + \left\{ R_p^{(c)} \times 10^{-3} \right\} \Delta I_{dp}^{(c)} \text{ (kV) from (4.231),} \tag{4.240}$$

$$0 = -\Delta I_d^{(c)} + \Delta I_{ds}^{(c)} + \Delta I_{dc}^{(c)} + \Delta I_{dp}^{(c)} \quad \text{(A) from (4.233).} \tag{4.241}$$

VSC AC side converter equations

$$\boldsymbol{0} = -\Delta \underset{\sim}{v}_{sRI}^{(c)} + K_C^{(c)} \Delta \underset{\sim}{v}_{RI}^{(d,c1)} - K_C^{(c)} \Delta \underset{\sim}{v}_{RI}^{(d,c2)} \quad \text{(pu) from (4.218),} \tag{4.242}$$

$$0 = (v_{sR_0}^{(c)}/V_{s_0}^{(c)})\Delta v_{sR}^{(c)} + (v_{sI_0}^{(c)}/V_{s_0}^{(c)})\Delta v_{sI}^{(c)} - \Delta V_s^{(c)} \quad \text{(pu) from (4.220),} \tag{4.243}$$

$$0 = -\left(v_{sI_0}^{(c)}/(V_{s_0}^{(c)})^2\right)\Delta v_{sR}^{(c)} + \left(v_{sR_0}^{(c)}/(V_{s_0}^{(c)})^2\right)\Delta v_{sI}^{(c)} - \Delta\theta_s^{(c)} \quad \text{(rad) from (4.220),} \tag{4.244}$$

$$0 = -\left(v_{sI_0}^{(n,cr)}/(V_{s_0}^{(n,cr)})^2\right)\Delta v_{sR}^{(n,cr)} + \left(v_{sR_0}^{(n,cr)}/(V_{s_0}^{(n,cr)})^2\right)\Delta v_{sI}^{(n,cr)} - \Delta\theta^{(cr)} \quad \text{(rad)} \tag{4.245}$$

from (4.196),

$$\begin{bmatrix} 0 \\ 0 \end{bmatrix} = \begin{bmatrix} v_{sR_0}^{(c)} & v_{sI_0}^{(c)} \\ v_{sI_0}^{(c)} & -v_{sR_0}^{(c)} \end{bmatrix} \begin{bmatrix} \Delta i_{sR}^{(c)} \\ \Delta i_{sI}^{(c)} \end{bmatrix} + \begin{bmatrix} i_{sR_0} & i_{sI_0} \\ -i_{sI_0} & i_{sR_0} \end{bmatrix} \begin{bmatrix} \Delta v_{sR}^{(c)} \\ \Delta v_{sI}^{(c)} \end{bmatrix} - \begin{bmatrix} \Delta P_s^{(c)} \\ \Delta Q_s^{(c)} \end{bmatrix} \quad \text{(pu)} \tag{4.246}$$

from (4.204) on page 163,

$$0 = \Delta\theta_s^{(c)} - \Delta\theta^{(cr)} - \Delta\alpha^{(c)} \quad \text{(rad) from (4.214),} \tag{4.247}$$

$$0 = \left(\frac{V_{s_0}^{(c)}}{V_{d_0}^{(c)}}\right)\Delta V_d^{(c)} - \Delta V_s^{(c)} + \left(\frac{V_{s_0}^{(c)}}{m_0^{(c)}}\right)\Delta m^{(c)} \quad \text{(pu) from (4.222).} \tag{4.248}$$

VSC terminal 1 / network bus interface

$$\boldsymbol{0} = -\Delta \underset{\sim}{i}_{sRI}^{(c)} + (1/K_C^{(c)})\Delta \underset{\sim}{i}_{RI}^{(d,c1)} \quad \text{(pu) from (4.223),} \tag{4.249}$$

$$0 = K_I^{(c1)}\Delta \underset{\sim}{i}_{RI}^{(d,c1)} - \Delta \underset{\sim}{i}_{RI}^{(n,c1)} \quad \text{(pu) from (4.224),} \tag{4.250}$$

$$0 = K_V^{(c1)}\Delta \underset{\sim}{v}_{RI}^{(d,c1)} - \Delta \underset{\sim}{v}_{RI}^{(n,c1)} \quad \text{(pu) from (4.189) pg. 160.} \tag{4.251}$$

VSC terminal 2 / network bus interface

Note that if the VSC is shunt connected equations (4.252) to (4.254) are omitted from the model as are variables $\Delta \underset{\sim}{i}_{RI}^{(d,c2)}$ and $\Delta \underset{\sim}{v}_{RI}^{(n,c2)}$, including the coefficients of $\Delta \underset{\sim}{v}_{RI}^{(n,c2)}$ in (4.242).

$$\boldsymbol{0} = \Delta \underset{\sim}{i}_{sRI}^{(c)} + (1/K_C^{(c)})\Delta \underset{\sim}{i}_{RI}^{(d,c2)} \quad \text{(pu) from (4.223),} \tag{4.252}$$

$$0 = K_I^{(c2)}\Delta \underset{\sim}{i}_{RI}^{(d,c2)} - \Delta \underset{\sim}{i}_{RI}^{(n,c2)} \quad \text{(pu) from (4.225),} \tag{4.253}$$

$$0 = K_V^{(c2)} \Delta \underset{\sim}{y}_{RI}^{(d,\,c2)} - \Delta \underset{\sim}{y}_{RI}^{(n,\,c2)} \quad \text{(pu) from (4.189) pg. 160.} \tag{4.254}$$

<u>Input to / Output from the DC side of the VSC</u>

$$0 = -\Delta I_{ds}^{(c)} + \Delta U_d^{(c)} \quad \text{(A)} \quad \text{and} \quad 0 = -\Delta V_d^{(c)} + \Delta y_d^{(c)} \quad \text{(kV) from (4.234),} \tag{4.255}$$

$$\text{or conversely,} \quad 0 = -\Delta V_d^{(c)} + \Delta U_d^{(c)} \quad \text{(kV)} \quad \text{and} \quad 0 = -\Delta I_{ds}^{(c)} + \Delta y_d^{(c)} \quad \text{(A).} \tag{4.256}$$

<u>VSC control inputs</u>

As mentioned earlier, in a flexible VSC model, it is desirable for the control inputs $\Delta u_1^{(c)}$ and $\Delta u_2^{(c)}$ to be constrained to two independent variables depending on the user's requirements. For detailed VSC modelling the inputs are set to the VSC phase angle and voltage modulation respectively:

$$0 = -\Delta\alpha^{(c)} + \Delta u_1^{(c)} \quad \text{(rad) from (4.235),} \tag{4.257}$$

$$0 = -\Delta m^{(c)} + \Delta u_2^{(c)} \quad \text{from (4.235),} \tag{4.258}$$

Examples of useful alternative constraints on the first control input are:

$$0 = -\Delta V_d^{(c)} + \Delta u_1^{(c)}, \quad \text{or} \quad 0 = -\Delta P_d^{(c)} + \Delta u_1^{(c)}, \quad \text{or} \quad 0 = -\Delta y_1^{(c)} + \Delta u_1^{(c)} \tag{4.259}$$

in which $\Delta y_1^{(c)}$ is a remote AC power system signal such as the power flow in a transmission element. Similarly, examples of alternative constraints on the second control input are:

$$0 = -\Delta Q_s^{(c)} + \Delta u_2^{(c)}, \quad \text{or} \quad 0 = -\Delta y_2^{(c)} + \Delta u_2^{(c)} \tag{4.260}$$

in which $\Delta y_2^{(c)}$ is a remote AC power system signal such as the reactive power flow in a transmission element or the voltage magnitude of remote bus.

The linearized VSC equations (4.236) to (4.260) can be assembled into a compact matrix form which is suitable for integration with the linearized DAEs of the power system as outlined in Section 4.4 and elaborated on in [12].

4.3.4 Simplified STATCOM model

The purpose of the STATCOM is similar to that of the SVC, that is, to control voltage by absorbing or generating reactive power. The STATCOM model in this section is based on a simplified version of the VSC model described in Section 4.3.3. The AC terminal *(c2)* of the VSC in Figure 4.14 on page 166 is connected to the zero voltage reference plane and the AC terminal *(c1)* is renamed to *(c)*. That is, the VSC at the heart of the STATCOM is shunt connected. The basic concepts on which this model are based are found in [25, 26] , for example. The simplified VSC model, which is depicted in Figure 4.15, makes the following idealizing assumptions.

- The VSC phase control system is sufficiently fast and accurate that the generated AC current phasor is assumed to always be in quadrature with the voltage phasor at the phase reference bus as depicted in Figure 4.15(b). This has the consequence that the VSC input $u_1 = \alpha^{(c)} = 0$ (see Section 4.3.3, page 170).

- The VSC fixed losses, $P_{sl}^{(c)}$, are negligible.

- The VSC DC capacitor is sufficiently large that the DC voltage is assumed to be constant during small-disturbances. Thus the VSC capacitor is depicted as a fixed DC source in Figure 4.15(a).

- A lossless reactor is connected between the converter AC terminal *(c)* and the phase reference bus as shown in Figure 4.15(a). This reactor is included in the model and therefore the VSC phase reference bus is AC terminal *(s)* in the simplified VSC model of the STATCOM.

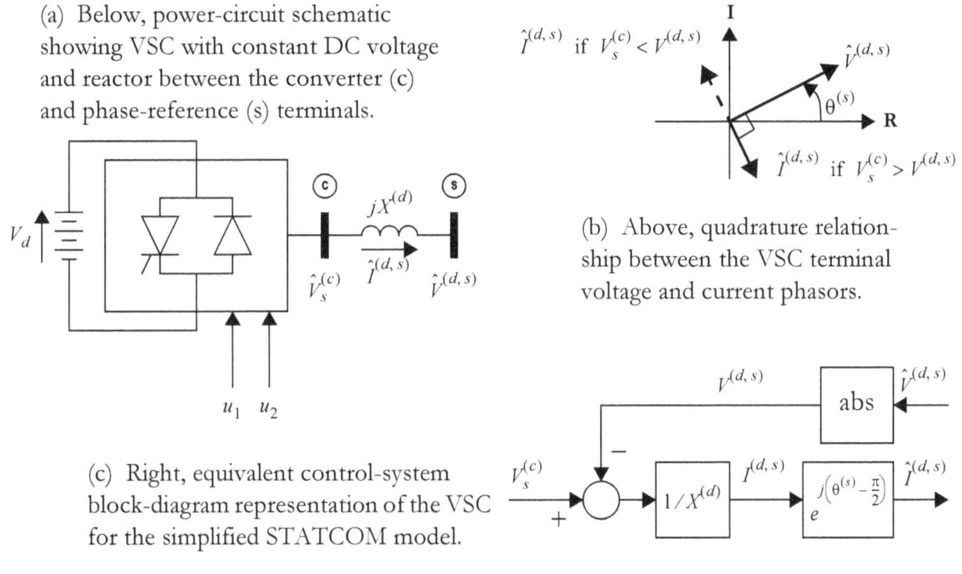

(a) Below, power-circuit schematic showing VSC with constant DC voltage and reactor between the converter (c) and phase-reference (s) terminals.

(b) Above, quadrature relationship between the VSC terminal voltage and current phasors.

(c) Right, equivalent control-system block-diagram representation of the VSC for the simplified STATCOM model.

Figure 4.15 Simplified VSC model for a STATCOM: (a) power-circuit schematic, (b) relationship between terminal voltage and current phasors, (c) equivalent control-system block-diagram of the VSC.

Based on the above assumptions the equivalent control-system block-diagram representation of the simplified VSC model for the STATCOM shown in Figure 4.15(c) is now formulated.

In accordance with Table 4.13 on page 160 the base values of voltage and apparent power for the AC terminal of the VSC are specified respectively as $V_{usb}^{(d,s)}$ (kV) and $S_{usb}^{(d,s)}$ (MVA). The corresponding quantities for the network bus to which the STATCOM VSC terminal *(s)* is connected are $V_{usb}^{(n,s)}$ (kV) and $S_{usb}^{(n,s)}$ (MVA).

The voltage phasor at the phase reference terminal *(s)* is,.

$$\hat{V}^{(d,s)} = V^{(d,s)} e^{j\theta^{(s)}} \text{ in per-unit on } V_{usb}^{(d,s)}, \tag{4.261}$$

and is regarded as an input to the VSC model derived from the voltage at the network bus to which the terminal is connected.

The voltage phasor at the converter terminal *(c)* is:

$$\hat{V}^{(d,c)} = \hat{V}_s^{(c)} = V_s^{(c)} e^{j\theta_s^{(c)}} = V_s^{(c)} e^{j(\theta^{(s)} + \alpha^{(c)})} = V_s^{(c)} e^{j\theta^{(s)}} \text{ in per-unit on } V_{usb}^{(d,s)}. \tag{4.262}$$

The equality of the phases of the voltages at the *(s)* and *(c)* terminals is derived from equation (4.214) on page 167 with the assumption, stated earlier, that the VSC phase-angle control input, $\alpha^{(c)}$, is zero.

Since it is assumed that the DC voltage is constant it follows from (4.213) on page 166 that the magnitude of the voltage at VSC AC terminal *(c)* is equal to the VSC modulation ratio multiplied by a constant gain factor, i.e.

$$V_s^{(c)} = V^{(d,c)} = \beta^{(c)} m^{(c)} \text{ where the constant } \beta^{(c)} = k_c^{(c)}(V_d^{(c)}/V_{usb}^{(d,c)}). \tag{4.263}$$

Thus, for modelling purposes, it is permissible and convenient to eliminate the modulation ratio and instead treat the VSC AC terminal voltage as the second control input signal to the VSC, i.e. $u_2 = V_s^{(c)}$.

The current output from the simplified VSC model is,

$$\hat{I}^{(d,s)} = \left(\frac{1}{jX^{(d)}}\right)(V_s^{(c)} - V^{(d,s)})e^{j\theta^{(s)}} = I^{(d,s)} e^{j\left(\theta^{(s)} - \frac{\pi}{2}\right)}, \tag{4.264}$$

where the magnitude of the injected current, which is in quadrature with the voltage, is

$$I^{(d,s)} = (V_s^{(c)} - V^{(d,s)})/X^{(d)}, \tag{4.265}$$

in per-unit on the VSC current base of $I_b^{(d,s)} = \dfrac{S_{usb}^{(d,s)}}{\sqrt{3}V_{usb}^{(d,s)}} \times 10^3$ (A).

Thus, equations (4.264) and (4.265) which represent the current injected by the simplified VSC model into the network, are equivalent to the control-system block-diagram in Figure 4.15(c).

A voltage control system can then be combined with the above simplified VSC model in order to regulate the voltage at the STATCOM terminal, *(s)*. This results in a simplified model of a STATCOM such as that depicted in Figure 4.16. In this model the STATCOM current droop feedback is explicitly represented by $V_{dr} = K_d I^{(d,s)}$. In common with the SVC, the STATCOM voltage control system usually incorporates current droop designed to reduce the voltage at the regulated bus in proportion to the current injected by the STATCOM into the network. Typically, the droop setting, K_d, is between 0.01 and 0.05 per-unit on the STATCOM base quantities. It is also possible to add a supplementary control signal to the AVR summing junction from, say, a Power Oscillation Damper (POD).

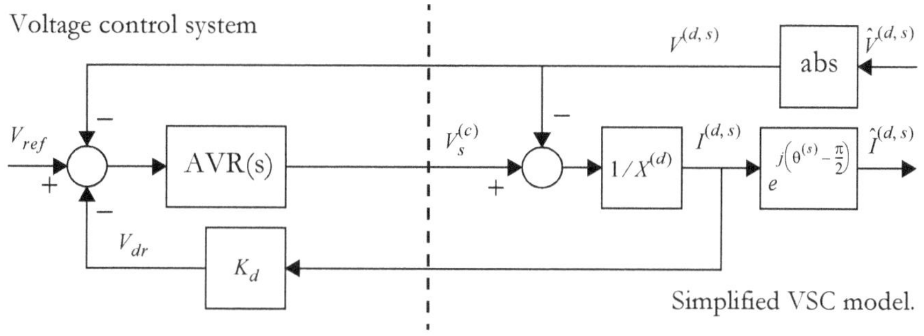

Figure 4.16 Simplified STATCOM model formed by combining the simplified VSC model in Figure 4.15(c) with a voltage control system.

The small-signal representation of the simplified VSC model is now derived. As described in Section 4.3.1 the initial steady-state values of the voltage $\underset{\sim}{v}_{RI_0}^{(d,s)}$ and current $\underset{\sim}{i}_{RI_0}^{(d,s)}$ components in the VSC per-unit system are obtained from that the initial steady-state values of $V_0^{(n,s)}$, $\theta^{(s)}$, $P_0^{(n,s)}$ and $Q_0^{(n,s)}$ obtained from the power flow solution. It is noted that in the power flow the STATCOM is assumed to be represented as a PV bus [1] in which, for consistency with the assumption that the STATCOM current is in quadrature with its terminal voltage, $P_0^{(n,s)} = 0$.

1. In power flow terminology the term "PV bus" denotes a bus at which the generated power and voltage magnitude are specified; the generated reactive power and voltage angle are the unknowns to be determined by the power flow solver.

With appropriate changes in notation equation (4.193) on page 161 is used to derive the STATCOM voltage magnitude $V^{(d,s)}$ from the voltage components $\underset{\sim}{v}_{RI}^{(d,s)}$ as

$$V^{(d,s)} = \sqrt{(v_R^{(d,s)})^2 + (v_I^{(d,s)})^2},$$

which is linearized about the operating point, in accordance with (4.195) on page 161, to yield:

$$\Delta V^{(d,s)} = \left\{ v_{R_0}^{(d,s)} / V_0^{(d,s)} \right\} \Delta v_R^{(d,s)} + \left\{ v_{I_0}^{(d,s)} / V_0^{(d,s)} \right\} \Delta v_I^{(d,s)}. \tag{4.266}$$

From (4.196) on page 161, with appropriate changes in notation, the perturbation in the STATCOM terminal voltage phase angle is:

$$\Delta\theta^{(s)} = -\left(\frac{v_{I_0}^{(d,s)}}{(V_0^{(d,s)})^2} \right) \Delta v_R^{(d,s)} + \left(\frac{v_{R_0}^{(d,s)}}{(V_0^{(d,s)})^2} \right) \Delta v_I^{(d,s)}. \tag{4.267}$$

The perturbation in the STATCOM current magnitude is obtained by linearizing equation (4.265) to yield,

$$\Delta I^{(d,s)} = (\Delta V_s^{(c)} - \Delta V^{(d,s)}) / X^{(d)}. \tag{4.268}$$

From equation (4.264) the STATCOM current phasor can be rewritten as,

$$\hat{I}^{(d,s)} = i_R^{(d,s)} + j i_I^{(d,s)} = I^{(d,s)} \left\{ \cos\left(\theta^{(s)} - \frac{\pi}{2}\right) + j\sin\left(\theta^{(s)} - \frac{\pi}{2}\right) \right\}, \tag{4.269}$$

from which it follows that

$$i_R^{(d,s)} = I^{(d,s)} \sin\theta^{(s)} \quad \text{and} \quad i_I^{(d,s)} = -I^{(d,s)} \cos\theta^{(s)}. \tag{4.270}$$

The perturbations in the current components output from the STATCOM terminal are derived by linearizing the preceding equations at the operating point to yield,

$$\Delta i_R^{(d,s)} = \sin\theta_0^{(s)} \Delta I^{(d,s)} - i_{I_0}^{(d,s)} \Delta\theta^{(s)} \quad \text{and} \quad \Delta i_I^{(d,s)} = -\cos\theta_0^{(s)} \Delta I^{(d,s)} + i_{R_0}^{(d,s)} \Delta\theta^{(s)}. \tag{4.271}$$

From (4.204) on page 163, with appropriate changes in notation, it follows that the perturbation in the reactive power output from the STATCOM is:

$$\Delta Q^{(d,s)} = \begin{bmatrix} v_{I_0}^{(d,s)} & -v_{R_0}^{(d,s)} \end{bmatrix} \Delta \underset{\sim}{i}_{RI}^{(d,s)} + \begin{bmatrix} -i_{I_0}^{(d,s)} & i_{R_0}^{(d,s)} \end{bmatrix} \Delta \underset{\sim}{v}_{RI}^{(d,s)}. \tag{4.272}$$

In accordance with equations (4.189) and (4.190) on page 160 the perturbations in the STATCOM terminal voltage and current components, in per-unit on the base values of the network bus to which the STATCOM is connected, are respectively,

$$\Delta \underset{\sim}{v}_{RI}^{(n,s)} = K_V^{(s)} \Delta \underset{\sim}{v}_{RI}^{(d,s)} \quad \text{and} \quad \Delta \underset{\sim}{I}_{RI}^{(n,s)} = K_I^{(s)} \Delta \underset{\sim}{i}_{RI}^{(d,s)}. \tag{4.273}$$

As for the SVC model, the linearized equations (4.266) to (4.268) and (4.271) to (4.273) for the simplified STATCOM model can be rewritten in the following matrix form suitable for interconnection with the transmission network model. Note that the model does not have any state-variables. The linearized equations are of the form:

$$\begin{bmatrix} 0 \\ 0 \end{bmatrix} = \begin{bmatrix} J_{gz} & J_{gi} & J_{gv} \\ J_{iz} & J_{ii} & J_{iv} \end{bmatrix} \begin{bmatrix} \Delta \underset{\sim}{z} \\ \Delta \underset{\sim}{i}_{RI}^{(n)} \\ \Delta \underset{\sim}{v}_{RI}^{(n)} \end{bmatrix} + \begin{bmatrix} J_{gu} \\ J_{iu} \end{bmatrix} \Delta \underset{\sim}{u} \quad \text{in which} \tag{4.274}$$

$$\Delta \underset{\sim}{z} = \begin{bmatrix} \Delta i_R^{(d,s)} & \Delta i_I^{(d,s)} & \Delta v_R^{(d,s)} & \Delta v_I^{(d,s)} & \Delta V^{(d,s)} & \Delta \theta^{(s)} & \Delta I^{(d,s)} & \Delta Q^{(d,s)} \end{bmatrix}^T \text{ and } \Delta \underset{\sim}{u} = \Delta V_s^{(c)}.$$

Note that these equations do not include the model of the STATCOM voltage regulator or any supplementary control system – they represent only the simplified VSC component of the STATCOM.

4.3.5 Modelling of HVDC Transmission Systems

Two models of HVDC transmission systems are considered with the general structure in Figure 4.17. The first is based on Voltage Sourced Converters (VSCXs) and the second on line-commutated Thyristor Controlled Converters (TCCXs). The HVDC transmission system may comprise overhead lines or cables; the system can be either mono-polar or bipolar. For both the rectifier and inverter it assumed that the number of bridges in series, N_B, is the same as the number of converter transformers operating in parallel on the AC side.

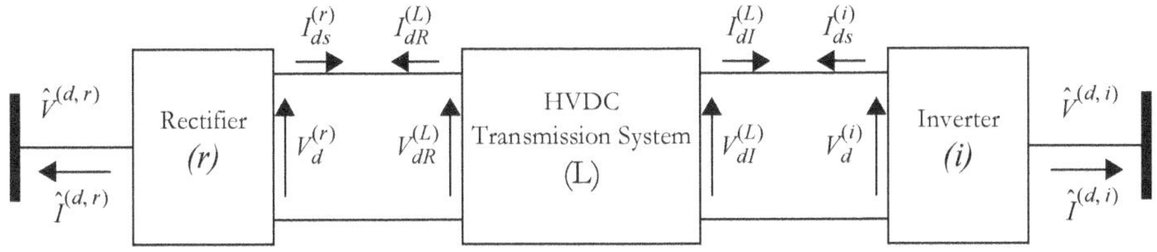

Figure 4.17 Structure of HVDC system model showing the interface between the DC side components. The directions of positive voltage and current-flow at the external interfaces of the components are indicated by the arrows.

Because the models of both VSCX and TCCX systems employ the same model of the HVDC link, the model of the link is analysed first. Models of the VSCX and TCCX systems are then considered in turn.

The HVDC models described in the following sections assume that the following information is provided by the power flow solution: the steady-state voltages, the real and reactive power flows at the AC terminals of the rectifier and inverter, and the initial steady-state values of the DC voltages of each converter. All plant on the AC side of the converter terminals, e.g. three-winding transformers, reactors, is assumed to be modelled in the power flow analysis. The nature of the interface between the AC terminals of the rectifier and inverter and the buses to which they are connected in the AC system is shown in Figure 4.12.

4.3.6 Model of a distributed-parameter HVDC transmission line or cable

In the small-signal model of a long, HVDC transmission line or cable it is not possible to represent the strictly distributed-parameter nature of the circuit. Let us assume that the dynamics of the line is adequately represented by N_T T-sections, of which the kth section is shown in Figure 4.18. Clearly, the greater the number of sections the closer to a distributed-parameter system the model becomes. Let:

R_L - total link resistance in (ohm),
L_L - total link inductance in (Henry),
C_L - total link capacitance (Farad),
R_{CL} - total cable sheath resistance (ohm).

The parameters of each T-section in SI units are therefore:
$L = L_L/(2\, N_T)$, $R = R_L/(2\, N_T)$, $C = C_L/N_T$, $R_C = R_{CL}/N_T$.

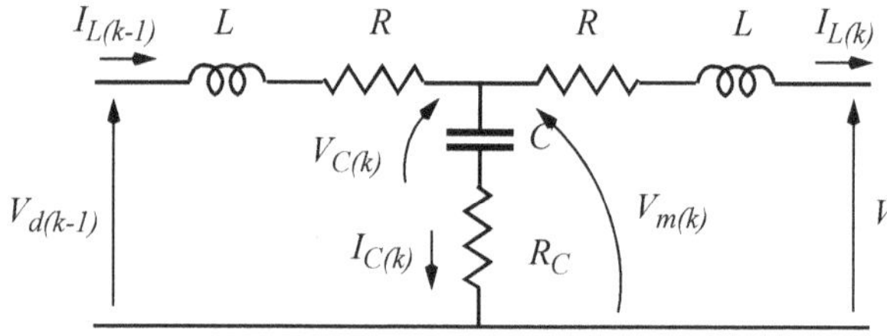

Figure 4.18 The kth T-section of a long, HVDC transmission line or cable.

The circuit diagram of the HVDC line / cable depicted in Figure 4.19 comprises the series connection of N_T of the above T-sections. Note that:

1. The series resistance R_r (ohm) and inductance L_r (H) connected to the rectifier end of the link *(R)* includes the resistance $R_r^{(r)}$ (ohm) and inductance $L_r^{(r)}$ (H) of the series smoothing reactor, if any, connected to the rectifier, i.e.

$$R_r = R + R_r^{(r)} \quad \text{and} \quad L_r = L + L_r^{(r)}. \tag{4.275}$$

2. Similarly, the series RL branch connected to the inverter node *(I)* includes the series smoothing reactor, if any, connected to the inverter, i.e.

$$R_i = R + R_r^{(i)} \quad \text{and} \quad L_i = L + L_r^{(i)}. \tag{4.276}$$

3. The parameters of the internal series RL elements are $2R$ and $2L$ because the right arm of the kth T-section is connected in series with the left arm of the $(k+1)^{th}$ T-section.

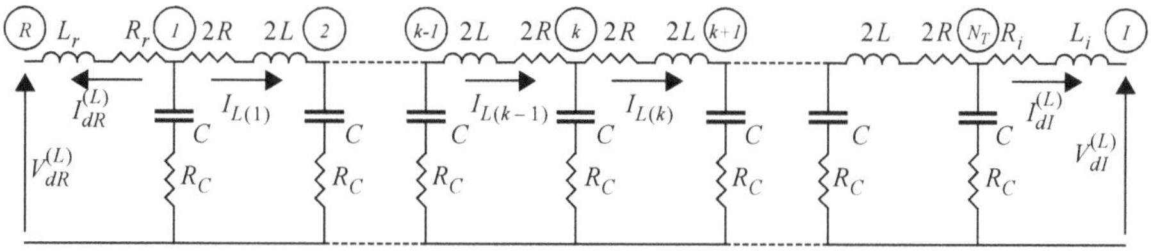

Figure 4.19 Circuit diagram of a HVDC link model comprising the series connection of N_T of the T-sections in Figure 4.18 and the rectifier and inverter smoothing reactors.

The rectifier- and inverter-end voltages, $V_{dR}^{(L)}$ and $V_{dI}^{(L)}$, are treated as model inputs. The rectifier- and inverter-end currents, $I_{dR}^{(L)} = -I_{L(0)}$ and $I_{dI}^{(L)} = I_{L(N_T+1)}$, are outputs from the link model which are to be input to the devices connected to the respective DC terminals. The state-equations for the inductor currents are written first, followed by the state-equations for the capacitor voltages and finally the algebraic nodal voltage and current equations. All currents are in Amperes (A) and all voltages are in Volts (V) - except $V_{dR}^{(L)}$ and $V_{dI}^{(L)}$ which are in kV.

Inductor current state-equations

$$pI_{L(0)} = -(R_r/L_r)I_{L(0)} + (V_{dR}^{(L)} \times 10^3 - V_{m(1)})/L_r, \tag{4.277}$$

$$pI_{L(k-1)} = -(R/L)I_{L(k-1)} + (V_{m(k-1)} - V_{m(k)})/(2L), \quad k = 2, \ldots, N_T, \tag{4.278}$$

$$pI_{L(N_T)} = -(R_i/L_i)I_{L(N_T)} + (V_{m(N_T)} - V_{dI}^{(L)} \times 10^3)/L_i. \tag{4.279}$$

Capacitor voltage state-equations

$$pV_{C(k)} = I_{C(k)}/C, \quad k = 1, \ldots, N_T. \tag{4.280}$$

Nodal voltage and current equations

$$0 = V_{C(k)} - V_{m(k)} + R_C I_{C(k)}, \quad k = 1, \ldots, N_T, \tag{4.281}$$

$$0 = -I_{L(k-1)} + I_{L(k)} + I_{C(k)}, \quad k = 1, \ldots, N_T. \tag{4.282}$$

Note that the equations (4.277) to (4.282) are linear. Each variable can therefore be replaced by its perturbed value at the steady-state value, e.g. the variable $V_{dR}^{(L)}$ is replaced by $\Delta V_{dR}^{(L)}$ at the steady-state value $V_{dR_0}^{(L)}$. Given the initial steady-state values of the rectifier- and inverter-end voltages, the initial values of the other variables are obtained by solving the aforementioned set of linear equations with the rates of change in the inductor currents and capacitor voltages set to zero. The initial values of the rectifier and inverter end currents are given by:

$$I_{dI_0} = -I_{dR_0} = (V_{dR_0} - V_{dI_0})/R_L \times 10^3 \text{ A.} \tag{4.283}$$

Simplified modelling of the HVDC transmission line / cable is appropriate when the link is short or when detailed modelling such as that described above is otherwise unnecessary or infeasible. For example, the link may be represented as a series RL branch incorporating the rectifier and inverter smoothing reactors, or simply as a series resistance.

4.3.7 Model of HVDC transmission with Voltage Sourced Converters (VSCX)

In the model of a VSCX the rectifier *(r)* in Figure 4.17 is represented by a shunt connected VSC model as described in Section 4.3.3. The linearized equations for the rectifier are listed in Section 4.3.3.1 in which the superscript *(c)* is replaced by *(r)*. The linearized model of the inverter is similarly represented - but with *(c)* replaced by *(i)*. For both converters the DC input and output are the perturbations in the DC current and voltage respectively, i.e. equation (4.255) on page 172 applies to both converters. Since the converters are shunt connected the equations and variables associated with the second terminal, *(r2)* and *(i2)*, of the respective converters are omitted from the model equations as described in Section 4.3.3. The HVDC link is represented by the equations (4.277) to (4.282). The interconnections between the linearized equations for the converters and HVDC link are represented by the following linear constraint equations:

$$0 = I_{ds}^{(r)} + I_{dR}^{(L)} \text{ (A)},\ 0 = I_{ds}^{(i)} + I_{dI}^{(L)} \text{ (A)},\ 0 = V_{d}^{(r)} - V_{dR}^{(L)} \text{ (kV)},\ 0 = V_{d}^{(i)} - V_{dI}^{(L)} \text{ (kV)} \tag{4.284}$$

Typically, one of the converters is used to control the power transferred by the link and the other converter is used to control its DC voltage. Each converter is normally used to control either its reactive power output or the AC voltage of an adjacent bus. The control systems are specific to the application and are omitted from the VSCX model.

The linearized model equations of the VSCX system can be written in the following form which is suitable for integration with the linearized DAEs of the power system as described in Section 4.4. The superscript *(T)* denotes that the quantity is associated with the integrated VSCX system.

$$\begin{bmatrix} p\Delta \underset{\sim}{x}^{(T)} \\ 0 \\ 0 \end{bmatrix} = \begin{bmatrix} J_{fx}^{(T)} & J_{fz}^{(T)} & 0 \\ J_{gx}^{(T)} & J_{gz}^{(T)} & J_{gi}^{(T)} \\ 0 & J_{iz}^{(T)} & J_{ii}^{(T)} \end{bmatrix} \begin{bmatrix} \Delta \underset{\sim}{x}^{(T)} \\ \Delta \underset{\sim}{z}^{(T)} \\ \Delta \underset{\sim}{i}_{RI}^{(n,T)} \end{bmatrix} + \begin{bmatrix} 0 \\ J_{gv}^{(T)} \\ 0 \end{bmatrix} \Delta \underset{\sim}{v}_{RI}^{(n,T)} + \begin{bmatrix} 0 \\ J_{gu_c}^{(T)} \\ 0 \end{bmatrix} \Delta \underset{\sim}{u}_{c}^{(T)} \tag{4.285}$$

The state ($\Delta \underset{\sim}{x}^{(T)}$), algebraic ($\Delta \underset{\sim}{z}^{(T)}$) and control-input, ($\Delta \underset{\sim}{u}_c^{(T)}$) variables in the preceding equation combine the corresponding variables from the HVDC link model and the rectifier and inverter models. The current components injected by the VSCX into the AC buses - to which the rectifier and inverter AC terminals are connected - are $\Delta \underset{\sim}{i}_{RI}^{(n,T)}$ and the associated bus voltages are $\Delta \underset{\sim}{v}_{RI}^{(n,T)}$. The latter current and voltage quantities are in per unit on the base values of the buses to which the converter terminals are connected.

4.3.8 Model of HVDC transmission with Voltage Commutated Converters

A small-signal model of a HVDC transmission system with voltage-commutated thyristor-controlled converters is now presented. The resulting TCCX model has the structure shown in Figure 4.17. A general purpose converter model, depicted in Figure 4.20, is formulated on the basis of [10, 25, 29, 30, 31, 32, 33]; it is then used in a modular fashion to represent the converter at one end of the link operating in rectifier mode and at the other end as an inverter. The distributed parameter model of the HVDC link in Section 4.3.5 is used to connect the rectifier and inverter.

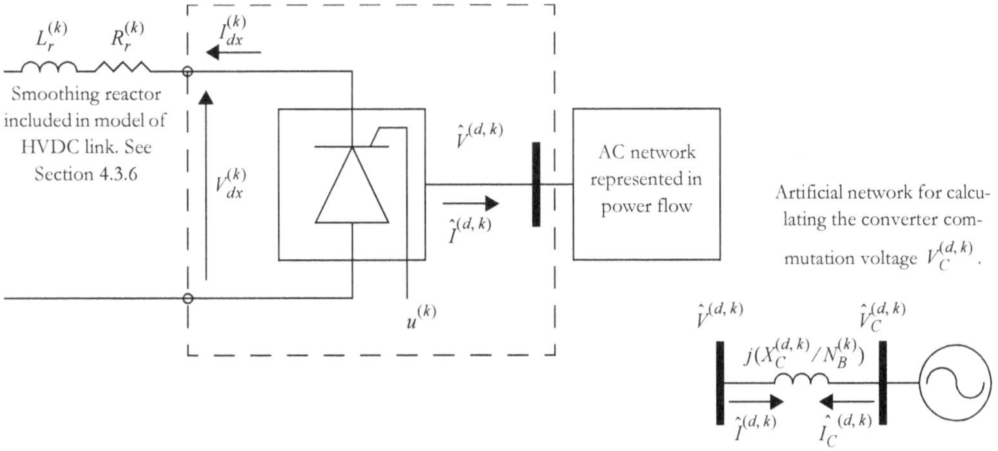

Figure 4.20 Voltage-commutated thyristor-controlled converter. The artificial network for calculating the converter commutation voltage is also shown.

In the following the converter identifier indicated by the superscript *(k)* is substituted either with *(r)* for the rectifier or *(i)* for the inverter. The AC terminal quantities are defined in per-unit on the device base values as specified in Table 4.13 on page 160. The DC side quantities are in SI units as specified below.

As mentioned in Section 4.3.5, the converter is assumed to comprise $N_B^{(k)}$ identical bridges connected in series on the DC side and in parallel on the AC side.

4.3.8.1 Internal DC voltage and current variables

For the purpose of developing the voltage-commutated thyristor-controlled converter model, internal values of the DC current, $I_{dx}^{(k)}$ (A), and DC voltage, $V_{dx}^{(k)}$ (kV), are employed. In Section 4.3.8.9 the relationships are established between these internal DC quantities and their corresponding external interface values shown in the general HVDC link model of Figure 4.17.

It is important to note that the DC current *must* flow in the direction of the thyristor valves and by definition this is the positive direction of $I_{dx}^{(k)}$. Consequently $I_{dx}^{(k)}$ is positive whether the converter is operating as a rectifier or as an inverter. The direction of positive DC voltage $V_{dx}^{(k)}$ is defined to be in the positive direction of $I_{dx}^{(k)}$. Thus, as will be seen in the following, $V_{dx}^{(k)}$ is positive if the converter is operating as a rectifier and is negative when operating as an inverter.

4.3.8.2 Relationship between the AC and DC quantities of the converter

In the analysis of voltage-commutated converters it is assumed that an ideal, sinusoidal, three-phase source is connected to the converter through a reactance, which is referred to as the commutating reactance, $X_C^{(d,k)}$ per bridge. The AC voltage source is referred to as the commutating voltage. Assuming a thyristor firing-angle delay, $\alpha^{(k)}$ (rad), and an overlap angle, $\mu^{(k)}$ (rad), due to the commutating reactance, it is shown in the analysis of [29] that the average DC voltage at the converter is,

$$V_{dx}^{(k)} = V_{d0}^{(k)} \cos\alpha^{(k)} - (R_C^{(k)} \times 10^{-3}) I_{dx}^{(k)}, \quad \text{(kV)}, \tag{4.286}$$

in which the DC current is $I_{dx}^{(k)}$ (A) and where the no-delay, no-load DC output voltage $V_{d0}^{(k)}$ (kV) is given in terms of the magnitude of the commutating voltage magnitude, $V_C^{(d,k)}$ (per-unit on $V_{usb}^{(d,k)}$ kV):

$$V_{d0}^{(k)} = c_{vd}^{(k)} V_C^{(d,k)} \quad \text{(kV), where} \quad c_{vd}^{(k)} = \left(\frac{3}{\pi}\sqrt{2}\right)(N_B^{(k)} V_{usb}^{(d,k)}) \quad \text{(kV), and} \tag{4.287}$$

$$R_C^{(k)} = c_{rd}^{(k)} X_C^{(d,k)} \text{ (ohm), in which } c_{rd}^{(k)} = \left(\frac{3}{\pi}\right) N_B^{(k)} \left(\frac{(V_{usb}^{(d,k)})^2}{S_{usb}^{(d,k)}}\right) \text{ (ohm).} \tag{4.288}$$

The so-called "commutation resistance", $R_C^{(k)}$, is an artefact that accounts for the DC voltage drop due to commutation overlap; it does not generate power losses.

By substituting for $V_{d0}^{(k)}$ from (4.287) into equation (4.286) the DC voltage is obtained in terms of the commutating voltage magnitude, the DC current, firing-angle delay and commutation resistance as follows:

$$V_{dx}^{(k)} = c_{vd}^{(k)} V_C^{(d,k)} \cos(\alpha^{(k)}) - (R_C^{(k)} \times 10^{-3}) I_{dx}^{(k)} \text{ (kV).} \tag{4.289}$$

In using this converter model the value of the commutation reactance must be selected, ideally it is the Thévenin impedance looking into the AC system from the converter AC terminals. The use of the commutation bus is therefore an artefact for the calculation the DC voltage, $V_{dx}^{(k)}$. Importantly (i) the transformers and associated components connected to the AC terminals of the converter are represented in the AC network model; (ii) the commutation reactance and commutation bus are not included in the AC network model.

The converter switching losses are assumed to be fixed and are represented by a user supplied quantity $P_{sl}^{(d,k)}$ pu on $S_{usb}^{(d,k)}$. Thus, due to the conservation of energy, the DC power output from the converter is related to the AC power output $P^{(d,k)}$ by:

$$P_d^{(k)} = -(P^{(d,k)} + P_{sl}^{(d,k)}) S_{usb}^{(d,k)} = V_{dx}^{(k)} I_{dx}^{(k)} \times 10^3 \text{ MW.} \tag{4.290}$$

As mentioned earlier, it is assumed that the initial steady-state values of the following quantities are obtained from the power flow solution: (i) the real and reactive power output from the converter AC terminal, $P_0^{(n,k)}$ and $Q_0^{(n,k)}$ respectively; (ii) the voltage magnitude and angle at the AC terminal, $V_0^{(n,k)}$ and $\theta_0^{(n,k)}$ respectively; and (iii) the DC voltage $V_{dx_0}^{(k)}$. The initial steady-state values $P_0^{(d,k)}$, $Q_0^{(d,k)}$, $\underset{\sim}{v}_{RI_0}^{(d,k)}$ and $\underset{\sim}{i}_{RI_0}^{(d,k)}$, are calculated from the power flow solution as described in Section 4.3.1. The initial steady-state values of the DC power and current are then obtained from (4.290) as follows:

$$P_{d_0}^{(k)} = -(P_0^{(d,k)} + P_{sl}^{(d,k)}) S_{usb}^{(d,k)} \text{ (MW) and } I_{dx_0}^{(k)} = P_{d_0}^{(k)} / V_{dx_0}^{(k)} \times 10^3 \text{ (A).} \tag{4.291}$$

The perturbation in the DC power output of the converter is now obtained by linearizing equation (4.290) about the above operating point values.

$$\Delta P_d^{(k)} = -S_{usb}^{(d,k)} \Delta P^{(d,k)} = (V_{dx_0}^{(k)} \times 10^3) \Delta I_{dx}^{(k)} + (I_{dx_0}^{(k)} \times 10^3) \Delta V_{dx}^{(k)} . \quad (4.292)$$

4.3.8.3 Perturbations in commutation voltage in terms of converter terminal quantities

From Figure 4.20 the commutating voltage phasor is obtained from the conditions at the converter AC terminals as follows.

$$\hat{V}_C^{(d,k)} = V_C^{(d,k)} e^{j\theta_C^{(k)}} = v_{CR}^{(d,k)} + j v_{CI}^{(d,k)} = \hat{V}^{(d,k)} - j\left(\frac{X_C^{(d,k)}}{N_B^{(k)}}\right)\hat{I}^{(d,k)} \text{ pu on } V_{usb}^{(d,k)}, \quad (4.293)$$

from which,

$$\begin{aligned} \underset{\sim}{v}_{CRI}^{(d,k)} &= \left[v_{CR}^{(d,k)} \; v_{CI}^{(d,k)}\right]^T = V_C^{(d,k)}\left[\cos(\theta_C^{(k)}) \; \sin(\theta_C^{(k)})\right]^T \\ &= \left[\left\{v_R^{(d,k)} + \left(\frac{X_C^{(d,k)}}{N_B^{(k)}}\right) i_I^{(d,k)}\right\} \; \left\{v_I^{(d,k)} - \left(\frac{X_C^{(d,k)}}{N_B^{(k)}}\right) i_R^{(d,k)}\right\}\right]^T \end{aligned} \quad (4.294)$$

The initial steady-state values $\underset{\sim}{v}_{RI_0}^{(d,k)}$ and $\underset{\sim}{i}_{RI_0}^{(d,k)}$ calculated in the previous section are then substituted in (4.294) to calculate the initial values of the commutating voltage components, $\underset{\sim}{v}_{CRI_0}^{(k)}$.

Based on (4.193) on page 161, the commutating voltage magnitude $V_C^{(d,k)}$ is:

$$V_C^{(d,k)} = \sqrt{(v_{CR}^{(d,k)})^2 + (v_{CI}^{(d,k)})^2} = \sqrt{(\underset{\sim}{v}_{CRI}^{(d,k)})^T \underset{\sim}{v}_{CRI}^{(d,k)}}, \quad (4.295)$$

which is linearized according to (4.195) to yield the following expression for the perturbation in the commutation voltage about its initial steady-state value,

$$\Delta V_C^{(d,k)} = \left(\frac{1}{V_{C_0}^{(d,k)}}\right)(\underset{\sim}{v}_{CRI_0}^{(d,k)})^T \Delta \underset{\sim}{v}_{CRI}^{(d,k)} \text{ (pu)}, \quad (4.296)$$

where $V_{C_0}^{(d,k)}$ is obtained by substituting $\underset{\sim}{v}_{CRI_0}^{(d,k)}$ in (4.295).

The initial steady-state firing-angle delay $\alpha_0^{(k)}$ is obtained by back substituting into (4.289), $V_{dx_0}^{(k)}$ from the power flow solution, $I_{dx_0}^{(k)}$ from (4.291) and $V_{C_0}^{(d,k)}$ (calculated above), to give:

$$\alpha_0^{(k)} = \operatorname{acos}\left(\frac{V_{dx_0}^{(k)} + (R_C^{(k)} \times 10^{-3}) I_{dx_0}^{(k)}}{c_{vd}^{(k)} V_{C_0}^{(d,k)}}\right). \quad (4.297)$$

The perturbation in the DC voltage about its initial steady-state value is obtained by linearizing equation (4.289) to yield:

$$\Delta V_{dx}^{(k)} = [c_{vd}^{(k)}\cos(\alpha_0^{(k)})]\Delta V_C^{(d,k)} - [c_{vd}^{(k)} V_{C_0}^{(d,k)}\sin(\alpha_0^{(k)})]\Delta\alpha^{(k)} - [R_C^{(k)} \times 10^{-3}]\Delta I_{dx}^{(k)} . \quad (4.298)$$

4.3.8.4 Perturbation in real and reactive power supplied from the commutation bus

The current output from the converter terminal, $\hat{I}^{(d,k)}$, is equal and opposite to that output from the commutation bus, $\hat{I}_C^{(d,k)}$, i.e.

$$\hat{I}_C^{(d,k)} = -\hat{I}^{(d,k)} . \quad (4.299)$$

The real and reactive power supplied from the commutation bus is:

$$P_C^{(d,k)} + jQ_C^{(d,k)} = \hat{V}_C^{(d,k)}(\hat{I}_C^{(d,k)})^* . \quad (4.300)$$

Substituting in the preceding equation for $\hat{V}_C^{(d,k)}$ in terms of the converter terminal voltage and current from (4.293) and then substituting for $\hat{I}_C^{(d,k)}$ from (4.299) yields:

$$\begin{aligned} P_C^{(d,k)} + jQ_C^{(d,k)} &= -\left(\hat{V}^{(d,k)} - j\left(\frac{X_C^{(d,k)}}{N_B^{(k)}}\right)\hat{I}^{(d,k)}\right)(\hat{I}^{(d,k)})^* \\ &= -\hat{V}^{(d,k)}(\hat{I}^{(d,k)})^* + j\left(\frac{X_C^{(d,k)}}{N_B^{(k)}}\right)(I^{(d,k)})^2 \\ &= -P^{(d,k)} - j\left(Q^{(d,k)} - \left(\frac{X_C^{(d,k)}}{N_B^{(k)}}\right)(I^{(d,k)})^2\right) \end{aligned}$$

From the preceding equation the real and reactive power output from the commutation bus is expressed in terms of the real and reactive power output from the converter AC terminal, $P^{(d,k)}$ and $Q^{(d,k)}$, and the magnitude of the AC converter current, $I^{(d,k)}$, i.e.:

$$P_C^{(d,k)} = -P^{(d,k)} \quad \text{and} \quad Q_C^{(d,k)} = -\left(Q^{(d,k)} - \left(\frac{X_C^{(d,k)}}{N_B^{(k)}}\right)(I^{(d,k)})^2\right) . \quad (4.301)$$

The initial steady-state value $I_0^{(d,k)}$ is obtained from (4.201) on page 162 using the previously determined current components $i_{RI_0}^{(d,k)}$. Then $P_{C_0}^{(d,k)}$ and $Q_{C_0}^{(d,k)}$ are derived by substituting the previously determined values of $P_0^{(d,k)}$, $Q_0^{(d,k)}$ and $I_0^{(d,k)}$ in (4.301).

Let $\phi^{(k)}$ (rad) be the angle by which the fundamental-frequency commutation bus current phasor $\hat{I}_C^{(d,k)}$ lags the rms phase-neutral commutation voltage phasor $\hat{V}_C^{(d,k)}$. The factor $\cos(\phi^{(k)})$ is the power factor - or displacement factor - of the fundamental waveforms and is a function of the converter firing-delay and extinction-delay angles, $\alpha^{(k)}$ (rad) and $\delta^{(k)}$ (rad), respectively.

An approximate value of the power factor is

$$\cos(\phi^{(k)}) = (\cos(\alpha^{(k)}) + \cos(\delta^{(k)}))/2. \tag{4.302}$$

However, an exact expression for the power factor angle is provided in [29]:

$$\tan(\phi^{(k)}) = T^{(k)} = \frac{G^{(k)}}{H^{(k)}} = \frac{2(\delta^{(k)} - \alpha^{(k)}) + \sin(2\alpha^{(k)}) - \sin(2\delta^{(k)})}{\cos(2\alpha^{(k)}) - \cos(2\delta^{(k)})}. \tag{4.303}$$

The reactive power output from the commutation bus is related to the real power output and the power-factor angle, $\phi^{(k)}$, by:

$$Q_C^{(d,k)} = P_C^{(d,k)}\tan(\phi^{(k)}) = P_C^{(d,k)}T^{(k)}. \tag{4.304}$$

Equations (4.303) and (4.304) are now used to calculate the initial steady-state value of the extinction-delay angle. The initial steady-state values $P_{C_0}^{(d,k)}$ and $Q_{C_0}^{(d,k)}$ were determined earlier in this section and $\alpha_0^{(k)}$ was determined in (4.297). Thus, from (4.304) it follows that $\phi_0^{(k)} = \mathrm{atan2}(Q_{C_0}^{(d,k)}, P_{C_0}^{(d,k)})$ [1]. Substituting for the known values of $T_0^{(k)} = \tan(\phi_0^{(k)})$ and $\alpha_0^{(k)}$ in (4.303) results in the following non-linear equation for the initial value of the extinction-delay angle:

$$\delta_0^{(k)} = \left\{c_0 + \sin(2\delta_0^{(k)}) - T_0^{(k)}\cos(2\delta_0^{(k)})\right\}/2, \text{ where} \tag{4.305}$$

$$c_0 = 2\alpha_0^{(k)} - \sin(2\alpha_0^{(k)}) + T_0^{(k)}\cos(2\alpha_0^{(k)}) \text{ is known.}$$

---

1. Note that the atan2 function is used to ensure that the power-factor angle is located in the correct quadrant. This is important when the converter is operating as an inverter. See the footnote on page 126 for the definition of atan2.

The preceding equation can be solved using any one of a number of methods including a fixed-point iteration method or the Newton-Raphson method with an initial estimate of $\delta_0^{(k)}$ determined from (4.302).

Let $\mu^{(k)}$ (rad) be the overlap angle, then

$$\mu^{(k)} = \delta^{(k)} - \alpha^{(k)}. \tag{4.306}$$

It is noted that inconsistencies can arise between the steady-state solution of converter variables in the power flow and the initial conditions determined for the purpose of linearizing the converter model. Such inconsistencies are frequently attributable to the way in which the power-factor angle is expressed as a function of the thyristor firing and extinction angles.

The linearized form of (4.303) is:

$$\begin{aligned}\Delta T^{(k)} &= (1/H_0^{(k)})\Delta G^{(k)} - (T_0^{(k)}/H_0^{(k)})\Delta H^{(k)} \\ &= (c_{t\alpha_0}^{(k)}\Delta\alpha^{(k)} + c_{t\delta_0}^{(k)}\Delta\delta^{(k)})\end{aligned} \tag{4.307}$$

$$\begin{aligned}\text{where}\quad c_{t\alpha_0}^{(k)} &= (2/H_0^{(k)})\left\{\cos(2\alpha_0^{(k)}) + T_0^{(k)}\sin(2\alpha_0^{(k)}) - 1\right\}, \\ \text{and}\quad c_{t\delta_0}^{(k)} &= -(2/H_0^{(k)})\left\{\cos(2\delta_0^{(k)}) + T_0^{(k)}\sin(2\delta_0^{(k)}) - 1\right\}.\end{aligned} \tag{4.308}$$

The perturbation in the reactive power output from the commutation bus is thus obtained by linearizing equation (4.304) about the operating point to give:

$$\Delta Q_C^{(d,k)} = T_0^{(k)}\Delta P_C^{(d,k)} + P_{C_0}^{(d,k)}\Delta T^{(k)}. \tag{4.309}$$

4.3.8.5 Perturbations in converter AC current and apparent power output

From equations (4.290) and (4.301) the following expressions for the real and reactive power output from the converter terminals are found:

$$P^{(d,k)} = -(P_d^{(k)}/S_{usb}^{(d,k)} + P_{sl}^{(d,k)}) \text{ and } Q^{(d,k)} = -Q_C^{(d,k)} + \left(\frac{X_C^{(d,k)}}{N_B^{(k)}}\right)(I^{(d,k)})^2. \tag{4.310}$$

The perturbations in these quantities about the operating point are obtained by linearizing the preceding equations to yield:

$$\Delta P^{(d,k)} = -(1/S_{usb}^{(d,k)})\Delta P_d^{(k)} \text{ and } \Delta Q^{(d,k)} = -\Delta Q_C^{(d,k)} + 2\left(\frac{X_C^{(d,k)} I_0^{(d,k)}}{N_B^{(k)}}\right)\Delta I^{(d,k)}. \tag{4.311}$$

The perturbations in the AC converter current components about the operating point are obtained in terms of the perturbation in the converter apparent power output ($\Delta \underset{\sim}{S}^{(d,k)} = \begin{bmatrix}\Delta P^{(d,k)} & \Delta Q^{(d,k)}\end{bmatrix}^T$) and the perturbations in the AC terminal voltage components by application of equations (4.199) and (4.200) on page 162 to yield:

$$\Delta \underset{\sim}{i}_{RI}^{(d,k)} = \boldsymbol{J}_{is}^{(d,k)} \Delta \underset{\sim}{S}^{(d,k)} + \boldsymbol{J}_{iv}^{(d,k)} \Delta \underset{\sim}{v}_{RI}^{(d,k)} \quad \text{(pu), in which} \tag{4.312}$$

$$\boldsymbol{J}_{is}^{(d,k)} = \left(\frac{1}{V_0^{(d,k)}}\right)^2 \begin{bmatrix} v_{R_0}^{(d,k)} & v_{I_0}^{(d,k)} \\ v_{I_0}^{(d,k)} & -v_{R_0}^{(d,k)} \end{bmatrix}, \text{ and}$$

$$\boldsymbol{J}_{iv}^{(d,k)} = \left(\frac{1}{V_0^{(d,k)}}\right)^2 \begin{bmatrix} (P_0^{(d,k)} - 2i_{R_0}^{(d,k)} v_{R_0}^{(d,k)}) & (Q_0^{(d,k)} - 2i_{R_0}^{(d,k)} v_{I_0}^{(d,k)}) \\ -(Q_0^{(d,k)} - 2i_{I_0}^{(d,k)} v_{R_0}^{(d,k)}) & (P_0^{(d,k)} - 2i_{I_0}^{(d,k)} v_{I_0}^{(d,k)}) \end{bmatrix}. \tag{4.313}$$

The perturbations in the converter AC current required in equation (4.311) are obtained by applying equation (4.202) on page 162 to the converter AC terminal *(k)* as follows:

$$\Delta I^{(d,k)} = (1/I_0^{(d,k)})(\underset{\sim}{i}_{RI_0}^{(d,k)})^T \Delta \underset{\sim}{i}_{RI}^{(d,k)} \quad \text{(pu)}. \tag{4.314}$$

4.3.8.6 Modifications of the general converter model for inverter operation

The voltage-commutated thyristor-controlled converter model developed above is general in that it applies whether the converter is operating as a rectifier or inverter. When the converter is operated as an inverter the firing-angle delay $\alpha^{(k)}$ must be greater than $\pi/2$ rad in order to produce a DC voltage that opposes the flow of the DC current through the thyristor valves. The DC voltage produced by the rectifier forces the DC current through the inverter valves against the opposing inverter DC voltage. When describing the operation of the inverter it is common practice to refer to the firing-angle advance, $\beta^{(k)}$ (rad), and the extinction-angle advance, $\gamma^{(k)}$ (rad) which are related to the corresponding delay angles by:

$$\beta^{(k)} = \pi - \alpha^{(k)} \quad \text{and} \quad \gamma^{(k)} = \pi - \delta^{(k)}. \tag{4.315}$$

Thus, although it is mathematically unnecessary, in the case of inverter operation the above equations of the advance angles are added to the general converter equations in accordance with conventional practice.

4.3.8.7 Converter control input

The converter model has a single control input signal denoted by $u^{(k)}$. From a detailed modelling perspective this input is usually the thyristor firing-angle delay, $\alpha^{(k)}$. However, it is also common when representing the inverter control system to employ either the firing-angle advance, $\beta^{(k)}$, or the extinction-angle advance, $\gamma^{(k)}$, as the control signal. Thus, one of the following constraints on the converter input signal is typically employed for detailed modelling of the converter controls:

$$u^{(k)} = \alpha^{(k)} \text{ (rad) or } u^{(k)} = \beta^{(k)} \text{ (rad) or } u^{(k)} = \gamma^{(k)} \text{ (rad)}. \tag{4.316}$$

It is often desirable to provide a simplified functional representation of the converter control systems. For example, if the converter is equipped with a fast acting control system whose objective is to maintain constant DC power flow then the control input $u^{(k)}$ can be set to $P_d^{(k)}$ rather than $\alpha^{(k)}$.

4.3.8.8 Summary of the linearized converter algebraic equations

A consolidated list is presented below of the linearized algebraic equations for the voltage-commutated thyristor-controlled converter model. The initial steady-state operating condition is determined from the power flow solution as described in the preceding sections.

DC side equations

$$0 = [c_{vd}^{(k)}\cos(\alpha_0^{(k)})]\Delta V_C^{(d,k)} - [c_{vd}^{(k)} V_{C_0}^{(d,k)}\sin(\alpha_0^{(k)})]\Delta\alpha^{(k)} - [R_C^{(k)} \times 10^{-3}]\Delta I_{dx}^{(k)} - \Delta V_{dx}^{(k)} \tag{4.317}$$

from (4.298),

$$0 = (V_{dx_0}^{(k)} \times 10^3)\Delta I_{dx}^{(k)} + (I_{dx_0}^{(k)} \times 10^3)\Delta V_{dx}^{(k)} - \Delta P_d^{(k)} \quad \text{from (4.292)}. \tag{4.318}$$

Commutation bus equations

$$0 = \left[\left\{\Delta v_R^{(d,k)} + \left(\frac{X_C^{(d,k)}}{N_B^{(k)}}\right)\Delta i_I^{(d,k)}\right\} \quad \left\{\Delta v_I^{(d,k)} - \left(\frac{X_C^{(d,k)}}{N_B^{(k)}}\right)\Delta i_R^{(d,k)}\right\}\right]^T - \Delta \underset{\sim}{v}_{CRI}^{(d,k)} \tag{4.319}$$

from (4.294),

$$0 = \left(\frac{1}{V_{C_0}^{(d,k)}}\right)(\underset{\sim}{v}_{CRI_0}^{(d,k)})^T \Delta \underset{\sim}{v}_{CRI}^{(d,k)} - \Delta V_C^{(d,k)} \quad \text{from (4.296)}, \tag{4.320}$$

$$0 = \Delta P^{(d,k)} + \Delta P_C^{(d,k)} \quad \text{from (4.301)}, \tag{4.321}$$

$$0 = c_{t\alpha_0}^{(k)}\Delta\alpha^{(k)} + c_{t\delta_0}^{(k)}\Delta\delta^{(k)} - \Delta T^{(k)} \tag{4.322}$$

from (4.307) and (4.308) where $T^{(k)} = \tan(\phi^{(k)})$;

$$0 = T_0^{(k)}\Delta P_C^{(d,k)} + P_{C_0}^{(d,k)}\Delta T^{(k)} - \Delta Q_C^{(d,k)} \quad \text{from (4.309)}. \tag{4.323}$$

Converter AC terminal equations

$$0 = \Delta P_d^{(k)} + S_{usb}^{(d,k)} \Delta P^{(d,k)} \quad \text{from (4.311)}, \tag{4.324}$$

$$0 = -\Delta Q_C^{(d,k)} + 2\left(\frac{X_C^{(d,k)} I_0^{(d,k)}}{N_B^{(k)}}\right)\Delta I^{(d,k)} - \Delta Q^{(d,k)} \quad \text{from (4.311)}, \tag{4.325}$$

$$0 = \boldsymbol{J}_{is}^{(d,k)} \Delta \underset{\sim}{S}^{(d,k)} + \boldsymbol{J}_{iv}^{(d,k)} \Delta \underset{\sim}{v}_{RI}^{(d,k)} - \Delta \underset{\sim}{i}_{RI}^{(d,k)} \quad \text{from (4.312) and (4.313)}, \tag{4.326}$$

$$0 = (1/I_0^{(d,k)})(\underset{\sim}{i}_{RI_0}^{(d,k)})^T \Delta \underset{\sim}{i}_{RI}^{(d,k)} - \Delta I^{(d,k)} \quad \text{from (4.314)}. \tag{4.327}$$

Advance angle equations

$$0 = \Delta\alpha^{(k)} + \Delta\beta^{(k)} \quad \text{and} \quad 0 = \Delta\delta^{(k)} + \Delta\gamma^{(k)} \quad \text{from (4.315)}. \tag{4.328}$$

Input control signal equation

As mentioned in Section 4.3.8.7 it is desirable in a flexible converter model for the control input signal to be constrained to an independent variable depending of the user's specific requirements. For detailed modelling the input control signal will typically be defined by one of the following equations:

$$0 = \Delta\alpha^{(k)} - \Delta u^{(k)} \quad \text{or} \quad 0 = \Delta\beta^{(k)} - \Delta u^{(k)} \quad \text{or} \quad 0 = \Delta\gamma^{(k)} - \Delta u^{(k)}, \quad \text{from (4.316)}. \tag{4.329}$$

Alternatively if the functional behaviour of the converter control system is to be represented then the input is constrained to some other user-selected system variable $\Delta y_c^{(k)}$ such that:

$$0 = \Delta y_c^{(k)} - \Delta u^{(k)}. \tag{4.330}$$

4.3.8.9 Relationship between the internal model and external interface values of the DC voltage and current in the general HVDC transmission model.

As described in Section 4.3.8.1 the positive direction of the DC current variable $I_{dx}^{(k)}$ used in the internal formulation of the converter model is in the direction of the thyristor valves. The positive direction of the DC voltage variable $V_{dx}^{(k)}$ used in the formulation of the model is in the positive direction of $I_{dx}^{(k)}$. The rectifier *(r)* and inverter *(i)* define external interface variables for the DC voltage and current for the respective converters in the general model of the HVDC transmission system shown in Figure 4.17 on page 177. Figure 4.21 shows the relationship between the internal and external interface DC current and voltage variables for two scenarios:

1. The DC current flow is from the rectifier to inverter for which the relationships between the internal and external variables are:

$$\text{for the rectifier:} \quad I_{ds}^{(r)} = I_{dx}^{(r)} \quad \text{and} \quad V_{d}^{(r)} = V_{dx}^{(r)}, \tag{4.331}$$

$$\text{for the inverter:} \quad I_{ds}^{(i)} = -I_{dx}^{(i)} \quad \text{and} \quad V_{d}^{(i)} = -V_{dx}^{(i)}. \tag{4.332}$$

2. The DC current flow is from the inverter to rectifier for which the relationships between the internal and external variables are:

$$\text{for the rectifier:} \quad I_{ds}^{(r)} = -I_{dx}^{(r)} \quad \text{and} \quad V_{d}^{(r)} = -V_{dx}^{(r)}, \tag{4.333}$$

$$\text{for the inverter:} \quad I_{ds}^{(i)} = I_{dx}^{(i)} \quad \text{and} \quad V_{d}^{(i)} = V_{dx}^{(i)}. \tag{4.334}$$

Of course, in both cases the DC power flow is from the rectifier to the inverter; in the first case the DC current and power flow in the same direction, whereas in the second case the direction of DC power flow is opposite to that of the DC current

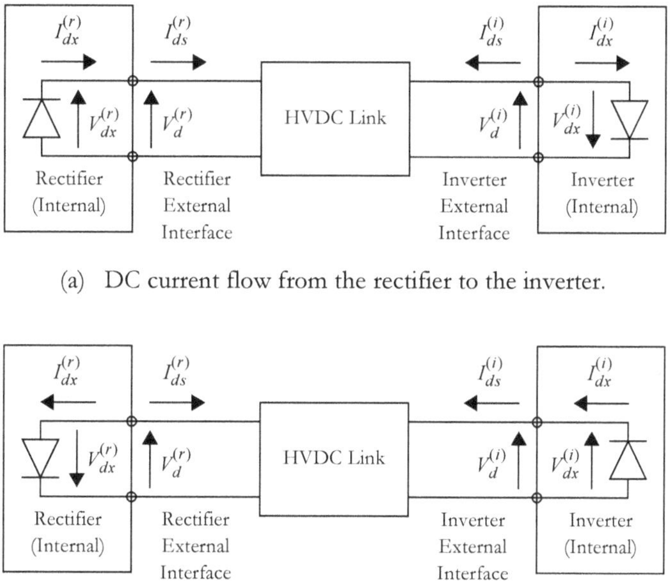

(a) DC current flow from the rectifier to the inverter.

(b) DC current flow from the inverter to the rectifier.

Figure 4.21 Relationship between converter internal and external DC variables.

4.3.8.10 Assembly of the TCCX model

In the small-signal model of a TCCX the rectifier *(r)* in Figure 4.17 is represented by a voltage-commutated thyristor-controlled converter model in accordance with the linearized equations listed in Section 4.3.8.8. In the rectifier model the superscript *(k)* in the general equations is replaced by *(r)*. The relationship between the internal values of the rectifier DC voltage and current - and the corresponding external interface values of these quantities - is defined by either equation (4.331) or (4.333) depending on the direction of DC current flow. The linearized model of the inverter is similarly represented but with *(k)* replaced by *(i)*. The

HVDC transmission line/cable is represented by the equations (4.277) to (4.282) on page 179. The interconnections between the linearized equations for the converters and HVDC line/cable are represented by the following linear constraint equations:

$$0 = I_{ds}^{(r)} + I_{dR}^{(L)} \text{ (A)},\ 0 = I_{ds}^{(i)} + I_{dI}^{(L)} \text{ (A)},\ 0 = V_d^{(r)} - V_{dR}^{(L)} \text{ (kV)},\ 0 = V_d^{(i)} - V_{dI}^{(L)} \text{ (kV)}. \quad (4.335)$$

It is possible to implement a wide range of control strategies with this general purpose model. The principal control strategies are: (i) the rectifier operates in constant-current control and the inverter operates in constant extinction angle control, or (ii) the rectifier operates in constant firing-angle control and the inverter operates in constant-current control. Supplementary control strategies to achieve a range of objectives are possible, including regulation of power flow, regulation of frequency and damping control. The control systems are specific to the application and are omitted from the model.

The linearized model equations of the TCCX system can be written in the following form which is suitable for integration with the linearized DAEs of the power system as described in Section 4.4. The superscript *(T)* denotes that the quantity is associated with the integrated TCCX system.

$$\begin{bmatrix} p\Delta\underset{\sim}{x}^{(T)} \\ \boldsymbol{0} \\ \boldsymbol{0} \end{bmatrix} = \begin{bmatrix} \boldsymbol{J}_{fx}^{(T)} & \boldsymbol{J}_{fz}^{(T)} & \boldsymbol{0} \\ \boldsymbol{J}_{gx}^{(T)} & \boldsymbol{J}_{gz}^{(T)} & \boldsymbol{J}_{gi}^{(T)} \\ \boldsymbol{0} & \boldsymbol{J}_{iz}^{(T)} & \boldsymbol{J}_{ii}^{(T)} \end{bmatrix} \begin{bmatrix} \Delta\underset{\sim}{x}^{(T)} \\ \Delta\underset{\sim}{z}^{(T)} \\ \Delta\underset{\sim}{i}_{RI}^{(n,\,T)} \end{bmatrix} + \begin{bmatrix} \boldsymbol{0} \\ \boldsymbol{J}_{gv}^{(T)} \\ \boldsymbol{0} \end{bmatrix} \Delta\underset{\sim}{v}_{RI}^{(n,\,T)} + \begin{bmatrix} \boldsymbol{0} \\ \boldsymbol{J}_{gu_c}^{(T)} \\ \boldsymbol{0} \end{bmatrix} \Delta\underset{\sim}{u}_c^{(T)} \quad (4.336)$$

The state, the algebraic, and the control-input variables ($\Delta\underset{\sim}{x}^{(T)}$, $\Delta\underset{\sim}{z}^{(T)}$, $\Delta\underset{\sim}{u}_c^{(T)}$) in the preceding equation combine the corresponding variables from the rectifier and inverter models and the HVDC line/cable model. The current components injected by the TCCX into the AC buses to which the rectifier and inverter AC terminals are connected are $\Delta\underset{\sim}{i}_{RI}^{(n,\,T)}$ and the associated bus voltages are $\Delta\underset{\sim}{v}_{RI}^{(n,\,T)}$. These latter current and voltage quantities are in per-unit of the base values of the network buses to which the converter terminals are connected.

4.3.9 Thyristor Controlled Series Capacitor (TCSC)

A model for the TCSC suitable for small-signal rotor-angle stability analysis is shown in Figure 4.22.

It is assumed that under steady-state conditions the TCSC is represented in the power flow as a series susceptance, $b_0^{(n)}$, between buses *j* and *k* as shown in Figure 4.22(a). The superscript *(n)* denotes that the susceptance is in per-unit on the network base quantities. The steady-state voltages at buses *j* and *k*, $\hat{V}_{j_0} = \hat{V}_0^{(n,\,1)}$ and $\hat{V}_{k_0} = \hat{V}_0^{(n,\,2)}$ respectively, are also obtained from the power flow solution.

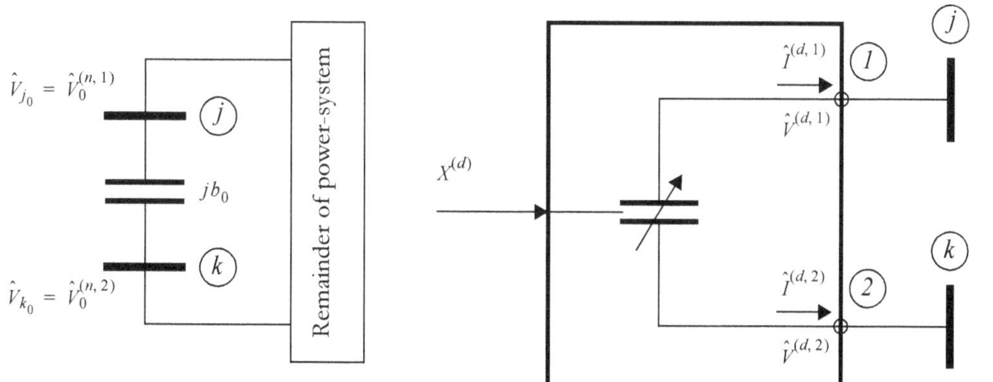

(a) Steady-state representation of TCSC in power-flow. (b) Dynamic representation of TCSC.

Figure 4.22 Representation of the TCSC (a) under steady-state operating conditions in the power flow, and (b) as a dynamic device with controllable series reactance.

Thus in the complex network nodal current equations in (4.337) below the TCSC is represented by the admittance $y_{jk} = y_{kj} = jb_0^{(n)}$ of the branch between buses *j* and *k*.

$$\begin{bmatrix} \hat{I}_1 \\ | \\ \hat{I}_j \\ \hat{I}_k \\ | \end{bmatrix} = \begin{bmatrix} Y_{11} & \cdots & -Y_{1j} & -Y_{1k} & \cdots \\ | \quad | & & | & | & | \\ -Y_{j1} & \cdots & [Y_{j1} + \ldots + Y_{jk} + \ldots] & -Y_{jk} & \cdots \\ -Y_{k1} & \cdots & -Y_{kj} & [Y_{k1} + \ldots + Y_{kj} + \ldots] & \cdots \\ | \quad | & & | & | & | \end{bmatrix} \cdot \begin{bmatrix} \hat{V}_1 \\ | \\ \hat{V}_j \\ \hat{V}_k \\ | \end{bmatrix} \tag{4.337}$$

where $\hat{V}_m$ and $\hat{I}_m$, $m = 1, \ldots, N$, are respectively the voltage and current injected into bus *m*, each in per-unit on the network base quantities.

In order to represent the TCSC in a dynamic model of the system, those terms associated with the steady-state representation of the TCSC are deleted from the nodal admittance matrix. At bus *j*, these are respectively the series element $-y_{jk}$ and term y_{jk} in the self admittance term Y_{jj}. The same approach is adopted at bus *k* for the terms $-y_{kj}$ and y_{kj}. In the dynamic model depicted in Figure 4.22(b) the series branch is replaced by equivalent current injections from terminals *1* and *2* of the TCSC into the network buses *j* and *k*. It is assumed that the control input is the reactance $X^{(d)}$ of the series capacitor. The superscript *(d)* denotes that the quantity is in per-unit on the TCSC base quantities.

The methodology outlined below can be adapted to derive the relationships of greater complexity for the current flows through a series branch whose impedance (or admittance

$y = g + jb$) is to be controlled dynamically; there may also be shunt elements at terminals *1* and *2* in Figure 4.22(a).

It is assumed that the following network base quantities are specified for buses *j* and *k* to which terminals *1* & *2* of the TCSC are connected:

$$V_{usb}^{(n)} = V_{usb}^{(n,1)} = V_{usb}^{(n,2)} \text{ (kV) and } S_{usb}^{(n)} = S_{usb}^{(n,1)} = S_{usb}^{(n,2)} \text{ (MVA)}. \tag{4.338}$$

Correspondingly, it is assumed that the TCSC device base quantities are:

$$V_{usb}^{(d)} = V_{usb}^{(d,1)} = V_{usb}^{(d,2)} \text{ (kV) and } S_{usb}^{(d)} = S_{usb}^{(d,1)} = S_{usb}^{(d,2)} \text{ (MVA)}. \tag{4.339}$$

The base value of admittance in the TCSC per-unit system is $Y_b^{(d)} = S_{usb}^{(d)}/(V_{usb}^{(d)})^2$ ohm and in the network per-unit system it is $Y_b^{(n)} = S_{usb}^{(n)}/(V_{usb}^{(n)})^2$ ohm. Thus, the susceptance in the network and TCSC per-unit systems are related by

$$b^{(n)} = K_Y b^{(d)}, \tag{4.340}$$

in which $K_Y = K_S/K_V^2$ and where K_V and K_S are defined in accordance with (4.189) and (4.190) on page 160. Thus, $b_0^{(d)} = b_0^{(n)}/K_Y$.

Let $\hat{I}^{(d)} = i_R^{(d)} + ji_I^{(d)}$ be the phasor current flow from terminal *2* to *1* of the TCSC and let

$$\hat{V}^{(d)} = v_R^{(d)} + jv_I^{(d)} = \hat{V}^{(d,2)} - \hat{V}^{(d,1)} = (v_R^{(d,2)} - v_R^{(d,1)}) + j(v_I^{(d,2)} - v_I^{(d,1)}) \tag{4.341}$$

be the *voltage difference* between the terminals of the TCSC. The TCSC terminal currents are:

$$\hat{I}^{(d,1)} = i_R^{(d,1)} + ji_I^{(d,1)} = \hat{I}^{(d)} \text{ and } \hat{I}^{(d,2)} = i_R^{(d,2)} + ji_I^{(d,2)} = -\hat{I}^{(d)}. \tag{4.342}$$

The TCSC current/voltage characteristic is $\hat{I}^{(d)} = (jb^{(d)})\hat{V}^{(d)} = -b^{(d)}v_I^{(d)} + jb^{(d)}v_R^{(d)}$ from which it follows that:

$$i_R^{(d)} = -b^{(d)}v_I^{(d)} \text{ and } i_I^{(d)} = b^{(d)}v_R^{(d)}. \tag{4.343}$$

The initial steady-state value of the TCSC voltage difference is $\hat{V}_0^{(d)} = (\hat{V}_0^{(n,2)} - \hat{V}_0^{(n,1)})/K_V$.

The TCSC control input is the series reactance rather than the series susceptance. The two quantities are related by $b^{(d)} = -1/X^{(d)}$ for which the linearized form is:

$$\Delta b^{(d)} = (1/X_0^{(d)})^2 \Delta X^{(d)} = (b_0^{(d)})^2 \Delta X^{(d)}. \tag{4.344}$$

The perturbation in the TCSC current components about their initial steady-state values are found by linearizing (4.343) and substituting for $\Delta b^{(d)}$ from (4.344) to yield:

$$\Delta i_R^{(d)} = -b_0^{(d)}\Delta v_I^{(d)} - v_{I_0}^{(d)}(b_0^{(d)})^2\Delta X^{(d)} \text{ and } \Delta i_I^{(d)} = b_0^{(d)}\Delta v_R^{(d)} + v_{R_0}^{(d)}(b_0^{(d)})^2\Delta X^{(d)}. \quad (4.345)$$

The linearized form of the relationships between the TCSC current $\Delta \underset{\sim}{i}_{RI}^{(d)} = \left[\Delta i_R^{(d)} \ \Delta i_I^{(d)}\right]^T$ and its terminal currents are found from (4.342) to be,

$$\Delta \underset{\sim}{i}_{RI}^{(d,1)} = \Delta \underset{\sim}{i}_{RI}^{(d)} \text{ and } \Delta \underset{\sim}{i}_{RI}^{(d,2)} = -\Delta \underset{\sim}{i}_{RI}^{(d)} \quad (4.346)$$

The linearized form of the equations describing the interconnection between the device terminals and the network buses to which they are connected are:

$$\Delta \underset{\sim}{i}_{RI}^{(n,1)} = K_I \Delta \underset{\sim}{i}_{RI}^{(d,1)}, \quad \Delta \underset{\sim}{i}_{RI}^{(n,2)} = K_I \Delta \underset{\sim}{i}_{RI}^{(d,2)} \text{ for the current and} \quad (4.347)$$

$$\Delta \underset{\sim}{v}_{RI}^{(n,1)} = K_V \Delta \underset{\sim}{v}_{RI}^{(d,1)}, \quad \Delta \underset{\sim}{v}_{RI}^{(n,2)} = K_V \Delta \underset{\sim}{v}_{RI}^{(d,2)} \text{ for the voltages.} \quad (4.348)$$

The linearized equations (4.345) to (4.348) are sufficient to represent the TCSC. However, it is also desirable to provide supplementary output equations for perturbations in the following quantities at one or both of the TCSC terminals:

- the voltage magnitude and angle by application of (4.195) and (4.196) on page 161;
- the real and reactive power flow by application of (4.204) on page 163;
- the magnitude of the current flow by application of (4.202) on page 162.

The TCSC linearized model equations (4.345) to (4.348) together with the supplementary output equations can be written in a form suitable for integration with the linearized DAEs of the power system as described in Section 4.4.

4.4 Linearized power system model

A fundamental assumption in this work is that during steady-state operation all generators and loads produce balanced three-phase fundamental-frequency voltages and currents. It is assumed the interconnecting AC network is three-phase and balanced. Consequently, for steady-state modelling purposes, a fundamental-frequency positive phase sequence representation of the generators, loads and AC transmission system is employed. In this representation the balanced three-phase system voltages and currents are represented by fundamental-frequency positive phase sequence stationary complex phasors. For the purpose of small-disturbance modelling the stationary assumption is relaxed to permit small perturbations in the current and voltage phasor magnitudes and phases. The resulting concept of quasi-stationary, fundamental-frequency, positive phase sequence phasors [34] forms the basis for rotor-angle small-signal stability analysis of power systems in this section.

A high-level mathematical description of how the linearized differential and algebraic equations (DAEs) of the interconnected system are assembled from (i) the DAEs of each of the devices and their associated control systems, and (ii) the network nodal admittance equations which are provided below. The resulting equations have a modular and extremely sparse structure which must be exploited to ensure computationally efficient analysis of large power systems.

4.4.1 General form of the linearized DAEs for a device and its controls

It is assumed that the ith device has an ordered list of n_{t_i} AC terminals which are connected to a corresponding list of network buses $b_i(k)$, $k = 1, \ldots, n_{t_i}$. It is also assumed that the device and its controls do not have inputs from any other device. (Note that the equations of devices that are interconnected through their control systems are combined to form a single super-device). The ith device and its associated controls is represented by a set of linearized DAEs of the following form:

$$
\begin{aligned}
p\Delta \underset{\sim}{x}_i &= \boldsymbol{J}_{fx}^{(i)}\Delta \underset{\sim}{x}_i + \boldsymbol{J}_{fz}^{(i)}\Delta \underset{\sim}{z}_i + \sum_{l=1}^{n_{t_i}} \left(\boldsymbol{J}_{fi}^{(i,\, b_i(l))}\Delta \underset{\sim}{i}_{RI}^{(i,\, b_i(l))} + \boldsymbol{J}_{fv}^{(i,\, b_i(l))}\Delta \underset{\sim}{v}_{RI}^{(b_i(l))}\right) + \boldsymbol{J}_{fu}^{(i)}\Delta \underset{\sim}{u}_i \\
\boldsymbol{0} &= \boldsymbol{J}_{gx}^{(i)}\Delta \underset{\sim}{x}_i + \boldsymbol{J}_{gz}^{(i)}\Delta \underset{\sim}{z}_i + \sum_{l=1}^{n_{t_i}} \left(\boldsymbol{J}_{gi}^{(i,\, b_i(l))}\Delta \underset{\sim}{i}_{RI}^{(i,\, b_i(l))} + \boldsymbol{J}_{gv}^{(i,\, b_i(l))}\Delta \underset{\sim}{v}_{RI}^{(b_i(l))}\right) + \boldsymbol{J}_{gu}^{(i)}\Delta \underset{\sim}{u}_i \\
\boldsymbol{0} &= \boldsymbol{J}_{ix}^{(i,\, b_i(m))}\Delta \underset{\sim}{x}_i + \boldsymbol{J}_{iz}^{(i,\, b_i(m))}\Delta \underset{\sim}{z}_i + \ldots \\
&\quad \sum_{l=1}^{n_{t_i}} \left(\boldsymbol{J}_{ii}^{(i,\, b_i(m),\, b_i(l))}\Delta \underset{\sim}{i}_{RI}^{(i,\, b_i(l))} + \boldsymbol{J}_{iv}^{(i,\, b_i(m),\, b_i(l))}\Delta \underset{\sim}{v}_{RI}^{(b_i(l))}\right) + \ldots \\
&\quad \boldsymbol{J}_{iu}^{(i,\, b_i(m))}\Delta \underset{\sim}{u}_i \quad \text{for } m = 1, \ldots, n_{t_i}
\end{aligned}
\tag{4.349}
$$

in which $\Delta \underset{\sim}{x}_i$, $\Delta \underset{\sim}{z}_i$ and $\Delta \underset{\sim}{u}_i$ are respectively vectors of n_{x_i} states, n_{z_i} algebraic variables and n_{u_i} external input-variables, associated with the ith device. Moreover, $\Delta \underset{\sim}{i}_{RI}^{(i,\, b_i(l))} = \left[\Delta i_R^{(i,\, b_i(l))} \quad \Delta i_I^{(i,\, b_i(l))}\right]^T$ are the real and imaginary components of the current injected by the lth terminal of the ith device into bus $b_i(l)$, $l = 1, \ldots, n_{t_i}$ and $\Delta \underset{\sim}{v}_{RI}^{(b_i(l))} = \left[\Delta v_R^{(b_i(l))} \quad \Delta v_I^{(b_i(l))}\right]^T$ are the real and imaginary components of the voltage at bus

$b_i(l)$, $l = 1, \ldots, n_{t_i}$. The voltage and current components are in the synchronously-rotating network frame of reference, and are each in per-unit on their respective network base quantities. The constant coefficient matrices $\boldsymbol{J}_{fx}^{(i)}$, $\boldsymbol{J}_{fz}^{(i)}$, etc. are typically sparse and depending on the device some of the coefficient matrices may be zero.

4.4.2 General form of the network nodal current equations

It is assumed that bus k is connected to a list d_k of n_{d_k} current injecting dynamic devices and to a list c_k of n_{c_k} immediately adjacent buses through series admittance elements. Each element of d_k contains the identifier of both the device and the terminal within the device which is connected to the bus. Most buses in large sparse networks do not have any dynamic devices connected to them and so in most cases d_k is empty. Applying Kirchoff's Current Law to bus k results in the following nodal current equations, one for the real component of the current and the other for the imaginary component.

$$0 = \boldsymbol{Y}_{kk}\Delta \underset{\sim}{v}_{RI}^{(k)} + \sum_{l=1}^{n_{c_k}} \boldsymbol{Y}_{kl}\Delta \underset{\sim}{v}_{RI}^{(c_k(l))} - \sum_{i=1}^{n_{d_k}} \Delta \underset{\sim}{i}_{RI}^{(d_k(i),\,k)} , \tag{4.350}$$

$$\text{in which } \boldsymbol{Y}_{kk} = \begin{bmatrix} G_{kk} & -B_{kk} \\ B_{kk} & G_{kk} \end{bmatrix} \text{ and } \boldsymbol{Y}_{kl} = -\begin{bmatrix} G_{kl} & -B_{kl} \\ B_{kl} & G_{kl} \end{bmatrix} \tag{4.351}$$

correspond respectively to (i) the sum of all admittance elements connected to bus k, and (ii) the negated total series admittance between buses k and l (i.e. admittances of parallel branches between two nodes are summed). These equations are sparse in the sense that they involve the voltages at a very small subset of the buses in the network.

The network nodal current equations for all of the buses $k = 1, \ldots, n_b$ are now expressed in the following matrix form in which the buses connected to dynamic devices are partitioned from the internal passive buses. It is emphasised that computationally sparse matrix storage and analysis methods are used in which only the non-zero admittance blocks in each nodal current equation are stored and analysed. Furthermore, for many purposes distinguishing between the dynamic and passive buses is unnecessary. The network nodal current equations are:

$$\begin{bmatrix} \boldsymbol{0} \\ \boldsymbol{0} \end{bmatrix} = \begin{bmatrix} \boldsymbol{Y}_{dd} & \boldsymbol{Y}_{dp} \\ \boldsymbol{Y}_{pd} & \boldsymbol{Y}_{pp} \end{bmatrix} \begin{bmatrix} \Delta \underset{\sim}{v}_{RI}^{(d)} \\ \Delta \underset{\sim}{v}_{RI}^{(p)} \end{bmatrix} - \begin{bmatrix} \boldsymbol{J}_{di} \\ \boldsymbol{0} \end{bmatrix} \Delta \underset{\sim}{i}_{RI}, \tag{4.352}$$

in which:

$$\Delta \underset{\sim}{v}_{RI}^{(d)} = \begin{bmatrix} \Delta \underset{\sim}{v}_{RI}^{(1)} \\ \dots \\ \Delta \underset{\sim}{v}_{RI}^{(n_{db})} \end{bmatrix} \text{ and } \Delta \underset{\sim}{v}_{RI}^{(p)}$$

are respectively vectors of voltage components (i) of the n_{db} buses to which dynamic devices *(d)* are connected; and (ii) the remaining set of passive buses *(p)*. Furthermore,

$$\Delta \underset{\sim}{i}_{RI} = \begin{bmatrix} \Delta \underset{\sim}{i}_{RI}^{(1)} \\ \dots \\ \Delta \underset{\sim}{i}_{RI}^{(n_d)} \end{bmatrix}$$

is the vector of the current components injected by the n_d dynamic devices into the buses to which they are connected. For the ith device the current injection vector is composed of n_{t_i} elements, one for each of its AC terminals, such that:

$$\Delta \underset{\sim}{i}_{RI}^{(i)} = \begin{bmatrix} \Delta \underset{\sim}{i}_{RI}^{(i,\, b_i(1))} \\ \dots \\ \Delta \underset{\sim}{i}_{RI}^{(i,\, b_i(n_{t_i}))} \end{bmatrix}. \tag{4.353}$$

4.4.3 General form of the linearized DAEs of the interconnected power system

The equations for each dynamic device with the form in (4.349) are interconnected through the network nodal current equations (4.352) to yield the following general form of the linearized DAEs of the interconnected power system.

$$\begin{bmatrix} p\Delta \underset{\sim}{x} \\ 0 \\ 0 \\ 0 \\ 0 \\ 0 \end{bmatrix} = \begin{bmatrix} J_{fx} & J_{fz} & J_{fi} & J_{fv} & 0 \\ J_{gx} & J_{gz} & J_{gi} & J_{gv} & 0 \\ J_{ix} & J_{iz} & J_{ii} & J_{iv} & 0 \\ 0 & 0 & -J_{di} & Y_{dd} & Y_{dp} \\ 0 & 0 & 0 & Y_{pd} & Y_{pp} \end{bmatrix} \begin{bmatrix} \Delta \underset{\sim}{x} \\ \Delta \underset{\sim}{z} \\ \Delta \underset{\sim}{i}_{RI} \\ \Delta \underset{\sim}{v}_{RI}^{(d)} \\ \Delta \underset{\sim}{v}_{RI}^{(p)} \end{bmatrix} + \begin{bmatrix} J_{fu} \\ J_{gu} \\ J_{iu} \\ 0 \\ 0 \\ 0 \end{bmatrix} \Delta \underset{\sim}{u} \tag{4.354}$$

In equation (4.354) $\Delta \underset{\sim}{x} = \begin{bmatrix} \Delta \underset{\sim}{x}_1 \\ \dots \\ \Delta \underset{\sim}{x}_{n_d} \end{bmatrix}$, $\Delta \underset{\sim}{z} = \begin{bmatrix} \Delta \underset{\sim}{z}_1 \\ \dots \\ \Delta \underset{\sim}{z}_{n_d} \end{bmatrix}$, $\Delta \underset{\sim}{u} = \begin{bmatrix} \Delta \underset{\sim}{u}_1 \\ \dots \\ \Delta \underset{\sim}{u}_{n_d} \end{bmatrix}$ are respectively the system state-variables, the internal algebraic-variables and the external system input variables of all the dynamic devices connected to the system. The coefficient matrix $\boldsymbol{J}_{fx}$ has the following block diagonal structure:

$$\boldsymbol{J}_{fx} = \mathfrak{D}(\boldsymbol{J}_{fx}^{(1)}, \ldots, \boldsymbol{J}_{fx}^{(n_d)}), \tag{4.355}$$

and the coefficient matrices $\boldsymbol{J}_{fz}$, $\boldsymbol{J}_{fu}$, $\boldsymbol{J}_{gx}$, $\boldsymbol{J}_{gz}$ and $\boldsymbol{J}_{gu}$ similarly have a block diagonal structure. The structure of the coefficient matrices associated with the current and voltage components such as $\boldsymbol{J}_{fi}$, $\boldsymbol{J}_{gv}$ and $\boldsymbol{J}_{iz}$ depend on the relative ordering of the dynamic devices and the buses to which they are connected. In any event, for large systems, these matrices are also very sparse.

4.4.4 Example demonstrating the structure of the linearized DAEs

For illustrative purposes the structure of the linearized DAEs of the small power system in Figure 4.23 is given in Figure 4.24. The system model in Figure 4.23 incorporates a device (1) with two terminals connected to two buses (1 and 2); buses 1 and 2 are each connected to two devices (1 and 2) and (1 and 3) respectively; and finally the bus (3) is connected to one dynamic device (4). This interconnection of dynamic devices and buses represents the range of possibilities usually encountered in practice.

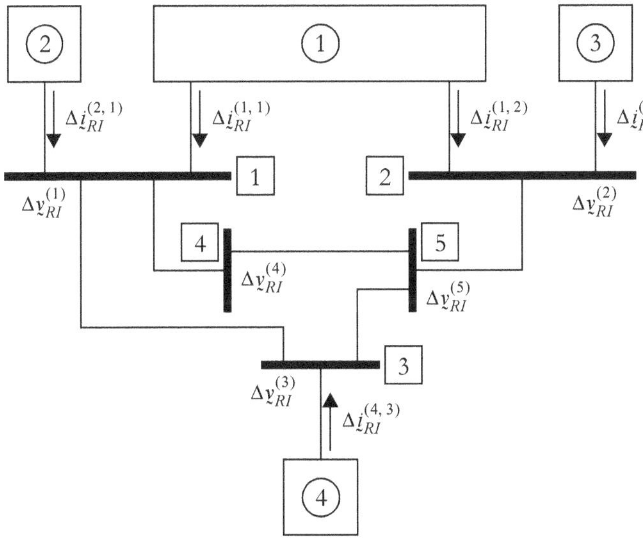

Figure 4.23 Example network to illustrate the modular and sparse structure of the linearized DAEs of the interconnected power system.

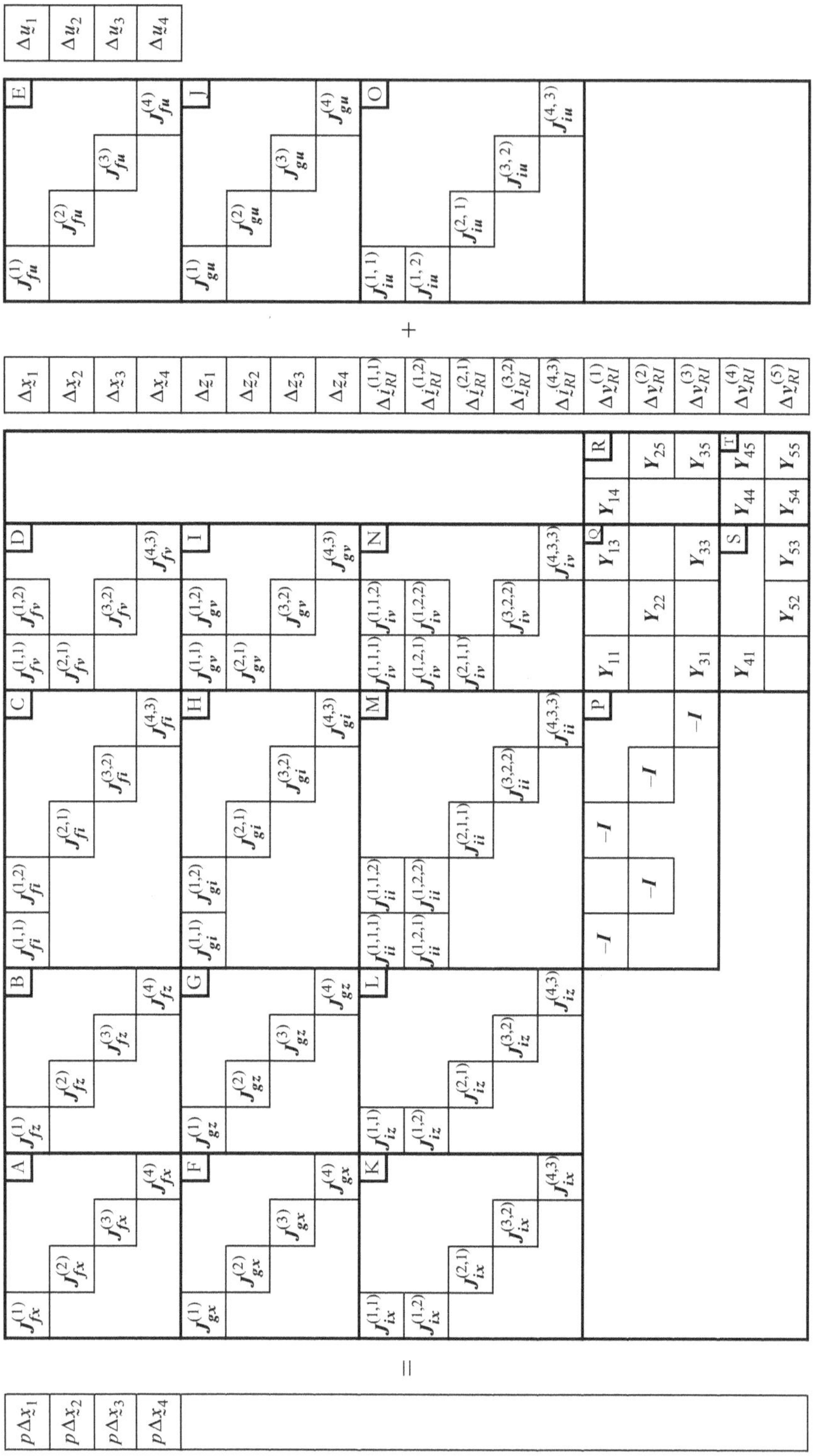

Figure 4.24 Structure of the linearized DAEs of the system in Figure 4.23.

In Figure 4.23 blank space indicates the corresponding matrix entries are zeros. The matrix block labelled 'A' corresponds to the sub-matrix $\boldsymbol{J}_{fx}$ in equation (4.354), block 'B' corresponds to $\boldsymbol{J}_{fz}$, etc.

4.5 Load models

4.5.1 Types of load models

In large scale power system stability studies loads are typically aggregated at bulk supply substations. In rotor-angle stability studies a static representation of loads is commonly employed, as reported in a recent international survey [37]. In such a representation the real and reactive power consumed by the load at a point in time is dependent on the voltage and frequency at the same instant [35, 36]. Composite load models, such as those described in [38], which incorporate an equivalent representation of the distribution network as well as a variety of dynamic and static load model components are not considered in this book. The following is a general representation of a static load model.

$$
\begin{aligned}
P &= \frac{P_0}{a_s}\left[a_1\left(\frac{V}{V_0}\right)^{m_{p1}} + a_2\left(\frac{V}{V_0}\right)^{m_{p2}} + a_3\left(\frac{V}{V_0}\right)^{m_{p3}}\right](1 + a_f(f-f_0)) \\
Q &= \frac{Q_0}{b_s}\left[b_1\left(\frac{V}{V_0}\right)^{m_{q1}} + b_2\left(\frac{V}{V_0}\right)^{m_{q2}} + b_3\left(\frac{V}{V_0}\right)^{m_{q3}}\right](1 + b_f(f-f_0))
\end{aligned}
\tag{4.356}
$$

in which P and Q are the real and reactive power consumed by the load in per-unit on the base MVA of the system; V is the terminal voltage of the load in per-unit of the base voltage of the bus to which the load is connected; and f is the frequency of the load bus voltage in per-unit of the system base frequency. P_0, Q_0, V_0 and $f_0 = 1$ are the corresponding initial steady-state values. The load model parameters are a_i, m_{pi}, b_i, m_{qi} for $i = 1, \ldots, 3$, a_f and b_f. Furthermore, in (4.356)

$$
a_s = \sum_{i=1}^{3} a_i \text{ and } b_s = \sum_{i=1}^{3} b_i . \tag{4.357}
$$

The formulation in (4.356) can be used to represent a range of commonly employed load models such as the ZIP (composite constant impedance, current and power) representation and the exponential load model.

Three basic types of static loads can be represented by the following:

1. A constant impedance load: $a_1 = 1$ *and* $a_2 = a_3 = 0$, $m_{p1} = 2$; $b_1 = 1$ *and* $b_2 = b_3 = 0$, $m_{q1} = 2$, $a_f = b_f = 0$;

2. A constant current load: $a_1 = a_3 = 0$ *and* $a_2 = 1$, $m_{p2} = 1$; $b_1 = b_3 = 0$ *and* $b_2 = 1$, $m_{q2} = 1$, $a_f = b_f = 0$;

3. A constant power load: $a_1 = a_2 = 0$ *and* $a_3 = 1$, $m_{p3} = 0$; $b_1 = b_2 = 0$ *and* $b_3 = 1$, $m_{q3} = 0$, $a_f = b_f = 0$.

Note that the load can be any linear combination of the above and the type of the real and reactive parts may differ. For example, a load may have a constant-current real component and a constant-impedance reactive component; this combination is the most commonly used static load model in the absence of any further information or measurement on the load characteristics of the system under study [35]. The frequency dependence of loads is not modelled.

Work, such as [38, 40], shows that loads may have a significant impact on the damping of rotor modes and therefore accurate dynamic modelling, whenever possible, of loads is highly desirable in both transient and small-signal stability studies.

4.5.2 Linearized load models

For the purposes of small signal analysis, equations (4.356) are readily linearized about the steady-state operating point to yield:

$$\Delta P = C_{pv}\Delta V + C_{pf}\Delta f \quad \text{and} \quad \Delta Q = C_{qv}\Delta V + C_{qf}\Delta f, \tag{4.358}$$

$$\text{in which } C_{pv} = \left(\frac{P_0}{V_0}\right)n_p, \quad C_{pf} = P_0 a_f, \text{ and } n_p = \left(\frac{1}{a_s}\right)\sum_{i=1}^{3}(a_i m_{pi}) \tag{4.359}$$

and C_{qv}, n_q and C_{qf} are calculated analogously.

The above small-signal model of a load is connected to the network by relating the real and reactive power, voltage magnitude and bus voltage frequency to the voltage components, $\underset{\sim}{v}_{RI}$, of the bus to which the load is connected and the current components, $\underset{\sim}{i}_{RI}$, injected by the load into the bus. The reference for the voltage and current components is the synchronously rotating network frame of reference; the voltage and current are each in per-unit on their respective network base quantities. It is assumed that the initial values of the real and reactive power, P_0 and Q_0, and the voltage magnitude and angle, V_0 and θ_0 are given by the power flow solution. From these initial values the following are derived:

$$v_{R_0} = V_0\cos(\theta_0), \quad v_{I_0} = V_0\sin(\theta_0). \tag{4.360}$$

The voltage magnitude is $V = \sqrt{v_R^2 + v_I^2}$ which is linearized about the steady-state operating point to yield:

$$\Delta V = \left(\frac{v_{R_0}}{V_0}\right)\Delta v_R + \left(\frac{v_{I_0}}{V_0}\right)\Delta v_I. \tag{4.361}$$

The perturbation in the frequency of the load bus frequency is given by:

$$\Delta f = \left(\frac{1}{\omega_b}\right)\frac{d\Delta\theta}{dt}, \tag{4.362}$$

in which ω_b is the base value of system frequency in rad/s and $\Delta\theta$ is the perturbation in the bus voltage angle in radians and t is time in seconds. It is important to recognise that $\Delta\theta$ is not a state-variable and therefore that rate-of-change of bus voltage angle is approximated by means of a highpass (i.e. washout) filter with a very short time constant, for which $T_f = \alpha_f\left(\frac{2\pi}{\omega_b}\right)$ with $\alpha_f = 0.1$ is a reasonable choice[1]. Thus, in the s-domain the bus frequency perturbation is approximated by:

$$\Delta f(s) = \left(\frac{1}{\omega_b}\right)\frac{s}{1+sT_f}\Delta\theta(s). \tag{4.363}$$

Transformation of the preceding equation to the time-domain results in the following state- and algebraic equation:

$$\begin{aligned} p\Delta x_f &= -\left(\frac{1}{T_f}\right)\Delta x_f - \left(\frac{1}{\omega_b T_f^2}\right)\Delta\theta \\ \Delta f &= \Delta x_f + \left(\frac{1}{\omega_b T_f}\right)\Delta\theta \end{aligned} \tag{4.364}$$

The bus voltage angle is $\theta = \operatorname{atan2}(v_I, v_R)$ which upon linearization yields the following expression for $\Delta\theta$:

$$\Delta\theta = -\left(\frac{v_{I_0}}{V_0^2}\right)\Delta v_R + \left(\frac{v_{R_0}}{V_0^2}\right)\Delta v_I. \tag{4.365}$$

The current injected by the load into the bus to which it is connected is given by:

$$(i_R + ji_I) = -\left(\frac{P+jQ}{v_R+jv_I}\right)^* = -\left(\frac{1}{V^2}\right)((Pv_R + Qv_I) + j(Pv_I - Qv_R)). \tag{4.366}$$

By linearizing the real and imaginary components of the preceding equation about the initial steady-state operating point the following equation for the perturbation of the current components injected by the load into the network are obtained:

1. Some time domain analysis programs utilize integration algorithms which require integration time-steps to be shorter than the shortest model time-constant. For use in such programs a larger value of α_f may be required.

$$\Delta \underset{\sim}{i}_{RI} = -\boldsymbol{J}_{iz}\Delta \underset{\sim}{z} - \boldsymbol{J}_{iv}\Delta \underset{\sim}{v}_{RI}, \tag{4.367}$$

$$\text{in which } \Delta \underset{\sim}{z} = \left[\Delta P\ \Delta Q\ \Delta V\ \Delta f\ \Delta\theta\right]^T, \tag{4.368}$$

$$\boldsymbol{J}_{iz} = \left(\frac{1}{V_0^2}\right)\begin{bmatrix} v_{R_0} & v_{I_0} & 2V_0 i_{R_0} & 0 & 0 \\ v_{I_0} & -v_{R_0} & 2V_0 i_{I_0} & 0 & 0 \end{bmatrix} \text{ and } \boldsymbol{J}_{iv} = \left(\frac{1}{V_0^2}\right)\begin{bmatrix} P_0 & Q_0 \\ -Q_0 & P_0 \end{bmatrix}. \tag{4.369}$$

The initial steady-state values of the current components are obtained from (4.366). The linearized state- and algebraic-equations (4.358), (4.361), (4.364), (4.365) and (4.367) representing the load are now amalgamated into the following matrix equation which is in the general form of the device equations in (4.349) on page 196.

$$\begin{bmatrix} p\Delta x_f \\ \boldsymbol{0} \\ \boldsymbol{0} \end{bmatrix} = \begin{bmatrix} \boldsymbol{J}_{fx} & \boldsymbol{J}_{fz} & \boldsymbol{0} & \boldsymbol{0} \\ \boldsymbol{J}_{gx} & \boldsymbol{J}_{gz} & \boldsymbol{0} & \boldsymbol{J}_{gv} \\ \boldsymbol{0} & -\boldsymbol{J}_{iz} & -\boldsymbol{I} & -\boldsymbol{J}_{iv} \end{bmatrix}\begin{bmatrix} \Delta x_f \\ \Delta \underset{\sim}{z} \\ \Delta \underset{\sim}{i}_{RI} \\ \Delta \underset{\sim}{v}_{RI} \end{bmatrix} \tag{4.370}$$

$$\text{in which } \boldsymbol{J}_{fx} = -\left(\frac{1}{T_f}\right), \quad \boldsymbol{J}_{fz} = \begin{bmatrix} 0 & 0 & 0 & 0 & -\left(\frac{1}{\omega_b T_f^2}\right) \end{bmatrix}, \quad \boldsymbol{J}_{gx} = \begin{bmatrix} 0 & 0 & 0 & 1 & 0 \end{bmatrix}^T,$$

$$\boldsymbol{J}_{gz} = \begin{bmatrix} -1 & 0 & C_{pv} & C_{pf} & 0 \\ 0 & -1 & C_{qv} & C_{qf} & 0 \\ 0 & 0 & -1 & 0 & 0 \\ 0 & 0 & 0 & -1 & \left(\frac{1}{\omega_b T_f}\right) \\ 0 & 0 & 0 & 0 & -1 \end{bmatrix}, \text{ and } \boldsymbol{J}_{gv} = \begin{bmatrix} 0 & 0 \\ 0 & 0 \\ \left(\frac{v_{R_0}}{V_0}\right) & \left(\frac{v_{I_0}}{V_0}\right) \\ 0 & 0 \\ -\left(\frac{v_{I_0}}{V_0^2}\right) & \left(\frac{v_{R_0}}{V_0^2}\right) \end{bmatrix}; \tag{4.371}$$

$\boldsymbol{J}_{iz}$ and $\boldsymbol{J}_{iv}$ are defined in (4.369).

If the frequency dependence of loads is neglected then equation (4.370) is simplified by omitting the state-variable Δx_f, the algebraic-variables Δf and $\Delta\theta$, and the associated equations. In this case it is also possible to omit the explicit equations for the load and instead implicitly include the effect of the voltage dependence of the load by modifying the self-admittance of the bus to which the load is connected as follows. It can be shown that the perturbation of the current injected by the load is given by:

$$\Delta \underset{\sim}{i}_{RI} = -\boldsymbol{Y}_L \Delta \underset{\sim}{v}_{RI}. \tag{4.372}$$

$$\text{in which } \boldsymbol{Y}_L = \left(\frac{1}{V_0^2}\right)\begin{bmatrix} \left(P_0 + c_R\left(\frac{v_{R_0}}{V_0}\right)\right) & \left(Q_0 + c_R\left(\frac{v_{I_0}}{V_0}\right)\right) \\ -\left(Q_0 - c_I\left(\frac{v_{R_0}}{V_0}\right)\right) & \left(P_0 + c_I\left(\frac{v_{I_0}}{V_0}\right)\right) \end{bmatrix} \text{ where} \tag{4.373}$$

$$c_R = P_0\left(\frac{v_{R_0}}{V_0}\right)(n_p - 2) + Q_0\left(\frac{v_{I_0}}{V_0}\right)(n_q - 2) \text{ and}$$

$$c_I = P_0\left(\frac{v_{I_0}}{V_0}\right)(n_p - 2) - Q_0\left(\frac{v_{R_0}}{V_0}\right)(n_q - 2)\,. \tag{4.374}$$

4.6 References

[1] Siemens PTI PSS®E, developed and distributed by "Siemens Power Technologies International (PTI)", See URL (correct at 4 March 2015): http://w3.siemens.com/smartgrid/global/en/products-systems-solutions/software-solutions/planning-data-management-software/pti/pages/power-technologies-international-(pti).aspx

[2] GE PSLF™, developed and distributed by "General Electric Company", See URL (correct at 4 March 2015): http://www.geenergyconsulting.com/practice-area/software-products/pslf

[3] W. G. Heffron and R. A. Phillips, "Effect of a Modern Amplidyne Voltage Regulator on Underexcited Operation of Large Turbine Generators [includes discussion]," *Power Apparatus and Systems, part iii. Transactions of the American Institute of Electrical Engineers,* vol. 71, 1952.

[4] F. P. de Mello and C. Concordia, "Concepts of Synchronous Machine Stability as Affected by Excitation Control," *Power Apparatus and Systems, IEEE Transactions on,* vol. PAS-88, pp. 316-329, 1969.

[5] A. E. Blondel, C. O. Mailloux and C. A. Adams, *Synchronous Motors and Converters: Theory and Methods of Calculation and Testing.* McGraw-Hill Book Company, 1913.

[6] R. H. Park, "Two-reaction theory of synchronous machines generalized method of analysis-part I," *American Institute of Electrical Engineers, Transactions of the*, vol. 48, pp. 716-727, 1929.

[7] P. C. Krause, F. Nozari, T. L. Skvarenina and D. W. Olive, "The Theory of Neglecting Stator Transients," *Power Apparatus and Systems, IEEE Transactions on,* vol. PAS-98, pp. 141-148, 1979.

[8] A. W. Rankin, "Per-unit impedances of synchronous machines," *American Institute of Electrical Engineers, Transactions of the*, vol. 64, pp. 569-573, 1945.

[9] A. W. Rankin, "Per-Unit Impedances of Synchronous Machines - II," *American Institute of Electrical Engineers, Transactions of the*, vol. 64, pp. 839-841, 1945.

[10] P. Kundur, *Power system stability and control.* New York: McGraw-Hill, 1994.

[11] "IEEE Recommended Practice for Excitation System Models for Power System Stability Studies," *IEEE Std 421.5-2005 (Revision of IEEE Std 421.5-1992),* pp. 1-85, 2006.

[12] D.J. Vowles, M.J. Gibbard, *Small-Signal Modelling for the Analysis of Rotor-Angle Stability and Control of Large Power Systems*, AUPress In preparation.

[13] *Determination of synchronous machine stability study constants,* vol. 1-3, Electric Power Research Institute, 1980.

[14] IEEE Guide: Test Procedures for Synchronous Machines Part I-- Acceptance and Performance Testing Part II-Test Procedures and Parameter Determination for Dynamic Analysis," *IEEE Std 115-2010.*

[15] IEEE Guide for Synchronous Generator Modeling Practices and Applications in Power System Stability Analyses, *IEEE Std 1110-2002 (Revision of IEEE Std 1110-1991)*, pp. 1-72, 2003.

[16] F. P. de Mello and L. H. Hannett, "Validation of Synchronous Machine Models and Derivation of Model Parameters from Tests," *Power Apparatus and Systems, IEEE Transactions on,* vol. PAS-100, pp. 662-672, 1981.

[17] I. M. Canay, "Causes of Discrepancies on Calculation of Rotor Quantities and Exact Equivalent Diagrams of the Synchronous Machine," *Power Apparatus and Systems, IEEE Transactions on,* vol. PAS-88, pp. 1114-1120, 1969.

[18] I. M. Canay, "Modelling of alternating-current machines having multiple rotor circuits," *Energy Conversion, IEEE Transactions on*, vol. 8, pp. 280-296, 1993.

[19] G. Shackshaft and P. B. Henser, et. al. "Model of generator saturation for use in power-system studies," *Electrical Engineers, Proceedings of the Institution of,* vol. 126, pp. 759-763, 1979.

[20] C. Concordia, *Synchronous Machines: Theory and Performance.* John Wiley & Sons, Inc., New York, 1951.

[21] B. Adkins and R. G. Harley, *The General Theory of Alternating Current Machines: Applications to Practical Problems.* Chapman and Hall, Ltd., London, 1975.

[22] S. D. Umans, J. A. Mallick and G. L. Wilson, "Modeling of Solid Rotor Turbogenerators Part I: Theory and Techniques," *Power Apparatus and Systems, IEEE Transactions on,* vol. PAS-97, pp. 269-277, 1978.

[23] J. A. Mallick, G. L. Wilson and S. D. Umans, "Modeling of Solid Rotor Turbogenerators Part II: Example of Model Derivation and Use in Digital Simulation," *Power Apparatus and Systems, IEEE Transactions on,* vol. PAS-97, pp. 278-291, 1978.

[24] Website for the electronic version of this book: http://www.adelaide.edu.au/press/titles/small-signal/

[25] N. G. Hingorani and L. Gyugyi, *Understanding FACTS: Concepts and Technology of Flexible AC Transmission Systems*, IEEE Press, New York, 2000.

[26] X. P. Zhang, C. Rehtanz, and B. Pal, *Flexible AC Transmission Systems: Modelling and Control,* Springer-Verlag, 2012.

[27] Y. H. Song and A. T. Johns, Editors, *Flexible ac transmission systems (FACTS)*, The Institution of Electrical Engineers, Stevenage, UK, 1999.

[28] S. Arabi and P. Kundur, "A versatile FACTS device model for powerflow and stability simulations", *Power Systems, IEEE Transactions on*, vol. 11, pp. 1944-1950, 1996.

[29] E.W. Kimbark, *Direct Current Transmission*, Wiley-Interscience, New York, 1971.

[30] E. Uhlmann, *Power Transmission by Direct Current*, Springer-Verlag, Berlin/Heidelberg, 1975.

[31] J. Arrillaga, *High Voltage Direct Current Transmission*, 2nd Edition The Institution of Electrical Engineers, Stevenage, UK, 1998.

[32] J. Arrillaga and B. Smith, *AC-DC Power System Analysis*, The Institution of Electrical Engineers, Stevenage, UK, 1998.

[33] S. Arabi, G. J. Rogers, D. Y. Wong, P. Kundur, and M. G. Lauby, "Small signal stability program analysis of SVC and HVDC in AC power systems," *Power Systems, IEEE Transactions on*, vol. 6, pp. 1147-1153, 1991.

[34] M. Ilić and J. Zaborszky, *Dynamics and Control of Large Electric Power Systems*, John Wiley and Sons, Inc., 2000.

[35] IEEE Task Force on Load Representation for Dynamic Performance, "Load Representation for Dynamic Performance Analysis," *Power Systems, IEEE Transactions on*, vol. 8, pp. 472-482, 1993.

[36] IEEE Task Force on Load Representation for Dynamic Performance, "Standard Models for Power Flow and Dynamic Performance Simulation," *Power Systems, IEEE Transactions on*, vol. 10, pp. 1302-1313, 1995.

[37] J. V. Milanovic, K. Yamashita, S. Martinez Villanueva, S. Z. Djokic, and L. M. Korunovic, "International Industry Practice on Power System Load Modeling," *Power Systems, IEEE Transactions on*, vol. 28, pp. 3038-3046, 2013.

[38] D. Kosterev, A. Meklin, J. Undrill, B. Lesieutre, W. Price, D. Chassin, R. Bravo, and S. Yang, "Load modeling in power system studies: WECC progress update," in *Power and Energy Society General Meeting - Conversion and Delivery of Electrical Energy in the 21st Century, 2008 IEEE*, 2008, pp. 1-8.

[39] "Load representation for dynamic performance analysis [of power systems]," *Power Systems, IEEE Transactions on,* vol. 8, pp. 472-482, 1993.

[40] I. A. Hiskens and J. V. Milanovic, "Load modelling in studies of power system damping," *Power Systems, IEEE Transactions on,* vol. 10, pp. 1781-1788, 1995.

Appendix 4–I Linearization of the classical parameter model of the generator.

The objective is to develop the linearized equations for the generator model in terms of the classically-defined standard parameters in the form described in Section 4.2.13.4. The formulation proceeds assuming that there are three rotor windings in each axis after which it is demonstrated that the equations are readily reformulated for models with fewer rotor windings.

The rotor d- and q-axis state-variables are respectively:

$$\underset{\sim}{\varphi}_{cd} = \begin{bmatrix} \varphi_d' & \varphi_{1d} & \varphi_{2d} \end{bmatrix}^T \text{ and } \underset{\sim}{\varphi}_{cq} = \begin{bmatrix} \varphi_q' & \varphi_{1q} & \varphi_{2q} \end{bmatrix}^T \tag{4.375}$$

The algebraic-variables associated with the d-axis rotor windings are grouped as follows:

$$\underset{\sim}{z}_{fd} = I_{fd},\ \underset{\sim}{z}_{1d} = \begin{bmatrix} i_{1d} & \varphi_{s1d} & \varphi_d'' \end{bmatrix}^T \text{ and } \underset{\sim}{z}_{2d} = \begin{bmatrix} i_{2d} & \varphi_{s2d} & \varphi_d''' \end{bmatrix}^T. \tag{4.376}$$

Similarly for the q-axis rotor windings:

$$\underset{\sim}{z}_{1q} = I_{1q},\ \underset{\sim}{z}_{2q} = \begin{bmatrix} i_{2q} & \varphi_{s2q} & \varphi_q'' \end{bmatrix}^T \text{ and } \underset{\sim}{z}_{3q} = \begin{bmatrix} i_{3q} & \varphi_{s3q} & \varphi_q''' \end{bmatrix}^T \tag{4.377}$$

Based on the above groupings of state- and algebraic-variables the d-axis rotor-winding equations in (4.133) to (4.142) on page 147 are linearized about the initial operating condition derived in Section 4.2.13.3 as follows.

$$\begin{bmatrix} p\Delta\underset{\sim}{\varphi}_{cd} \\ 0 \\ \mathit{0} \\ \mathit{0} \end{bmatrix} = \begin{bmatrix} \mathit{0} & \mathbf{c}_{\varphi fd} & \mathbf{C}_{\varphi 1d} & \mathbf{C}_{\varphi 2d} \\ \mathbf{c}_{f\varphi d} & -1 & \mathbf{c}_{f1d} & \mathbf{c}_{f2d} \\ \mathbf{C}_{1\varphi d} & \mathit{0} & \mathbf{C}_{11d} & \mathbf{C}_{12d} \\ \mathbf{C}_{2\varphi d} & \mathit{0} & \mathbf{C}_{21d} & \mathbf{C}_{22d} \end{bmatrix} \begin{bmatrix} \Delta\underset{\sim}{\varphi}_{cd} \\ \Delta\underset{\sim}{z}_{fd} \\ \Delta\underset{\sim}{z}_{1d} \\ \Delta\underset{\sim}{z}_{2d} \end{bmatrix} + \begin{bmatrix} \mathit{0} \\ c_{fid} \\ \mathbf{c}_{1id} \\ \mathbf{c}_{2id} \end{bmatrix} \Delta i_d + \begin{bmatrix} \mathit{0} \\ 1 \\ \mathit{0} \\ \mathit{0} \end{bmatrix} \Delta I_{sd} + \begin{bmatrix} \mathbf{b}_{ced} \\ 0 \\ \mathit{0} \\ \mathit{0} \end{bmatrix} \Delta E_{fd} \tag{4.378}$$

in which the coefficient sub-matrices are:

$$\mathbf{c}_{\varphi fd} = \begin{bmatrix} -\frac{1}{T_{d0}'} & 0 & 0 \end{bmatrix}^T,\ \mathbf{C}_{\varphi 1d} = \begin{bmatrix} 0 & 0 & 0 \\ 0 & \frac{1}{T_{d0}''} & 0 \\ 0 & 0 & 0 \end{bmatrix},\ \mathbf{C}_{\varphi 2d} = \begin{bmatrix} 0 & 0 & 0 \\ 0 & 0 & 0 \\ 0 & \frac{1}{T_{d0}'''} & 0 \end{bmatrix} \tag{4.379}$$

$$c_{fid} = (L_d - L_d'),\ \mathbf{c}_{1id} = \begin{bmatrix} 0 & -(L_d' - L_l) & 0 \end{bmatrix}^T,\ \mathbf{c}_{2id} = \begin{bmatrix} 0 & -(L_d'' - L_l) & 0 \end{bmatrix}^T \tag{4.380}$$

$$c_{f\varphi d} = \begin{bmatrix} 1 & 0 & 0 \end{bmatrix}, \; C_{1\varphi d} = \begin{bmatrix} 0 & 0 & 0 \\ 1 & -1 & 0 \\ \left(\dfrac{L_d'' - L_l}{L_d' - L_l}\right) & \left(\dfrac{L_d' - L_d''}{L_d' - L_l}\right) & 0 \end{bmatrix}, \; C_{2\varphi d} = \begin{bmatrix} 0 & 0 & 0 \\ 0 & 0 & -1 \\ 0 & 0 & \left(\dfrac{L_d'' - L_d'''}{L_d'' - L_l}\right) \end{bmatrix} \tag{4.381}$$

$$c_{f1d} = \begin{bmatrix} -c_{fid} & 0 & 0 \end{bmatrix}, \; c_{f2d} = c_{f1d} \tag{4.382}$$

$$C_{11d} = \begin{bmatrix} -1 & -\left(\dfrac{L_d' - L_d''}{(L_d' - L_l)^2}\right) & 0 \\ 0 & -1 & 0 \\ 0 & 0 & -1 \end{bmatrix}, \; C_{12d} = \begin{bmatrix} -c_{1id} & 0 & 0 \end{bmatrix} \tag{4.383}$$

$$C_{21d} = \begin{bmatrix} 0 & 0 & 0 \\ 0 & 0 & 1 \\ 0 & 0 & \left(\dfrac{L_d''' - L_l}{L_d'' - L_l}\right) \end{bmatrix}, \; C_{22d} = \begin{bmatrix} -1 & -\left(\dfrac{L_d'' - L_d'''}{(L_d'' - L_l)^2}\right) & 0 \\ 0 & -1 & 0 \\ 0 & 0 & -1 \end{bmatrix} \tag{4.384}$$

$$\text{and } b_{ced} = \begin{bmatrix} \dfrac{1}{T_{d0}'} & 0 & 0 \end{bmatrix}^T \tag{4.385}$$

The d-axis rotor-winding algebraic variables z_{rd} are defined as:

$$z_{rd} = \begin{bmatrix} z_{fd} \\ z_{1d} \\ z_{2d} \end{bmatrix} \tag{4.386}$$

The coefficient sub-matrices in (4.379) to (4.384) are combined to form the following consolidated set of sub-matrices:

$$C_{\varphi zd} = \begin{bmatrix} c_{\varphi fd} & C_{\varphi 1d} & C_{\varphi 2d} \end{bmatrix}, \; C_{z\varphi d} = \begin{bmatrix} c_{f\varphi d} \\ C_{1\varphi d} \\ C_{2\varphi d} \end{bmatrix}, \; C_{zzd} = \begin{bmatrix} -1 & c_{f1d} & c_{f2d} \\ 0 & C_{11d} & C_{12d} \\ 0 & C_{21d} & C_{22d} \end{bmatrix} \text{ and}$$

$$c_{zid} = \begin{bmatrix} c_{fid} \\ c_{1id} \\ c_{2id} \end{bmatrix} \tag{4.387}$$

and the following sub-matrix is defined:

$$c_{zsd} = \begin{bmatrix} 1 \\ 0 \\ 0 \end{bmatrix} \tag{4.388}$$

The consolidated vector of algebraic variables in (4.386) and associated consolidated coefficient sub-matrices in equations (4.387) and (4.388) are substituted in the *d*-axis rotor-winding equations in (4.378) to yield the following compact formulation:

$$\begin{bmatrix} p\Delta\underset{\sim}{\varphi}_{cd} \\ 0 \end{bmatrix} = \begin{bmatrix} 0 & C_{\varphi zd} \\ C_{z\varphi d} & C_{zzd} \end{bmatrix} \begin{bmatrix} \Delta\underset{\sim}{\varphi}_{cd} \\ \Delta\underset{\sim}{z}_{rd} \end{bmatrix} + \begin{bmatrix} 0 \\ c_{zid} \end{bmatrix} \Delta i_d + \begin{bmatrix} 0 \\ c_{zsd} \end{bmatrix} \Delta I_{sd} + \begin{bmatrix} b_{ced} \\ 0 \end{bmatrix} \Delta E_{fd} \tag{4.389}$$

The *q*-axis rotor-winding equations (4.143) to (4.152) on page 148 following compact formulation of the linearized *q*-axis rotor-winding equations is similarly developed to yield,

$$\begin{bmatrix} p\Delta\underset{\sim}{\varphi}_{cq} \\ 0 \end{bmatrix} = \begin{bmatrix} 0 & C_{\varphi zq} \\ C_{z\varphi q} & C_{zzq} \end{bmatrix} \begin{bmatrix} \Delta\underset{\sim}{\varphi}_{cq} \\ \Delta\underset{\sim}{z}_{rq} \end{bmatrix} + \begin{bmatrix} 0 \\ c_{ziq} \end{bmatrix} \Delta i_q + \begin{bmatrix} 0 \\ c_{zsq} \end{bmatrix} \Delta I_{sq} , \tag{4.390}$$

in which the following variable and coefficient sub-matrix definitions apply:

$$\underset{\sim}{z}_{rq} = \begin{bmatrix} \underset{\sim}{z}_{1q} \\ \underset{\sim}{z}_{2q} \\ \underset{\sim}{z}_{3q} \end{bmatrix}, \tag{4.391}$$

$$C_{\varphi zq} = \begin{bmatrix} c_{\varphi 1q} & C_{\varphi 2q} & C_{\varphi 3q} \end{bmatrix}, \quad C_{z\varphi q} = \begin{bmatrix} c_{1\varphi q} \\ C_{2\varphi q} \\ C_{3\varphi q} \end{bmatrix}, \quad C_{zzq} = \begin{bmatrix} -1 & c_{12q} & c_{13q} \\ 0 & C_{22q} & C_{23q} \\ 0 & C_{32q} & C_{33q} \end{bmatrix},$$

$$c_{ziq} = \begin{bmatrix} c_{1iq} \\ c_{2iq} \\ c_{3iq} \end{bmatrix} \text{ and } c_{zsq} = \begin{bmatrix} 1 \\ 0 \\ 0 \end{bmatrix}. \tag{4.392}$$

in which, analogously with the *d*-axis sub-matrix definitions in equations (4.379) to (4.384), the following *q*-axis sub-matrix definitions are obtained from the *q*-axis rotor-equations.

$$c_{\varphi 1q} = \begin{bmatrix} -\frac{1}{T_{q0}'} & 0 & 0 \end{bmatrix}^T, \quad C_{\varphi 2q} = \begin{bmatrix} 0 & 0 & 0 \\ 0 & \frac{1}{T_{q0}''} & 0 \\ 0 & 0 & 0 \end{bmatrix}, \quad C_{\varphi 3q} = \begin{bmatrix} 0 & 0 & 0 \\ 0 & 0 & 0 \\ 0 & \frac{1}{T_{q0}'''} & 0 \end{bmatrix}, \tag{4.393}$$

$$c_{1iq} = (L_q - L_q'), \quad c_{2iq} = \begin{bmatrix} 0 & -(L_q' - L_l) & 0 \end{bmatrix}^T, \quad c_{3iq} = \begin{bmatrix} 0 & -(L_q'' - L_l) & 0 \end{bmatrix}^T, \tag{4.394}$$

$$c_{1\varphi q} = \begin{bmatrix} 1 & 0 & 0 \end{bmatrix},\ C_{2\varphi q} = \begin{bmatrix} 0 & 0 & 0 \\ 1 & -1 & 0 \\ \left(\frac{L_q'' - L_l}{L_q' - L_l}\right) & \left(\frac{L_q' - L_q''}{L_q' - L_l}\right) & 0 \end{bmatrix},\ C_{3\varphi q} = \begin{bmatrix} 0 & 0 & 0 \\ 0 & 0 & -1 \\ 0 & 0 & \left(\frac{L_q'' - L_q'''}{L_q'' - L_l}\right) \end{bmatrix}, \quad (4.395)$$

$$c_{12q} = \begin{bmatrix} -c_{1iq} & 0 & 0 \end{bmatrix},\ c_{13q} = c_{12q}, \quad (4.396)$$

$$C_{22q} = \begin{bmatrix} -1 - \left(\frac{L_q' - L_q''}{(L_q' - L_l)^2}\right) & 0 \\ 0 & -1 & 0 \\ 0 & 0 & -1 \end{bmatrix},\ C_{23q} = \begin{bmatrix} -c_{q2i} & 0 & 0 \end{bmatrix}, \quad (4.397)$$

$$C_{32q} = \begin{bmatrix} 0 & 0 & 0 \\ 0 & 0 & 1 \\ 0 & 0 & \left(\frac{L_q''' - L_l}{L_q'' - L_l}\right) \end{bmatrix},\ C_{33q} = \begin{bmatrix} -1 - \left(\frac{L_q'' - L_q'''}{(L_q'' - L_l)^2}\right) & 0 \\ 0 & -1 & 0 \\ 0 & 0 & -1 \end{bmatrix}. \quad (4.398)$$

The d- and q-axis rotor-winding equations (4.389) and (4.390) are now combined to give,

$$\begin{bmatrix} p\Delta\underset{\sim}{\varphi}_{cdq} \\ 0 \end{bmatrix} = \begin{bmatrix} 0 & C_{\varphi zdq} \\ C_{z\varphi dq} & C_{zzdq} \end{bmatrix} \begin{bmatrix} \Delta\underset{\sim}{\varphi}_{cdq} \\ \Delta\underset{\sim}{z}_{rdq} \end{bmatrix} + \begin{bmatrix} 0 \\ C_{zidq} \end{bmatrix} \Delta\underset{\sim}{i}_{dq} + \begin{bmatrix} 0 \\ C_{zsdq} \end{bmatrix} \Delta\underset{\sim}{I}_{sdq} + \begin{bmatrix} b_{ce} \\ 0 \end{bmatrix} \Delta E_{fd}. \quad (4.399)$$

The following variable and coefficient matrix definitions apply to the preceding equation.

$$\underset{\sim}{\varphi}_{cdq} = \begin{bmatrix} \underset{\sim}{\varphi}_{cd} \\ \underset{\sim}{\varphi}_{cq} \end{bmatrix},\ \underset{\sim}{z}_{rdq} = \begin{bmatrix} \underset{\sim}{z}_{rd} \\ \underset{\sim}{z}_{rq} \end{bmatrix},\ \underset{\sim}{i}_{dq} = \begin{bmatrix} i_d \\ i_q \end{bmatrix} \text{ and } \underset{\sim}{I}_{sdq} = \begin{bmatrix} I_{sd} \\ I_{sq} \end{bmatrix} \quad (4.400)$$

$$C_{\varphi zdq} = \mathfrak{D}(C_{\varphi zd}, C_{\varphi zq}),\ C_{z\varphi dq} = \mathfrak{D}(C_{z\varphi d}, C_{z\varphi q}),\ C_{zzdq} = \mathfrak{D}(C_{zzd}, C_{zzq}) \quad (4.401)$$

$$C_{zidq} = \mathfrak{D}(c_{zid}, c_{ziq}),\ C_{zsdq} = \mathfrak{D}(c_{zsd}, c_{zsq}) \text{ and } b_{ce} = \begin{bmatrix} b_{ced} \\ 0 \end{bmatrix} \quad (4.402)$$

In order to readily generalize the formulation of the equations to generator models with fewer than three rotor windings in each of the axes it, is convenient to define the kth-transient flux-linkages in terms of the rotor winding state- and algebraic variables as follows. The coefficient matrices in the following equation change depending on the number of rotor windings:

$$\Delta\underset{\sim}{\varphi}^{k}_{dq} = \begin{bmatrix} C_{k\varphi dq} & C_{kzdq} \end{bmatrix} \begin{bmatrix} \Delta\underset{\sim}{\varphi}_{cdq} \\ \Delta\underset{\sim}{z}_{rdq} \end{bmatrix} \tag{4.403}$$

in which $C_{k\varphi dq} = \mathfrak{D}(c_{k\varphi d}, c_{k\varphi q})$ and $C_{kzdq} = \mathfrak{D}(c_{kzd}, c_{kzq})$ and where (4.404)

$$c_{kzd} = \begin{bmatrix} 0 & c_{k1d} & c_{k2d} \end{bmatrix}, \; c_{kzq} = \begin{bmatrix} 0 & c_{k2q} & c_{k3q} \end{bmatrix}. \tag{4.405}$$

For the case of three rotor-windings in each axis the constituent matrices have the following values:

$$c_{k\varphi d} = c_{k\varphi q} = c_{k1d} = c_{k2q} = \boldsymbol{0} \text{ and } c_{k2d} = c_{k3q} = \begin{bmatrix} 0 & 0 & 1 \end{bmatrix}. \tag{4.406}$$

The d- and q-axis voltage equations (4.158) and (4.159) on page 149 are consolidated to give:

$$\Delta\underset{\sim}{v}_{dq} = \omega_0 W_{dq} \Delta\underset{\sim}{\varphi}^{k}_{dq} - Z^{k}_{dq} \Delta\underset{\sim}{i}_{dq} \tag{4.407}$$

$$\text{in which } W_{dq} = \begin{bmatrix} 0 & -1 \\ 1 & 0 \end{bmatrix} \text{ and } Z^{k}_{dq} = \begin{bmatrix} r_s & -\omega_0 L^{k}_{q} \\ \omega_0 L^{k}_{d} & r_s \end{bmatrix}. \tag{4.408}$$

Three alternative methods of representing the saturation level have been presented: (i) the resultant airgap flux-linkages (φ_{ag}) in Section 4.2.13.2.3; (ii) the resultant k^{th}-transient flux-linkages (φ^{k}) in Section 4.2.13.2.1; and (iii) the transient d-axis flux-linkages (φ_d') in Section 4.2.13.2.2. For each of these methods the perturbations in the saturation demagnetization currents can be expressed as follows. The definitions of the coefficient matrices in the equation are given in Table 4.14 for each of the three methods for representing the saturation level.

$$\Delta\underset{\sim}{I}_{sdq} = C_{s\varphi dq} \Delta\underset{\sim}{\varphi}_{cdq} + C_{skdq} \Delta\underset{\sim}{\varphi}^{k}_{dq} + C_{sidq} \Delta\underset{\sim}{i}_{dq} \tag{4.409}$$

In Table 4.14 the saturation characteristics for the d- and q-axes are combined into a single diagonal matrix,

$$S_{dq}(\varphi_m) = \mathfrak{D}(S_d(\varphi_m), S_q(\varphi_m)) \tag{4.410}$$

where φ_m is the saturation level indicator for the method chosen to represent the effects of magnetic saturation. The saturation demagnetizing current components in the d- and q-axes in terms of the selected saturation level indicator φ_m are, in matrix form,

$$\underset{\sim}{I}_{sdq} = S_{dq}(\varphi_m) \underset{\sim}{\varphi}_{mdq} \tag{4.411}$$

in which $\underset{\sim}{\varphi}_{mdq} = \begin{bmatrix} \varphi_{md} & \varphi_{mq} \end{bmatrix}^T$ is the vector of the d- and q-axis components of the saturation level indicator.

Table 4.14 Coefficient matrices of the linearized, saturation demagnetizing current equations for the three methods of representing the saturation level.

#	φ_m	Coefficient matrices $C_{s\varphi dq}$, C_{skdq} and C_{sidq}
1	φ_{ag}	See Section 4.2.13.2.3, $\varphi_m = \varphi_{ag}$ and $\underset{\sim}{\varphi}_{mdq} = \begin{bmatrix}\varphi_{ad} & \varphi_{aq}\end{bmatrix}^T$ $\boldsymbol{C_{s\varphi dq}} = \boldsymbol{0}$ $\boldsymbol{C_{skdq}} = \boldsymbol{S_{dq}}(\varphi_{ag_0}) + \frac{1}{\varphi_{ag_0}}\left(\left(\frac{\partial \boldsymbol{S_{dq}}(\varphi_{ag})}{\partial \varphi_{ag}}\right)_0 \underset{\sim}{\varphi}_{adq_0}\underset{\sim}{\varphi}_{adq_0}^T\right)$ and $\boldsymbol{C_{sidq}} = -\boldsymbol{C_{skdq}}\boldsymbol{L_{dqai}}$ in which $\underset{\sim}{\varphi}_{adq} = \underset{\sim}{\varphi}_{dq}^k - \boldsymbol{L_{aidq}}\underset{\sim}{i}_{dq}$ where $\boldsymbol{L_{aidq}} = \mathfrak{D}((L_d^k - L_l), (L_q^k - L_l))$ and $\varphi_{ag} = \sqrt{(\underset{\sim}{\varphi}_{adq})^T(\underset{\sim}{\varphi}_{adq})}$
2	φ^k	See Section 4.2.13.2.1, $\varphi_m = \varphi^k$ and $\underset{\sim}{\varphi}_{mdq} = \begin{bmatrix}\varphi_d^k & \varphi_q^k\end{bmatrix}^T$ $\boldsymbol{C_{s\varphi dq}} = \boldsymbol{0}$ $\boldsymbol{C_{skdq}} = \boldsymbol{S_{dq}}(\varphi_0^k) + \frac{1}{\varphi_0^k}\left(\left(\frac{\partial \boldsymbol{S_{dq}}(\varphi^k)}{\partial \varphi^k}\right)_0 \underset{\sim}{\varphi}_{dq_0}^k\left(\underset{\sim}{\varphi}_{dq_0}^k\right)^T\right)$ and $\boldsymbol{C_{sidq}} = \boldsymbol{0}$ in which $\boldsymbol{S_{dq}}(\varphi^k) = \mathfrak{D}(S_d(\varphi^k), S_q(\varphi^k))$ & $\varphi^k = \sqrt{(\underset{\sim}{\varphi}_{dq}^k)^T(\underset{\sim}{\varphi}_{dq}^k)}$
3	φ_d'	See Section 4.2.13.2.2, $\varphi_m = \varphi_d'$ and $\underset{\sim}{\varphi}_{mdq} = \begin{bmatrix}\varphi_d' & 0\end{bmatrix}^T$ $\boldsymbol{C_{s\varphi dq}} = \boldsymbol{S_{dq}}(\varphi_{d_0}') + \left(\left(\frac{\partial \boldsymbol{S_{dq}}(\varphi_d')}{\partial \varphi_d'}\right)_0 \underset{\sim}{\varphi}_{cdq_0}\right)$, $\boldsymbol{C_{skdq}} = \boldsymbol{0}$ and $\boldsymbol{C_{sidq}} = \boldsymbol{0}$ in which $\boldsymbol{S_{dq}}(\varphi_d') = \mathfrak{D}\left(\begin{bmatrix}S_d(\varphi_d') & 0 & 0\end{bmatrix}, \boldsymbol{0}\right)$

Equations (4.403), (4.407) and (4.409) are incorporated with the rotor-winding equations in (4.399) to yield the following system of DAEs:

$$\begin{bmatrix} p\Delta\underset{\sim}{\varphi}_{cdq} \\ \boldsymbol{0} \\ \boldsymbol{0} \\ \boldsymbol{0} \\ \boldsymbol{0} \end{bmatrix} = \begin{bmatrix} \boldsymbol{0} & \boldsymbol{C_{\varphi zdq}} & \boldsymbol{0} & \boldsymbol{0} & \boldsymbol{0} & \boldsymbol{0} \\ \boldsymbol{C_{z\varphi dq}} & \boldsymbol{C_{zzdq}} & \boldsymbol{0} & \boldsymbol{C_{zsdq}} & \boldsymbol{C_{zidq}} & \boldsymbol{0} \\ \boldsymbol{C_{k\varphi dq}} & \boldsymbol{C_{kzdq}} & -\boldsymbol{I} & \boldsymbol{0} & \boldsymbol{0} & \boldsymbol{0} \\ \boldsymbol{C_{s\varphi dq}} & \boldsymbol{0} & \boldsymbol{C_{skdq}} & -\boldsymbol{I} & \boldsymbol{C_{sidq}} & \boldsymbol{0} \\ \boldsymbol{0} & \boldsymbol{0} & \omega_0\boldsymbol{W_{dq}} & \boldsymbol{0} & -\boldsymbol{Z_{dq}^k} & -\boldsymbol{I} \end{bmatrix} \begin{bmatrix} \Delta\underset{\sim}{\varphi}_{cdq} \\ \Delta\underset{\sim}{z}_{rdq} \\ \Delta\underset{\sim}{\varphi}_{dq}^k \\ \Delta\underset{\sim}{I}_{sdq} \\ \Delta\underset{\sim}{i}_{dq} \\ \Delta\underset{\sim}{v}_{dq} \end{bmatrix} + \begin{bmatrix} \boldsymbol{b_{ce}} \\ \boldsymbol{0} \\ \boldsymbol{0} \\ \boldsymbol{0} \\ \boldsymbol{0} \end{bmatrix} \Delta E_{fd} \qquad (4.412)$$

In order to simplify the structure of the above equations it is convenient to define the following consolidated vector of algebraic variables:

$$\Delta \underset{\sim}{z}_e = \begin{bmatrix} \Delta \underset{\sim}{z}_{rdq} \\ \Delta \underset{\sim}{\varphi}^k_{dq} \\ \Delta \underset{\sim}{I}_{sdq} \end{bmatrix} \tag{4.413}$$

and the associated consolidated matrix coefficients:

$$\boldsymbol{C}_{ce} = \begin{bmatrix} \boldsymbol{0} & \boldsymbol{C}_{\varphi z dq} & \boldsymbol{0} & \boldsymbol{0} \end{bmatrix}, \; \boldsymbol{C}_{ec} = \begin{bmatrix} \boldsymbol{C}_{z\varphi dq} \\ \boldsymbol{C}_{k\varphi dq} \\ \boldsymbol{C}_{s\varphi dq} \end{bmatrix}, \; \boldsymbol{C}_{ee} = \begin{bmatrix} \boldsymbol{C}_{zzdq} & \boldsymbol{0} & \boldsymbol{C}_{zsdq} \\ \boldsymbol{C}_{kzdq} & -\boldsymbol{I} & \boldsymbol{0} \\ \boldsymbol{0} & \boldsymbol{C}_{skdq} & -\boldsymbol{I} \end{bmatrix}, \; \boldsymbol{C}_{ei} = \begin{bmatrix} \boldsymbol{C}_{zidq} \\ \boldsymbol{0} \\ \boldsymbol{C}_{sidq} \end{bmatrix} \text{ and}$$

$$\boldsymbol{C}_{ve} = \begin{bmatrix} \boldsymbol{0} & \omega_0 \boldsymbol{W}_{dq} & \boldsymbol{0} \end{bmatrix}. \tag{4.414}$$

Substitution of the quantities in (4.413) and (4.414) into (4.412) results in the following compact form of the linearized electromagnetic equations:

$$\begin{bmatrix} p\Delta \underset{\sim}{\varphi}_{cdq} \\ \boldsymbol{0} \\ \boldsymbol{0} \end{bmatrix} = \begin{bmatrix} \boldsymbol{0} & \boldsymbol{C}_{ce} & \boldsymbol{0} & \boldsymbol{0} \\ \boldsymbol{C}_{ec} & \boldsymbol{C}_{ee} & \boldsymbol{C}_{ei} & \boldsymbol{0} \\ \boldsymbol{0} & \boldsymbol{C}_{ve} & -\boldsymbol{Z}^k_{dq} & -\boldsymbol{I} \end{bmatrix} \begin{bmatrix} \Delta \underset{\sim}{\varphi}_{rdq} \\ \Delta \underset{\sim}{z}_e \\ \Delta \underset{\sim}{i}_{dq} \\ \Delta \underset{\sim}{v}_{dq} \end{bmatrix} + \begin{bmatrix} \boldsymbol{b}_{re} \\ \boldsymbol{0} \\ \boldsymbol{0} \end{bmatrix} \Delta E_{fd} \tag{4.415}$$

The algebraic variables $\Delta \underset{\sim}{z}_e$ are now eliminated from (4.415) to yield the generator electromagnetic equations in the required form:

$$\begin{bmatrix} p\Delta \underset{\sim}{\varphi}_{cdq} \\ \boldsymbol{0} \end{bmatrix} = \begin{bmatrix} \boldsymbol{A}_r & \boldsymbol{B}_{ri} & \boldsymbol{0} \\ \boldsymbol{C}_{vr} & \boldsymbol{D}_{vi} & -\boldsymbol{I} \end{bmatrix} \begin{bmatrix} \Delta \underset{\sim}{\varphi}_{cdq} \\ \Delta \underset{\sim}{i}_{dq} \\ \Delta \underset{\sim}{v}_{dq} \end{bmatrix} + \begin{bmatrix} \boldsymbol{b}_{re} \\ \boldsymbol{0} \end{bmatrix} \Delta E_{fd} \tag{4.416}$$

$$\text{in which } \boldsymbol{A}_r = \boldsymbol{C}_{ce}\boldsymbol{J}_{ec}, \; \boldsymbol{B}_{ri} = \boldsymbol{C}_{ce}\boldsymbol{J}_{ei}, \; \boldsymbol{C}_{vr} = \boldsymbol{C}_{ve}\boldsymbol{J}_{ec} \text{ and } \boldsymbol{D}_{vi} = \boldsymbol{C}_{ve}\boldsymbol{J}_{ei} - \boldsymbol{Z}^k_{dq} \tag{4.417}$$

$$\text{and where } \boldsymbol{J}_{ec} = -\boldsymbol{C}^{-1}_{ee}\boldsymbol{C}_{ec} \text{ and } \boldsymbol{J}_{ei} = -\boldsymbol{C}^{-1}_{ee}\boldsymbol{C}_{ei} \tag{4.418}$$

The perturbations in the algebraic variables are given by:

$$\Delta \underset{\sim}{z}_e = \boldsymbol{J}_{ec}\Delta \underset{\sim}{\varphi}_{cdq} + \boldsymbol{J}_{ei}\Delta \underset{\sim}{i}_{dq} \tag{4.419}$$

As outlined in Section 4.2.13.4 the electromagnetic equations in (4.416) are combined with the rotor equations of motion and the network interface equations as in (4.117) on page 133. However, $\Delta \underset{\sim}{x}_r$ and $\Delta \underset{\sim}{u}_m$ are redefined as follows:

$$\Delta \underset{\sim}{x}_r = \begin{bmatrix} \Delta \underset{\sim}{\varphi}_{cd} \\ \Delta \underset{\sim}{\varphi}_{cq} \end{bmatrix} \text{ and } \Delta \underset{\sim}{u}_m = \begin{bmatrix} \Delta E_{fd} \\ \Delta T_m \end{bmatrix}. \tag{4.420}$$

Appendix 4–II Forms of the equations of motion of the rotors of a generating unit

App. 4–II.1 Introduction

In the literature various forms of the equations of motion are employed, often depending on the application. However, it is important to understand the nature of any approximations used and their relevance to the application. When attempting to simulate small-signal events using a large-signal (transient stability) software, it has been found that the damping of variables does not match that derived from a small-signal analysis software. The reason may be associated with the form of the shaft equations employed in the large-signal simulation software. Consequently, several forms of the large-signal equations in per-unit form are derived in this appendix, followed by the associated small-signal versions.

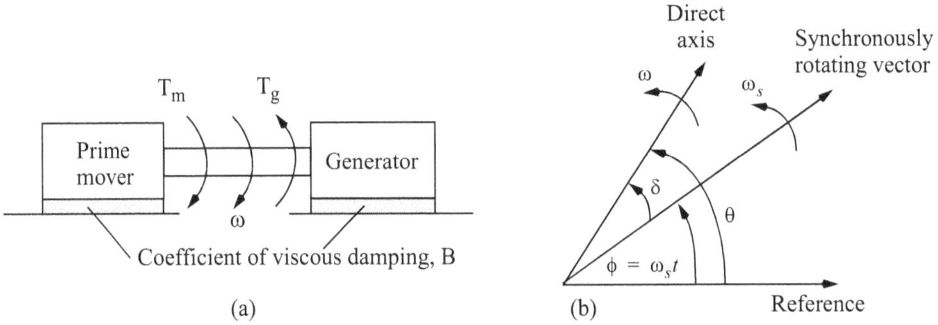

Figure 4.25 (a) Rotating mechanical system, prime mover and generator
(b) Rotating speed vectors and associated angles

For a given generating unit consider the mechanical system of Figure 4.25(a) rotating at an angular velocity $\omega(t)$ electrical rad/s (the term 'speed' is synonymous). Let us assume (i) a two-pole generator (i.e. $n_{pp} = 1$), (ii) the total moment of inertia of the rotating system is J (kg-m^2), (iii) the effective coefficient of viscous friction for small disturbances in speed in the vicinity of synchronous speed is B (N-m/rad/s), and (iv) the shaft is infinitely stiff. The equation of motion of the shaft *in SI units*, due to a net accelerating torque $T_a(t)$ (N-m) acting on it, is:

$$J\frac{d}{dt}\omega(t) + B(\omega(t) - \omega_s) = T_m(t) - T_g(t) = T_a(t), \tag{4.421}$$

where $T_m(t)$ is the prime mover torque (N-m); $T_g(t)$ is the airgap torque of electro-magnetic origin developed by the generator; time is in seconds.

Under perturbed, stable conditions the instantaneous angular velocity ($\omega(t)$) of the unit varies about the synchronously rotating speed reference (ω_s, rad/s) of the system. As shown in Figure 4.25(b) the rotor angle $\theta(t)$ of the unit with respect to a stationary system reference frame is:

$$\theta(t) = \omega_s t + \delta(t) \text{ (rad)},$$

where $\delta(t)$ (rad) is the rotor angle of the unit with respect to the synchronously rotating speed reference. The rotational speed of the shaft is thus:

$$\omega(t) = \frac{d\theta(t)}{dt} = \omega_s + \frac{d\delta(t)}{dt} \text{ (rad/s), or } \frac{d\delta(t)}{dt} = \omega(t) - \omega_s. \tag{4.422}$$

For clarity at the present, let us denote per-unit quantities by the subscript p and let base speed be ω_b rad/s, thus the per unit speed is

$$\omega_p(t) = \omega(t)/\omega_b, \text{ and } \frac{d\omega(t)}{dt} = \omega_b \cdot \frac{d\omega_p(t)}{dt}. \tag{4.423}$$

Since we are considering a two pole machine base electrical and mechanical speeds are identical. Let us define the per-unit synchronous speed as,

$$\omega_0 = \omega_s/\omega_b. \tag{4.424}$$

Note that normally, but not necessarily, $\omega_b = \omega_s$ rad/s.

Let us define base power S_b (VA) and base torque T_b (N-m) by the relationship:

$$S_b = \omega_b T_b \text{ (see Tables 4.2-4.4 on page 102).} \tag{4.425}$$

Dropping the time dependency of the variables and dividing (4.422) through by ω_b, we find that in per-unit

$$\frac{d\delta}{dt} = \omega_b(\omega_p - \omega_0) \text{ and } \frac{d\omega_p}{dt} = \frac{1}{\omega_b} \cdot \frac{d^2\delta}{dt^2}. \tag{4.426}$$

Let us now consider several forms of the per-unit shaft equations.

App. 4–II.2 Shaft equations expressed in terms of per-unit angular speed and torques

Dividing through (4.421) by the base torque T_b of the generating unit, and expressing the equation in terms of per unit speed ω_p based on (4.423), we find

$$\frac{\omega_b J}{T_b} \cdot \frac{d\omega_p}{dt} + \frac{B\omega_b}{T_b} \cdot (\omega_p - \omega_0) = \frac{T_m}{T_b} - \frac{T_g}{T_b} = T_{mp} - T_{gp} \text{ per unit.} \tag{4.427}$$

Firstly, let us rearrange in (4.427) the term, $\frac{\omega_b J}{T_b} = 2\left[\frac{\omega_b^2 J}{2S_b}\right] = 2H$, where

$$H = \left[\frac{\omega_b^2 J}{2}\right] / S_b. \tag{4.428}$$

H is defined as the inertia constant of the unit, and is the ratio of the stored energy in Joules of the rotating system at base speed to the base apparent power, S_b, of the unit in VA. The rotating system normally consists of the prime mover, and the generator and exciter.

Secondly, consider in (4.427) the term, $(B\omega_b)/T_b = D$ where D is a damping torque coefficient. It is the damping torque for a speed difference equal to base speed per unit of the base torque.

Thirdly, the terms on the right hand side of (4.427) are the per unit mechanical and airgap electro-mechanical torques, T_{mp} and T_{gp} respectively.

The per-unit equations of motion of the rotor resulting from (4.426) and (4.427), respectively, are

$$\frac{d\delta}{dt} = \omega_b \cdot (\omega_p - \omega_0) \quad \text{and} \quad 2H \cdot \frac{d\omega_p}{dt} + D \cdot (\omega_p - \omega_0) = T_{mp} - T_{gp}, \tag{4.429}$$

which are the equations of motion listed in equations (4.56) and (4.57) on page 114.

Note that there are no approximations made in the derivation of the above equations. However, the relationship between damping torque and speed perturbations about synchronous speed, characterized by the coefficient D, is generally unknown at any operating condition. Typically, the damping torque coefficient D is small, and being unknown, it is often set to zero or some low value. Importantly, damping effects of electromagnetic origin are due to losses in the resistances of the damper windings of the generator models. Such losses should therefore not be accounted for in damping torque coefficient of the shaft acceleration equation.

The per-unit equations of rotor motion given by (4.429) are linear in the speed and torque variables. These variable can therefore be replaced by the perturbed variables to form the set of linearized equations, i.e.

$$\frac{d\Delta\delta}{dt} = \omega_b \cdot \Delta\omega_p \quad \text{and} \quad 2H \cdot \frac{d\Delta\omega_p}{dt} + D \cdot \Delta\omega_p = \Delta T_{mp} - \Delta T_{gp}; \tag{4.430}$$

these are the linearized equations of motion listed in (4.58) and (4.59) on page 114.

App. 4–II.3 Per-unit shaft acceleration equation in terms of rotor-speed and power

In a rotating system, the power, torque and speed of rotation - all in SI units - are related through the product non-linearity

$$P = \omega T \ \text{(W)}. \tag{4.431}$$

The mechanical power delivered to the shaft by the turbine and the power transferred across the airgap of the machine are respectively,

$$P_m = \omega T_m \ \text{(W)} \quad \text{and} \quad P_g = \omega T_g \ \text{(W)}. \tag{4.432}$$

Note that because the shaft is assumed to be rigid the mechanical and electrical rotor-speeds are identical, observing that in this analysis a two pole machine is assumed.

Based on equation (4.432) it can be shown that the per-unit shaft acceleration equation can be expressed in terms of the per-unit accelerating power $P_{ap} = (P_{mp} - P_{gp})$, i.e.

$$2H\frac{d\omega_p}{dt} + D(\omega_p - \omega_0) = (P_{mp} - P_{gp})/\omega_p \tag{4.433}$$

for which the linearized form is,

$$2H\frac{d\Delta\omega_p}{dt} + D\Delta\omega_p = (\Delta P_{mp} - \Delta P_{gp})/\omega_0. \tag{4.434}$$

Furthermore, it can be shown that in the linearized acceleration equations the terms in the torque perturbations in (4.430) and in terms of the power perturbations (4.434) are exactly equivalent, i.e. $\Delta T_{mp} - \Delta T_{gp} = (\Delta P_{mp} - \Delta P_{gp})/\omega_0$.

It is important to note that no approximations have been made in the formulation of this non-linear equation. Specifically, perturbations in the rotor-speed in calculating the relationship between the accelerating torque and power have been retained.

As mentioned earlier normally, but not necessarily, synchronous speed ω_s rad/s is equal to the base speed ω_b rad/s in which case $\omega_0 = \omega_s/\omega_b = 1$ per-unit. *If this is the case it follows that the per-unit perturbations in torque and power are identical.*

App. 4–II.4 Shaft acceleration equation neglecting speed perturbations in the torque/power relationship

It is commonly assumed, for the purpose of calculating the generator *d*- and *q*-axis stator voltages, that the perturbations of the rotor-speed from synchronous speed are negligible. As stated in Sections 4.2.2 and 4.2.5 this assumption is adhered to in this book, i.e. the airgap power is related to the airgap torque by:

$$P_{gp} = \omega_0 T_{gp} \quad \text{(repeat of (4.52) on page 113).}$$

Given the above approximation it is *essential* that the relationship between the mechanical power and torque also neglect perturbations in the rotor-speed, that is,

$$P_{mp} = \omega_0 T_{mp} . \tag{4.435}$$

The consequence of ignoring the above approximation but rather setting $P_{mp} = \omega_p T_{mp}$ is discussed in Appendix 4–II.5.

With the assumption that perturbations in the rotor-speed are negligible for the purpose of calculating the shaft accelerating power, it is shown that the linearized shaft equations expressed in terms of torque perturbations (4.430) are identical to the equations expressed in terms of power perturbations(4.434).

App. 4–II.5 A common misunderstanding in calculating the accelerating torque and power

The consequences of a misunderstanding that sometimes occurs in the formulation of the shaft acceleration equation are now discussed. The misunderstanding is that the rotor speed perturbations are neglected in the relationship between the airgap torque and power (as discussed above) but erroneously the speed perturbations are retained in the relationship between the mechanical power and torque. That is to say, the error is to define the per-unit mechanical and airgap power differently as follows:

$$\text{Error:} \quad P_{mp} = \omega_p T_{mp} \text{ and } P_{gp} = \omega_0 T_{gp} . \tag{4.436}$$

From the preceding equation the following inconsistent expression for the accelerating torque is derived,

$$T_{mp} - T_{gp} = P_{mp}/\omega_p - P_{gp}/\omega_0 , \tag{4.437}$$

which when substituted into the per-unit acceleration equation (4.429) results in,

$$\text{Error:} \quad 2H \cdot \frac{d\omega_p}{dt} + D \cdot (\omega_p - \omega_0) = P_{mp}/\omega_p - P_{gp}/\omega_0 . \tag{4.438}$$

Linearizing the above inconsistent expression for the acceleration equation about the operating point yields,

$$2H \cdot \frac{d\Delta\omega_p}{dt} + D \cdot \Delta\omega_p = \Delta P_{mp}/\omega_{p_0} - (P_{mp_0}/\omega_{p_0}^2)\Delta\omega_p - \Delta P_{gp}/\omega_0$$

which, upon substitution of $\omega_{p_0} = \omega_0$, results in the following incorrect form of the linearized acceleration equation.

$$2H \cdot \frac{d\Delta\omega_p}{dt} + (D + P_{mp_0}/\omega_0^2) \cdot \Delta\omega_p = (\Delta P_{mp} - \Delta P_{gp})/\omega_0 = (\Delta T_{mp} - \Delta T_{gp}) . \qquad (4.439)$$

Comparison of the correct formulation of the linearized acceleration in (4.430) with the above expression (4.439) shows that the effect of the misunderstanding is erroneously to increase the damping coefficient of the generator model by P_{mp_0}/ω_0^2.

Chapter 5

Concepts in the tuning of power system stabilizers for a single machine system

5.1 Introduction

Although this chapter is concerned with the application of a power system stabilizer (PSS) to a single-machine system, the concepts for the most part are applicable to multi-machine systems: such applications will be discussed in Chapters 9 and 10. Various important aspects of the tuning of the PSS can therefore considered in some detail because the analysis involves a simple system only.

The reasons for the wide-spread deployment of PSSs in power systems today are twofold, (i) to stabilize the unstable electro-mechanical modes in the system, (ii) to ensure that there is an adequate margin of stability for these modes over a wide range of operating conditions and contingencies, that is, the electro-mechanical modes are adequately damped. Some systems, such as the Eastern Australian grid, would be unstable without the use of both PSSs and stabilizers installed on certain FACTS devices.

A marginally stable electro-mechanical mode is oscillatory in nature and is very lightly damped. The frequency of rotor oscillations is typically between 1.5 to 15 rad/s, and the 5% settling time may be many tens of seconds. Typically a mode of a lengthy duration would not satisfy the system operator's criterion for modal damping. A stable mode is said to be 'positively' damped, whereas an unstable mode is referred to as being 'negatively' damped.

With the growth of power systems, and the need to transmit power over long distances by means of high-voltage transmission lines, the problems of instability following a major fault or disturbance have increased. Instability in such events is typically the result of a generator falling out of step due to insufficient synchronizing torques being available to hold generators in synchronism. In order to increase the synchronizing torques between generators, high-gain fast-acting excitation systems were developed with the objective of increasing field flux linkages rapidly during and following the fault. However, such high-gain excitation systems may introduce negative damping on certain electro-mechanical modes.

In linear control systems design, rate feedback is employed not only to stabilize an unstable system but also to enhance the system's damping performance. A PSS that uses generator speed (i.e. the rate of change of rotor angle) as a stabilizing signal is such a rate-feedback controller. However, to introduce on the shaft of the generator a torque of electromagnetic origin *that is purely a damping torque* requires that the compensation transfer function provided in the PSS is properly designed. 'Pure' damping occurs when the induced electrical torque is in phase with speed; *this is an essential function of the PSS*. Such a torque *opposes* a change in rotor-speed.

The main role of the PSS is to provide damping of the electro-mechanical modes for *small disturbances* on the system. Therefore, in order to analyse the dynamic performance of the system and to tune stabilizers, the non-linear equations describing the dynamic behaviour of the generator and system are linearized about a steady-state operating condition. As outlined in Section 2.1.1 *a set of linear equations* in terms of the *new set of perturbed variables* results. A significant consequence is that the powerful methods and techniques provided by linear control systems theory become available both for the analysis of dynamic performance and for controller design.

This chapter is concerned with illustrating the concepts associated with the design and tuning of a PSS using small-signal analysis techniques. The performance of the system under large-magnitude disturbances - such as fault conditions as mentioned above - is treated briefly in Chapter 10. However it should be mentioned that, following clearance of a fault, the system may appear to be stable following the second or subsequent swings in the rotor angles but becomes unstable as the steady-state is seemingly approached. Such instability is due to the existence of one or more unstable electro-mechanical modes in the post-fault operating condition. For example, transmission lines may have been switched out of service in order to clear the fault and instability is a consequence of network voltages falling during the post fault period. The PSSs must be designed to ensure small-signal stable operation in the steady-state that follows the worst-case contingencies.

The benefits of small-signal analysis in complementing large-signal (or transient stability) analysis are described in Section 10.9.1.1.

The paper by de Mello and Concordia in 1969 provided the basis for the design of many Power System Stabilizers in operation today [1]. Based on the concept of damping torques developed on the generator shaft, a technique is presented in the paper for the design and tuning of a speed-input PSS for a single-machine infinite-bus (SMIB) system. The PSS transfer function is designed to provide phase compensation for the transfer function between the voltage reference of the AVR and the electrical torque. Ideally, any perturbations in shaft speed produce pure damping torques on the shaft.

In a set of papers by Larsen and Swann in 1981 [2] the concepts in [1] were extended and applied to the tuning of PSSs and their tuning on site. Firstly, frequency response measurements between the voltage reference of the AVR and the terminal voltage yield a transfer function which, because the speed perturbations are assumed negligible, is equivalent to the phase response between the voltage reference and electrical torque. The transfer function is called the generator (*G*), excitation system (E) and power system (P) transfer function, GEP(s). Secondly, the PSS compensation is then designed to offset the phase-lags in GEP(s) by means of phase-lead transfer function blocks. Finally, the PSS gain is raised until prolonged oscillations are observed, i.e. the generator is on the brink of modal instability. The PSS gain is then set to $1/3^{rd}$ of the limiting value, providing a gain margin of 3:1 or 10 dB. This approach to PSS tuning is considered in more detail in Chapter 6.

An alternative approach which is applicable to single- and multi-machine systems is based on the Method of Residues and is also described in Chapter 6. Some of the features of the these methods, and approaches to the tuning and implementation of PSSs, are described in an IEEE Tutorial Course [3].

Using the damping torque concepts developed for the single machine system [1], [2] and [4], a procedure called the P-Vr approach for the tuning of PSSs in multi-machine systems was proposed by Gibbard in 1988 [5]. This procedure is in part an extension of the GEP Method, however, *specific and meaningful information is derived concerning both the phase compensation for, and the gain setting of, the PSS.* (See Section 1.2, item 3)

The P-Vr tuning procedure is described in this chapter for a single machine system to illustrate in some detail the concepts in the tuning of PSSs, however - as stated earlier - the procedure is readily applied to multi-machine systems and is the subject of later chapters.

The literature on PSSs, and their associated stabilizing signals, for single- and multi-machine systems is fairly extensive and is discussed in more detail in Chapter 8, 'Types of Power System Stabilizers'. The purpose of this chapter, however, is to provide an understanding of the fundamentals of PSS design and tuning.

In this chapter the preliminary tuning of a PSS will be based on the Heffron and Phillips model of the SMIB system [6]. This model, being fourth order, is amenable to simple analysis and thus it is possible to derive simple, closed-form solutions for certain transfer functions and torque-related expressions which are applicable to a range of operating conditions. *These features provide a more detailed insight and understanding of the relevant concepts.* However, to partially bridge the gap between the low-order Heffron and Phillips model and the practical models used in multi-machine systems, the tuning of a PSS for a higher-order machine model will be employed later in the chapter. For higher-order models it is not practical to derive similar closed-form solutions and therefore other approaches must be employed. In addition, a somewhat more realistic representation of the external system will be adopted so that system quantities, such as voltage levels on busbars, lie within normally acceptable ranges.

Some of the earlier material in this Chapter, Section 5.2 to 5.7, is also covered in the IEEE Tutorial Course [3].

In this and subsequent chapters we will employ the term "*range of operating conditions*". It is assumed that, for the subsequent analysis, a set of steady-state conditions are selected *which encompass those conditions for which the stabilizer is to be tuned.* The latter conditions should include normal operation and contingencies such as line and generator outages, etc. By judiciously selecting the encompassing set should result in a reduction of the number of conditions that need to be studied and result in an acceptable stabilizer design.

In the following sections the excitation system [1], which includes the automatic voltage regulator (AVR), is modelled by a simple first-order transfer function. In practice for the tuning of the PSS an accurate model of the excitation system and associated parameters are required. The excitation system should be properly tuned and the model validated by measurements. Because the excitation system is in the PSS control loop the resulting performance of the PSS is likely to be poor if the that model is inadequate.

5.2 Heffron and Phillips' Model of single machine - infinite bus system

De Mello and Concordia based their analysis on a linearized model, developed by Heffron and Phillips [6], of a single machine connected to an infinite bus through an external impedance, $r_e + jx_e$, representing the sum of the impedances of the generator transformer and the Thévenin equivalent impedance of the system. The Heffron and Phillips model includes a third-order representation of the generator and a first-order model of the excitation system as shown in Figure 5.1. The constants K_1 to K_6 are defined in [1] and are given in Appendix 5–I.1.

1. According to [7] the excitation system is comprised of that "equipment providing field current for a synchronous machine, including all power, regulating, control, and protective elements". The regulation of terminal voltage is a function of the AVR.

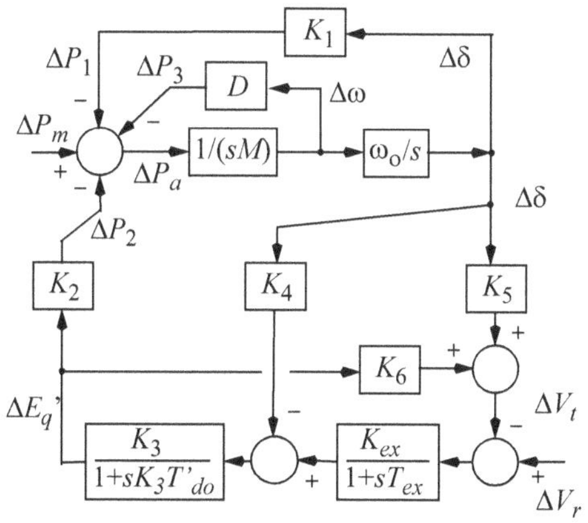

Figure 5.1 Heffron and Phillips' Model of a SMIB system.

All variables are in the Laplace (or s) domain, although much of the later analysis is conducted in the frequency domain with $s = j\omega_f$, where ω_f is the frequency in rad/s. The per-unit perturbations in the variables are defined below:

ΔP_1 Electrical torque, a function of rotor angle
ΔP_2 Torque of electro-magnetic origin
ΔP_3 Damping torque associated with windage, friction, and losses in the damper windings
ΔP_a Accelerating torque acting on the shaft of the generator-turbine unit
ΔP_m Prime mover torque
$\Delta E_q'$ Voltage proportional to direct axis flux linkages
ΔV_t Terminal voltage
ΔV_r Reference voltage
$\Delta\omega$ Rotor speed
$\Delta\delta$ Rotor angle

5.3 Synchronizing and damping torques acting on the rotor of a synchronous generator

In the context of a linearized model of a single-machine infinite-bus system, de Mello and Concordia [1] developed the concept of a complex torque $\Delta P(s)$ of electro-magnetic origin acting on the shaft of a generator. For reasons discussed shortly, the component of this torque in-phase with speed was called a *damping torque* and that component in-phase with rotor angle was called a *synchronizing torque*. Both these torques are *braking* torques, that is, they act to oppose changes in speed or rotor angle, respectively.

In the context of the Heffron and Phillips' model of Figure 5.1, let us consider the torque of electro-magnetic origin, ΔP_0, developed on the rotor of the generator. If the perturbations in prime-mover torque are negligible, $\Delta P_m = 0$, the sum of the electrical torques acting on the rotor can be expressed as

$$\Delta P_0(s) = \Delta P_a = \Delta P_1(s) + \Delta P_2(s) + \Delta P_3(s), \tag{5.1}$$

$$\text{or } \Delta P_0(s) = K_1 \Delta\delta(s) + K_2 \Delta E_q'(s) + D\Delta\omega(s). \tag{5.2}$$

Based on (4.58) the relation between the rotor angle (rad) and the per-unit speed can be expressed as

$$\Delta\delta(s) = \omega_0 \Delta\omega(s)/s, \tag{5.3}$$

in the s-domain. The dependency on the Laplace operator 's', i.e. (s), will not be shown when the variables are clearly in the s-domain.

The terms in (5.2), other than $K_2 \Delta E_q'$, are expressed in terms of $\Delta\omega$ and $\Delta\delta$. Let the transfer function between $\Delta\omega$ and $\Delta E_q'$ be $(\Delta E_q')/\Delta\omega$, and let that between $\Delta\delta$ and $\Delta E_q'$ be $\Delta E_q'/\Delta\delta$ where, based on (5.3),

$$\frac{\Delta E_q'}{\Delta\delta} = \frac{s}{\omega_0} \cdot \frac{\Delta E_q'}{\Delta\omega}. \tag{5.4}$$

The component of electrical torque $K_2 \Delta E_q'(s)$ in (5.2) can be split into components, a real component in phase with rotor angle and a real component in phase with speed. The expression for $\Delta P_0(s)$ can then be written in the form:

$$\Delta P_0 = [K_1 + K_2 \Re(\Delta E_q'/\Delta\delta)]\Delta\delta + [D + K_2 \Re(\Delta E_q'/\Delta\omega)]\Delta\omega. \tag{5.5}$$

The paths in the Heffron and Phillips' model encompassed by transfer function $\Delta E_q'/\Delta\delta$ are shown in Figure 5.2. This transfer function can be found by block diagram manipulation to be

$$\frac{\Delta E_q'}{\Delta\delta} = \frac{-K_3[K_4(1 + sT_{ex}) + K_5 K_{ex}]}{s^2 K_3 T_{ex} T'_{d0} + s(T_{ex} + K_3 T'_{d0}) + (1 + K_3 K_6 K_{ex})}. \tag{5.6}$$

Referring to (5.5), let us define the coefficients of $\Delta\delta$ and $\Delta\omega$ as $k_s(s)$ and $k_d(s)$ respectively, i.e.

$$k_s = K_1 + K_2 \Re(\Delta E_q'/\Delta\delta) \quad \text{and} \quad k_d = D + K_2 \Re(\Delta E_q'/\Delta\omega), \tag{5.7}$$

$$\text{hence (5.5) becomes} \qquad \Delta P_0 = k_d \Delta\omega + k_s \Delta\delta. \tag{5.8}$$

The terms $k_d(s)\Delta\omega(s)$ and $k_s(s)\Delta\delta(s)$ in (5.8) are damping and synchronizing torques, respectively. Consequently $k_d(s)$ is known as a *damping torque coefficient* and $k_s(s)$ as a *synchronizing torque coefficient.*

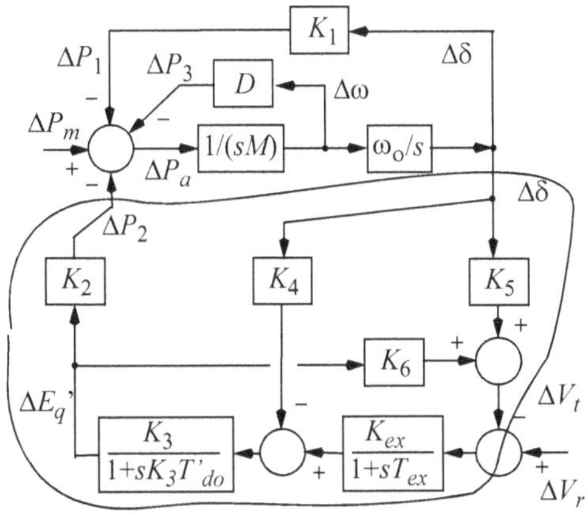

Figure 5.2 Evaluation of the transfer function $\Delta E_q'/\Delta\delta$ in the Heffron and Phillips' model.

As mentioned earlier, for analysis in the frequency domain, let $s = j\omega_f$ where ω_f is the frequency in rad/s. From the shaft relation (5.3), we note that in the frequency domain the rotor angle and speed are related by $\Delta\delta(j\omega_f) = -j(\omega_0/\omega_f)\Delta\omega(j\omega_f)$. Equation (5.8) can thus be written in the form:

$$\Delta P_0(j\omega_f) = k_d\Delta\omega(j\omega_f) - j(\omega_0/\omega_f)k_s\Delta\omega(j\omega_f) = (a+jb)\Delta\omega(j\omega_f), \tag{5.9}$$

where $a = k_d$ and $b = -(\omega_0/\omega_f)k_s$ are *torque coefficients* and are respectively the real and imaginary parts of the transfer function $\Delta P_0(j\omega_f)/\Delta\omega(j\omega_f)$. Thus any *positive torque coefficient in phase with speed* produces a positive *damping* torque on the machine shaft. Correspondingly, any *positive torque coefficient in quadrature lagging on speed* implies a positive *synchronizing* torque.

A useful expression for each of the torque coefficients can be derived from (5.9), i.e.

$$k_d = \Re\left\{\frac{\Delta P_0(j\omega_f)}{\Delta\omega(j\omega_f)}\right\} \quad \text{and} \quad k_s = -\Im\left\{\frac{\omega_f}{\omega_0}\cdot\frac{\Delta P_0(j\omega_f)}{\Delta\omega(j\omega_f)}\right\}. \tag{5.10}$$

Later in Section 5.10.2 reference is made to the phrase '*disabling the shaft dynamics of the machine*' for the purpose of calculating the torque coefficients. The shaft dynamics are disabled if the speed signal is completely isolated, for analysis purposes, from the accelerating torque acting on the shaft. This achieved by opening the output of the block $1/(sM)$ in Figure 5.2 [1]. We can now treat the speed signal $\Delta\omega$ as an input signal to the transfer function $\Delta P_0/\Delta\omega$ in (5.10), and hence calculate the torque coefficients.

1. Since $M = 2H$, setting the machine inertia constant, H, to infinity serves the same purpose [4].

5.4 The role of the Power System Stabilizer - some simple concepts

Before further examining the Heffron and Phillips' model in detail, and in order to provide some insight, some simple concepts about the role, function and effects of a PSS can be demonstrated by simplifying the Heffron and Phillips' model shown in Figure 5.1. Let us assume that (i) there are no perturbations in the reference voltage V_r, (ii) the exciter and open-circuit time constants, T_{ex} and $K_3 T'_{d0}$ (with $K_3 \approx 0.3$), respectively, are very short, and (iii) any disturbances to the system occur through mechanical torque perturbations, ΔP_m. The model of the generator in Figure 5.1 can then be simplified to the second-order model shown in Figure 5.3 in which the torques acting of the shafts and the resulting speed and angular deviations are shown [1].

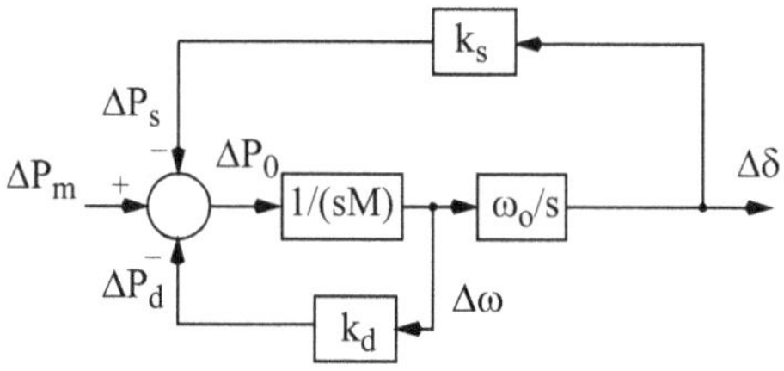

Figure 5.3 A very simplified model of the generator in a SMIB system.

In Figure 5.3 the torque of electro-magnetic origin ΔP_0 acting on the rotor is the sum of the synchronizing and damping torques, ΔP_s and ΔP_d, respectively, i.e.

$$\Delta P_0(s) = \Delta P_s + \Delta P_d = k_s \Delta\delta + k_d \Delta\omega . \qquad \text{((5.8) repeated)}$$

The transfer function between the mechanical torque perturbation and the rotor angle response of the generator shaft is

$$\frac{\Delta\delta}{\Delta P_m} = \frac{\omega_0/(2H)}{s^2 + k_d/(2H)s + \omega_0 k_s/(2H)} . \tag{5.11}$$

Typically the damping of such a system is light and its response to a step input is oscillatory. Thus there are a pair of complex poles that lie at:

$$s_1, s_2 = -\frac{k_d}{4H} \pm j\frac{1}{2}\sqrt{\frac{4\omega_0 k_s}{(2H)} - \left(\frac{k_d}{2H}\right)^2} = -\frac{k_d}{4H} \pm j\sqrt{\frac{\omega_0 k_s}{2H}} , \tag{5.12}$$

---

1. The model in Figure 5.3 has the same form as that which is obtained with a classical generator model. In the latter it is assumed that the voltage proportional to d-axis flux-linkages are constant during the study period (i.e. $\Delta E_q' = 0$).

when $k_d \ll 2\sqrt{\omega_0 k_s \cdot 2H}$. The frequency of the damped rotor oscillations (the imaginary part of the pole) is $\omega_d \approx \sqrt{\omega_0 k_s/(2H)}$. From a consideration of the pole locations we can conclude that as the damping torque coefficient k_d is increased (hypothetically) from some initial value:

- The poles shift more-or-less directly to the left [1] in the complex *s*-plane at a constant frequency of oscillation as long as the associated damping ratio is less than about 0.2 - 0.3.

- The *damping constant*, defined in Section 2.8.2.1, is $\alpha = k_d/(4H)$, i.e. it is inversely proportional to *H*, the inertia constant [2]. For example, consider two generating units with identical per-unit parameters on machine base but their inertia constants are in the ratio of 2:1, say; of the two, the lighter unit is the better damped.

- The *frequency of damped oscillation* is proportional to the square root of ratio $\omega_0 k_s/(2H)$. Thus, if switching lines out of service halves the synchronizing torque coefficient - or if the inertia constant of a unit is doubled - the frequency of oscillation is decreased by about 30% (to $1/\sqrt{2}$).

- If the damping ratio exceeds ~0.2 as k_d is increased, the trajectory of the complex poles move along a semi-circle of constant radius in the complex *s*-plane (see Figure 2.8).

- If the value of $-\alpha = k_d/(4H)$ is positive (i.e. the damping torque coefficient k_d is negative) then, according to (5.12), a pair of poles lie in the right-half of the complex *s*-plane; the system is therefore unstable.

Let us assume that we can add a feedback loop from rotor speed $\Delta\omega$ to the torque signal ΔP_{dp} - as shown in Figure 5.4 - such that $\Delta P_{dp} = k\,\Delta\omega$. It is clear that increasing the gain *k* has the same effect as increasing the *damping torque coefficient* k_d, that is, enhancing the damping of rotor oscillations. A PSS is a device that ideally induces on the rotor a torque of electro-magnetic origin proportional to speed perturbations. The 'ideal' PSS gain *k*, which is a *damping torque coefficient,* we shall call the *damping gain* of the PSS. The 'ideal' PSS will produce a direct left-shift in the rotor mode, as manifested in (5.12), from $-k_d/(4H)$ to $-(k+k_d)/(4H)$. The gaol in the tuning of a practical PSS is to achieve the same result, the damping gain of the PSS being adjusted to meet the specifications on damping for the rotor modes of oscillation.

---

1. By 'direct left-shift' is implied that the eigenvalue / mode shift is $-\alpha \pm j0$, $\alpha \geq 0$.
2. Some additional significance of this result is derived from the analysis in Section 13.2.2.

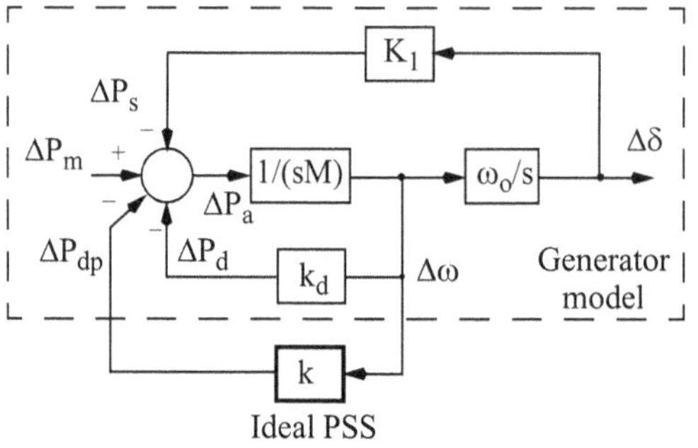

Figure 5.4 The ideal PSS introduces a pure damping torque ΔP_{dp} on the rotor of the generator.

As noted Section 5.1, the PSS feeds back a signal proportional to speed, i.e. the rate of change of position, a technique known as "rate feedback". The technique is commonly used in conventional control system design to control speed of a rotating device [1], to improve the damping - or to stabilize - a closed-loop system. Under ideal steady-state conditions the perturbations in speed are zero in a stable system - and thus rate feedback acts only when the system is disturbed in some way, e.g. by noise - which is always present.

Note that speed appears to be the ideal stabilizing signal because the damping torque induced on the shaft of the generator by the associated PSS is related to speed through a simple gain. Moreover, this gain being a damping torque coefficient has practical significance, e.g. a damping gain in the range 20 - 30 pu on machine MVA rating is a moderate damping gain setting for a speed-PSS. Practical forms of the "speed-PSS", such as the "integral-of-accelerating-power PSS", are discussed in Chapter 8.

In practice the 'ideal' speed-PSS, consisting simply of a damping gain k, is replaced by a transfer function $H_{PSS}(s) = kG(s)$, where $G(s)$ consists of a compensating transfer function and relevant filters. In this chapter the design of the transfer function $kG(s)$ is outlined for a SMIB system to illustrate a number of concepts which are also applicable to multi-machine systems.

5.5 The inherent synchronizing and damping torques in a SMIB system

Prior to attempting to discuss or tune the PSS for a generator it is instructive to ascertain the levels of the inherent damping of the rotor modes. One approach is to calculate the eigenvalues of the system. A second approach is to determine the natural or inherent damping torque coefficient over the range of rotor frequencies for the system (although in this case

1. In order to control speed the feedback signal (from the 'ideal PSS') must negated at the summing junction as shown in Figure 5.4.

of a SMIB system there is only one mode of rotor oscillation). In order to provide a complementary point of view, the second approach will be adopted.

Based on the Heffron and Phillips' model of Figure 5.1 and equations (5.7) and (5.8) the synchronizing and damping torque coefficients can be determined for the SMIB system. In the absence of a PSS, k_s and k_d in (5.7) are the synchronizing and damping torque coefficients *inherently* produced by the generator. Using a simple example let us calculate these torque coefficients to examine not only their values in the vicinity of the frequency of the single rotor mode but also how they vary with the generator loading.

5.5.1 Example 5.1

For present purposes a third-order model of the generator and a first-order model of the excitation system are used. The unit is connected to an infinite bus through transformer and transmission line reactances representing the external system.

The parameters of a SMIB system and the steady-state operating conditions are:

Generator: $D = 0$, $H = 3.0$ MWs/MVA, $r_a = 0$, $x_d = 1.9$ pu, $x_q = 1.8$ pu, $x'_d = 0.30$ pu, $T'_{do} = 6.5$ s, rating 250 MVA.
Exciter: $K_{ex} = 200$ pu, $T_{ex} = 0.02$ s.
Transformer and line reactance: $x_t = 0.15$ pu and $x_L = 0.225$ pu, respectively.
The generator is under closed-loop voltage control, terminal voltage $V_t = 1.0$ pu.
Operating Conditions: System frequency = 50 Hz. Rated real power output is $P = 0.9$ pu, and reactive power outputs are $Q = -0.20, 0.0, 0.2$ and 0.4 pu.
The machine and system base is 250 MVA. The constants K_1 to K_6 are defined in [1] and Appendix 5–I.1.

Note. For present purposes it is assumed that this SMIB system is a representation of a generator and transmission system within a multi-machine system in which the electro-mechanical modes may range from 1.5 - 15 rad/s. While there is a single rotor mode in this example, the PSS is to be tuned for the latter frequency range.

For each operating condition at 0.9 pu real power output the rotor angle, the terminal voltage angle, the infinite-bus voltage, the K-constants, the eigenvalues for the rotor mode, and the inherent synchronizing and damping torque coefficients are given in Table 5.1. The eigen-analysis reveals that all the real parts of the rotor mode are positive and thus the system is unstable for the range of reactive power outputs of the generator; the frequency of the unstable rotor modes is between 7.9 and 9.3 rad/s. The damping and synchronizing torque coefficients for all operating conditions are given at a frequency of 8.69 rad/s, a value close to the mid-range of the frequencies of oscillation of the rotor modes. The negative damping torque coefficients at 8.69 rad/s (-16.9 to -2.3 pu) are consistent with the unstable rotor modes. Relative to the damping torque coefficients the variation of the synchronizing torque coefficients (1.6 to 1.2 pu) is much less over the range of operating conditions.

Table 5.1 *K*-constants, rotor modes and inherent torque coefficients k_d, k_s, for $P = 0.9$ pu,

Q pu	Rotor Angle deg.	Angle Term. Voltage deg.	Infinite Bus Voltage pu	K_1	K_2	K_3	K_4	K_5	K_6	Eigenvalues, rotor mode	k_d at 8.69 rad/s	k_s at 8.69 rad/s
-0.2	85.9	17.4	1.127	1.355	1.665	0.297	2.664	-0.121	0.204	$1.15 \pm j\,9.23$	-16.9	1.62
0.0	77.0	18.6	1.055	1.289	1.523	0.297	2.437	-0.072	0.292	$0.514 \pm j\,8.70$	-6.56	1.44
0.2	70.0	20.0	0.985	1.202	1.371	0.297	2.194	-0.051	0.357	$0.267 \pm j\,8.25$	-3.22	1.29
0.4	64.9	21.6	0.915	1.122	1.227	0.297	1.964	-0.048	0.404	$0.198 \pm j\,7.94$	-2.34	1.20

For later reference it is instructive to consider the frequency response of the transfer function for the inherent torque coefficients. Referring to Figure 5.2 and (5.6), the transfer function is

$$\left.\frac{\Delta P_a}{\Delta\omega}\right|_{\Delta P_m = 0} = \frac{\Delta P_1 + \Delta P_2}{\Delta\omega}$$

$$= \frac{\omega_0}{s}\left[K_1 - \frac{K_2K_3[K_4(1+sT_{ex}) + K_5K_{ex}]}{s^2K_3T_{ex}T'_{d0} + s(T_{ex} + K_3T'_{d0}) + (1 + K_3K_6K_{ex})}\right] \tag{5.13}$$

where ΔP_a is the accelerating torque. An examination of (5.13) reveals that at both low and high frequencies ($s \to j0$ or $s \to j\infty$) the transfer function rolls off at 20 dB/decade at a phase angle of $-90°$, i.e. the damping torque coefficients at these frequencies are zero. There is phase variation from $-90°$ in the intermediate frequency range as shown in the responses in Figure 5.5(a).

Note from Figure 5.5(a) the damping torque coefficient, k_d = *gain* x cosine(*phase angle*), is negative for all the selected outputs; the associated inherent damping torques in Figure 5.5(b) are therefore destabilizing. This result is consistent with those revealed in Table 5.1.

Clearly for this system the degree of instability (as revealed by both the *negative* inherent damping torque coefficient and the real part of the eigenvalue of the rotor mode) increases as the generator operates at increasingly leading power factors. The potential for small-signal instability is a characteristic of generator operation at leading power factors. A device called an under-excitation limiter is normally fitted to a machine to prevent the steady-state operating point from drifting too far into the leading power factor region.

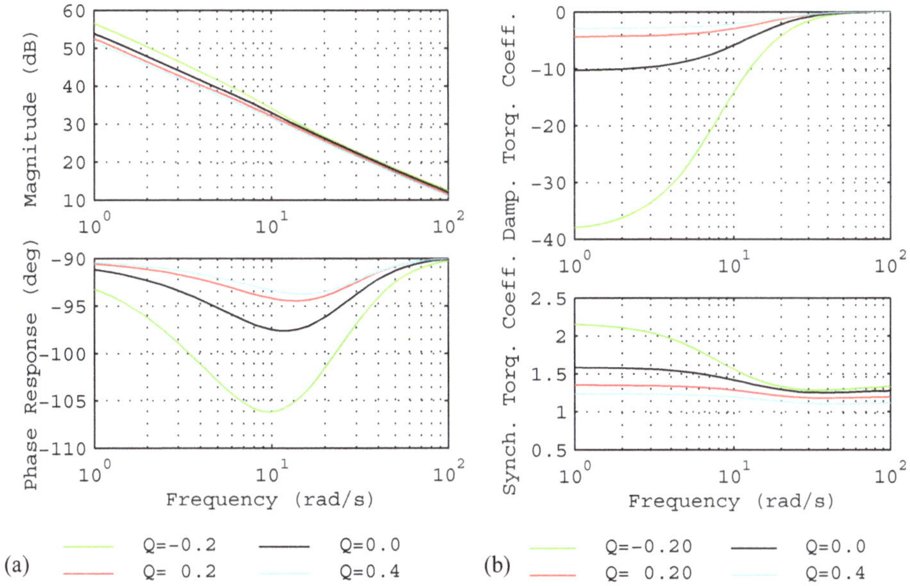

Figure 5.5 (a) Frequency response of the transfer function $\Delta P_a/\Delta\omega$ for the inherent torque coefficients for real power output $P = 0.9$ pu as reactive power Q (pu) varied. (b) The associated damping and synchronizing torque coefficients.

We have examined only a few operating conditions at rated real power output of 0.9 pu. For present purposes, however, this simple set of studies reveals that a PSS - when installed - should possess a damping gain, i.e. a positive damping torque coefficient, greater than $k = 16.9$ pu to ensure stability under the most onerous operating condition, $P = 0.9$ and $Q = -0.2$ pu.

Note from Table 5.1 that the voltage at the infinite-bus end of the high voltage lines is outside a practical range of 0.95 to 1.05 pu for three of the four operating conditions. Later in Section 5.10.4 - when a more practical system is employed - we will ensure that a more appropriate set of operating conditions is studied.

5.6 Effect of the excitation system gain on stability

In practice it is common for excitation system gains to lie in the range of 200 to 400 pu [1], or more. High gains may be desirable to satisfy requirements for steady-state voltage regulation, or to boost generator field flux linkages during the fault period thereby enhancing syn-

1. For example, subject to certain conditions, a requirement in a set of Rules [8] requires that generators must have an excitation control system that regulates voltage at an agreed location to within 0.5% of the set-point. This implies an excitation system gain of at least 200 pu.

chronizing torques between generation in the early post-fault period. In the case of small-signal dynamic performance, the effect of the excitation system gain on damping torques, and therefore stability, is analysed in detail in [1].

From Table 5.1 it is observed that the system is unstable for the range of steady-state operating conditions covered in the studies. Because the excitation system gain of 200 pu in this application is considered fairly high, it is instructive to assess the effect of lower and higher gains on the rotor modes.

In Figure 5.6 are shown the loci of the rotor mode as the excitation system gain is increased from zero to 300 pu for two of the operating conditions included in Table 5.1. Without a PSS this SMIB system is stable only at very low excitation system gains, i.e. less than 30 pu. As the power factor becomes less lagging, i.e. at $Q = 0$, the damping of the rotor mode tends to degrade further as the gain increases.

In this simple model of the generator and excitation system the higher-order dynamics are ignored. Such a model may be satisfactory at low excitation system gains, but at high gains the effect of the unmodelled dynamics is to degrade stability. Care therefore should be taken when analysing simplified low-order models in high-gain excitation systems.

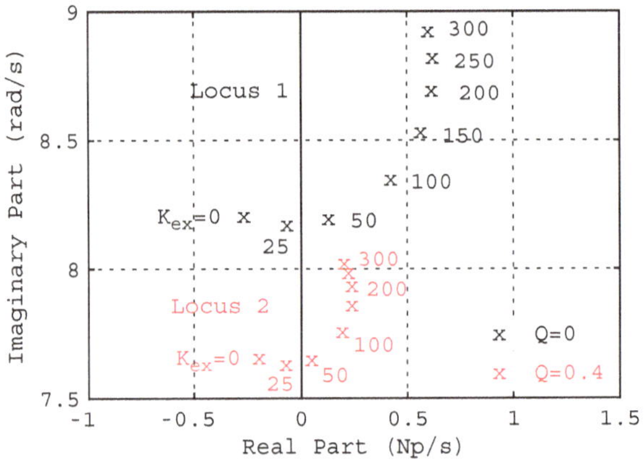

Figure 5.6 Variation of the rotor mode with increasing excitation system gain K_{ex} for two steady-state operating conditions, $P = 0.9$ and $Q = 0$, $Q = 0.4$ pu. No PSS in service.

5.7 Effect of an idealized PSS on stability

In Section 5.4 it was noted for the very simplified model of the SMIB system that the left-shift in the rotor mode due to an idealized speed-PSS is $k/(4H)$, where k is the damping gain of the PSS and H is the inertia constant. Somewhat analogous to the treatment in Section 5.4 of the effect of the PSS on the rotor mode, let us consider the performance of an idealized PSS on the SMIB system represented by the Heffron and Phillips model of Figure 5.1 on page 227.

Because the element D in Figure 5.1 is a pure damping gain, $D = \Delta P_{dp}/\Delta\omega$, it can be replaced - or augmented - by an idealized speed-PSS of gain k. Setting $k = D = 20$ pu and by calculating the poles of the closed-loop transfer function $\Delta V_t(s)/\Delta V_r(s)$ ((5.69) Appendix 5–I.2), the effect on the rotor mode of such a PSS can be assessed.

Let us consider in Table 5.1 on page 234 the steady-state condition $P = 0.9$, $Q = 0$ pu for which the rotor mode is $0.514 \pm j8.70$ with no PSS in service. Based on the above-mentioned calculation the rotor mode with the idealized PSS is $-1.040 \pm j8.57$. The mode shift due to the action of the idealized PSS is $-1.554 \mp j0.13$ which, for practical purposes, is a direct left-shift of 1.554 Np/s. This is in good agreement with the shift of $k/(4H) = 1.667$ Np/s for an inertia constant of $H = 3$ MWs/MVA. The result will be reviewed later when a practical PSS is employed with the Heffron and Phillips model of the SMIB system.

5.8 Tuning concepts for a speed-PSS for a SMIB system

The simple Heffron and Phillips model for the linearized SMIB system, shown in Figure 5.1 on page 227, will used to illustrate the concepts employed in the tuning of a PSS based on the so-called 'P-Vr' method.

As foreshadowed in Section 5.4, the form of the transfer function of the speed-PSS is

$$H_{PSS}(s) = kG(s) \quad \text{where} \quad G(s) = G_c(s) \cdot G_W(s) \cdot G_{LP}(s); \tag{5.14}$$

the right-hand-side transfer functions are defined below. Let us now consider a five-step procedure for the tuning of the PSS transfer function $H_{PSS}(s)$.

1. Determine the compensating transfer function $G_c(s)$ such that, over a selected range of modal frequencies, a torque of electro-magnetic origin proportional to speed is induced by the PSS on the shaft of the generator.

2. Select the value of the damping gain k.

3. Select the parameters of the PSS washout filter $G_W(s)$. This filter blocks steady-state offsets (i.e. DC signals) and significantly attenuates low-frequency signals below the range of rotor modal frequencies. Ideally, its transfer function over the range of rotor modal frequencies is $1\angle 0°$.

4. Select the parameters of the low-pass filter $G_{LP}(s)$ that significantly attenuates high-frequency signals above the range of rotor modal frequencies. In the range its transfer function is ideally $1\angle 0°$. The filter may also attenuate the higher-frequency shaft torsional modes to ensure that they are not excited by the PSS. If a true-speed input PSS is employed, which is rare in practice, it may require specialized filters to attenuate the torsional modes.

5. Implement the PSS tuning, assess its performance, and confirm the validity of the design over an encompassing range of operating conditions.

These steps are discussed in the following sections.

5.8.1 Determination of compensating transfer function

The determination of compensating transfer function $G_c(s)$ is the first step in the PSS tuning procedure outlined above.

The output of the speed-PSS is injected at the summing junction of the voltage reference and terminal voltage signals - or at some point in the AVR of low signal-power level. The PSS forms a feedback loop as shown in dashed lines in Figure 5.7. Note that the sign on the output signal of the PSS is (maybe unexpectedly) $+\Delta V_s$. However, comparing Figure 5.7 with Figure 5.4, we note that the negation already occurs at the output ΔP_2 of the internal path $\Delta P_2 <\!-\!\!-\Delta\omega$ in Figure 5.7 and hence a positive sign must arise the input summing junction.

From Steps 3 and 4 above it follows that, over the range of frequencies of the electro-mechanical modes (i.e. 1.5 to 15 rad/s), the product of the washout and the low-pass filter transfer functions $G_W(s)G_{LP}(s)$ should be close to $1\angle 0°$. Under these conditions the PSS transfer function of (5.14) is of the ideal form

$$H_{PSS}(s) = kG_c(s), \tag{5.15}$$

where k is the desired damping torque coefficient (or PSS damping gain), and $G_c(s)$ is a compensating transfer function.

From an examination of Figure 5.1 on page 227 the equation for the torque of electromagnetic origin is, in (5.1),

$$\Delta P_0(s) = \Delta P_1(s) + \Delta P_2(s) + \Delta P_3(s), \quad \Delta P_m = 0.$$

The action of the speed-input PSS is to induce a torque of electromagnetic origin proportional to speed on the shaft of the generator through the electro-magnetic components of torque ΔP_1 and ΔP_2, shown in Figure 5.7. The system being linear, the principle of superposition can be employed to derive expressions for each of the torque components in terms of the relevant variables, and then combine them appropriately.

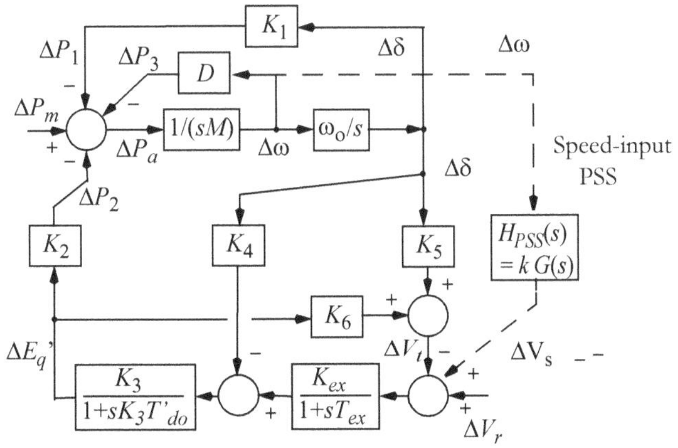

Figure 5.7 Inclusion of a speed-stabilized PSS in the SMIB model.

The paths in the Heffron and Phillips' model encompassed by transfer function $\Delta P_2/\Delta\delta$ are shown by the solid line in Figure 5.8(a). In the absence of perturbations in reference voltage ΔV_r (or stabilizing signal ΔV_s) the latter transfer function can be shown by block diagram manipulation to be:

$$:\left.\frac{\Delta P_2}{\Delta\delta}\right|_{\Delta V_r = 0} = -\left(\frac{(K_2K_3)[K_4(1+sT_{ex})+K_5K_{ex}]}{s^2K_3T_{ex}T'_{d0}+s(T_{ex}+K_3T'_{d0})+(1+K_3K_6K_{ex})}\right) \tag{5.16}$$

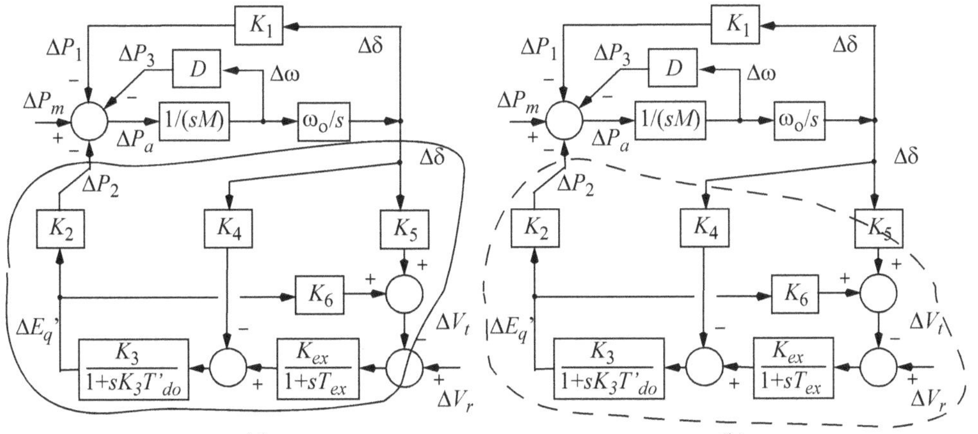

(a) (b)

Figure 5.8 Evaluation of transfer functions,
(a) $(\Delta P_2/\Delta\delta)|_{\Delta V_r = 0}$ (b) $(\Delta P_2/\Delta V_r)|_{\Delta\omega = 0}$

The transfer functions (5.16) and $\Delta P_1/\Delta\delta = K_1$ can be combined into a single transfer function, $H_{P\delta}(s)$, i.e.

$$H_{P\delta}(s) = \left.\frac{\Delta P_\delta}{\Delta\delta}\right|_{\Delta V_r = 0} = K_1 + \left.\frac{\Delta P_2}{\Delta\delta}\right|_{\Delta V_r = 0}. \tag{5.17}$$

Let us now examine the path in Figure 5.8(b) from the voltage reference ΔV_r to the torque component ΔP_2. The blocks enclosed by the dashed line form the transfer function [1]:

$$H_{PVr}(s) = \left.\frac{\Delta P_2}{\Delta V_r}\right|_{\Delta\omega = 0} = \frac{K_2 K_3 K_{ex}}{s^2 K_3 T_{ex} T'_{d0} + s(T_{ex} + K_3 T'_{d0}) + (1 + K_3 K_6 K_{ex})}. \tag{5.18}$$

A form of this transfer function that will be useful later is:

$$H_{PVr}(s) = \frac{k_c}{1 + c_1 s + c_2 s^2}. \tag{5.19}$$

It has been emphasized that Figure 5.8(b) and (5.18) represent the transfer function from the voltage reference to a component of the torque of electromagnetic origin for no perturbations in the steady-state rotation of the rotor, i.e. $\Delta\omega = \Delta\delta = 0$; this, in effect, implies that the *shaft dynamics are disabled.* Importantly, the transfer function $H_{PVr}(s)$, which is calculated *with the shaft dynamics disabled,* will be referred to in the following chapters as the *P-Vr transfer function* or the *P-Vr characteristic of the generator.*

A useful modified form of Figure 5.7 is represented by the block diagram of Figure 5.9. The P-Vr transfer function $H_{PVr}(s)$ is formed as $(\Delta P_2/\Delta V_r)|_{\Delta Vs = 0}$ or $(\Delta P_2/\Delta V_s)|_{\Delta Vr = 0}$. The blocks in Figure 5.9 representing the dynamic behaviour of the shaft are shown as $J(s) = 1/(sM)$ and $N(s) = \omega_o/s$. Turbine / governor action can be included - if desired - in the block $H_{GOV}(s)$. The transfer function of the speed-PSS is represented in its ideal form as

$$H_{PSS}(s) = \frac{\Delta V_s}{\Delta\omega} = kG_c(s). \tag{5.20}$$

Observe that in Figure 5.9 there are three distinct feedback paths, namely, the path through the PSS and the transfer function $H_{PVr}(s)$, the path through the rotor angle, and the path through the governor and turbine. We shall find the formation of the first two of the separate paths is useful and revealing in the analysis of both single- and multi-machine systems.

---

1. This transfer function is related to the GEP(*s*) function as explained in Section 6.4

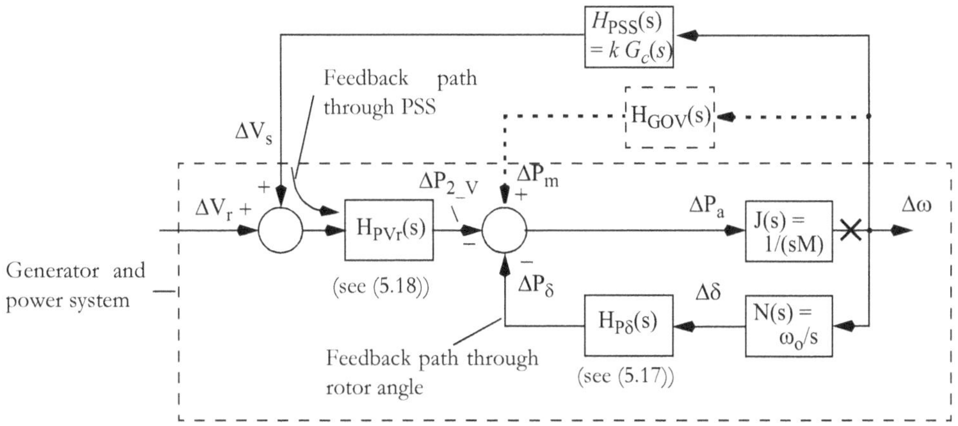

Figure 5.9 Modified form of Heffron and Phillips' model including PSS and turbine/governor transfer functions. (**X** is the point at which the speed path is opened to disable the shaft dynamics.)

From Figure 5.9 we note that the output of block $H_{PVr}(s)$ comprises the superposition of two components $\Delta P_2 = H_{PVr}[\Delta V_r + \Delta V_s]$. For no disturbances at the reference input,

$$\begin{aligned} \Delta P_2\big|_{\Delta V_r = 0} &= \frac{\Delta P_2}{\Delta V_s} \cdot \frac{\Delta V_s}{\Delta\omega} \cdot \Delta\omega \\ &= H_{PVr}(s) \cdot kG_c(s) \cdot \Delta\omega \end{aligned} \tag{5.21}$$

when each input-output relation is replaced by its transfer function. By definition, the term $H_{PVr}(s) \cdot kG_c(s)$ in (5.21) is a *damping torque coefficient* (see Section 5.3). It is clear that a torque of electromagnetic origin in phase with speed is induced by the PSS if this coefficient is set to the scalar PSS damping gain, *k*, i.e.

$$H_{PVr}(s) \cdot kG_c(s) = k, \tag{5.22}$$

$$\text{thus} \quad G_c(s) = 1/H_{PVr}(s). \tag{5.23}$$

Note that $G_c(s)$ is chosen to compensate for the transfer function $H_{PVr}(s)$. Following the substitution of (5.23) in (5.20), the basic transfer function for the speed-PSS becomes

$$H_{PSS}(s) = kG_c(s) = k/H_{PVr}(s). \tag{5.24}$$

Let us determine, for the Heffron and Phillips' model, the compensating transfer function to be implemented by the PSS. Substitution of (5.18) in (5.23) yields:

$$G_c(s) = \frac{s^2 K_3 T_{ex} T'_{d0} + s(T_{ex} + K_3 T'_{d0}) + (1 + K_3 K_6 K_{ex})}{K_2 K_3 K_{ex}}. \tag{5.25}$$

In an alternative form this compensating transfer function can be expressed as:

$$G_c(s) = \frac{(1+K_3K_6K_{ex})}{K_2K_3K_{ex}} \cdot \left[1 + s \cdot \frac{T_{ex} + K_3T'_{d0}}{1+K_3K_6K_{ex}} + s^2 \cdot \frac{K_3T_{ex}T'_{d0}}{1+K_3K_6K_{ex}}\right]. \tag{5.26}$$

As has now been established for the single-machine system, we will find that in the multi-machine case the compensation provided by the PSS is simply the inverse of the P-Vr transfer function $H_{PVr}(s)$; it will form the *basis for the tuning of PSSs* in the following sections and chapters. Note:

- We are proposing to cancel the set of poles in the transfer function $\Delta P_2/\Delta V_r$ of (5.18) with a corresponding set of zeros; this is allowable as long as the poles of (5.18) are stable.

- The transfer function $G_c(s)$ in (5.25) is not proper (see definition in Section 2.2), however, this is remedied later.

The basis for the calculation of the P-Vr transfer function can be established from Figure 5.9 with the PSS and governor blocks removed from the diagram. The torque of electromagnetic origin is $\Delta P_{2-V} + \Delta P_\delta$. If, in the figure, the speed output from the block *J*(*s*) is opened at **X** then the speed signal - and thus the shaft dynamics - are disabled ($\Delta\omega = \Delta\delta = 0$). Under this condition the output of the block $H_{P\delta}(s)$, ΔP_δ, is zero and thus the P-Vr transfer function can be calculated directly from the transfer function $(\Delta P_{2-V}/\Delta V_r)|_{\Delta\omega = 0}$.

The inherent damping and synchronizing torques of the generator are supplied through the block $H_{P\delta}(s)$. As was revealed in Example 5.1, Section 5.5.1, the inherent damping torques may augment or degrade the damping torque induced by the PSS over the frequency range of the rotor modes of interest.

For this simple SMIB system, with a third-order model of the machine, the compensating transfer function $G_c(s)$ of the PSS can be calculated *directly* from (5.26) for a given operating condition. We shall see in Chapter 10 for higher-order generator models and multi-machine systems that such a direct closed-form calculation is not practical and alternative methods will be employed for determining the P-Vr characteristic and the associated compensating transfer function $G_c(s)$. However, for present purposes, the P-Vr transfer function is calculated directly from (5.18); this is illustrated in the following example.

5.8.2 The nature of the P-Vr characteristic

In a theoretical analysis [10] it is shown that the P-Vr transfer function of a machine consists of two components. The first component depends on the parameters of the generator and its AVR/exciter and is independent of the external system. On the other hand, the second component depends on both the parameters of the generator and the dynamics of all other generators in the system; however, this second term is dominated by the Thévenin equivalent impedance seen from the terminals of the generator looking into the rest of the system. This result provides a theoretical basis for the observation in [5] that the P-Vr transfer function is relatively robust to changes in the system operating conditions. Typically the frequency response of the P-Vr transfer function for both the gain (for real power outputs greater than 0.7 pu) and the phase shift - in particular - do not vary appreciably over a wide range of operating conditions and system configurations. Consequently, PSSs designed based on the synthesized P-Vr transfer function are also robust over a wide range of operating conditions.

It must be emphasized for future reference: *The PSS must be tuned to be robust* [1] *to a full range of N and N-1 operating conditions. For this purpose it is necessary to select a set of operating conditions which encompass, and therefore include, the range of conditions.* Be examining the bordering conditions *this approach reduces the number of cases for which the P-Vr characteristics must be evaluated.*

5.8.3 Example 5.2: Evaluate the P-Vr characteristics of the generator and determine the PSS compensating transfer function.

The P-Vr characteristics for the generator in the SMIB system of Example 5.1 are to be calculated using the P-Vr transfer function (5.18) when the real power output is $P = 0.9$ pu and the reactive power outputs are $Q = -0.2$, 0.0, 0.2 and 0.4 pu. The P-Vr frequency response plots are calculated using the machine and system parameters listed in Example 5.1, together with the *K*-constants found in Table 5.1 on page 234.

The P-Vr frequency response plots which are shown in Figure 5.10 cover the range of modal frequencies of 7.9 to 9.3 rad/s for the four operating conditions, Cases A to D.

Based on a mid-range modal frequency of 8.7 rad/s [2] the phase of the P-Vr characteristics are, from Figure 5.10, -58.9°, -47.7°, -41.3° and -37.5° for values of $Q = -0.2$, 0, 0.2 and 0.4 pu, respectively. We will select the P-Vr characteristic for Case B, $Q = 0$, as its phase of -47.7° lies close to the middle of the band of characteristics at the modal frequency, assuming for the present that the PSS is to be tuned for a real power output of $P = 0.9$ pu. This means the phase of the characteristic for *any other value* of Q, $-0.2 < Q < 0.4$, will be within about 11° of that of the selected P-Vr characteristic. We shall find later, for a range of more practical operating conditions, the latter phase variation is typically within ±5° to ±10° of the selected characteristic [3].

---

1. See item 3 of Section 1.2
2. This is the frequency of the unstable rotor mode for $P = 0.9$, $Q = 0$.
3. A wider spread exceeding ±10° may not be unusual depending on the generator parameters and if the range of leading and lagging power outputs of the machine is wide. The rated reactive power range of the generator may be wide but not necessarily achievable due to voltage constraints, transformer tap ranges, etc.

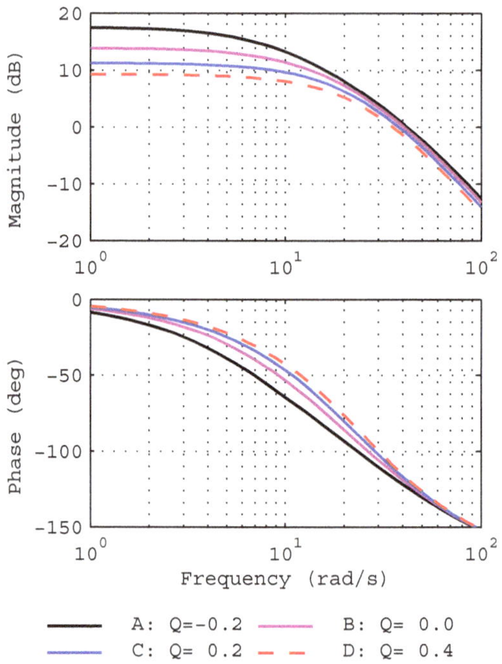

Figure 5.10 P-Vr characteristics of SMIB System for $P = 0.9$, $Q = -0.2$, 0.0, 0.2 and 0.4 pu.

The steady-state or static gain of the P-Vr transfer function is, from (5.18), $K_2K_3K_{ex}/(1 + K_3K_6K_{ex})$. The static gain range is within ±4.5 dB of the gain of the P-Vr characteristic for Case B, $Q = 0$. Later, in Chapter 10 for the multi-machine system, we shall find that over the range of operating conditions the gain variation is typically within ±3 dB of the selected characteristic. However, in this example the range of reactive power outputs in Cases A and D may be considered to represent the more extreme conditions and may be weighted accordingly. We shall refer to a selected or representative P-Vr characteristic such as Case B as the "Design Case".

We have now determined the P-Vr characteristic for which the compensating transfer function of the PSS is to be evaluated. Substitution in (5.26) of the relevant parameters given in Example 5.1, together with the K-constants for $Q = 0$ from Table 5.1 on page 234, yields the compensating transfer function:

$$G_c(s) = 0.20263(1 + s0.10638 + s^2 0.0021058)\,. \tag{5.27}$$

As noted earlier the above transfer function is not proper; this will be remedied in Section 5.8.6. It has also been noted that the required form of the PSS transfer function is $kG_c(s)$. We will now determine the damping gain k.

5.8.4 Determination of the damping gain *k* of the PSS

The determination of the damping gain *k* is the second step in the PSS tuning procedure outlined in Section 5.8.

In Section 5.4 and in the associated Figure 5.4 it is shown that an ideal PSS introduced a damping torque of electro-magnetic origin proportional to speed, $\Delta P_{dp} = k\,\Delta\omega$, where *k* is both a damping torque coefficient and the damping gain of the ideal PSS. Likewise, we find from consideration both of the loop through the PSS in Figure 5.9 and of equations (5.21) - (5.23) that a torque proportional to speed is produced, i.e.

$$\Delta P_{2-V}\big|_{\Delta Vr = 0} = H_{PSS}(s) \cdot H_{PVr}(s)\Delta\omega = kG_c(s) \cdot (1/G_c(s)) \cdot \Delta\omega = k\,\Delta\omega\,. \qquad (5.28)$$

The damping gain *k* of the PSS should be large enough to:

- swamp any inherent negative damping torques over the range of rotor modes for the range of operating conditions,
- ensure that the most lightly-damped rotor mode satisfies the criteria for system damping.

As pointed out earlier, a damping gain of 20 to 30 pu on machine rating is considered to be a moderate gain.

5.8.5 Example 5.3. Calculation of the damping gain setting for the PSS

For illustrative purposes it will be assumed that, for the most poorly damped rotor mode, the damping performance criterion as represented by its 2% settling time should be better than 8 s; this implies that the real part of the rotor mode should be more negative than $-\alpha = -4/8 = -0.5$ Np/s (see Section 2.8.2.1)

According to Table 5.1 on page 234, without a PSS the most poorly damped mode is that for the operating condition $P = 0.9$, $Q = -0.2$, i.e. $1.146 \pm j9.23$. To achieve the specified damping target of -0.5 Np/s the PSS must cause a left shift in the mode of at least $-\Delta\alpha = -0.5 - 1.146 = -1.646$ Np/s. From Section 5.4 and (5.12) we note, for a *rough* guide, that for the very simplified model of the SMIB system the left-shift ($\Delta\alpha$) in the mode associated with a damping gain k_d is $\Delta\alpha = k_d/(4H)$. For an inertia constant $H = 3$ the value of k_d required to achieve the specified left shift is $k_d = 4H\Delta\alpha = 4 \times 3 \times 1.646 = 19.75$ pu. Thus, we anticipate that a moderate PSS damping gain of $k = 20$ pu will achieve the specified damping performance. Moreover, the value for the damping gain of 20 pu will swamp the inherent damping torque coefficient of -16.9 pu for $Q = -0.2$ pu.

5.8.6 Washout and low-pass filters

5.8.6.1 The washout filter

The selection of parameters of the washout filter is the third step listed in the PSS tuning procedure in Section 5.8.

A washout filter is incorporated in the PSS to ensure that steady-state or slow changes in system frequency - and thus shaft speed - do not offset, in the steady state, the terminal voltage of the generator from its reference value. The transfer function of the washout filter is of the form:

$$G_W(s) = \left.\frac{sT_W}{1+sT_W}\right|_{s=j\omega_f}. \tag{5.29}$$

The frequency response of the filter is shown in Figure 5.11(a). Note from (5.29) that as $\omega_f \to 0$, $|G_W(j\omega_f)| \to 0$, i.e. steady-state offsets or d.c. signal levels are blocked. Signals below the corner frequency $1/T_W$ are attenuated, and the phase shift introduced by the filter tends to 90° as $\omega_f \to 0$. For frequencies much greater than the corner, $G_W(j\omega_f) \to 1\angle 0$, i.e. unity gain. A basis for the selection of the washout time constant T_W is to place its corner frequency (i.e. $1/T_W$ rad/s) about a decade below the lowest frequency of the rotor modes of oscillation, normally an inter-area mode. At a frequency a decade above the corner frequency the phase lead introduced by the washout filter is about $5°$, i.e. almost negligible for tuning purposes. In particular, this basis for the selection of the washout time constant mitigates against any excessive phase lead at low inter-area modal frequencies.

Two washout filters in cascade are often employed in practice to block slow ramp-like changes in system frequency that occur in system operation. The washout filter will be discussed more fully in Chapter 8.

5.8.6.2 The low-pass filter

The selection of the parameters of the low-pass filter is the fourth step in the PSS tuning procedure of Section 5.8.

Low-pass filters are added to attenuate high-frequency signals that would otherwise be amplified by the PSS, and to ensure that the PSS transfer function is proper. Attenuation of shaft torsional mode components is typically accomplished by specialized filtering of the speed input signal. The simplest form of the filter is

$$G_{LP}(s) = \frac{1}{(1+sT_1)(1+sT_2)\ldots(1+sT_p)}. \tag{5.30}$$

The corner frequencies associated with the time constants T_1, T_2, ... are usually placed a decade above the highest frequency of the rotor modes of oscillation, normally a local area or an inter-machine mode of oscillation. The order p of the filter is selected to provide adequate high-frequency attenuation and / or to ensure a proper PSS transfer function. The frequency response plot of a second-order low-pass filter $G_{LP}(s) = 1/(1+s/\omega_c)^2$ is shown in Figure 5.11(b).

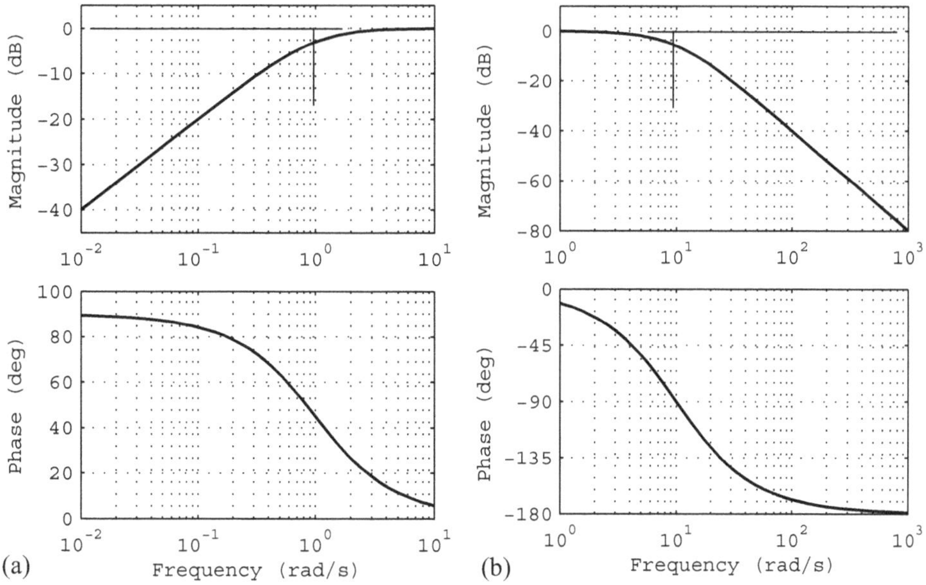

Figure 5.11 Frequency response of (a) a first-order washout filter and (b) a second-order low-pass filter. The corner frequencies are normalized to 1 and 10 rad/s, respectively. For another corner frequency, ω_c rad/s, scale frequency axes to ω_c or $\omega_c/10$.

Assume that the two corner frequencies of the second-order low-pass filter are set to 200 rad/s [1]. Based on Figure 5.11, a decade below the corner, i.e. at 20 rad/s, the phase lag is $11.4°$. Ideally, for tuning purposes the phase lags due to the low-pass corners should be small to negligible over the range of the modes of rotor oscillation, typically 1.5 to 15 rad/s. Of course, in a specific application the relevant modal frequency range might be much less. Often, in practice, the maximum corner frequency of the low-pass filters may be limited by the PSS manufacturer to values in the range from about 20 to 50 rad/s.

Note that torsional modes may not be insignificant and must be attenuated. The frequencies of these modes may be as low as 8 Hz (50 rad/s). For example, if 40 dB attenuation at 50 rad/s is required, the corner frequency of a second-order filter is 5 rad/s. Such a filter introduces a phase lag of 90° at 5 rad/s - possibly in the mid-range of modal frequencies; more-

1. For the purposes of this study the corner frequencies of 200 rad/s are chosen so that the effects of the low-pass filter over the modal frequency range are minimal.

over, the damping gain of the PSS amplifies the torsional modes if insufficiently attenuated. This conflict between the requirements of the PSS design and attenuation of the torsional frequencies is overcome either by the adoption of specialized filtering of the PSS speed-input signal [1] or by using alternative PSS structures such as the 'integral-of-accelerating-power' PSS; this is discussed in more detail in Chapter 8.

5.9 Implementation of the PSS in a SMIB System

The implementation of the PSS design, assessment of its performance, and confirmation of the validity of the design is the fifth step listed in the PSS tuning procedure of Section 5.8.

5.9.1 The transfer function of the PSS

In summary, it has been noted that the following are among the requirements for the implementation of a practical PSS:

- A washout filter (5.29) is required to eliminate the offset resulting from the steady-state level of the input signal to the PSS. Its response to frequencies in the range associated with rotor modes should ideally be $1\angle 0°$.
- The elemental PSS transfer function in (5.26) is not proper; the selection of a low-pass filter (5.30) of an appropriate order can overcome this deficiency. The response of the filter to frequencies in the range associated with rotor modes should ideally be $1\angle 0°$.
- The PSS must not excite the torsional modes of the rotors of the generator - prime-mover system. Depending on the magnitude and frequency of the torsional modes the higher-order low-pass filter having an order 2 or greater may serve this purpose - though separate specialized filtering of the PSS speed-input signal may be required.
- The damping gain setting k (pu on generator rating) is selected to have a moderate value 20 - 30 pu and / or to satisfy the criterion for system damping performance - though very high damping gain settings should be avoided.

The practical form of the PSS should include or account for the above set of requirements over the range of modal frequencies. Two general forms of the speed-PSS, $kG(s)$, which includes the washout and low-pass filters, therefore become

$$H_{PSS}(s) = kG(s) = k\left(\frac{1}{H_{PVr}(s)}\right) \cdot G_W(s) \cdot G_{LP}(s), \tag{5.31}$$

$$\text{or} \quad H_{PSS}(s) = kG(s) = kG_c(s) \cdot G_W(s) \cdot G_{LP}(s), \tag{5.32}$$

where the filter transfer functions are given by (5.29) and (5.30), respectively.

1. 'Notch filters' are used for the same purpose [9].

The compensating transfer function $G_c(s)$ of the speed-PSS (5.26) is of the form:

$$G_c(s) = \frac{1 + c_1 s + c_2 s^2}{k_c}. \tag{5.33}$$

Substitution of the latter equation and the filter transfer functions in (5.32) yields an equation containing all the relevant parameters for the practical PSS for this single machine case, i.e.

$$H_{PSS}(s) = kG(s) = k \cdot \frac{sT_W}{1 + sT_W} \cdot \frac{1}{k_c} \cdot \frac{1 + c_1 s + c_2 s^2}{(1 + sT_1)(1 + sT_2)}. \tag{5.34}$$

It is important to note that, in the context of (5.34), the gain k has been referred to as the '*damping gain*' of the PSS. If the washout filter is ignored, the 'DC' gain of the PSS is k/k_c; *conventionally this is referred to as the 'PSS Gain'*. The choice of the parameters and the implementation of the practical PSS is now examined.

5.9.2 Example 5.4. The dynamic performance of the speed-PSS

In Examples 5.1 to 5.3 a SMIB system has been analysed for the operating conditions P = 0.9 pu, Q = -0.2, 0.0, 0.2 and 0.4 pu. The single rotor mode of oscillation to which the criterion for damping must be applied is given in Table 5.1; these modes lie in the range 7.9 to 9.3 rad/s and all are unstable. The components of the tuning procedure will be now be reviewed.

The PSS compensating transfer function. This was evaluated in Example 5.2. Because the PSS is a fixed parameter device, i.e. its parameters are not changed with changes in operating conditions, the compensating transfer function is based on a P-Vr characteristic that best represented the phase responses of the characteristics over the set of operating conditions. The appropriate P-Vr characteristic was shown to be that for the operating condition $P = 0.9, Q = 0$ (Case B); the associated compensating transfer function is given by (5.27), i.e.

$$G_c(s) = (1 + s0.10638 + s^2 0.0021058)/4.935\,.$$

The washout filter. The corner frequency of the filter is placed a decade or more below the lowest rotor modal frequency, 7.9 rad/s. For a washout time constant of T_W = 5 s the filter's corner frequency is 0.2 rad/s, a value that satisfies the latter requirement. The frequency response of this washout filter is shown in Figure 5.11(a). (Note that a washout time-constant of 5 s will ensure that the PSS contributes a pure damping torque at a frequency of about 10/5 = 2 rad/s.)

The low-pass filter. The corner frequency of the filter should ideally be a decade or more above the highest rotor modal frequency, 9.3 rad/s. The selection of a second-order filter with corners at 200 rad/s results in a small phase lag of 5° at 9.3 rad/s; this selection ensures that the PSS transfer function is proper.

(If connected to multi-machine system, and if modal frequencies other than that at 9 rad/s and other issues are not of concern, the washout and low-pass filter time constants could be modified to 2 and 0.01 s, say.)

The PSS transfer function. On insertion into (5.34) of the parameters determined above, together with the damping gain of $k = 20$ pu on machine rating, the practical PSS transfer function becomes

$$H_{PSS}(s) = 20 \cdot \frac{s5}{1+s5} \cdot \frac{1}{4.935} \cdot \frac{(1+s0.1064+s^2 0.002106)}{(1+s0.005)(1+s0.005)} . \tag{5.35}$$

A plot of the PSS transfer function is shown in Figure 5.12 in which the following features are observed. (i) Below the range of modal frequencies, 1-15 rad/s, the washout filter becomes effective. (ii) Over the range of modal frequencies the responses are the mirror image of those of the P-Vr characteristic for Design Case B, Figure 5.10. The phase lag of $\sim 45°$ in the P-Vr characteristic at the modal frequency of 8-9 rad/s is cancelled by the phase lead introduced by the PSS. (iii) Over the range of modal frequencies the PSS gain is close to $20/4.935$, or 12.2 dB. (iv) Above the modal frequency range the second-order low-pass filter becomes effective. As mentioned earlier, to reduce the high frequency gain the corner frequencies of the low-pass filter should be reduced.

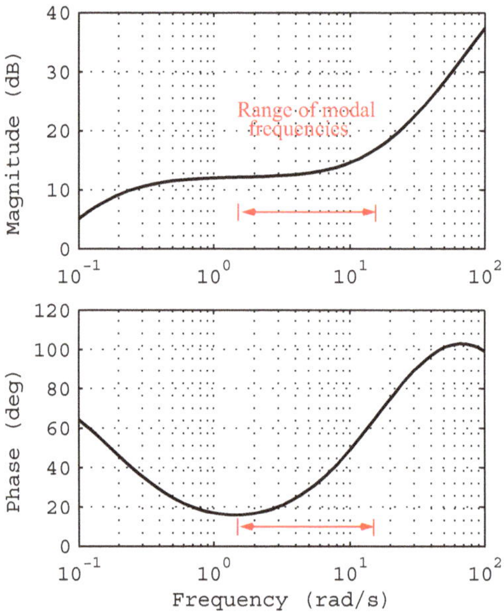

Figure 5.12 Frequency response plot of PSS transfer function, (5.35). Over the modal frequency range the responses are the mirror image of those of the P-Vr characteristic for Design Case B

Note that in the transfer function of the compensation in (5.35), i.e.

$$G_{Hc}(s) = \frac{1}{4.935} \cdot \frac{(1 + s0.1064 + s^2 0.002106)}{(1 + s0.005)(1 + s0.005)},$$

(i) the zeros are real and lie at -12.5 and -38.0, and (ii) the low-pass filter time constants (5 ms) may be too short to implement in a PSS in practice [1]. However, the 5 ms time constants are retained in the present analysis so that their effect over the range of modal frequencies, 1.5 - 15 rad/s (~0.25 - 2.5 Hz), is minimal.

The rotor modes of the SMIB system with the PSS in service can now be calculated assuming that the SMIB and PSS parameters are those provided in Example 5.1 and (5.35). The resulting rotor modes and the associated shifts in the modes for the selected operating conditions are listed in Table 5.2; the same information is displayed in Figure 5.13 on page 252.

Table 5.2 Effect on rotor modes with PSS in service (P = 0.9 pu), k = 20 pu.

Case / Q pu	Eigenvalues, rotor mode		Mode shift
	with PSS out of service	with PSS in service	
A / -0.2	$1.148 \pm j9.23$	$-0.684 + j9.54$	$-1.832 + j0.305$
B / 0.0	$0.514 \pm j8.70$	$-1.072 + j8.67$	$-1.586 - j0.025$
C / 0.2	$0.267 \pm j8.25$	$-1.008 + j8.08$	$-1.275 - j0.166$
D / 0.4	$0.198 \pm j7.94$	$-0.836 + j7.75$	$-1.034 - j0.193$

The table reveals that, with the PSS in service, the 2% settling time of the rotor mode for all operating conditions is shorter than the criterion of 8 s, i.e. all modes are better damped than a mode with a real part of -0.5 Np/s. The PSS design therefore satisfies this performance specification.

Note that the left-shift in modes varies over a range of generator reactive output. The reasons for this and the variations in modal frequencies are now examined.

5.9.3 Analysis of the variation in the mode shifts over the range of operating conditions

Upon an examination of Table 5.2 and Figure 5.13 the following questions arise: Why does the extent of the left shift of the modes increase as the reactive power output of the generator changes from 0.4 lagging to 0.2 leading? Bearing in mind that the PSS tuning is based on the Design Case B (P = 0.9, Q = 0 pu), why does the modal frequency increase when the reactive power output decreases from that for the Case B, and decrease as the reactive out-

1. The realization of an alternative transfer function to overcome both this and the high gain at higher frequencies is considered in an example in Section 5.12.

put increases from that for the design case? The following addresses these and some other questions.

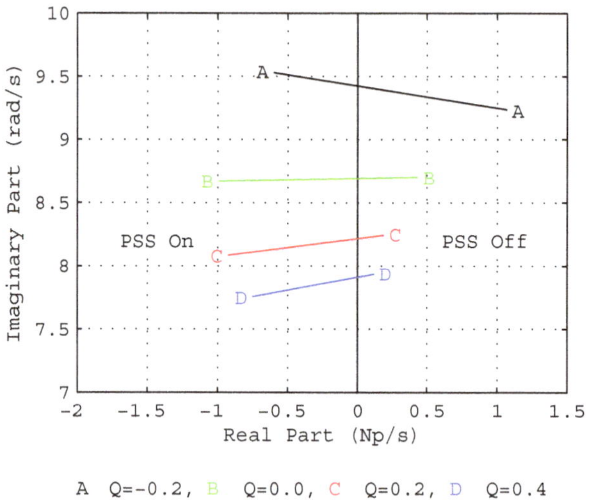

Figure 5.13 Rotor modes of oscillation for the PSS in and out of service for operating conditions $P = 0.9$ pu: $Q = -0.2$ (A), 0 (B), 0.2 (C) and 0.4 (D) pu; damping gain $k = 20$ pu.

In Example 5.3 of Section 5.8.5 it was pointed out that, for the very simplified SMIB system of Figure 5.4, an estimate of the left-shifts of the rotor modes due to the action of the ideal PSS is $\alpha = k_d/(4H) = 1.667$. Because the left-shifts of the rotor modes in Table 5.2 differ significantly from the latter value, let us investigate the reasons for these discrepancies.

For a SMIB system it will be shown in Chapter 13 that, due to an increment in the damping gain Δk of the PSS, the shift in the complex rotor mode λ_h is given by

$$\Delta\lambda_h = -\frac{\rho(\lambda_h)}{2H} \cdot H_{PVr}(\lambda_h) \cdot G_c(\lambda_h) \cdot \Delta k, \tag{5.36}$$

where $\rho(\lambda_h)$ is the complex participation factor of the generator's speed state in the mode λ_h, evaluated with the PSS in service with the damping-gain setting, $k = k_o$. [Equation (5.36) is derived from (13.8)]. The left-shift in mode h is $-\Re\{\Delta\lambda_h\}$. Ideally, the compensating transfer function of the PSS is given by (5.23), i.e. $G_c(\lambda_h) = 1/H_{PVr}(\lambda_h)$, thus (5.36) becomes

$$\Delta\lambda_h = -\frac{\rho(\lambda_h)}{2H} \cdot \Delta k. \tag{5.37}$$

If the participation factor $\rho(\lambda_h)$ of the generator is complex, the mode shift

$$\Delta\lambda_h = -[\rho_r(\lambda_h) + j\rho_i(\lambda_h)]\Delta k/(2H), \tag{5.38}$$

is complex. However, for cases in which the participation of the generator is relatively high the participation factor $\rho(\lambda_h)$ is real or almost real, the mode shift $\Delta\lambda_h$ is directly to the left in the complex s-plane [1]; this left-shift is a prime aim of stabilization using PSSs.

Note that equations (5.23), (5.36) and (5.37) strictly apply at the complex rotor mode $\lambda_h = \alpha_h \pm j\omega_h$. However, because the modes are relatively lightly damped, with damping ratios typically less than 0.15 to 0.20, it is assumed that *conventional frequency response analysis* can be applied with $s = \lambda_h \approx j\omega_h$. The affect of this assumption will be assessed below.

Strictly, equation (5.37) applies only to the case when the compensating transfer function $G_c(s)$ is based on the P-Vr characteristic for the operating condition selected to be the Design Case, i.e. $P = 0.90$, $Q = 0$ as determined in Example 5.2. However, it is illustrative to examine the effect on the mode shift of employing the P-Vr characteristic at some other operating condition p than that for the Design Case, and attempt to account for the contributions to any differences in the respective mode shifts. For the Design Case let $G_{co}(s) = G_c(s)$. The P-Vr characteristic at operating condition p is given by,

$$H_{PVrp}(s) = \frac{k_{cp}}{1 + c_{1p}s + c_{2p}s^2}. \tag{5.39}$$

Substituting for $G_{co}(s)$ and $H_{PVrp}(s)$ from (5.33) and (5.39), respectively, in (5.36) the mode shift for operating condition p is:

$$\Delta\lambda_p = \left[-\frac{\rho_p(s)}{2H}\bigg|_{s=\lambda_h}\right] \cdot \frac{k_{cp}}{(1 + c_{1p}s + c_{2p}s^2)} \cdot \frac{(1 + c_{1o}s + c_{2o}s^2)}{k_{co}} \cdot G_W(s) \cdot G_{LP}(s) \cdot \Delta k, \tag{5.40}$$

where subscript 'o' refers to the coefficients in the compensating transfer function which is implemented in the PSS. In Section 3.10 it is shown that the participation factor is a function of the eigenvectors and thus it must be calculated for the mode $\lambda_{hp} = \alpha_{hp} \pm j\omega_{hp}$. The other terms in (5.40) will be calculated for $s = \lambda_{hp}$ and for the frequency $s \approx j\omega_{hp}$. The expression for the mode shift in (5.40) reduces to:

$$\Delta\lambda_{hp} \approx \left[\frac{|\rho_p(s)|}{2H}\right] \cdot \frac{K_p(s)}{K_o(s)} \cdot \Delta k \cdot K_{filt} \;\angle\{180° + \Phi_p(\lambda_{hp}) + \phi_p(s) - \phi_o(s) + \phi_{filt}(s)\}, \tag{5.41}$$

where the gains K_o, K_p and the phase shifts ϕ_o, ϕ_p are calculated from the P-Vr characteristics (5.18); Φ_p is the phase angle of the participation factor at λ_{hp}. The gain and phase contributed by the combination of the washout and low-pass filters are K_{filt} and ϕ_{filt}.

---

1. For this case of a single rotor mode $\rho(\lambda_h) \approx 0.5$ in (5.37).

Based on (5.41), for the modal frequency $s = \lambda_h$ and the frequency $s = j\omega_h$, we can calculate and account for all contributions to the mode shift and provide an estimation of that shift. This then allows us to:

- compare the estimated value with the actual mode shift (which employs the complex modal frequency λ_h);
- examine the contributions to the mode shift for operating conditions other than the condition used as basis for tuning purposes, and hence
- gain an understanding of the nature of the complex mode shifts observed in Figure 5.13 in terms of the P-Vr characteristics for the various operating conditions versus the condition used as the basis for tuning purposes.

For each operating condition in Table 5.3 the estimated mode shifts evaluated from (5.41) and their components using $s = \lambda_{hp}$ are compared to those shifts which are calculated with $s = j\omega_{hp}$. The operating conditions and closed-loop rotor modes are those listed in Table 5.2. In columns 4 to 10, Table 5.3, are incorporated the components or elements that determine not only the nature and extent of the left-shift but also the change in the modal frequency due to the action of the PSS. Note, however, the incremental gain Δk in (5.41) assumes a value of 20 pu based on the damping gain setting, $k_o = 0$ (because the PSS is initially out of service). The significance of using what may be considered to be large value of incremental gain and its effect on the participation factor is discussed below.

From Table 5.3 the following are preliminary insights into the PSS design are derived:

1. For this example and for the operating condition *p*, the estimated left-shifts in column 10 based on P-Vr characteristic for *p* agree within one percent for two methods of calculation, $s = \lambda_h$ and $s = j\omega_h$. For tuning purposes the use of the conventional frequency response method in the analysis does not lead to significant errors. For damping ratios of the rotor modes exceeding 0.15 to 0.2 the accuracy of the calculation may decrease significantly and should be verified.

2. Consider in Figure 5.13 and Table 5.2 a selected operating condition - say Case C - with the PSS off; the mode for this condition is $0.267 \pm j8.25$. When the PSS is in service with a damping gain of 20 pu the mode is shifted by $-1.275 - j0.166$ to $-1.008 + j8.08$. With a small imaginary component in the mode shift, *the compensating transfer function $G_c(\lambda_h)$ has effected a more-or-less pure left shift of the mode* while the *damping gain has determined the extent of the shift.* For the selected operating condition the components in columns 3 to 9 in Table 5.3 which contribute to the mode shift do not alter significantly as the damping gain is increased. Therefore, as implied by (5.36), $\Delta\lambda_h \propto \Delta k$, i.e. *the incremental mode shift is proportional to the increment in*

damping gain; it will be noted in Figure 10.26 that this applies to relatively large changes in gain of 5 to 20 pu.

Table 5.3 Components of estimated mode shift (5.41), $s = \lambda_h$ or $s = j\omega_h$, $P = 0.9$, Q pu

Case / Q pu (P = 0.9 pu)	Modal Response ($s = \lambda_h$), or Freq. Response ($s = j\omega_h$)	Average particip-ation factor $p_{av}(\lambda)$ mag, phase (abs), (°)	Washout Filter mag (abs), phase °	Low-pass filter mag(abs), phase °	Product $H_{PVrp}(s) \cdot G_{co}(s)$ mag, phase (abs), (°)	Estimated Mode Shift using P-Vr for operating condition p	Actual Mode Shift based on P-Vr for Design Case B (see Table 5.2)
col.1	col.2	col.3	cols.4-5	cols.6-7	cols.8-9	col.10	col.11
A / -0.2	λ_h	0.450, 5.7	1.00, 1.2	1.00, -5.5	1.240, -12.1	$-1.838 + j0.347$	$-1.832 + j0.305$
	$j\omega_h$		1.00, 1.2	1.00, -5.5	1.241, -11.2	$-1.831 + j0.315$	
B / 0.0	λ_h	0.475, 4.7	1.00, 1.3	1.01, -5.0	1.000, 0	$-1.601 - j0.028$	$-1.586 - j0.025$
	$j\omega_h$		1.00, 1.3	1.00, -5.0	1.000, 0	$-1.579 - j0.028$	
C / 0.2	λ_h	0.483, 4.2	1.00, 1.4	1.01, -4.6	0.792, 7.0	$-1.276 - j0.179$	$-1.275 - j0.166$
	$j\omega_h$		1.00, 1.4	1.00, -4.6	0.797, 6.2	$-1.268 - j0.161$	
D / 0.4	λ_h	0.484, 3.5	1.00, 1.4	1.01, -4.4	0.644, 10.7	$-1.028 - j0.204$	$-1.034 - j0.193$
	$j\omega_h$		1.00, 1.5	1.00, -4.4	0.651, 9.7	$-1.030 - j0.186$	

Eigenvalues, rotor modes, PSS in service (see Table 5.2)

λ_h for Cases A: $-0.684 + j9.54$; B: $-1.072 + j8.67$; C: $-1.008 + j8.08$; D: $-0.836 + j7.75$.
$\omega_h = Im(\lambda_h)$.

Note: H=3.0 MWs/MVA; $\Delta k = 20$ pu on machine MVA rating; *K*-constants from Table 5.1 on page 234; $(K_p/K_o)\angle(\phi_p - \phi_o)$ represents $H_{PVrp}(s)G_{co}(s)$ at $s = \lambda_h$ or $s = j\omega_h$.

Consider now the contributions to the complex mode shift as revealed in the columns of Table 5.3.

3. The participation factor is non-linearly dependent on the complex modal frequency, λ_h, through its eigenvectors, $p_{hk} = w_{hk}\, v_{kh}$. The participation factors in column 3 are therefore taken as the average of the complex participation factors for the PSS in- and out-of-service. For case A the latter two factors are, respectively, $0.463\angle 11.4°$ at a damping gain of $k = 20$ pu and $0.442\angle{-0.2°}$ at $k = 0$ pu. Because the participation factors change as the gain k increases from 0 to 20 pu, the average value is assumed to be a best estimate for present purposes. Note from (5.38) that *a*

negative phase angle for the participation factor leads to an increase the modal frequency ω_h, while a positive angle produces a decrease in frequency. The latter observations explain in part the shapes of the eigenvalue plots in Figure 5.13 for the range of operating conditions. Note the magnitude of the participation factors are less than 0.5; the reasons for the slightly lower values are explained in Chapter 9.

4. The phase angles introduced by the washout and low-pass filters in columns 5 and 7 contribute to the change in modal frequency, i.e. *a negative phase angle so introduced produces an increase in modal frequency and vice-versa.*

5. For the P-Vr characteristic on which the PSS compensation is based (i.e. Case B) the product $H_{PVro}(\lambda_h) \cdot G_{co}(\lambda_h) = 1\angle 0°$ (see columns 8-9). As shown in Figure 5.10 on page 244 the phase of the P-Vr characteristic $H_{PVrp}(\lambda_h)$ for Case A (Q = -0.2) lags (is more negative than) that of the selected P-Vr (case B, Q = 0). Consequently the phase of the product $H_{PVrp}(\lambda_h) \cdot G_{co}(\lambda_h)$ in (5.36) is negative, thus leading to an increase in modal frequency. For Cases C and D (Q = 0.2, 0.4) there is a decrease in modal frequency corresponding to their P-Vr phase characteristics being more positive than that of Case B at the modal frequency.

6. The low-frequency gain of the P-Vr characteristic (Figure 5.10 on page 244) for Case A being greater than that for the design Case B results in the magnitude of the product $H_{PVrp}(\lambda_h) \cdot G_{co}(\lambda_h)$ in (5.36) being greater than 1. *The magnitude of the resulting mode shift is thus greater for those P-Vr characteristics whose low frequency gains are greater than that of the design Case B and vice-versa.*

7. For the Design Case B the *actual* mode shift (column 11) at the modal frequency $s = \lambda_h$ should ideally be purely real. However, the non-zero phase shifts in the washout and low-pass filters as well as the participation factor introduce a small imaginary component into the actual mode shift.

8. The main factors which cause the left-shift to deviate from the ideal, or from that for the Design Case, are highlighted in (5.36), i.e. the variations with operating condition of both the participation factor from $0.5\angle 0°$ and the product $H_{PVrp}(\lambda_h) \cdot G_{co}(\lambda_h)$ from $1.0\angle 0°$.

Note that the observations and concepts introduced in the items above are particularly relevant to the design of PSSs in multi-machine systems that are considered in Section 10.4.

For implementing the compensation based on the P-Vr characteristic selected for the tuning of PSSs, it may be that the manufacturer of the PSS has not provided sufficient number of blocks to supply the phase lead required at the modal frequency λ_h. The phase of the product $H_{PVr}(\lambda_h) \cdot G(\lambda_h)$ (of unity magnitude) will then be negative (due to the net phase lag) and will result in an increase in the frequency of the mode shift according to (5.36). If the net phase lag is large, say $30°$, the increase in modal frequency is equal to $\sin 30° = 0.5$ times the magnitude of the mode shift; correspondingly the left-shift is reduced to $\cos 30° = 0.87$ times the magnitude.

5.10 Tuning of a PSS for a higher-order generator model in a SMIB system

In the previous sections of this chapter there are several noteworthy features:

- The tuning of the PSS was based on a *closed-form* relation for the compensating transfer function of the PSS, given by (5.26), for the third-order generator model and its first-order excitation system. In practice such an expression is not easily derived for a higher-order generator, particularly in a multi-machine system.

- Because of the simplicity of the SMIB system model the P-Vr characteristics and other relations can be presented as closed-form expressions in an in-depth analysis. Insightful information about the process of PSS tuning and its performance can then follow.

- The operating conditions in the SMIB system, in which the external system is represented by a simple series impedance, resulted in bus voltage levels outside the range of 0.95 to 1.05 pu for the more extreme operating conditions as shown in Table 5.1. Results are more meaningful if practical operation within voltage limits is observed; this is the case in the following analysis for a much wider range of conditions including line outages.

- The real power output of the generator was confined to its rated value. The performance of the system with the PSS in service at rated and at lower real power outputs is also of interest.

- For *practical purposes* the phase characteristics of the P-Vr about the design case were revealed to be more-or-less invariant over a range of operating conditions. The variation in the P-Vr gain characteristics was as much as ±4 dB, this value however depends on the choice of the Design Case P-Vr characteristic. Do these observations apply to higher-order generator models with operational constraints applied?

In order to introduce further operational and modelling considerations, the P-Vr characteristics will thus be based on both a sixth-order generator model and its excitation system and a wider, practical and encompassing set of normal and contingency operating conditions. In normal practice, higher-order generator models and higher-order excitation systems are used in the simulation of the dynamic performance of power systems.

The purpose of the following sections is to examine again (i) the synchronizing and damping torque coefficients of the sixth-order generator and its excitation system in a SMIB context, (ii) its P-Vr characteristics, and (iii) the tuning of the PSS. *The procedure will form the basis for the tuning of PSSs in multi-machine power systems*. In addition, it is of interest to examine the effect of the d- and q-axis windings of the generator on the synchronizing and damping torque coefficients.

The more practical SMIB power system and the model of the generating unit are described in Section 5.10.1. The tuning of the associated PSS is illustrated by means of an example in which both normal and line-outage conditions are considered for a range of generator real and reactive power outputs. The P-Vr characteristics are calculated and the performance of the PSS - the tuning of which is based on these characteristics - is then examined. From such an examination the implications for the tuning of PSSs in a multi-machine system are assessed.

5.10.1 The power system model

A 'more practical' SMIB system now considered consists of the generator connected to an infinite bus through a step-up transformer and a pair of transmission lines as shown in Figure 5.14. However, in this case the transformer is fitted with a tap changer, shunt capacitance is included in the model of the transmission lines, and a constant impedance load is connected to the high voltage terminals of the generator transformer. Post-fault contingencies are represented by one or both of the circuits 'a' and 'b' being out-of-service. This arrangement represents more closely a practical configuration in the vicinity of a generating station that feeds into a large system. Clearly, it not intended to model a tightly-meshed system of generating stations, loads and interconnecting transmission lines.

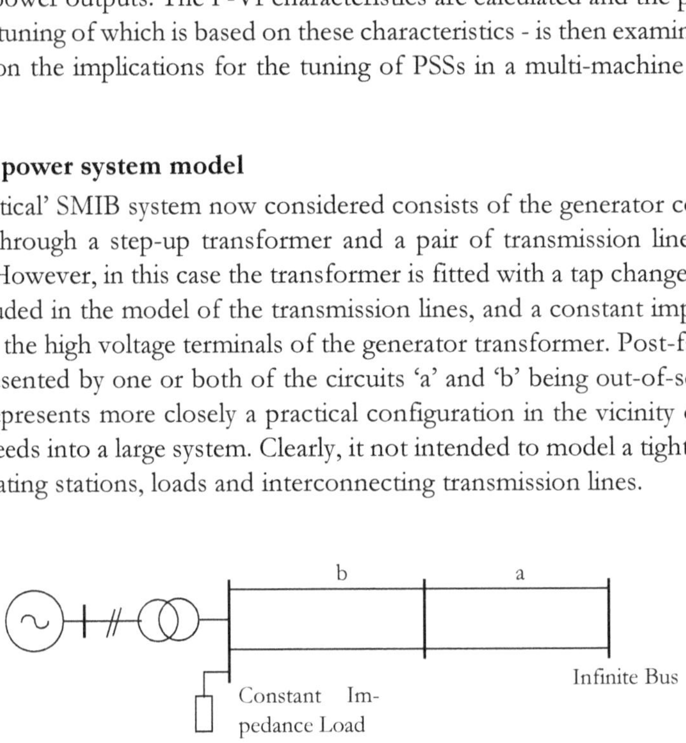

Figure 5.14 A 'more practical' SMIB system.

For a given set of machine and system parameters, the real and reactive power outputs (P, Q) of the generator together with the complex power of the load are selected as input quantities together with a terminal voltage setting of 1 pu.

The Thévenin equivalent at the generator terminals of the system between the infinite bus and the generator terminals is calculated. The model of the generating unit, together with the Thévenin-equivalent voltage source and the associated impedance, form the model of the SMIB system.

For convenience, the coupled-circuit form of the sixth-order generator model described in [9] is adopted for the analysis in the remainder of this chapter. Closed-loop control of the generator terminal voltage is again implemented by means of a simple first-order model of the excitation system.

The linearized model of the generator and excitation system used in this application is also that described in [9]. The equations for the sixth-order generator in a SMIB system, together with those of the excitation system, are given for completeness in Appendix 5–I.3. Because a model of the stabilizer is required in a later section, the equations for the complete generator-excitation system-PSS are also included in the Appendix. These equations are in state-space form, rather than the transfer function form employed in the earlier sections of this chapter.

5.10.2 Calculation of the synchronizing and damping torque coefficients

In Section 5.3 it was noted that the synchronizing and damping torque coefficients can be calculated from the transfer function $\Delta P_0/\Delta\omega$ given by (5.10) with the shaft dynamics disabled. In the context of the state equations given by (3.9), this implies removing the row and column associated with the speed state in the $\boldsymbol{A}$ matrix and treating shaft speed $\Delta\omega$ as an input signal. The latter column then becomes a column associated with shaft speed in the input matrix, $\boldsymbol{B}$. The column associated with the speed output in the output matrix $\boldsymbol{C}$ is also moved to the column of the $\boldsymbol{D}$ matrix corresponding to the speed input; the torque of electro-magnetic origin ΔP_0 remains an output signal.

For illustrative purposes and simplicity, assume that (i) a set of third-order state equations describes the dynamic performance of the system, (ii) $x_2(t)$ is the speed state $\Delta\omega$, and (iii) $y(t)$ is the output ΔP_0. The state equations are:

$$\begin{bmatrix}\dot{x}_1(t)\\ \dot{x}_2(t)\\ \dot{x}_3(t)\end{bmatrix} = \begin{bmatrix}a_{11} & a_{12} & a_{13}\\ a_{21} & a_{22} & a_{23}\\ a_{31} & a_{32} & a_{33}\end{bmatrix}\cdot\begin{bmatrix}x_1(t)\\ x_2(t)\\ x_3(t)\end{bmatrix} + \begin{bmatrix}b_1\\ b_2\\ b_3\end{bmatrix}\cdot u(t),\quad y(t) = \begin{bmatrix}c_{11} & c_{12} & c_{13}\end{bmatrix}\cdot\begin{bmatrix}x_1(t)\\ x_2(t)\\ x_3(t)\end{bmatrix}. \tag{5.42}$$

Setting the input $u(t)$ to zero and eliminating the rows of the **A**, **B** & **C** matrices associated with the speed state as well as the corresponding column in the **A** matrix, we find:

$$\begin{bmatrix}\dot{x}_1(t)\\ \dot{x}_3(t)\end{bmatrix} = \begin{bmatrix}a_{11} & a_{13}\\ a_{31} & a_{33}\end{bmatrix}\cdot\begin{bmatrix}x_1(t)\\ x_3(t)\end{bmatrix} + \begin{bmatrix}a_{12}\\ a_{32}\end{bmatrix}\cdot\begin{bmatrix}x_2(t)\end{bmatrix},\quad y(t) = \begin{bmatrix}c_{11} & c_{13}\end{bmatrix}\cdot\begin{bmatrix}x_1(t)\\ x_3(t)\end{bmatrix} + c_{12}\cdot x_2(t). \tag{5.43}$$

with $x_2(t)$ being the speed input signal. From the modified state equations (5.43) the frequency response $\Delta P_0(j\omega_f)/\Delta\omega(j\omega_f)$ is calculated. The calculation of the frequency responses of the torque coefficients is outlined in Section 5.3 and is given by:

$$k_d = \Re\left\{\frac{\Delta P_0(j\omega_f)}{\Delta\omega(j\omega_f)}\right\} \quad\text{and}\quad k_s = -\Im\left\{\frac{\omega_f}{\omega_0}\cdot\frac{\Delta P_0(j\omega_f)}{\Delta\omega(j\omega_f)}\right\} \qquad \text{[(5.10), repeated]}.$$

5.10.3 Calculation of the P-Vr characteristics for a SMIB system with high-order generator models

As mentioned in the Introduction, Section 5.1, the theoretical basis for the use of the P-Vr characteristic for the tuning of PSSs is given in [10]; in this paper the application of P-Vr characteristic in PSS designs, described in earlier papers [5] and [11], is confirmed. The analysis in [10] is applied to multi-machine power systems - of which the SMIB system is a special case; the results of this analysis are reviewed in the multi-machine context in Section 9.4.

The tuning of the speed-PSS described in Section 5.8 is based on determining the compensating transfer function of PSS, $G_c(s)$, given by (5.23), i.e.

$$G_c(s) = 1/H_{PVr}(s).$$

In the case of the Heffron and Phillips' model of the SMIB system a closed-form expression for the second-order P-Vr transfer function, $H_{PVr}(s)$ - and thus $G_c(s)$, was derived. For the seventh-order generator-excitation system model, the P-Vr characteristic is sixth-order (the shaft dynamics being disabled). For this transfer function, and *for multi-machine systems in particular*, the derivation of a closed-form solution for $G_c(s)$ is not only tedious but is unnecessary. Unnecessary, because

- the calculation of the P-Vr frequency responses is straight-forward and is based on a set of state equations relating the torque of electromagnetic origin to the reference voltage with the speed state disabled;
- from the set of P-Vr frequency responses - such as those in Figure 5.10 - the P-Vr characteristic is selected that best represents the family of such characteristics over the range of operating conditions;
- typically, only a low-order representation of the P-Vr characteristic is required in order to provide a suitable PSS compensating function, $G_c(s)$;
- it is usually relatively easy to synthesize a low-order transfer function representation of the selected P-Vr characteristic.

Because the P-Vr transfer function is that from the voltage reference input to the torque of electro-magnetic origin as output *with the shaft dynamics disabled*, its calculation is similar to that for the torque coefficients in (5.42) and (5.43). The torque of electromagnetic origin ΔP_0 remains an output signal. Retaining the third-order system of (5.42) for illustrative purposes for this case, we note that $y(t)$ remains the torque of electromagnetic origin ΔP_0 and $x_2(t)$ the speed state $\Delta\omega$, however, the input $u(t)$ is now the voltage reference signal, ΔV_r. On elimination of the speed state, (5.42) reduces to:

$$\begin{bmatrix} \dot{x}_1(t) \\ \dot{x}_3(t) \end{bmatrix} = \begin{bmatrix} a_{11} & a_{13} \\ a_{31} & a_{33} \end{bmatrix} \cdot \begin{bmatrix} x_1(t) \\ x_3(t) \end{bmatrix} + \begin{bmatrix} b_1 \\ b_3 \end{bmatrix} \cdot u(t), \quad y(t) = \begin{bmatrix} c_{11} & c_{13} \end{bmatrix} \cdot \begin{bmatrix} x_1(t) \\ x_3(t) \end{bmatrix}. \tag{5.44}$$

The P-Vr characteristic $\Delta P_0(j\omega_f)/\Delta V_r(j\omega_f)$ is calculated from the state equations, modified as illustrated by (5.44).

5.10.4 Example 5.5: Tuning and analysis of the performance of the PSS for the higher-order generator model

5.10.4.1 Parameters of the SMIB system and generator

The generator is connected to the infinite bus through a step-up transformer and two 330 kV transmission lines; the lines are connected to a common bus at their midpoint as shown in Figure 5.14. Nominal system frequency is 50 Hz.

Parameters:

- Transmission lines. For each of the four 330 kV line sections, length 290 km, the parameters are given on the generator MVA rating and base of 500 MVA. The series impedance each line is $Z_L = 0.0225 + j0.225$ pu, shunt susceptance $b = 0.11$ pu. For the outage of transmission line section 'a' in Figure 5.14 the series impedance from the generator HV bus to the infinite bus is $Z_L = 0.03375 + j0.3375$ pu, the associated shunt susceptance is $b = 0.33$ pu. For the outage of two lines, both sections 'a' and 'b' are out of service and $Z_L = 0.045 + j0.450$ pu, $b = 0.22$ pu.

- Transformer. Series impedance is $Z_t = j0.15$ pu on 500 MVA. The off-nominal tap setting *t* is the per-unit turns ratio (= 1 pu at nominal taps); the taps are assumed to be on the low-voltage side and adjustable in steps of 1%. (In practice, taps are located on the high voltage side of the transformer, however, the convention in [9] is adopted.)

- Constant shunt admittance load located at the station's high voltage bus. The per unit complex load power at 1 pu voltage is $P_o = 0.09$, $Q_o = 0, 0.02, 0.03, 0.04$ (or $Q_o = 0.29$ pu when a shunt reactor is brought into service).

- Generator. Rating 500 MVA.

$D = 0$, $H = 3.0$ MWs/MVA, $r_a = 0$, $x_d = 1.9$ pu, $x_q = 1.8$ pu, $x'_d = 0.30$ pu, $x_l = 0.20$ pu, $x'_q = 0.55$ pu, $x''_d = 0.26$ pu, $x''_q = 0.26$ pu, $T'_{do} = 6.5$ s, $T'_{qo} = 1.4$ s, $T''_{do} = 0.035$ s, $T''_{qo} = 0.04$ s, on the generator rating.

The per-unit parameters on generator rating for the corresponding sixth-order coupled-circuit model are:

$X_d = 1.9$, $X_{ad} = 1.7$, $R_{fd} = 0.00088453$, $X_{fd} = 0.106250$, $R_{1d} = 0.0227364$, $X_{1d} = 0.15$, $X_q = 1.8$, $X_{aq} = 1.6$, $R_{1q} = 0.0046564$, $X_{1q} = 0.4480$, $R_{2q} = 0.0336146$, $X_{2q} = 0.0724138$.

Excitation system: $K_{ex} = 200$ pu, $T_{ex} = 0.02$ s.

The state-equation model for the generator, which is under closed-loop voltage control, is given in Appendix 5–I.3.1.

The ranges of operating conditions for the generator are $P = 0.9$ and 0.7 pu, $-0.2 \leq Q \leq 0.4$ pu, and $P = 0.1$, 0.3, 0.5, 0.7 and 0.9 pu at unity power factor. Normal conditions and one line and two lines out-of-service are analyzed. All conditions are subject to the 330 kV bus-bar voltages lying in the range 0.95 to 1.05 pu. *Note that the above selection of operating conditions encompasses not only a wide range of normal and outage conditions but also reactive power outputs at 0.9 and 0.7 pu real power.*

5.10.4.2 Steady-state and dynamic performance of the system, no PSS in service

The steady-state voltages and rotor angles, together with the associated eigenvalues, are calculated for the range of operating conditions using the system and generating-unit parameters listed above. The relevant results are given in Table 5.4 for selected operating conditions.

Without a PSS at a generator real-power output of 0.9 pu (450 MW), the system is unstable for all the selected operating conditions. For all real power outputs and a given system configuration (e.g. a single line outage) shown in Table 5.4 it is significant that:

- the rotor angle increases as the power factor becomes more leading - and can exceed 90°;
- correspondingly, the stability of the system degrades.

Furthermore, at a selected complex power output (e.g. $S = 0.9 + j0.3$ pu), stability degrades with the increased series impedance of the transmission system associated with the outages of line sections 'a' then 'a' and 'b'.

5.10.4.3 Inherent synchronizing and damping torques coefficients, 6th order generator model

As was the case for the third-order Heffron and Phillips' model in Section 5.3, it is again revealing to examine the inherent synchronizing and damping torque coefficients for the rotor before the tuning of the PSS is considered. In this case the generator terminal voltage is controlled by a first-order model of the excitation system.

Table 5.4 Steady-state operating conditions and eigenvalues for the SMIB system; no PSS in service (P, Q in pu on 500MVA base)

Case	Generator Output, P, Q pu	Lines out of service	Load pu Po, Qo #	Trans-former taps %	Rotor angle $\delta°$	Voltage pu ($\angle°$)			Eigenvalues, Rotor Mode
						Gen.Ter-minals	Gen. HV Bus	Inf. Bus	
A	0.9, -0.1	none	0.09, 0	0	80.7	$1\angle17.6$	1.024	1.020	$0.773 \pm j9.16$
B	0.9, 0	none	0.09, 0.04	-1	76.0	$1\angle17.7$	1.019	1.002	$0.552 \pm j9.12$
C	0.9, 0.2	none	0.09, 0.04	-4	67.6	$1\angle17.6$	1.021	0.959	$0.261 \pm j9.02$
D	0.9, 0.4	none	0.09, 0.29	-6	60.8	$1\angle17.6$	1.015	0.967	$0.113 \pm j8.98$
E	0.9, -0.1	one	0.09, 0	+5	87.7	$1\angle24.5$	0.978	1.017	$0.936 \pm j8.48$
F	0.9, 0.3	one	0.09, 0.03	-8	68.2	$1\angle21.7$	1.053	0.952	$0.245 \pm j8.44$
G	0.9, -0.07	two	0.09, 0.03	+6	91.7	$1\angle30.1$	0.965	1.049	$0.927 \pm j7.98$
H	0.9, 0.3	two	0.09, 0.03	-7	74.0	$1\angle27.6$	1.041	0.951	$0.322 \pm j7.84$
J	0.7, -0.2	none	0.09, 0.04	0	76.2	$1\angle13.1$	1.035	1.049	$0.323 \pm j9.06$
K	0.7, 0	none	0.09, 0.04	-3	64.6	$1\angle13.0$	1.036	1.004	$-0.065 \pm j8.92$
L	0.7, 0.2	none	0.09, 0.04	-5	55.9	$1\angle13.1$	1.029	0.954	$-0.198 \pm j8.76$
M	0.7, -0.2	one	0.09, 0.03	+5	81.3	$1\angle18.2$	0.990	1.044	$0.493 \pm j8.41$
N	0.7, 0.2	one	0.09, 0.03	-6	59.6	$1\angle16.8$	1.040	0.950	$-0.096 \pm j8.19$
# Load at 1 pu voltage on generator HV bus									

The modified state equations for the calculations of the inherent torque coefficients are derived as outlined in Section 5.10.2. From these equations the frequency response of the synchronizing and damping torque coefficients are calculated, that is, from the transfer function $\Delta P_0(j\omega_f)/\Delta\omega(j\omega_f)$.

With reference to Figure 5.15 and Table 5.4 it is noted that the inherent damping torque coefficient is markedly negative for many of the cases over the range of the rotor modal frequencies, 7.8 to 9.2 rad/s. This observation applies at the more heavily stressed conditions, particularly (i) as the series impedance of the transmission system increases with the line outages, and (ii) at the more leading power factors. The least damped condition is that for case G with a coefficient of -13 pu, only slightly poorer than that for case A. Note that: (i) as the damping torque coefficients become more negative for the cases shown in the figures so the degree of instability increases for the corresponding cases in Table 5.4; (ii) at the lower inter-area modal frequencies, potentially 1.5 to 6 rad/s, the inherent damping torque coefficients tend to be negative. Such degradations in damping must be remedied by the action of the PSS.

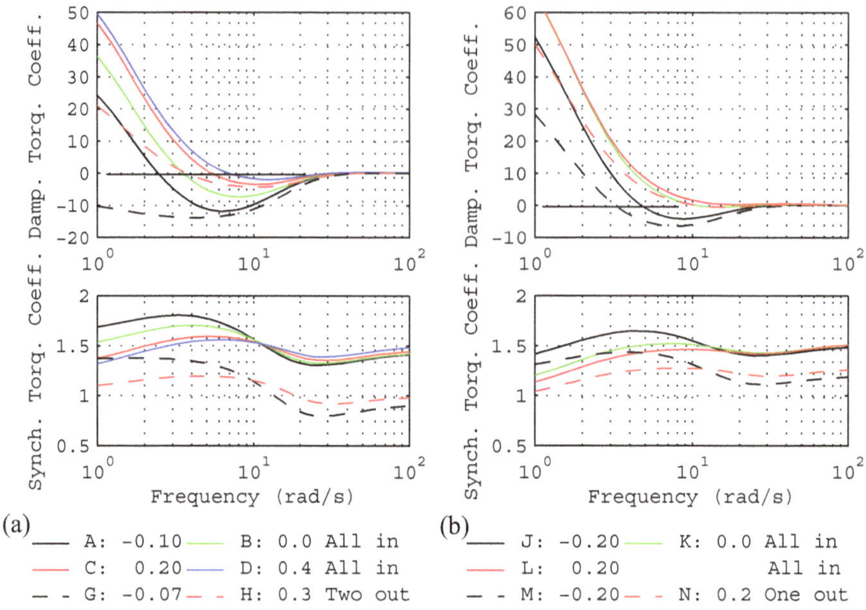

Figure 5.15 Inherent synchronizing and damping torque coefficients for SMIB system at a generator real power output (a) $P = 0.9$ pu, and (b) $P = 0.7$ pu. Reactive power output varies between 0.2 pu leading and 0.4 pu lagging with all lines in service, or the outage of one or two lines.

As observed in Section 5.5.1 the damping gain k of the PSS - when installed - should be greater than $k = 13$ pu to ensure stability under the most onerous operating condition, i.e. in Case G: $P = 0.9$, $Q = -0.07$ pu, with two lines out of service.

Note from Table 5.4 that the *magnitudes* of damping ratios of the rotor modes are generally less than 0.1. The torque coefficients are therefore again calculated using conventional frequency response methods with $s = j\omega_f$.

5.10.5 The P-Vr characteristics for a SMIB system with a 6th order generator model

The basis of the P-Vr characteristics and their calculation are discussed in Section 5.10.3.

The P-Vr characteristics are shown in Figure 5.16 for generator real power outputs of 0.9 and 0.7 pu. These characteristics mainly relate to the more extreme ends of the range of reactive power outputs and system configurations listed in Table 5.4.

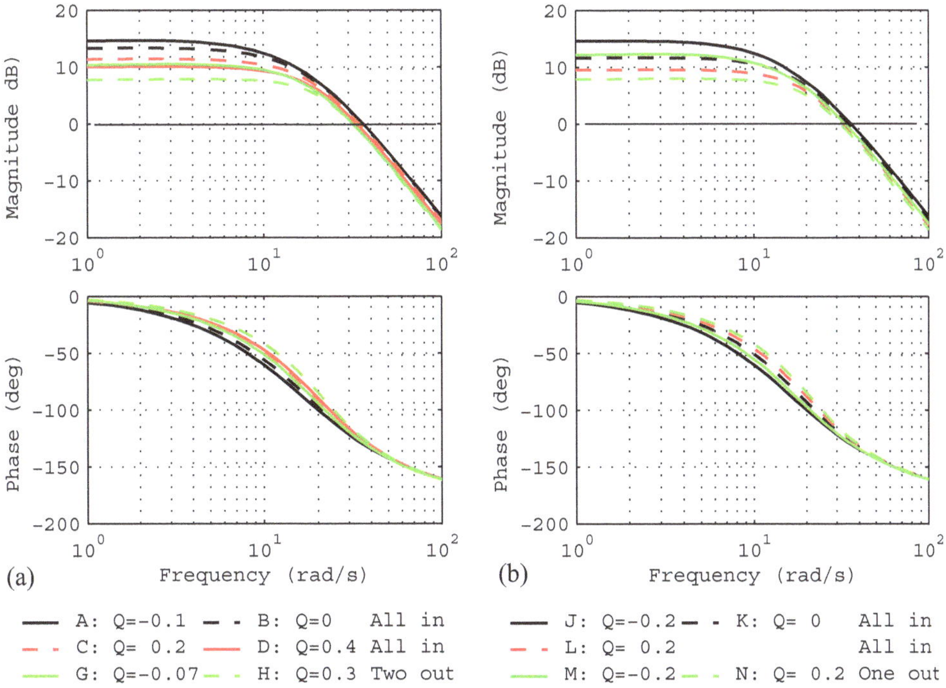

Figure 5.16 P-Vr characteristics for the SMIB system at generator real power outputs (a) $P = 0.9$ pu, and (b) $P = 0.7$ pu; all lines in service, or outage of one or two lines.

Note that, for constant real power outputs, the low-frequency gains of the P-Vr characteristics increase as the reactive power changes from lagging to leading.

As is the case for the lower-order generator model of Section 5.8, the P-Vr characteristics in the latter figures form the basis for the tuning of the speed-PSS.

5.10.6 Tuning a speed-PSS for a SMIB system with a 6th order generator model

The same form $kG(s)$ of the transfer function of the speed-PSS is employed as in Section 5.8. The tuning of the PSS transfer function is considered in the same five steps.

Step 1 of Section 5.8.
The determination of the compensating transfer function $G_c(s)$ such that, over a selected range of modal frequencies, a torque of electro-magnetic origin proportional to speed is induced by the PSS on the rotor of the generator.

5.10.6.1 Selection of the P-Vr characteristic for compensating transfer function $G_c(s)$

For the selected set of operating conditions listed in Table 5.4 the frequencies of the rotor modes cover a range from 7.8 to 9.2 rad/s. Based on a mid-range modal frequency of about

8.7 rad/s, the phase of the P-Vr characteristics for $P = 0.9$ pu varies over the range $-53.5°$ (Case A) to $-35.8°$ (Cases J & H). Let us choose the P-Vr characteristic for $P = 0.9$, $Q = 0.2$ pu (Case C) because its phase of $-43.5°$ lies close to the middle of the band of characteristics at the modal frequency. The phase of any characteristic associated with the operating conditions listed in Table 5.4 is thus within $\pm 10°$ of that of the selected characteristic. Case C will be called the 'Design Case'.

At low frequencies (1 rad/s), the extent of the range of gains of the P-Vr characteristics shown in Figure 5.16(a) is from 7.8 dB (Cases H) to 14.6 dB (Case A). These differ from the selected P-Vr characteristic (Case C) by -3.9 and +2.9 dB, respectively.

An inspection of the P-Vr characteristics for $P = 0.7$ pu in Figure 5.16(b) reveals that these characteristics are suitably represented by Design Case C.

5.10.6.2 *Synthesis of the compensating transfer function $G_c(s)$*

We wish to synthesize the P-Vr transfer function, $H_{PVrS}(s)$, for Design Case C in order to form the PSS compensating transfer function $G_c(s)$ by means of (5.23). In order to provide some useful insights the process is illustrated using some basic concepts in frequency response analysis.

Imagine, on the P-Vr characteristic of Case C in Figure 5.16(a), the straight-line asymptotes as $\omega_f \to 0$ and as $\omega_f \to \infty$ are drawn on the magnitude plot for the Design Case C. These asymptotes roll off at 0 and -40dB/decade, respectively, and intersect at 19 rad/s. Because the slope of the magnitude plot monotonically decreases with increasing frequency it is assumed that there are no zeros in the transfer function (or, if any exist, they are cancelled by closely-located poles). Furthermore, the phase response tends to $-180°$ as $\omega_f \to \infty$. A second-order form of the synthesized transfer function $H_{PVrS}(s)$ possessing a pair of complex poles is thus assumed, i.e.

$$H_{PVrS}(s) = k_{co}/(s^2 + 2\xi\omega_n s + \omega_n^2) \quad \text{or} \quad H_{PVrS}(s) = k_c/\left(1 + \frac{2\xi s}{\omega_n} + \left[\frac{s}{\omega_n}\right]^2\right), \tag{5.45}$$

where $k_c = k_{co}/\omega_n^2$. The frequency response characteristics for such a transfer function are shown in Figure 2.16. From a comparison of the P-Vr response for Case C in Figure 5.16(a) with that of the latter figure it is observed that:

- The intersection of the asymptotes occurs at the undamped natural frequency $\omega_n = 19$ rad/s; the associated magnitude response in Figure 2.16 is 6 dB down at this frequency. Assume that the damping ratio $\xi = 1$; the corresponding gain from the P-Vr plot is about 4 dB down. The damping ratio is thus less than 1.0 and therefore the transfer function possesses a pair of complex poles.

- There is no resonance apparent in the magnitude response, i.e. the damping ratio ξ is greater than 0.707.

A trial-and-error method, a curve fitting process, or some other convenient method can now be adopted to determine the damping ratio ξ. From the eigenvalues of the ***A*** matrix for the P-Vr characteristic in (5.45) it may be possible for this simple SMIB system to isolate a pair of poles for the second-order transfer function of (5.45). These are found to be located at $-16 \pm j10.2$ - from which $\xi = 0.85$ and $\omega_n = 19$ rad/s. The low-frequency gain of the P-Vr characteristic for Case C in Figure 5.16(a) is 11.7 dB or 3.84 pu. Inserting the relevant values into (5.45), the synthesized P-Vr transfer function for Case C becomes:

$$H_{PVrS}(s) = 3.84/(1 + 0.0895s + 0.00277s^2). \tag{5.46}$$

Thus, based on (5.23), *Step 1* for the determination of the compensating transfer function $G_c(s)$ can be completed, i.e.

$$G_c(s) = 1/(H_{PVrS}(s)) = (1 + 0.0895s + 0.00277s^2)/3.84. \tag{5.47}$$

Steps 2, 3 and 4 of Section 5.8:
These steps cover the selection of: the value of the damping gain *k*; the parameters of the PSS washout filter; and the high-pass filter parameters that ensure $G_c(s)$ is proper. It is assumed that attenuation of the torsional modes is not required.

As stated earlier a practical form of the speed-PSS for the SMIB system includes the damping gain *k*, together with the washout and low-pass filters, as explained in Section 5.9. Incorporating the PSS compensating transfer function $1/H_{PVr}(s)$ from (5.45), the practical PSS transfer function described in (5.34) becomes:

$$H_{PSS}(s) = kG(s) = k \cdot \frac{sT_W}{1 + sT_W} \cdot \frac{1}{k_c} \cdot \frac{1 + \frac{2\xi s}{\omega_n} + \left[\frac{s}{\omega_n}\right]^2}{(1 + sT_1)(1 + sT_2)}. \tag{5.48}$$

For the range of modal frequencies of rotor oscillation shown in Table 5.4 (i.e. 7.8 - 9.2 rad/s) an examination of Figure 5.15 for the damping torque coefficients reveals that a damping gain of about 13 pu is required to overcome the inherent negative damping. A value of $k = 20$ pu is adopted on a trial basis. The same parameters as in Section 5.9 are adopted for the washout and low-pass filters because the range of modal frequencies is essentially the same. With the insertion of the latter parameters together with those from (5.47) into (5.48) the transfer function of the speed-PSS for the SMIB system becomes:

$$H_{PSS}(s) = 20 \cdot \frac{s5}{1 + s5} \cdot \frac{1}{3.84} \cdot \frac{1 + s0.0895 + s^2 0.00277}{(1 + s0.005)(1 + s0.005)}. \tag{5.49}$$

Note that (i) the zeros in the transfer function of the compensation are complex and (ii) the low-pass filter time constants of 5 ms are very short. These matters are considered in more detail in Section 5.12.

Prior to considering *Step 5* in the tuning procedure of Section 5.8 (i.e. the assessment if the dynamic performance of the PSS) it is instructive to view the frequency characteristic of the damping torque induced by the PSS. *Step 5* is discussed in Section 5.11.

5.10.6.3 Damping and synchronizing torque coefficients induced by the PSS

Let us assess the coefficients of the synchronizing and damping torques induced on the generator through the action of the PSS. Not only does this serve as a check and partial validation of the analysis and tuning of the PSS represented by (5.49), but it also provides additional insights into the action of the PSS.

Referring to Figure 5.9 on page 241, the procedure for examining the synchronizing and damping torques developed through the action of the PSS involves disabling the shaft dynamics of the machine, injecting a speed perturbation into the PSS+excitation-system+machine loop only, and calculating the complex torque ΔP_a, i.e.

$$\Delta P_a\big|_{\Delta\delta,\ \Delta P_m = 0} = H_{PVr}(s) \cdot H_{PSS}(s) \cdot \Delta\omega\,. \tag{5.50}$$

Note that the inherent torque coefficients associated with the path $\Delta\omega \rightarrow \Delta P_\delta$ are excluded in formulation of (5.50).

The product $H_{PVr}(s) \cdot H_{PSS}(s)$ in (5.50) is by definition a complex torque coefficient, $\Gamma(s)$. This coefficient can be expressed in terms of the PSS compensating transfer function $G_c(s)$ and the washout and low-pass filters ($G_W(s)$ and $G_{LP}(s)$) by:

$$\Gamma(s) = H_{PVr}(s) \cdot G(s) \cdot k = H_{PVr}(s) \cdot G_c(s) \cdot G_W(s) \cdot G_{LP}(s) \cdot k\,. \tag{5.51}$$

For the operating condition which forms PSS Design Case C, the compensating transfer function of the PSS is ideally given by (5.23), i.e. $G_c(s) = 1/H_{PVr}(s)$, thus (5.51) becomes

$$\Gamma(s) = G_W(s) \cdot G_{LP}(s) \cdot k\,. \tag{5.52}$$

Assuming the transfer functions of the washout and low-pass filters are each real numbers, i.e. $1\angle 0°$, at the modal frequency, equation (5.52) represents the already-stated purpose of the speed-PSS, i.e. it should induce a damping torque coefficient equal to k as outlined in Section 5.8.1.

For operating conditions other than the Design Case, (5.51) becomes:

$$\Gamma_p(s) = H_{PVrp}(s) \cdot G_{co}(s) \cdot G_W(s) \cdot G_{LP}(s) \cdot k\,, \tag{5.53}$$

where the subscripts '*p*' refer to the P-Vr transfer function which changes with operating condition; subscripts '*o*' refer to the compensating transfer function which is implemented in the PSS. At the modal frequency $s = \lambda_h \approx j\omega_h$ the transfer functions of the washout and low-pass filters are close to $1\angle 0°$, thus (5.53) becomes

$$\Gamma_p(j\omega_h) = \frac{K_p(\omega_h)}{K_o(\omega_h)} \cdot k \;\angle\{\phi_p(\omega_h) - \phi_o(\omega_h)\}, \tag{5.54}$$

where K_p, ϕ_p and K_o, ϕ_o are defined following (5.41).

As shown in Section 5.8.3 and Section 5.10.6.1, at the modal frequency ω_h the P-Vr phase responses typically lie in a narrow band of some $\pm 5° - \pm 10°$ about that of the Design Case. The torque coefficient in (5.54) is essentially a real number, i.e. a damping torque coefficient, given by

$$\Gamma_p(\omega_h) = \frac{K_p(\omega_h)}{K_o(\omega_h)} \cdot k. \tag{5.55}$$

The significance of the above result is that the damping torque coefficient induced by the PSS at any operating condition and the modal frequency ω_h is the damping gain k modified by the ratio of the P-Vr gain at the operating condition to that of the Design Case.

Firstly, consider now the calculation of the torque coefficients induced through the PSS path only in Figure 5.9 on page 241. The complex coefficient $\Gamma(s)$ is expressed in (5.51) in terms of the synthesized and PSS transfer functions $H_{PVrS}(s)$ and $H_{PSS}(s)$, given by (5.46) and (5.49), respectively. For the SMIB system the frequency responses of the synchronizing and damping components are derived from $\Gamma(s)$ using the result from (5.10) as employed in Section 5.10.2.

In Figure 5.17 are shown the synchronizing and damping torque coefficients induced by the PSS for two operating conditions, the Design Case C (P = 0.9, Q = 0.2 pu), and the most poorly-damped condition Case G (P = 0.9, Q = -0.07 pu, two lines out-of service). The details of the steady-state operating conditions for these cases are given in Table 5.4.

Two important features of the responses of the damping torque coefficients are seen in the figure. Firstly, they are more-or-less flat over the range of modal frequencies of interest, 7 to 9 rad/s. Secondly, the damping torque coefficient of 20 pu required by the design is achieved in the Design Case C, however, for Case G a value slightly less results. In Figure 5.16(a) the respective P-Vr gains at a modal frequency in the vicinity of 7 to 9 rad/s are 3.7 and 3.3 pu. According to (5.55) the damping torque coefficient induced by the PSS for Case G is $(3.3/3.7) \times 20 = 17.8$, a value which closely agrees with that found from Figure 5.17.

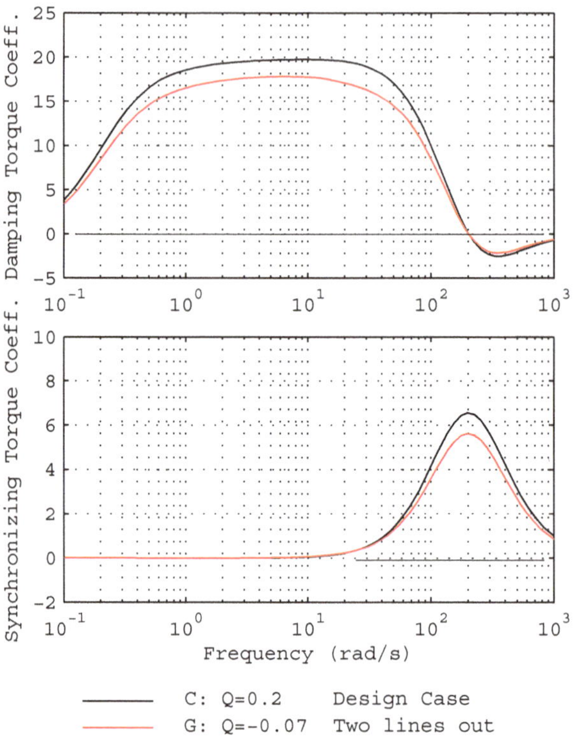

Figure 5.17 Coefficients of the per unit generator synchronizing and damping torques induced through the PSS path (i.e. excludes the inherent torques); $P = 0.9$ pu.

Secondly, consider inherent torque coefficients that are induced through the rotor angle path $\Delta\omega \rightarrow \Delta P_\delta$ in Figure 5.9 on page 241. The frequency response $\Delta P_\delta(j\omega_f)/\Delta\omega(j\omega_f)$ for Cases C and G are shown in Figure 5.18 and possess some interesting features. Firstly, the magnitude responses roll off fairly consistently at 20 dB/decade and the phase responses vary about an angle of $-90°$. This is consistent with integration in the transfer function $\Delta\delta/\Delta\omega = \omega_o/(j\omega_f)$ (as revealed in (5.3) and Figure 5.9). Secondly, the damping torque coefficient is related to the cosine of the phase angle and is therefore positive when the phase angles are more positive than $-90°$.

However, as foreshadowed in Section 5.5.1, the positive contribution of the PSS to the generator torque coefficients will be reduced if the inherent torque coefficients are negative. As established in Section 5.10.4.3 and Figure 5.15 the inherent damping torque coefficients are mainly negative at the modal frequencies listed in Table 5.4 on page 263; it is therefore necessary to examine the net effect on the torque coefficients of the PSS and rotor-angle paths.

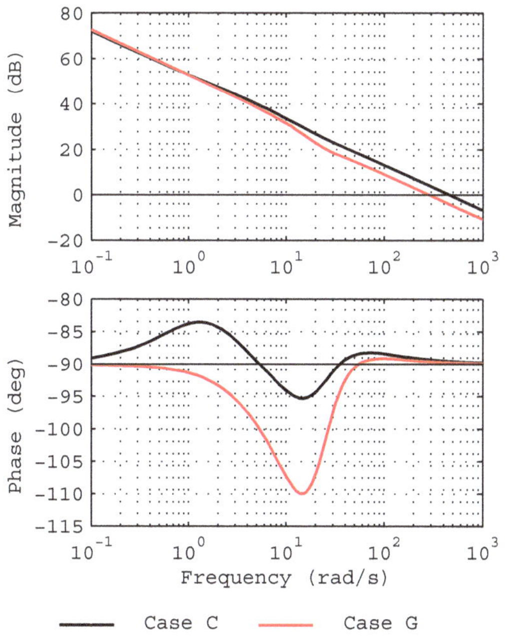

Figure 5.18 Frequency response of the inherent torque coefficients for Case C (the PSS Design Case); $P = 0.9$, $Q = 0.2$ pu, and Case G (Outage of two lines); $P = 0.9$, $Q = -0.07$pu.

In Figure 5.19(a) and (b) the frequency responses of the inherent and PSS-induced torques coefficients are shown separately and summed together. Figure 5.19(a) applies to the Design Case C and Figure 5.19(b) to the 'worst case' operating condition, Case G. The responses of the PSS-induced torques coefficients in the two figures are the same as those in Figure 5.17. However, due to the very different nature of the inherent torque coefficients in the two cases, the combined responses are markedly different. While the combined responses are 'somewhat' flat over the modal frequency range 7 to 9 rad/s, the values of the sum of the damping torque coefficients are significantly different, i.e. about 17 and 5 pu in Figure 5.19(a) and (b), respectively. Although these coefficients are both positive, the eigenvalue of the rotor mode for the Design Case will be characterized by a significantly greater shift into the left-half s-plane than that for Case G. This is to be examined in Section 5.11 in assessing the performance of the PSS over the range of operating conditions.

Due to the inclusion of the damper windings to represent rotor eddy-current losses in the model of the sixth-order generator the frequency response characteristics of the inherent damping torque coefficients are seen in Figure 5.19 to vary greatly over the selected range of operating conditions. In Case C at low frequencies (less than 1 rad/s) the damping torque coefficient is significant, being of the order of 50 to 60 pu. For case G, on the other hand, the damping torque coefficient is negative and destabilizing.

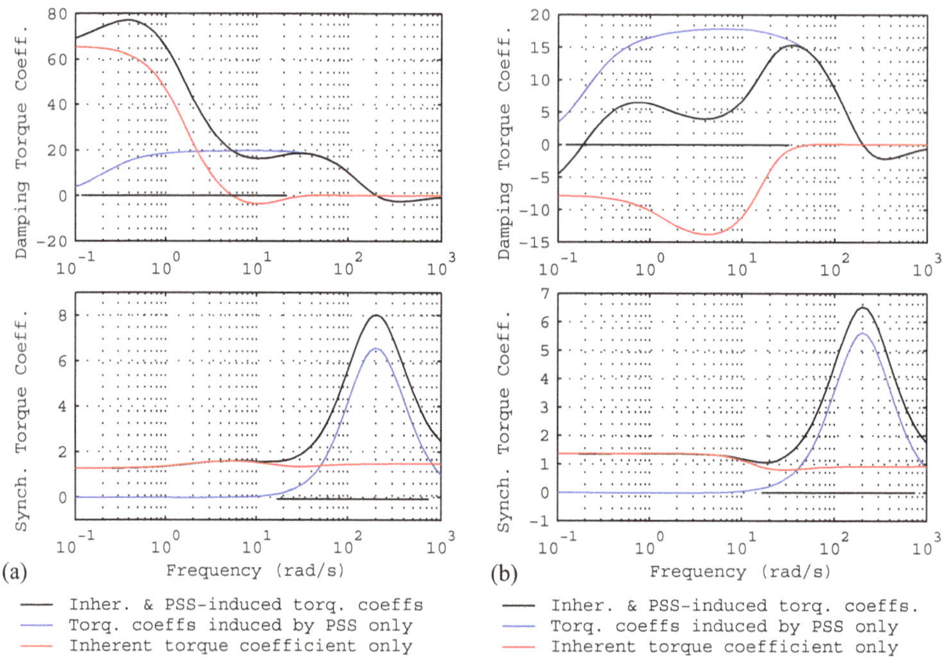

Figure 5.19 Components of the torque coefficients induced on the generator for the operating condition (a) Case C (the PSS Design Case); P = 0.9, Q = 0.2 pu, (b) a 'worst case' operating condition, Case G (outage, two lines); P = 0.9, Q = - 0.07 pu.

As an aside, it is of interest to ascertain what are the components of, or the contributions to, the inherent torque coefficients by individual rotor windings given a set of generator and system parameters and selected operating conditions. With shaft dynamics disabled the inherent torque and its components are calculated from (5.77) in Appendix 5–I.3.2, i.e.

$$\Delta P_0 = K_1\Delta\delta + K_2\Delta\psi_{fd} + K_{21}\Delta\psi_{d1} + K_{22}\Delta\psi_{q1} + K_{23}\Delta\psi_{q2},$$

the rotor angle and flux states being given in terms of the input speed signal by (5.76). The coefficient K_1 is a synchronizing torque coefficient whereas K_2, K_{21}, K_{22} and K_{23} are torque coefficients due to flux linkages associated with the field, direct-axis and the two quadrature-axis windings. The inherent synchronizing and damping torque coefficients are shown in Figure 5.20 for two operating conditions with all lines in service, P = 0.9 pu and (a) Q = 0.2, and (b) Q = -0.07 pu (outage, two lines); these conditions are the same as in Figure 5.19.

In the leading power factor operating condition (Q = -0.07 pu, Figure 5.20(b)) the contributions to the damping torque coefficients by the q-axis windings are negligible, however, in Figure 5.20(a) for the lagging power case (Q = 0.2 pu) they are significant. The field and d-axis windings tend to be the dominant contributors to the damping torque coefficients.

While it is possible to assign to each of the damping windings a contribution to the torque coefficients, concerns over the validity of the results are likely to arise due to questions relating to the accuracy of both the model and the parameters values attributed to the windings. The frequency dependence of the damping contribution of the damping windings emphasizes the importance of employing higher-order generator models to adequately represent the damping performance of power systems.

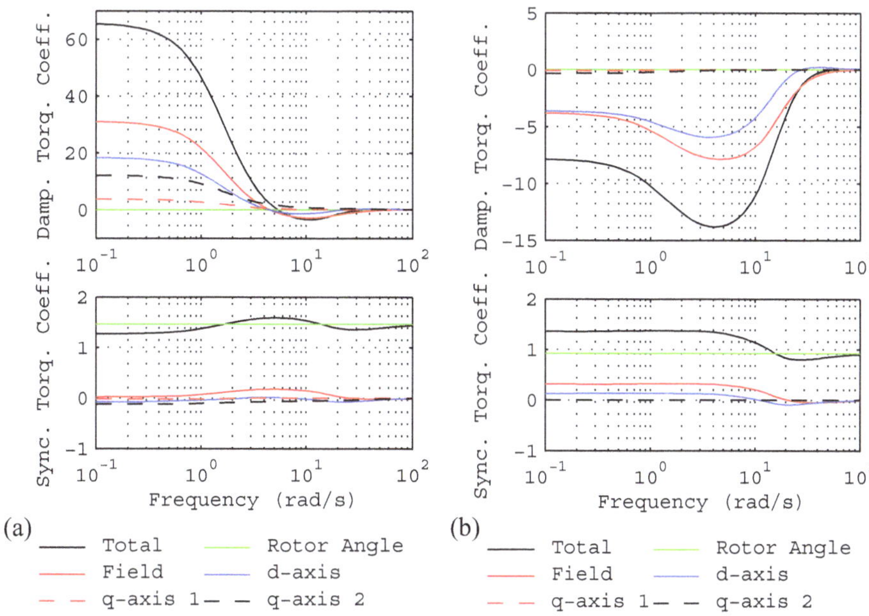

Figure 5.20 Components of the inherent synchronizing and damping torque coefficients for SMIB system for generator real power output (a) $P = 0.9$, $Q = 0.2$ pu, all lines in service, and (b) $P = 0.9$, $Q = -0.07$ pu, outage of two lines.

5.11 Performance of the PSS for a higher-order generator model

Step 5 in the tuning procedure outlined in Section 5.8 is 'the confirmation of the validity of the design, and assessment of the performance of the PSS'.

The improvements in the damping of the rotor modes of oscillation due to the operation of the PSS are revealed in Table 5.5 and Figure 5.21.

Table 5.5 Effect on rotor modes with PSS in service; 6th order generator model. PSS damping gain k = 20 pu on machine rating.

Case	Generator Output P, Q pu	Lines out of service	Eigenvalues, rotor mode		Mode shift
			with PSS out of service	with PSS in service	
A	0.9, -0.1	none	$0.773 \pm j9.16$	$-1.156 \pm j9.51$	$-1.929 \pm j0.35$
B	0.9, 0	none	$0.552 \pm j9.12$	$-1.271 \pm j9.31$	$-1.823 \pm j0.20$
C	0.9, 0.2	none	$0.261 \pm j9.02$	$-1.338 \pm j9.03$	$-1.599 \pm j0.02$
D	0.9, 0.4	none	$0.113 \pm j8.98$	$-1.305 \pm j8.93$	$-1.418 \mp j0.05$
E	0.9, -0.1	one	$0.936 \pm j8.48$	$-0.632 \pm j8.64$	$-1.568 \pm j0.16$
F	0.9, 0.3	one	$0.245 \pm j8.44$	$-1.042 \pm j8.34$	$-1.687 \mp j0.10$
G	0.9, -0.07	two	$0.927 \pm j7.98$	$-0.409 \pm j8.04$	$-1.336 \pm j0.06$
H	0.9, 0.3	two	$0.322 \pm j7.84$	$-0.774 \pm j7.72$	$-1.096 \mp j0.11$
J	0.7, -0.2	none	$0.323 \pm j9.06$	$-1.727 \pm j9.46$	$-2.050 \pm j0.40$
K	0.7, 0	none	$-0.065 \pm j8.92$	$-1.770 \pm j8.94$	$-1.705 \pm j0.02$
L	0.7, 0.2	none	$-0.198 \pm j8.76$	$-1.579 \pm j8.66$	$-1.381 \mp j0.10$
M	0.7, -0.2	one	$0.493 \pm j8.41$	$-1.171 \pm j8.57$	$-1.664 \pm j0.15$
N	0.7, 0.2	one	$-0.096 \pm j8.18$	$-1.240 \pm j8.03$	$-1.144 \mp j0.15$

It is informative to compare these results for the 6th order generator model with those in Figure 5.13 on page 252 and Table 5.2 for the 3rd order model. The left-shift of the mode in Table 5.5 for Design Case C is 1.60 Np/s, which agrees well both with that the Design Case B in Table 5.2 and with the predicted left-shift $k_d/(4H)$ = 1.67 Np/s for a SMIB system in Section 5.9.3. As expected, the shift in the associated frequency (ω_h = 0.02 rad/s) in Design Case C is minimal for the reasons discussed in the latter section.

The same explanations that are given in Section 5.9.3 for the nature and magnitude for the mode shifts are also applicable to the 6th order generator in a SMIB system. In the results for the latter, however, there are some additional issues to consider; these are:

- At a lower real power output, i.e. $P = 0.7$ pu, and a selected reactive power output, the rotor mode when the PSS is either in- or out-of-service is better damped than for $P = 0.9$ pu, e.g. comparing Cases B and K in Table 5.5 for all lines in service. This is to be expected as the corresponding steady-state rotor angle is smaller (see Table 5.4 on page 263).

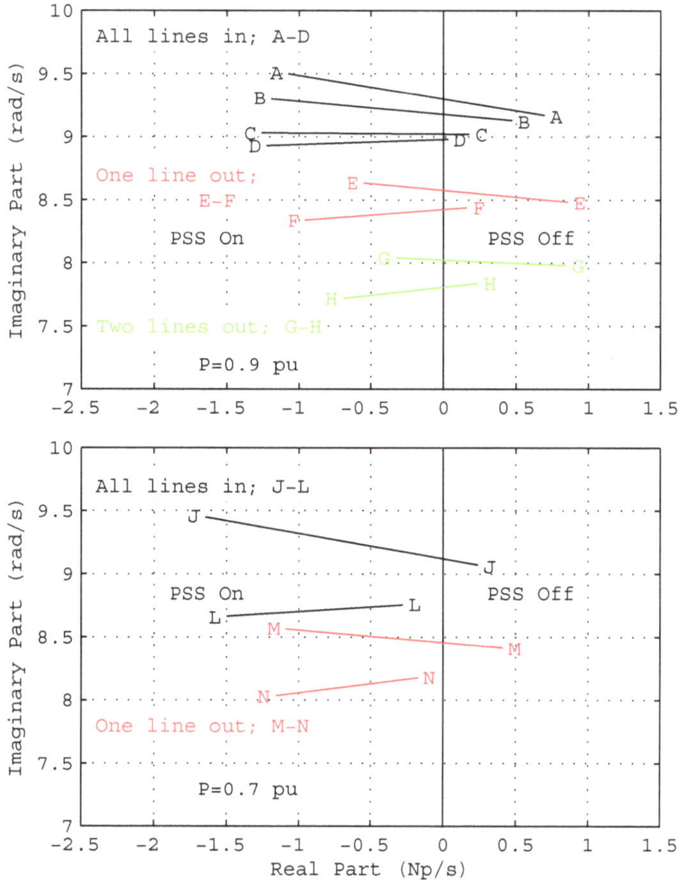

A P=0.9 Q=-0.1, B Q=0.0, C Q=0.2, D Q=0.4
E P=0.9 Q=-0.1, F Q=0.3; G P=0.9 Q=-0.1, H Q=0.3
J P=0.7 Q=-0.2, L Q=0.2; M P=0.7 Q=-0.2, N Q=0.2

Figure 5.21 Rotor mode for the PSS on and off for the set of operating conditions given in Table 5.5, Cases A to N. The Design Case is Case C. The upper plot is for $P = 0.9$ pu, the lower for $P = 0.7$ pu.

- The modal frequencies decrease as one and then two lines are taken out of service - with the result that the series impedance of the transmission system increases. This matter was observed in Section 5.4 based on a simple analysis of the factors that determine the real and imaginary parts of the rotor mode.

- It is observed in Figures 5.13 and 5.21 that, with PSSs off, the most poorly damped rotor mode is associated with the most leading power-factor condition. However, a characteristic of the P-Vr tuning approach is that the gain in the P-Vr characteristics is greatest for this condition. As illustrated in Figure 5.21 the net result with the PSS in service is that the left-shift in the rotor mode is greatest for the leading power factor condition - a desirable and beneficial outcome.

The discussion so far has considered real power outputs at or near rated values. The question arises: how does the PSS perform at lower values of real power outputs? To illustrate the answer, the P-Vr characteristics and the rotor modes of oscillation are also calculated at real power outputs at $P = 0.9$, 0.7, 0.5, 0.3 and 0.1 pu, all at unity power factor. The P-Vr characteristics are shown in Figure 5.22 together with that for the Design Case C ($P = 0.9$, $Q = 0.2$ pu). The associated rotor modes with and without the PSS in service are listed in Table 5.6.

Table 5.6 Shifts in rotor modes at lower real power outputs with PSS in service. PSS damping gain $k = 20$ pu on machine rating.

Case	Generator Output, pu P, Q pu	Lines out of service	Eigenvalues, rotor mode		Mode shift
			with PSS out of service	with PSS in service	
B	0.9, 0	none	$0.552 \pm j9.12$	$-1.271 \pm j9.31$	$-1.823 \pm j0.20$
K	0.7, 0	none	$-0.065 \pm j8.92$	$-1.770 \pm j8.94$	$-1.705 \pm j0.02$
R	0.5, 0	none	$-0.539 \pm j8.50$	$-1.874 \pm j8.36$	$-1.335 \mp j0.14$
S	0.3, 0	none	$-0.779 \pm j8.04$	$-1.610 \pm j7.89$	$-0.831 \mp j0.15$
T	0.1, 0	none	$-0.816 \pm j7.77$	$-1.121 \pm j7.70$	$-0.305 \mp j0.07$
C (Ref)	0.9, 0.2	none	$0.261 \pm j9.02$	$-1.338 \pm j9.03$	$-1.599 \pm j0.02$

Based on Table 5.6 let us consider the performance of the system with the PSS out-of-service as the real power output is decreased at unity power factor. The rotor mode is stable for real power outputs of 0.7 pu or less. With further reduction in real power the damping improves; for power outputs less than 0.5 pu the damping performance criterion stated in Section 5.8.5 is satisfied, namely, that the real part of the rotor mode should be more negative than −0.5 Np/s. In fact, at the lower power levels in this example, PSS action may not be required - but this depends on the encompassing range of operating conditions.

For real power outputs between 0.9 and 0.5 pu, Figure 5.22 reveals that the magnitude and phase responses in the P-Vr characteristic lie in a relatively narrow band, the centre of which is the design characteristic, Case C. The shifts in the rotor modes due to the action of the PSS are listed in Table 5.6. The explanation for the extent of the mode shifts in the power range is similar to that given in Section 5.9.3. For real power outputs less than 0.5 pu the left shift of the mode reduces significantly with reduction in power output. This is due to the gain of the associated P-Vr characteristics reducing significantly below that of the Design Case characteristic [1]. Nevertheless, the damping of the rotor mode for lower power levels below 0.5 down to 0.1 pu is markedly enhanced. Typically, in practice, the PSS may be switched out of service for real power levels less than ~0.3 pu.

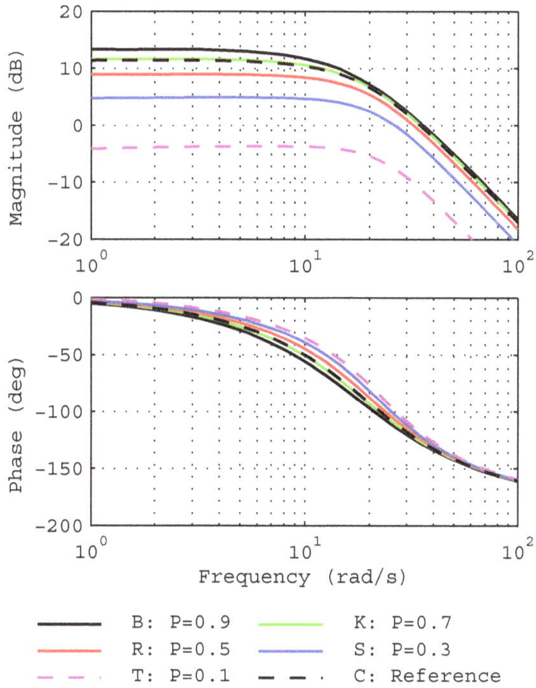

Figure 5.22 P-Vr characteristics for the SMIB system for real power outputs from 0.1 to 0.9 pu at unity power factor.

While the effect of the variation of the real power output on the P-Vr characteristic is noted, information concerning the effect of reactive power output on the characteristic may be deduced from Figure 5.16 on page 265. From the figure for the system in this example and loading conditions (P = 0.9 & 0.7 pu) it can be deduced that, at constant real power output, (i) the magnitude responses of the P-Vr transfer function typically lie within a band of ±2 dB about the Design Case C, (ii) the phase responses lie within a band of ±10° over the range of modal frequency. It may also be observed that the magnitude responses at constant real power decrease as the power factors change from maximum leading to maximum lagging. The two effects on the P-Vr characteristics of the variations in real and reactive outputs are explained in the context of multi-machine systems in Chapters 9 and 10. It is interesting to examine the relation between the P-Vr characteristics of Figure 5.16 and the plots in Figure 5.21.

1. The reason for the nature of the P-Vr gain variation is discussed in Section 9.4.1.

Due to the basis of selection of the synthesized P-Vr for Design Case the PSS transfer function is robust - for practical purposes - to changes in operating conditions at the higher values of generator real power outputs, together with variations in reactive power, was pointed out in Section 5.9.3 [1]. The results of further studies as revealed in Figures 5.16 and 5.22, also demonstrate the robustness of the PSS design [2] to wide variations in operating conditions. The two effects on the P-Vr characteristics of the variations in real and reactive outputs are explained in the context of multi-machine systems in Section 9.4.1; the robustness of fixed-parameter PSSs and an associated theoretical basis for robustness are discussed further.

5.12 Alternative form of PSS compensation transfer function

In Section 5.10.6.2 the compensation transfer function for the PSS included a pair of complex zeros to represent the inverse of the P-Vr transfer function. Referring to (5.49) and ignoring the gain $k_c = 3.84$ for present purposes, the relevant transfer function is

$$G_c(s) = \frac{1 + s0.0895 + s^2 0.00277}{(1 + s0.005)(1 + s0.005)}. \tag{5.56}$$

The frequency response plot for this PSS transfer function is shown in Figure 5.23.

A practical PSS may not be capable of accepting complex zeros, moreover, time constants of 5 ms may be too short to implement in either a digital or analog PSS. Alternatively, let us assume the transfer function may be represented by a set of lead blocks of the form:

$$G_c(s) = \prod_{i=1}^{k} \frac{(1 + sT_{nk})}{(1 + sT_{dk})}, \text{ in which the zeros are real and } T_{nk} > T_{dk}. \tag{5.57}$$

The alternative transfer function in (5.57) is required not only to increase the values of the time constants T_{dk} in the denominator of the transfer function in (5.56) but also to provide the phase lead determined from the P-Vr characteristic over the range of modal frequencies, 1.5 - 15 rad/s. A third-order lead transfer function is found which closely matches the frequency response over the desired modal frequency range as shown in Figure 5.23:

$$G_c(s) = \left[\frac{1 + 0.0438s}{1 + 0.0158s}\right]^3 \approx \left[\frac{1 + s/22.8}{1 + s/63.3}\right]^3. \tag{5.58}$$

1. For operating conditions, P = 0.9 and 0.7 pu, the magnitude of the P-Vr transfer functions consistently lie within a band of ± 2 dB, that is, by factors of 1.26 and 0.79 (see Figure 5.16). In the case of the SMIB system this provides confidence that for practical purposes the left-shifts of the rotor mode over the encompassing range of operating conditions will lie in the range 1.26:1 and 0.79:1.
2. See item 3 of Section 1.2

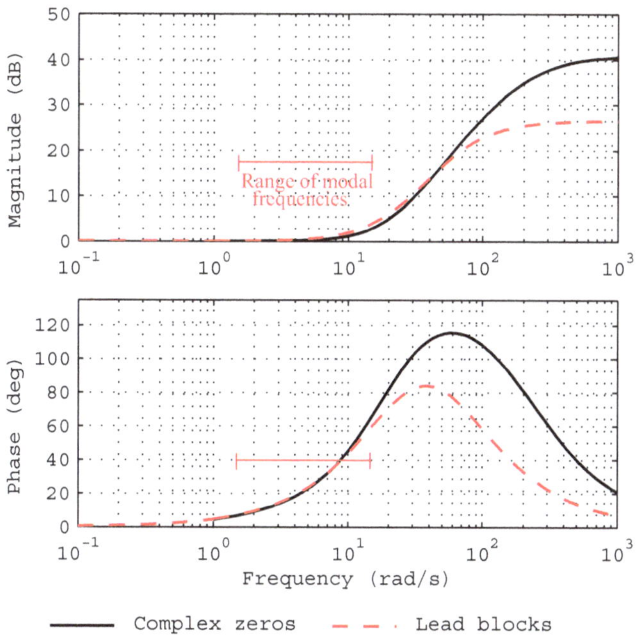

Figure 5.23 Frequency response of the PSS second-order transfer function when its zeros are complex or, alternatively, when the transfer function is implemented by three lead blocks.

The maximum deviation of the latter frequency response from that of that derived from (5.56) is [0.8 dB, 1.3°] at 10 rad/s and [1.1 dB, 5.9°] at 15 rad/s. Although the denominator time constants are increased from 5 ms to 16 ms, a further increase in these time constant may impact deleteriously on the desired frequency response of the compensation in the range of modal frequencies. It is also noted in Figure 5.23 that the gain at high frequencies in the alternative compensation is reduced for torsional frequencies above 100 rad/s (~16 Hz). The concern that the high frequency gain in the PSS might excite such modes is overcome in the 'integral-of-accelerating-power PSS' by means of a pre-filter, a purpose of which is to significantly attenuate the torsional modes (see Section 8.5).

5.13 Tuning an electric power-PSS based on the P-Vr approach

In [14] the authors highlight an electric-power based PSS that does not require the implementation of a phase-lead transfer-function network. The power-based PSS may have some attraction (i) in cases where only the damping of local modes is of concern; (ii) in a PSS for which the number of blocks may be restricted [1]. It is sometimes the case too that a unit is provided with a power input PSS even in situations where this would not have been the preferred choice. The P-Vr approach of Sections 5.8.1 to 5.8.5 can also be applied to the tuning

1. An alternative approach to tuning a power-based PSS is described in Section 8.3.

of this type of PSS with the advantages revealed earlier, namely that the procedure is systematic and the value of the PSS gain is a meaningful quantity - it is also the damping gain defined in Section 5.4.

In the following the transfer function of the power-PSS is derived and assumes that the number of blocks in the PSS is restricted. The analysis commences with equation (5.21) which is employed in development of the P-Vr-based speed-PSS in Section 5.8.1. The latter equation is based on an ideal PSS developing a damping torque proportional to rotor speed and takes the form:

$$\left.\frac{\Delta P_2}{\Delta\omega}\right|_{\Delta V_r = 0|sdd} = k_\omega = \left.\frac{\Delta P_2}{\Delta V_s}\right|_{sdd} \cdot \frac{\Delta V_s}{\Delta\omega} \text{ pu,} \tag{5.59}$$

assuming the shaft dynamics are disabled (*sdd*) on all generators; k_ω is a damping torque coefficient. (It is seen from Figure 5.9 that $\Delta P_2/\Delta V_r$ and $\Delta P_2/\Delta V_s$ are the same transfer functions.)

The term $\left.\frac{\Delta P_2}{\Delta V_s}\right|_{sdd}$ is identified as the P-Vr characteristic $H_{PVr}(s)$ and $\frac{\Delta V_s}{\Delta\omega}$ the transfer function of the speed-PSS; k_ω is also the damping gain, $G_{c\omega}(s)$ is the compensation transfer function of the speed-PSS. That is:

$$\Delta P_2|_{\Delta V_r = 0|sdd} = H_{PVr}(s) \cdot k_\omega G_{c\omega}(s) \cdot \Delta\omega \quad \text{((5.21) repeated)} \tag{5.60}$$

Based (4.59) the equation of rotor motion derived for the case when the perturbations in mechanical power are zero is

$$2Hs\Delta\omega = -\Delta P_e = -\Delta P_2 \text{ pu or } \Delta\omega = -\frac{1}{2Hs} \cdot \Delta P_2 . \tag{5.61}$$

The substitution for $\Delta\omega$ from (5.61) in (5.59) yields:

$$\left.\frac{\Delta P_2}{\Delta\omega}\right|_{\Delta V_r = 0|sdd} = 2Hs \cdot \left.\frac{\Delta P_2}{\Delta V_s}\right|_{sdd} \cdot (-1)\frac{\Delta V_s}{\Delta P_2} . \tag{5.62}$$

In the above equation the term $\left.\frac{\Delta P_2}{\Delta V_s}\right|_{sdd}$ may be identified as $H_{PVr}(s)$, the P-Vr transfer function, and $\frac{\Delta V_s}{\Delta P_2}$ identified as the transfer function of the power-PSS.

Strictly, over the range of modal frequencies of interest we require $\left.\frac{\Delta P_2}{\Delta\omega}\right|_{\Delta V_r = 0|sdd} = k_\omega$.

Furthermore, in order to attenuate low and high frequencies responses outside the range, washout and the low-pass filters are added, i.e.

$$\left.\frac{\Delta P_2}{\Delta\omega}\right|_{\Delta V_r = 0|sdd} = k_\omega \cdot \left(\frac{sT_1}{1+sT_1}\right) \cdot \left(\frac{sT_2}{1+sT_2}\right) \cdot \left(\frac{1}{1+sT_{LP}}\right). \tag{5.63}$$

By equating (5.62) and (5.63), the transfer function of the power PSS is determined, namely:

$$PSS_{power} = \frac{\Delta V_s}{\Delta P_2} = -k_\omega \cdot \left(\frac{sT_1}{1+sT_1}\right) \cdot \left(\frac{sT_2}{1+sT_2}\right) \cdot \left(\frac{1}{1+sT_{LP}}\right) \cdot \left[1 \Big/ \left(2Hs \cdot \left.\frac{\Delta P_2}{\Delta V_s}\right|_{sdd}\right)\right]. \tag{5.64}$$

For the Design Case (e.g. Case C in Section 5.10.6) the P-Vr transfer function $\left.\frac{\Delta P_2}{\Delta V_s}\right|_{sdd}$ is calculated as in Sections 5.8.1 to 5.8.3. The 'extended P-Vr transfer function', $2Hs \cdot \left.\frac{\Delta P_2}{\Delta V_s}\right|_{sdd}$, is then derived by curve-fitting. Let us assume that synthesized transfer function is of the simple form $As/(1+Bs)$. The resulting transfer function of the fixed-parameter power-PSS can be then be formed, i.e.:

$$PSS_{power}(s) = -k_\omega \cdot \left(\frac{sT_2}{1+sT_2}\right) \cdot \left(\frac{T_1/A}{1+sT_1}\right) \cdot \left(\frac{1+sB}{1+sT_{LP}}\right). \tag{5.65}$$

Note that, because of negation in the power-PSS path of (5.64), the signal ΔV_s is fed into the summing junction with a positive sign [1].

5.13.1 Example 5.6: Tuning of a power-based PSS

Case C (the 'Design Case'), and the more extreme cases, Cases A, G and H in the SMIB system of Section 5.10.5 and Table 5.4 are used to demonstrate the tuning of a power-PSS.

The P-Vr characteristic for the Design Case C, shown in Figure 5.24(a), is combined with the transfer function $2Hs$ to form the 'extended P-Vr', XPVr(s), $2Hs \cdot \left.\frac{\Delta P_2}{\Delta V_s}\right|_{sdd}$; the P-Vr and the XPVr are also shown in Figure 5.24(a). The synthesized transfer function of XPVr(s) - which results from curve fitting - is shown in the same figure and found to be

$$\text{XPVr(s)} = As/(1+Bs) = 27.1s/(1+0.108s). \tag{5.66}$$

1. A similar phenomenon occurs with the speed-PSS. See Section 5.8.1.

Assuming the damping gain is $k_\omega = 20$ pu on generator rating, and that the time constants of the washout and low-pass filters are 5 s and 0.01 s respectively, the transfer function of the power-PSS based on Design Case C is:

$$PSS_{power} = -20 \cdot \left(\frac{s5}{1+s5}\right) \cdot \left(\frac{5/27.1}{1+s5}\right) \cdot \left(\frac{1+s0.108}{1+s0.01}\right). \tag{5.67}$$

For comparison, the transfer-function of speed-PSS derived in Section 5.10.6.2 is

$$H_{PSS}(s) = 20 \cdot \frac{s5}{1+s5} \cdot \frac{1}{3.84} \cdot \frac{1+s0.0895+s^2 0.00277}{(1+s0.005)(1+s0.005)}. \quad \text{((5.49) repeated)} \tag{5.68}$$

The eigenvalues for the four cases with the speed- and power-PSSs out- and in-service are calculated; these are listed in Table 5.7.

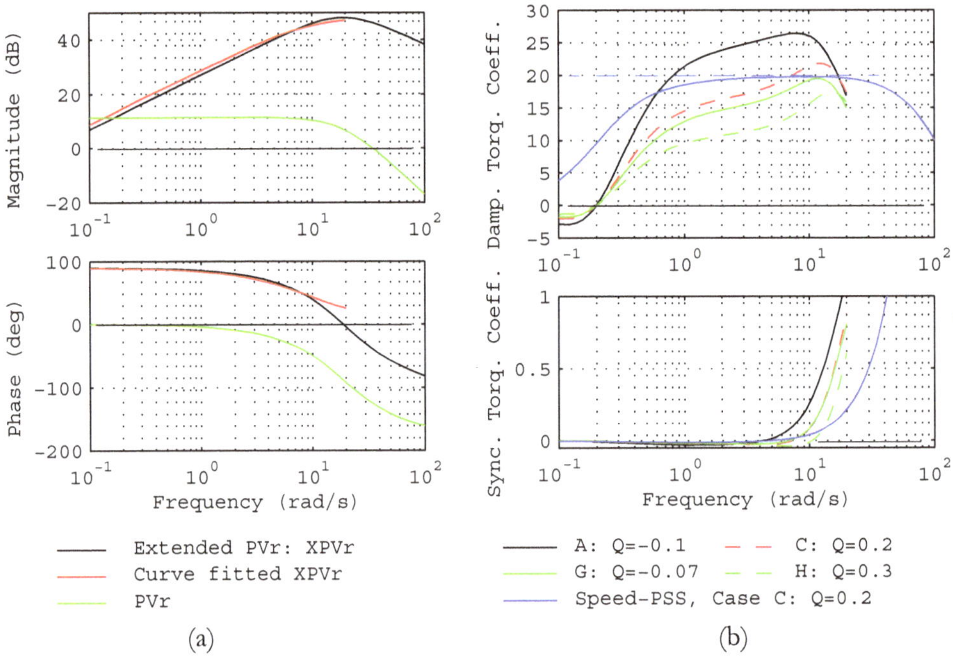

Figure 5.24

(a) Case C: P-Vr, the extended P-Vr (XPVr), and the synthesized characteristic of XPVr(s)
(b) Synchronizing and damping torque coefficients induced by the *power-PSS*.
Cases A and C: All lines in service; Cases G and H: Two lines out. For comparison the torque coefficients of the *speed-PSS* for Case C from Figure 5.17 are also plotted.

Table 5.7 Rotor modes with speed- and power-PSS in service; 6th order generator model. PSS damping gain $k_\omega = 20$ pu on machine rating.,

Case	Generator Output P, Q pu	Lines out of service	Eigenvalues, rotor mode		
			with PSS out of service	with speed-PSS in service	with power-PSS in service
A	0.9, -0.1	none	$0.773 \pm j9.16$	$-1.156 \pm j9.51$	$-1.443 \pm j9.53$
C	0.9, 0.2	none	$0.261 \pm j9.02$	$-1.338 \pm j9.03$	$-1.524 \pm j8.95$
G	0.9, -0.07	two	$0.927 \pm j7.98$	$-0.409 \pm j8.04$	$-0.465 \pm j7.98$
H	0.9, 0.3	two	$0.322 \pm j7.84$	$-0.774 \pm j7.72$	$-0.809 \pm j7.65$

Note in Table 5.7 that the rotor mode shifts for both types of PSS are directly to the left. However, it is observed that the left-shifts due to the power-PSS are greater than those of the speed-PSS. The differences arise due to (i) mismatches between the XPVr characteristic based on Case C and the fitted transfer-function XPVR(s), (ii) differences between the P-Vr characteristics of Design Case C and the other (more extreme) cases as seen in Figure 5.16.

In the case of the speed-PSS the induced damping torque coefficients in the PSS path, shown in Figure 5.17 for Cases C and G, are 20 and 18 pu over a modal frequency range 2 to 15 rad/s. The question arises: what are the corresponding characteristics of the damping torque coefficients for the power-PSS?

The characteristics of the synchronizing and damping torque coefficients for the power-PSS are illustrated in Figure 5.24(b) for the four cases when the damping gain is 20 pu. At the rotor modal frequencies of 7.5 to 9.5 rad/s the damping torque coefficients range from 14 to 26 pu. The reasons for the differences between the damping torque coefficients for Case C and the other cases are as explained above for the greater rotor-mode shifts by the power-PSS. The torque coefficients for a speed-PSS based on Case C are shown in Figure 5.24(b) for comparison.

However, in a multi-machine scenario with inter-area modal frequencies above 2 rad/s, damping torque coefficients exceeding 11 pu are potentially induced by the power-PSS for the worst-case operating condition (H). Adjusting the damping gain setting from 20 pu will raise or lower the damping torque coefficients and likewise modify the left-shifts of the rotor mode.

Note that: (i) In synthesizing the power-PSS difficulties may occur in finding a simple curve-fitted transfer function for the XPVr due to the shape of P-Vr and the modal frequencies of interest. (ii) As will be discussed in Chapter 8, power-PSSs may cause terminal voltage and reactive power swings due to changes and ramping of mechanical power [15]. Reducing the washout time constant from 5 s in the latter example may alleviate this problem.

5.14 Summary: P-Vr approach to the tuning of a fixed-parameter PSS

The P-Vr approach provides a systematic and consistent method for tuning PSSs for a selected set of encompassing operating conditions. It possesses the following features.

1. The aim of the PSS tuning procedure is to introduce on the generator shaft a damping torque (a torque proportional to machine speed); this causes the modes of rotor oscillation to be shifted to the left in the complex s-plane.

2. The PSS compensation transfer function $G_c(s)$ is tuned to achieve a direct *left-shift* in the complex s-plane of the relevant modes of rotor oscillation.

3. The damping gain k (pu on machine rating) of the PSS determines the *extent* of the left-shift.

4. The damping torque coefficient contributions induced by the PSS can be designed to be constant ('flat') over a desired range of local- and inter-area modal frequencies (e.g. as in Figure 5.17). The damping gain should be selected to ensure that the damping torque coefficients swamp any inherent negative contributions by the generator over the range of operating conditions.

5. If the above features are realized the PSS transfer function $kG_c(s)$ is said to be robust (see item 3, Section 1.2).

The analysis of the P-Vr approach to the tuning of fixed-parameter speed-PSS and its implications for a single-machine infinite-bus system have demonstrated the following.

1. For practical purposes, the *phase response* of the P-Vr is more-or-less invariant over an encompassing range of operating conditions. Similarly, at the higher real power outputs, typically 0.5 to 1 pu of rated power, the magnitude response retains its shape and consistently lies in a band of ±2 dB of the Design Case P-Vr characteristic. The PSS based upon the generator's P-Vr characteristics is robust because it induces positive damping torque coefficients on the shaft of the generator over: (i) the defined range of modal frequencies, (ii) the encompassing range of N and N-1 operating conditions, (iii) a range of leading and lagging power factors.

2. At higher levels of generator real power output the magnitude and phase of the P-Vr frequency response characteristics lie in relatively narrow bands for a wide range of operating conditions. This permits the selection of a Design Case whose magnitude and phase response are within the band.

3. In the following chapters it is recommended that system studies be conducted over a "wide range of operating conditions". The selection of the "Design Case" should be representative of the P-Vr characteristics for a set of operating conditions which encompasses the extreme range of operation, including the most leading and lagging reactive power output conditions. That is, by a *prudent* choice of an encompassing,

representative set of *N* and *N*-1 operating conditions *a fewer number of studies may be required* to determine the selection of the "Design Case".

4. The selection of the Design Case may be biased by experience, e.g. power factors less than 0.98 leading are very unlikely to occur.

5. The formulation of a PSS 'damping gain' (in pu on machine MVA rating) has the advantages, (i) it is the damping torque coefficient induced on the generator over the design range of modal frequencies; (ii) it is a meaningful number - a moderate damping gain is 20-30 pu, high values exceed 30 to 50 pu. The term 'PSS gain' lacks meaning unless it is clearly defined on such a basis.

6. As the operating conditions change the *inherent* damping torque coefficients can vary from significant negative to large positive values over the modal frequency range of concern. By adjustment of the damping gain such negative torques must be swamped by the positive damping coefficients induced by the PSS.

7. An alternative PSS tuning approach is the GEP(s) method (see Chapter 6). It employs the phase characteristics of the P-Vr method but provides no information or guidance on the selection of the PSS gains. The extension of the analytical approach to the GEP(s) to include the additional information available from the P-Vr method is a simple further step.

8. It is shown in Section 5.10.6.2 that the compensation transfer function for the speed-PSS may include a pair of complex zeros to represent the inverse of the P-Vr transfer function. A practical PSS may not be capable of accepting complex zeros, however, it is shown the transfer function may be represented by a set of lead blocks of the form $(1 + sT_{nk})/(1 + sT_{dk})$, in which the zeros are real and $T_{nk} > T_{dk}$.

9. In the tuning of speed-PSSs a trade-off may need to be found between the modal frequency range of interest, the possible high gain of the PSS compensation transfer function at higher frequencies, the low-pass filter parameters, and the attenuation of the torsional modes. Use of notch filters [9], or employing integral-of-accelerating-power PSSs (Section 8.5), may overcome some of these issues. In the case of hydro-generators no adverse interaction between the generator and the network at the torsional modal frequencies have been reported [9].

10. A comprehensive evaluation of PSS performance must include consideration of the response of the system to large disturbances (i.e. transient stability analysis). This aspect has been omitted from this chapter which has focussed on fundamental concepts but is examined briefly in Section 10.9. The settings of limiters on the output of the PSS are not considered.

11. It is shown in Section 5.13 that the P-Vr approach for tuning a speed-PSS can be adapted to the tuning of a power-based PSS [1]. The same advantages of the speed-PSS apply to power-PSS; for example, the procedure is systematic, the value of the PSS gain is a meaningful quantity - it is also a damping gain. An advantage of this

power-PSS is that it does not require the implementation of a phase-lead transfer-function network.The power-PSS may be more relevant, say, to cases for which a simple PSS is required for damping a local mode, when the number of blocks provided for implementing the PSS is limited, a speed-stabilizing signal is not readily available, etc.

Other benefits of the P-Vr approach will become evident when (i) the GEP and the Method of Residues are discussed in Chapter 6, and (ii) multi-machine systems are analysed in Chapter 10.

5.15 References

[1] F. P. de Mello and C. Concordia, "Concepts of Synchronous Machine Stability as Affected by Excitation Control, "*Power Apparatus and Systems, IEEE Transactions on*, vol. PAS-88, pp. 316-329, 1969.

[2] E.V. Larsen and D.A. Swann, "Applying Power System Stabilizers: Part I-III", *Power Apparatus and Systems, IEEE Transactions on*, vol. PAS-100, pp. 3017-3046, 1981.

[3] IEEE Power & Energy Society, *IEEE Tutorial Course: Power System Stabilization Via Excitation Control*, 09TP250, 2009.

[4] F. P. de Mello, J. S. Czuba, P. A. Rushe, and J. R. Willis, "Developments in application of stabilizing measures through excitation control," in *CIGRE*, 1986, 38–05.

[5] M. J. Gibbard, "Co-ordinated design of multimachine power system stabilisers based on damping torque concepts," *IEE Proceedings C Generation, Transmission and Distribution*, vol. 135, pp. 276-284, 1988.

[6] W. G. Heffron and R. A. Phillips, "Effect of a Modern Amplidyne Voltage Regulator on Underexcited Operation of Large Turbine Generators," *Power Apparatus and Systems, IEEE Transactions on, Part III. Transactions of the American Institute of Electrical Engineers*, vol. 71, pp. 692-697, 1952.

[7] IEEE Standard 421.1, *IEEE Standard Definitions for Excitation Systems for Synchronous Machines,* 2007.

[8] Australian Energy Market Commission, National Electricity Rules, Vers. 55, March 2013.

[9] P. Kundur, *Power System Stability and Control,* McGraw-Hill, Inc., 1994.

[10] D.M. Lam and H. Yee, "A Study of Frequency Responses of Generator Electrical Torques For Power System Stabilizer Design", *Power Systems, IEEE Transactions on*, vol. 13, pp. 1136-1142, 1998.

---

1. An alternative approach to tuning a power-based PSS is described in Section 8.3.

[11] M. J. Gibbard, "Robust design of fixed-parameter power system stabilisers over a wide range of operating conditions," *Power Systems, IEEE Transactions on*, vol. 6, pp. 794-800, 1991.

[12] D. Simfukwe and B. C. Pal, "Robust and Low Order Power Oscillation Damper Design Through Polynomial Control," *Power Systems, IEEE Transactions on*, vol. 28, pp. 1599-1608, 2013.

[13] D. D. Simfukwe, B. C. Pal, R. A. Jabr, and N. Martins, "Robust and low-order design of flexible ac transmission systems and power system stabilisers for oscillation damping," *Generation, Transmission & Distribution, IET*, vol. 6, pp. 445-452, 2012.

[14] G. R. Berube, L. M. Hajagos, and R. Beaulieu, "Practical utility experience with application of power system stabilizers," in *Power Engineering Society Summer Meeting, 1999. IEEE*, 1999, pp. 104-109 vol. 1.

[15] J.C.R. Ferraz, N. Martins, N. Zeni Jr, J.M.C. Soares, G.N. Taranto, "Adverse Increase in Generator Terminal Voltage and Reactive Power Transients Caused by Power System Stabilizers," *Proceedings of IEEE Power Engineering Society Winter Meeting*, 2002.

Appendix 5–I

App. 5–I.1 K-coefficients, Heffron and Phillips Model of SMIB System

The steady-state values of variables are designated with the subscript ‘0’. All quantities are in per-unit except angles are in degrees. Computation of the initial conditions is based on solution of the SMIB model equations given in the Appendix of [1].

V_b	Source voltage of the Thévenin equivalent of the system connected at the generator terminals.
e_{t0}	Generator terminal voltage.
e_{d0}, e_{q0}	Direct and quadrature components of the terminal voltage.
i_{d0}, i_{q0}	Direct and quadrature components of the generator current.
E_{q0}	Voltage behind quadrature-axis reactance of the generator.
δ_0	Angle between the quadrature axis of the generator and the infinite bus.
r_e, x_e	Resistance and reactance of the Thévenin equivalent of system connected at the generator terminals.

$$A = r_e^2 + (x_e + X'_d) \cdot (X_q + x_e)$$

$$K_1 = (E_{q0} V_b [r_e \sin\delta_0 + (x_e + X'_d)\cos\delta_0])/A + \ldots$$
$$i_{q0} V_b (X_q - X'_d)[(x_e + X_q)\sin\delta_0 - r_e \cos\delta_0]/A$$

$$K_2 = r_e E_{q0}/A + i_{q0}[1 + (x_e + X_q)(X_q - X'_d)/A]$$

$$K_3 = [1 + (x_e + X_q)(X_d - X'_d)/A]^{-1}$$

$$K_4 = (V_b (X_d - X'_d)[(x_e + X_q)\sin\delta_0 - r_e \cos\delta_0])/A$$

$$K_5 = V_b (e_{d0}/e_{t0}) X_q [r_e \sin\delta_0 + (x_e + X'_d)\cos\delta_0]/A + \ldots$$
$$V_b (e_{q0}/e_{t0}) X'_d [r_e \cos\delta_0 - (x_e + X_q)\sin\delta_0]/A$$

$$K_6 = (e_{q0}/e_{t0})[1 - X'_d (x_e + X_q)/A] + r_e X_q (e_{d0}/e_{t0})/A$$

App. 5–I.2 Transfer function of the SMIB system with closed-loop control of terminal voltage

Based on Figure 5.1 for the Heffron and Phillips model for the SMIB system with closed-loop voltage control, the transfer function for the terminal voltage response to perturbations in reference voltage is:

$$\frac{\Delta V_t}{\Delta V_r} = \frac{s^2 MK_3K_6 + sDK_3K_6 + \omega_0 K_3(K_1K_6 - K_2K_5)}{s^4 a_4 + s^3 a_3 + s^2 a_2 + sa_1 + a_0}, \tag{5.69}$$

where the damping torque coefficient $D = 0$.

$$a_4 = c_3 T'_{do} T_{ex},\quad a_3 = c_3 T'_{do} + c_2 T_{ex},\quad a_2 = c_2 + T_{ex} c_1 + MK_3K_6K_{ex},$$

$$a_1 = c_1 + \omega_0 T_{ex} c_0 + DK_3K_6K_{ex},\quad a_0 = \omega_0[c_0 + K_3K_{ex}(K_1K_6 - K_2K_5)],$$

$$c_3 = MK_3,\quad c_2 = M + DK_3T'_{do},\quad c_1 = \omega_0 K_1K_3T'_{do},\quad c_0 = K_1 - K_2K_3K_4.$$

The K-coefficients are listed in Appendix 5–I.1

App. 5–I.3 Model of the 6th-order generator and excitation system

App. 5–I.3.1 State-space model

In Section 12.6 of [9] the state-space model of a 6th-order generator model, having one and two damper windings on the direct and quadrature axes, respectively, is developed. It is of the form:

$$\dot{x}(t) = A \cdot \boldsymbol{x}(t) + B \cdot \boldsymbol{u}(t),\quad y(t) = C \cdot \boldsymbol{x}(t) + D \cdot \boldsymbol{u}(t).\ \text{Thus,}$$

$$\begin{bmatrix} \Delta\dot{\omega}_r \\ \Delta\dot{\delta} \\ \Delta\dot{\psi}_{fd} \\ \Delta\dot{\psi}_{1d} \\ \Delta\dot{\psi}_{1q} \\ \Delta\dot{\psi}_{2q} \end{bmatrix} = \begin{bmatrix} a_{11} & a_{12} & a_{13} & a_{14} & a_{15} & a_{16} \\ a_{21} & 0 & 0 & 0 & 0 & 0 \\ 0 & a_{32} & a_{33} & a_{34} & a_{35} & a_{36} \\ 0 & a_{42} & a_{43} & a_{44} & a_{45} & a_{46} \\ 0 & a_{52} & a_{53} & a_{54} & a_{55} & a_{56} \\ 0 & a_{62} & a_{63} & a_{64} & a_{65} & a_{66} \end{bmatrix} \begin{bmatrix} \Delta\omega_r \\ \Delta\delta \\ \Delta\psi_{fd} \\ \Delta\psi_{1d} \\ \Delta\psi_{1q} \\ \Delta\psi_{2q} \end{bmatrix} + \begin{bmatrix} b_{11} & 0 \\ 0 & 0 \\ 0 & b_{32} \\ 0 & 0 \\ 0 & 0 \\ 0 & 0 \end{bmatrix} \begin{bmatrix} \Delta P_m \\ \Delta E_{fd} \end{bmatrix}. \tag{5.70}$$

The elements of the ***A***- and ***B***-matrices are defined in Section 12.6 of [9].

Elements a_{12} to a_{16} are functions of K_1, K_2, K_{21}, K_{22} and K_{23}, respectively (see equations 12.171 to 12.183, [9]).

The elements a_{3i}, a_{4i}, a_{5i}, a_{6i} are functions of the generator parameters. Note, in equation 12.179 of [9]

$$a_{44} = -\frac{\omega_0 R_{1d}}{L_{1d}}\left(1 - \frac{L''_{ads}}{L_{1d}} + m_3 L''_{ads}\right).$$

The terminal voltage is given by equation 12.185 of [9]:

$$\Delta V_t = K_5 \Delta\delta + K_6 \Delta\psi_{fd} + K_{61}\Delta\psi_{1d} + K_{62}\Delta\psi_{1q} + K_{63}\Delta\psi_{2q}, \tag{5.71}$$

where $K_5 = \frac{e_{d0}}{E_{t0}}\{-R_a m_1 + L_l n_1 + n_1 L''_{aqs}\} + \frac{e_{q0}}{E_{t0}}\{-L_l m_1 - R_a n_1 + m_1 L''_{ads}\}$,

$$K_6 = \frac{e_{d0}}{E_{t0}}\{-R_a m_2 + L_l n_2 + n_2 L''_{aqs}\} + \frac{e_{q0}}{E_{t0}}\left\{-L_l m_2 - R_a n_2 + L''_{ads}\left(\frac{1}{L_{fd}} - m_2\right)\right\},$$

$$K_{61} = \frac{e_{d0}}{E_{t0}}\{-R_a m_3 + L_l n_3 + n_3 L''_{aqs}\} + \frac{e_{q0}}{E_{t0}}\left\{-L_l m_3 - R_a n_3 + L''_{ads}\left(\frac{1}{L_{1d}} - m_3\right)\right\},$$

$$K_{62} = \frac{e_{d0}}{E_{t0}}\left\{-R_a m_4 + L_l n_4 - L''_{aqs}\left(\frac{1}{L_{1q}} - n_4\right)\right\} + \frac{e_{q0}}{E_{t0}}\{-L_l m_4 - R_a n_4 - m_4 L''_{ads}\},$$

$$K_{63} = \frac{e_{d0}}{E_{t0}}\left\{-R_a m_5 + L_l n_5 - L''_{aqs}\left(\frac{1}{L_{2q}} - n_5\right)\right\} + \frac{e_{q0}}{E_{t0}}\{-L_l m_5 - R_a n_5 - m_5 L''_{ads}\}.$$

The coefficients $m_1 \ldots m_5$, $n_1 \ldots n_5$ are defined in Section 12.6 of [9]. The transfer function of a first-order excitation system of the form shown in Figure 5.7 on page 239 is

$$\Delta E_{fd} = \frac{K_{ex}}{1 + sT_{ex}}(\Delta V_r - \Delta V_t + \Delta V_s), \tag{5.72}$$

where ΔE_{fd} is the field voltage (pu), and ΔV_s in the output from a PSS, if fitted.

The time-domain form of (5.72) is

$$\Delta \dot{E}_{fd} = -\frac{1}{T_{ex}}\Delta E_{fd} + \frac{K_{ex}}{T_{ex}}(\Delta V_r - \Delta V_t + \Delta V_s), \text{ or}$$

$$\Delta \dot{E}_{fd} = a_{77}\Delta E_{fd} + a_{72}\Delta\delta + a_{73}\Delta\psi_{fd} + a_{74}\Delta\psi_{d1} + a_{75}\Delta\psi_{q1} + a_{76}\Delta\psi_{q2} + b_{72}(\Delta V_r + \Delta V_s) \tag{5.73}$$

where $a_{71} = 0$, $a_{72} = -K_5 K_{ex}/T_{ex}$, $a_{73} = -K_6 K_{ex}/T_{ex}$, $a_{74} = -K_{61} K_{ex}/T_{ex}$,

$$a_{75} = -K_{62} K_{ex}/T_{ex},\ a_{76} = -K_{63} K_{ex}/T_{ex},\ a_{77} = -1/T_{ex},\ b_{72} = K_{ex}/T_{ex}. \tag{5.74}$$

Augmenting the state-space model (5.72) with (5.73), the equation for the model becomes:

$$\begin{bmatrix} \Delta\dot{\omega}_r \\ \Delta\dot{\delta} \\ \Delta\dot{\psi}_{fd} \\ \Delta\dot{\psi}_{1d} \\ \Delta\dot{\psi}_{1q} \\ \Delta\dot{\psi}_{2q} \\ \Delta\dot{E}_{fd} \end{bmatrix} = \begin{bmatrix} a_{11} & a_{12} & a_{13} & a_{14} & a_{15} & a_{16} & 0 \\ a_{21} & 0 & 0 & 0 & 0 & 0 & 0 \\ 0 & a_{32} & a_{33} & a_{34} & a_{35} & a_{36} & a_{37} \\ 0 & a_{42} & a_{43} & a_{44} & a_{45} & a_{46} & 0 \\ 0 & a_{52} & a_{53} & a_{54} & a_{55} & a_{56} & 0 \\ 0 & a_{62} & a_{63} & a_{64} & a_{65} & a_{66} & 0 \\ 0 & a_{72} & a_{73} & a_{74} & a_{75} & a_{76} & a_{77} \end{bmatrix} \begin{bmatrix} \Delta\omega_r \\ \Delta\delta \\ \Delta\psi_{fd} \\ \Delta\psi_{1d} \\ \Delta\psi_{1q} \\ \Delta\psi_{2q} \\ \Delta E_{fd} \end{bmatrix} + \begin{bmatrix} b_{11} & 0 \\ 0 & 0 \\ 0 & 0 \\ 0 & 0 \\ 0 & 0 \\ 0 & 0 \\ 0 & b_{72} \end{bmatrix} \begin{bmatrix} \Delta P_m \\ \Delta V_r \end{bmatrix} \tag{5.75}$$

Note, the element a_{37} assumes the value of $b_{32} = \omega_0 R_{fd}/L_{adu}$ in (5.70).

App. 5–I.3.2 Calculation of the inherent torque coefficients.

In Section 5.3 it was noted for the SMIB system that the synchronizing and damping torque coefficients can be calculated from the transfer function $\Delta P_0(s)/\Delta\omega(s)$ given by (5.9) on page 229 with the shaft dynamics disabled. For the 6th-order generator model a new set of ABCD matrices is formed by (ii) eliminating the first row in (5.75) and (ii) expressing the electrical torque ΔP_0 as output and the speed signal $\Delta\omega_r$ as input, i.e.

$$\begin{bmatrix} \Delta\dot{\delta} \\ \Delta\dot{\psi}_{fd} \\ \Delta\dot{\psi}_{1d} \\ \Delta\dot{\psi}_{1q} \\ \Delta\dot{\psi}_{2q} \\ \Delta\dot{E}_{fd} \end{bmatrix} = \begin{bmatrix} 0 & 0 & 0 & 0 & 0 & 0 \\ a_{32} & a_{33} & a_{34} & a_{35} & a_{36} & a_{37} \\ a_{42} & a_{43} & a_{44} & a_{45} & a_{46} & 0 \\ a_{52} & a_{53} & a_{54} & a_{55} & a_{56} & 0 \\ a_{62} & a_{63} & a_{64} & a_{65} & a_{66} & 0 \\ a_{72} & a_{73} & a_{74} & a_{75} & a_{76} & a_{77} \end{bmatrix} \begin{bmatrix} \Delta\delta \\ \Delta\psi_{fd} \\ \Delta\psi_{1d} \\ \Delta\psi_{1q} \\ \Delta\psi_{2q} \\ \Delta E_{fd} \end{bmatrix} + \begin{bmatrix} a_{21} \\ 0 \\ 0 \\ 0 \\ 0 \\ 0 \\ 0 \end{bmatrix} \begin{bmatrix} \Delta\omega_r \end{bmatrix} \tag{5.76}$$

The expression for torque, ΔP_0, is given by Equation 12.170 in [9]:

$$\Delta P_0 = K_1 \Delta\delta + K_2 \Delta\psi_{fd} + K_{21} \Delta\psi_{d1} + K_{22} \Delta\psi_{q1} + K_{23} \Delta\psi_{q2}. \tag{5.77}$$

These K-coefficients are listed in Equation 12.171 of [9].

Chapter 6

Tuning of PSSs using methods based on Residues and the GEP transfer function

6.1 Introduction

In Section 5.8 the P-Vr method for the tuning of the PSS for a generator in a single-machine infinite-bus (SMIB) system is described. Several other methods, which will be shown to be somewhat related to the P-Vr method, are described in the literature. Two other methods will be discussed here, the first is based on Transfer-Function Residues, the second on the so-called GEP Method. The P-Vr method, the Method of Residues and the GEP Method are reconciled for a practical, multi-machine system in [1]. However, for illustrative purposes in this chapter we will examine only the application of the Residues and GEP Methods to a generator in a SMIB system.

The background to the Method of Residues is provided in [2] and its application to PSSs is illustrated in Appendix A of [3]. The method is also used in practice for the design of Power Oscillation Dampers (PODs) which are fitted to FACTS devices such as SVCs, typically to enhance the damping of inter-area modes. The design of PODs using the Method of Residues is described in [4], however, this topic is considered in more detail in Chapter 11.

6.2 Method of Residues

6.2.1 Theoretical basis for the Method

The theoretical basis, calculation and significance of the residues of a transfer function are discussed earlier in Section 2.5. In essence, for a set of distinct poles r_i is the residue of the pole at $s = p_i$. A transfer function $G(s)$ is described by its partial fraction expansion equation (2.14), or by

$$G(s) = \frac{P(s)}{Q(s)} = \frac{r_1}{s-p_1} + \frac{r_2}{s-p_2} + \ldots + \frac{r_i}{s-p_i} + \ldots + \frac{r_n}{s-p_n}. \tag{6.1}$$

The derivation of residues from the state equations is outlined in Section 3.7.

Consider a SMIB system for which a PSS is to be designed and installed. The transfer function from the reference voltage input to the speed output signal of the generator is $G_S(s) = \Delta\omega/\Delta V_{ref}$. The PSS, with transfer function $F(s)$, is a speed-input PSS (although other stabilizing signals can be employed). When operating in closed-loop the PSS output is connected to the AVR summing junction, as shown in Figure 6.1.

It is emphasized that the following simple approach to the determination of the compensation transfer function of the PSS is based on the change of the rotor mode of oscillation *when the PSS feedback path is switched from open to closed loop.*

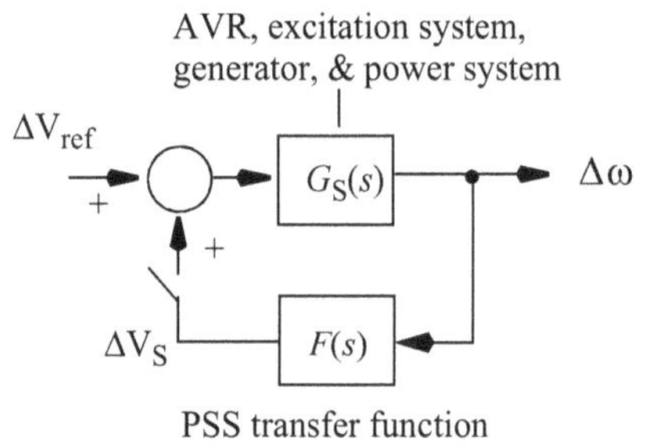

Figure 6.1 SMIB system $G_S(s)$ and PSS transfer function $F(s)$ on open loop.

Let the PSS transfer function be:

$$F(s) = k_R H(s) = k_R G_c(s) \cdot G_W(s) \cdot G_{LP}(s), \tag{6.2}$$

where the transfer function $G_c(s)$ of the PSS in this application is designed to provide the appropriate phase compensation and is assumed to consist of *m* lead or lag blocks of the form [1]:

1. This form is used in the determination of the order *m* and time constants T_n and T_d in Appendix 6–I.1.

$$G_c(s) = \left[\frac{1 + T_n s}{1 + T_d s}\right]^m . \tag{6.3}$$

The PSS gain setting in (6.2) is k_R in pu on device base (note, this is not the 'damping gain' associated with the P-Vr method). The washout and low-pass filter transfer functions, $G_W(s)$ and $G_{LP}(s)$, are given by (5.29) and (5.30), respectively. It is assumed that the values of the time constants in the latter two transfer functions have been appropriately selected (see Section 5.8.6). The objective of the tuning procedure is to determine the values k_R, T_n, T_d and m that satisfy the relevant requirements on damping.

Note that in Figure 6.1 positive feedback is assumed for the following analysis. The transfer function of the SMIB system and PSS when the loop is closed is therefore:

$$W(s) = \frac{G_S(s)}{1 - G_S(s)F(s)} = \frac{G_S(s)}{1 - k_R G_S(s)H(s)} \quad \text{using (6.2).}$$

The poles of the closed-loop transfer function are derived from its characteristic equation:

$$1 - G_S(s)F(s) = 0 ; \tag{6.4}$$

these poles are also the eigenvalues of the system of $W(s)$ [1]. Let us evaluate the shift $\Delta\lambda_h$ in the pole (eigenvalue) λ_h resulting from the closure of the feedback loop. Assume the plant and system, $G_S(s)$, is excited by the eigenvalue λ_h when on open loop, i.e. from (6.1),

$$G_S(\lambda_h) = r_h/(s - \lambda_h),$$

where r_h is the residue of the eigenvalue λ_h of the forward-loop transfer function, $G_S(\lambda_h)$. The associated characteristic equation is, from (6.4),

$$1 - \frac{r_h}{s - \lambda_h} F(\lambda_h) = 0, \text{ or } s = \lambda_h + k_R r_h H(\lambda_h), \tag{6.5}$$

noting that $F(s) = k_R H(s)$ in (6.2). Suppose the pole of the closed-loop system is shifted by a small amount $\Delta\lambda_h$ from the open-loop pole λ_h. The root of the new characteristic equation is thus $s = (\lambda_h + \Delta\lambda_h)$ and (6.5) becomes:

$$(\lambda_h + \Delta\lambda_h) - \lambda_h - k_R r_h H(\lambda_h + \Delta\lambda_h) = 0 . \tag{6.6}$$

If the mode shift is 'small' then the transfer-function $H(s)$ in the neighbourhood of $s = \lambda_h$ can be represented by the first-order Taylor series expansion:

$$H(\lambda_h + \Delta\lambda_h) = H(\lambda_h) + \left(\left.\frac{\partial H(s)}{\partial s}\right|_{s = \lambda_h}\right)\Delta\lambda_h . \tag{6.7}$$

1. See Sections 3.7and 2.5

Substitution of (6.7) in (6.6) leads to

$$\Delta\lambda_h = \frac{k_R r_h H(\lambda_h)}{1 - r_h k_R \dfrac{\partial H(\lambda_h)}{\partial \lambda_h}} \; ; \tag{6.8}$$

equations for the evaluation of $\dfrac{\partial H(\lambda_h)}{\partial \lambda_h}$ are given in Appendix 6–I.2, (6.25) and (6.26).

If the gain k_R is chosen such that

$$\left| r_h k_R \frac{\partial H(\lambda_h)}{\partial \lambda_h} \right| \ll 1 \, , \tag{6.9}$$

then (6.8) reduces to:

$$\Delta\lambda_h \approx k_R r_h H(\lambda_h) . \tag{6.10}$$

The result in (6.10) is significant for the design of the PSS compensation. As illustrated in Figure 6.2 the residue of the eigenvalue λ_h of $G_S(s)$ is a complex number, $|r_h| \angle\theta_h{}^\circ$, where $\theta_h = \arg\{r_h\}$. In order for the mode shift $\Delta\lambda_h$ in (6.10) to be $\pm 180°$, i.e. a direct left-shift of λ_h in the complex *s*-plane is required,

$$\arg\{r_h H(\lambda_h)\} = \pm 180° . \tag{6.11}$$

The compensation angle ϕ provided by the PSS thus must be

$$\phi = \arg\{H(\lambda_h)\} = \pm 180 - \arg\{r_h\} \quad (°) . \tag{6.12}$$

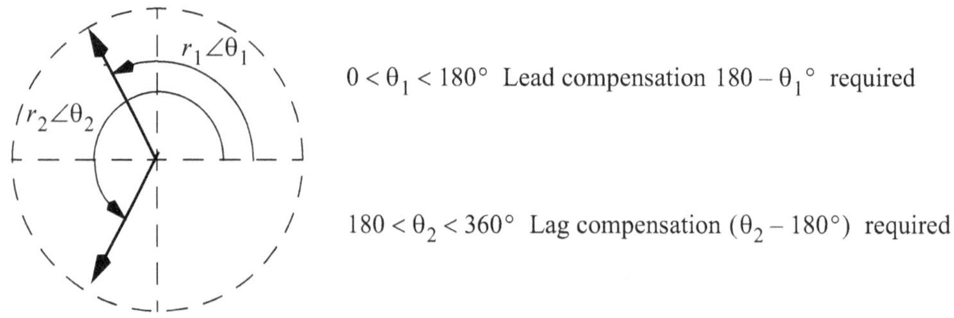

Figure 6.2 Examples of phase compensation required for possible residues of $G_S(\lambda_h)$.

Based on (6.10) the following is a procedure for determining the PSS parameters.

To effect a left shift in the rotor mode of interest, $\Delta\lambda_h$, choose the parameters T_n, T_d for the *m* lead or lag compensator blocks, (6.3), such that $\arg\{r_h H(\lambda_h)\} = \pm 180°$.

In order to determine a nominal value of an upper limit on the gain k_R such that the approximation for the mode shift in (6.10) is acceptable, let us define a nominal limit k_{Rm} based on (6.9) as

$$\left| r_h k_{Rm} \frac{\partial H(\lambda_h)}{\partial \lambda_h} \right| = 1, \text{ or} \tag{6.13}$$

$$k_{Rm} = \frac{1}{\left| r_h \frac{\partial H(\lambda_h)}{\partial \lambda_h} \right|} \text{ pu.} \tag{6.14}$$

An acceptable gain might lie in the range $0 < k_R < 0.1 k_{Rm}$, say, but will depend on the nature of the problem.

It was shown in Section 3.7 that the residue for mode h is the product of an 'observability measure' ($\hat{\boldsymbol{c}}_{mh}$) and a 'controllability measure' ($\hat{\boldsymbol{b}}_{hq}$) given by

$$r_{mq}^{h} = \hat{\boldsymbol{c}}_{mh} \cdot \hat{\boldsymbol{b}}_{hq} . \tag{6.15}$$

Ideally for a selected stabilizing signal these measures should reflect both good observability of the stabilizing signal and good controllability of the output variable.

From (6.10) it is noted that (i) the larger the magnitude of the residue, the greater is the mode shift; (ii) for a robust design it is desirable that the magnitude and phase of the residue remain more-or-less unchanged over the range of operating conditions; (iii) the real and imaginary components of the open-loop mode λ_h may also vary over a range of operating conditions in practice.

6.3 Tuning a speed-PSS using the Method of Residues

6.3.1 Calculation of the compensation transfer function of the PSS

As an example consider the tuning of a true-speed PSS founded on the Method of Residues. The design is based on a selected set of operating conditions for the SMIB system described in Table 5.4 on page 263 in which the rotor modes with the PSSs out of service are also listed. To represent a range of operating conditions four normal conditions are selected (Cases A, B, C, D) together with cases in which one and two lines are out of service (E, F and G, H respectively).

For the transfer function from the voltage reference input signal to speed-output perturbations the residues, which are calculated using a small-signal dynamics software package, are displayed in Figure 6.3. The residues lie in a relatively narrow phase-band of approximately $21°$ with a spread in magnitude between 0.19 and 0.31 units.

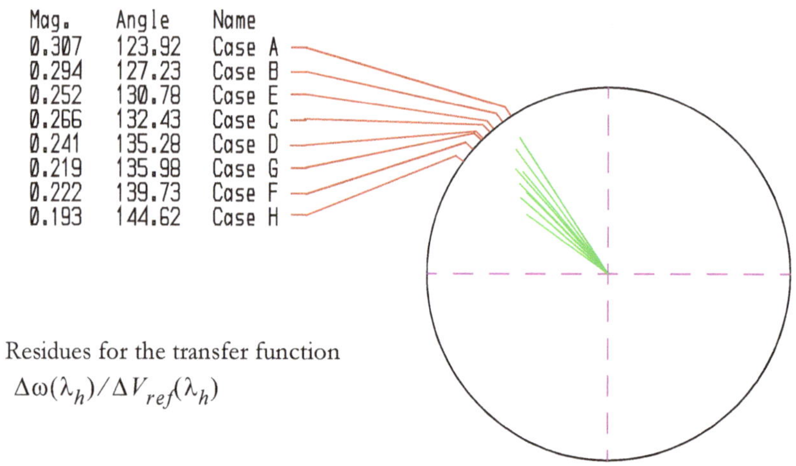

Figure 6.3 Polar plot of residues for selected cases A to H for the SMIB system. The residue for Case C is selected as representing the group.

It is now necessary to select from the group in Figure 6.3 a representative residue on which the compensation is to be based. If the residue of maximum amplitude is selected (e.g. that for Case A in the figure) the resulting value of nominal limit k_{Rm} would be lower than if some other residue were selected (according to (6.14)). The decision depends on the application or may be determined by the system criteria which specify the minimum level for damping (say) for the outage of two lines (Cases G and H here). However, as in Section 5.10.6.1, the design Case C is again selected to facilitate a comparison of the performance between the PSSs based on the Residues and P-Vr methods. From the polar plot in Figure 6.3 the residue for Case C, $0.266\angle 132.4°$, is selected as it fairly well represents the group in amplitude and phase. The PSS is thus required to provide phase-lead compensation of $180° - 132.4° = 47.6°$.

From Table 5.4 for Case C the modal frequency is $\lambda_C = 0.261 \pm j9.02$. Compensation for the phase shift introduced by the washout and low-pass filters at this complex modal frequency is also required.

Let us assume a single lead transfer function, and washout and low-pass filters with time constants of $T_W = 5$ s and $T_{LP} = 0.0125$ s, respectively, are employed. The PSS of (6.2) thus takes the form:

$$F(s) = k_R H(s) = k_R \left[\frac{sT_W}{1 + sT_W}\right]\left[\frac{1 + T_n s}{1 + T_d s}\right]\left[\frac{1}{1 + T_{LP} s}\right]. \tag{6.16}$$

At the complex modal frequency a net phase lag of 5.27° is introduced by the washout and low pass filters; the total phase compensation required is therefore $47.57 + 5.27 = 52.84°$ at the rotor mode $s = \lambda_C = 0.261 \pm j9.02$. Employing the algorithm outlined in

Appendix 6–I.1 an iterative procedure is used to calculate the parameters of the lead transfer function for the rotor mode. The resulting time constants are T_n = 0.359 and T_d = 0.038 s.

It was suggested earlier that an acceptable gain might lie in the range $0 < k_R < 0.1k_{Rm}$, where the nominal upper gain limit k_{Rm} is determined by (6.14). Substitution in the latter equation with values $r_h = 0.266\angle 132.4°$ and $\left.\frac{\partial H(s)}{\partial s}\right|_{s=\lambda_h} = 0.171 - j0.221$ yields k_{Rm} = 13.4 pu on machine MVA rating. Let us assume an acceptable gain range for k_R of 0 to 1.34 pu.

6.3.2 Design Case C. Performance of the PSS with increasing PSS gain

Using the PSS transfer function given by (6.16) let us estimate the values of the rotor mode as the gain k_R is increased from zero (open loop) to 10% of k_{Rm}, i.e 1.34 pu.

Three values of the mode are calculated for each value of k_R, that is, (i) an approximate value of the mode based on (6.10), (ii) a corrected value based on (6.8), and (iii) an eigenvalue calculated using the software package Mudpack [5]. The results of these calculations are shown in Table 6.1.

Table 6.1 Rotor modes as PSS gain k_R is increased from zero to 10% of k_{Rm}.

Gain k_R (% of k_{Rm})	Approximate *		Corrected+		Eigenanalysis		Difference EigAnal-Approx.		Difference EigAnal-Corr'd	
	Real	Imag	Real	Imag	Real	Imag	Real	Imag	Real	Imag
0	0.261	9.018	0.261	9.018	0.261	9.018	-	-	-	-
1	0.148	9.020	0.148	9.019	0.147	9.016	-0.001	-0.004	-0.001	-0.003
2	0.035	9.021	0.034	9.016	0.030	9.012	-0.005	-0.009	-0.004	-0.004
3	-0.078	9.021	-0.080	9.011	-0.089	9.004	-0.011	-0.017	-0.009	-0.007
4	-0.192	9.021	-0.194	9.003	-0.210	8.991	-0.018	-0.030	-0.016	-0.012
5	-0.305	9.021	-0.308	8.993	-0.334	8.974	-0.029	-0.047	-0.026	-0.019
6	-0.418	9.025	-0.422	8.981	-0.459	8.953	-0.041	-0.072	-0.037	-0.028
7	-0.531	9.022	-0.537	8.966	-0.587	8.925	-0.056	-0.097	-0.050	-0.041
8	-0.644	9.022	-0.651	8.949	-0.716	8.892	-0.072	-0.130	-0.065	-0.057
9	-0.758	9.022	-0.765	8.930	-0.846	8.851	-0.088	-0.171	-0.081	-0.079
10#	-0.870	9.022	-0.879	8.909	-0.976	8.804	-0.106	-0.218	-0.097	-0.105

$0.1k_{Rm}$=1.34 pu * Eqn. (6.10) + Eqn. (6.8) EigAnal: Eigen-analysis

From the table it is observed that the real parts of the rotor modes based on the approximate and corrected estimated values start to differ from the eigenvalues by more than 10% for PSS gains k_R greater than 8% to 10% (i.e. 1.1 - 1.3 pu). It should be remembered that more accurate eigen-analysis takes into account the effect of other zeros and poles on the modal trajectory as k_R is increased whereas the approximate trajectory increases linearly. Engineering judgement based on system requirements is needed to decide on the value of the gain setting - and if it is appropriate for other operating conditions.

6.3.3 Significance of the gain k_R and the damping gain k

The gain k_R itself has no significance unless it can be related to the damping torque coefficient induced by the PSS, i.e. the damping gain k. Let us relate the results of the simulation study in Table 6.1 to that based on the P-Vr-based design in Chapter 5 for the same operating condition, Case C in Table 5.4. However, to facilitate the comparison, the following modifications are made to the PSS based on the P-Vr transfer function in (5.49) on page 267, (i) the low-pass filter time constants of the PSS transfer function are set to 0.0125 and 0.005 s, and (ii) its damping gain is set to k = 77.4% of 20 pu (15.5 pu); the washout filter time constant remains at 5 s. The latter gain setting results in the real part of the eigenvalue (the rotor mode) being equal to that produced by simulation for the tuning design based on Residues Method, i.e −0.976 Np/s with k_R = 10% of k_{Rm} (see column 6 of Table 6.1).

Referring to Figure 6.4 we note that for the PSS tuned based on the P-Vr Method the damping torque coefficient is essentially flat at ~15 pu over the frequency range 1 to 10 rad/s for the setting of the damping gain to 15.5 pu. For the PSS design based on Residues Method the damping torque coefficient equals that of the P-Vr method at 15 pu at a frequency of ~9.2 rad/s. However, in the residues-based method the damping torque coefficient is seen to vary markedly over the frequency range 2 to 12 rad/s, a variation which will be shown to be unsatisfactory if the SMIB system represents an approximate, reduced equivalent of a larger multi-machine system in which a range of electro-mechanical modal frequencies exist. Thus, in a multi-machine system to ensure a robust PSS design using the Residues Method it is necessary to take into account all of the electro-mechanical modes in which the generator participates. A reduced-order system equivalent with a range of modal frequencies is proposed in [6] to facilitate robust application of the Residues Method. Nevertheless, it is necessary to verify robustness by closely examining damping performance over an encompassing range of operating conditions.

6.4 Conclusions, Method of Residues

Based on the preceding analysis and the example it is noted that:

- In the Method of Residues tuning procedure, the determination of the PSS compensation parameters is based both on the value of the *complex* rotor mode on open loop and on the variation of the mode on closed loop as the gain k_R is incremented (as in Table 6.1). In both the P-Vr method and the GEP Method (which follows next) the analysis is based on frequency response, $s = j\omega_f$. Clearly an analysis based on *complex* modal frequencies in the Method of Residues is likely to be more accurate - all else being equal; this is discussed in Section 5.9.3 with reference to Table 5.3 on page 255.

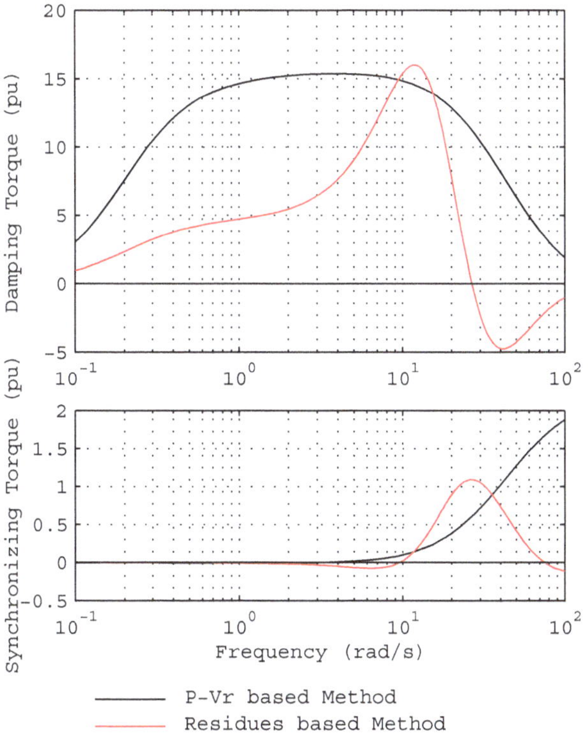

Figure 6.4 Case C: Synchronising and damping torque coefficients induced on the generator. PSS tuning is based on (i) P-Vr Method and (ii) Method of Residues.

- The relations (6.10) and (6.8) for the approximate and corrected mode shifts provide useful estimates of the mode shifts with increase in gain k_R. However, in the multi-machine case, eigen-analysis may reveal that the effect of system poles and zeros may cause the trajectory of the rotor mode to deviate from a direct left-shift at relatively low values of the gain k_R.

- In the example presented the Method of Residues can provide a basis for determining the PSS transfer function, and yields acceptable values of mode shifts with increasing PSS gain. However, the performance of the PSS needs to be validated over an encompassing range of operating conditions using a small-signal dynamics software package, particularly in the application of the method to PSSs in multi-machine systems [4], [10].

- In the P-Vr method, the damping gain k represents the damping torque coefficient induced by the PSS over a desired range of rotor modal frequencies. In the Method of Residues the value of the PSS gain k_R has no obvious significance.

- The P-Vr method yields inherently robust PSS designs (see item 3 of Section 1.2). Moreover, the PSS can be tuned to yield a more-or-less constant positive damping torque coefficient over a modal frequency range covering the low frequency inter-area modes to the higher frequency local-area and intra-plant modes (see Figure 6.4). In comparison, robust PSS design using the Residue Method requires accurate determination of the residues of all rotor modes in which the generator participates significantly.

6.5 The GEP Method

Larsen and Swann described in 1981 a practical procedure for tuning PSSs based on measurements taken in the field [7]. The procedure is based on the design approach of de Mello and Concordia [8]. The transfer function between the voltage reference input to the AVR and the electrical torque developed on the shaft is called the generator, excitation system and power system transfer function, GEP(s). GEP(s) can be shown to be proportional to the transfer function from voltage reference ($V_r(s)$) to *terminal voltage* ($V_t(s)$); the frequency response $V_t(j\omega_f)/V_r(j\omega_f)$ is relatively straight-forward to measure in the field. The compensation angle for the PSS transfer function is the negative of the phase shift of the measured frequency response. From this result, a compensating transfer function is synthesized for the PSS. A further test is performed to determine the gain setting of the PSS. This test consists of raising the gain until the onset of instability is observed; the PSS gain is then set to 1/3rd of this value - providing a gain margin of about 10 dB. Further developments of the GEP approach are reported in [6], [9].

The use of the field measured frequency-response for PSS design relies on the assumption that, because the generator is connected to a large power system, its speed remains more-or-less constant during the frequency response measurements. This is equivalent to assuming that the inertia constant of the unit is very large or, alternatively, the speed and angle perturbations are negligible. For example, based on the Heffron and Phillips model of a SMIB system under closed-loop voltage control in Figure 5.8 on page 239, the transfer functions for both the torque of electro-magnetic origin and the terminal voltage can be written with respect to the reference voltage perturbations as:

$$\frac{\Delta P_2(s)}{\Delta V_r(s)} = \frac{K_2 K_3 K_{ex}}{s^2 K_3 T_{ex} T'_{d0} + s(T_{ex} + K_3 T'_{d0}) + (1 + K_3 K_6 K_{ex})}, \quad \text{and} \tag{6.17}$$

$$\frac{\Delta V_t(s)}{\Delta V_r(s)} = \frac{K_6 K_3 K_{ex}}{s^2 K_3 T_{ex} T'_{d0} + s(T_{ex} + K_3 T'_{d0}) + (1 + K_3 K_6 K_{ex})}, \tag{6.18}$$

respectively; both these equations are independent of the shaft-dynamics. An inspection of the two transfer functions reveal that (i) they are simply related by the scalar ratio K_2/K_6 and, (ii) the phase responses of terminal voltage and the P-Vr characteristics are identical. The PSS tuned according to the GEP method must therefore introduce phase lead to compensate for the lagging phase characteristic of $\Delta V_t(j\omega_f)/\Delta V_r(j\omega_f)$. The phase response of the synthesized compensation therefore corresponds to the negation of the lagging phase characteristic.

6.6 Tuning a speed-PSS using the GEP Method

Let us consider the tuning of a true-speed PSS based on the GEP method for the sixth-order generator and SMIB system of Section 5.10.4.1. Because the measurements of the frequency responses when conducted in the field are likely to be at lower levels of real power output, operating conditions Cases R and K for the SMIB system listed in Table 5.6 on page 276 are used as examples. (For reasons of security, field measurements are likely to be carried out at lower values of real power output.) The generator is under closed-loop voltage control with power outputs of 0.5 and 0.7 pu at unity power factor; the associated rotor modes are $-0.539 \pm j8.50$ and $-0.065 \pm j8.92$, respectively.

The terminal voltage frequency responses are compared with those of their P-Vr characteristics in Figure 6.5. The presence of the resonance associated with the lightly damped rotor mode for Case K is particularly evident in the terminal voltage responses of the GEP frequency responses. If the GEP transfer function is calculated with all shaft dynamics disabled (GEPSDD), the resonances in Cases R and K are eliminated and the phase response of GEPSDD generally agrees closely with that of the P-Vr transfer function, as seen in Figure 6.6.

The P-Vr characteristic representing the synthesized GEPSDD phase responses is

$$(1 + s0.005)/(1 + s0.05)^2 .$$

Thus the associated transfer function of the PSS derived from the negation of the synthesized GEPSDD phase responses takes the form,

$$k_G \cdot \frac{5s}{1 + 5s} \cdot \left[\frac{1 + s0.05}{1 + s0.005}\right]^2 , \tag{6.19}$$

which includes a washout and a first-order a low-pass filter; let $k_G = 1$.

6.6.1 Example 2. Performance of the PSS based on Design Case C

For purposes of comparison with the Method of Residues the performance of the PSS based on the GEP approach is evaluated for Design Case C of Table 5.4, $P = 0.9$ $Q = 0.2$ pu using the PSS given by (6.19).

The damping torque coefficient for the PSS is calculated as in Section 6.3.3 for the Residue Method. The frequency response of the damping torque coefficient is shown in Figure 6.7 and reveals that the value of the coefficient at the modal frequency of 9.0 rad/s is 4.1 pu. Consequently the PSS gain $k_G = 1$ is equivalent to a damping gain of 4.1 pu in the vicinity of the modal frequency of rotor oscillations.

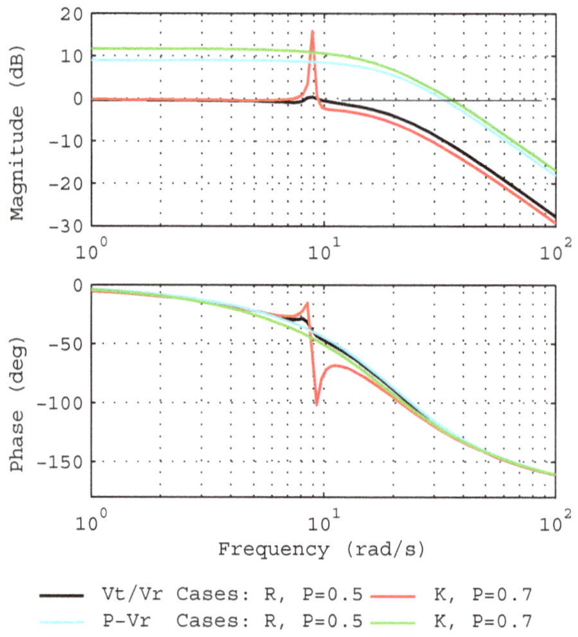

Figure 6.5 Frequency responses for the transfer function $\Delta V_t/\Delta V_r$ and the P-Vr characteristics for lower levels of real power output at unity power factor.

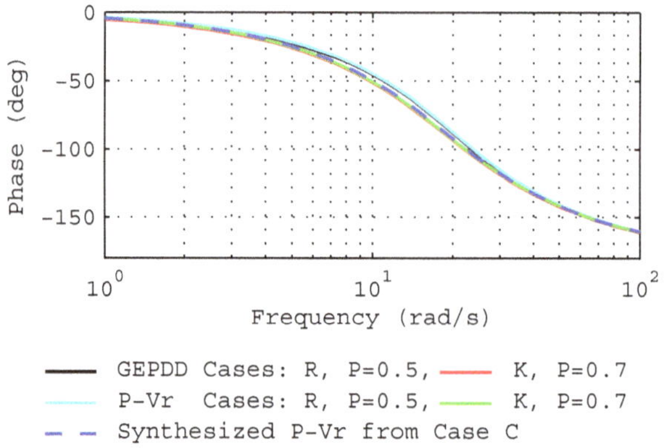

Figure 6.6 Comparison of the GEP phase response with shaft dynamics disabled (GEPS-DD) and those of the associated P-Vr characteristics. The synthesized phase response for the design Case C of Table 5.4 is also shown.

Note that in Figure 6.7 the damping torque coefficient has, relatively, a more level frequency response over the range 1.5-15rad/s than that in Figure 6.4 for the PSS design based on the Method of Residues. This is because the phase compensation of (6.19) reflects that of the P-Vr approach over the frequency range - rather than that designed at a single selected modal frequency.

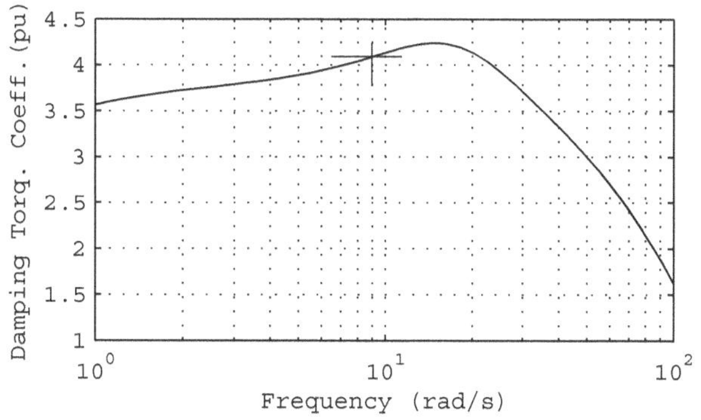

Figure 6.7 Case C: PSS-induced damping torque coefficient for a PSS designed according to the GEP method and with gain $k_G = 1$.

Let us compare the mode shifts based on the GEP Method and the P-Vr method. For $k_G = 4.88$ pu (equivalent to a PSS damping gain of 20 pu on machine MVA rating), the rotor mode using the GEP method is $-1.35 \pm j8.94$ for operating condition C. The mode shift associated with this gain is $-1.61 \mp j0.08$, a value which is close agreement with that derived based on the P-Vr method, i.e. $-1.60 \pm j0.02$ (refer to Table 5.5).

From a comparison of Figure 6.5 and Figure 6.6, the following observations are offered.

- As is to be expected, the magnitude responses of the conventional GEP and the P-Vr transfer functions differ. (Because the generator is under closed-loop voltage control the gain of the GEP transfer function tends to unity at low frequencies.)

- The conventional GEP phase response may be 'distorted' by a number of resonances associated with the rotor modal frequencies in a multi-machine system. (These should be significantly attenuated by the damping introduced by a properly-designed PSS). Such phase responses and the lack of appropriate magnitude information complicate the synthesis of the PSS transfer function. (It is usually a matter of finding a number of lead networks to obtain the desired lead compensation over the range of modal frequencies.)

- For measurements in the field, one would be reluctant to proceed with frequency response measurements approaching a resonance because of an uncertainty concerning its magnitude. The range of the frequencies measured may therefore be restricted to lower values when resonances are encountered. Such resonances may not be encountered at lower real power outputs from the generator.

- The magnitude responses for the P-Vr and the GEPSDD transfer functions differ by a constant gain value. There is close agreement between the P-Vr and GEPSDD phase responses for the sixth-order model of the generator. The GEPSDD phase response which is determined by analysis in this exercise is seen to provide a good 'smoothed' representation of that of the GEP.

6.7 Conclusions, GEP method

Based on the preceding analysis and Example 2 it is noted that:

1. In a single- or multi-machine system the field-measured GEP responses may be adulterated by resonances due to lightly damped local or intra-station modes. The range of the measured frequency responses may be curtailed as the size of the resonance under field conditions is typically unknown.

2. Field-measured GEPs can assist in the validation of the small-signal system model of the generator and system used in simulation-based PSS tuning methods. It should be noted that, for PSS tuning using analytical techniques, accurate models of the generator and excitation system are highly desirable.

3. By eliminating the resonances associated with the rotor modes of the conventional GEP transfer functions, the phase responses of the GEPSDD transfer functions determined by analysis provide a sounder basis for the determination of phase compensation required for PSS than by field measurements. However, for the evaluation of the GEPSDD one would need to have confidence in the accuracy of the model of the generator and system.

4. If, in the tuning of a PSS based on the conventional analysis of GEP(s), the shaft dynamics are disabled [1] then the same phase information is available in the frequency responses of both the GEP(s) and the P-Vr methods. However, the associated magnitude responses in the GEP(s) method are ignored. Because the P-Vr method provides guidelines for the settings of PSS damping gains, the concepts, analysis and results in Chapters 10, 12 -14 are lost.

5. Of the three methods which are discussed in this chapter and are complementary, the P-Vr frequency response provides continuous, consistent information over the range of rotor modal frequencies and encompassing operating conditions for both *magnitude* and phase; this simplifies considerably the synthesis of the PSS transfer function and its tuning.

---

1. The transfer function in the case is $\Delta P(s)/\Delta V_{ref}(s)$.

6. *The PSS damping gain associated with the P-Vr method has particular significance.* It provides a sound and proper basis for the systematic selection of PSS gains. Moreover, its significance as a damping torque coefficient is used in the theoretical and practical applications demonstrated in Chapters 12 to 14. It is also a meaningful quantity. For example, 20 pu damping gain on machine rating is a moderate gain value; neither the PSS gain k_R of (6.2), associated with the Residue Method, nor k_G (6.19) of the GEP Method, have any significant meaning when expressed on machine rating.

7. Because of the advantages listed above, and other merits, the P-Vr approach is employed in practice by a number of organizations for the tuning of PSSs in the multi-machine environment described in Chapter 10.

6.8 References

[1] M. J. Gibbard and D. J. Vowles, "Reconciliation of methods of compensation for PSSs in multimachine systems," *Power Systems, IEEE Transactions on*, vol. 19, pp. 463-472, 2004.

[2] F. L. Pagola, I. J. Perez-Arriaga and G. C. Verghese, "On sensitivities, residues and participations: applications to oscillatory stability analysis and control," *Power Systems, IEEE Transactions on*, vol. 4, pp. 278-285, 1989.

[3] CIGRE Technical Brochure no. 166 prepared by Task Force 38.02.16, "Impact of Interactions among Power System Controls", *published by CIGRE*, 2000.

[4] P. Pourbeik and M. J. Gibbard, "Damping and synchronizing torques induced on generators by FACTS stabilizers in multimachine power systems," *Power Systems, IEEE Transactions on*, vol. 11, pp. 1920-1925, 1996.

[5] D.J. Vowles and M.J. Gibbard, *Mudpack User Manual: Version 10S-03*, School of Electrical and Electronic Engineering, The University of Adelaide, July 2014.

[6] F. De Marco, N. Martins, and J. C. R. Ferrari, "An Automatic Method for Power System Stabilizers Phase Compensation Design," *Power Systems, IEEE Transactions on*, vol. 28, pp. 997-1007, 2013.

[7] E. V. Larsen and D. A. Swan, "Applying power system stabilizers: Part I – III," *Power Apparatus and Systems, IEEE Transactions on*, vol. PAS-100, pp. 3017–3046, June 1981.

[8] F. P. Mello and C. Concordia, "Concepts of Synchronous Machine Stability as Affected by Excitation Control," *Power Apparatus and Systems, IEEE Transactions on*, vol. PAS-88, pp. 316-329, 1969.

[9] P. Kundur, M. Klein, G. J. Rogers, and M. S. Zywno, "Application of power system stabilizers for enhancement of overall stability," *Power Systems, IEEE Transactions on*, vol. 4, pp. 614–626, 1989.

[10] P. Pourbeik, *Design and Coordination of Stabilisers for Generators and FACTS devices in Multimachine Power Systems*, PhD Thesis, The University of Adelaide, Australia, 1997.

Appendix 6–I

App. 6–I.1 Algorithm for the calculation of stabilizer parameters

The algorithm is based on the general form of the stabilizer transfer function $F(s)$

$$F(s) = k_{fds} \cdot H(s) = k_{fds} \cdot \left[\frac{T_W s}{1 + T_W s}\right]^w \cdot \left[\frac{1 + T_n s}{1 + T_d s}\right]^m \cdot \left[\frac{1}{1 + T_{LP} s}\right]^z, \tag{6.20}$$

which consists of the compensator, and the washout and low-pass filters [4], [10]. Lead compensation is to be designed such that the compensation angle ϕ provided by the stabilizer is $\phi = \arg\{H(\lambda_h)\} = \pm 180 - \arg\{r_h\}$ (°) at the selected complex frequency $s = \lambda = -\sigma + j\omega$. For this complex mode let us assume that the washout and low-pass filters introduce a phase lead of $\Phi°$; the maximum phase lead to be contributed at frequency λ by the compensator is then $\theta_m = \phi - \Phi$ (°). For a compensator consisting of m first-order lead blocks the phase lead to be contributed by each block is $\theta_{max} = \theta_m / m$. Let us therefore consider the first-order lead compensator described in Section 2.12.1.4 Example 10 [1].

$$G(s) = (1 + sT)/(1 + s\alpha T). \tag{6.21}$$

If the maximum phase lead that can be produced by the above compensator is assumed to be 60°, then the allowable range for α is $0.07 < \alpha < 1$. Substitution of the complex frequency $s = \lambda = -\sigma + j\omega$ in (6.21) results in:

$$G(\lambda) = \frac{[1 - \sigma T(1 + \alpha) + \alpha T^2(\sigma^2 + \omega^2)] + j\omega T(1 - \alpha)}{1 - 2\sigma\alpha T + \alpha^2 T^2(\sigma^2 + \omega^2)}.$$

The phase lead introduced by the compensator at $s = \lambda$ is equal to

$$\theta = \frac{\theta_m}{m} = \operatorname{atan}\left[\frac{\omega T(1 - \alpha)}{1 - \sigma T(1 + \alpha) + \alpha T^2(\sigma^2 + \omega^2)}\right]. \tag{6.22}$$

The desired maximum phase lead, θ_{max} occurs at $\omega = \omega_c$ where $d\theta/d\omega|_{\omega = \omega_c} = 0$. Noting that σ is a constant and applying the constraints $T \neq 0$ and $\alpha \neq 1$, the expression $d\theta/d\omega|_{\omega = \omega_c} = 0$ is solved to give

$$\omega_c = \sqrt{\frac{1}{\alpha T^2} - \frac{\sigma}{\alpha T} - \frac{\sigma}{T} + \sigma^2}. \tag{6.23}$$

---

1. Note that for $\sigma = 0$, i.e. $s = j\omega_f$, $\omega_c = 1/(\sqrt{\alpha}T)$ and $\theta_{max} = \operatorname{asin}[(1-\alpha)/(1+\alpha)]$.

Substitution of (6.23) into (6.22) yields an expression for the maximum phase lead:

$$\theta_{max} = -\text{atan}(Z) \text{ where } Z = \frac{\left(\sqrt{\frac{1-\sigma T-\alpha\sigma T+\sigma^2\alpha T^2}{\alpha T^2}}\right)\frac{T}{2}(\alpha-1)}{1-\sigma T-\alpha\sigma T+\sigma^2\alpha T^2}. \tag{6.24}$$

To design a first-order lead block to provide the required maximum phase lead θ_{max} at the modal frequency $\lambda_h = \sigma_h + j\omega_h$, the following procedure is proposed to solve for the compensator parameters α and T in (6.21).

1. Select m such that $\theta_{max} < 60°$ where $\theta_{max} = \theta_m/m$.

2. Calculate $\alpha_0 = \frac{1-\sin(\theta_{max})}{1+\sin(\theta_{max})}$ as an initial estimate of α.

3. Calculate $T_0 = 1/(\sqrt{\alpha_0}\omega_h)$ as an initial estimate of T.

4. Set (i) a tolerance level for the iterative calculations, e.g. $\varepsilon = 0.001$, and (ii) the counter to $k = 1$.

5. Given that $\omega_c = \omega_h$, solve for T using (6.23) in the form

   $$\alpha_{k-1}(\sigma_h^2 - \omega_h^2)T^2 - \sigma_h(1+\alpha_{k-1})T + 1 = 0,$$

   and choose the smallest positive value of T. Set $T_k = T$.

6. Solve (6.24) for α letting $X = (\tan\theta_{max})^2$, i.e. solve the equation

   $$[4T_k^2X\sigma_h^2 - 4T_kX\sigma_h - 1]\alpha^2 + [2+4X-4X\sigma_hT_k]\alpha - 1 = 0,$$

   and choose the smallest positive value of α. Set $\alpha_k = \alpha$.

7. If $(|\alpha_k - \alpha_{k-1}| < \varepsilon)$ and $(|T_k - T_{k-1}| < \varepsilon)$ then end, else go to 5.

The required first-order compensator parameters are $\alpha = \alpha_k$ and $T = T_k$.

Note that if the compensation angle is negative, i.e. $\theta_{max} < 0$, the parameters of a lag compensation transfer function are calculated.

App. 6–I.2 Calculation of the nominal upper limit of the range of stabilizer gains

A nominal measure of the upper limit of range of stabilizer gain is shown to be:

$$k_{Rm} = 1/\left|r_h \frac{\partial H(\lambda_h)}{\partial \lambda_h}\right| . \qquad \text{(6.14) (repeated)}$$

The residue r_h is specified as are the parameters of the washout and low-pass filters. Having selected the desired order m of the compensator its parameters are calculated using the above algorithm. It then remains to calculate $\frac{\partial H(s)}{\partial s}$ and evaluate it at the selected modal frequency $s = \lambda_h = -\sigma_h + j\omega_h$.

Assuming the three blocks in the transfer function are in forms such as those in (6.20), the derivative can be expressed in a general form:

$$\frac{\partial}{\partial s} J(s) = \frac{\partial}{\partial s}\left\{\left[\frac{a+sT_1}{1+sT_2}\right]^p\right\} = p \cdot \frac{(a+sT_1)^{p-1}(T_1 - aT_2)}{(1+sT_2)^{p+1}} . \qquad (6.25)$$

From the above expression the derivative of the three transfer functions in (6.20) can derived by setting the following quantities for the:

compensator $G_c(s)$, $a = 1,\ T_n = T_1,\ T_d = T_2,\ m = p;$

washout filter $G_W(s)$, $a = 0,\ T_w = T_1 = T_2,\ w = p;$

low-pass filter $G_{LP}(s)$, $a = 1,\ T_1 = 0,\ T_{LP} = T_2,\ z = p.$

Given $H(s)$ in (6.20) is of the form $H(s) = A(s)B(s)C(s)$, the expression for the derivative is:

$$\frac{\partial H(s)}{\partial s} = A(s)B(s)\frac{\partial}{\partial s}C(s) + B(s)C(s)\frac{\partial}{\partial s}A(s) + C(s)A(s)\frac{\partial}{\partial s}B(s) . \qquad (6.26)$$

Chapter 7

Introduction to the Tuning of Automatic Voltage Regulators

7.1 Introduction

7.1.1 Purposes

Given a model and the parameters of the generator and its exciter, there is little published in the literature describing the various methods for the tuning of automatic voltage regulators (AVRs) to achieve certain performance specifications for the generator off- and on-line.

An aim of this chapter is to introduce and provide an analytical basis for various tuning methodologies, which provide a set of parameters for the particular AVR model. Further analysis may depend on the type and form of the AVR supplied by a manufacturer. However, even for complex AVR structures, the proposed methodologies may provide an initial set of parameters based on a simplified model of the AVR. Subsequent fine-tuning, based on the complex structure, can then yield an appropriate final set of parameters.

It should be emphasized that the tuning methodologies considered here are based on the concept of transient gain reduction, though various other design approaches are employed [1]. Depending on the type of AVR, rate-feedback may also be used to essentially effect a similar behaviour as transient gain reduction. Furthermore, more modern systems which employ proportional-integral-derivative (PID) controls can be tuned to give a response akin to transient gain reduction. It is recognized that manufacturers of the equipment have their own, effective procedures for tuning. However, when tuning, it is important in a number of scenarios to account for the power system characteristics over an encompassing range of normal and outage conditions. The latter considerations are often of concern to the transmission service provider (TSP) who may be responsible for system security. It is therefore desirable that staff in such TSPs understand the relevant methodologies and can undertake or validate, if necessary, the tuning of AVRs.

A further objective in the description of the methodologies is to provide for young engineers an introductory and a reference text which not only covers the relevant control systems background but also highlights the power system requirements and performance.

7.1.2 Coverage of the topic

Because powerful methods of analysis are available in linear control systems theory, the tuning of AVRs is based on small-signal analysis and the linearized models of the power system and associated devices. The performance of the resulting tuned AVR, and the other elements of the power system, should then be subject to simulation studies for an appropriate set of large-signal disturbances over the range of operating conditions. In such simulations the limits on AVR and the exciter quantities, as well as saturation, should be modelled.

7.2 The excitation control system of a synchronous generator

The IEEE Standard 421.1 [2] defines the *excitation control system* (ECS) as the feedback control system that includes the synchronous generator and its excitation system. Essentially the excitation control system is the system which excites and controls the rotor field current of the generator and thus the term ECS includes the generator. The excitation control system as well as the *excitation system* (ES) are shown in the block diagram of Figure 7.1. The excitation system is defined as 'the equipment providing field current for a synchronous generator, including all power, regulating, control, and protective elements'. The main 'power element' is the exciter, however, the 'regulating, control, and protective elements' are referred to in the Standard as the 'synchronous machine regulator'.

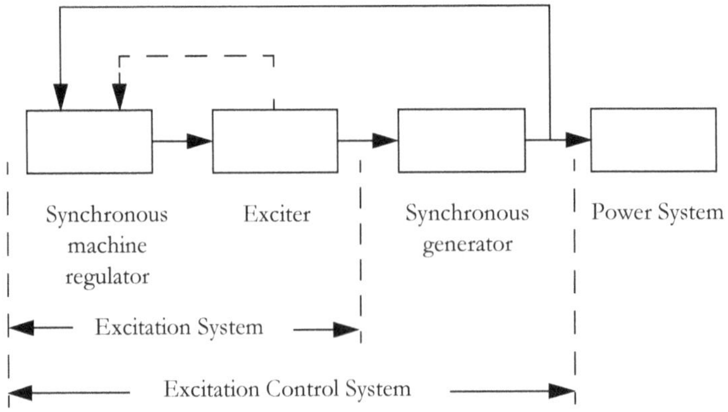

Figure 7.1 Block diagram of the excitation control system [2].

We shall refer to the 'synchronous machine regulator' as the 'automatic voltage regulator' or AVR [1]. We will assume the AVR comprises all the control elements and any lower-power, power-electronic devices which drive the input to the exciter.

IEEE Standard 421.2 [3] concerning aspects of the testing and evaluation of dynamic performance of excitation control systems is also of interest here.

A component of the AVR is the compensating control provided to ensure that the *excitation control system* satisfies certain steady-state and dynamic performance criteria for the unit off- and on-line. The main objective in 'AVR tuning' is to determine the parameters of the appropriate compensator which satisfy the criteria. The block diagram of the *excitation control system*, which forms the basis for the analysis which follows later, is shown in Figure 7.2. The element $K_A/(1 + sT_A)$ typically represents the simplified dynamics of the AVR power amplifier and the gain. However, this block diagram does not apply to rate-feedback compensation which is treated in Section 7.10.

When the generator is on-line the transfer function $G_{gen}(s)$ includes the dynamics associated with the external power system. When the unit is off-line it is assumed to be under closed-loop voltage control and operating at rated voltage at synchronous speed. It should be noted that the dynamic behaviour of the excitation system and generator may differ significantly when off- or on-line under closed-loop voltage control.

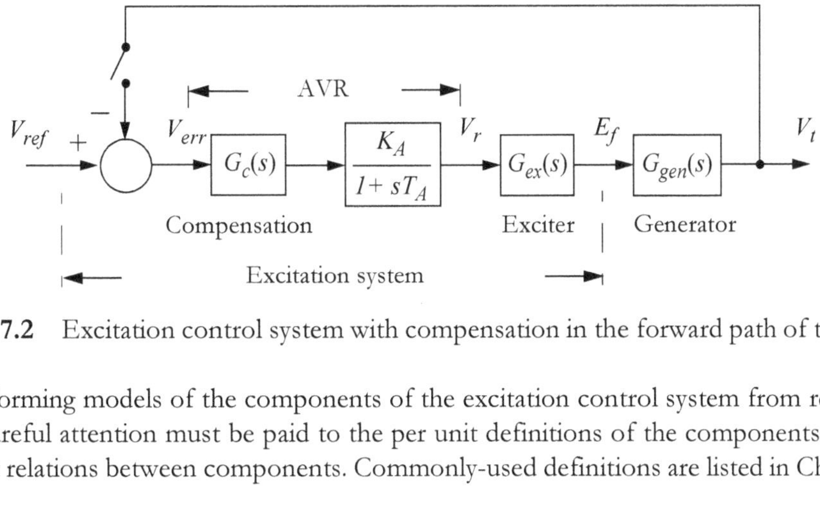

Figure 7.2 Excitation control system with compensation in the forward path of the AVR

When forming models of the components of the excitation control system from results of tests, careful attention must be paid to the per unit definitions of the components and the per unit relations between components. Commonly-used definitions are listed in Chapter 4.

---

1. The IEEE Standard 421.1 [2] refers to 'automatic voltage regulator' as 'a term often used to designate either the voltage regulator alone or the complete control system comprised of limiters, etc.'

7.3 Types of compensation and methods of analysis

A variety of types of compensation is employed in the AVR and a number of these will be discussed in the following sections. Each of these are analysed and, where appropriate, a design method or procedure is formulated. The aim in each method is to achieve a desired transient gain reduction over a selected range of modal frequencies. For each type, illustrative examples are provided of the tuning method to achieve specified dynamic performance criteria with the generator off- and on-line in a simple power system. The types of compensation considered are:

- *Transient gain reduction (TGR)*, the simpler type of compensation [1] (Section 7.6).
- *Proportional plus Integral plus Derivative (PID) compensation* (Section 7.7). A number of forms of PID compensation is analysed and procedures for the calculation of the parameters are proposed. In the case of a type of PID, called Type 2B here, a detailed analysis is undertaken to determine a suitable set of PID parameters for both normal and N-1 operating conditions for a remote, three-generator power station in which brushless AC exciters are installed. (Section 7.11).
- *Proportional plus Integral (PI) compensation* (Section 7.9). The concepts in the PI design procedure are simple, and follow on from the earlier sections. The extension of PI compensation to PID compensation using a series lead-lag block is illustrated and the equations for the parameter conversion to the PID structure are formulated.
- *Rate feedback of the AVR or exciter output (RFB)* (Section 7.10). The aim of the analysis is to determine the gain and time constant in the feedback transfer function that satisfy the performance specifications. A method for calculating the latter parameters is proposed which is based on a simple model of the excitation system. However, it is shown that the results can be applied to more complex systems which, for instance, include PI compensation in the forward path.

The small-signal performance of the system is analysed using software packages such as Matlab® [4] or Mudpack [5]. Several methods of linear system analysis such as frequency response, root-locus, step response and eigen-analysis are employed. The theory behind these linear system analysis techniques are described in references [8] and [9].

7.4 Steady-state and dynamic performance requirements on the generator and excitation system

In some types of excitation systems there is a requirement for high values of gain in the forward loop of the excitation system for closed-loop voltage control. Such gains are typically employed (i) to provide fast response of the generator terminal voltage to disturbances, (ii)

1. Transient gain reduction has not been used in some cases with fast-response, high gain, static excitations systems [6], [7].

to boost field flux linkages following a major disturbance in order to increase synchronizing power, and (iii) to satisfy requirements on the error in the terminal voltage in the steady-state ('zero frequency' and final equilibrium point).

At higher modal frequencies high gains in the forward loop of the closed-loop voltage control system are destabilizing. In order to provide a stable, robust system it is necessary to reduce, by compensation, the high forward loop gain (K_A) to a lower "transient gain" (K_T) at higher frequencies. This concept is illustrated in Figure 7.3.

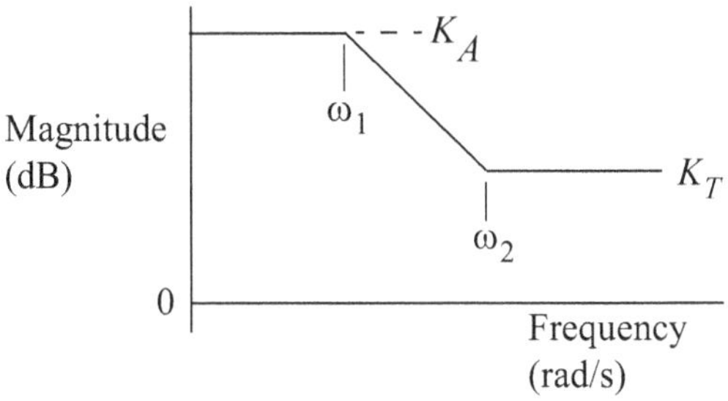

Figure 7.3 Straight-line approximation of the magnitude response of the compensation transfer function showing high gain at low frequency and transient gain reduction at higher frequencies.

In general, the basis of the approaches is (i) to establish a transient gain reduction over an appropriate range of modal frequencies, and (ii) to provide a well-damped response of the generator terminal voltage to a step change in reference voltage. Such a range of modal frequencies depends on the rating and location of the generator and its participation in the inter-area, local-area and intra-station modes. It will be assumed for present purposes that the range of modal frequencies is 1.5 to 15 rad/s. On the other hand, if local-area modes only were of concern the relevant frequency range might be 6 to 12 rad/s (see Section 1.5). The compensation provided should also satisfy the criteria for the dynamic performance of the generator off-line under closed-loop voltage control (this matter is treated in sections 7.6.3, 7.7.2.3 and 7.11.2.2).

The level of transient gain is determined in the first instance by the closed-loop performance of the generator and excitation system under voltage control when the unit is off-line and running at synchronous speed. In order that the terminal voltage response to a step change in reference voltage is well damped, it is suggested in [10] that the transient gain K_T should be approximately less than $T'_{do}/(2T_E)$. For example, if the generator's open-circuit time constant is $T'_{do} = 5$ s and the exciter time constant is $T_E = 0.1$s, the transient gain should be less than 25 pu on machine base. It is also stated in [10] that "... the closed-loop response of this voltage-component regulating-loop (i.e. ignoring the contribution from rotor angle) under load conditions is not materially different from the closed-loop response on open circuit". This implies that the approximate value $K_T < T'_{do}/(2T_E)$ is appropriate on-line. However, it also pointed out in the discussion in [10] that the performance with values of transient gain exceeding $T'_{do}/(2T_E)$ has proved satisfactory. Note that cases are illustrated in which the closed-loop terminal voltage response to a step change in reference voltage is well damped but the rotor modes are poorly damped. Such cases may required a coordinated tuning with the PSS installed on the generator.

The range of the transient gain is typically 25 to 50 pu on machine base. In the examples in this chapter a transient gain of 32 pu (~30 dB) is usually adopted.

In order to ascertain the characteristics of the closed-loop voltage control system of a generator, either off- or on-line, step changes in the voltage reference of up to 5% of rated terminal voltage are employed in practice. The size of the steps is such that the response is measurable above the noise level but not so large as to produce over-voltages of the exciter output. However, for the small-signal (i.e. linear) simulation studies in this book, a step size of 1% from 1 pu is a more convenient and meaningful measure.

Minimum performance requirements on generators and excitation systems are usually specified by the system operator in a code or a set of rules; the latter will be referred to as the 'Rules'. As an example - and as applied in this chapter - typical requirements that may be imposed on the generator and excitation system are:

- Under closed-loop voltage control the generating unit must be stable when off-line and on-line. For planning purposes, when the unit is on-line the halving time of any inter-regional or intra-regional rotor oscillations should be less than 5 s (e.g. [11]).

- Each excitation control system must provide continuous voltage regulation to within 0.5% of the selected set-point value at all operating points within generator capability, (e.g. [11]). This is interpreted as requiring that the effective DC gain of the terminal voltage control loop is at least 200 pu for the steady-state terminal voltage error to be less than 0.5% (see Section 2.10.1) (this is assuming a proportional-only control).

- With the generator on-line and under closed-loop voltage control, the settling time following a disturbance equivalent to a 5% step change in the measured generating-unit terminal voltage must be less than 5 s. This must be satisfied at all operating points within the generating unit capability. It is assumed here that the step change does not lead to the activation of limiters in the excitation system. The settling time is the time for the terminal voltage response to decay to within a prescribed percentage of the final steady-state level. (In Section 7.11 a 10% settling time employed; unless otherwise stated a 2% settling time is adopted in this book.)

- When the unit is under closed-loop voltage control and is running off-line at rated speed the corresponding settling time is 2.5 s, or less.

The objective of the analysis in this chapter is to determine the compensation that must provided by the AVR to satisfy the relevant Rules and, when appropriate, to align with excitation system models in the IEEE Standard 421.5 [12].

Other helpful background material is provided in [7] and in [13] to [15].

As mentioned earlier an objective of this chapter is to provide a theoretical background and some guidelines to the various approaches to AVR tuning. Each method of compensation that is analysed is followed by an illustrative example, typically for the generator off- and on-line under closed-loop voltage control. Because the requirements and performance specifications may vary from application to application, there is flexibility available for fine tuning of the controls based on any of the approaches.

The following discussions are based on small-signal analysis. In particular, Bode plots of the open-loop frequency response of a system will be employed to assess the stability and performance of the closed-loop system using the concepts of Phase Margin and the gain-cross-over frequency. The open-loop system should therefore have no poles or zeros in the right-half of the s-plane [8].

As emphasized earlier, the results obtained based on small-signal analysis should be reviewed in the context of appropriate and relevant large-signal (transient stability) studies.

In modelling excitation systems, limiting of the outputs is imposed on certain types of blocks, namely integrators, first-order blocks, lag-lead and lead-lag blocks. Use of an integrator, say with windup limiting, can result in additional phase shifts.Windup should not be an issue in strictly small-signal analysis but one should be aware of it occurring following large-signal disturbances. (See Appendix 7–I.4.)

7.5 A single-machine infinite-bus test system

In the examples to illustrate the application of the methods of compensation the single-machine infinite-bus (SMIB) system shown in Figure 7.4 is employed [1]. One line ('a') of a pair of parallel lines is out of service, e.g. for maintenance or following a fault on the line. The generator output at unity power factor is 0.4 pu power (a value above the range 0.1 to 0.3 pu at which the PSSs typically are switched into service). The load is modelled as constant impedance. For on-line analysis the generator transfer function block, referred to as $G_{gen}(s)$, accounts for the dynamics of both the generator and the system to which it is connected. The sixth-order generator parameters are listed in Appendix 7–I.1. Note that the model of the exciter is a simple first-order lag; a more detailed model is used for a brushless exciter in Section 7.8.

1. For the studies in Section 7.8 concerning the Type 2B PID, a different but more relevant power system is chosen.

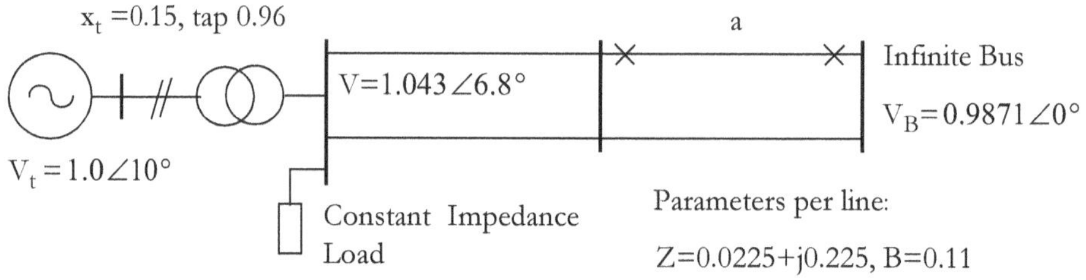

Figure 7.4 SMIB test system, outage of line 'a'; Case W.
Generator output P = 0.4, Q = 0 pu. Load: P = 0.02, Q = 0 pu at 1 pu voltage.
Parameters [1] are in pu on machine base (MVA). System frequency is 50 Hz.

Only this one operating condition (an N-1 condition) is analysed to illustrate some of the issues that may arise in the design process. Without a PSS installed on the generator in this case a lightly damped rotor mode is present. In practice an encompassing range of normal and N-1 operating conditions should be examined in order to determine the appropriate compensation. For the purposes of this chapter we will deem damping performance to be adequate if the halving time of the rotor mode of oscillation is shorter than 5 seconds [2]. This is consistent with the requirements in the Australian National Electricity Rules (NER) [11].

7.6 Transient Gain Reduction (TGR) Compensation

7.6.1 Introduction

In classical control terminology Transient Gain Reduction (TGR) is referred to as cascade lag-lead compensation; such compensation is incorporated in the control system of the AVR as shown in Figure 7.5.

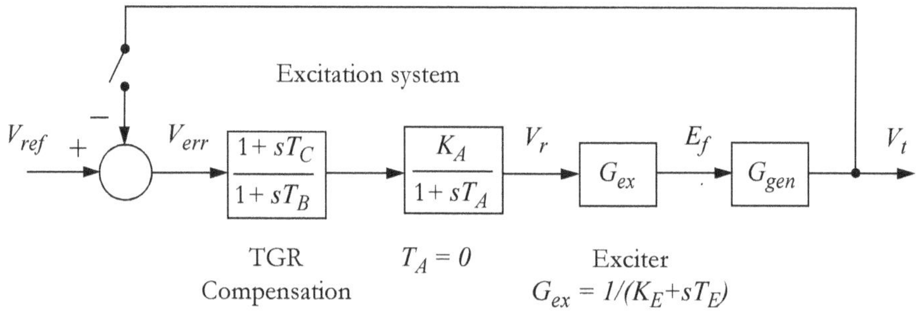

Figure 7.5 Excitation control system with Transient Gain Reduction

1. The generator and excitation system parameters which are given in Appendix 7–I.1.1 differ from those in Section 5.10.4. However, the transformer and line parameters are the same on the generator MVA base as those in the latter section.

2. The stability limit is derived from $1 \cdot e^{-a5} = 0.5$, i.e. $-a = -0.139$ Np/s.

The transfer functions for the TGR block and the associated classical lag-lead block are:

$$\frac{1+sT_C}{1+sT_B} \quad \text{or} \quad \frac{1+sT}{1+s\alpha T}, \text{ respectively, where } T_B > T_C \text{ or } \alpha > 1\,. \tag{7.1}$$

Let ω_f be the frequency of the exciting sinusoidal signal. In the frequency domain at high frequencies ($\omega_f T \gg 1$) the transfer functions in (7.1) reduce to T_C/T_B or $1/\alpha$. At such frequencies the gain of the AVR transfer function in the forward loop is

$$V_r(j\omega_f)/V_{err}(j\omega_f) = K_T = K_A \cdot T_C/T_B,$$

where K_T is the per-unit transient gain. TGR compensation thus provides a gain reduction $K_T/K_A = T_C/T_B$.

The concept of TGR is illustrated in Figure 7.3 on page 317 in which the corner frequencies are $\omega_1 = 1/T_B$ and $\omega_2 = 1/T_C$.

The classical design approach for determining the parameters T and α are considered in Section 2.12.1.5 and in texts on control system analysis [8], [9]. However, let us demonstrate a somewhat different approach bearing in mind that the relevant performance requirements of the generation and excitation system have to be satisfied for the generator on-line. In particular, the unit must operate stably at lower power levels before the PSS is switched on, say, at 0.3 pu power. Furthermore, the requirements should apply for the appropriate range of operating conditions, particularly at leading power factors, N-1 contingencies, etc.

A method for determining the parameters for TGR compensation and evaluating its performance is demonstrated in the following illustrative examples which consider both the on-line and off-line cases.

7.6.2 The performance of the generator and compensated excitation system on-line

7.6.2.1 Preliminary off-line considerations

When the unit is off-line (i.e. with the generator main breaker open and the unit running isolated from the power system) and under closed-loop voltage control the transfer function of the generator is assumed to be $G_{gen}(s) = 1/(1 + T'_{d0})$, $T'_{d0} = 5.0$ s. The simple transfer function of the exciter is $G_{ex}(s) = K_E/(1 + sT_E)$, where $K_E = 1.0$, $T_E = 0.1$ s. (For the on-line analysis the generator parameters are listed in Appendix 7–I.1.1.) For this unity feedback system, shown in Figure 7.5, the low frequency or 'DC' gain of the forward loop is thus K_A when off-line. In order for the voltage regulation to be better than 0.5% when the unit is off-line, the AVR gain $K_A = 250$ pu is selected; this results in a steady-state error of $1/(1 + 250) = 0.4\,\%$ - which is less than the specified value of 0.5%. (see Section 7.4).

7.6.2.2 On-line studies

The approach to the design of the TGR block is demonstrated for an excitation system with an AVR gain initially set to lower values of K_A without compensation. The initial study is based on the generator on-line connected to the external system shown in Figure 7.4. A study of the performance of the generator on-line without and with TGR compensation in its AVR will follow.

For $K_A = 1$ pu, $T_A = 0$ and without compensation the Bode Plot V_t/V_{ref} with the terminal-voltage feedback loop open is shown in Figure 7.6. If the forward-loop gain K_A is increased to 32 pu (~30 dB) without compensation, we note that the gain cross-over frequency (ω_{c0}) is 3.1 rad/s and the associated Phase Margin (PM) is 78°. With such a PM not only is the system stable when the voltage feedback loop is closed, but the time response of the terminal voltage to a step change in reference voltage will be over-damped. (A PM > 60° for the terminal-voltage feedback loop typically provides a well-damped response to disturbances in that loop. However, the response associated with the lightly-damped rotor mode at about 9 rad/s (evident in Figure 7.6) may be superimposed on the well-damped terminal voltage response to a step change in reference voltage.)

A transient gain $K_T = 32$ pu will be adopted in this and other examples. As stated in Section 7.4 it is suggested that the transient gain should be less than 25 pu for the latter time constants. However, it will be established that the selected value of transient gain is satisfactory for this N-1 operating condition.

The proposed AVR gain is $K_A = 250$ pu or 48 dB. Since the desired transient gain is $K_T = 32$, the TGR transfer-function must provide attenuation of $K_T/K_A = 32/250 = 0.128$ (-18 dB) at frequencies less than that of gain-crossover frequency ($\omega_{co} = 3.1$ rad/s). Furthermore, the PM of 78° with $K_T = 32$ pu should not be significantly reduced by the TGR transfer-function. As pointed out, such a value for the PM is likely to lead, under closed-loop voltage control with TGR, to an over-damped terminal voltage response to a step change in reference voltage.

Let us locate the upper corner $1/T_C$ of the TGR transfer function at a decade below ω_{co}. The effect of the corner at $1/T_C = \omega_{co}/10$ will cause the PM to be reduced by about 5°. Therefore $1/T_C = 0.31$ rad/s, or $T_C = 3.125$ s.

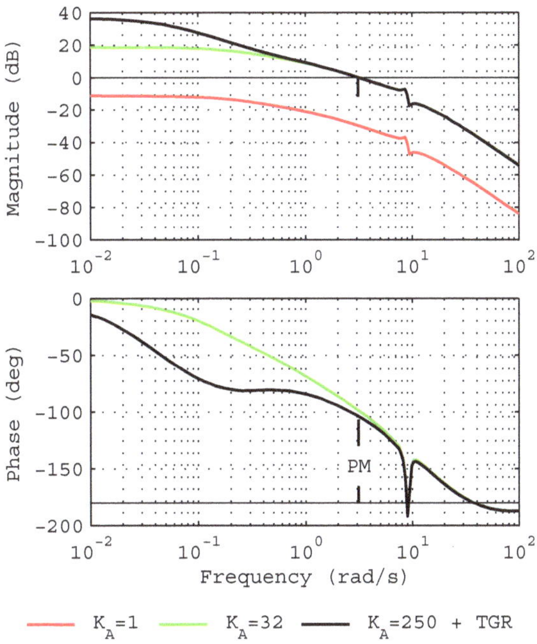

Figure 7.6 Unit on-line: Bode Plots of the open-loop terminal voltage for (i) no TGR compensation, $K_A = 1$ pu; (ii) no TGR, $K_A = 32$ pu; (iii) with TGR, $K_A = 250$ pu.

The attenuation provided by the TGR transfer function over the frequency range, from DC to high frequencies ($\omega T \gg 1$), is shown in Section 7.6.1 to be T_C/T_B or $1/\alpha$. The log-magnitude attenuation of the TGR transfer function is 18 dB, i.e. $20\log_{10}\alpha = 18$ dB, or $\alpha = 8$. The lower corner frequency of the TGR transfer function is then located at $1/(\alpha T) = 0.04$ rad/s, i.e. $T_B = \alpha T = 25$ s.

Thus, for $K_A = 250$ pu, the TGR transfer function is $\dfrac{1 + s3.125}{1 + s25.0}$. The Bode Plots for the case $K_A = 250$ pu (with TGR compensation), is shown in Figure 7.6.

The PM for the compensated case is $76°$ at 3.1 rad/s (a value close to that of $78°$ for the uncompensated case with $K_A = 32$ pu). Eigen-analysis conducted on the closed-loop system (with outage of line 'a') reveals that the rotor mode of oscillation is $-0.27 \pm j8.83$; the associated halving time is 2.55 s which satisfies the damping performance requirements. The closed-loop time responses for *perturbations* in the terminal voltage, exciter voltage and field current due to a step increase in the reference voltage of 0.01 pu (i.e. 1%) are shown in Figure 7.7.

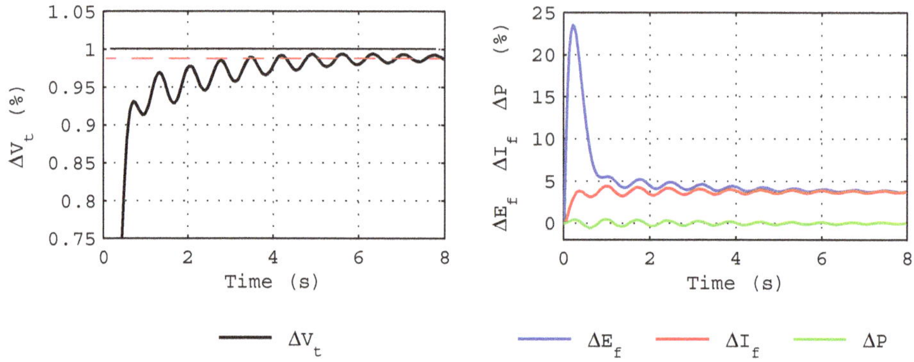

Figure 7.7 Unit on-line with closed-loop control of terminal voltage: *perturbations* in generator terminal voltage (V_t), field voltage (E_f) and current (I_f) for a step change in reference voltage from a steady-state value of 1.0 pu to 1.01 pu (1%).

Increasing the forward-loop gain, or increasing the time constant T_C, could reduce the PM from 76° to a lower value. This would improve the closed-loop terminal voltage response from over-damped to well damped. However, there may be concern that the compensated system lacks robustness, say, to ±6 dB variation in the loop gain. For example, it may be deduced from Figure 7.6 that a 6 dB increase in gain increases the gain-cross-over frequency to about 6 rad/s. This would not only result in a poorly-damped terminal voltage response to a step change in the reference-input but, due the proximity to the lightly-damped rotor mode, a damped oscillation of frequency 8.8 rad/s would be superimposed on it. Bearing in mind that the system is operating in a N-1 condition, the risk associated with the lack of robustness would need to be taken into consideration. It will be shown later it is necessary to examine a range of encompassing N and N-1 conditions to establish the validity of the selected compensation.

As will be illustrated in Appendix 7–I.5, it is desirable for the phase response shown in Figure 7.6 to be "flatter" in the vicinity of the gain cross-over frequency to ensure a phase margin of at least, say of 60°, for a ±6 dB variation in the loop gain.

7.6.3 The performance of the generator and compensated excitation system off-line

The performance of the generating unit when running off-line at rated speed and under closed-loop voltage control is now considered.

Typically, the only relevant generator parameter when off-line is its open-circuit time constant [1]. As stated earlier, the generator is modelled in Figure 7.5 by the simple transfer function $1/(1+sT'_{d0})$. The gain $K_A = 250$ pu.

The open-loop Bode Plots, V_t/V_{ref}, (i) with no compensation, and (ii) with the TGR parameters determined above for the on-line case, are shown in Figure 7.8. With no compensation, the phase margin (PM) of 26° at 21 rad/s suggests that under closed-loop voltage control the responses to step changes in voltage will be lightly damped and oscillatory. With compensation the phase margin PM is 60° at 5.5 rad/s, thus the closed-loop step response is adequately damped. The time responses of terminal voltage to a 1% step change in reference voltage are shown in Figure 7.9.

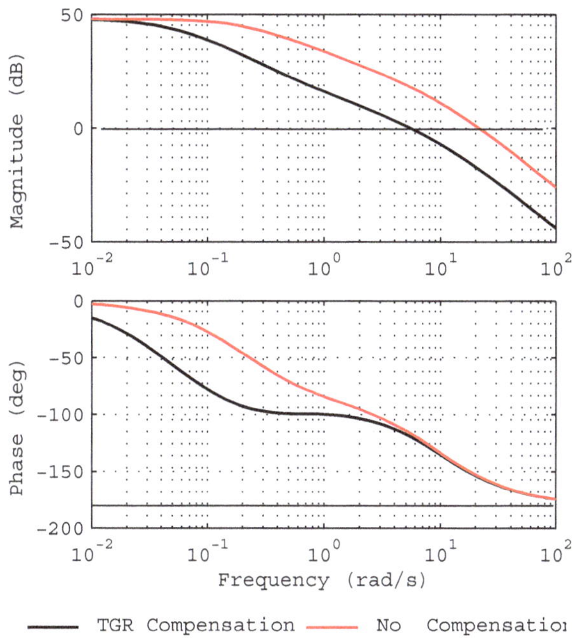

Figure 7.8 Unit off-line, $K_A = 250$ pu: Bode Plots of the open-loop terminal voltage V_t/V_{ref} with and without TGR (with TGR, PM=60° at 5.5 rad/s; without TGR, PM=26° at 21 rad/s).

1. The effects of generator saturation have been ignored. Saturation may result in a reduction of loop gain, at 1 pu terminal voltage, in the simple model of the generator.
In some types of excitation systems, such as brushless, it may be necessary to model the demagnetizing effects of the generator field current on the performance of the exciter and the reduction in generated field voltage due to loading of the excitation system rectifier [12].
These effects are taken into account in Section 7.11.

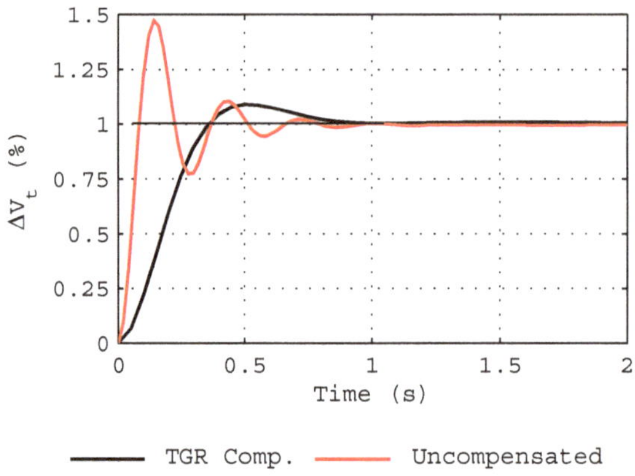

Figure 7.9 Unit off-line: Perturbation in terminal voltage on closed loop due to a step change in reference voltage from a steady-state value of 1.0 pu to 1.01 pu (1%) (with and without TGR compensation).

7.6.4 Comparison of performance of the excitation control system on- and off-line

The analysis of the performance of the generating unit on- and off-line is summarised in Table 7.1.

Table 7.1 Performance of the unit on- and off-line with TGR compensation

Performance Measures		On-line	Off-line
Phase Margin from Bode Plots K_A=250 pu	Uncompensated	32° at 13.2 rad/s	26° at 21 rad/s
	TGR compensation	76° at 3.1 rad/s	60° at 5.5 rad/s
Performance under closed-loop voltage control with TGR Compensation (K_A = 250 pu):			
Rotor mode of oscillation	Value	$-0.27 \pm j\,8.8$ [a]	-
	Halving time (s)	2.6	-
Voltage response to a step change of 0 to 1% in V_{ref}	Peak overshoot (% of step size)	none	9% at 0.55 s
	90% rise time (s)	0.6	0.3
	2% settling time (s)	4.3	0.8
Terminal voltage regulation	Steady-state error, (% of step size)	1.5	0.4
	Effective DC gain (pu)	67	250
Note a: Without compensation the on-line rotor mode is $-0.054 \pm j9.35$.			

The performance of the generating unit off-line satisfies the performance criteria. However, it is significant that, for the selected operating condition with the unit on-line, the effective steady-state ('DC') gain is reduced from 250 to 67 pu. As a result, the steady-state error exceeds the requirements of the Rules. Furthermore, due to the lower gain, the response of terminal voltage to a step change in voltage reference is somewhat sluggish (see Figure 7.7) - although the settling time satisfies the specification of 5 s.

Ideally, it is desirable to adjust the AVR gain of the unit so that the effective gain on-line is 200 pu or more, i.e. by a factor of 200/67 = 3, so that the requirements of the Rules are satisfied. The AVR gain must therefore be increased significantly; however, this increase must be attenuated at low frequencies by the same factor so that conditions in the vicinity of the gain cross-over frequency in the Bode Plot remain unchanged. The high-gain solution may be unacceptable. The resolution of this problem is (i) to review the design in the light of the Rules and the relevant operating conditions, (ii) to examine alternative methods of compensation - for example, PID compensation which is considered next.

7.7 PID compensation

In considering the use of Proportional plus Integral plus Derivative (PID) Compensation it should be emphasized that the approach adopted here is aimed specifically to its application in the tuning of AVRs. The design of PID controllers for other applications, such as process controls, is covered in texts such as [14], [15].

The structure of the PID Compensation for application to excitation control systems is shown in Figure 7.10. The block $K_G/(1+sT_G)$ in the AVR is a simple representation of other dynamic or control elements.

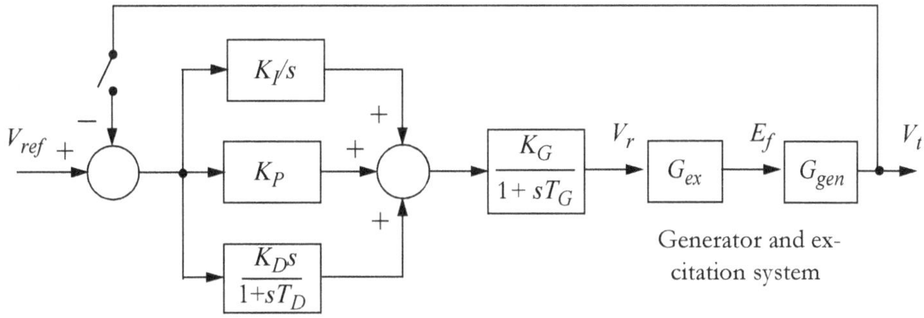

Figure 7.10 Block diagram of PID compensation.

When under closed-loop voltage control the purpose of the integrating block is to 'integrate out' any steady-state voltage error to zero by providing, in effect, an infinite steady-state gain. The voltage regulation is therefore zero or, in the steady state, the pu terminal voltage is equal to the pu reference voltage (see Section 2.10.1.2). This feature applies to both off-

and on-line conditions and therefore it is an attractive approach to compensation to ensure the relevant requirements of the Rules are always satisfied. As in Transient Gain Reduction a second aim for the compensator is to provide a relatively low transient gain, K_T, say 25 to 50 pu, in the forward path over the relevant frequencies of the rotor modes of oscillation.

The derivative action of the derivative block (i) speeds up what otherwise might be a sluggish response of the unit to disturbances in terminal voltage, (ii) boosts the speed of response of the field voltage following the application and clearance of a major fault, and (iii) limits the potentially high gain at high frequencies by means of the low-pass filter $1/(1+sT_D)$. During the immediate post-fault transient the associated increase in field flux linkages enhances the synchronizing power flow between the generator and the system - thereby enhancing stability.

Procedures for the general-purpose tuning of PID compensation is covered fairly extensively in the literature. However, a systematic approach to the design of PID compensation for application to the tuning of AVRs of synchronous generators is to be developed for which the theoretical background to the method is outlined in the following section. Thereafter, an example of the method is provided for comparison with the other methods of compensation described in other sections.

We will assume that the general form of the desired characteristics for PID compensation in the application to AVR tuning is summarized in the straight-line frequency response plot of the transfer function $G_c(s)$ in Figure 7.11.

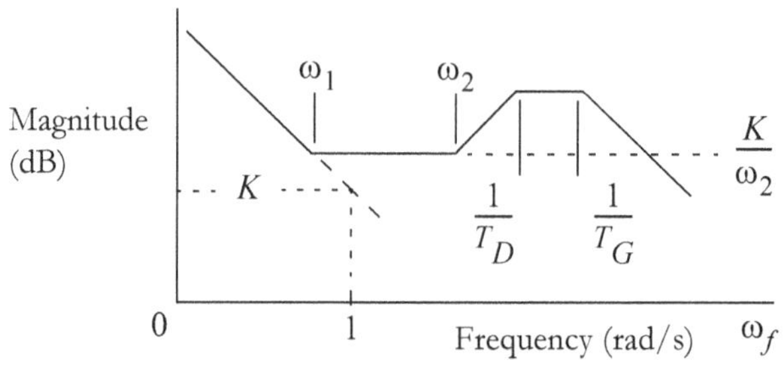

Figure 7.11 Straight-line approximation of the magnitude response of the PID transfer function $G_c(s)$.

7.7.1 PID Compensation: Theoretical Background

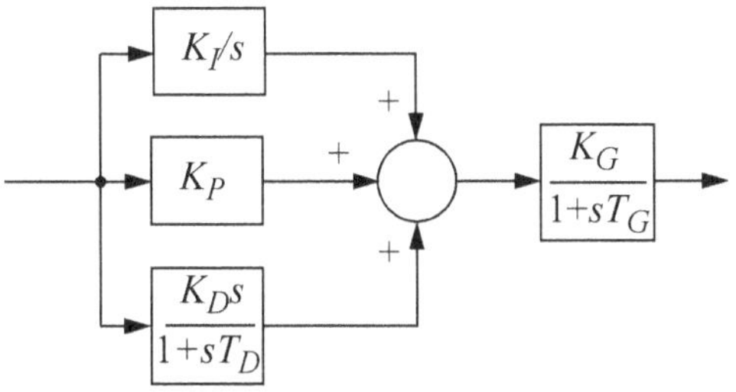

Figure 7.12 Block diagram of the practical PID compensator

The block diagram of a practical form of the PID compensator is shown in Figure 7.12. Its transfer function is:

$$G_c(s) = \left[\frac{K_I}{s} + K_P + \frac{sK_D}{1+sT_D}\right] \cdot \frac{K_G}{1+sT_G} . \tag{7.2}$$

It is convenient to consider a number of forms of this compensator both from the analysis and practical applications points of view. We will describe the forms that are to be analysed as PID Types 1 to 3 as follows:

Type 1: $T_D = 0$, $K_G = 1$, $T_G = 0$. The PID transfer function is

$$G_{c1}(s) = (K_I/s) + K_P + sK_D \tag{7.3}$$

Type 2A and 2B: $T_D > 0$, $K_G > 0$, $T_G = 0$. The PID transfer function is

$$G_{c2}(s) = \{(K_I/s) + K_P + [sK_D/(1+sT_D)]\} \cdot K_G . \tag{7.4}$$

Type 3: $T_D > 0$, $K_G > 0$, $T_G > 0$. The PID transfer function is described by (7.2) with $G_{c3}(s) = G_c(s)$.

While the forms of the transfer functions for Types 2A and 2B are identical, the locations of the upper corner frequencies ω_2 and $1/T_D$ with respect to the range of modal frequencies differ. This is explained in more detail in Section 7.7.1.2.

The analysis of the PID transfer function represented by (7.2) is somewhat complex. Initially however, we can obtain useful insights and information from an analysis based on the simple Type 1 PID transfer function of (7.3). We will return to Types 2 and 3 later.

7.7.1.1 Characteristics of Type 1 PID compensation

In Type 1 PID compensation it is assumed that the derivative time constant is zero, $T_D = 0$. Three convenient forms of the simple Type 1 PID are:

$$G_{c1}(s) = \frac{K_I}{s} + K_P + sK_D = \frac{K_D}{s} \cdot \left[s^2 + s\frac{K_P}{K_D} + \frac{K_I}{K_D}\right] = \frac{K_D s^2 + K_P s + K_I}{s}. \tag{7.5}$$

The zeros of this transfer function $(s = -\omega_1, -\omega_2)$ can be derived from the numerator of (7.5) when it is expressed in the following form:

$$(s + \omega_1)(s + \omega_2) = s^2 + s(\omega_1 + \omega_2) + \omega_1\omega_2 = 0. \tag{7.6}$$

Equating the coefficients of s in (7.6) with the numerator coefficients $s^2 + (K_P/K_D)s + (K_I/K_D)$ in (7.5), we find that the corner frequencies are

$$\omega_{1,2} = \frac{K_P}{2K_D} \pm \frac{1}{2K_D}\sqrt{K_P^2 - 4K_DK_I} = \frac{K_P}{2K_D}\left(1 \pm \sqrt{1 - \frac{4K_DK_I}{K_P^2}}\right), \text{ moreover,} \tag{7.7}$$

$$(\omega_1 + \omega_2) = K_P/K_D \text{ and } \omega_1\omega_2 = K_I/K_D. \tag{7.8}$$

In the sequel it is helpful to express (7.5) in terms of the zeros $(s = -\omega_1, -\omega_2)$:

$$G_{c1}(s) = \left(\frac{K_D}{s}\right)(s^2 + s(\omega_1 + \omega_2) + \omega_1\omega_2). \tag{7.9}$$

Note that if $K_P^2 \gg 4K_IK_D$ the corner frequencies reduce to

$$\omega_1 \approx \frac{K_I}{K_P}, \text{ and } \omega_2 \approx \frac{K_P}{K_D}\left(1 - \frac{K_DK_I}{K_P^2}\right) \approx \frac{K_P}{K_D}. \tag{7.10}$$

If the corner frequencies are well-spaced apart, say $\omega_2/\omega_1 > 10$, we observe that the frequency ω_1 is that associated with the corner of the transfer function $(K_I/s + K_P)$, likewise ω_2 with the corner of $(K_P + sK_D)$.

For the simple Type 1 model of the PID, the condition for the corner frequencies of (7.7) to be real is $K_P \geq 2\sqrt{K_DK_I}$. Thus the value of the proportional gain, K_{Pb}, at which the real zeros evolve from complex values is

$$K_{Pb} = 2\sqrt{K_DK_I}. \tag{7.11}$$

Substitution of this value of K_{Pb} in (7.7) yields the identical corner frequencies $\omega_{1,2} = \frac{K_{Pb}}{2K_D}$. According to (7.8) when $\omega_1 = \omega_2$ we find

$$\omega_{1,2} = \sqrt{K_I/K_D}\Big|_{K_P = K_{Pb}} \text{ rad/s,} \tag{7.12}$$

at which frequency the zeros are real for increasing values of K_P.

The frequency response of the transfer-function $G_{c1}(s)$ is obtained by setting $s = j\omega_f$ in (7.5) or equivalently in (7.9):

$$G_{c1}(j\omega_f) = K_P + j\left(K_D\omega_f - \frac{K_I}{\omega_f}\right) = K_D(\omega_1 + \omega_2) + j\left(\frac{K_D}{\omega_f}\right)(\omega_f^2 - \omega_1\omega_2). \tag{7.13}$$

For real values of the corner frequency the straight-line approximation of the magnitude response of the simple Type 1 PID transfer function (7.5) is shown in Figure 7.13.

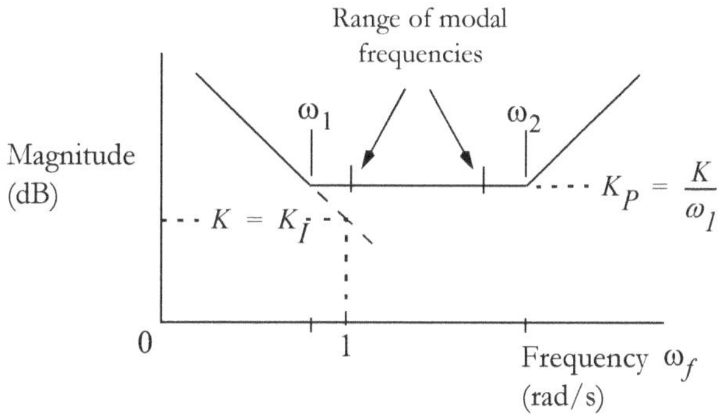

Figure 7.13 Straight-line approximation of the magnitude response of the simple Type 1 PID transfer function $G_{c1}(s)$ $(T_D = 0)$.

The desired transient gain reduction of the compensator is ideally established over the range of modal frequencies in the constant gain region of the straight-line response between the corner frequencies ω_1 and ω_2. At very low and at high frequencies the slope of the magnitude response is -20 and $+20$ dB/decade, respectively. In the actual magnitude response the gain is a minimum, $K_{min} = K_P$, at the geometric mean of the two corner frequencies [1], i.e. at a frequency of $\omega_{min} = \sqrt{\omega_1\omega_2}$.

Using the result of equation (7.8), the frequency at minimum gain is given by

$$\omega_{min} = \sqrt{K_I/K_D}, \tag{7.14}$$

and is independent of the value of the proportional gain, K_P. Again we note the values of $\omega_{1,2}$ from (7.12) and ω_{min} from (7.14) are identical.

Let us assume K_P is to be varied over a range of values consistent with an appropriate level of transient gain reduction. Substitution of $s = j\omega_{min} = j\sqrt{K_I/K_D}$ in (7.5) for $G_{c1}(s)$ yields a value of $G_{c1}(j\omega_{min}) = K_P$. In other words, the *minimum gain of the simple Type 1 PID over the frequency range is equal to the setting of the proportional gain, K_P*. As has been noted - in terms of the magnitude response of Figure 7.13 - the straight-line segment between ω_1 and ω_2 represents the gain K_P (dB). The value of the proportional gain K_P is thus close to the value of the transient gain, K_T, particularly when the corners are well-spaced apart.

The above analysis provides a procedure for AVR tuning when based on the simple Type 1 PID transfer function (7.5). The steps in the procedure are:

1. Select a value for the desired minimum transient gain K_P over the modal frequency range. Note that, since this is the minimum value between the corner frequencies, the effective gains closer to the corners will be somewhat higher.

2. Select the value for the lower corner frequency ω_1 which is typically close to the value K_I/K_P (see (7.10)); hence deduce a value for $K_I = \omega_1 K_P$.

3. In order to achieve the desired transient gain reduction, set ω_{min} to a value in the vicinity of the geometric mean of the selected modal frequency range, say $\omega_{min} = 3$ rad/s for a range of modal frequencies, 1.0 to 10 rad/s. From (7.14) $\omega_{min} = \sqrt{K_I/K_D}$; hence estimate $K_D = K_I/\omega_{min}^2$.

4. From (7.11) check that the value of K_P selected ensures that the corner frequencies assume real values. That is, ensure $K_P \geq 2\sqrt{K_D K_I}$.

5. Evaluate the actual corner frequencies of the simple Type 1 PID transfer function from (7.7), and plot the frequency responses.

6. Analyse the dynamic performance of the generator and excitation system when off-line and on-line over a range over steady-state operating conditions. Adjust the

---

1. Alternatively, for a transfer function of the form of (7.5), i.e. $(as^2 + bs + c)/s$, a simple analysis reveals the magnitude of the transfer function is a minimum at $\omega_{min} = \sqrt{c/a}$, $s = j\omega$. Furthermore, the phase angle traverses from $-90°$ at low frequencies to $+90°$ at high frequencies and passes through zero phase at ω_{min}. This result can also be determined easily from (7.13).

parameters of the PID to satisfy the performance specifications for the generating unit off- and on-line; repeat steps 1 to 6.

It should be noted that the gains, as determined and discussed in the context of this chapter, are in per-unit on the relevant generator and exciter field voltage and current base values, and generator stator base voltage. However, the gains of the AVR as identified in the actual software (or hardware) of the excitation system may be on quite different and varied bases depending on the manufacturer's scaling system in their control design. As such, care needs to be exercised in the field when translating gains to and from the actual settings in the controls and those which are used in simulation platforms such as that discussed here.

7.7.1.2 Characteristics of Type 2 PID compensation

An expanded form of the Type 2 PID transfer function of (7.4), in which $T_D > 0,\ K_G > 0,\ T_G = 0$, is

$$G_{c2}(s) = \left\{[s^2(K_D + K_P T_D) + s(K_P + K_I T_D) + K_I]/[s(1 + sT_D)]\right\} \cdot K_G. \tag{7.15}$$

The zeros of this transfer function are derived from the numerator of (7.15):

$$s^2 + s\left(\frac{K_P + K_I T_D}{K_D + K_P T_D}\right) + \frac{K_I}{K_D + K_P T_D} = 0. \tag{7.16}$$

This equation is of the form:

$$(s + \omega_1)(s + \omega_2) = s^2 + s(\omega_1 + \omega_2) + \omega_1\omega_2 = 0. \tag{7.17}$$

Equating the coefficients of s in (7.16) and (7.17), and solving for $\omega_{1,2}$, we find:

$$\begin{aligned}\omega_{1,2} &= \frac{K_P + K_I T_D}{2(K_D + K_P T_D)} \pm \frac{\sqrt{(K_P + K_I T_D)^2 - 4(K_D + K_P T_D)K_I}}{2(K_D + K_P T_D)} \\ &= \left(\frac{K_P + K_I T_D}{2(K_D + K_P T_D)}\right)\left(1 \pm \sqrt{1 - \frac{4(K_D + K_P T_D)K_I}{(K_P + K_I T_D)^2}}\right)\end{aligned} \tag{7.18}$$

$$\text{thus } (\omega_1 + \omega_2) = (K_P + K_I T_D)/(K_D + K_P T_D); \quad \omega_1\omega_2 = K_I/(K_D + K_P T_D). \tag{7.19}$$

From (7.18) it can be shown that the zeros of $G_{c2}(s)$, $\omega_{1,2}$, are real if

$$K_P \geq K_I T_D \pm 2\sqrt{K_D K_I}. \tag{7.20}$$

An alternative form of (7.15) - which accommodates real corner frequencies ω_1 and ω_2 - is

$$G_{c2}(s) = K \cdot \frac{(1 + s/\omega_1)}{s} \cdot \frac{(1 + s/\omega_2)}{(1 + sT_D)}, \tag{7.21}$$

where $K = K_I \cdot K_G$. The method for deriving the straight-line approximation of the frequency response of $G_{c2}(j\omega_f)$ based on (7.21) is outlined in Section 2.12 and texts on control system analysis. This response is shown in Figure 7.14 for what will be called Type 2A PID compensation in which

$$\omega_1 < \text{range of modal frequencies} < \omega_2 < (1/T_D).$$

Type 2B PID compensation, for which $\omega_1 < \omega_2 <$ range of modal frequencies, and $\omega_1 < \omega_2 < (1/T_D)$ will be considered in Section 7.8.

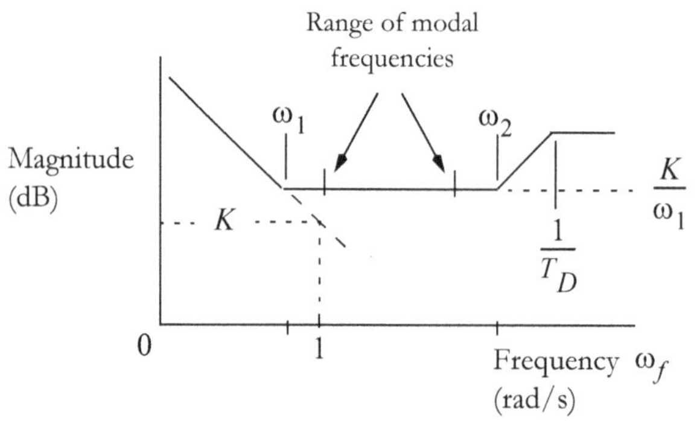

Figure 7.14 Straight-line approximation of the frequency response of the Type 2A PID compensator transfer function $G_{c2}(s)$.

In a Type 2A PID the desired transient gain reduction of the compensator is ideally established over the range of modal frequencies in the constant gain region of the straight-line response between the corner frequencies ω_1 and ω_2. For the Type 2A PID (with $K_G = 1$) the transient gain K_T is assumed to set to the value of K_P, the proportional gain for purposes of design. At low and high frequencies the slope of the response on either side of these frequencies is -20 and $+20$ dB/decade, respectively. The effect of the corner $1/T_D$ $(>\omega_2)$ results in the gain at high frequencies of the PID transfer function (7.15) tending to

$$K_G[(K_D/T_D) + K_P]\text{, as } s \to \infty. \tag{7.22}$$

This gain may be considered unacceptably high and appropriate adjustments to the parameters in (7.22) may be required.

The frequency at which the magnitude response of the PID transfer function (7.15) is a minimum involves some tedious analysis [1].

1. This involves finding the minimum of the magnitude of the PID transfer function (7.15) with $s = j\omega_f$ and making use of the Symbolic Maths Toolbox in MATLAB®.

For a non-zero value of T_D the corner frequency $1/T_D$ is typically a decade or more greater than ω_{min}. In this case the magnitude and phase contribution of the transfer function $1/(1+sT_D)$ to the frequency response at ω_{min} is reduced by a factor of more than 0.995 (–0.043 dB) and by an angle less than $5.7°$, respectively. As a result the minimum of the magnitude of the practical PID transfer function occurs at a slightly lower frequency than ω_{min} based on (7.14).

7.7.1.3 Characteristics of Type 3 PID compensation

Based on (7.2) and (7.15), the Type 3 PID Compensator can be expressed as

$$G_{c3}(s) = \frac{s^2(K_D + K_P T_D) + s(K_P + K_I T_D) + K_I}{s(1+sT_D)} \cdot \frac{K_G}{1+sT_G}, \tag{7.23}$$

which, like (7.21), is of the form:

$$G_{c3}(s) = K_I \cdot \frac{(1+s/\omega_1)}{s} \cdot \frac{(1+s/\omega_2)}{(1+sT_D)} \cdot \frac{K_G}{1+sT_G}, \tag{7.24}$$

where ω_1 and ω_2 are real values given by (7.18) and $T_D > 0$, $K_G > 0$, $T_G > 0$. Being three transfer function blocks in series, the frequency response analysis $G_{c3}(s)$ based on (7.24) can be conducted as described in Sections 2.7 and 2.12, and in texts on control system analysis.

With $T_G > 0$ in (7.24), and $1/T_G > 1/T_D$, the low-pass filter $1/(1+sT_G)$ may be employed to attenuate high frequency signals - such as noise - as illustrated in Figure 7.11 on page 328.

7.7.2 Tuning methodology for PID Compensation Types 1 and 2A

7.7.2.1 Tuning of simple Type 1 PID compensation

Let us adopt the following specifications as the starting point for the tuning of Type 1 PID compensation based on (7.5).

1. Over the range of local- and inter-area modal frequencies of 1.5 to 12 rad/s the effective transient gain, K_T, of the AVR / excitation system is to be 25 - 50 pu on machine base (i.e. 28.0 - 34.0 dB); a value of 32 pu will be selected.

2. The corner frequencies of PID transfer function are to lie outside the range of modal frequencies of 1 to 10 rad/s.

3. For a 0 to 1% step change at the voltage reference input of the closed-loop voltage control system, off-line or on-line, (a) the terminal voltage overshoot should be less than 7.5%, (b) the 90% rise time should be less than 1 s, and (c) the 2% settling time of the terminal voltage should be less than 5s.

4. The criterion for on-line system damping performance is that the real parts of all rotor modes should be less than -0.139 Np/s (a halving time of 5 s).

Let us follow the steps in Section 7.7.1.1 for the determination of the parameters of the simple PID transfer function.

1. Assume the proportional gain setting of the PID, K_P, is 32 pu; let $K_G = 1$.

2. Assume the lower corner frequency, $\omega_1 \approx K_I/K_P$, is 0.5 rad/s. The value of K_I is thus 16 pu/s.

3. Based on the straight-line magnitude response in Figure 7.13, select the value of frequency (rad/s) at which the magnitude of the simple PID is to be a minimum. For the range of modal frequencies, say, a value of $\omega_{min} = 3.2$ rad/s is close to the geometric mean ($\sqrt{10}$) of the specified modal frequency range. According to (7.14) the ratio $K_I/K_D = \omega_{min}^2 = 10.24$, thus selecting $K_I = 16$ yields $K_D = 1.56$ pu-s; let $K_D = 1.5$ pu-s (say) with the result that $\omega_{min} = 3.27$ rad/s.

4. From (7.11), the value of the proportional gain K_P at which the complex zeros of the Type 1 PID transfer function assume a real value is $K_{Pb} = 2\sqrt{K_D K_I} = 9.80$ pu. We also know that the value of gain at ω_{min} is the setting of K_P, $K_P > K_{Pb}$.

5. With $K_P = 32$ pu, $K_I = 16$ pu/s and $K_D = 1.5$ pu-s the corner frequencies of the Type 1 PID are calculated from (7.7). The values are $\omega_1 = 0.51$ and $\omega_2 = 20.8$ rad/s; both values are outside the specified range of modal frequencies.

It is instructive to derive the frequency response of the simple Type 1 PID compensator for a range of values of K_P from 0 to 44 pu, including the selected gain setting of 32 pu. The responses are shown in Figure 7.15. For $K_P = 32$ pu we note the value is almost constant over the selected modal frequency range, that is, for all intents and purposes *the value of the transient gain K_T is equal to the proportional gain K_P.*

The high-frequency gain (above the range of modal frequencies) may become excessive. To limit the gain the time constant T_D in the PID transfer function of (7.4) is set to a non-zero value. This is considered in the tuning of Type 2A PID compensation in the following section.

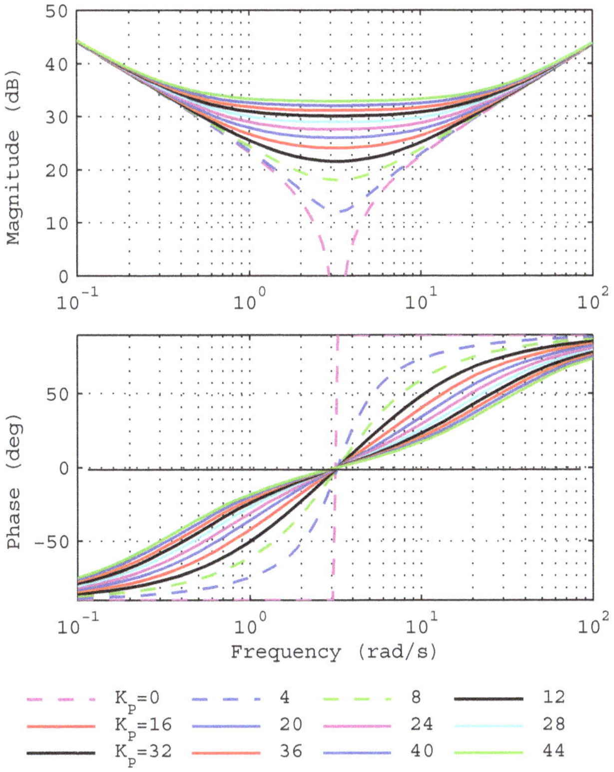

Figure 7.15 Frequency responses of the Type 1 PID for K_I = 16 pu/s, K_D = 1.5 pu-s, T_D = 0 s and values of K_P between 0 and 44 pu; the dashed lines apply to complex zeros of $G_{c1}(s)$.

7.7.2.2 Tuning of Type 2A PID compensation

An approach to the tuning of this compensation for the off- and on-line dynamic performance of the AVR is covered in some detail in Sections 7.7.2.2 to 7.7.2.4. Because a lightly damped rotor mode arises in the on-line case, an exploratory tuning of a PSS for the generator and system is considered.

As foreshadowed in Section 7.7.1.2 and Figure 7.14 the low-pass filter $1/(1+sT_D)$ is introduced to the Type 2A PID transfer function to limit its high-frequency gain.

The same specifications as for the Type 1 PID are assumed for the tuning of Type 2A PID compensation based on (7.4), however, an appropriate value of the time constant T_D must be determined. For the range of modal frequencies, 1.5 to 12 rad/s, a frequency of ω_{min} = 3.27 rad/s is nominally assumed at which the magnitude response of the Type 1 PID transfer function is a minimum; in fact, the minimum will be a lower value for the Type 2A PID.

Steps 1, 2 and 3 of the PID design procedure are the same as those listed in Section 7.7.2.1 for the Type 1 PID, which yielded $K_P = 32$ pu, $K_I = 16$ pu/s and $K_D = 1.5$ pu-s. The corner associated with T_D is selected to have a value greater than that of the upper corner frequency, $\omega_2 = 20.8$ rad/s, calculated in step 5 for the Type 1 PID. Let us consider three alternative values for T_D, namely $T_D = 0.0125,\ 0.025$ and 0.0375 s; the associated corner frequencies of 80, 40 and 26.7 rad/s, respectively.

With $T_D < 1/\omega_2$ and $K_G = 1$ the value of the proportional gain K_{Pb} at which the real corners evolve from complex values is derived from (7.20), i.e. $K_{Pb} = K_I T_D \pm 2\sqrt{K_D K_I}$ pu. For the range of values of T_D selected, the corner frequencies are real for values of the gain K_P greater than $K_{Pb} = 10.4$ pu.

The corner frequencies ω_1 and ω_2 of the Type 2A PID are calculated from (7.18); the values are given in Table 7.2 and are outside the specified range of modal frequencies, 1 to 10 rad/s.The frequencies at which the magnitude responses of the Type 2A PID transfer function (7.15) are a minimum, together with other relevant statistics, are also listed in Table 7.2.

The frequency responses of the PID for the selected values of T_D are shown in Figure 7.16.

Table 7.2 Characteristics of Type 2A PID (7.15)

T_D (s)	Corner Frequencies (rad/s)			At minimum of magnitude response*:			High frequency gain
	ω_1	ω_2	$1/T_D$	ω_{min} rad/s	Magnitude (pu)	Phase °	(pu) | (dB)
0	0.51	20.8	-	3.27	32	0	-
0.0125	0.51	16.4	80	2.92	32.2	-1.6	152 | 43.6
0.0250	0.51	13.6	40	2.73	32.3	-3.0	92 | 39.3
0.0375	0.51	11.6	26.7	2.58	32.5	-4.3	72 | 37.1

PID Parameters are $K_P = 32$ pu, $K_I = 16$ pu/s, $K_D = 1.5$ pu-s, $K_G = 1$ pu;
$T_D = 1/\omega_D$ is a variable parameter.
* Values read off frequency responses in Figure 7.16

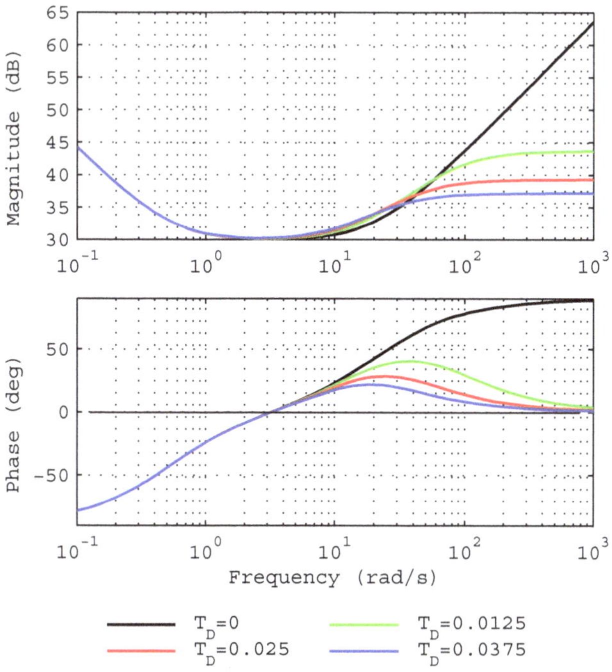

Figure 7.16 Frequency response of the Type 2A PID for a range of values of T_D. The PID parameters are $K_P = 32$ pu, $K_I = 16$ pu/s, $K_D = 1.5$ pu-s, $K_G = 1$ pu, T_D.

7.7.2.3 Dynamic performance of the Type 2A PID with the generating unit off-line

The following studies are based on the SMIB system in which one line is out of service as shown in Figure 7.4. The block diagram of PID compensation and generator are shown in Figure 7.10 on page 327; the gain K_G of the series block in the PID is set to unity and its time constant T_G is zero. The parameters of the sixth-order generator are listed in Appendix 7–I.1.1. When off-line, it is assumed - using first-order transfer functions - that the relevant generator and excitation parameters are $T'_{do} = 5.0$ s and $T_E = 0.1$ s, respectively.

The purpose of the following analysis is to establish whether the specifications for the off-line performance are satisfied. Firstly, from the open-loop frequency response plot of the PID plus the excitation system and generator, it is desirable to determine information on the stability of the closed-loop voltage control system and the nature of its dynamic performance. The Bode plot of the open-loop system, V_t/V_{ref}, is shown in Figure 7.17 for the range of values of T_D listed in Table 7.2.

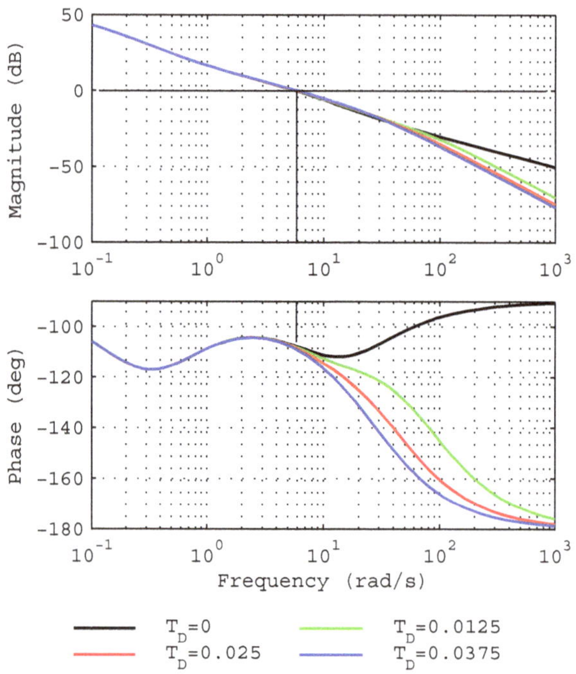

Figure 7.17 Unit off-line. Bode plot V_t/V_{ref} of the open-loop voltage control system. Type 2A PID parameters are $K_P = 32$ pu, $K_I = 16$ pu/s, $K_D = 1.5$ pu-s, $K_G = 1$ pu; T_D s.

From Figure 7.17 it is observed that at the gain-crossover frequency of 5.9 rad/s the phase margin is approximately 71° - 73° for all values of T_D. The generating unit when off-line is therefore stable under closed-loop voltage control and the response of terminal voltage to a step change in reference voltage is well damped. This is observed in Figure 7.18 which reveals that the value of the time constant T_D has little effect on the response.

Because the phase margin shown in Figure 7.17 for all values of T_D is adequate, it may be of interest to increase the gains of the PID by increasing the setting of K_G, say, by factors of two or three (6 and 9.5 dB respectively). By lowering the 0 dB axis in the Bode diagram of Figure 7.17 by 6 or 9.5 dB the resulting phase margins are shown in Table 7.3 to lie between 69° and 56°.

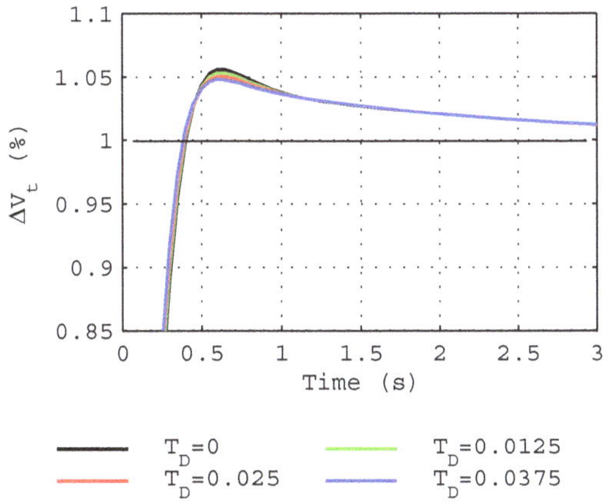

Figure 7.18 Unit off-line. Perturbations in terminal voltage due to a step change in reference voltage from a steady-state value of 1.0 pu to 1.01 pu (1%). The unit is under closed-loop voltage control for the specified range of values of T_D, $K_G = 1$. (Note scale range on y-axis).

Table 7.3 Phase Margins of the off-line generator for a range of settings of K_G and T_D

K_G (pu)	$T_D = 0.$ s		$T_D = 0.0125$ s		$T_D = 0.025$ s		$T_D = 0.0375$ s	
	PM°	ω_{gco}	PM°	ω_{gco}	PM°	ω_{gco}	PM°	ω_{gco}
1	73	5.7	72	5.7	72	5.8	71	5.9
2	69	9.9	67	10.2	65	10.5	62	10.8
3	68	13.4	65	14.1	60	14.6	56	14.9

PM: Phase Margin (°). ω_{gco}: Gain cross-over frequency (rad/s)

For the purpose of analysis of the performance of the unit on-line in the next section, it is desirable to choose an appropriate value of K_G and T_D. To ensure the closed-loop terminal voltage response to a disturbance is well-damped, let us choose from the table a value of phase margin better than (greater than) 65°. Moreover, with higher values of the gain K_G the transient gain and the high frequency gain of the PID may be too high in the particular application. Let us restrict the evaluation of the closed-loop terminal voltage responses to those for $K_G = 1$ and 2 pu. To complement the responses shown in Figure 7.18 for $K_G = 1$ pu, the closed-loop voltage responses to a step change in the reference voltage when the unit is off-line is shown in Figure 7.19 for $K_G = 2$ pu and the range of values of T_D.

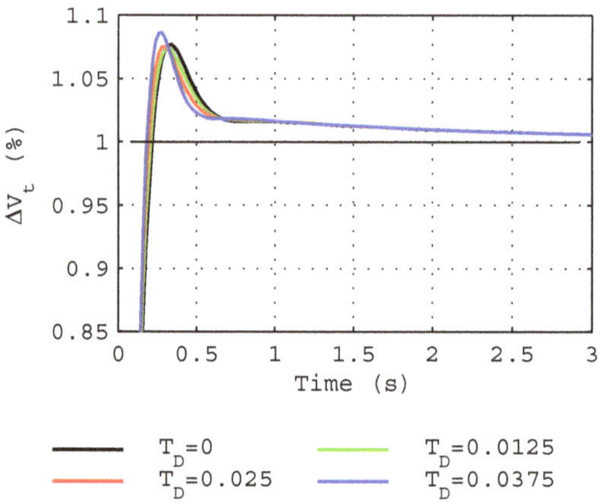

Figure 7.19 Unit off-line. Perturbations in terminal voltage from the initial steady-state value due to a step change from1.0 to 1.01 pu (1%) in the reference voltage under closed-loop voltage control for a range of values of T_D, $K_G = 2$ pu. (Note scale range on y-axis).

A comparison between the closed-loop step responses for generator off-line for the gains $K_G = 1$ and 2 pu is shown in Table 7.4 (step change of +0.01 pu (1%) in reference voltage).

Table 7.4 Characteristics of closed-loop, off-line, 0-0.01 pu step responses for $K_G = 1$ and 2 pu

K_G	Rise Time: 90% of step size (s)	Peak Overshoot (% on step size)	Time to Peak (s)	2% Settling Time (s) ##
1	~0.29 s for all values of T_D	5.6 to 4.6% for all values of T_D *	0.63 to 0.60 s for all values of T_D *	~2.1 s for all values of T_D
2	~0.16 s for all values of T_D	~7.5% for all values except for T_D=0.0375 s the value is 8.6%	0.33 to 0.30 s for all values - except for T_D=0.0375 s the value is 0.27s	< 0.62 s for all values of T_D

* Note: First to last values in the range are $T_D = 0$, 0.0125, 0.025, 0.0375 s
2% Settling-Time requirement is less than 5 s (see Section 7.7.2.1)

In this application, selecting a value of $T_D = 0.025$ s ensures that the overshoot of the terminal voltage to a step in the reference is less than 7.5% of the step size for both gain $K_G = 1$ and 2 pu; for higher values of T_D the peak overshoot increases rapidly. Moreover, for $K_G = 1$ and 2 pu we note from Table 7.3 that the phase margin for $T_D = 0.025$ s is better than or equal to the specified limiting value of 65°. We also note that the step responses are markedly faster for $K_G = 2$ than for $K_G = 1$ pu.

7.7.2.4 Dynamic performance of the Type 2A PID with the generating unit on-line
Selecting $T_D = 0.025$ s and gain $K_G = 1$ and 2 pu, let us consider the stability of the generating unit on-line and under closed-loop voltage control.

The SMIB test power system is shown in Figure 7.4. Let us consider two operating conditions,

- A line outage condition, Case W: generator output P = 0.4, Q = 0 pu (as in Figure 7.4 on page 320).

- All lines in service, Design Case C: generator output P = 0.9, Q = 0.2 pu [1].

As in the previous AVR tuning method (Section 7.6), Case W is analysed initially. The Bode plot of the open-loop transfer function V_t/V_{ref} for the generating unit on-line is displayed in Figure 7.20 together with the response when the PID is replaced by a simple gain element having the same value as the PID proportional gain ($K_P = 32$ pu). Over the selected range of rotor modal frequencies, 1 to 10 rad/s, the magnitude responses in the figure being in close agreement reveals that the transient gains are practically identical.

At the gain cross-over frequency of 3.2 rad/s the Phase Margin of 79° suggests that for $K_G = 1$ pu the closed-loop step response V_t/V_{ref} should be well-damped; however the resonance at ~9 rad/s in the Bode plot may result in a *rotor* oscillation of about that frequency being superimposed on the terminal voltage response. By 'raising' the magnitude plot of Figure 7.20 by 6 dB - for $K_G = 2$ pu - the Bode plot reveals that the Phase Margin is reduced to 68° at a gain cross-over frequency of 6.6 rad/s. The closed-loop terminal voltage response should again be well-damped, but the oscillatory response of the rotor mode should be accentuated as the gain cross-over frequency approaches the resonant frequency of ~9 rad/s. These results are demonstrated in the associated closed-loop step responses of Figure 7.21.

Because of oscillatory nature of the responses, as revealed in the closed-loop performance, it is clear that the rotor mode is lightly damped and that a PSS is required.

---

1. Case C is the 'Design Case' in Section 5.11. With all lines in service the steady-state power flow conditions for Case C are the same as those in Table 5.4. However, the generator and exciter parameters in Chapter 5 differ from those listed in the Appendix 7–I.1.1. Comparison of results between this and Chapter 5 may be misleading.

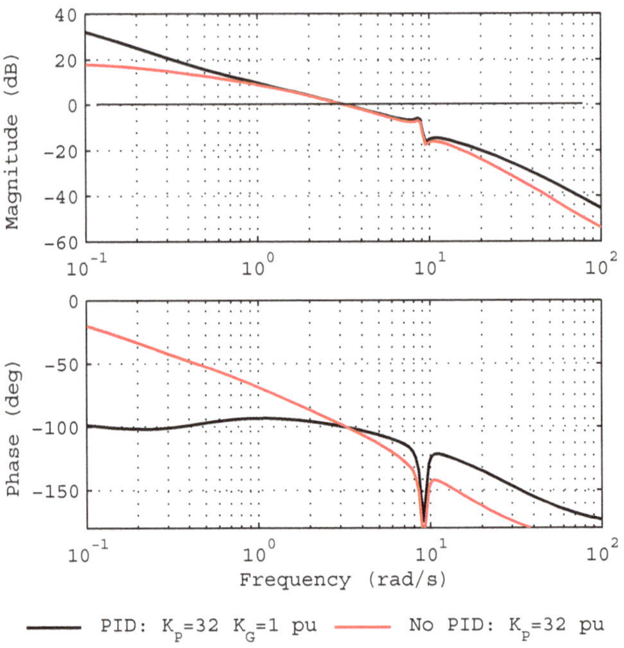

Figure 7.20 Case W: Unit on-line. Bode plot of V_t/V_{ref}.
PID parameters: K_P = 32 pu, K_I = 16 pu/s, K_D = 1.5 pu-s, K_G = 1 pu; T_D = 0.025 s.
For 'No PID' the AVR gain is 32 pu.

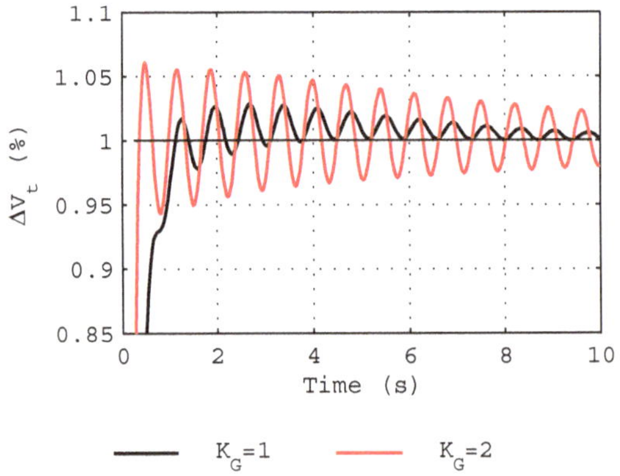

Figure 7.21 Case W: Unit on-line. Perturbations in terminal voltage due to a step change in reference voltage from a steady-state value of 1.0 pu to 1.01 pu (1%). Generator is under closed-loop control: gains K_G = 1 and 2.0 pu. (Note scale range on y-axis.)

7.7.2.5 Exploratory tuning of a PSS to improve the damping of the oscillatory mode

A PSS is typically switched into service at a lower value of real power output, say 0.1 to 0.3 pu. The procedure for tuning the PSS is the same as that outlined in Section 5.10.4. We will assume that, for this system, Case C has also been established to be the PSS 'design case' for the set of generator and excitation system parameters listed in Appendix 7–I.1.1.

The time-domain responses are faster for $K_G = 2.0$ than for $K_G = 1.0$ pu, however, for $K_G = 2.0$ pu the oscillatory response is less well damped. Because it may be necessary to select a value of K_G such that $1 \le K_G \le 2$ pu, the P-Vr characteristic is derived for the design case for the two values of K_G. These characteristics, together with the synthesized characteristics for Case C, are shown in Figure 7.22.

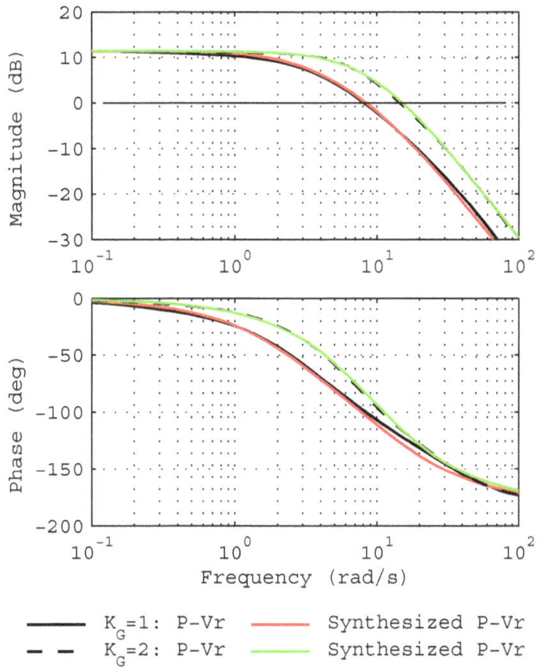

Figure 7.22 P-Vr characteristic for Case C for the set of PID parameters: $K_P = 32$ pu, $K_I = 16$ pu/s, $K_D = 1.5$ pu-s, $T_D = 0.025$ s and $K_G = 1.0$ or 2.0 pu. The synthesized P-Vr agrees closely with actual P-Vr plots.

Including the low-pass and washout filters, the resulting PSS transfer functions based on Case C are:

$$G_{PSS1} = \frac{k}{3.725} \cdot \frac{s5}{1+s5} \cdot \frac{(1+s0.357)(1+s0.0752)}{(1+s0.0067)(1+s0.0067)} \text{ for } K_G = 1.0 \text{ pu, and}$$

$$G_{PSS2} = \frac{k}{3.738} \cdot \frac{s5}{1+s5} \cdot \frac{(1+s0.143)(1+s0.0794)}{(1+s0.0067)(1+s0.0067)} \text{ for } K_G = 2.0 \text{ pu}, \tag{7.25}$$

respectively, with the damping gain k set to 20 pu on machine MVA rating [1].

With the PSS1 and PSS2 in service, the closed-loop step responses V_t/V_{ref} are replotted in Figure 7.23 for Cases C and W.

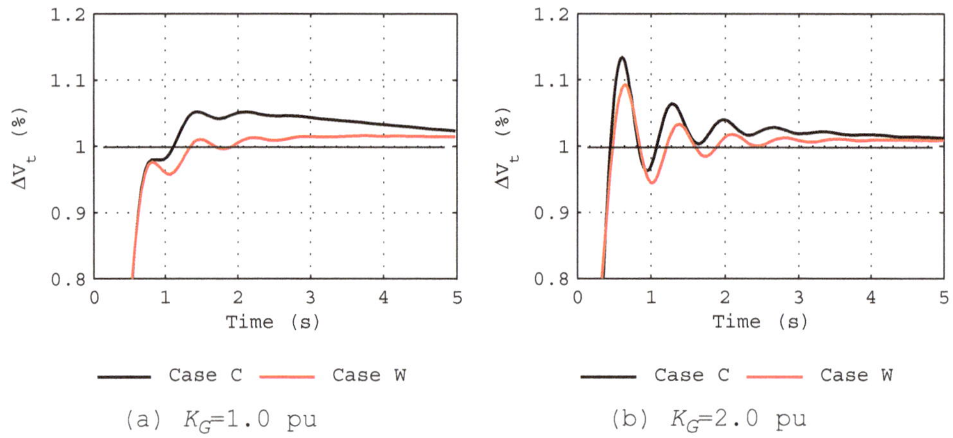

Figure 7.23 Cases C and W: Unit on-line. Perturbations in terminal voltage due to a step change in reference voltage from a steady-state value of 1.0 pu to 1.01 pu (1%). Generator is under closed-loop control with (a) PSS1, gain $K_G = 1$ pu in service and (b) PSS2 for $K_G = 2$. PSS damping gain is 20 pu on machine MVA base. (Note scale on y axis.)

The time-domain performance and modal characteristics of the closed-loop AVR, excitation system and generator on-line are summarized in Table 7.5 when K_G is set to 1 and 2 pu.

7.7.2.6 Concluding remarks: PID Type 2A Compensation based on Cases C and W

Based solely on the limited set of operating conditions considered in the tuning of the PID-based AVR it may be concluded that with a PSS in service the setting of the PID gain K_G should lie between 1 and 2 pu. The former and latter settings yield on-line terminal voltages responses to a step change, shown in Figure 7.23(a) and (b), which may considered sluggish and over-responsive, respectively. Depending on the actual performance specifications, further investigations should concentrate on values of K_G, which lies in the range $1.3 \le K_G \le 1.6$, say [2]. The rotor mode is well damped with the PSS in service and satisfies the damping criterion for rotor modes. Clearly there are a number of issues that should be

1. The low-pass filter time constants (6.7 ms) are very short. Such time constants should typically be 3 or more times the cycle time of the PSS processor to reduce phase errors at higher frequencies.

reviewed in the analysis of both the on- and off-line cases before a final set of parameters for the PID is chosen. Such issues are:

Table 7.5 Characteristics of closed-loop, on-line, step responses for $K_G = 1, 2$ pu.

Case No.	K_G pu	Time domain characteristics				Rotor modes		
		90% Rise Time (s)	Peak Over-shoot (%)	Time to Peak (s)	2% Sett-ling Time (s)	PSS in service	PSS out of service	Mode shift
C	1.0	0.63 s	5.2%	1.45 s	5.3 s	$-1.94 \pm j8.90$	$-0.05 \pm j9.08$	$-1.88 \mp j0.18$
	2.0	0.41 s	13.4%	0.61 s	3.5 s	$-1.48 \pm j9.13$	$0.33 \pm j9.15$	$-1.82 \mp j0.02$
W	1.0	0.63 s	1.0%	1.48 s	1.24 s	$-1.56 \pm j8.72$	$-0.24 \pm j8.87$	$-1.32 \mp j0.15$
	2.0	0.40 s	9.2%	0.65 s	1.49 s	$-1.39 \pm j8.78$	$-0.11 \pm j8.93$	$-1.28 \mp j0.15$
PID Parameters are K_P =32 pu, K_I =16 pu/s, K_D =1.5 pu-s, K_G; T_D =0.025 s PSS damping gain is 20 pu on machine MVA rating.								

- In practice, to ensure that the specifications are satisfied it is necessary to conduct a set of on-line studies for an encompassing range of normal and N-1 operating conditions.
- What is the maximum acceptable value for the high-frequency gain $(K_P+K_D/T_D)K_G$ (e.g. see Table 7.2 for Type 2A PID)?
- The choice of an acceptable range of transient gains ($K_P \cdot K_G$).

The example also reveals a basis for coordinating the tuning of the PID controls with the tuning the PSS when the on-line responses do not meet the specifications for the rotor modes.

7.8 Type 2B PID Compensation: Theory and Application to AVR tuning

The theoretical basis and calculation of the parameters of the Type 2B PID are presented in this section. However, because an application to the tuning of AVRs in a remote three-generator power station with brushless excitation systems is more detailed and complex it is covered at the end of this chapter in Section 7.11.

---

2. K_G should be less than 1.56 to satisfy the specified upper limit on the transient gain of $K_P \cdot K_G = 50$ pu.

7.8.1 Tuning of Type 2B PID compensation

In contrast to the Type 2A PID, in the Type 2B PID the range of modal frequencies lies above the corner ω_2 rad/s; this is illustrated in Figure 7.24.

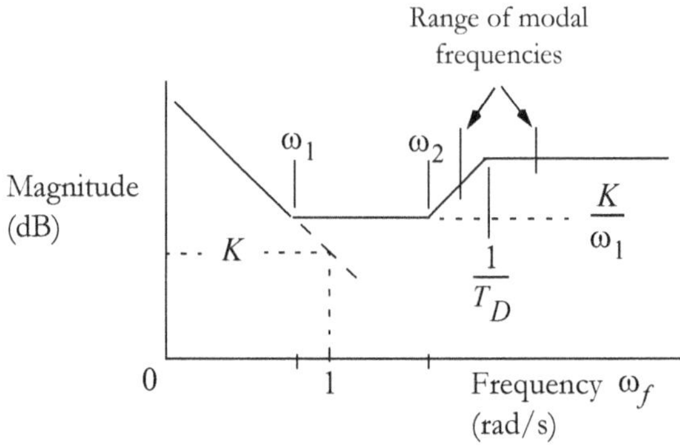

Figure 7.24 Straight-line approximation of the frequency response of the Type 2B PID compensator transfer function $G_{c2B}(s)$; range of modal frequencies > ω_2.

The use of the Type 2B PID compensator may occur in cases such as when the PID is required to contribute phase lead at low frequencies. In such a case the generator and exciter time constants, T'_{do} and T_{ex}, may be relatively long. Phase lead in the compensator is provided at the lower corner frequencies ω_1 and ω_2; integration ensures the steady-state error between the reference and terminal voltages is integrated out.

Equations (7.15) to (7.22) are applied to Type 2B PID. The transient gain K_{TB}, which ideally applies over - or above - the range of modal frequencies is, from (7.22),

$$K_{TB} = K_G[(K_D/T_D) + K_P]\,, \text{ as } s \to \infty\,. \tag{7.26}$$

Note that $K_{TB} > K_P \cdot K_G$. Furthermore, the corner frequencies ω_1 and ω_2 in Figure 7.24 are given by (7.19), i.e. $\omega_1\omega_2 = K_I/(K_D + K_P T_D)$. Assuming $\omega_1 \approx K_I/K_P$, the upper corner is then

$$\omega_2 \approx 1/(T_D + K_D/K_P)\,, \tag{7.27}$$

$$\text{or } K_D = K_P[(1/\omega_2) - T_D]\,. \tag{7.28}$$

Substituting (7.28) in (7.26) we find

$$K_{TB}/(K_P \cdot K_G) = \omega_D/\omega_2 \quad \text{where } \omega_D = 1/T_D\,. \tag{7.29}$$

Equations (7.28) and (7.29) form the basis for the Type 2B PID design.

7.8.2 Example: Evaluation of Type 2B PID parameters.

This example demonstrates the application of the above results to the tuning of the AVRs for a more difficult case of a remote, base-load, three-machine power station supplying energy via 132 kV lines to a high voltage grid. The purpose of this example - and its continuation in Section 7.11 - is

- to illustrate the determination of the PID parameters which satisfy certain performance specifications over a wide range of N and N-1 operating conditions;
- to examine a systematic and structured method for the selection of PID parameters which are robust over the range of operating conditions;
- to linearize the non-linear model of the brushless exciter and account for the variation of its small-signal parameters with the steady-state operating conditions;
- to establish the requirements for software for automating and expediting the calculations in the design process for application in practical cases.

In this application the tuning of the AVRs is more complex because the time constants T'_{d0} and T_E of the generator and brushless exciter are relatively long and the only tunable parameters in the AVR are those of the PID; the tuning is covered in some detail in Section 7.11. However, for this application the calculation of the characteristics of a relevant set of candidate PID parameters are required and are therefore examined in the following section.

7.8.2.1 Frequency response characteristics of Type 2B PIDs

The range of modal frequencies is known to be 4 to 7 rad/s. Assume that over this frequency range an effective value of $K_T \approx 32$ pu is required when $K_G = 1$ pu. According to (7.26) and Figure 7.24 the desired transient gain K_{TB} must be somewhat higher, say, 40 - 70 pu; the values of K_{TB}, ω_1, ω_2 and ω_D are subject to the condition: $\omega_1 < \omega_2 < \omega_D$.

The PID parameters are the calculated based on (7.28) and (7.29) using the following relationships:

$$T_D = 1/\omega_D; \quad K_P = \omega_2 K_{TB}/\omega_D; \quad K_D = (K_{TB} - K_P)/\omega_D; \quad K_I = K_P\omega_1. \qquad (7.30)$$

For several sets of values for K_{TB}, ω_1 , ω_2 and ω_D the PID Type 2B parameters are derived using the above algorithm and are listed in Table 7.6. The associated frequency responses, which are shown in Figure 7.25, demonstrate the effect of modifying the parameters in the vicinity of 1 rad/s.

It is noted from Figure 7.25 or Table 7.6:

- Over the frequency range 4 to 7 rad/s the gain is close to the desired value of transient gain, $K_T \approx 32$ pu.

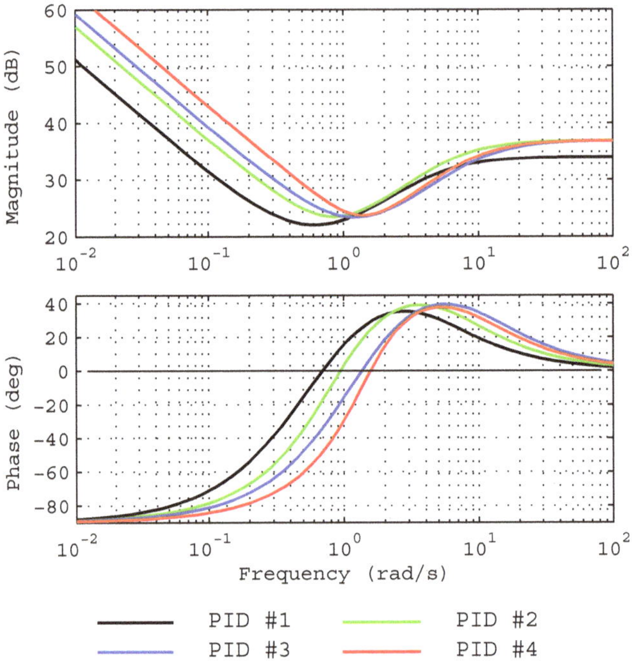

Figure 7.25 Frequency response plots of four Type 2B PIDs with the parameters in Table 7.6.

Table 7.6 Calculated parameters for PID Type 2B

Para-meter Set No.	Selected quantities				Calculated PID parameters			
	K_{TB}	ω_1 (rad/s)	ω_2 (rad/s)	ω_D (rad/s)	K_P pu	K_I (pu/s)	K_D (pu s)	T_D (s)
1	50	0.30	1.2	5	12	3.6	7.60	0.200
2	70	0.50	1.4	7	14	7.0	8.00	0.143
3	70	0.65	2.2	11	14	9.1	5.09	0.0909
4	70	1.0	1.9	9.5	14	14	5.89	0.1053

- The minimum value of gain is about 23 dB (i.e. $K_P = 14$ pu) at frequencies less than ω_2.
- With the higher values of the corner frequency ω_2, Figure 7.25 shows that PID Sets 2, 3 and 4 provide additional phase lead in the range of modal frequencies 3-8 rad/s. For PID type 2 compensation the transfer function $(1 + s/\omega_2)/(1 + sT_D)$ in (7.21) is

a lead-lag element, thus by increasing ω_D with respect to ω_2 the peak phase-lead is increased.

Although K_{TB}, ω_1, ω_2 and ω_D are selected to calculate the PID parameters, equations (7.30) can be rearranged to calculate the parameters based on some other choice, e.g. K_P, K_{TB}, ω_1 and ω_2, where $K_P < K_{TB}$.

The example of the detailed tuning of the AVRs in the remote three-generator power plant is described in Section 7.11. Other methods of AVR tuning are considered in the following Sections 7.9 and 7.10.

7.9 Proportional plus Integral Compensation

7.9.1 Simple PI Compensation

The structure of the Proportional plus Integral Compensation (PI) in the AVR is shown in Figure 7.26. We shall refer to this as Simple PI Compensation.

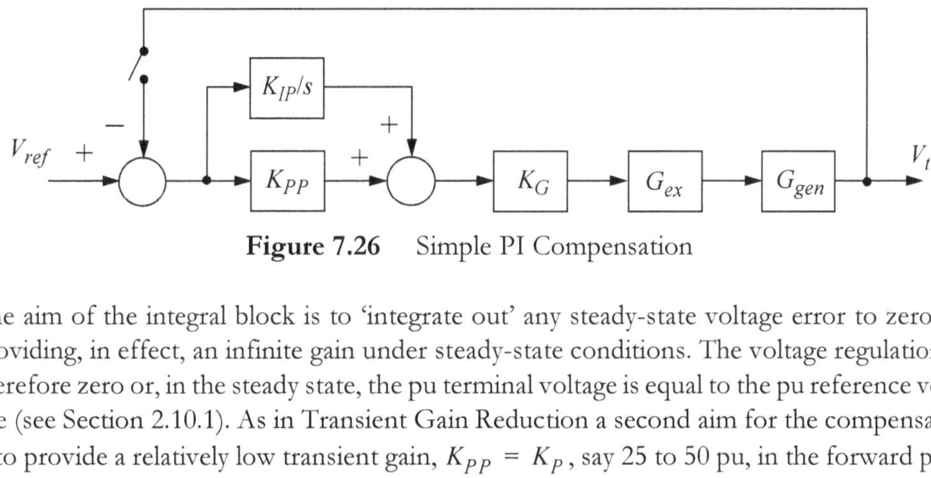

Figure 7.26 Simple PI Compensation

The aim of the integral block is to 'integrate out' any steady-state voltage error to zero by providing, in effect, an infinite gain under steady-state conditions. The voltage regulation is therefore zero or, in the steady state, the pu terminal voltage is equal to the pu reference voltage (see Section 2.10.1). As in Transient Gain Reduction a second aim for the compensator is to provide a relatively low transient gain, $K_{PP} = K_P$, say 25 to 50 pu, in the forward path over the frequencies of the rotor modes of oscillation. Let $K_G = 1$.

The form of the simple PI compensator transfer function is:

$$G_c(s) = K_{IP}/s + K_{PP} = K_{IP}(1 + T_1 s)/s \text{, where } T_1 = 1/\omega_1 = K_{PP}/K_{IP}. \quad (7.31)$$

Let $K_{PP} = 32$ pu be the desired transient gain, let's place the corner frequency $\omega_1 = K_{IP}/K_{PP}$ about a decade below the lowest frequency mode, say 5 rad/s, and let $\omega_1 = 0.5$ rad/s. The frequency response of the Simple PI Compensator with the resulting integrator gain, $K_{IP} = K_{PP}\omega_1 = 16$ pu/s, is shown in Figure 7.27. The compensator transfer function is thus:

$$G_{PI}(s) = 16(1+2s)/s\,. \tag{7.32}$$

An alternative implementation of PI compensation using positive feedback is described in Appendix 7–I.3.

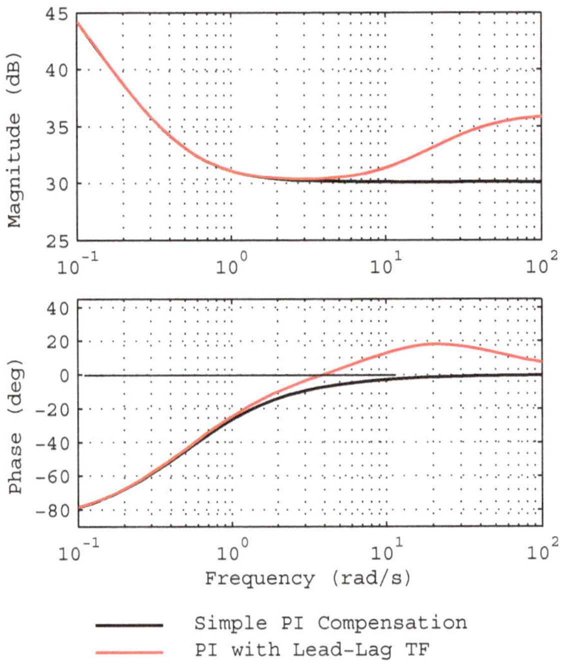

Figure 7.27 Frequency response plots of (i) Simple PI Compensation; (ii) Simple PI Compensation with a series lead-lag transfer function block

7.9.2 Conversion to a PID Compensator with an additional lead-lag block

For a number of reasons it may be desirable to boost the gain of the AVR at higher frequencies, say, to improve the rate of response of the field voltage during the fault interval or in the immediate post-fault period. This can be achieved by inserting a lead-lag block in series with the PI block; the maximum phase lead of the lead-lag should occur at, or above, the highest local-mode frequency. With this block it is also possible to improve the phase response of the compensator at the higher modal frequencies. In Section 2.12.1.4 and [8], [9] the lead-lag compensator is described; a form of the transfer function is $G_{LD}(s) = (1+sT_2)/(1+s\alpha T_2)$, where the values for α are less than unity.

For application with the simple PI compensator in Section 7.9.1, assume (i) the frequency at which the maximum phase lead in the transfer function $G_{LD}(j\omega_f)$ occurs is $\omega_m = 20$ rad/s and (ii) the high frequency gain is to be boosted by a factor of two (6 dB). The former requires that $\omega_m = 1/(\sqrt{\alpha}T_2) = 20$ rad/s and the factor $(1/\alpha) = 2$; hence the time constant of lead-lag block is $T_2 = 0.07071$ s. The frequency response of the Simple PI Compensator modified by the series lead-lag block is shown in Figure 7.27. This response is similar to that shown for a Type 2A PID in Figure 7.16 for $T_D>0$.

The PI with the series lead-lag compensator illustrated above can be converted to a PID form. Letting $T_D = \alpha T_2$ s, the transfer function of the PI plus lead-lag compensator is

$$G_c = \frac{K_{IP}(1+sT_1)}{s} \cdot \frac{(1+sT_2)}{1+sT_D} = \frac{K_{IP}T_1T_2s^2 + K_{IP}(T_1+T_2)s + K_{IP}}{s(1+sT_D)}. \tag{7.33}$$

This equation is identical in form to (7.15) for Type 2 PID with $K_G=1$; let us equate coefficients of the powers of *s* in the numerators of (7.33) and (7.15). We find:

$$\begin{aligned} K_I &= K_{IP} \\ K_P &= K_I(T_1+T_2-T_D) \quad \text{for} \quad T_1 > T_2 > T_D. \\ K_D &= K_I T_1 T_2 - K_P T_D \end{aligned} \tag{7.34}$$

These equations are solved sequentially for K_I, K_P and K_D.

In the above example the parameters for the PI of (7.31) and the series lead-lag compensator of (7.33) are:

$K_{IP} = 16$ pu-s, $T_I = 2$ s, $T_2 = 0.07071$ s, $T_D = 0.03536$ s.

By substitution of these values in (7.34) we find the equivalent PID parameters are:

$K_I = 16$ pu-s, $K_P = 32.57$ pu, $K_D = 1.111$pu/s, $T_D = 0.03536$ s.

The significance of the conversion to PID parameters is the following. It may be convenient or simpler to determine the parameters of a PI and series lead-lag compensator rather than to directly determine a set of PID parameters. If it is necessary to frame the compensation in the form of a PID controller the conversion is readily calculated based on (7.34).

Note. There are further modifications possible to the frequency response of the Simple PI Compensator to meet particular requirements. For example: The addition of a series first-order low-pass block $1/(1+s0.01)$ would provide high-frequency roll-off of -20 dB/decade on magnitude at frequencies above 100 rad/s. However there would be some reduction in the phase lead over the range of rotor modes e.g. about 5 to 6 degrees at 10 rad/s.

7.10 Rate feedback compensation

7.10.1 Method of analysis

The block diagram of the generator and excitation system with rate feedback (RFB) of the exciter voltage are shown in Figure 7.28. For the determination of the rate feedback gain K_F and associated time constant T_F of the feedback loop, both frequency response and root locus methods may be employed.

Compensation based on rate feedback employs either the output voltage of the exciter (which is also the voltage input to the field of the generator) or a signal related to exciter field current (if field voltage is inaccessible). As in the case of TGR tuning, the aim of the compensation is to derive for the Excitation System (ES) a desired transient gain reduction over the range of modal frequencies. This is achieved by appropriately determining the values of the feedback gain and time constant, K_F and T_F, respectively.

In the types of analyses of ESs considered, a simple low-order model of the ES is employed. It will then be shown that the analyses can be extended to account for more complex dynamic systems.

The tuning of the ES shown in Figure 7.28 will be illustrated first. This will be followed by an examination of the dynamic performance of the generator off-line when it is under closed-loop voltage control with the tuned ES. Finally, the associated performance of the generator when on-line is assessed over a range of operating conditions to ascertain if the performance specifications are satisfied.

7.10.2 Tuning of the Excitation System (ES)

As evident from Figure 7.28 the ES comprises only the closed loop formed by the AVR, the exciter and the rate-feedback path, the rate-feedback transfer function being:

$$G_F(s) = (sK_F)/(1+sT_F). \tag{7.35}$$

Note that the block diagram in Figure 7.28 applies to the cases with the generator off-line as well as on-line. In the latter case the block G_{gen} includes both the generator and the system to which it is connected.

In the case of a brushless excitation system the output voltage of the exciter is not accessible for measurement. In this case a signal proportional to the *exciter* field current, typically designated V_{fe}, is used as the signal for rate feedback. The exciter field current is closely related to the field voltage of the main generator (i.e. E_f). To cover the cases of the exciter output voltage being available or unavailable for feedback, a generalised approach is adopted in which a voltage V pu, not defined, is the feedback signal - as shown in Figure 7.29. For the purposes of initially illustrating a procedure, a simple first-order system is assumed for the forward loop where the gain K and the time constant T represent those of the AVR, or the

AVR plus exciter, (referred to as the 'Plant'). It will be demonstrated that this simple system can then be modified to include additional dynamics in the forward path.

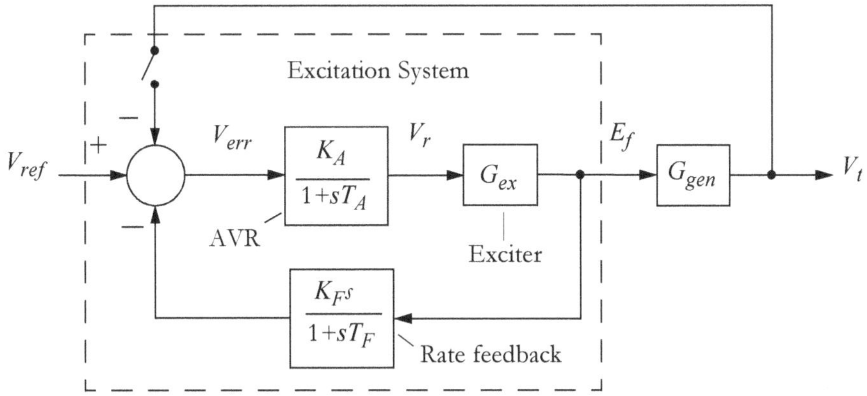

Figure 7.28 Generator and Excitation System with field-voltage feedback compensation. The block G_{gen} accounts for the generator (and system) model when off-line (and on-line).

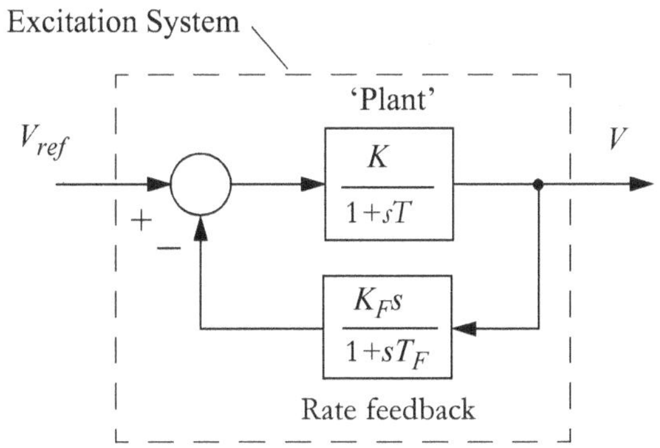

Figure 7.29 The Excitation System is shown as a simple closed-loop system with rate feedback.

The closed-loop transfer function of the Excitation System shown in Figure 7.29 is

$$\frac{V}{V_{ref}} = \frac{\frac{K}{T \cdot T_F} \cdot (1 + sT_F)}{s^2 + \left(\frac{T + T_F + KK_F}{T \cdot T_F}\right)s + \frac{1}{T \cdot T_F}}. \tag{7.36}$$

This transfer function can be expressed as

$$G_{CL}(s) = \frac{V}{V_{ref}} = K \cdot \frac{(1 + sT_F)}{(1 + s/\omega_1)(1 + s/\omega_2)}, \text{ in which} \tag{7.37}$$

$$\omega_1\omega_2 = 1/(T \cdot T_F) \text{ and } (\omega_1 + \omega_2) = (T + T_F + KK_F)/(T \cdot T_F) \tag{7.38}$$

The following analysis is mainly based on frequency response methods, however, the root locus technique is used to demonstrate the fine-tuning of the rate feedback parameters which is applicable to higher order excitation systems.

7.10.3 Rate feedback compensation using Frequency Response Methods.

The closed-loop transfer function (7.37) has one zero at $s = -\omega_F = -1/T_F$, associated with the feedback path, and two poles at $s = -\omega_1, -\omega_2$. In order to obtain a more-or-less constant transient gain reduction over the range of modal frequencies, say 1 to 10 rad/s, the straight-line frequency response for the magnitude of the transfer function $G_{CL}(j\omega_f)$ should have the form shown in Figure 7.30.

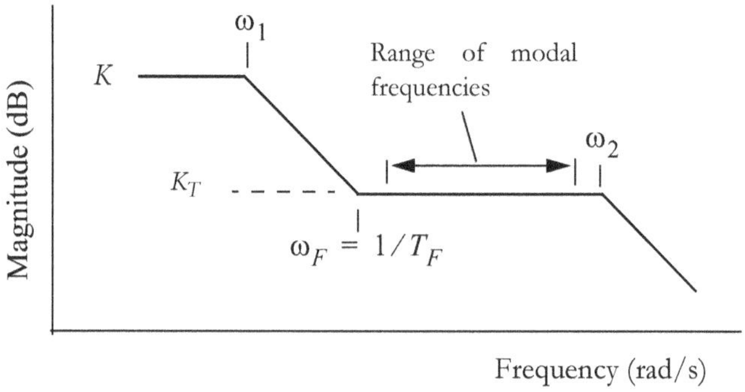

Figure 7.30 Desired form of the straight-line magnitude response of the transfer function $|G_{CL}(j\omega_f)|$ for the closed-loop, rate-feedback control system.

From the desired form of the transfer function shown in Figure 7.30, the corner frequency ω_2 should ideally be greater than, say, $10\omega_F$ to cover the desired range of modal frequencies. Over that range the transient gain is K_T. The lower corner ω_1 typically should be less than $\omega_F/10$ (to minimise the effect of the corner on the modal frequency range). Assume that the transfer function in (7.37) can be divided into two blocks in series, say,

$$G_{c1}(s) = K \cdot \frac{(1 + sT_F)}{(1 + s/\omega_1)} \text{ and } G_{c2}(s) = \frac{1}{1 + s/\omega_2}. \tag{7.39}$$

At frequencies greater than $1/T_F$, $G_{c1}(s) \rightarrow K_T = K \cdot T_F\omega_1$ as $s \rightarrow \infty$, where K_T is the desired transient gain. Thus,

$$\omega_1 = K_T/(K \cdot T_F). \tag{7.40}$$

It is noted that $G_{c2}(j\omega_f) \rightarrow 1$ when $\omega_f < \omega_2/10$. Moreover, from (7.38):

$$\omega_1\omega_2 = 1/(T \cdot T_F) \text{ which, from (7.40), implies } \omega_2 = K/(K_T \cdot T). \tag{7.41}$$

Note that the upper corner frequency ω_2 is independent of the rate feedback parameters, K_F and T_F. Following substitution of (7.40) and (7.41) in (7.38), we find

$$T_F = \frac{K_T}{K} \cdot T + \frac{K \cdot K_T}{K - K_T} \cdot K_F. \tag{7.42}$$

If a value of the rate feedback gain, K_F, is selected the associated value of the time constant T_F can be determined from (7.42), together with the lower corner frequency ω_1 from (7.40). In this case the lower corner frequency cannot be specified.

However, according to (7.40), $\omega_1 = K_T/(K \cdot T_F)$, either ω_1 or T_F can be selected, given values of K and K_T. An approach based on the frequency response of Figure 7.30 suggests that the corner frequency $1/T_F$ is a more meaningful quantity to select than the rate gain K_F. Thus, given the value of T_F, K_F can then calculated from a rearranged form of (7.42), i.e.

$$K_F = \frac{1}{K}\left\{\left(\frac{K}{K_T} - 1\right)T_F - \left(1 - \frac{K_T}{K}\right)T\right\} \approx \left(\frac{K - K_T}{K \cdot K_T}\right)T_F \quad \text{if } T \text{ is small.} \tag{7.43}$$

Note that, if T is small, (7.43) can be expressed as

$$K_T \approx K\Big/\left(1 + \frac{K_F \cdot K}{T_F}\right) = \frac{T_F}{K_F}, \text{ if } (K_F \cdot K)/T_F \gg 1. \tag{7.44}$$

Thus, given the values of K_F and T_F, the ratio T_F/K_F is an estimate of the upper limit on the value of the transient gain K_T.

An examination of the frequency response plot of (7.37) for $G_{CL}(j\omega_f)$ will reveal whether the desired transient gain is more-or-less achieved over the range of modal frequencies.

In the above analysis a simple model has been employed for the ES. In practice, the models of both the AVR and the exciter may be of higher order. In some cases it may be possible to adapt the above analyses to satisfy such systems. However, in other cases it is possible to represent the excitation control system by low-order models in order to determine initial set of values of the rate feedback parameters, K_F and T_F. These parameters can then be fine-tuned using the more complex models and the associated frequency responses.

In (7.42) all quantities except K_F and T_F, the rate feedback parameters, are either known or selectable. Using frequency response techniques an analysis of two cases will be considered: Case 1 is based on the configuration in Figure 7.29; in Case 2 transient gain reduction or PI compensation is included in the forward path of the AVR. Furthermore, it will been shown that AVR and exciter transfer functions possessing additional dynamics can be incorporated in the analyses. In all cases a check should be made that the limits imposed on the ES by conditions such as (7.40) and (7.41) are valid

7.10.3.1 Case 1: No TGR or PI compensation in forward loop of the AVR

Three illustrative examples based on the rate feedback of the exciter output are described for Case 1. In Example 1 below the parameters of the rate-feedback block are determined subject to certain specifications. In Example 2 the effects of additional dynamics in the forward loop of the AVR are investigated. Finally, in Example 3 the significance of the rate-feedback parameters employed in the Sample Data for the AC2A model of the excitation system [12] is assessed against those values calculated by the approach adopted in Section 7.10.3.

7.10.3.1.1 Example 1, Case 1. Simple excitation system

Assuming rate feedback of the exciter output voltage E_f, the parameters of the ES in Figures 7.28 and 7.29 are $K_A = K = 250$ pu, and the exciter gain and time constant are $K_E = 1$ and $T_E = T = 0.1$ s, respectively; the time constant T_A is assumed negligible. The specification for the transient gain is $K_T = 32$ pu (30.1 ±3 dB) over the modal frequency range 1 to 10 rad/s.

The desired form of the straight-line frequency response of the magnitude of the closed-loop excitation control system is shown in Figure 7.30. Referring to the latter figure, let us assume the feedback time constant of is $T_F = 2$ s. At the associated corner frequency $\omega_F = 0.5$ rad/s we know, based on Section 2.12.1.3, that the transient gain K_T is close to 30.1 dB +3 dB.

From (7.40), the lower corner frequency is $\omega_1 = K_T/(K_A T_F) = 0.064$ rad/s. Similarly, based on (7.41) the upper corner frequency is $\omega_2 = K_A/(K_T T_E) = 78.1$ rad/s; at this frequency the transient gain is close to 30.1 dB -3 dB. For $T_F = 2$ s the rate-feedback gain $K_F = 0.0541$ pu-s is calculated from (7.43). The resulting exciter frequency response is shown in Figure 7.31 together with those for several smaller values of T_F and associated values of the gain K_F. The closed-loop responses reveal that for values of $T_F > 1.2$ s the transient gain requirements are satisfied.

The relevant characteristics of the responses are summarised in Table 7.7.

Table 7.7 Characteristics of frequency responses for varying rate-feedback time constant T_F

Ex. 1, Case 1 Study No.	T_F (s)	ω_F (rad/s)	ω_1 (rad/s)	ω_2 (rad/s)	K_F (pu-s)	See Note 1.
1	0.747	1.34	0.171	78.13	0.020	34.4 dB −44.3°
2	1.222	0.818	0.105	78.13	0.0329	32.3 dB −34.0°
3	2.00	0.500	0.064	78.13	0.0541	31.1 dB −23.5°

ES parameters: $K = K_A = 250$ pu, $K_T = 32$ pu, $T = T_E = 0.1$ s.
Note 1. Magnitude and phase of the responses are calculated at 1.0 rad/s.

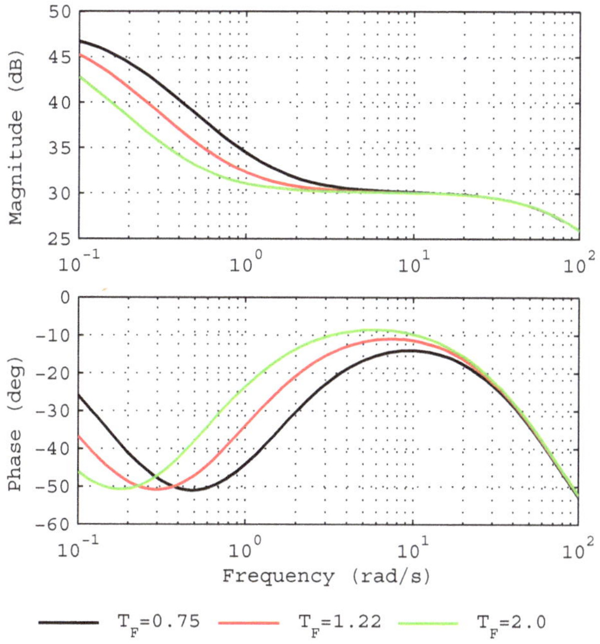

Figure 7.31 Example 1, Case 1: Closed-loop frequency responses V/V_{ref} for the simplified rate-feedback excitation system model in Figure 7.29 for increasing values of the time constant T_F (s) in the rate-feedback block.

7.10.3.1.2 Example 2, Case 1. Higher-order excitation system

It has been pointed out that, in practice, the models of both the AVR and the exciter may be of higher order. Referring to Figures 7.28 and 7.29 let us assume the AVR and exciter transfer functions, G_A and G_{ex}, possess additional dynamics and are of the forms:

$$G_A = \frac{K_A}{1+sT_A} \text{ and } G_{ex} = \frac{1}{(1+sT_E)} \cdot \frac{1+sT_2}{(1+sT_1)}\text{, respectively,}$$

where $T_A = 0.05$ s, $T_1 = 0.02$ s, $T_2 = 0.01$ s. Assume that the rate feedback parameters are those for Study 2 in Table 7.7, namely, $T_F = 1.22$ s and $K_F = 0.033$ pu-s, and $K = K_A = 250$ pu, $K_T = 32$ pu, and $T = T_E = 0.1$ s. The above analysis is based on $T = 0.1$ s and, because the time constants T_A, T_1 and T_2 of the additional elements are shorter than that of the exciter, the corners associated with additional dynamics lie at higher frequencies than $1/T_E$. The application of frequency response approach for Case 1 is therefore valid. The frequency response with the additional faster time constants is shown in Figure 7.32, and is compared with the frequency response if the faster dynamics are ignored.

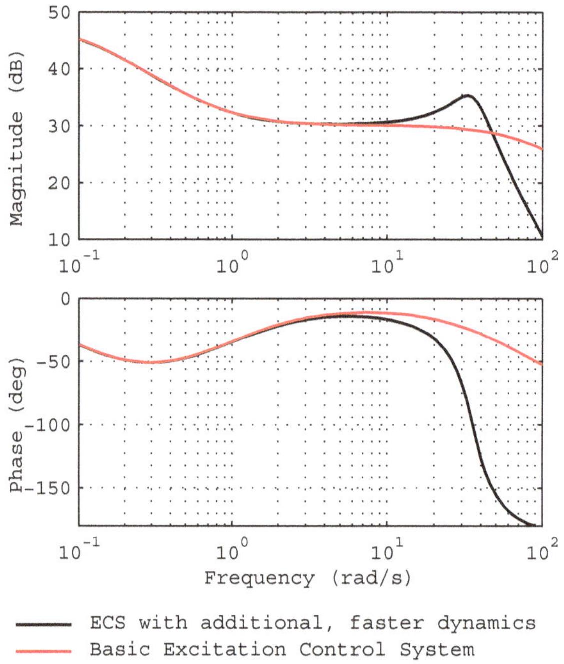

Figure 7.32 Comparison of the closed-loop frequency responses of an excitation system having additional, faster dynamics than that for the simplified system in Figure 7.29. $T_F = 1.22$ s and $K_F = 0.033$ pu-s.

In this case, over the range of modal frequencies of interest, the additional faster dynamics have little effect on the frequency response of the simplified excitation system which has been tuned ignoring the effects of faster dynamics. The exciter time constant of 0.1 s has been employed in the example is long. Typically, in practice, the exciter time may be less that 0.05 s; such values cause the upper corner $\omega_2 = K_A/(K_T T_E)$ to increase in frequency. The magnitude and phase responses, based on the frequency response for Case 1, therefore tend to be 'flatter' over a range beyond the 1 to 10 rad/s range of modal frequencies.

An aside: Say the exciter time constant is $T_E = 0.05$ pu and in the forward loop there is a first-order lag with a time constant $T_l = 0.25$ s (corner frequency 4 rad/s) which lies in the modal frequency range of interest. In this event the pole at −4 Np/s can be cancelled - for practical purposes - and the pole shifted to say, −25 Np/s.

7.10.3.1.3 Example 3, Case 1: Application to excitation system model AC2A

The approach adopted in Example 1 can be adapted to excitation system (ES) models in IEEE Standard 421.5 [12]. Type AC2A ES represents a field-controlled alternator-rectifier exciter system (a brushless ES). A small-signal model of the AC2A is shown in Figure 7.33 in which, for present purposes, a simple model $1/(K_E + sT_E)$ is used for the exciter.

The following set of sample data for the Type AC2A model is provided in [12]:
$K_A = 400$ pu, $T_A = 0.01$ s, $T_B = T_C = 0$ s, $K_B = 25$ pu, $K_H = 1$ pu, $K_E = 1.0$ pu, $T_E = 0.6$ s, $K_F = 0.03$ pu-s, $T_F = 1.0$ s.

By block diagram manipulation of Figure 7.33 with $K_E = 1.0$ pu, it can be shown that

$$\frac{\Delta E_{FD}}{\Delta V_A} = \frac{K_B/(1 + K_H K_B)}{1 + [sT_E/(1 + K_H K_B)]}.$$

The above transfer functions reveals that the exciter time constant is reduced by a factor $1/(1 + K_H K_B)$, and consequently the speed of response of the exciter is increased. The AC2A model of the excitation system can then be expressed in the form shown in Figure 7.29 with $T = T_E/(1 + K_H K_B)$ and $K = K_A K_B/(1 + K_H K_B)$, assuming $T_A = 0$.

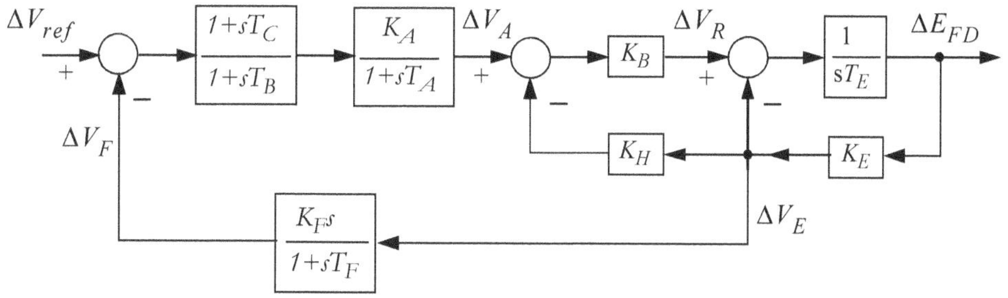

Figure 7.33 Small-signal model of the AC2A ES with a simplified exciter model ($T_A = 0$)

In this example the forward loop of the ES, $\Delta E_{FD}/\Delta V_A$, introduces a corner at $1/T = 43.3$ rad/s lying above the selected range of modal frequencies 1 - 10 rad/s. The approach developed in Section 7.10.3 is thus valid since this corner lies above the upper end of the modal frequency range.

Let us consider the following two studies:

1. assume the desired transient gain is $K_T = 32$ pu. Given $T_F = 1.0$ s calculate the associated rate-feedback gains K_F based on (7.43);

2. calculate the effective transient gain K_T associated with the data supplied in [12] for the AC2A excitation system model.

In Study 2, equation (7.43) is solved for K_T, i.e.

$$K_T^2 T - K_T(KT + T_F K + K^2 K_F) + T_F K^2 = 0 . \tag{7.45}$$

The results are summarised in Studies 1 and 2 of Table 7.8.

Table 7.8 Comparison of the closed-loop parameters for AC2A [12] in Studies 1 & 2

Study No.	T_F (s)	K_T (pu)	ω_1 (rad/s)	ω_2 (rad/s)	K_F (pu-s)	Basis of calculation	Comment
1	1.0	32.0	0.083	52.1	0.0286*	Based on frequency response analysis, Case 1	K_F calculated from (7.43) with $K_T = 32$ pu
2	1.0	30.6*	0.080	54.46	0.030	Based on data set for exciter AC2A [12]	K_T calculated from (7.45) with $K_F = 0.03$ pu
* Calculated value, given the selected value of K_T or K_F. ω_1, $1/T_F$ and ω_2 are the corner frequencies of the closed-loop transfer function (7.37).							

In Study 2 it is of interest to note that, for the AC2A ES with the rate-feedback time constant $T_F = 1$ s and gain $K_F = 0.03$ pu, the calculated transient gain is $K_T = 30.6$ pu. These values are close to those in Study 1 when the transient gain of K_T is set to 32 pu and the calculated rate-feedback gain is $K_F = 0.0286$ pu, i.e. the AC2A has an inherent transient gain close to that which has been adopted in this chapter. This observation is confirmed in the frequency response plots of Figure 7.34; moreover, the transient gain is more-or-less constant over the selected modal frequency range, 1 to 10 rad/s.

It is noted that the combination of $K_F = 0.03$ pu-s and $T_F = 1.0$ s is commonly used in the sample data sets for a variety of ESs in IEEE Standard 421.5 [12]. According to (7.44), for latter values the effective transient gain is $K_T \approx T_F/K_F = 33.3$ pu.

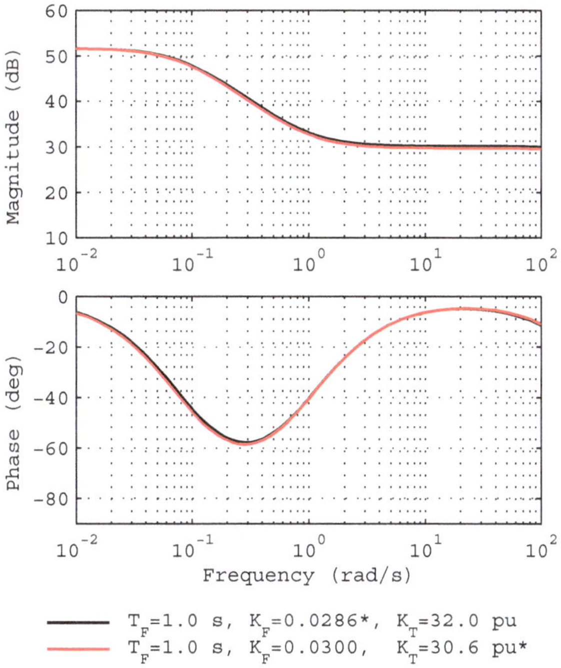

Figure 7.34 Frequency responses V/V_{ref} of the closed-loop rate-feedback ES for Studies 1 and 2 of Table 7.8. * Calculated values when either K_T or K_F is the specified quantity.

7.10.3.2 Case 2: Rate feedback with TGR or PI compensation in forward loop of the AVR

In Case 1, previously considered in Section 7.10.3.1, transient gain reduction or PI compensation is omitted from the forward loop.

TGR or PI compensation may be employed in conjunction with rate feedback of the exciter voltage or AVR output (the 'Plant' output) - as shown in general form in Figure 7.35(a).

The objective of the analysis is to derive a constant transient gain K_T at frequencies in the modal frequency range, say 1 to 10 rad/s. This implies that all corner frequencies in the transfer function of the ES should lie outside of the latter range.

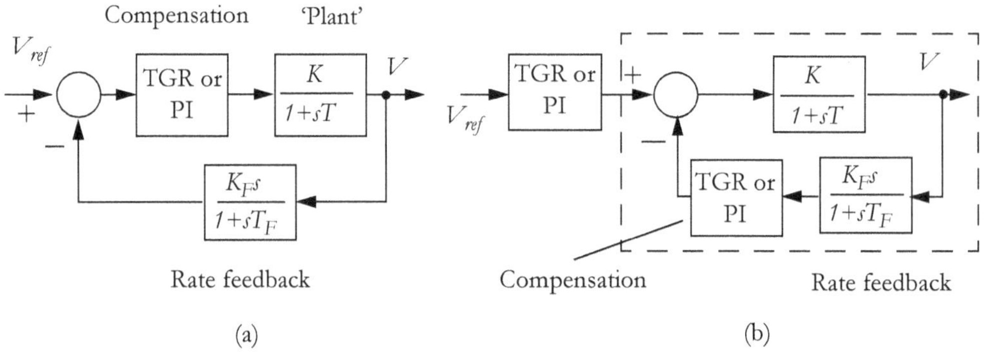

Figure 7.35 Compensation with rate feedback of the exciter output voltage E_f or the equivalent exciter field current (V_{fe})
(a) General form of the simple block diagram; (b) Equivalent form.

Based on the equivalent form of the block diagram in Figure 7.35(b), the following analysis considers the behaviour of system in the lower and higher frequency ranges. The lower range includes the corner frequencies of the TGR or PI compensation together with the corner frequency $1/T_F$ of the rate feedback block. The higher frequencies range which exceeds 10 rad/s includes plant corner $1/T$ - and possibly additional higher corner frequencies We require $1/T_F \ll 1/T$ - assuming $1/T_F$ is the highest corner in the lower frequency range - and $1/T > 10$ rad/s. Ideally, the magnitude of the transient gain K_T should then be constant over the range of modal frequencies.

Consider the limits as $\omega_f \rightarrow \infty$, $s = j\omega_f$ of the following transfer functions whose corners lie in the lower frequency range:

TGR transfer function (7.1) $\quad (1 + sT_C)/(1 + sT_B) \rightarrow T_C/T_B$;

PI transfer function (7.31) $\quad K_{PP} + (K_{IP}/s) \rightarrow K_{PP}$;

Rate feedback transfer function (7.35) $\quad sK_F/(1 + sT_F) \rightarrow K_F/T_F$; (7.46)

The plant transfer function when $\omega_f < 1/(10T)$ is $\quad K/(1 + j\omega_f T) \rightarrow K$.

The upper corner frequency of the TGR, $1/T_C$, and the corner frequency of the PI, $1/T_{PI} = K_{IP}/K_{PP}$, are such that they are less than $1/T_F$.

Let $K_C = T_C/T_B$ or let $K_C = K_{PP}$ (these are the high frequency gains of the TGR or PI transfer functions in (7.46), respectively).

Under the condition that,

$$\omega_f > 1/T_F > 1/T_C \text{ or } \omega_f > 1/T_F > 1/T_{PI}, \tag{7.47}$$

the transfer function for the systems shown in Figure 7.35 is

$$W(j\omega_f) = \frac{V}{V_{ref}} = K_C \cdot \left[\frac{K}{(1 + KK_CK_F/T_F)} \cdot \frac{1}{1 + (j\omega_f T)/(1 + KK_CK_F/T_F)}\right]. \tag{7.48}$$

The gain of the transfer function (7.48) represents the transient gain K_T over the range of frequencies of interest:

$$\text{i.e.,} \quad K_T = K_C \cdot \frac{K}{(1 + KK_CK_F/T_F)}, \quad \text{or} \quad \frac{K_F}{T_F} = \frac{1}{K_T} - \frac{1}{KK_C}. \tag{7.49}$$

Note the corner frequency of the transfer function $W(j\omega_f)$ in (7.48) lies at a value greater than $1/T$. The application of these results is considered in Example 4.

7.10.3.2.1 Example 4, Case 2: Transient gain reduction or PI compensation with rate feedback

The application of rate feedback with either TGR or PI compensation in a closed-loop control system is shown in Figure 7.35(a). The following parameters are provided for the compensation and the plant, i.e.:

- TGR: $T_C = 5$ s, $T_B = 12.5$ s, upper corner of TGR is $1/T_C = 0.2$ rad/s;
- PI: $K_{PP} = 0.4$ pu, $K_{IP} = 0.08$ pu/s, corner of PI is $K_{IP} / K_{PP} = 0.2$ rad/s.
- Plant: $K = 250$ pu, $T = 0.05$ s.
- For both forms of compensation: transient gain $K_T = 32.0$ pu. Assume $T_F = 2.0$ s.

Condition (7.47) is valid for both types of compensator. Calculate the values of the feedback gain K_F for each compensation.

Based on (7.49), the rate feedback gains for the respective compensator types and for the selected value of the feedback time constant are:

$$\text{TGR:} \quad K_F = \left(\frac{1}{32} - \frac{1}{250 \times 0.4}\right) \times 2 = 0.0425 \text{ pu} \quad \text{with } K_C = T_C/T_B = 0.4 \text{ pu;}$$

$$\text{PI:} \quad K_F = \left(\frac{1}{32} - \frac{1}{250 \times 0.4}\right) \times 2 = 0.0425 \text{ pu} \quad \text{with } K_C = K_{PP} = 0.4 \text{ pu.}$$

For the system of Figure 7.35 the frequency responses for the cases of TGR and PI compensation are shown in Figure 7.36. For comparison, the response of the high-frequency transfer-function model of the system, $W(j\omega_f)$ (7.48), $\omega_f \gg (1/T_F) = 0.5$ rad/s, is also displayed. All three magnitude responses satisfy the transient gain requirement of 30 dB (32 pu) over the selected range of modal frequencies. Because the (upper) corners of the TGR and PI compensation and the feedback time constant are the same in both cases, their phase responses are almost identical for frequencies greater than 0.5 rad/s ($= 1/T_F$). The upper corner frequencies of all three responses lie at $[1 + (KK_CK_F)/T_F]/T = 63$ rad/s, a value greater than $1/T = 20$ rad/s. For the purposes of comparison note that the parameters of the TGR and PI compensation have been chosen such that the product $K.K_C$ in (7.49) is the same in each case.

This example demonstrates that the relationships in (7.49) provide a basis for determining the parameters for rate feedback analysis when coupled with other compensation functions.

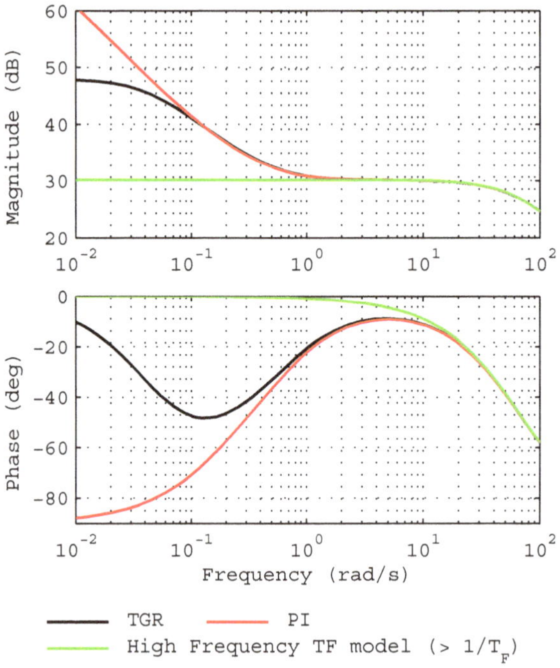

Figure 7.36 Frequency responses for the system of Figure 7.35(a): TGR or PI compensation in the forward loop and, for comparison, the high-frequency transfer-function model, $W(j\omega_f)$ (7.48), valid for $\omega_f \gg (1/T_F) = 0.5$ rad/s.

7.10.3.3 Case 3. Effect of other or additional dynamics in the forward loop

The analysis of rate feedback compensation has been based on the excitation system (ES) configurations of Figures 7.29 and 7.35, (i) without and with TGR or PI compensation and

(ii) when the corner frequency of the first-order 'plant' transfer function, $K/(1+sT)$ in the ES, lies above the range of modal frequencies. It may be that the 'plant' transfer function is of higher order than one, and/or that the additional pole(s) lie in the range of modal frequencies, perhaps associated with higher-order models of linearized excitation systems. Such additional dynamics in the forward loop can be accommodated in the analysis which has been developed in Example 5, below.

7.10.3.3.1 Example 5, Case 3. TGR or PI compensation with rate feedback and additional dynamics in the forward loop

Let us consider Example 4 which includes transient gain reduction or PI compensation with rate feedback. Let us assume that there are additional blocks which introduce poles at -2.5, -20 and -50 Np/s in the 'plant'. The pole at -2.5 is associated with a corner frequency of 2.5 rad/s. Unfortunately, this corner lies in the modal frequency range of 1 - 10 rad/s over which a transient gain of 30 dB is required. Moreover, this pole being associated with a 'plant' parameter may vary somewhat with the plant loading between -2.3 and -2.6 Np/s.

An approach which is adopted for this scenario is to 'cancel' the pole at -2.5 Np/s and shift it to a higher frequency beyond the modal frequency range using the lead-lag transfer function, say $(1+s0.42)/(1+s0.042)$, with corners at 2.38 and 23.8 rad/s [1]. The corners of the modified dynamics all lie above or at that of the 'plant' corner frequency of $1/T = 20$ rad/s and thus the identical design used in Case 2 above is employed, i.e. the parameters of the TGR, PI and rate feedback parameters are the same as in Example 4. The transfer function of the block associated with the 'plant' in Figure 7.35(a) therefore takes the form:

$$\frac{250}{(1+0.05s)} \cdot \frac{1}{(1+0.4s)} \cdot \frac{(1+0.42s)}{(1+0.042s)} \cdot \frac{1}{(1+0.05s)(1+0.02s)}.$$

The frequency response of the closed-loop 'plant' with TGR or PI compensation and rate feedback, with and without the additional dynamics, is shown in Figure 7.37 on page 368.

Figure 7.37 reveals that the magnitude plots with and without additional dynamics agree closely over the range 1-10 rad/s. However, the phase plots start to diverge only at 2 - 3 rad/s and at 10 rad/s there is an additional phase lag of about 20° due to the additional dynamics. The performance of the closed-loop 'plant' can be improved with further fine-tuning.

1. Note: there is not complete cancellation of the pole at -2.5 with the zero at -2.38. For a disturbance to the system the magnitude of the response associated with the almost cancelled pole at -2.5 should be small.

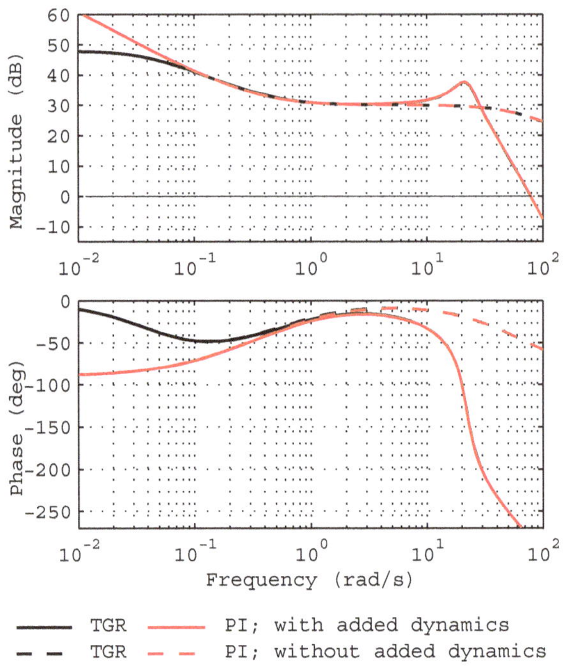

Figure 7.37 Frequency responses for the closed-loop 'plant' of Figure 7.35 for TGR or PI compensation in the forward loop and rate feedback. With additional dynamics the poles lie at -2.5, -20, -23.8, -50, the zero at -2.38 Np/s.

7.10.4 Rate feedback compensation using the Root Locus Method

What is the significance of the root locus method for the purpose of determining the rate-feedback parameters in addition to frequency response techniques? Firstly, if the AVR and/ or the exciter models are of higher order or differ from the form assumed in (7.36) / (7.37), it may be possible to derive initial estimates only of the rate-feedback parameters K_F and T_F from the analysis of Section 7.10.3. Secondly, the robustness or sensitivity of the damping of the poorly-damped closed-loop poles to changes in a parameter value can be assessed The use of a combination of the root locus method and frequency response techniques to fine-tune the estimated parameters may then yield an acceptable set of parameter values.

The following is an unconventional application of the well-known root locus method described in [8] or [9]. The basis for this application of the method to determine how the closed-loop poles of the off-line generating unit under closed-loop voltage control vary as a parameter such as K_F in the feedback path is varied from zero to infinity.

The block diagram of Figure 7.28 is manipulated in several steps into a form that is amenable to determination of the gain K_F using the root locus method, i.e. the gain K_F appears in the forward path of the open-loop system. The desired form of the open/closed loop system is that shown in Figure 7.38. In this form the generator and its voltage control loop become a feedback loop about the AVR and exciter. Note that the closed-loop poles of the transfer functions V_t/V_{ref} and E_f/V_{dum} are identical, however the zeros in the two transfer functions differ. As the terminal voltage reference V_{ref}, shown in Figure 7.28 on page 355, is not relevant to this scenario it is ignored in the root-locus analysis. This is now in the classical form of a closed-loop system for root locus analysis using Matlab®, the gain k being varied over the range $0 \to \infty$. (Note that for the unit on-line a root locus analysis can be conducted using a power system small-signal software package. A succession of eigen-analyses is performed as the gain k in the block diagram of Figure 7.38 is varied over an appropriate range).

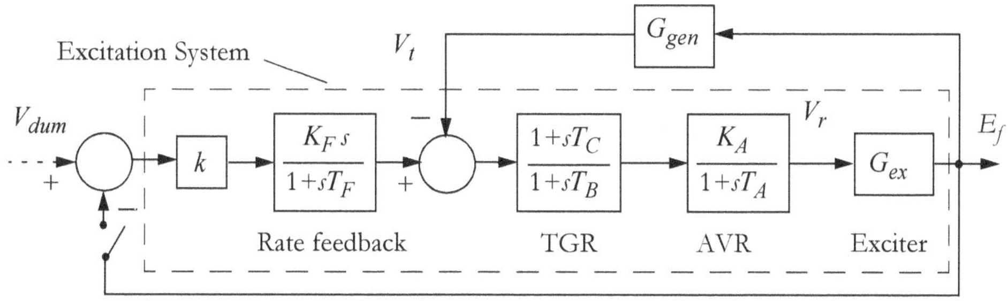

Figure 7.38 Block diagram of a generating unit off-line. An equivalent form of Figure 7.28 for root-locus analysis with a variable gain k.

7.10.4.1 Example 6. Root Locus Method. Application to the generating unit off-line

For the analysis of the off-line performance of the generating unit the same parameters are adopted as in Section 7.10.3.2.1 for TGR compensation with rate feedback of the exciter voltage.

Generator and exciter parameters: $T'_{d0} = 5$ s, $K_E = 1.0$ pu, $T_E = 0.10$ s;
AVR parameters: $K_A = 250.0$ pu, $T_A = 0.05$ s;
Transient gain reduction parameters: $T_C = 5.0$ s, $T_B = 12.5$ s;
Tuning of the feedback block parameters yielded values of $K_F = 0.0425$ pu-s and $T_F = 2.0$ s.

With reference to Figure 7.35, the exciter transfer function $G_{ex}(s)$ is included in the 'plant' transfer function; the feedback signal is the field-voltage perturbation. Because the condition (7.47), $1/T_F \ll \omega_f \ll 1/T_E < 1/T_A$, applies to this scenario, the same values of K_F and T_F apply to this example. Of interest is the effect on the damping of the closed-loop system of changes in K_F. This can be determined from the plot of the root loci shown in Figure 7.39 as the gain k in Figure 7.38 is varied $0 \to \infty$ with K_F set to 0.0425 pu-s. The associated closed-loop poles for $k = 1$, i.e. $k.K_F = 0.0425$, are marked on the plot and are all well damped.

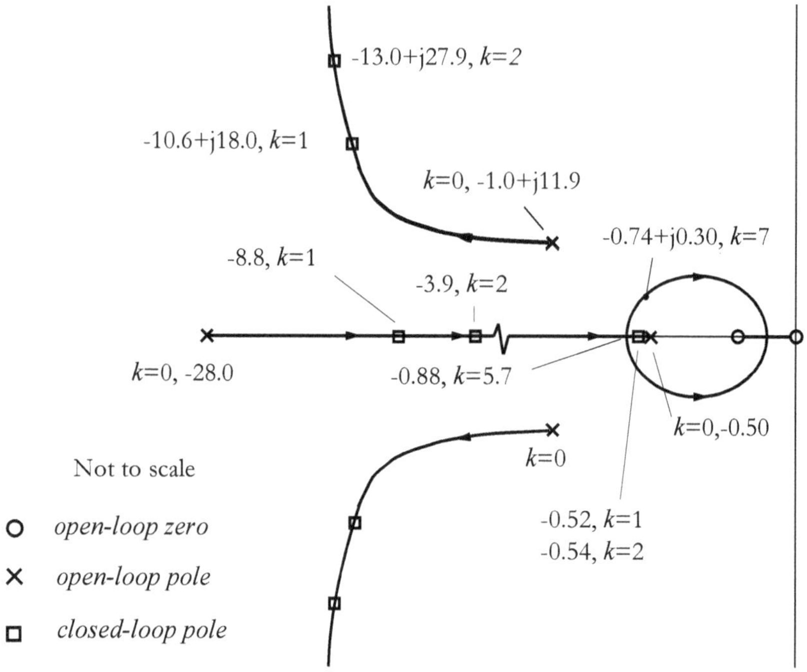

Figure 7.39 Root locus plot for the exciter-voltage feedback gain $k.K_F$, $K_F = 0.0425$ pu-s and $T_F = 2.0$ s. Loci of the closed-loop poles start at the open-loop poles and terminate at the finite or infinite zeros as gain k is varied $0 \rightarrow \infty$.

The closed-loop poles for an increase in the rate-feedback gain of 6 dB ($k = 2$) are also marked on the locus plot; the dominant closed-loop pole at -0.52 Np/s ($k = 1$) is not significantly affected. Furthermore, it is observed that the damping of the dominant pole improves for an increase in gain of 5.7 times (15 dB) before the damping commences to decrease (the damping constant changes from -0.52 Np/s to -0.88 Np/s). For a decrease in gain, say to $k = 1/5.7 = 0.18$, the closed-loop pole lies just to the left of the open-loop pole at -0.50 Np/s; the dominant pole of the off-line generator under closed-loop voltage control is therefore robust to gain variations of ± 15 dB.

It is noted that the corner frequency $1/T_E = 10$ rad/s is a value at the upper end of the modal frequency range. Nevertheless, a frequency response plot of the ES shows that transient gain is held constant at 30 dB from 1.5 to 20 rad/s.

The performance of the generator on-line under closed-loop voltage control with rate feedback compensation is very similar to those studied for TGR and PID compensation. The similarity is a result of selecting the transient gain to be the same ($K_T = 32$ pu) over the selected modal frequency, 1-10 rad/s, in all cases.

7.11 Tuning of AVRs with Type 2B PID compensation in a three-generator system

This section is a continuation of the example described in Section 7.8. Its purpose is

- to demonstrate, in some detail, the more complex tuning of the AVRs;
- to satisfy the dynamic and steady-state performance specifications over a wide range of normal and line-outage operating conditions;
- to analyse the performance
  - of a generator operating off-line at rated speed and under closed-loop voltage control;
  - of one, two or three machines on-line at part and at rated real power output for a range of reactive power generation;
- to include models of the non-linear and the linearized brushless excitation system and to determine the variation of parameters of the linearized model with operating conditions.

7.11.1 The three-generator, 132 kV power system

A power station containing three identical generators, each rated 50 MW 0.85 power factor, is connected by double-circuit 132 kV lines to a high voltage system, represented by an infinite bus, as shown in Figure 7.40.

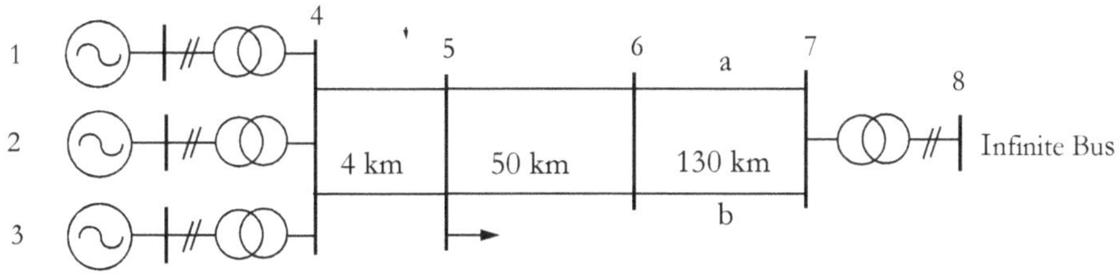

Figure 7.40 The three-generator, 132 kV power system

The parameters of the 5th order, salient-pole generator rated 58.8 MVA and its exciter are listed in Appendix 7–I.1.2. The exciter is an AC generator with a rotating rectifier and is represented by an AC8B Excitation System Model [12] shown in Appendix 7–I.2, Figure 7.49.

Note that the following applies only to the analysis associated with the system shown in Figure 7.40 and the associated AVRs with Type 2B PID compensation.

The PID parameters are to be determined assuming $K_G = 1$, $T_G = 0$.

Transmission line parameters in per unit / 100 km on system base (100 MVA) are:
Z = 0.0632+j0.2347, b = 0.0484.

The load at bus 5 is 50 MW, 10 MVAr lag when it is 'on', zero when 'off'. The most onerous system contingency is the outage of the 130 km line 'a'.

The transformer parameters are:
for each generator: Z = j0.20 pu on 100 MVA, tap range ±10 %.
for the transformer at the Infinite Bus: Z = j0.05 pu on 100 MVA, tap range ±10 %.

The range of operating conditions is summarised in Table 7.9.

For each set of study cases C*1 to C*5 and C*6 to C*10, the real power output of each generator is maintained constant for the five reactive power outputs between 25 Mvar lagging to 20 Mvar leading. That is:

- Output of each generator: 50 MW at: 25, 12.5 Mvar lag, 0 Mvar, and 10, 20 Mvar lead;
- Output of each generator: 25 MW at: 25, 12.5 Mvar lag, 0 Mvar, and 10, 20 Mvar lead;
- Number of generators on-line: one, two or three. Units are equally loaded; unequal loadings are not considered in these studies.

Table 7.9 Power system operating conditions

No.of Units	Power (MW) *	Cases	Load#	Line 'a'	Cases	Load#	Line 'a'
		Mvar: 25 lag to 20 lead	in or out	in or out	Mvar: 25 lag to 20 lead	in or out	in or out
One	50	C01-C05	in	in	C06-C10	in	out
	50	C11-C15	out	in	C16-C20	out	out
	25	C21-C25	out	in	-	-	-
Two	50	C41-C45	in	in	C46-C50	in	out
	50	C51-C55	out	in	C56-C60	out	out
	25	C61-C65	out	in	-	-	-
Three	50	C71-C75	in	in	C76-C80	in	out
	50	C81-C85	out	in	C86-C90	out	out
	25	C91-C95	out	in	-	-	-

* Power output per generator
\# Load: 50 MW 10 Mvar. Line 'a' in or out of service.

The features of this generator-brushless-exciter and power system are: (i) the lines are long with a surge impedance loading (SIL) of 45 MW; (ii) at rated output of the station the lines are heavily loaded (about 1.7xSIL); (iii) with the outage of a line the loading on the second circuit is about 3.4xSIL; (iv) the open-circuit time constants of the generator and exciter are relatively long; (v) the only adjustable parameters in the AVR are those of the PID and the gain K_G; (vi) for planning purposes the halving time [1] of any rotor modes should be less than 5 s. An implication of items (iv) and (v) is that the PID must introduce adequate phase lead at lower frequencies, i.e. about 1 to 4 rad/s. This not only ensures stability but also satisfies a requirement that, for a small step-change in reference voltage, the settling time of the terminal voltage response to lie within a band of ±10 % of its final value in less than 5 s when the generator is on-line; when off-line the 10% settling time is 2.5 s.

To ensure that the tuning of the PID covers a range of operating conditions, the 75 generating/operating conditions shown in Table 7.9 are examined. However, certain system conditions are not credible because 132 kV bus voltages are outside the range of 95-108%, or taps are at their limiting positions; several cases - such as C71 and C90 at maximum lagging or leading reactive power output - are therefore discarded. The terminal voltage of each generator is maintained at 1 pu.

7.11.2 The frequency response characteristics of the brushless exciter and generator

The closed-loop terminal voltage control system of each generator is shown in the block diagram in Figure 7.41; note that the generator model accounts for the effects of the external system when the unit is on-line. Because the models of both the generator and the exciter are non-linear, the parameters of the linearized model will change with conditions at the generator terminals. In order to establish suitable parameters for the excitation control system it is necessary to determine the variation of the generator-exciter characteristics with terminal conditions.

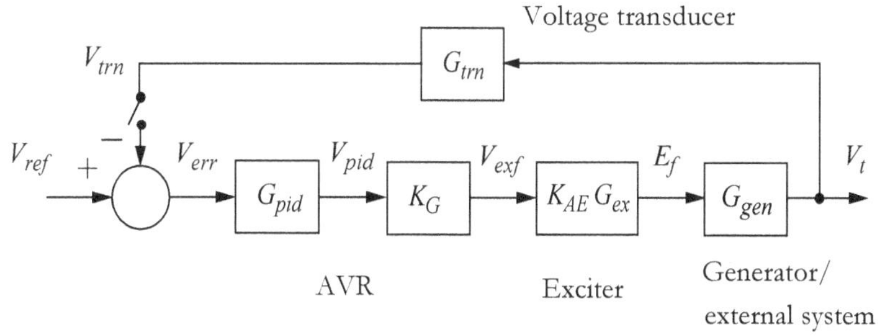

Figure 7.41 Terminal voltage control system. The gain K_{AE} accounts for the per unit system of the excitation control system which includes a brushless exciter.

Over a range of terminal conditions such a characteristic is the frequency response of the generator-exciter transfer function, as measured between the exciter field voltage as input

1. See definition in Section 10.2.2.

and the transducer voltage output, $V_{trn}(j\omega_f)/V_{exf}(j\omega_f)$. The use of the latter transfer function is particularly pertinent to brushless excitation systems in which the exciter output voltage is not accessible for measurement.

A 5th order model of the generator and a non-linear exciter model are available in a small-signal power system dynamic performance package. Such models are automatically linearized at each operating condition by the software. The non-linear and linearized exciter model are shown in Figure 7.49 and 7.50 of Appendix 7–I.2.

Based on (i) the power system of Figure 7.40, (ii) the system and device parameters given in Section 7.11.1, the frequency responses of relevant blocks in the voltage control loop are calculated for selected operating conditions. The set of frequency responses for the generator and exciter, V_{trn}/V_{exf}, are shown in Table 7.42 when either one, two or three generators are on-line; the output of a generator is 25 or 50 MW at 1 pu terminal voltage.

For normal and N-1 operation of this system, operation at lagging power factors is more likely to occur. The selection of the PID parameters may be influenced accordingly.

It is noted from Figure 7.42 that, for feasible cases C01 to C95 the gain in the generator/exciter frequency responses in the region of 1.0 rad/s varies within ±6 dB, and the phase varies by about $\pm 25°$. The variations in the responses over the frequency range are due not only to the range of steady-state conditions at the generator terminals but also to the associated parameter values in the small-signal model of the exciter. An example of the exciter parameters and the steady-state field voltage is illustrated in Table 7.10 of Appendix 7–I.2 for operating conditions C16 to C20 in which a single machine is on-line.

The significance of the phase variation is the following. Let us assume that when the PID is added to the forward loop the gain-cross-over frequency of the Bode plot of $V_{trn}(j\omega_f)/V_{ref}(j\omega_f)$ occurs at 1 rad/s. The gain variation in the generator/exciter frequency responses at 1 rad/s is small but the phase variation remains at about $\pm 25°$. This will result in a similar variation in the phase margin over the range of operating conditions with implications for both stability and transient response to a step change in reference voltage. The Bode plots suggest that, when the units are under closed-loop voltage control, the greater phase lags in the leading power factor cases (i) are not conducive to stability, and (ii) result in the terminal voltage response to step changes in reference voltage being less-well or poorly damped.

To determine an appropriate set of PID parameters for the range of operating conditions let us base the analysis on a condition in the middle of the band of phase variations, say Case C17, in which the output of a single generator is 50MW, 12.5 Mvar lagging; the line ‘a’ is out of service and the load is disconnected.

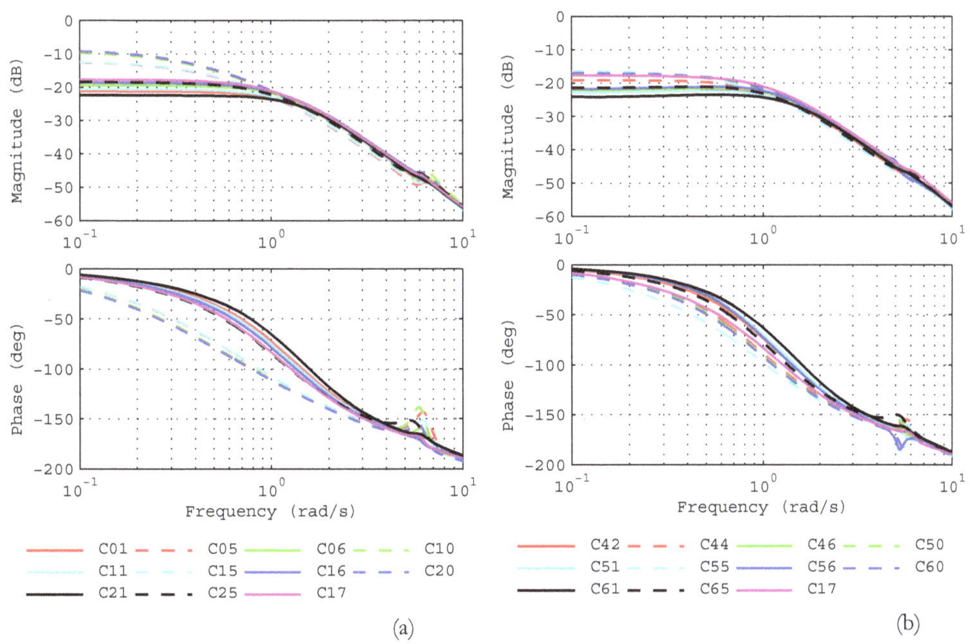

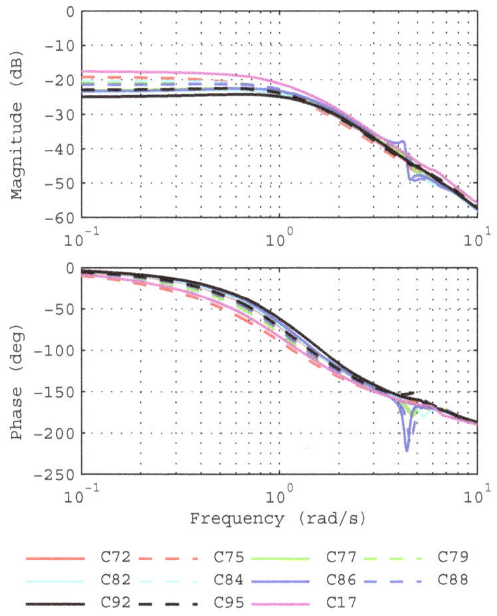

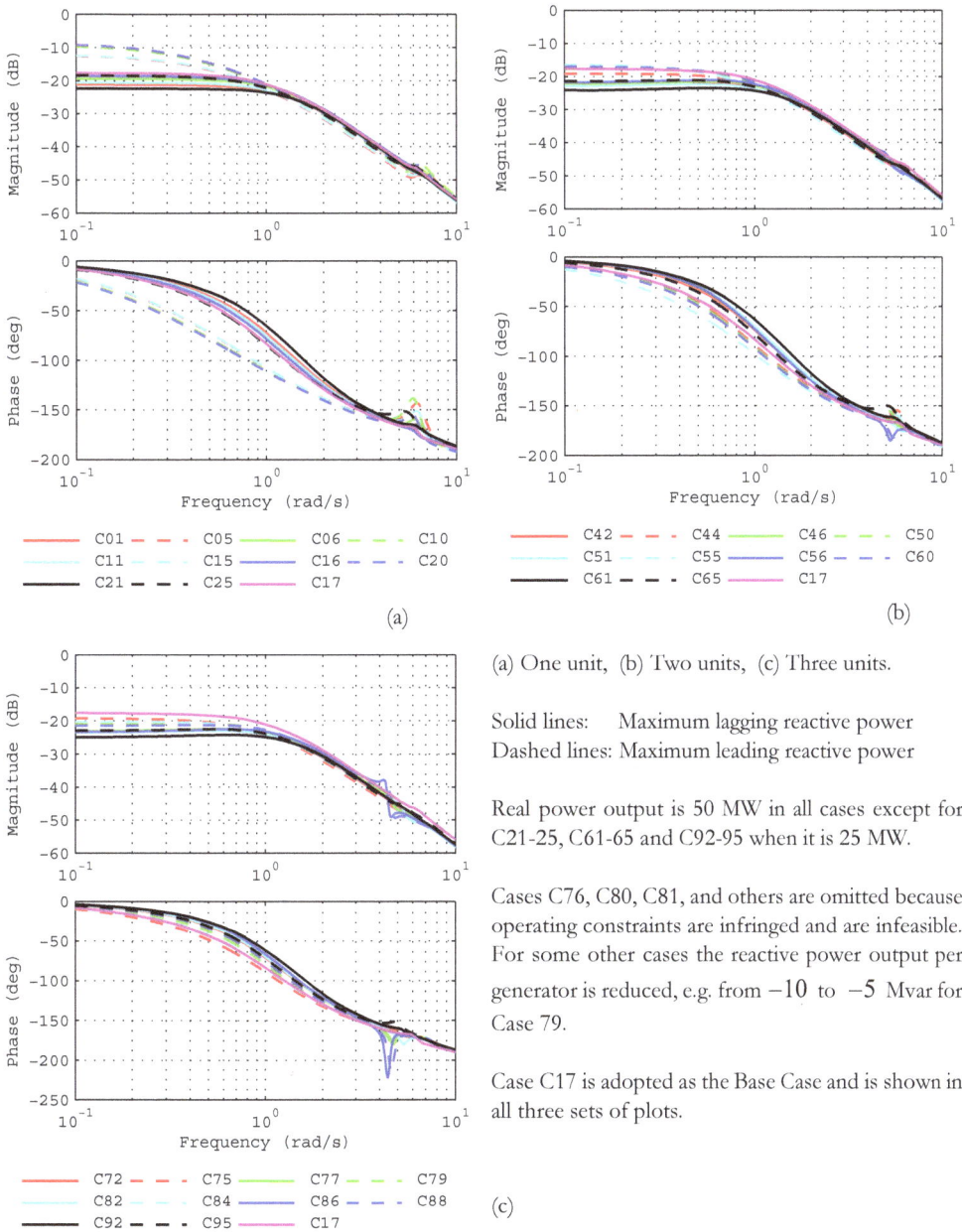

(a) One unit, (b) Two units, (c) Three units.

Solid lines: Maximum lagging reactive power
Dashed lines: Maximum leading reactive power

Real power output is 50 MW in all cases except for C21-25, C61-65 and C92-95 when it is 25 MW.

Cases C76, C80, C81, and others are omitted because operating constraints are infringed and are infeasible. For some other cases the reactive power output per generator is reduced, e.g. from -10 to -5 Mvar for Case 79.

Case C17 is adopted as the Base Case and is shown in all three sets of plots.

(c)

Figure 7.42 Envelopes of frequency responses between the generator terminal voltage transducer and the exciter field voltage, V_{trn}/V_{exf}, for the feasible range of operating conditions shown in Table 7.9

7.11.2.1 Calculation of the PID Type 2B parameters

In order to establish a basis for the compensation to be provided by the PID, let us consider for Case C17 the frequency responses of the generator and exciter, $V_{trn}(j\omega_f)/V_{exf}(j\omega_f)$, and the AVR, $V_{exf}/V_{err} = PID(j\omega_f)$. The PID parameters selected for trial are those for set No. 2 in Table 7.6 on page 350. These two responses are shown in Figure 7.43 together with the phase response of the open-loop transfer function V_{trn}/V_{ref}.

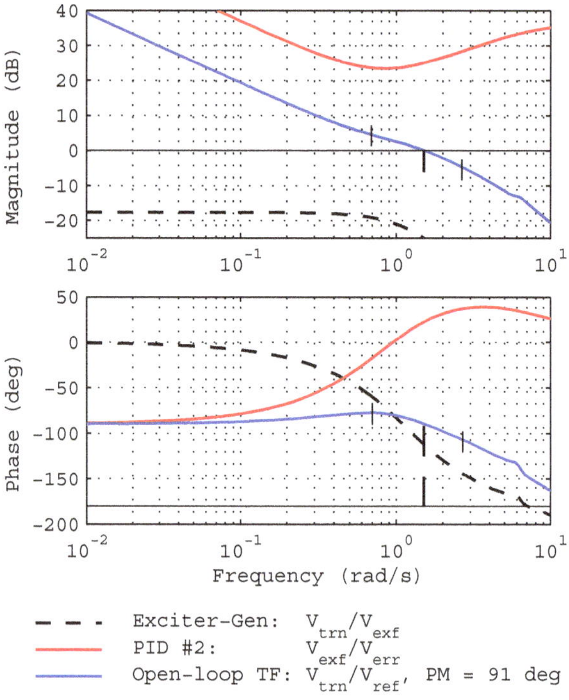

Figure 7.43 Case C17 (one unit): Frequency responses of the component transfer functions in the open-loop system including the PID parameter Set No. 2 (see Table 7.6 on page 350), i.e. K_P = 14 pu, K_I = 7.0 pu/s, K_D = 8.0 pu-s, T_D = 0.143 s, K_G = 1.0. Gain cross-over frequency of the open-loop transfer function is 1.51 rad/s.

From the open-loop transfer function in Figure 7.43 it is noted that (i) the gain-cross-over frequency occurs in the range 0.7 - 2.5 rad/s for which the possible variations in the loop gain lie in the range ±4.5 dB; (ii) the phase margin varies from 103° to 75° over the same frequency range.

In determining appropriate PID parameters the phase margin should be more-or-less constant about the gain-cross-over frequency for robustness to gain variations. Selecting a phase margin of 65°, say, ensures the closed-loop response of terminal voltage to a step change in reference voltage is not significantly over-damped (for large values of the phase margin) or under-damped (for small values of the phase margin). In the case of higher values of loop gain associated with the gain-cross-over frequency exceeding 3 rad/s we note that the closed-loop step response of terminal voltage is likely to contain a damped oscillatory component due to the electro-mechanical modal resonance at 5 to 6 rad/s.

We will therefore examine an approach to derive a more of less constant phase margin of 65° over an appropriate frequency range for Case 17. The 'phase matching' method which achieves this objective is described in Appendix 7–I.5 in which it is shown in Figure 7.55 that the parameter set No. 4 for PID Type 2B in Table 7.6 provides the required phase margin.

The significance of the analysis of the phase margin for the Base Case 17 is revealed in its effect on the terminal voltage response of the closed-loop system due to a step change in reference voltage for the full set of operating conditions. As shown in Figure 7.44 the response of the system incorporating PID Set No. 2 (phase margin 96°) is well damped. However, with PID set 4 (phase margin 66°) a satisfactory, suitably-damped response results. Moreover, the settling-time requirement that the response lies within 10% of its final value within 5 s is satisfied with both PID Sets 2 and 4.

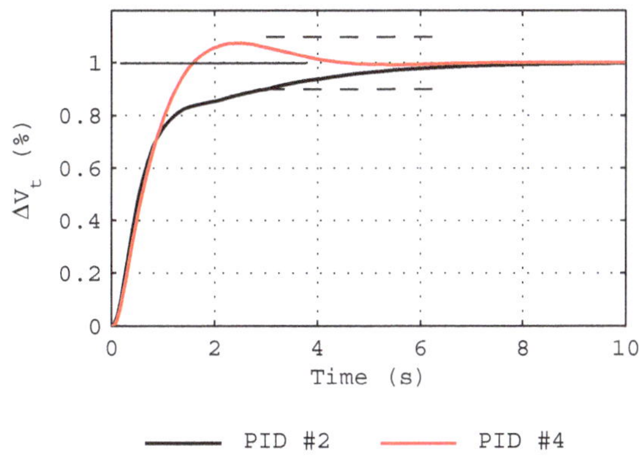

Figure 7.44 Case C17. Single unit only on-line. Perturbations in terminal voltage (V_t) due to a step change in reference voltage from a steady-state value of 1.0 pu to 1.01 pu (1%). PID parameter Sets 2 and 4, Table 7.6. The ±10 % band about the final value is also shown.

The dynamic performance of the single generator with PID Set No. 4, parameter values $K_P = 14$ pu, $K_I = 14$ pu/s, $K_D = 5.89$ pu-s and $T_D = 0.1053$ s, appears satisfactory. The application of this PID set to all the feasible operating cases and conditions for a generator off-line and one, two and three units on-line is now examined.

7.11.2.2 Dynamic performance of a unit off-line under closed-loop terminal voltage control

When the generator is operating off-line at rated speed and under closed-loop voltage control it is required to satisfy the relevant performance specifications. For example, such specifications may require that the measured terminal voltage settles within 10% of the final value in less than 5 s for a step change of 1% in the terminal voltage (see Section 7.4). in the analysis.When the generator is off-line in the following analysis the 10% settling time in terminal voltage for a 1% change in reference voltage is 2.5 s.

In previous examples simple models of the exciter and the generator have been employed, i.e. $1/(K_E + sT_E)$ and $1/(1 + sT'_{d0})$. However, (i) at 1 pu voltage the small-signal gain of the generator is determined by the slope of the saturation curve and is less than unity; (ii) the perturbations in generator field current modulates the generator field voltage by two mechanisms represented in the model of the exciter in Appendix 7–I.2, Figure 7.50. The mechanisms are (i) the effect of the demagnetization term, K_{DE}, and (ii) the non-linear reduction in rectifier average output voltage with increase in the rectifier load, i.e the generator field current. The latter mechanism is represented by the value of the gain K_{CE} and the associated mode of operation of the rectifier.

The linearized model of the off-line, fifth-order salient-pole generator and the exciter are formed automatically. The off-line unit operates a rated voltage and speed. The other elements in the voltage control loop are PID Set No. 4 (see Appendix 7–I.5), the per unitizing gain K_{AE} and the terminal voltage transducer, time constant T_{trn}. The generator and exciter parameters are given in Appendix 7–I.1.2. The Bode Plot of the open voltage-control loop, $V_{trn}(j\omega_f)/V_{ref}(j\omega_f)$, and the associated closed-loop response in generator terminal voltage due to a +1% step in the reference voltage, are shown respectively in (a) and (b) of Figure 7.45.

The closed-loop step response is adequately damped, as predicted by the Bode plot, and satisfies the performance specification.

7.11.2.3 Dynamic performance over a range of operating conditions; one, two and three units on-line based on PID parameter Set No. 4.

The open-loop frequency responses for one, two and three units on-line are examined to derive information on both the damping of the voltage control loop and the stability of the power system under closed-loop conditions. The margins of rotor angle stability under closed-loop conditions are also examined, assuming for planning purposes a 5 s halving time for the dominant mode. Finally, the closed-loop responses of the generator terminal voltage to step changes in its reference voltage are assessed to determine if the requirement that the response lies within 10% of its final value within 5 s is satisfied over the range of operating conditions.

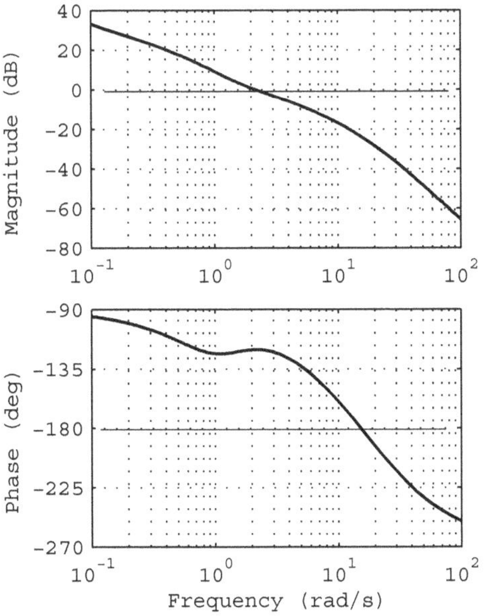

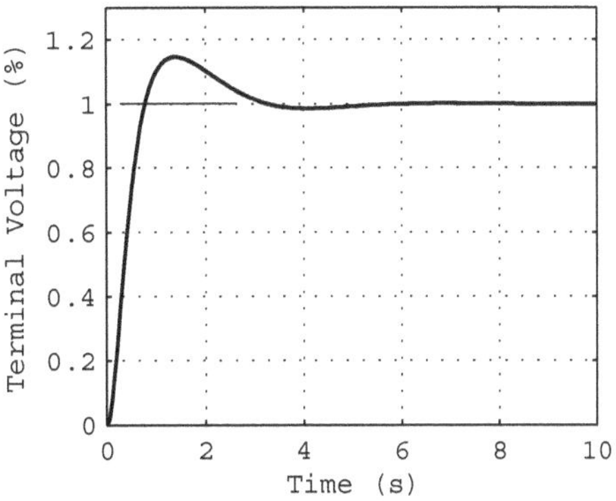

(b) Perturbation in generator terminal voltage due to a step change in reference voltage from a steady-state value of 1.0 pu to 1.01 pu (1%).
10% Settling time < 2.5 s.

(a) Bode Plot: Phase Margin 60 deg at 2.2 rad/s.

Figure 7.45 Generator off-line, operating at rated speed under terminal voltage control with PID Set No. 4. (a) Open-loop Bode Plot. (b) Perturbation in closed-loop terminal voltage step-response.

7.11.2.3.1 Open-loop frequency responses; one, two and three units on-line.

In the following Bode plots for generator #1 the *terminal voltage feedback path is open on that generator, but is closed on the other generators when more than one unit is on-line.* The Bode plots are shown in Figure 7.46 for the cases when one, two or three generators are on-line. When all machines are under closed-loop voltage control these open-loop plots should reveal the nature of (i) the damping in the voltage control loop on generator #1, (ii) the stability of the system, and (iii) the terminal voltage response of generator #1 to a step in its reference voltage. One can equally well apply the above analysis to unit #2 or #3 instead.

Because the Phase Margins derived from the Bode plots in Figure 7.46 are all positive the system is stable over the range of operating conditions. However, the Phase Margins are much less than the desired value of $65°$ at higher values of leading reactive power output, e.g. for C20 the PM is $44°$ at 1.26 rad/s. Thus under leading power factor operation and closed-loop voltage control the system damping is degraded. However, in the cases of one, two or three generators on-line at rated real power output it should be noted that the higher values of leading reactive power output are unlikely to arise in practice. In such cases the reactive power import to the system at the infinite bus is somewhat greater than that absorbed by the generator. For example, in case C55 the output of two units is 100 MW -40 Mvar and the reactive import from the infinite bus is 64 Mvar. Similarly in case C15 for one generator, output 50 MW -20 Mvar, 15 Mvar is imported from the system. Such reactive flows from the real power sink to the real power source are unwarranted and uneconomic - especially for the outage of line 'a'. Thermal limits of transmission lines and transformers, which may be relevant under outage conditions, have been ignored. Clearly, under leading power factor operation the reactive power absorbed by the generators must be limited. Such limits would need to be determined by further studies.

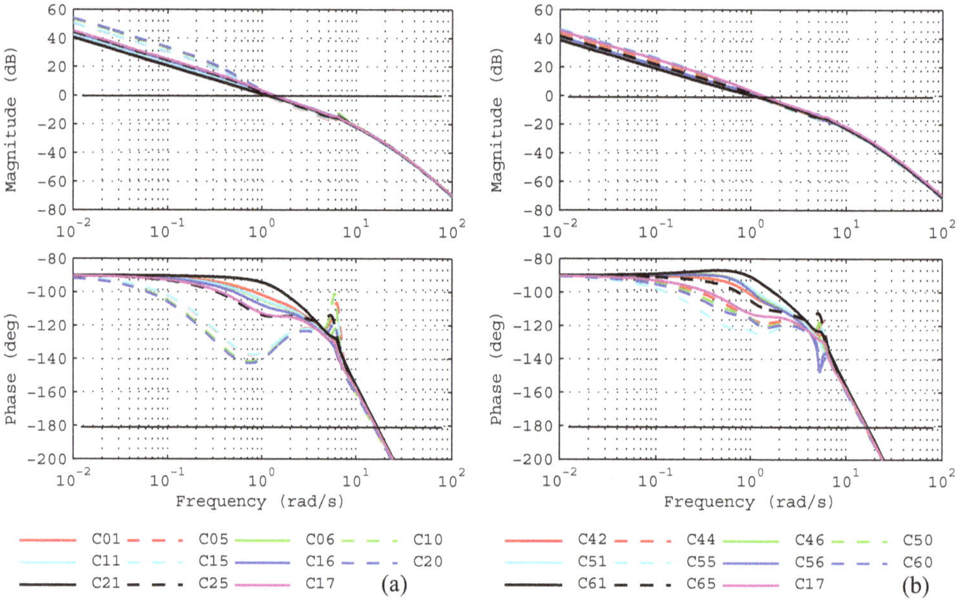

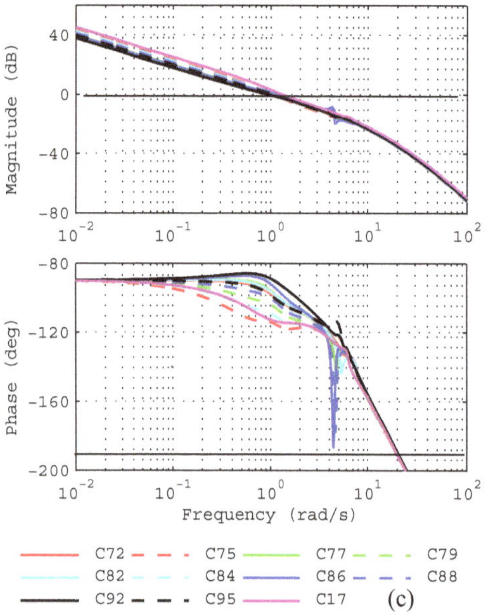

(a) One unit, (b) Two units, (c) Three units.
PID Set No.4 installed on all generators

Solid lines: Maximum lagging reactive power output.
Dashed lines: Maximum leading reactive power output.

Unit real power output is 50 MW in all cases except for C21-25, C61-65 and C92-95 when it is 25 MW.

Cases C76, C80, C81, and others are omitted because operating constraints are infringed and are infeasible. For some other cases the reactive power output per generator is reduced, e.g. from -10 to -5 Mvar for Case 79.

Case C17 is adopted as the Base Case and is shown in all three sets of plots.

Figure 7.46 All feasible Cases, C01 to C95: Bode plots $V_{trn}(j\omega_f)/V_{ref}(j\omega_f)$ of the terminal voltage when the feedback path is open on generator #1 with PID Set No. 4.

7.11.2.3.2 System eigenvalues when generators are under closed-loop voltage control

While the Phase Margins derived from the Bode plots show that the system is stable, an examination of the eigenvalues for the rotor modes reveals the degree of stability of these modes. The most onerous conditions most likely to yield rotor angle instability are the cases for which line 'a' in Figure 7.40 is out of service and the load at bus 5 is off, i.e. the rated output of the station is carried over line 'b'. The local and inter-machine modes for one or more units on-line are seen in Figure 7.47.

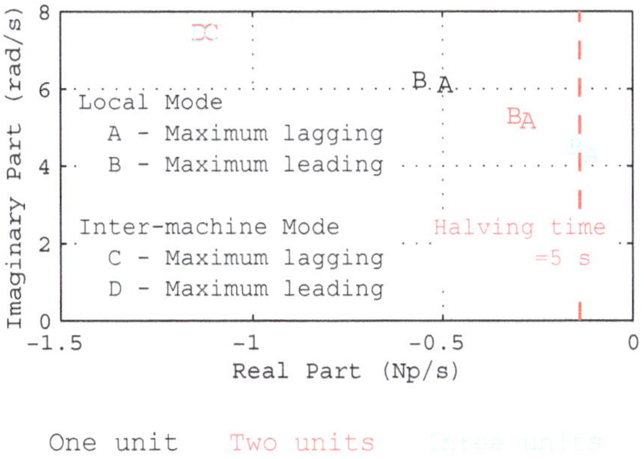

Figure 7.47 Eigenvalues for one, two and three units on-line. Line 'a' is out-of-service and the load at bus 5 is off. Conditions are shown for feasible maximum lagging and leading reactive power outputs at rated real power. PID Set No. 4 installed on all generators.

As shown in Figure 7.47, when three machines are on-line at rated real power output the 5 s halving time is breached, or nearly breached (cases C86-C88). To provide an adequate margin of stability for the most onerous condition it is therefore necessary to install power system stabilizers on the generators. (This is not considered here.)

7.11.2.3.3 Step responses for the range of feasible operating conditions; one, two or three units on-line.

Based on the PID parameter Set 4 in Table 7.6, let us determine the terminal voltage response of the closed-loop system to a +1% step change in reference voltage of generator #1 over the range of operating conditions C01 to C95 considered in Figure 7.42.

The perturbations in the generator # 1 terminal voltage from its initial steady-state value are shown in Figure 7.48. For each of the Case sets in Table 7.9, e.g. C01 - C05, C50 - C60, only the maximum feasible lagging and the maximum leading reactive power cases are plotted. We observe the following.

1. All Cases C01 to C95 satisfy the terminal voltage settling-time criterion. As intended, the choice of Case C17 as the base case results in a satisfactory set of responses. The

overshoot of the terminal voltage response at leading power factors may be of concern when one unit is on-line.

2. In Figure 7.46(a) the phase of the extreme Case C21 (25 MW 25 Mvar lag) is some 20° less than that of Case C17 at a gain cross-over frequency of ~1 rad/s. Consequently the phase margin for Case C21 is likely to be 65+20 = 85°, a value which results in an over-damped response - as is evident in Figure 7.48(a). The converse argument applies to Case C20 (50 MW 20 Mvar lead), i.e. the resulting step response is lightly damped.

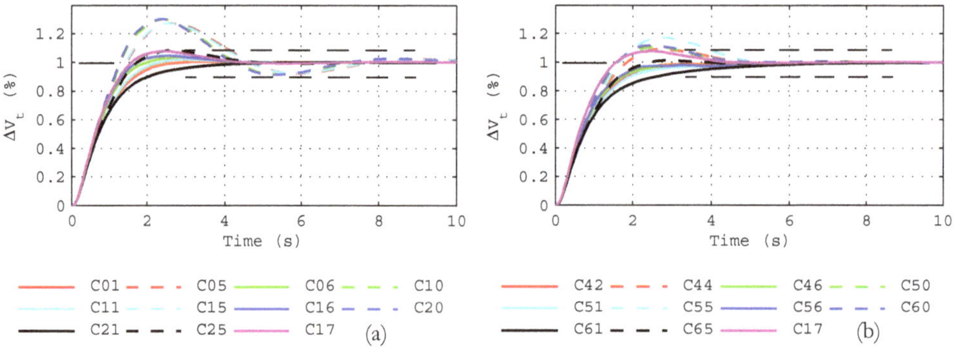

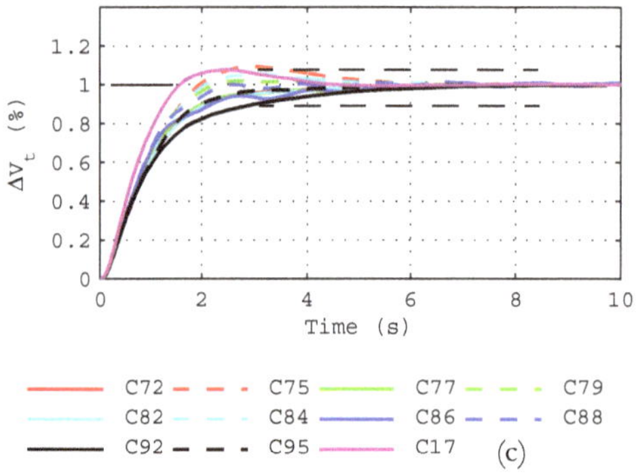

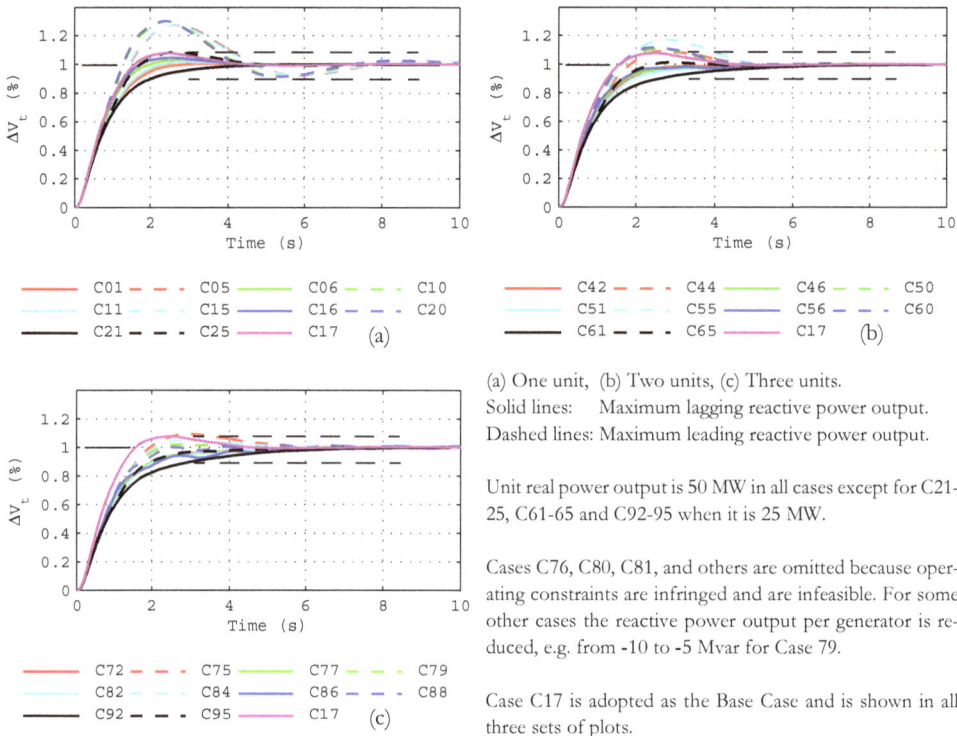

(a) One unit, (b) Two units, (c) Three units.
Solid lines: Maximum lagging reactive power output.
Dashed lines: Maximum leading reactive power output.

Unit real power output is 50 MW in all cases except for C21-25, C61-65 and C92-95 when it is 25 MW.

Cases C76, C80, C81, and others are omitted because operating constraints are infringed and are infeasible. For some other cases the reactive power output per generator is reduced, e.g. from -10 to -5 Mvar for Case 79.

Case C17 is adopted as the Base Case and is shown in all three sets of plots.

Figure 7.48 Closed-loop operation. The perturbations are shown in the terminal voltage of generator #1 due to a step change in reference voltage from a steady-state value of 1.0 pu to 1.01 pu (1%) for the feasible operating conditions. Responses lie within the 10% of the final value of the step amplitude in less than 5 s.

The lack of damping at leading power factors, highlighted in Section 7.11.2.3.1, results in the excessive over-shoot of the terminal voltage responses when one or two generators are on-line. The otherwise satisfactory small-signal performance of the three-generator power system based on parameter set No. 4 for PID Type 2 compensation (Table 7.6) is demonstrated in Figure 7.48 for N and N-1 conditions. Studies examining the provision of PSSs for the generators, the limiting of reactive power absorption by the generators, and the performance of the system for major disturbances would be undertaken in practice, but are beyond the scope of this chapter.

7.12 Summary, Chapter 7

Suppliers of AVRs and excitation systems may have developed tuning techniques which are peculiar to the structure and characteristics of their equipment. The aim of this chapter, however, is to *introduce* some of the basic concepts in the tuning of AVRs which may help others understand some of the approaches which could be used - as well as the relevant control system theory which underpins the analysis. The analysis of the various approaches to tuning are complemented by examples to demonstrate the design procedure and the performance of the type of compensation employed.

The concept of transient gain, which is the effective gain of the excitation system over a selected range of modal frequencies, forms the basis for the various types of compensation. Compensation such as Transient Gain Reduction (TGR) fulfils this objective when high gain excitation systems are required, however the steady-state difference between the desired and the actual terminal voltage following a disturbance may be greater than that specified. The use of PI compensation provides infinite gain at zero frequency and ensures zero error in the terminal voltage in the steady state.

The application of rate feedback of exciter voltage - or AVR output if exciter voltage is not accessible for measurement - is also studied.

To boost the speed of response of the generator field voltage following the occurrence of a disturbance, for example, PID compensation is employed. Assuming a general form of PID compensation given by

$$G_c(s) = \left[\frac{K_I}{s} + K_P + \frac{sK_D}{1+sT_D}\right] \cdot \frac{K_G}{1+sT_G}, \quad \text{((7.2) repeated)}$$

three types are analysed. Type 1: $T_D = 0$, $T_G = 0$, $K_G = 1$,
Type 2A or B: $T_D > 0$, $T_G = 0$, $K_G > 0$, and Type 3: $T_D > 0$, $T_G > 0$, $K_G > 0$.

To determine the parameters in (7.2), the user typically specifies the desired corner frequencies and the transient and high frequency gains required. Except for Type 2B PIDs, the modal frequencies and transient gain lie in a modal frequency range ω_1 to ω_2 .Type 2B PID compensation can be applied to cases in which the PID is required to contribute phase lead at low frequencies. Such circumstances can occur, for example, when the generator and exciter time constants, T'_{do} and T_E, are relatively long.

A detailed example is given of the application of PID type 2B compensation to a remote, three generator system over a range of operating conditions when one, two and three units are in operation. A comprehensive set of studies of the frequency response characteristics of the brushless-exciter-generator system for one or more machines on-line reveals that the closed-loop step responses of terminal voltage are likely to range from poorly to heavily damped. An emphasis in the studies has been to determine a basis for the evaluation of a suitable set of PID 2B parameters to satisfy the dynamic performance specifications over a wide range of operating conditions. The analysis is based on an operating condition which is chosen that 'best' represents those frequency response characteristics over the set of operating conditions.

The aim of new technique called the 'phase matching', explained in Appendix 7–I.5, is to improve the robustness of the generator controls to variations in the gain of the voltage control loop. The studies employing this technique have illustrated the importance of obtaining good models and parameters for both exciter and generator, preferably validated by test.

7.13 References

[1] IEEE Power & Energy Society, *IEEE Tutorial Course: Power System Stabilization Via Excitation Control*, 09TP250, 2009.

[2] IEEE Standard 421.1, *IEEE Standard Definitions for Excitation Systems for Synchronous Machines*, 2007.

[3] IEEE Standard 421.2, *IEEE Guide for Identification, Testing, and Evaluation of the Dynamic Performance of Excitation Control Systems*, 1990.

[4] The Mathworks, Inc. (2010, 23 August 2010). *MATLAB®*. Available: www.mathworks.com.

[5] D.J. Vowles and M.J. Gibbard, *Mudpack User Manual: Version 10S-03*, School of Electrical and Electronic Engineering, The University of Adelaide, July 2014.

[6] P. Kundur, M. Klein, G. J. Rogers, and M. S. Zywno, "Application of power system stabilizers for enhancement of overall system stability," *Power Systems, IEEE Transactions on*, vol. 4, pp. 614-626, 1989.

[7] P. Kundur, *Power System Stability and Control,* McGraw-Hill, Inc., 1994.

[8] Gene Franklin, J.D. Powell and Abbas Emami-Naeini, *Feedback Control of Dynamic Systems,* 6th edition, Pearson Higher Education Inc., 2009.

[9] R.C. Dorf and R.H. Bishop, *Modern Control Systems,* 11th edition, Prentice Hall, 2008.

[10] F.P. de Mello and C. Concordia, "Concepts of Synchronous Machine Stability as Affected by Excitation Control", *IEEE Trans., vol. PAS-88*, pp. 316-329, April 1969.

[11] Australian Energy Market Commission, *National Electricity Rules,* Version 55, March 2013, www.aemc.gov.au

[12] "IEEE Recommended Practice for Excitation System Models for Power System Stability Studies," IEEE Std. 421.5-2005 (Revision of IEEE Std. 421.5-1992), pp. 0_1-85, 2006.

[13] R.J. Koessler, "Techniques for Tuning Excitation System Parameters", *IEEE Trans. on Energy Conversion*, Vol. 3, No. 4, December 1988.

[14] F.G. Shinskey, *Process-Control Systems, Applications, Design and Tuning.* 4th Edition 1996, McGraw-Hill, New York.

[15] K. Astrom and T. Hagglund, *PID Controllers: Theory, Design and Tuning.* Second Edition 1995, Instrument Society of America, Research Triangle Park, North Carolina.

Appendix 7–I

App. 7–I.1 Generator and exciter parameters

App. 7–I.1.1 Parameters for the 6th order generator and a simple exciter

Note. These models are only used in the application of all types of compensation except that in Section 7.8 and 7.11 for Type 2B PIDs.

Generator model: 6th order, classical. Saturation ignored. Values in pu on machine MVA base (the MVA base is stated in the application). (These parameters are the same as those used for generator TPS_4 in Table 10.23 on page 527). System frequency is 50 Hz.

$D = 0$, $H = 2.6$ s, $r_a = 0$, $x_d = 2.3$, $x_q = 1.7$, $x'_d = 0.30$,

$T'_{d0} = 5.0$ s, $x_l = 0.2$, $x_q' = 0.40$, $T_{q0}' = 2.0$ s, $x''_q = 0.25$, $T''_{q0} = 0.25$ s,

$x_d'' = 0.25$, $T_{d0}'' = 0.03$ s.

Exciter: Simple linear first-order lag model: $K_E = 1.0$ pu, $T_E = 0.1$ s.

App. 7–I.1.2 Parameters for the 5th order salient-pole generator and a brushless AC exciter

Note. These models are only used in Section 7.8 and 7.11 for the application of the analysis of Type 2B PIDs.

Generator model: 5th order, salient-pole generator; saturation is included. All values are in per unit on machine rating (58.8 MVA) unless otherwise stated. System frequency is 50 Hz.

$D = 0$, $H = 5.5$ s, $r_a = 0$, $x_d = 1.5$, $x_q = 0.7$, $x'_d = 0.22$,

$T'_{d0} = 8.0$ s, $x_l = 0.10$, $x''_q = 0.16$, $T''_{q0} = 0.12$ s, $x''_d = 0.16$, $T''_{d0} = 0.04$ s,

$S_d(1.0) = 0.15$, $S_d(1.2) = 0.45$.

The exciter is an AC generator with a rotating rectifier and is represented by an AC8B Excitation System Model [12] and is shown in Figure 7.49. Its parameters are:

$K_E = 1.0$ $T_E = 0.7$ s, $K_{CE} = 0.1$, $K_{DE} = 1.25$, $K_{AE} = 1.75$, $T_A = 0$.

The terminal voltage transducer is represented by a first-order lag block, $T_{trn} = 20$ ms.

App. 7–I.2 Models of the brushless AC exciter

The model of the brushless AC exciter is shown in Figure 7.49.

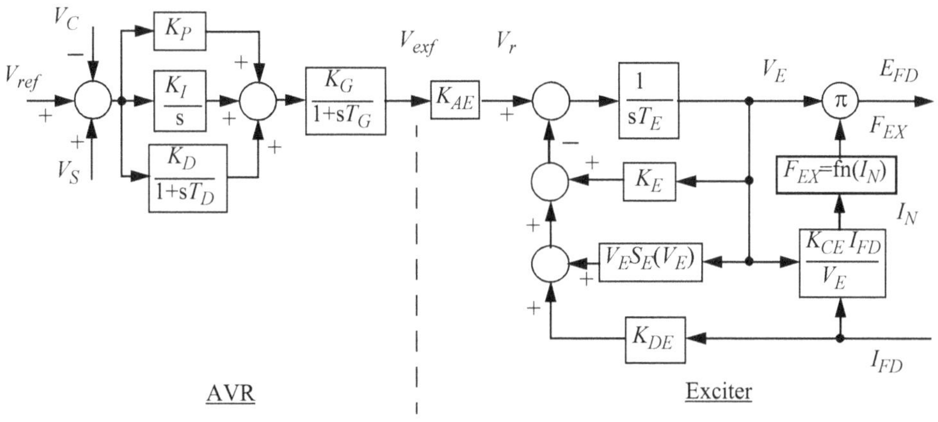

Figure 7.49 The brushless AC exciter is based on the AC8B Excitation System Model [12]

The rectifier regulation modes are expressed by the three equations:

$$F_{EX} = f(I_N) = \begin{cases} (1.0 - 0.577 I_N) & I_N \le 0.433 \\ \sqrt{0.75 - I_N^2} & 0.433 < I_N < 0.75 \\ 1.732(1.0 - I_N) & 0.75 \le I_N \le 1.0 \end{cases}$$

The AVR comprises the Type 2B PID ($T_G = 0$) in which the gain K_G may represent power amplification. The gain K_{AE}, included in the exciter model, is a factor which accounts for the per unitization of exciter and generator quantities.

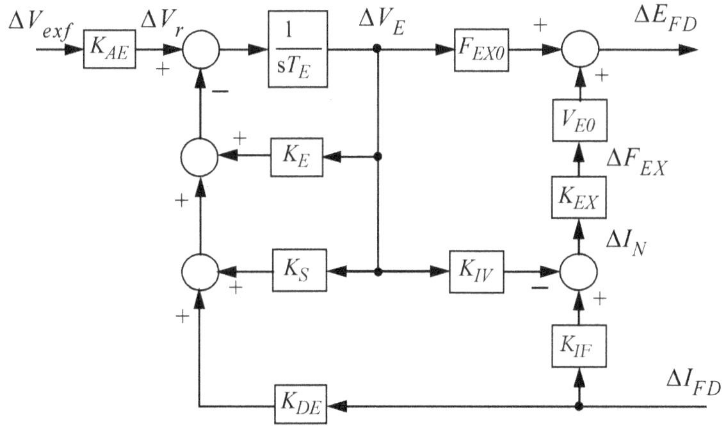

Figure 7.50 Small-signal model of the AC8B Excitation System Model

The parameters in the model are defined as follows:

$$K_{EX} = \begin{cases} -0.577 & I_{NO} \le 0.433 \\ -\left(\dfrac{I_{NO}}{\sqrt{0.75 - I_{NO}^2}}\right), & 0.433 < I_{NO} < 0.75, \\ -1.732 & 0.75 \le I_{NO} \le 1.0 \end{cases} \quad \left(K_{IV} = \frac{K_{CE} \cdot I_{FD0}}{V_{E0}^2}\right), \text{ and } K_{IF} = \frac{K_{CE}}{V_{E0}}.$$

The gain K_S is related to the saturation function of the exciter and is dependent on the initial steady-state value of the field voltage V_{E0}. It is given by the following expression:

$$K_S = S_E(V_{E0}) + \left(V_{E0} \times \left.\frac{\partial S_E(V_E)}{\partial V_E}\right|_{V_E = V_{E0}} \right).$$

Saturation in the exciter is assumed to be negligible under steady-state operating conditions and thus $K_S = 0$.

As an example, the values of the parameters of the linearized exciter model for cases C16 - C20 are provided in Table 7.10. The parameters of the exciter model are listed in Appendix 7–I.1.2.

Table 7.10 Parameters of the small-signal model of the exciter

Case	E_{F0}	K_{IF}	K_{IV}	V_{E0}
C16	2.16	0.044	0.041	2.26
C17	1.92	0.049	0.046	2.03
C18	1.61	0.059	0.056	1.70
C19	1.37	0.069	0.065	1.45
C20	1.20	0.079	0.075	1.27
K_{EX} = -0.577, F_{EX0} = 0.945, K_S = 0				

App. 7–I.3 PI Compensation using positive feedback

A simple positive feedback implementation can be employed for PI Compensation. The diagram of the associated control blocks is shown in Figure 7.51.

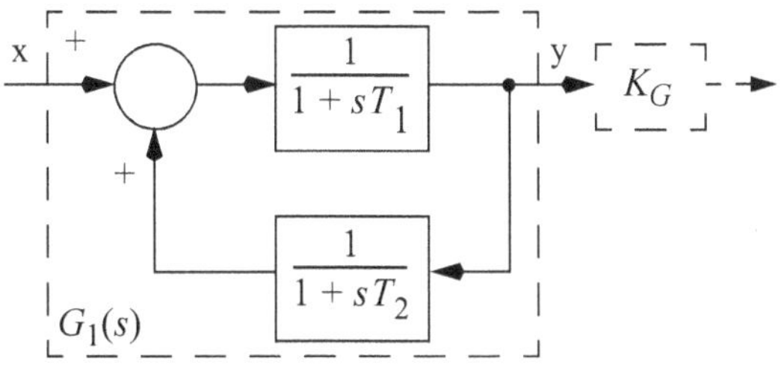

Figure 7.51 PI Compensator

The transfer function of the system from x to y, i.e. excluding the gain K_G, is:

$$\begin{aligned} G_1(s) &= \frac{1+sT_2}{s(sT_1T_2+T_1+T_2)} \\ &\approx \frac{(1+sT_2)}{T_2 s(1+sT_1)}, \quad \text{for } T_2 \gg T_1 \end{aligned} \tag{7.50}$$

where, for the approximation $T_2 \gg T_1$, the low and high frequency corners are $1/T_2$ and $1/T_1$ rad/s, respectively. These corners should be respectively about a decade or more below and above the extremes of the range of frequencies of rotor oscillations.

Equation (7.50) can be rearranged into the following form representative of the PI structure:

$$\begin{aligned} G_1(s) &= \frac{1}{T_1+T_2}\cdot\frac{1}{s}+\left(\frac{T_2}{T_1+T_2}\right)^2\frac{1}{1+s[(T_1T_2)/(T_1+T_2)]} \\ &\approx \frac{1}{T_2}\cdot\frac{1}{s}+\frac{1}{1+sT_1}, \quad \text{for } T_2 \gg T_1 \end{aligned}$$

For the compensator $G_1(s)$ the effective integrator gain is $\approx 1/T_2$ and the proportional gain is ≈ 1 over the range of rotor frequencies. The respective gains of the compensator $K_G G_1(s)$ become $\approx K_G/T_2$ and $\approx K_G$ if $K_G \neq 1$.

A plot of frequency responses of $G_1(s)$ is shown in Figure 7.52 for a range of values of T_2. As explained below, the value of T_1 is such that the phase angle approaches zero degrees in the mid-range of rotor frequencies, e.g. 4 rad/s.

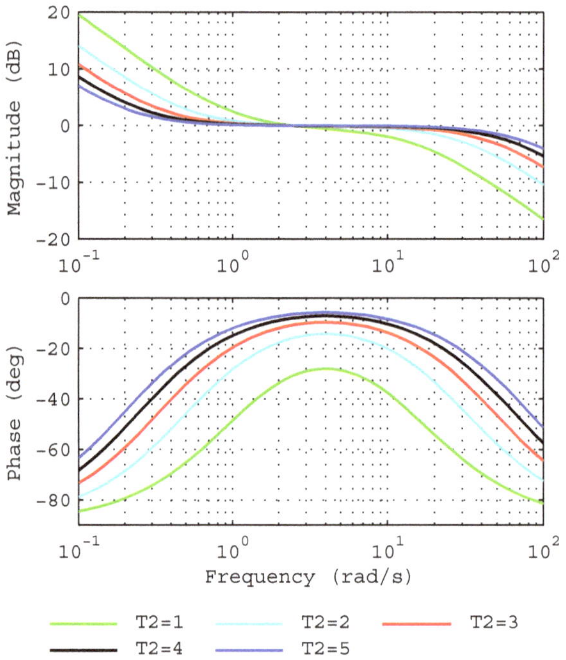

Figure 7.52 Frequency responses of the PI compensation using positive feedback for values of T_2 from 1 to 5 s; associated values of T_1 are such that the maximum phase angle is at 4 rad/s.

For design of the parameters of the PI compensator it may be desirable to place the phase angle characteristic such that phase is close to zero degrees over the range of rotor modal frequencies. From the figure we note that we can choose a frequency at which the phase angle is a maximum. This frequency, ω_{mx}, occurs at the geometric mean of the corner frequencies of the exact transfer function (7.50), i.e.

$$\omega_{mx} = \frac{1}{T_2}\sqrt{\frac{T_1 + T_2}{T_1}}. \tag{7.51}$$

Thus, given T_2 and ω_{mx}, the value of T_1 can be derived from (7.51) to yield:

$$T_1 = T_2/(T_2^2\omega_{mx}^2 - 1) \approx 1/(T_2\omega_{mx}^2) \text{ for } T_2 \gg T_1. \tag{7.52}$$

Based on (7.50) and (7.51), the value of the phase characteristic $G_1(j\omega_f)$ when $\omega_f = \omega_{mx}$ is

$$\begin{aligned} \angle G_1\big|_{\omega_{mx}} &= \operatorname{atan}[\omega_{mx}T_2] - 90° - \operatorname{atan}[(\omega_{mx}T_1T_2)/(T_1 + T_2)] \\ &= -2\operatorname{atan}[\sqrt{T_1/(T_1 + T_2)}] \end{aligned} \tag{7.53}$$

Figure 7.52 is based on the selection $\omega_{mx} = 4$ rad/s with T_2 being varied from 1 to 5 s; the corresponding values of T_1 and $\angle G_1(j\omega_f)$ at ω_{mx} are calculated from (7.52) and (7.53), respectively.

App. 7–I.4 Integrator Wind-up Limiting

Two types of limiter, anti-windup [1] and windup, are encountered in excitation system models. Examples of these types are shown in Figure 7.53 in the case of a simple integrator. The upper and lower limits are *UL* and *LL*, respectively.

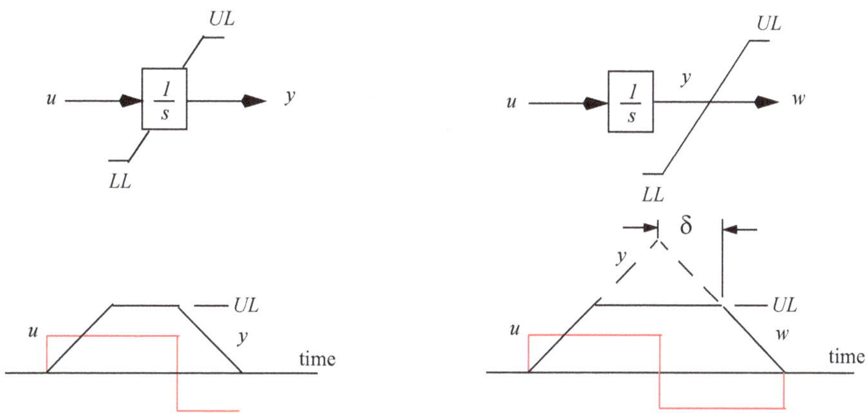

If $UL \geq y \geq LL$ then $dy/dt = u$
If $y \geq UL$ and $u \geq 0$ then set $dy/dt = 0$
If $y \leq LL$ and $u \leq 0$ then set $dy/dt = 0$

(a) Integrator with anti-windup limiting

If $UL \geq y \geq LL$ then $w = y$
If $y > UL$ then $w = UL$
If $y < LL$ then $w = LL$

(b) Integrator with windup limiting

Figure 7.53 Integrator with (a) anti-windup limiting, and (b) windup limiting.

The operation of the two types of limiters are illustrated in principle in Figure 7.53, (a) and (b). In illustration (b), with windup limiting, the output of the integrator $y(t)$ continues to increase once the limit *UL* is reached but starts to decrease only when the input $u(t)$ changes sign. Limiting ceases only when the output y falls below *UL*. With anti-windup limiting, however, it ceases limiting as soon as the input changes sign. The advantage of anti-windup limiting is that it eliminates the time delay δ between sign reversal and wind-down to *UL* that occurs in windup limiting.

Anti-windup and windup limiting occur in other types of transfer function blocks incorporating lead-lag and PI compensation for example (see [7], [12]).

---

1. Anti-windup limiting is also known as "non-windup" limiting ([7], [12]).

App. 7–I.5 A 'phase-matching' method for constant phase margin over an appropriate frequency range

Consider the phase responses of $V_{trn}(j\omega_f)/V_{exf}(j\omega_f)$ for Case C17 and the parameter Set No. 2 for PID Type 2B in Table 7.6; the responses are shown in Figure 7.54.

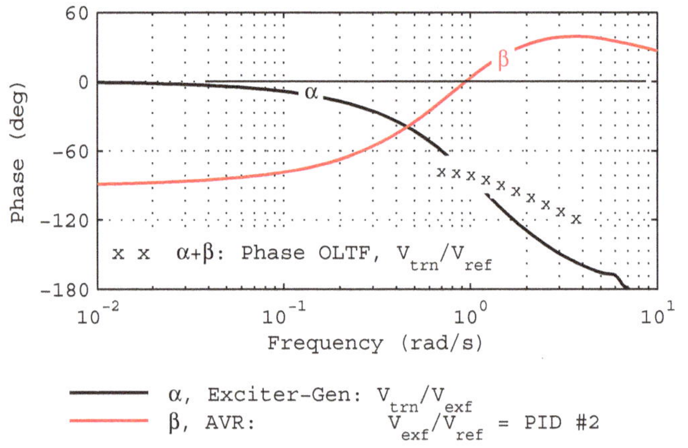

Figure 7.54 Case C17: Frequency responses of the phase of the component transfer functions of the open-loop system. The phase response of the open-loop transfer function with PID Set No. 2 is shown by x-x-x.

Let $\alpha(j\omega_f)$ and $\beta(j\omega_f)$ (deg.) be the phase responses of the generator-exciter and the PID with parameter Set No. 2, respectively, as shown in Figure 7.54. The phase of the open-loop transfer function is $\alpha + \beta$ shown by x-x-x in the figure. Depending on the location of the gain-crossover in the range 0.7 - 2.5 rad/s, the phase margin is the difference between the open-loop phase response $(\alpha + \beta)$ and $-180°$. At the gain-crossover-frequency, $\omega_f = \omega_c$, the phase margin is

$$PM = \alpha_c + \beta_c - (-180°) = \alpha_c + \beta_c + 180° . \tag{7.54}$$

At low frequencies $\alpha = 0°$ and, for the PID, $\beta = -90°$. Let

$$\beta' = \beta + 90° \tag{7.55}$$

so that at low frequencies both $\beta' = 0°$ and $\alpha = 0°$.

Assume the desired phase margin is PM_{des}, e.g. 65°. The required values of β', based on (7.54) and (7.55), are

$$\beta'_{des} = PM_{des} - \alpha - 90° \tag{7.56}$$

(If the actual value of β' is greater than the desired value of β'_{des} then the phase margin is greater (i.e more stable) than the desired phase margin PM_{des}; and vice-a-versa.)

Thus, in order to match the phase margin with the desired phase margin it is necessary to find the PID frequency response, β', that closely matches the line β'_{des} over the potential range of gain-crossover-frequencies. Let

$$\alpha' = -\alpha \text{, thus} \tag{7.57}$$

$$\beta'_{des} = PM_{des} + \alpha' - 90° . \tag{7.58}$$

In order to illustrate a design procedure based on (7.58) let us consider the following steps.

1. Given a selected system operating condition, choose (i) a set of parameters for a Type 2B PID as in Table 7.6, (ii) the design case C17 for the generator-exciter transfer function $V_{trn}(j\omega_f)/V_{exf}(j\omega_f)$ (see Figure 7.42), and (iii) set PM_{des} to $65°$, say.

2. Plot (i) α', the negated phase angle of the transfer function $V_{trn}(j\omega_f)/V_{exf}(j\omega_f)$ for the selected operating condition, (ii) β', the phase angle β of the Type 2B PID advanced by $90°$, and (iii) the line showing where the response of β' must lie with respect to the plot of α' to satisfy the Phase Margin requirement, $\beta'_{des} = PM_{des} + \alpha' - 90° = \alpha' - 25°$. The plots of α' and $\beta'_{des} = \alpha' - 25°$ is shown by 'x x x' in Figure 7.55

3. Based on the plot in Step 1 adjust the PID parameters systematically so that the desired phase margin is satisfied, i.e. plots of β' and β'_{des} match closely - or overlap -over the desired frequency range.

4. Check that the resulting PID satisfies the system performance criteria over the range of operating conditions in which one or more units are on-line.

Let us consider the determination of the PID parameters based on the above steps.

For Step 1 the system operating condition Case C17 and a set of parameters have already been selected for the analysis associated with Figure 7.54. The parameters are those in Set No. 2, Table 7.6, $K_P = 14$ pu, $K_I = 7.0$ pu/s, $K_D = 8.0$ pu-s, $T_D = 0.143$ s, $K_G = 1.0$. Let us base our analysis in this step on this set of PID parameters and Case C17.

The plots of α' and β' associated with the transfer function $V_{trn}(j\omega_f)/V_{exf}(j\omega_f)$ for Case C17 and the PID parameter set, respectively, are shown in Figure 7.55. Also shown is a plot (x x x) along which the angle β' of the desired PID must lie in order for the open-loop transfer function V_{trn}/V_{ref} to have the desired phase margin (assuming for this study that the gain-cross-over frequency for the resulting open-loop transfer function (OLTF) lies in the range 0.7 to 2.5 rad/s).

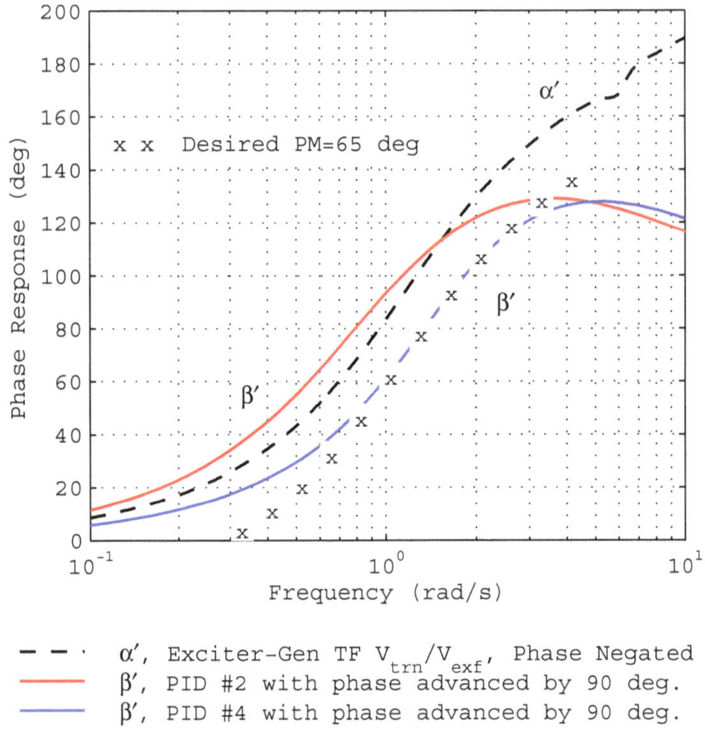

Figure 7.55 Plots of α' for exciter-generator transfer function (TF) for Case C17, and β' of the phase-advanced PID TF together with the plot of $\beta'_{des} = \alpha' - 25°$ (x x x) which represents the desired location of the phase plot of the PID TF.

It is clear from Figure 7.55 that PID parameter Set No. 2 produces excessive phase lead and therefore the phase margin of the OLTF is greater than the desired value $PM_{des} = 65°$. Referring to Table 7.6 or Figure 7.25 it is seen that, by increasing the values of the corner frequencies ω_1, ω_2 and ω_D for the PID sets, the plot of β' in Figure 7.55 approaches the desired phase margin plot. For PID Set No. 4 with parameters $K_P = 14$ pu, $K_I = 14$ pu/s, $K_D = 5.89$ pu-s, $T_D = 0.105$ s, $K_G = 1.0$, the plot of β' coincides with desired phase margin plot for gain-cross-over frequencies in the range 0.9 to 2.5 rad/s.

Based on the PID parameter Sets 2 and 4, the composite OLTF for Case C17 is plotted in Figure 7.56. From this Bode plot it is observed that (i) the gain cross-over frequencies for the two sets are 1.5 and 1.3 rad/s, respectively, (ii) with parameter Set 4 the phase margin is close to the desired value of 65°. The phase margin variations for PID Sets 2 and 4 for a loop-gain variation of ±6 dB are shown in Table 7.11.

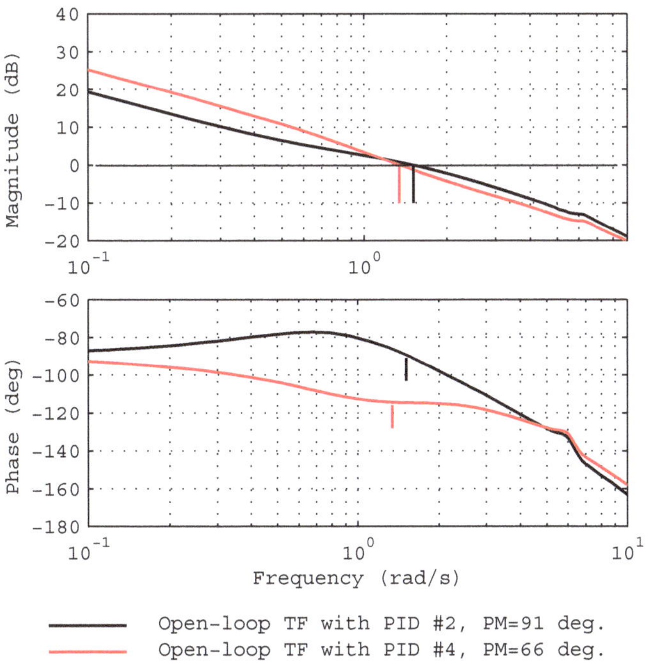

Figure 7.56 Case C17. Bode Plots of the OLTF comprising $V_{trn}(j\omega_f)/V_{ref}(j\omega_f)$ and PID Sets 2 or 4.

Table 7.11 Case C17. Robustness: Phase Margin variation for loop-gain change of ±6 dB

Set	Gain change (dB)	Phase Margin at frequency of ...		Gain change (dB)	Phase Margin at frequency of ...	
		PM (deg)	Frequency (rad/s)		PM (deg)	Frequency (rad/s)
2	–6	102	0.54	+6	69	3.0
4	–6	70	0.80	+6	64	2.4
Desired phase margin is 65°.						

From Figure 7.56 and Table 7.11, it is evident that the phase margin variation of about 6° associated with PID Set No. 4 implies it is robust to variation in the loop gain; this is revealed by the relatively small changes in its phase in the figure. In Set 2, however, not only is the phase margin variation considerably more but the phase margin exceeds the desired value of 65° over the gain variation of ±6 dB.

Chapter 8

Types of Power System Stabilizers

8.1 Introduction

In Chapter 5 a speed-PSS based on the P-Vr design approach is described. The purpose of this chapter is to describe in detail the theoretical basis for some of the widely deployed types of PSSs and the associated practical implications. For some other types of PSSs, including the multi-path, multi-band PSS developed by Hydro-Québec, only a brief overview is provided. Furthermore, the details of a number of other types of PSSs and their development are omitted from this book, for example: delta-omega stabilizers (without and with torsional filters) [1]; the use of notch filters to attenuate the first torsional mode [2]; the application of the coordinated AVR/PSS, called the "Desensitized Four Loops Regulator" [3].

The input to the PSS in Chapter 5 is assumed to be the 'true' rotor speed as measured directly by a high-fidelity tacho-generator, a toothed wheel, or some other device mounted on the shaft of the turbine-generator unit. In practice there may be physical difficulties in positioning any such device on the shaft as well as locating it to minimize the introduction of the torsional modes of the shaft into the speed signal. Moreover, other difficulties such as noise, lateral shaft movement (runout or 'wobble' [4]) in vertical units, may present themselves. In this chapter, however, synthesized speed perturbations, which are assumed to accurately represent the true rotor speed perturbations, are used as the input to the PSS. This means that the same basis and procedure as that outlined in Chapter 5 can be employed for the design and tuning of the PSS.

The major factor in the selection of a stabilizing signal for input to the PSS is the requirement that the modes of concern, which may be the local-, inter-area, and possibly the intra-station modes, must be observable by the signal over a wide range of operating conditions. Typically, perturbations in rotor speed, the electric power output, and the frequency at the generator terminals are the commonly-used local signals.

Various types of pre-filters are in use which convert one or more signals derived from variable(s) other than speed into a synthesized speed signal. Such variables are electric power, bus-voltage angle, frequency, terminal voltage and current; some manufacturers develop a 'speed' signal from such variables using various techniques. Lack of fidelity and resolution of the synthesized speed signal in representing the 'true' rotor speed are factors that result in degradation in the performance of the PSS when the design is implemented in practice. This chapter considers the design of the pre-filters which synthesize a speed signal, and highlights some issues which may be detrimental to the performance of the resulting pre-filter and speed-PSS.

The pre-filters which are discussed in the following sections employ as input signals:

- the electric power output of the generator,
- frequency (or the deviation of the frequency from its nominal value) at the generator terminals [5], and
- electric power and a 'speed' deviation signal in the widely-used 'integral-of-accelerating-power' pre-filter [7], [8], [9].

Some of the practical issues concerning different types of PSSs, field testing and other aspects are covered in [10], [11]. In practice, the engineer who is responsible for tuning the PSS does not often have the ability to influence the selection of the type of PSS. This chapter is intended to provide the reader with approaches to tuning PSSs in circumstances where 'ideal' performance is not possible because the most appropriate PSS may not have been specified or provided for the application.

PSS analysis and design procedures are based on linearized models for which the inputs are the perturbations of the above signals from their initial steady-state values.

Frequency is also derived by some manufacturers from voltage and current measurements at the generator terminals. The analysis in this chapter concerns only that derived from the rate of change of a bus voltage-angle.

In the design of PSSs attention must be paid to reducing the effects of the torsional modes of the turbine-generator unit on its dynamic performance [5].

Though the following concerns the small-signal analysis of PSS types, it should be borne in mind that it is necessary to limit the input to the PSS such that limiting occurs ahead of limiting at its output. This concept also applies to controllers other than PSSs.

Because the characteristics of washout filters may affect the performance of the pre-filter and PSS significantly, the time- and frequency-domain responses of a single washout filter and of two identical washouts in series are next examined, but in more detail than earlier in Chapter 5.

8.2 Dynamic characteristics of washout filters

8.2.1 Time-domain responses

In Section 5.8.6.1 the washout filter is introduced with the purpose of eliminating any steady-state offset, or DC level, in the input signal to a PSS. In this section, in addition to those of the single washout filter, the dynamic characteristics of two identical washout filters in series are examined and a comparison made with the dynamic performance of a single washout.

The transfer functions of one and two washout filters having a washout time constant of T_W (seconds) are, respectively:

$$G_{1W}(s) = \frac{sT_W}{1 + sT_W}, \text{ and} \tag{8.1}$$

$$G_{2W}(s) = \frac{s^2 T_W^2}{(1 + sT_W)^2}. \tag{8.2}$$

In analog terms, the analysis assumes a low impedance source drives the filters which then feed into a high impedance sink. Expressions for the time-domain responses of each of the filters to a step input of A_0 units and a ramp input of R_0 units/s are shown in Table 8.1.

Based on the definition of settling times in Section 2.8, the time-domain response of the single washout filter to a step input decays to zero with a 2% settling time of $4T_W$ s. However, for a ramp input the response of the single filter tends to a finite value $R_0 T_W$ - also with a settling time $4T_W$ s. Consequently, for a PSS having electrical power as the stabilizing signal, and with the input being a slow ramp in electrical power, the single washout filter produces a potentially undesirable offset in the terminal voltage of the generator. For the single washout filter the forms of the step and ramp responses are illustrated in Figure 8.1 for washout time constants of 4 and 8 s.

Table 8.1 Analytical expressions for responses of washout filters to step and ramp inputs.

Input signal at time $t(0+)$	Output responses	
	One washout filter	Two washout filters in series
Step, A_0 units	$Y_{1WS} = A_0 e^{-t/T_W}$ $(Y_{1WS} \rightarrow 0$ as $t \rightarrow \infty)$	$Y_{2WS} = A_0 e^{-t/T_W}(1 - t/T_W)$ $(Y_{2WS} \rightarrow 0$ as $t \rightarrow \infty)$
Ramp, R_0 units/s	$Y_{1WR} = R_0 T_W\left(1 - e^{-t/T_W}\right)$ $(Y_{1WR} \rightarrow R_0 T_W$ as $t \rightarrow \infty)$	$Y_{2WR} = R_0 t e^{-t/T_W}$ $(Y_{2WR} \rightarrow 0$ as $t \rightarrow \infty)$

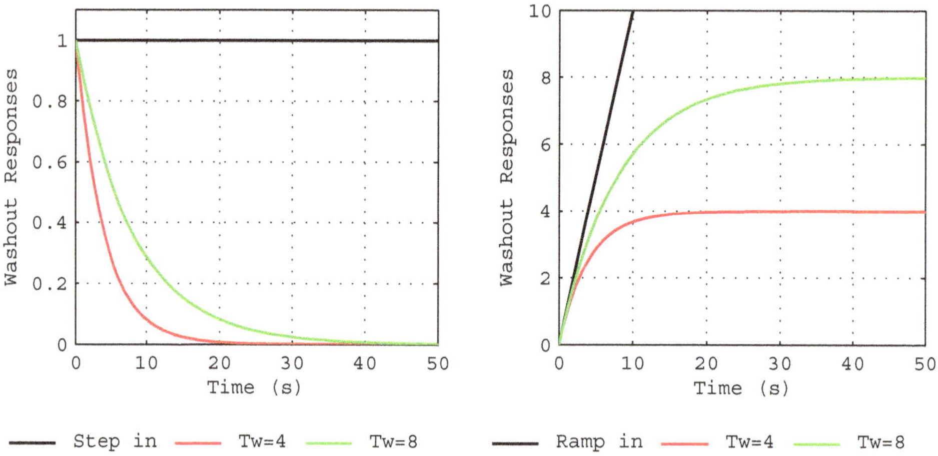

Figure 8.1 Responses of a single washout filter to step and ramp inputs of 1 unit and 1 unit/s, respectively, for washout time constants of 4 and 8 s.

For the case of two washout filters in series the following time-domain characteristics are of interest.

1. The responses of two identical washout filters in series to a step input of 1 unit are shown in Figure 8.2 for values of the washout time constant of 4 and 8 s. For a positive step input the response decays from the initial value A_0, passes through zero at time T_W s and under-shoots by a value $-A_0 e^{-2} = -0.135A_0$ at $2T_W$ s. It then decays to within $-0.02A_0$ of zero after approximately $5.5T_W$ s.

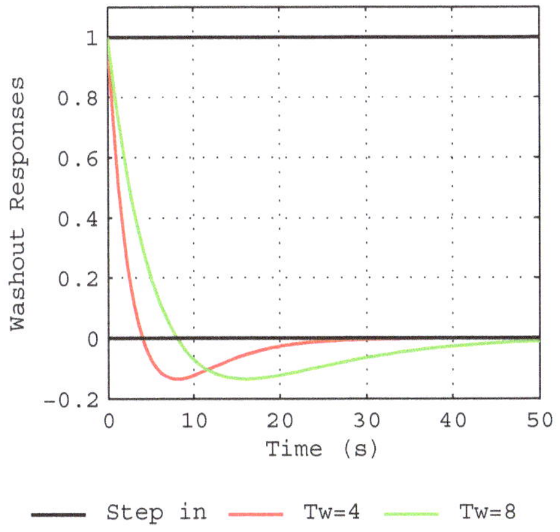

Figure 8.2 Responses of two identical washout filters in series to a step input of 1 unit for washout time constant values of 4 and 8 s.

2. For a positive ramp input the time domain response reaches a maximum value of $R_0 T_W e^{-1} = 0.368 R_0 T_W$ at time T_W s. It then decays to zero. Notice that the maximum value of the response depends on the ramp rate and the value of the washout time constant.

The responses to a ramp input of two identical washout filters in series is of particular interest in the discussion of the 'integral-of-accelerating-power' PSS considered in Section 8.5. Accordingly, the responses of two such filters to a ramp of 1 unit/s are shown in Figure 8.3 for a range of values of the washout time constant from 1 to 10 s.

In Figures 8.1 to 8.3 the time-domain characteristics listed in Table 8.1 are clearly illustrated.

8.2.2 Frequency-domain responses

The nature of the frequency response of a single washout filter, and its role in the dynamic performance of speed-PSSs, are discussed in Section 5.8.6.1. Because the application of two washouts is of interest in this chapter the frequency response of two identical washouts in series, time constant T_W, is shown in Figure 8.4. The response is normalised to a corner frequency of 1 rad/s (i.e. $T_W = 1$ s). For example, if $T_W = 5$ s the associated corner frequency is 0.2 rad/s, the magnitude and phase of the response at say 0.02 rad/s (as read off Figure 8.4 at $0.02/0.2 = 0.1$ rad/s) are then -40 dB and $169°$, respectively.

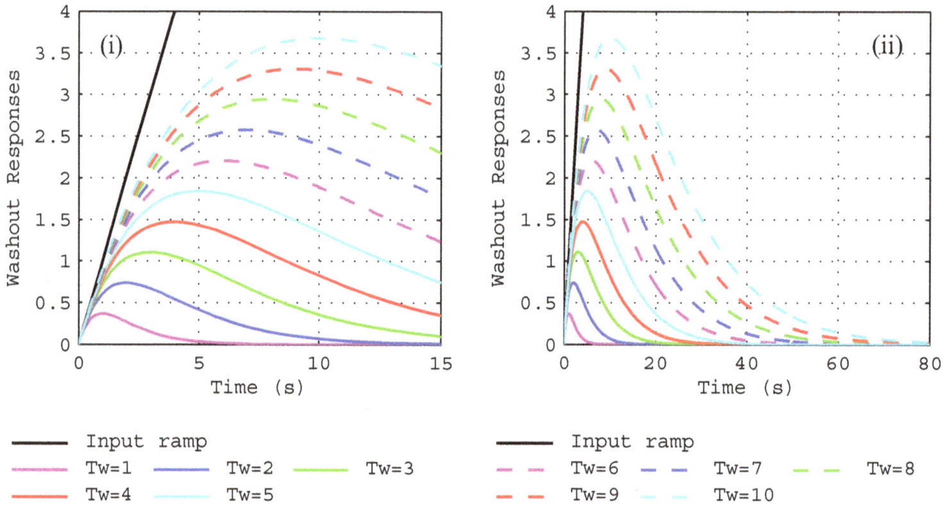

Figure 8.3 Responses of two identical washout filters in series to a ramp input of $R_0 = 1$ unit/s as the washout time constant T_W is varied from 1 to 10 s. Time-frames: (i) 0-15 s, (ii) 0-80 s. Solid lines T_W 1-5 s; dashed lines T_W 6-10 s. Peak occurs at T_W s.

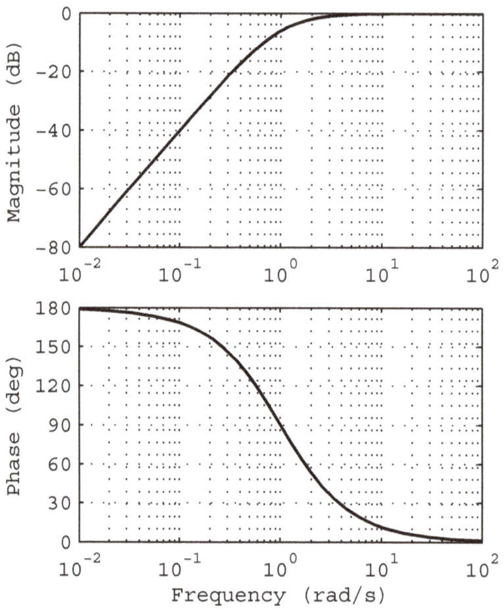

Figure 8.4 Frequency response for two identical washouts filters in series normalised to a corner frequency of 1 rad/s. (For a single washout filter, halve all vertical-axis quantities.)

8.2.3 Comparison of dynamic performance between a single and two washout filters.

Let us compare the features of a single washout filter with two identical washouts in series.

Consider a power system in which the lowest inter-area modal frequency is 2 rad/s. For the purpose of the design of the associated PSS, let us assume that if a single washout filter is employed the corner frequency of the washout would be 0.2 rad/s, say, a decade below the modal frequency. The time constant of the single filter is $T_{1W} = 5$ s; the phase lead introduced by the filter at the modal frequency is $5.7°$. To introduce the same phase lead at the modal frequency for two identical washouts the corner frequency of each should be 0.1 rad/s, i.e. $T_{2W} = 10$ s. Based on these assumptions a comparison of dynamic performance is summarized in Table 8.2.

Table 8.2 Characteristics of a single washout and two identical washout filters in series

	One washout filter	Two washout filters in series
Corner frequency, T_W	0.2 rad/s, 5 s	0.1 rad/s, 10 s
Phase lead introduced at 2 rad/s	5.7°	5.7°
<u>Step response of 1 unit</u>: Settling time Under-shoot Final value	 $4T_W = 20$ s - 0	 $\sim 5.4T_W = 54$ s $-0.135A_0 = -0.135$ at $2T_W$ s 0
<u>Ramp response of 1 unit/s</u>: Settling time Peak value Final value	 $4T_W = 20$ s Peak is the final value $T_W R_0 = 5$	 By calculation $0.368R_0T_W = 3.68$ 0

Some observations on the characteristics of the washout filters of Table 8.2 are listed below.

1. A reduction in time constants for both a single washout and two washouts in series improves their dynamic performance through lower settling times.

2. The performance of the single washout filter in Table 8.2 is superior to that of two washouts, except that the ramp response of the single washout filter tends to a finite value. As mentioned in Section 8.2.1, in the case of an electrical power PSS this characteristic can produce an offset in generator terminal voltage and reactive power output when a ramp in electrical power output occurs.

3. As noted, the time-domain performance of two identical washouts in series can be improved by reducing the time constant. However, such a reduction is a compro-

mise with the increase in phase lead at the lower modal frequencies. In the case of a reduction in the time constant, say from 10 s to 5 s, the phase lead at the modal frequency of 2 rad/s is increased from $5.7°$ to $11.4°$. If desired, the increased phase lead so introduced by the two washouts can be compensated for in the tuning of the PSS main compensation blocks.

4. If the inter-area modes are not of concern, the washout filter time constants can likewise be determined based on the relatively higher frequency of the local-area mode(s).

8.3 Performance of a PSS with electric power as the stabilizing signal.

8.3.1 Transfer function and parameters of the electric power pre-filter.

It has been common practice to use electrical power perturbations as a stabilizing signal. The analysis and implementation of the associated PSS is simplified if the electrical power perturbations are converted to speed perturbations by means of a pre-filter. The transfer function of this pre-filter is now discussed.

The equation of motion of the rotor of a synchronous generator for small-signal disturbances is given in the Laplace domain by,

$$\Delta P_m - \Delta P_e = 2Hs\Delta\omega + D\Delta\omega, \quad \text{all quantities in per unit.} \tag{8.3}$$

This equation has been a basis for analysis in Chapters 4 and 5. In (8.3) ΔP_m and ΔP_e are the perturbations in mechanical and electrical torques (or powers), respectively, acting on the shaft; $\Delta\omega$ is the perturbation in rotor speed.

If we assume that perturbations in mechanical power and damping torques $D\Delta\omega$ acting on the rotor are negligible, then (8.3) reduces to:

$$\Delta\omega_S = -\frac{1}{2Hs}\Delta P_e, \tag{8.4}$$

where $\Delta\omega_S$ is a speed signal synthesized from electrical power, and therefore can be employed as a stabilizing signal as long as the assumptions stated above are justified. The structure of the PSS becomes that shown in Figure 8.5. (The negative sign at the summing junction for ΔV_S reflects the inherent negation in (8.4)).

As shown in the figure the pseudo-integrator, or pre-filter, used in a practical PSS to replace the ideal integrator in (8.4) is given by the transfer function:

$$\Delta\omega_s = \frac{T_H/(2H)}{1 + sT_H}\Delta P_e. \tag{8.5}$$

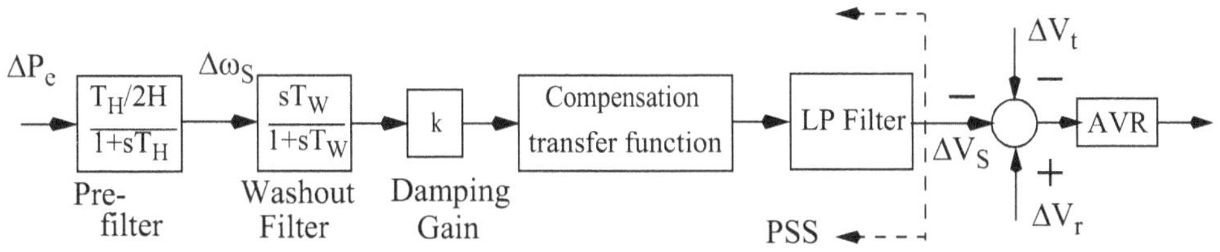

Figure 8.5 Structure of the PSS with electric power as the stabilizing signal. Note the negative sign of the PSS output signal ΔV_S at the summing junction to the AVR.

So that transfer function of the low-pass filter $T_H/(1+sT_H)$ acts as an integrator $1/s$ over the range of modal frequencies it is required that its corner frequency at $\omega_{fc} = 1/T_H$ rad/s should be a decade or more below the lowest (inter-area) modal frequency. With this choice of T_H, the gain of the filter rolls off at -20 dB/decade over the range of modal frequencies and its associated phase angle is approximately -90 deg. This is the case for an ideal integrator (see Section 2.12.1.2).

For example, assume the lowest (inter-area) modal frequency is 2 rad/s; the corner frequency should ideally be 0.2 rad/s or less. For a values of T_H of 5.0 and 7.5 s the corner frequencies are respectively 0.2 and 0.133 rad/s; Table 8.3 shows that for these values of T_H the frequency response of the associated pseudo-integrator agrees well with that of the ideal integrator at and above 2 rad/s. While it is common to set $T_H = T_W$ a higher value of T_H (say $T_H = 7.5$ s when $T_W = 5$ s) is sometimes used in practice.

Table 8.3 Responses of the ideal integrator and the pseudo-integrator at lower frequencies.

ω_f rad/s	Ideal: $s = \dfrac{1}{j\omega_f}$	$T_H/(1+sT_H)$	
		$T_H = 5$ s	$T_H = 7.5$ s
1	$1.0\angle{-90°}$	$0.9806\angle{-78.69°}$	$0.9912\angle{-82.41°}$
2	$0.50\angle{-90°}$	$0.4975\angle{-84.29°}$	$0.4989\angle{-86.19°}$
4	$0.25\angle{-90°}$	$0.2497\angle{-87.14°}$	$0.2499\angle{-88.09°}$

Because a synthesized speed signal is derived from the electrical power output using the pre-filter of (8.5), the design of the compensating transfer function of the PSS follows the procedure based on a speed-stabilizing signal as outlined in Section 5.10.6. The compensating transfer function is the same as for the speed PSS given in (5.49).

The rapid attenuation of the electric power signal with frequency ω_f is noted in Table 8.3. A feature of the use of electric power perturbations as a stabilizing signal is that the torsional oscillations which occur on the shafts of generating units are heavily attenuated. This topic is discussed later in Section 8.5.3.

Several cautionary comments follow.

1. The electrical power output of the generator will closely follow any ramping of the mechanical power output of the turbine. Depending on the ramp rate and the time constant of the washout filter(s), there may be a significant deviation in the associated PSS output. As a result of this signal being injected into the excitation system, there could be unacceptable variations in terminal voltage and hence in the reactive power output of the generator [13]. This problem is ameliorated by providing an appropriate pre-filter, such as in the Delta-P-omega stabilizer [14], or employing an 'integral-of-accelerating-power' PSS, to be discussed in Section 8.5.

2. Care should be taken to ensure that negative feedback of the PSS output signal is applied at the AVR summing junction.

3. Prior to purchase due care should be taken to ensure that the power input PSS provides for the synthesis of a rotor-speed signal from the electrical-power input.

8.3.2 Dynamic performance of a speed-PSS with an electric power pre-filter.

A PSS designed for a speed-stabilizing signal used with an electric power pre-filter forms the basis for the assessment of the dynamic performance of the integrated stabilizer. Five operating conditions for the sixth-order generator-SMIB system, shown in Table 5.6, are used to illustrate the performance of the PSS. Its performance is compared to that of the PSS which uses "true" rotor speed for the same fives cases.

A single washout filter with time constant of 5 s is selected in Section 5.10.6.2 for the five Cases. The associated corner frequency of 0.2 rad/s is more than a decade below the single rotor mode of oscillation (~ 9 rad/s) [1]. The inertia constant of the unit is 3 MWs/MVA. As revealed in Table 8.3 a suitable time constant for the pseudo-integrator is 7.5 s. The transfer function for the electric power pre-filter is thus

$$\frac{T_H/(2H)}{1+sT_H} = \frac{1.25}{1+7.5s}. \tag{8.6}$$

A comparison of the modes resulting from the use of an electric power pre-filter that synthesizes a rotor speed signal with those produced by a true rotor speed PSS is shown in Table 8.4. Because there is close agreement in the values of the modes, it is concluded that the pre-filter accurately synthesizes rotor speed perturbation with the caveat that slow variations in mechanical power may cause variations in the reactive output of the generator.

---

1. The corner frequency of 0.2 rad/s is a decade below any potential inter-area modal frequencies of 2 rad/s if the SMIB system represents a generator in a multi-machine system.

Table 8.4 Comparison of modes for Speed and Electric Power PSSs*

Case	Generator Output, P, Q pu	Rotor mode		
		with PSS out of service	true-speed PSS in service*	Speed PSS with Electric Power pre-filter
A	0.9, -0.1	$0.773 \pm j9.16$	$-1.156 \pm j9.51$	$-1.163 \pm j9.48$
B	0.9, 0	$0.552 \pm j9.12$	$-1.271 \pm j9.31$	$-1.275 \pm j9.29$
C	0.9, 0.2	$0.261 \pm j9.02$	$-1.338 \pm j9.03$	$-1.339 \pm j9.01$
D	0.9, 0.4	$0.113 \pm j8.98$	$-1.305 \pm j8.93$	$-1.305 \pm j8.91$
G#	0.9, -0.07	$0.927 \pm j7.98$	$-0.409 \pm j8.04$	$-0.409 \pm j8.02$

* Results for the true-speed PSS are given in Table 5.4
# Two lines are out of service in Case G. All lines are in service in Cases A - D.

8.4 Performance of a PSS with bus-frequency as the stabilizing signal.

The frequency of the generator terminal voltage is used by some manufacturers as a PSS stabilizing signal on the basis that bus frequency closely represents rotor speed perturbations in magnitude and phase. The following analysis applies only to frequency signal derived from the voltage angle. The frequency ω_{freq} is the rate of change of the terminal-voltage angle, α (rad), thus

$$\omega_{freq} = (1/\omega_0)(d\alpha/dt) \text{ pu of system frequency;} \tag{8.7}$$

the associated transfer function is

$$\omega_{freq}(s) = (1/\omega_0)s\alpha; \tag{8.8}$$

where $\omega_0 = 2\pi f_0$, f_0 being system frequency (Hz). Once again, the angular perturbations are converted to a pseudo-speed signal by means of pseudo-differentiation, pseudo-differentiation being employed *to limit the gain and noise amplification at high frequencies* associated with pure differentiation in (8.7). Moreover the torsional modes, if present in the terminal voltage, are amplified. Equation (8.8) yields the transfer function of the bus-frequency pre-filter:

$$\frac{\Delta\omega_{freq}(s)}{\Delta\alpha(s)} = (1/\omega_0)\frac{s}{(1+sT_F)}. \tag{8.9}$$

The time-constant T_F incorporates the phase-lag inherent in the measurement of the bus-voltage angle or bus-frequency. It is thus a property of the measurement transducer rather than being a tunable or selectable parameter.

In order that the transfer function $s/(1+sT_F)$ acts as a differentiator over the range of modal frequencies,

1. its low-frequency response should (ideally) pass through the 0 dB axis at 1 rad/s and roll-up at 20 dB/decade with an associated phase angle of 90°;

2. its corner frequency at $\omega_{fc} = 1/T_F$ rad/s should ideally be a decade or more above the highest modal frequency (typically a local-area mode). However the gain introduced by the transfer function at higher frequencies may be destabilizing.

Note that the frequency response of the pre-filter of (8.9) can be deduced from that of the single washout filter in Figure 8.4 by rearranging the pre-filter transfer function into the form
$$\frac{sT_F}{(1+sT_F)}\cdot\left[\frac{(1/\omega_0)}{T_F}\right].$$

Although the pre-filter transfer function of (8.9) synthesizes a speed signal from the derivative of bus-angular perturbations (frequency), the question arises how well does bus frequency represent the actual rotor speed perturbations in magnitude and phase? Let us examine the performance of a PSS equipped with a bus-frequency stabilizing signal.

8.4.1 Dynamic performance of a speed-PSS with a bus-frequency pre-filter

The bus-frequency pre-filter delivers a synthesized speed signal to a PSS whose design is based on a true rotor-speed stabilizing signal. This so-called bus-frequency PSS forms the basis for the assessment of the dynamic performance of the integrated PSS. Once again, the five Cases for the sixth-order generator-SMIB system, listed in Table 5.5, are used to investigate the performance of the pre-filter.

The pre-filter is assumed to be of the form given in (8.9); its parameters are determined as follows.

Assuming the upper modal frequency is 10 rad/s, ideally the corner frequency of the pre-filter should be set a decade higher, at 100 rad/s. However, due to the higher gains introduced at higher frequencies the choice of a corner frequency of 75 rad/s may be considered to be a suitable compromise; thus $T_F = 0.0133$ s. Nominal system frequency is 50 Hz, $\omega_0 = 100\pi$ rad/s. The combined transfer function of the pre-filter and the speed-PSS of (5.49), is thus

$$H(s) = \frac{\Delta V_s(s)}{\Delta\alpha(s)} = \left[\frac{[1/(2\pi f_0)]s}{(1+s0.0133)}\right]\cdot 20\cdot\frac{s5}{1+s5}\cdot\frac{1}{3.84}\cdot\frac{1+s0.0895+s^2 0.00277}{(1+s0.005)(1+s0.005)}. \quad (8.10)$$

The mode shift associated with both the original speed-PSS and the bus-frequency PSS of (8.10) are shown in Table 8.5; the damping gain is 20 pu for both PSSs.

Table 8.5 Comparison of real parts of mode shifts for speed and bus-frequency PSSs, damping gains 20 pu.

Case	Gen. Out-put. P, Q pu	Rotor mode ------ ## Excitation system mode for bus-frequency PSS ##			Rotor mode shifts, PSSs in service		Ratio*	
		PSSs off	true-speed PSS in service	bus-frequency PSS in service ------ ##	true-speed PSS	bus-frequency PSS	c	1/c
A	0.9, −0.1	0.77 ± j9.2 -	− 1.16 ± j9.5 -	− 0.46 ± j9.5 2.73 ± j80.0	− 1.93 ± j0.35 -	−1.24± j0.30 -	1.56 -	0.64 -
B	0.9, 0	0.55 ± j9.1 -	− 1.27 ± j9.3 -	− 0.60 ± j9.3 − 0.97 ± j80.9	− 1.82 ± j0.20 -	−1.15± j0.17 -	1.59 -	0.63 -
C	0.9, 0.2	0.26 ± j9.0 -	− 1.34 ± j9.0 -	− 0.71 ± j9.1 − 9.2 ± j81.9	− 1.60 ± j0.02 -	−0.97± j0.03 -	1.64 -	0.61 -
D	0.9, 0.4	0.11 ± j9.0 -	− 1.31 ± j8.9 -	− 0.73 ± j9.0 − 17.7 ± j82.0	− 1.42 ∓ j0.05 -	− 0.85 ∓ j0.01 -	1.68 -	0.60 -
G#	0.9, −0.07	0.93 ± j8.0 -	− 0.41 ± j8.0 -	− 0.24 ± j8.1 16.3± j76.4	− 1.34 ± j0.06 -	−1.17± j0.14 -	1.14 -	0.88 -

* Ratio: c= $\Re e$ *(true-speed PSS mode-shift)* / $\Re e$ *(bus-frequency PSS mode-shift)*

Note: Results for speed-PSS for a SMIB system are given in Table 5.5

Two lines are out of service in Case G. All lines are in service in Cases A - D

In column 5 the upper and lower quantities are the rotor mode and an excitation system mode, respectively

From the table the following is observed:

1. In Cases A and G a high-frequency mode associated with the excitation system and q-axis variables is unstable as the generator power factor becomes leading.
2. The real parts of the mode shifts for the bus-frequency PSS are degraded significantly (by a factor of 1/c) with respect to the speed-PSS.
3. Therefore, because the differences in the imaginary parts of the two sets of mode shifts are negligible, there appears to be a reduction in the loop gain of the PSS-SMIB system when employing rate of change of angle of the generator terminal voltage as the stabilizing-signal source.

Let us consider these observations commencing with no.1 above.

8.4.1.1 Stability of the closed-loop system for Case A

Let us examine the open-loop frequency response for the system of Case A, a leading power factor condition, listed in Table 8.5. The open-loop transfer function is $\Delta V_S/\Delta V_{ref}$, where ΔV_S and ΔV_{ref} are the output of the PSS and the AVR reference voltage, respectively; it includes the combined transfer function of the pre-filter and the speed-PSS given by (8.10). A block diagram of the transfer function of the open-loop system $G(s)H(s)$ is shown in Figure 8.6(a) and the associated frequency responses are given in Figure 8.6 (b).

As the open-loop system possesses one unstable pole-pair at $2.73 \pm j80.0$, the stability of the closed-loop system can be determined from the Nyquist Criterion based on the open-loop system $-G(s)H(s)$ [1]. In the case of Figure 8.6(b) (i) it can be shown that for closed-loop stability the gain at high frequencies must be less than unity (0 dB), and thus must be attenuated. This is achieved by changing the two time-constants of the low-pass filter of the PSS in (8.10) from 0.005 to 0.01 s. The associated response of the open-loop transfer function shown in Figure 8.6 (b)-(ii) results in a stable closed-loop system with poles at $-0.433 \pm j9.75$ and $-12.0 \pm j40.1$ for the rotor and exciter modes, respectively. Further studies are required to mitigate against instability for higher gains at high frequencies over the range of operating conditions.

8.4.2 Degradation in damping with the bus-frequency pre-filter

Based on Table 8.5 the improvement in the damping-constant of the rotor mode due to the frequency-PSS is substantially less than for the speed-PSS, although there is negligible change in the modal frequency for both PSSs. This suggests that the use of bus-frequency, derived from bus voltage-angle, results in a reduction in the loop-gain in the path through the machine and PSS. Consider the simple system shown in Figure 8.7. The voltages and angles are E and δ internal to the generator and V_t and α at its terminals; the voltage at the infinite bus is $V_b\angle 0°$. The generator internal reactance is x and that of the equivalent external circuit is x_e. (One might speculate that, for a simple system such as this, the perturbations in α are roughly related to those in δ by a factor $x_e/(x_e + x)$ - if the angles are not large.)

1. Because the PSS output is not negated at the summing junction of the AVR, the conventional open-loop transfer $G(s)H(s)$ must be negated for application of the Nyquist Criterion.

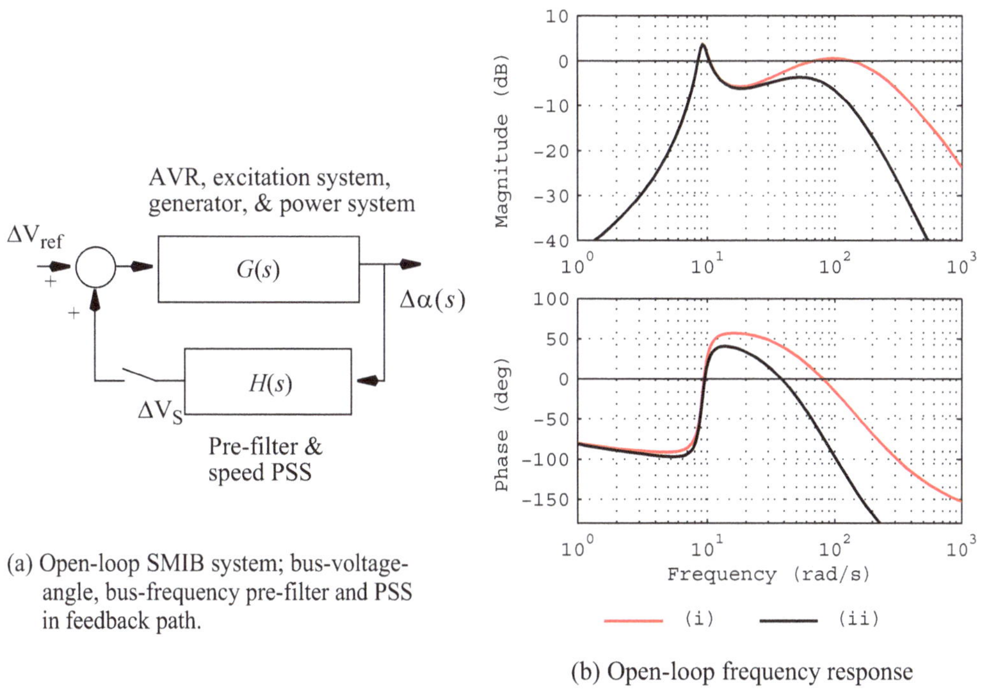

Figure 8.6 (a) Open-loop system $\Delta V_S / \Delta V_{ref}$. (b) Open-loop frequency response for Case A: (i) pre-filter and PSS transfer function given by (8.10) (the unstable mode in the open-loop system is $2.73 \pm j80.0$). (ii) the closed-loop system is stable with modification of the parameters of the low-pass filter.

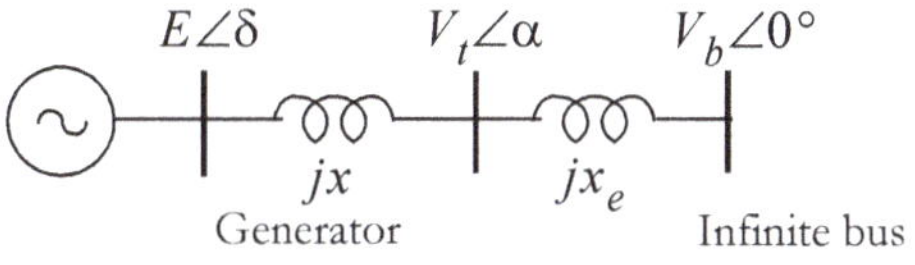

Figure 8.7 A simple SMIB system

It can be shown that for small perturbations

$$\Delta\delta = \left(1 + \frac{\cos\alpha_0}{k_0 \cos(\delta_0 - \alpha_0)}\right)\Delta\alpha \quad \text{where } k_0 = \left(\frac{x_e}{x}\right) \cdot \left(\frac{E}{V_b}\right). \tag{8.11}$$

Since perturbations in both rotor speed $\Delta\omega$ and bus frequency $\Delta\omega_{freq}$ are related to $\Delta\delta$ and $\Delta\alpha$ by equations of the form of (8.7), then

$$\Delta\omega = \left(1 + \frac{\cos\alpha_0}{k_0\cos(\delta_0 - \alpha_0)}\right)\Delta\omega_{freq} = c \cdot \Delta\omega_{freq}. \tag{8.12}$$

The coefficient c is greater than unity for transmission angles $\alpha_0 < 90°$. Because $\Delta\omega_{freq} < \Delta\omega$ the PSS loop-gain is, in effect, reduced by a factor of $1/c$.

As an example, assume for a pseudo steady-state condition that the internal voltage E is the voltage proportional to rotor flux linkages, E', and the reactance x is the transient reactance x'_d. If $x'_d = 0.3$, $x_e = 0.375$, then for $V_t = 1.0$,

$$E' = 1.036,\ V_b = 1.055,\ \delta_0 = 33.8°,\ \alpha_0 = 18.7°\ \text{, and}$$

$$k_0 = 1.23 \text{ and } c = 1.80.$$

Thus, using bus frequency as the stabilizing signal rather than the true rotor speed, the PSS loop-gain is inherently reduced, in this case by a factor of $1/c = 0.56$. (Note, as surmised above, the ratio $x_e/(x_e + x)$ is about 0.55.)

The above somewhat simplistic example reveals the order of the magnitude of the reduction in the PSS loop-gain for a pseudo steady-state condition. This example raises the question: what is the effect of the closed-loop dynamics and a more accurate generator model on the gain reduction?

Consider Case B of Table 8.5 on page 409 the rotor mode is $-0.60 \pm j9.30$ when the bus-frequency PSS is in service. Let us evaluate the frequency responses at 9.3 rad/s of both the true rotor speed ($\Delta\omega$) and the bus frequency $\Delta\omega_{freq}$ for perturbations in reference voltage. The ratio of the true rotor speed to the bus frequency at the modal frequency is 1.57; this ratio agrees well with the value of $c = 1.59$ in the table. The phase difference between the true and synthesized speeds is approximately $0°$ when the phase lag introduced by the corner $1/T_F$ in the pre-filter (8.9) is accounted for. Thus, for practical purposes, the true speed and the synthesized speeds are essentially in phase. We conclude that for the cases analysed the use of the bus-angle perturbations as the input signal to the PSS results in a gain reduction in the machine - PSS loop. Moreover, there is a significant reduction in the mode shift for the single rotor mode. Thus in the multi-machine context at the lower inter-area frequencies, in which the generator may participate, are there marked reductions in loop gain - and therefore reductions in the associated mode shifts due to the use of this type of PSS?

An analysis of the performance of the PSS over the range of normal and contingency conditions, such as that in Table 8.5, suggests that the effective attenuation in gain associated with the bus-frequency PSS is roughly $x_e/(x_e + x)$. As in Case G (Table 8.5) when the external impedance (jx_e) is increased the attenuation is significantly reduced. Accordingly, a judicious increase in PSS gain is required in order to provide a performance similar to that of a speed-input PSS over the encompassing range of operating conditions.

Note: In the event of significant transients that lead to sudden changes in bus-voltage angle, e.g. a line fault followed by the tripping of the circuit, the synthesized rotor speed derived from the bus-voltage angle will not necessarily be representative of the true rotor speed until the resulting large-amplitude oscillations have markedly decayed.

8.5 Performance of the "Integral-of-accelerating-power" PSS

8.5.1 Introduction

A third category of the PSS models listed in the IEEE Standard [15] is the integral-of-accelerating-power PSS (IAP PSS) and is referred to in the Standard as PSS2B. As shown in Figure 8.8, the IAP PSS consists of two main components, a pre-filter which develops a synthesized speed signal $\Delta\omega_S$ and a conventional PSS the design of which is based on the rotor-speed stabilizing signal discussed in Chapter 5.

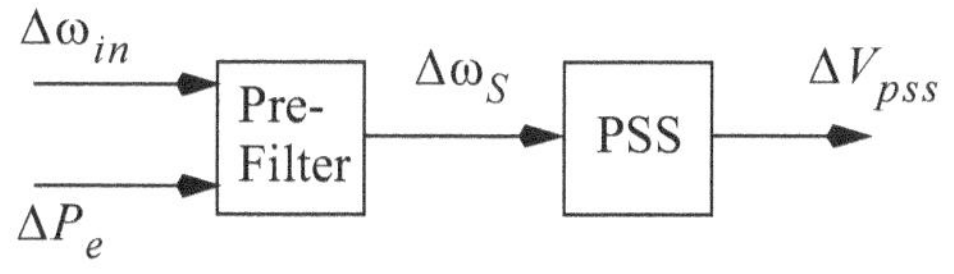

Figure 8.8 Components of the integral-of-accelerating-power PSS

The inputs to the pre-filter are 'speed' and electric power signals, $\Delta\omega_{in}$ and ΔP_e. The 'speed' signal may be derived in a number of ways, for example from

- frequency of the generator terminal voltage;
- speed of the rotor measured by tacho-generator, a toothed wheel mounted on the shaft. etc.;
- filtered values of the instantaneous three-phase voltages and currents, and processing of these signals.

These speed signals contain not only the inter-area and other rotor modes but also the torsional modes of the turbine-generator-exciter unit. In order that the latter modes are not excited by the PSS, the torsional modes must be significantly attenuated; this is one of the roles of the IAP pre-filter.

If the mechanical power output of the turbine - be it hydro, gas or steam - is changing, the electric power output of the generator will follow it closely, particularly as the mechanical power changes occur relatively slowly under normal operating conditions. As observed in Section 8.3, changing the mechanical - and hence electrical - power input to the PSS can perturb the terminal voltage of the generator, possibly causing undesirable swings in its reactive power output. Assuming the variation in mechanical power is a ramp, the pre-filter incorporates a 'ramp tracking filter' which tracks a ramp ideally with zero tracking error - and thereby offsets the ramp in the electrical power input; this is a second role of the IAP pre-filter.

Let us consider the influence of the torsional modes, the ramping of mechanical power, and the characteristics of the ramp tracking filter.

8.5.2 Torsional modes introduced by the speed stabilizing signal

A generating unit, in the case of a steam turbine, may consist of high pressure, intermediate and low-pressure stages, the generator and an exciter. The lumped masses are connected by shafts whose torsional stiffness is finite. As is illustrated in Chapter 9 for a linear spring-mass system, the rotating masses similarly exhibit modal frequencies and damping dependent on the inertia of the masses and the stiffness of the interconnecting shafts [5], [12].

Since the mechanical stiffness of the shaft components is at least an order of magnitude higher than the effective electro-mechanical coupling between the generator and the power system, the entire rotating mass of the mechanical shaft of a large turbo-generator is more-or-less uniformly subject to the power system's inter- and local-area modes of frequency 1.5 to 15 rad/s. The first torsional mode for large steam turbine units can be as low as 8 Hz (50 rad/s) [5], [6]. Depending on the mode shape of the particular torsional mode, a shaft-speed transducer that is located in a region of the shaft that closely corresponds to a peak of the torsional oscillations (an anti-node of the mode shape) can result in a significant component of the torsional mode in the speed signal. One way to avoid this problem is to locate the speed transducer at a node of the modal shape [5]; this, however, is not always practical since in some cases the node may lie inside a turbine stage.

It will be assumed in the analysis that the input speed signal to the pre-filter, $\Delta\omega_C(t)$, comprises the 'true' rotor speed component, $\Delta\omega_{in}(t)$, 'corrupted' by the first and higher torsional modes (as well as noise), $\Delta\Omega_t(t)$, i.e.

$$\Delta\omega_C(t) = \Delta\omega_{in}(t) + \Delta\Omega_t(t). \tag{8.13}$$

8.5.3 The electric power signal supplied to the pre-filter

The electric power signal input to the pre-filter is a filtered representation of the instantaneous electric power. The filtering process typically introduces a very small phase shift over the range of electro-mechanical modal frequencies and consequently the input power signal closely follows the low frequency perturbations in power associated with the local- and inter-area modes. Furthermore, as mentioned earlier, if the mechanical power output of the turbine is ramped, say, in the relatively slow process of generation despatch, i.e. changing power output from one level to another, the electrical power output of the generator will closely follow the mechanical power.

If there is any component of the torsional modes in the electric power signal, depending on how it is calculated, the component - being of significantly higher frequency than the rotor mode - will be significantly attenuated by the integration in the pre-filter.

8.5.4 The Ramp Tacking Filter (RTF)

The RTF is a low-pass filter of the form,

$$F(s) = \left[\frac{1 + sT_8}{(1 + sT_9)^M}\right]^N, \text{ where } T_8 = M \cdot T_9. \tag{8.14}$$

The RTF serves a number of purposes. Firstly, it tracks a ramp signal at its input with zero tracking error. Secondly, it significantly attenuates signals at frequencies above the corner frequency $1/T_9$. Thirdly, as will be demonstrated, it passes the low frequency perturbations associated with mechanical power changes with negligible attenuation. The frequency responses for two typical sets of parameter values for the RTF are shown in Figure 8.9. (It should be noted that the tracking feature of the RTF is defeated if T_8 deviates markedly from $T_8 = M \cdot T_9$, for example if $T_8 = 0$).

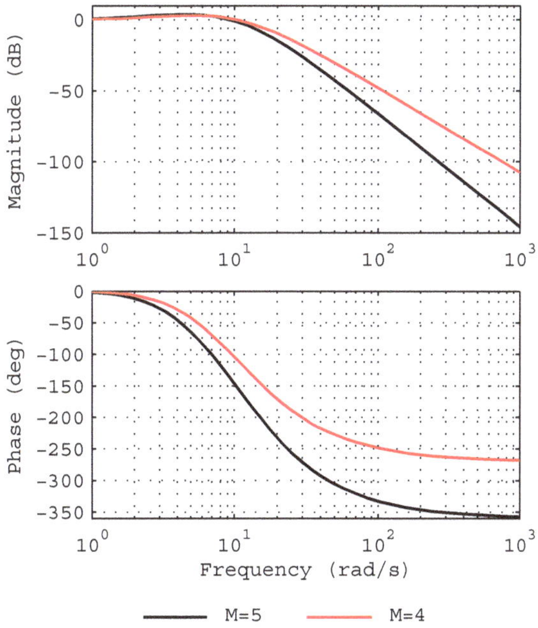

Figure 8.9 Frequency responses of the Ramp Tracking Filter for $N = 1$, $T_9 = 0.1$, $T_8 = M \cdot T_9$, $M = 5$ or $M = 4$.

It is clear from the plots that a torsional mode, say the first at 15 Hz (~95 rad/s), is attenuated by 50 dB or more. The parameters commonly used for the RTF are $N = 1$, $M = 5$ and $T_9 = 0.1$. The value of the time constant T_9 more-or-less determines the corner frequency of the RTF. If for example, the variations in mechanical power output are slow, and the torsional modes possess low frequency components, it may be desirable to reduce the value of T_9, and/or set $N = 2$.

While the RTF tracks a ramp input signal $U_R(t) = R_0 t$ with zero tracking error, it tracks a signal which is the integral of a ramp, i.e. a parabola $U_P(t) = 0.5 R_0 t^2$, with a constant tracking error. The ramp-tracking characteristics of the filter are analysed in Appendix 8–I.2. from which it can be shown that the steady-state tracking error to a parabolic input is finite, i.e. $e_{fss} = 10 R_0 T_9^2 = 0.1 R_0$ when $T_9 = 0.1$ and $N = 1$, $M = 5$.

8.6 Conceptual explanation of the action of the pre-filter in the IAP PSS

We will consider the action of the pre-filter in two steps, firstly without washout filters and then considering their effects.

8.6.1 Action of the pre-filter, no washout filters

As explained in Section 8.5.2 the input speed signal to the pre-filter, $\Delta\omega_C(t)$, is assumed to comprise the 'true' rotor speed component, $\Delta\omega_{in}(t)$, 'corrupted' by a torsional component and high frequency noise, $\Delta\Omega_t(t)$, i.e.

$$\Delta\omega_C(t) = \Delta\omega_{in}(t) + \Delta\Omega_t(t) \qquad \text{(8.13) repeated.}$$

Similarly, an input to the pre-filter is the perturbation in the electric power output of the generator $\Delta P_e(t)$, which closely follows the ramping of mechanical power but also contains perturbations in the associated local and inter-area modes.

For the purpose of explaining the conceptual basis of the pre-filter, let us assume that the basic structure of the pre-filter is that shown in Figure 8.10 (omitting the washout filters).

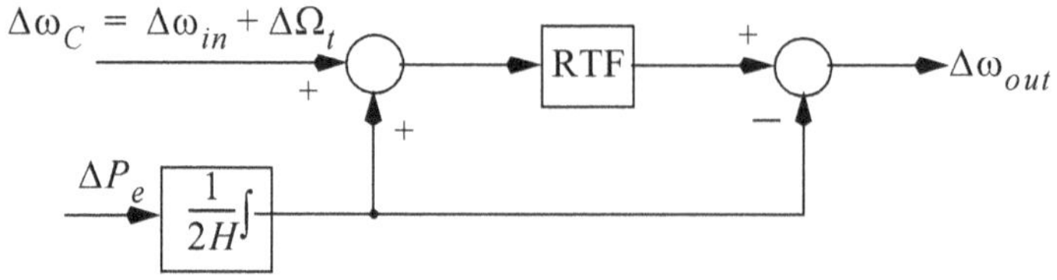

Figure 8.10 The basic structure of the pre-filter

In general, the relationship between the accelerating power (or torque) acting on the shaft and instantaneous speed is given by the shaft equation (4.64) on page 115.

$$\Delta P_m(t) - \Delta P_e(t) = 2H\frac{d}{dt}\Delta\omega_{in}(t) \text{ pu,} \tag{8.15}$$

where $\Delta P_m(t)$ is the perturbation in mechanical power output of the turbine and includes a ramp change in turbine power. Rearranging (8.15) and integrating the resulting expression, we can express the integral of the electrical power signal, *IPE*, as:

$$IPE = \frac{1}{2H}\int\Delta P_e(t)dt = \frac{1}{2H}\int\Delta P_m(t)dt - \Delta\omega_{in}(t). \tag{8.16}$$

Each term in (8.16) has the dimensions of speed (pu). Consequently, on the basis of (8.16), the output of the integration of the electrical power signal, $IPE = \frac{1}{2H}\int\Delta P_e(t)dt$, contains information not only on the mechanical power ramp and perturbations but also the *'true'* rotor speed $\Delta\omega_{in}(t)$. As mentioned, if any torsional modes, which typically exceed 8 Hz (50 rad/s), are present in the electrical power signal they are heavily attenuated through the integrator transfer function $\Delta P_e(s)/(2Hs)$, i.e. 50 dB at 50 rad/s for H=3 MWs/MVA.

Let us combine the signal *IPE* with the input speed signal $\Delta\omega_C(t)$ of (8.13), as shown diagrammatically in Figure 8.11(i). A signal *IPM* results:

$$IPM = IPE + \Delta\omega_C(t) = \left\{\frac{1}{2H}\int\Delta P_m(t)dt - \Delta\omega_{in}(t)\right\} + \{\Delta\omega_{in}(t) + \Delta\Omega_t(t)\}, \tag{8.17}$$

$$\text{i.e. } IPM = \frac{1}{2H}\int\Delta P_m(t)dt + \Delta\Omega_t(t). \tag{8.18}$$

Note that *IPM* contains only the perturbations in mechanical power and the torsional modes, *the true rotor speed signals* $\Delta\omega_{in}(t)$ *in* (8.17) *having been cancelled out;* this cancellation is an essential feature of the IAP pre-filter.

As shown in Figure 8.11(ii), the signal *IPM* is passed through the RTF. By judicious selection of the parameters of the RTF it will attenuate significantly the higher-frequency torsional modes and track the integral of the mechanical power ramp-changes with negligible tracking error [1]. An analysis of these features of the RTF are given in Appendix 8–I. The output of the RTF therefore contains the integral of mechanical power, $V_{rtf} = \frac{1}{2H}\int\Delta P_m(t)dt$ [2], the levels of the torsional modes having been attenuated significantly.

---

1. Strictly-speaking, because of the ideal integrator in the basic pre-filter structure shown in Figures 8.10 and 8.11, the tracking error of the RTF to a ramp in mechanical power is non-zero. As explained in Section 8.6.2.2 this error is very small, and is zero when there are one or more washout filters ahead of the integrator.
2. Note that the slow changes in the integral of mechanical power are not attenuated by the RTF (see its frequency response in Figure 8.9).

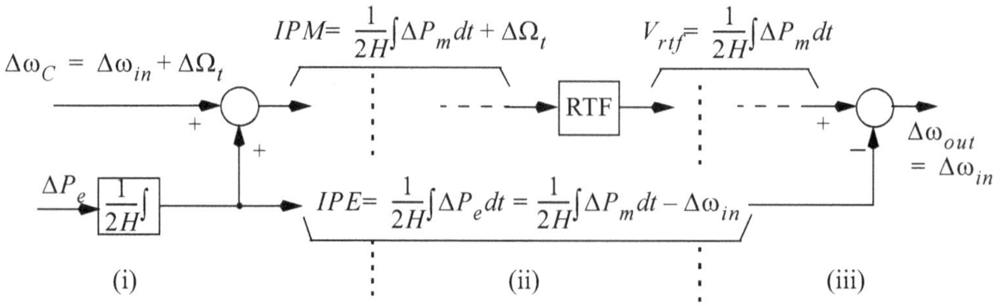

Figure 8.11 The action of the pre-filter. (i) The implementation of (8.13) and (8.16). (ii) The ramp tracking filter attenuates the torsional modes $\Delta\Omega_t(t)$ in the speed input and tracks the integral of the mechanical changes with negligible steady-state error. (iii) The signals containing the integral of the mechanical changes are cancelled out at the summing junction and the true speed signal $\Delta\omega_{out} = \Delta\omega_{in}$ is synthesized.

Finally, as shown in Figure 8.11(iii), the negated signal *IPE* is combined with the output of the RTF at the summing junction. The component $\frac{1}{2H}\int \Delta P_m(t)dt$ present in each signal is cancelled out resulting in the output of the pre-filter being the required 'true' rotor speed, $\Delta\omega_{in}(s)$.

To compensate for a difference in the levels of the speed signal in the speed-signal path from that derived from electric power [1], the gain k_s is provided as shown in Figure 8.12. (For example, this adjustment may be required if an attenuated speed signal is derived from bus frequency, see Section 8.4.2). Furthermore, to eliminate any steady-state levels in the electrical power and speed inputs, $\Delta P_e(t)$ and $\Delta\omega_C(t)$, two washout filters are added to each input; this completes the block diagram of the IAP pre-filter. (The effect on the synthesized speed signal $\Delta\omega_{out}$, say, of having two washouts in the speed input and one in the electric power input path is discussed briefly in the later Section 8.6.4.2.)

We know that the RTF follows a ramp input at its terminals with zero steady-state error e_{fss} between its input and output. In practice there are washout filters and an integrator between the mechanical ramp input and the input to the RTF. The input to the RTF may no longer be a ramp, how does this affect the steady-state error?

---

1. The degradation in performance of the PSS in such a case is illustrated in Figure 8.20(i).

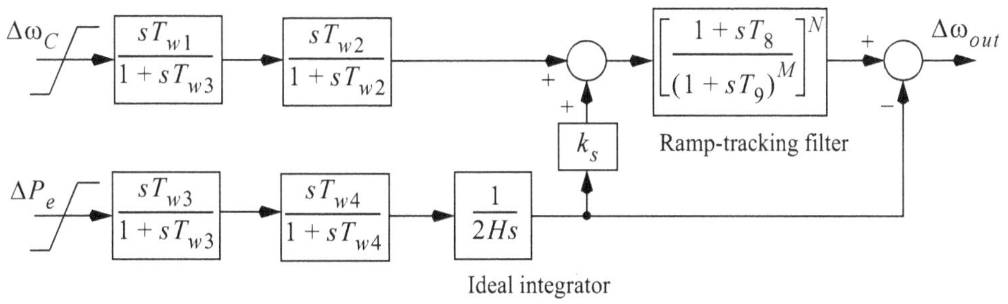

Figure 8.12 Block diagram of the prefilter for the IAP PSS. The gain k_s is set to unity in the following analysis.

8.6.2 Effect of the washout filters and integrators on the performance of the pre-filter

In the previous section the conceptual action of the pre-filter without washout filters was analysed; let us now consider their effect on the tracking of the RTF and the dynamic performance of the pre-filter.

In Section 8.2 the dynamic characteristics of one or two washout filters are analysed in their own right. However, as a diversion, let us (i) assume the speed and torsional signals are negligible and (ii) examine the steady-state and dynamic performance only of the path associated with the electric power input, namely the washout filters, the integrator and the RTF. This path is shown in Figure 8.13. *Note that a fictitious test input signal $U(s)$ is used for the purposes of this analysis and is a step, ramp, parabolic or cubic function of time only.* We will also consider two cases when the integrator in the pre-filter is represented as an ideal or as a pseudo-integrator; the latter is referred to as the 'practical' integrator. In essence, in this analysis the performance of the RTF to a particular set of characteristics of the mechanical power output is being studied.

It has been emphasized that the component $[1/(2H)] \cdot \int \Delta P_m(t) dt$ in the signal IPM should pass through the RTF with zero following error so that it cancels (ideally) the same component in the signal IPE when the mechanical power is ramped. Several questions arise. Due to the action of the washouts and the integrator, does the output of the RTF still follow its input with zero steady-state error when that input is no longer a ramp? For example, consider the output of the washout filters in Figure 8.13. Does the RTF track with zero error other mechanical power inputs, e.g. $a_n t^n$, $n > 1$?

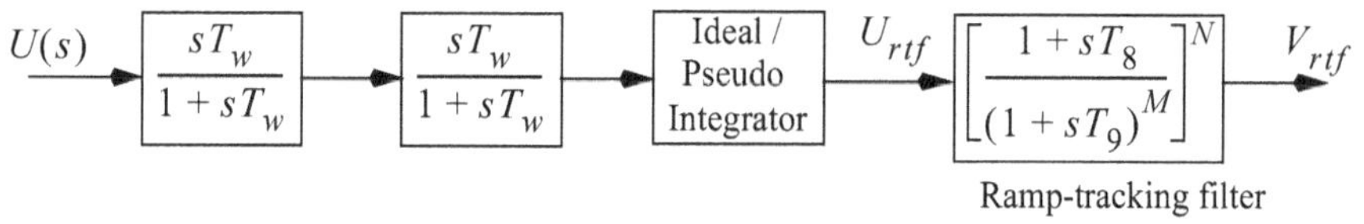

Figure 8.13 Path between the electric power input and the RTF output for a test input $U(s)$ which replaces the electric power signal.

Note that, because the blocks in the path of Figure 8.13 represent linear elements, the principle of superposition permits the performance of this sub-system to be analysed independently of the rest of the pre-filter. The behaviour of this sub-system also reflects its behaviour when it is incorporated in the complete prefilter.

8.6.2.1 Dynamic response of the isolated path of Figure 8.13 to a ramp input

Useful insight is provided by examining both the dynamic and steady-state responses at the input and output of the RTF as well the tracking errors for a ramp in mechanical power. We are concerned only with the path of Figure 8.13.

Let us now demonstrate the nature of the response of the RTF for a ramp of rate $R_0 = 0.0075$ pu/s in mechanical power output. For the current and later applications the parameters of the complete pre-filter of Figure 8.12 are given below:

- Washout filters: $T_{w1} = T_{w2} = T_{w3} = T_{w4} = 7.5$ s, assuming the lowest (inter-area) modal frequency is 1.5 to 2 rad/s (only T_{w3} and T_{w4} in Figure 8.12 are relevant to the signal path under study);
- Integrator: H=3 MWs/MVA; Pseudo-integrator (as derived in Section 8.3.1): $T_H = 7.5$ s;
- RTF: $N = 1$, $M = 5$, $T_9 = 0.1$ s, $T_8 = MT_9 = 0.5$ s. (The selection of $T_9 = 0.1$ s is mentioned in Section 8.5.4).

It is shown in Figure 8.14 (a) it is noted that, for the ideal integrator, the output of the RTF does not track the ramp in mechanical power but tends to a constant value $T_W^2 R_0/(2H) = 0.0703$ in the steady state. Furthermore, the output of the RTF $V_{rtf}(t)$ tracks its input $U_{rtf}(t)$ with negligible error which, as shown in Figure 8.14 (b), tends to zero in the steady state. For the pseudo-integrator, however, it is observed in Figure 8.14(a) that the output of the RTF follows the ramp in mechanical power with zero following error in the steady-state (i.e. after some 50 s). This is because the pseudo-integrator ceases to act as an integrator and becomes a low pass filter at low frequencies.

For a ramp in mechanical power it is also noted in Figure 8.14 (a) that zero tracking error between the input and output of the RTF is achieved for both the ideal and the pseudo- integrator.

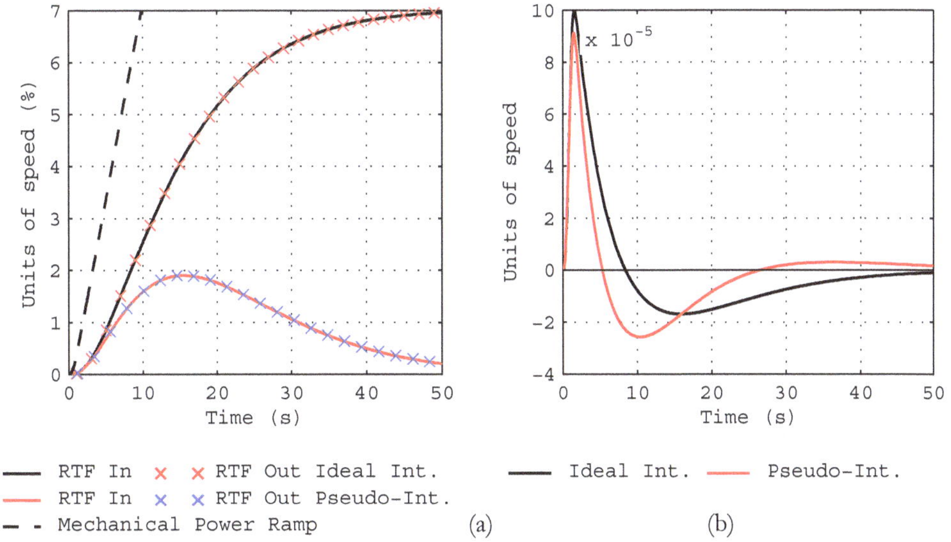

Figure 8.14 Responses to a ramp in mechanical power for ideal and pseudo-integrators with two washout filters in the isolated path of Figure 8.13. The plots show (a) the input and output responses of the RTF U_{rtf}, V_{rtf}, and (b) that the error across the RTF, $U_{rtf}(t) - V_{rtf}(t)$, in the responses is very small and tends to zero in the steady state.

(To avoid a discontinuity at time zero in Figure 8.14 (a) and (b), the initial slope of the mechanical power output is varied in parabolic fashion from zero to the ramp rate of 0.0075 pu/s at 1 s.)

8.6.2.2 The steady-state tracking - and tracking errors - of the RTF

For the RTF with the parameters given in Figure 8.9 it is known that its output V_{rtf} tracks a ramp change at its input U_{rtf} with zero steady state error. However, for a parabolic input to the RTF its output tracks the input with a constant following error after any initial transients have decayed away.

Let us examine the behaviour of the isolated path of Figure 8.13 in more detail. Firstly, for the sake of completeness, it is of interest to ascertain the performance of the RTF not only for the four types of mechanical power change $U(t)$, but also the effects of none, one and two washout filters on the tracking errors. Secondly, consideration is given to the effects of the ideal and pseudo-integrators, the transfer function of the latter being $T_H/[2H(1+sT_H)]$, (8.5). Of interest are not only the steady-state values of the input to the

RTF but also how closely the output of the RTF tracks the input to the RTF. Consequently, in Appendix 8–I.2 expressions are derived which analyse the nature of the tracking error for power changes of a general form $U(t) = a_n t^n$.[1]. The results are summarised in Table 8.6.

The upper value in each row of the table is the *steady-state input to the* RTF, (not the power changes at the input, $\Delta U(s)$). The steady-state input is

$$U_{rtf}\big|_{ss} = \lim_{t \to \infty} U_{rtf}(t) = \lim_{s \to 0} sU_{rtf}(s)$$.[2].

The lower value is the steady-state tracking error of the RTF, i.e. the difference between the steady-state input to the RTF and its output, i.e. $e_{fss} = U_{rtf}\big|_{ss} - V_{rtf}\big|_{ss}$. Note that:

- $x \to \infty$ means the quantity increases indefinitely with time.
- When the both the mechanical power and the input to the RTF are increasing indefinitely with time the tracking error may be zero or finite (e.g. columns 5 to 8, parabolic input).

Although the tracking error is zero for a ramp applied directly to the RTF (column 1), when a ramp is applied to an ideal integrator in the path the tracking error is non-zero (Table 8.6, ramp, col. 4). The conceptual discussion in Section 8.6 surrounding Figure 8.11, in which there is an ideal integrator in the path, is based on the assumption that the tracking error is zero. However, it can be shown that this error is small even for fast ramps. In practice of course, there are one or more washout filters in the power-signal path in which case the tracking error of the RTF is zero.

In summary, the practical case is the replacement of the ideal integrator by the pseudo-integrator of (8.5) with one or two washout filters in the electric power input path. As noted in Table 8.6 - and analysed in Appendix 8–I.2 - the steady-state tracking errors of the RTF are zero if a pseudo-integrator is employed when the mechanical power input is a step, ramp, or parabola.

---

1. The expressions are for the input to the RTF and the tracking error between RTF input and output. For the ideal integrator these are (8.29) and (8.32), respectively; for the pseudo-integrator they are (8.33) and(8.34).
2. Final Value Theorem. See Section 2.10.

Table 8.6 Steady-state input to the RTF $U_{rtf}|_{ss}$ and the tracking errors following changes in mechanical power

Type of mechanical power change $U(t)$, $t \geq 0$		Input to RTF / Tracking error	Input to RTF, $U_{rtf}\|_{ss}$, output $V_{rtf}\|_{ss}$ and tracking error of the RTF, $e_{fss} = U_{rtf}\|_{ss} - V_{rtf}\|_{ss}$ as $t \to \infty$, for mechanical power changes $U(t)$ applied to the path in Figure 8.13:							Notes
			RTF only.(	Ideal integrator & RTF (no washouts)	Pseudo-integrator & RTF (no washouts)	One washout, ideal integrator & RTF	One washout, pseudo-integrator & RTF	Two washouts, ideal integrator & RTF	Two washouts, pseudo-integrator & RTF	
1	**2**	**3**	**4**	**5**	**6**	**7**	**8**	**9**	**10**	
Step, $n = 0$	R_0	$U_{rtf}\|_{ss}$	R_0	$\to \infty$	$K_{0P}R_0$	$K_{1I}R_0$	0	0	0	$K_G = 10T_9^2$,
		e_{fss}	0	0	0	0	0	0	0	$K_{0I} = 1/(2H)$
Ramp, $n = 1$	$R_0 t$	$U_{rtf}\|_{ss}$	$\to \infty$	$\to \infty$	$\to \infty$	$\to \infty$	$K_{1P}R_0$	$K_{2I}R_0$	0	$K_{0P} = T_H/(2H)$,
		e_{fss}	0	$K_{0I}K_G R_0$	0	0	0	0	0	$K_{1I} = T_W/(2H)$,
Parabola, $n = 2$	$R_0 t^2/2$	$U_{rtf}\|_{ss}$	$\to \infty$	$\to \infty$	$\to \infty$	$\to \infty$	$\to \infty$	$\to \infty$	$K_{2P}R_0$	$K_{1P} = T_W T_H/(2H)$
		e_{fss}	$R_0 K_G$	$\to \infty$	$K_{0P}K_G R_0$	$K_{1I}K_G R_0$	0	0	0	
Cubic, $n = 3$	$R_0 t^3/6$	$U_{rtf}\|_{ss}$	$\to \infty$	$\to \infty$	$\to \infty$	$\to \infty$	$\to \infty$	$\to \infty$	$\to \infty$	$K_{2I} = T_W^2/(2H)$
		e_{fss}	$\to \infty$	$\to \infty$	$\to \infty$	$\to \infty$	$K_{1P}K_G R_0$	$K_{2I}K_G R_0$	0	$K_{2P} = T_W^2 T_H/(2H)$

RTF: N=1, M=5; Inertia constant H (MWs/MVA); Integrators: Ideal (I) $1/(2Hs)$, Pseudo (P) $T_H/[2H(1 + sT_H)]$

8.6.3 Dynamic performance of the complete pre-filter

The SMIB system, Case C, described in Sections 5.10 and 5.11 will be used to investigate the performance of the pre-filter using the parameters provided in Section 8.6.2.1. In Table 5.5 the input to the PSS is rotor speed; the rotor mode of oscillation is $-1.34 \pm j9.03$. Using an IAP PSS with the pre-filter parameters of Section 8.6.2.1 together with the SMIB PSS parameters (derived in Section 5.10.6), the value of the rotor mode is virtually unchanged at $-1.32 \pm j8.97$.

For illustrative purposes the following three disturbances are applied to the generating unit:

- A ramp increase in mechanical power input is 0.45 pu per minute, or 0.0075 pu/s, over a period of 20 s. (This rate is exaggerated to highlight certain features in the responses.) To avoid a discontinuity at time zero, the initial slope of the mechanical power output is varied in parabolic fashion from zero to the ramp rate of 0.0075 pu/s at 1 s.

- A relatively small step increase of 5% in terminal voltage reference at 4 s, followed by a step decrease of 5% at 12 s.

- An exaggerated, sustained torsional mode of 12 Hz (75.4 rad/s) and peak amplitude 0.25%, commencing at 12 s.

The simultaneous application of an increasing ramp, and the step change in voltage, should reveal how the pre-filter discriminates between the changes in mechanical power input and disassociated electrical power perturbations, oscillatory in nature, resulting from the change in reference voltage. While responses to small changes in mechanical power at the ramp rate specified are amenable to analysis using a small-signal model of the SMIB, the change in mechanical power of 0.45 pu per minute over a period of 20 s is not small. Although it is inconsistent to mix small- and large-signal analyses, the important issue here is the assessment the performance of the pre-filter which is a linear element. Moreover, using the small-signal model of the SMIB system *provides to the pre-filter the electric power and rotor speed signals inputs of the correct relative amplitudes and phase.* Again, for the purposes of illustration, the amplitude of the sustained torsional mode is exaggerated and is large, being of the same order of amplitude as the speed perturbations resulting from the step in reference voltage.

In Figure 8.15 the variable names and their locations in the pre-filter are defined for use in subsequent figures. Variable names *IPE* and *IPM* are defined earlier in (8.16) and (8.18) respectively. The 'true' rotor speed at the input is $\Delta\omega_{in}$; $\Delta\Omega_t$ represents the torsional modes present; $\Delta\omega_{out}$ is the speed output of the pre-filter (and ideally is equal to the 'true' speed input $\Delta\omega_{in}$). The output of the second speed washout filter is $\Delta\omega_{W2}$ and that of the second electric-power washout filter is ΔP_{W2}; ΔV_{rtf} is the output signal of the RTF.

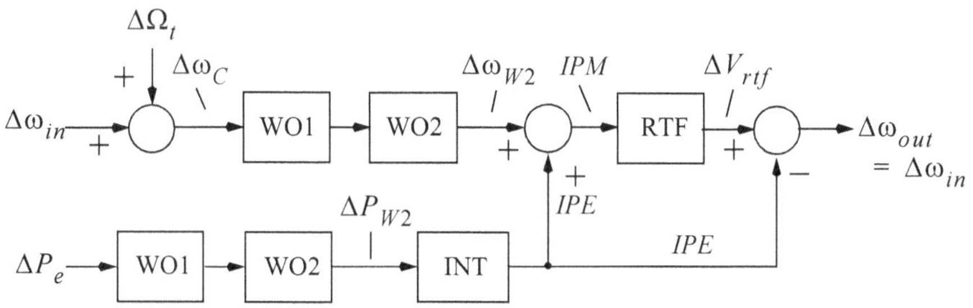

Figure 8.15 The variable names and their locations in the pre-filter are defined for use in Figures 8.16 to 8.18.

For the three disturbances the responses of the variables in the pre-filter are shown in Figures 8.16 to 8.18. The left- and right-hand plots in each figure show the relevant responses when torsional modes are absent or present, respectively. So that the responses to the changes in reference voltage are clearly discernible, the damping gain of the PSS in Case C, Section 5.10.6, is reduced from 20 to 10 pu.

From Figure 8.16 the following are noted:

- In (a)-(i) the nature and timing of two of the input disturbances are shown.
- In (a)-(ii) the decaying oscillatory responses in true speed $\Delta\omega_{in}$ due to the step changes in reference voltage are observed; the output of the second speed washout filter $\Delta\omega_{W2}$ (not shown) is identical for practical purposes.
- In (a)-(ii), as predicted by (8.16), the output of the pseudo-integrator (*IPE*) contains both the oscillatory rotor speed component and a component associated with the ramp in mechanical power. Importantly, *it is observed that the true speed component is eliminated from the signal IPM which is input to the* RTF.
- However, in (b)-(i) the signal *IPM* at the input to the RTF contains *a component associated with the ramp in mechanical power as well as the torsional mode,* $\Delta\Omega_t$. As mentioned, the true speed component seen in *IPE* is absent from *IPM.*

In (b)-(ii) is shown $\Delta\omega_C$, the torsional mode modulated by the true speed component.

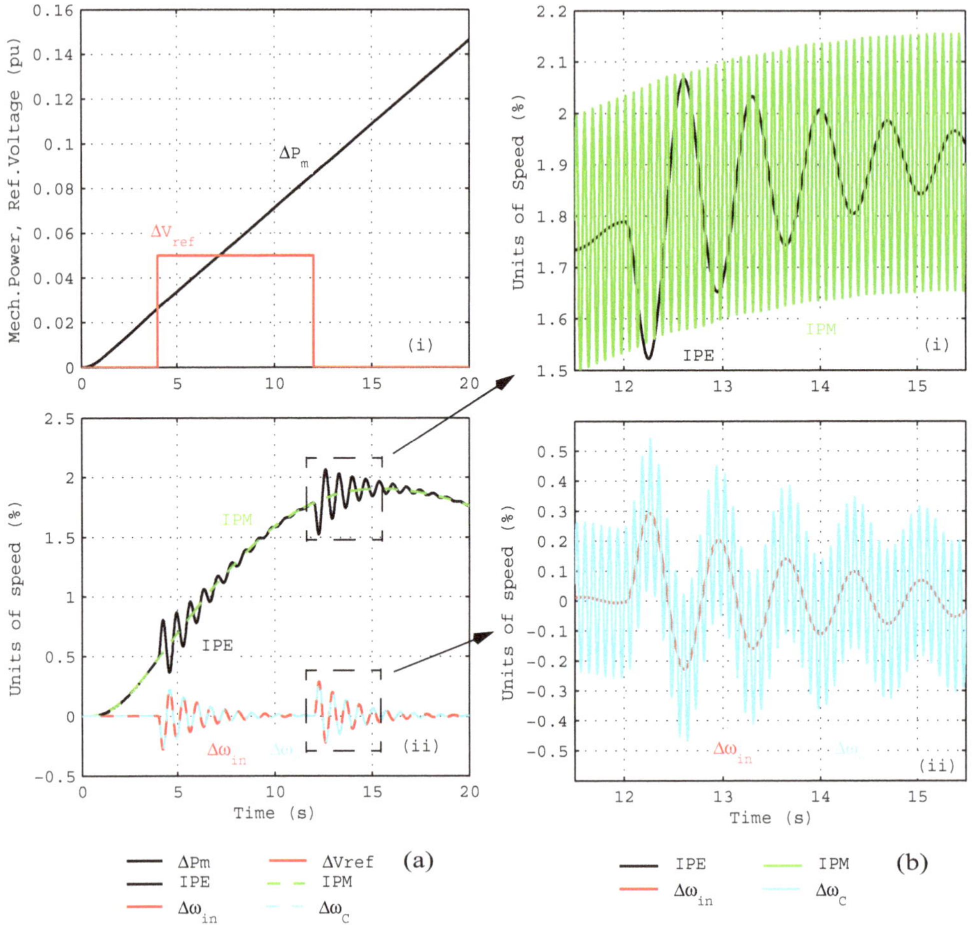

Figure 8.16

(a) Torsional mode absent

(i) Disturbances: Ramp in mechanical power ΔP_m and step changes in reference voltage, ΔV_{ref}

(ii) Inputs: $\Delta\omega_{in}$ & $\Delta\omega_C$. Responses of internal variables *IPE* & *IPM*

(b) Torsional mode present

(i) Responses of internal variables *IPE* & *IPM*

(ii) Inputs: $\Delta\omega_{in}$ & $\Delta\omega_C$

Consider Figure 8.17 in which are shown the responses of internal and external variables.

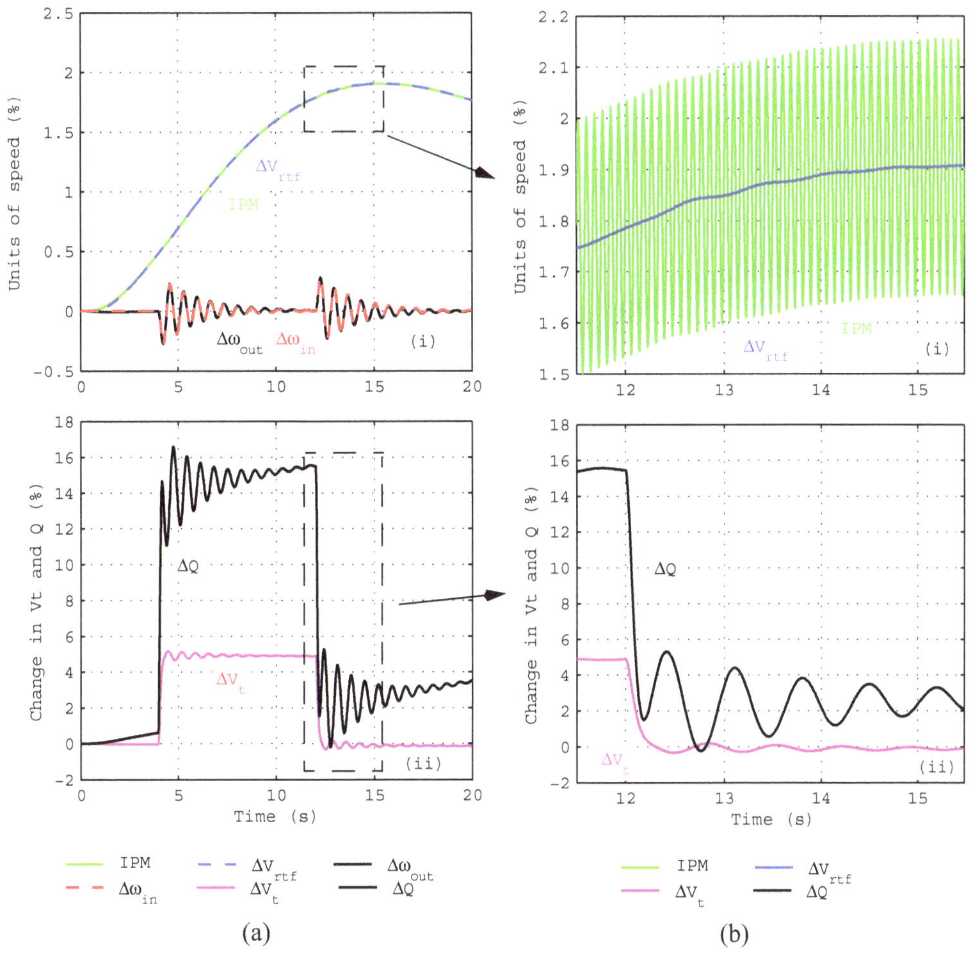

Figure 8.17

(a) Torsional mode absent	(b) Torsional mode present
(i) Responses of internal variables *IPM* & ΔV_{rtf}; input $\Delta\omega_{in}$ & output $\Delta\omega_{out}$	(i) Responses of internal variables *IPM* & ΔV_{rtf}

(ii) Responses: terminal voltage ΔV_t and reactive power ΔQ

In Figure 8.17 it is observed that:

- in (a)-(i) the output of the RTF, ΔV_{rtf}, follows the mechanical power component related to the input signal to the RTF, *IPM*, *with zero tracking error*;

- in (a)-(i) *the speed output signal* $\Delta\omega_{out}$ *from the pre-filter is identical to the 'true' speed input signal* $\Delta\omega_{in}$; the associated rotor mode is clearly evident in the terminal voltage and reactive power responses in (a)-(ii).

- in (b)-(i) the *torsional mode present at the input to the RTF, IPM, is not evident in the heavily attenuated output of the RTF,* ΔV_{rtf}.

In considering Figure 8.17(a)-(ii), it should be remembered that, as the electrical power output increases while following the mechanical power ramp, the reactive output of the generator will also ramp in order to supply the additional I^2X losses. Moreover, from the figure it is noted that there is also a step increase/decrease in reactive power output associated with the step changes in terminal voltage; this is superimposed on the reactive power ramp. In Figure 8.17(b)-(ii) there is no evidence of the heavily attenuated torsional mode in terminal voltage and reactive power responses.

The output of the second washout filter in the electrical power signal path ΔP_{W2} is displayed in Figure 8.18(a)-(ii), together with the output of the pseudo-integrator (*IPE*). The effect of the mechanical ramp change can be observed in both signals.

The responses of the speed output signal from the pre-filter $\Delta\omega_{out}$ and associated response of the PSS ΔV_{pss} are seen in Figure 8.18(a)-(ii). Note that there is negligible off-set in both these signals from their zero values. Therefore, as a consequence, the offset in the output of the pseudo-integrator (*IPE*) *due the ramping of mechanical power will not be manifested as an offset either in the PSS output, the terminal voltage, nor in the reactive power output of the unit.* When the torsional mode is present, due to amplification by the PSS, there is evidence of the attenuated torsional mode in the PSS output in the expanded display of Figure 8.18(b)-(ii). Bear in mind, however, the amplitude of the torsional mode, seen in Figure 8.18(b)-(i), and the ramp rate of mechanical power have been exaggerated for illustrative purposes.

Figures 8.16 to 8.18 confirm that, due to the action of a properly designed pre-filter, the effects of neither the ramping of the mechanical power output of the turbine, nor of torsional oscillations, are manifested in the output of the PSS. Furthermore, the swinging of terminal voltage and reactive power output due to ramping of power is not observed.

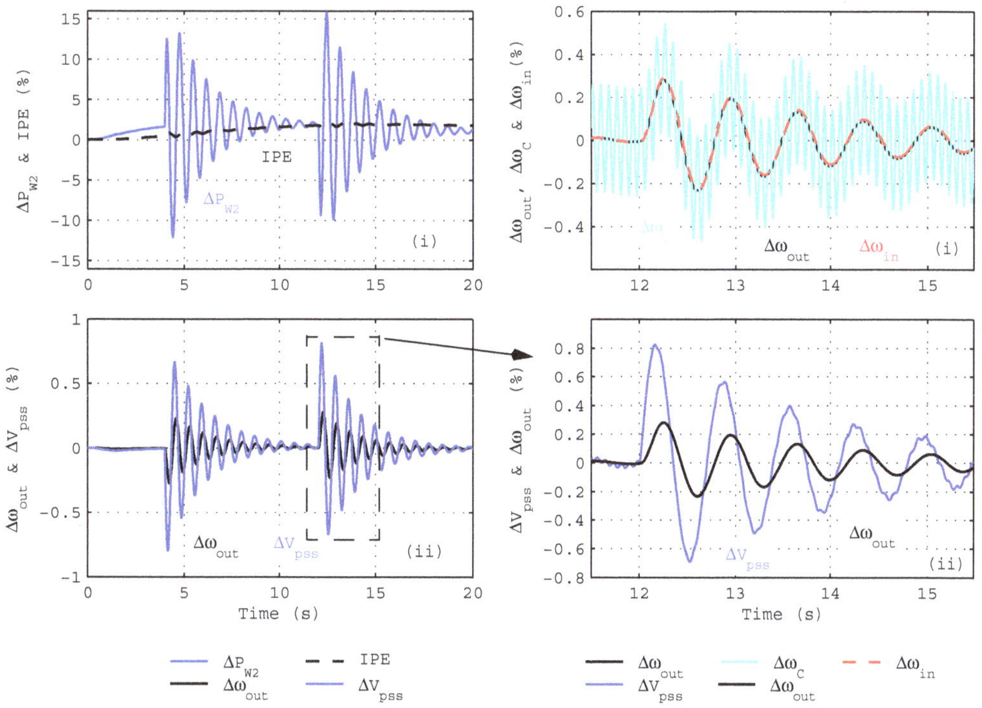

(a) (b)

Figure 8.18

(a) Torsional mode absent

(i) Responses: Washout ΔP_{W2} and *IPE*

(b) Torsional mode present

(i) Inputs: $\Delta\omega_{in}$ & $\Delta\omega_C$
Output: $\Delta\omega_{out}$

(ii) Outputs: $\Delta\omega_{out}$ & ΔV_{pss}

8.6.4 Potential causes of degradation in performance of the pre-filter of the IAP PSS

Degradation in the performance of the pre-filter may be attributable to a number of causes. Several of these are now examined.

8.6.4.1 Effects of non-ideal pre-processing the speed input signal to the pre-filter.

Any pre-processing of the speed input signal may result in incomplete cancellation of the speed signal at the input to the RTF; complete cancellation is seen as an essential feature of the pre-filter.

In Section 8.5.1 it is pointed out that the speed signal may be derived from a number of sources, including the true rotor speed which itself may be subject to some form of processing prior to injection to the pre-filter of the PSS. In the case of a 'speed' signal derived from bus-frequency the signal may be subject to attenuation as established in Section 8.4.2 For illustrative purposes it will now be assumed that the true rotor speed signal $\Delta\omega_{in}$ is processed through a first-order pre-processing filter prior to input to the PSS pre-filter.

Let the transfer function $G_A(s)$ of the speed pre-processing filter of the true rotor speed signal be

$$G_A(s) = (\Delta\omega_A/\Delta\omega_{in}) = A/(1+sT_A). \tag{8.19}$$

With this transfer function the effects of attenuation - or gain - and phase shift on the output speed signal of the PSS pre-filter, $\Delta\omega_S(=\Delta\omega_{out})$ are to be analysed. The output of the speed pre-processing filter is $\Delta\omega_A$, A and T_A are the gain and time constant. The relevant elements of the PSS pre-filter which includes the speed pre-processing filter are shown in Figure 8.19. Perturbations in mechanical power output and the torsional mode are assumed to be absent; according to (8.4) the true rotor speed is

$$-\Delta\omega_{in} = \frac{1}{2Hs}\Delta P_e. \tag{8.20}$$

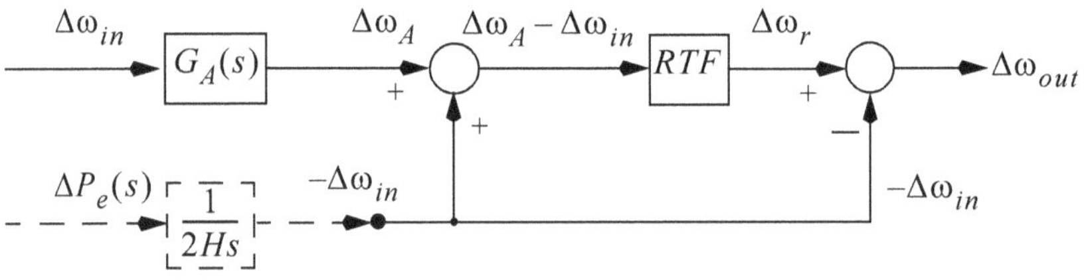

Figure 8.19 Signals in the IAP pre-filter assuming non-ideal pre-processing of the speed input signal through a transfer function $G_A(s)$ ($\Delta P_m = 0$).

Based on (8.19) and Figure 8.19 it can be shown that the output of the pre-filter is:

$$\Delta\omega_{out} = \left[\frac{(A-1)-sT_A}{1+sT_A}\cdot RTF(s)+1\right]\Delta\omega_{in}. \tag{8.21}$$

Clearly, at low frequencies $\Delta\omega_{out} \to A\Delta\omega_{in}$ and at high frequencies $\Delta\omega_{out} \to \Delta\omega_{in}$. Over the frequency range typically of interest the responses, or distortion factors $(\Delta\omega_{out}/\Delta\omega_{in})$ in the true speed, are shown in Figure 8.20 for a range of values of A and time constants T_A. The parameters of the RTF of (8.14) are $N = 1$, $M = 5$ and $T_9 = 0.1$ s.

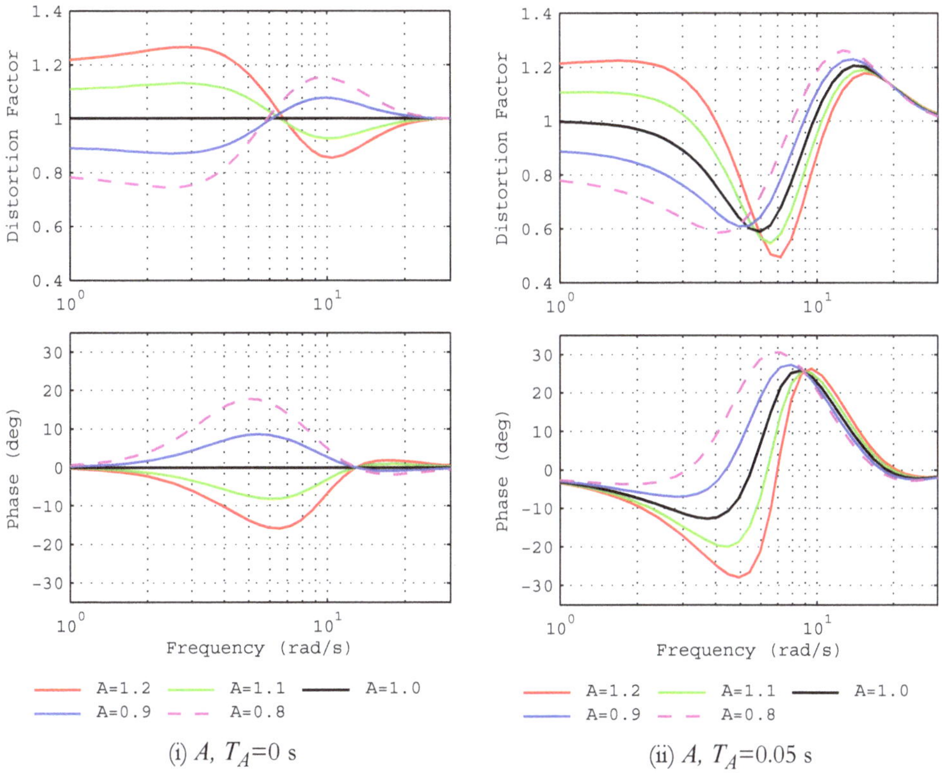

Figure 8.20 Distortion factors, $\Delta\omega_{out}/\Delta\omega_{in}$, due non-ideal pre-processing of the true speed input signal to the pre-filter. Values A: 0.8 to 1.2; (i) $T_A = 0$, (ii) $T_A = 0.05$ s.

Of concern in the figure are the effects of the amplitude and the phase shift on the pre-processed speed signal $\Delta\omega_A$ over the range of frequencies of the rotor modes, 1.5 to 15 rad/s, and their deviation from the ideal response of $1\angle 0°$. Although the range of values of A and T_A employed in Figure 8.20 may be considered somewhat extreme, the results imply that appropriate care is required in the pre-processing of the speed input signal to the pre-filter. These results show that depending on how the speed-input signal to the pre-filter is derived in practice, significant distortion in both gain and phase of the synthesised speed signal can occur.

Various methods can be employed for calculating the electric power. Any pre-processing filters which are employed in the electric power input signals paths may also result in incomplete cancellation of the speed signal at the input to the RTF. This would likewise result in distortion of the speed output of the PSS pre-filter. The effects of any pre-processing of input signals to the PSS pre-filter should therefore be examined to assess if they degrade the performance of the PSS.

Consider the case of a bus-frequency stabilizing input, the associated pseudo-speed signal ω_{freq} being derived from the rate of change of terminal voltage angle as in (8.9). The degradation in the amplitude of this signal is discussed in Section 8.4.2. The effect of such degradation on the output of prefilter $\Delta\omega_{out}$ is illustrated in Figure 8.20(i). Not only is the amplitude of $\Delta\omega_{out}$ modified but also is its phase- which could introduce an additional phase lag in the PSS over the modal frequency range of interest.

8.6.4.2 One or two washout filters in the electrical-power and speed paths?
Recall that a washout filter is introduced with the purpose of eliminating any steady-state offsets, or DC levels, in the input signal, as well as blocking very slow changes in the input. It is thus necessary to include at least one washout filter in each path of the pre-filter.

The effect on the response of the RTF of one or two washout filters in the electrical-power path has been examined in Section 8.6.2. The performance requirements for the pre-filter may thus determine the number of washout filters in this path.

What are the effects of choosing a different number of washout filters in the speed and electric power paths? The following requirements must be satisfied:

- When considering the presence of the local- and inter-area modes in each of the two signal paths, the frequency response of both one or two washout filters should be ideally, or close to, $1\angle 0°$ over the range of modal frequencies. This requirement dictates the value of the washout time constant, T_W.

- The frequency response of a pseudo-integrator in the electrical-power path should be ideally, or close to, $[1/(2H\omega_f)]\ \angle{-90°}$ over the range of modal frequencies. This requirement determines the time constant of the pseudo-integrator.

If there are different numbers of washout filters in the speed and power paths an imprecise cancellation of the true rotor-speed at the input to the RTF occurs under perturbed conditions. It is therefore desirable that the same number of washouts be employed in both input paths.

8.6.4.3 Effect on the synthesized speed signal of setting the RTF time constant T_8 to zero.
As in earlier sections, the SMIB system Case C, described in Section 5.10 will be used to investigate the performance of the pre-filter when the time constant T_8 is set to zero.

In order for the RTF to follow a ramp with zero steady-state error a requirement is that $T_8 = MT_9$ in the RTF transfer function of (8.14). Setting T_8 to zero turns the RTF into a simple low-pass filter of order M if $N = 1$. With this setting and for a ramp in mechanical power the output of the RTF follows the input signal IPM with non-zero error. In Figure 8.21, and comparing it with Figure 8.18(a)-(ii), this error is seen to manifest itself not only in the synthesized speed signal at the output of the pre-filter, but also in the PSS output. Consequently there is an associated undesirable swing of the generator terminal voltage and reactive power output during ramping of the mechanical power.

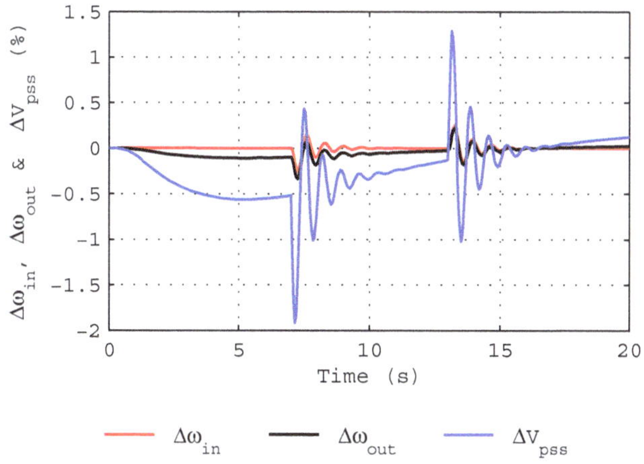

Figure 8.21 Deviation in the synthesized speed output ($\Delta\omega_{out}$) from the speed input ($\Delta\omega_{in}$), and the consequent effect on the PSS output signal (ΔV_{pss}), due to setting $T_8 = 0$ in the RTF. (Compare these responses with those in Figure 8.18(a)-(ii).)

8.7 The Multi-Band Power System Stabilizer

A fourth category of the PSS models listed in the IEEE Standard [15] is called the Multi-Band PSS (MB-PSS), PSS4B, which was first developed by Hydro-Québec and is in operation on the Hydro-Québec system [16], [17].

A block diagram of the MB-PSS structure is shown in Figure 8.22. It is noted that this PSS has three separate tunable paths, unlike the integral-of-accelerating-power PSS, PSS2B [15], which only has a single such path. The objective of the MB-PSS structure is to isolate and focus the PSS tuning in three frequency bands which account for three phenomena:

- 0.05 - 0.2 Hz (~0.3 - 1.2 rad/s): very slow oscillations associated with the common or global modes on a system [1];
- 0.2 - 1 Hz (~1.2 - 6 rad/s): low frequency, inter-area modes of rotor oscillation;
- 1 - 4 Hz (~6 - 25 rad/s): higher frequency, local-area and intra-plant modes of rotor oscillation.

---

1. Note that it is important not to confuse the *global mode* with low-frequency modes sometimes observed with hydro-turbines, for example. The latter modes may be associated with governor - water column interactions and are *localized* phenomena [19].

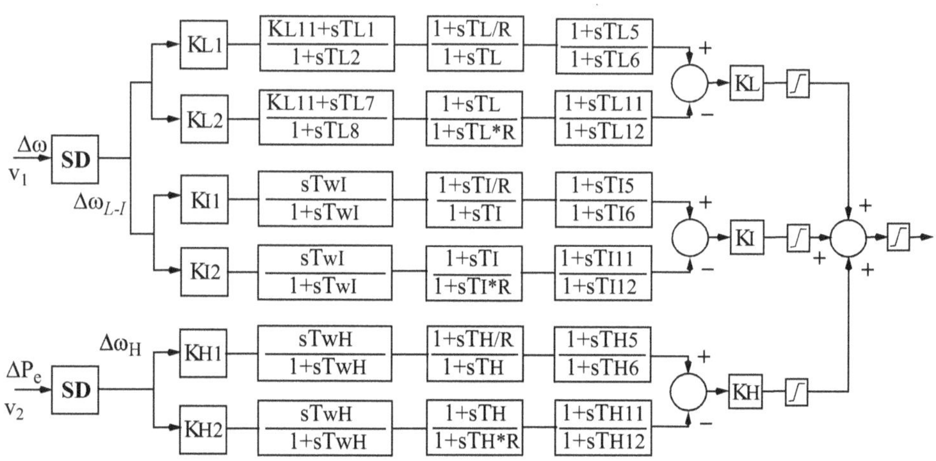

Figure 8.22 Multi-Band PSS. SD: Speed Transducer. (See [16], [17] for details).

In particular, the low frequency band is introduced to provide damping for very low frequency phenomena encountered on isolated systems [1], particularly the so-called global mode in such a system. It is stated in [17] that the MB-PSS and the integral-of-accelerating-power PSS "... can be tuned to achieve quite similar performance in the local, intra-unit and torsional modes ... since they both use an electric power signal to capture the high frequency dynamics. However, having many more degrees of freedom available to modulate its phase lead over a wide frequency range allows the MB-PSS to better balance its performance in inter-area modes from 0.1 to 0.8 Hz" (0.6 to 5 rad/s).

Low frequency oscillations have been observed, for example in hydro-systems: 0.63 rad/s between the Northwest and Southwest power systems in the US [18]; 0.31 to 0.50 rad/s on the Colombian system [19]. Oscillations lying in the intermediate range, associated with vortex instability in hydro machines, are reported to be less than 0.5 Hz (3 rad/s) [20], and about 1 Hz (6 rad/s) [21].

The speed signal $\Delta\omega_H$, input to the high frequency band, is derived from the measured generator electrical power output. A separate internal frequency transducer supplies a speed signal $\Delta\omega_{L-I}$ to the low and intermediate frequency bands. Washout filters are provided in the intermediate and high frequency bands; torsional (notch) filters may be incorporated in the PSS structure. In each of the three bands is a differential filter arrangement; it is of interest to understand the characteristics of such a filter. An analysis of a simplified form of the filter, shown in Figure 8.23, is conducted in Appendix 8–I.3.

1. Systems may be isolated because there are no synchronous links to neighbouring systems.

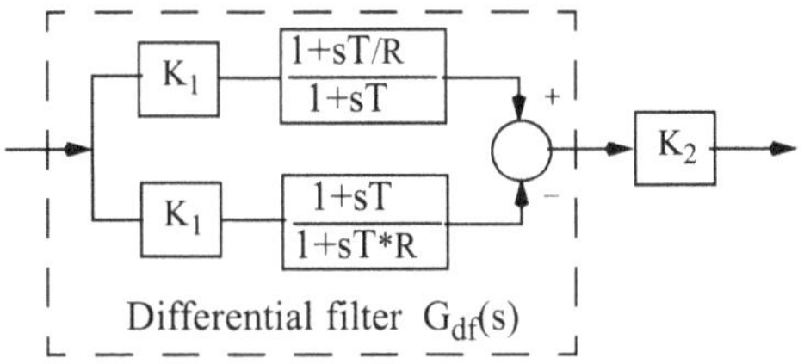

Figure 8.23 Filter *G*(*s*)

The analysis reveals the filter takes the form of the well-known Q-filter,

$$G(s) = \frac{K_2(2\xi/\omega_m)s}{1 + (2\xi/\omega_m)s + (s/\omega_m)^2}, \tag{8.22}$$

where ω_m rad/s is the frequency at which the frequency response is at its maximum value K_2. The frequency response of (8.22) with variation in damping ratio ξ is given Figure 2.21.

It is of interest to examine the nature of the frequency response of the MB-PSS omitting washout filters, speed transducers, and torsional (notch) filters. Let the gains and centre frequencies of the three bands, evaluated in Figure 5 of [16] be K_L = 5.0 pu, F_L = 0.04 Hz; K_I = 25.0 pu, F_I = 0.70 Hz; K_H = 120 pu, F_H = 8.0 Hz; respectively. The frequency responses of three bands and the output of the MB-PSS are shown in Figure 8.24; they agree closely with Figures 5 and 6 in [16].

In [17], a detailed comparison is provided on a test system between the designs of the MB-PSS (PSS4B) and the integral-of-accelerating-power PSS (PSS2B). For the MB-PSS it is found that, by separating out the low frequency and the higher frequency bands (each of which have their own limits and wash-out filters), the lower-frequency band limits and wash-out can be adjusted independently of the higher frequency bands to account for islanding and large frequency deviations.

Figure 8.24 reveals that, for the selected parameter values, the phase response varies between 35 and 60 degrees leading. That is, the phase response is relatively level over the range of 0.1 to 25 rad/s (0.02 to 4 Hz) in this case. However, the MB-PSS gain varies over a wide range. Interestingly, this approach contrasts with that of the P-Vr method (Section 5.8.1) in which the PSS transfer function attempts to account for the inherent gain and phase characteristic of the particular generator - on which the PSS is installed - over a relevant range of modal frequencies (e.g. see Figure 5.16) and an encompassing set of operating conditions.

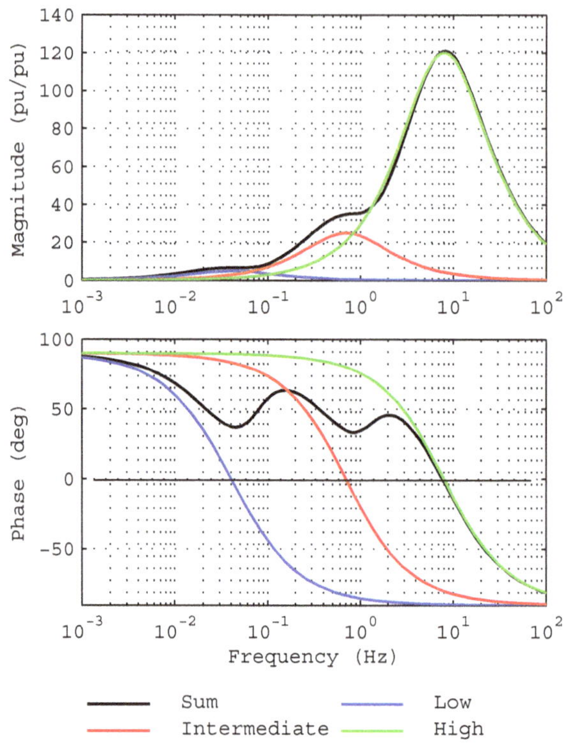

Figure 8.24 Frequency responses of the MB-PSS assuming a common speed signal input to the three differential filters in Figure 8.22

A number of methods for the tuning the MB-PSS has been offered. For example, the parameters of the MB-PSS are selected by adjusting the centre frequency and gain of each band so as to achieve the nearly flat phase response between 30 and 50 degrees over the range of frequencies, say, 0.05 Hz and 3 Hz (0.3 to 20 rad/s) in order to cover the global and intra-station modes. Other approaches, including optimization techniques, are proposed in [22], [23], [24] and [25].

8.8 Concluding remarks

In Chapter 5 a PSS based on the P-Vr design approach is described; it assumes a 'true' speed stabilizing signal is available. By 'true' speed is implied that the signal faithfully represents the generator speed in magnitude and phase, torsional oscillations being negligible. In this chapter electric-power and bus-frequency based pre-filters are employed to yield a *synthesized* speed signal for input to a P-Vr based speed-PSS. A similar objective applies to the pre-filter for the integral-of-accelerating-power PSS but overcomes some of the disadvantages of the previous two pre-filters.

The frequency of the generator terminal voltage is used as a PSS stabilizing signal on the basis that bus frequency closely represents rotor speed perturbations in magnitude and phase. Although it may be synthesized from terminal voltages, bus frequency is assumed to be derived from the rate of change of bus voltage-angle α, i.e. $\omega_{freq} = (1/\omega_0)(d\alpha/dt)$ pu of system frequency. Using the latter signal as the stabilizing signal for a 'true' speed-PSS is shown to reduce the effective damping gain of the PSS (by as much as 40% in the cases studied). However, the damping gain of the PSS can be increased to compensate for the gain reduction. Because differentiation of a signal occurs, care should be taken to provide adequate attenuation at high frequencies (i) to reduce noise, and (ii) to eliminate a possible source of instability - as is demonstrated in an example. In the signal processing for this and other forms of bus-frequency transducers, care should be taken to avoid the introduction of phase shifts which may degrade the design of the PSS unless they can be accounted for. The effect on bus frequency of large, sudden disturbances at the generator terminals should be examined.

The performance of the electric power PSS is shown to be close to that of designed for the conventional 'true' speed-PSS. However, in comparison with a 'true' speed stabilizing signal which in practice may contain torsional modes, the advantage of this pre-filter is that it significantly attenuates these modes in its output speed signal. However, the conventional PSS has the disadvantage that ramping of the mechanical power output of the prime mover causes variations in the terminal voltage and reactive power output of the generator. This problem can be ameliorated by use of an integral-of-accelerating-power PSS.

The integral-of-accelerating-power (IAP) pre-filter generates the speed signal for a PSS designed for a 'true' speed-stabilizing signal based on the P-Vr approach. A detailed analysis of the pre-filter for the IAP PSS is conducted and demonstrates the role and effects of the ramp tracking filter (RTF), and of the washout filters and the integrator in the power input path. It is shown that the RTF itself consists of a unity feedback system with two integrations in its forward path and therefore it tracks a ramp input at its input with zero error in the steady state. However, depending on the number of washout filters, the type of integrator, and the characteristics of the mechanical power output, the steady-state tracking errors may be finite but are small. Because the effective operation of the pre-filter relies on the cancellation of the speed signal at the output of the integrator by the input speed signal, care must be taken to ensure the fidelity - in amplitude and phase - of the speed input signal to the pre-filter. If the latter signal lacks fidelity with respect to the 'true' speed, the performance of the PSS may be markedly degraded.

PSS2B or PSS4B?

In considering the application of the multi-band and integral-of-accelerating-power PSSs the following few items may be pertinent.

In comparison to the integral-of-accelerating-power PSS (PSS2B) the feature of the Multi-Band PSS (PSS4B) is its ability to damp low-frequency and common-mode oscillations [17].

The sensitivity of the output of the latter PSS to ramping of mechanical power and slow system frequency drift is likely to be low because
(i) the gain in the low frequency band is relatively low (20-25% of the high frequency gain),
(ii) the corner frequencies of the washout filters in the intermediate band are 1 rad/s,
(iii) the high frequency speed signal transducer, in effect, has a washout corner frequency of about 1.2 rad/s (0.2 Hz)
Consequently, variations in reactive power are likely to be small.

For large steam units with the first torsional mode being about 8-10 Hz, notch filters may be required for the PSS4B. However, in the case of an integral-of-accelerating-power PSS with a ramp tracking filter having the characteristics shown in Figure 8.9, the attenuation of torsional frequencies at 8-10 Hz (50-60 rad/s) is 50 dB or more; notch filters may not be needed.

Thus, in generalizing, it is necessary to consider carefully - among other factors - the system characteristics as well of those of the generating units in order to specify the system damping performance requirements over the low to high range of modal frequencies. Following such an investigation it may then be possible to select the required PSS structure.

8.9 References

[1] P. Kundur, D. C. Lee, and H. M. Zein El-Din, "Power System Stabilizers for Thermal Units: Analytical Techniques and On-Site Validation," *Power Apparatus and Systems, IEEE Transactions on*, vol. PAS-100, pp. 81-95, 1981.

[2] W. Watson and M. E. Coultes, "Static Exciter Stabilizing Signals on Large Generators - Mechanical Problems," *Power Apparatus and Systems, IEEE Transactions on*, vol. PAS-92, pp. 204-211, 1973.

[3] H. Bourles, S. Peres, T. Margotin, and M. P. Houry, "Analysis and design of a robust coordinated AVR/PSS," *Power Systems, IEEE Transactions on*, vol. 13, pp. 568-575, 1998.

[4] Keay, F.W. and South, W.H., "Design of a Power System Stabilizer Sensing Frequency Deviation", *Power Apparatus and Systems, IEEE Transactions on*, vol. PAS-90, Issue: 2, March 1971, pp. 707 - 713.

[5] P. Kundur, *Power system stability and control.* New York: McGraw-Hill, 1994.

[6] P. Pourbeik, D. G. Ramey, N. Abi-Samra, D. Brooks, and A. Gaikwad, "Vulnerability of Large Steam Turbine Generators to Torsional Interactions During Electrical Grid Disturbances," *Power Systems, IEEE Transactions on*, vol. 22, pp. 1250-1258, 2007.

[7] A. Murdoch, S. Venkataraman, R. A. Lawson, and W. R. Pearson, "Integral of accelerating power type PSS. I. Theory, design, and tuning methodology," *Energy Conversion, IEEE Transactions on*, vol. 14, pp. 1658-1663, 1999.

[8] A. Murdoch, S. Venkataraman, and R. A. Lawson, "Integral of accelerating power type PSS. II. Field testing and performance verification," *Energy Conversion, IEEE Transactions on*, vol. 14, pp. 1664-1672, 1999.

[9] P. M. Paiva, J. M. Soares, N. Zeni, Jr., and F. H. Pons, "Extensive PSS use in large systems: the Argentinian case," in *Power Engineering Society Summer Meeting, 1999.* IEEE, 1999, pp. 68-75 vol.1.

[10] "IEEE Tutorial Course, Power System Stabilization via Excitation Control", IEEE Special Publication 09TP250,2009.

[11] G. R. Berube, L. M. Hajagos, and R. Beaulieu, "Practical utility experience with application of power system stabilizers," in *Power Engineering Society Summer Meeting, 1999.* IEEE, 1999, pp. 104-109 vol.1.

[12] P. M. Anderson, B. L. Agrawal and J. E. Van Ness, *Subsynchronous Resonance in Power Systems*, New York, IEEE Press.

[13] J.C.R. Ferraz, N. Martins, N. Zeni Jr, J.M.C. Soares, G.N. Taranto, "Adverse Increase in Generator Terminal Voltage and Reactive Power Transients Caused by Power System Stabilizers," *Proceedings of IEEE Power Engineering Society Winter Meeting*, 2002.

[14] D. C. Lee, R. E. Beaulieu, and J. R. R. Service, "A Power System Stabilizer Using Speed and Electrical Power Inputs-Design and Field Experience," *Power Apparatus and Systems, IEEE Transactions on*, vol. PAS-100, pp. 4151-4157, 1981.

[15] "IEEE Recommended Practice for Excitation System Models for Power System Stability Studies," *IEEE Std 421.5-2005 (Revision of IEEE Std 421.5-1992)*, pp. 0_1-85, 2006.

[16] R. Grondin, I. Kamwa, G. Trudel, L. Gerin-Lajoie, and J. Taborda, "Modeling and closed-loop validation of a new PSS concept, the multi-band PSS," in *Power Engineering Society General Meeting, 2003*, IEEE, 2003, p. 1809 Vol. 3.

[17] I. Kamwa, R. Grondin, and G. Trudel, "IEEE PSS2B versus PSS4B: the limits of performance of modern power system stabilizers," *Power Systems, IEEE Transactions on*, vol. 20, pp. 903-915, 2005.

[18] F. R. Schleif, G. E. Martin, and R. R. Angell, "Damping of System Oscillations with a Hydrogenerating Unit," *Power Apparatus and Systems, IEEE Transactions on*, vol. PAS-86, pp. 438-442, 1967.

[19] H. V. Pico, J. D. McCalley, A. Angel, R. Leon, and N. J. Castrillon, "Analysis of Very Low Frequency Oscillations in Hydro-Dominant Power Systems Using Multi-Unit Modeling," *Power Systems, IEEE Transactions on*, vol. 27, pp. 1906-1915, 2012.

[20] N. Martins, A. A. Barbosa, J. C. R. Ferraz, M. G. Dos Santos, A. L. B. Bergamo, C. S. Yung, V. R. Oliveira, and N. J. P. Macedo, "Retuning stabilizers for the north-

south Brazilian interconnection," in *Power Engineering Society Summer Meeting, 1999.* IEEE, 1999, pp. 58-67 vol.1.

[21] K.A. Lance, R. L.Bolden, G. Sheard, "Performance of generators with static excitation on a large network". *Proceedings of the Annual Engineering Conference.* Hobart, 1975, pp. 226-35.

[22] J. B. Simo, I. Kamwa, G. Trudel, and S. A. Tahan, "Validation of a new modal performance measure for flexible controllers design," *Power Systems, IEEE Transactions on*, vol. 11, pp. 819-826, 1996.

[23] I. Kamwa, G. Trudel, and L. Gerin-Lajoie, "Robust design and coordination of multiple damping controllers using nonlinear constrained optimization," in *Power Industry Computer Applications, 1999. PICA '99.* Proceedings of the 21st 1999 IEEE International Conference, 1999, pp. 87-94.

[24] L. Gerin-Lajoie, D. Lefebvre, M. Racine, L. Soulieres, and I. Kamwa, "Hydro-Québec experience with PSS tuning," in *Power Engineering Society Summer Meeting, 1999.* IEEE, 1999, pp. 88-95 vol.1.

[25] A. Khodabakhshian, R. Hemmati and M. Moazzami, "Multi-band power system stabilizer design by using CPCE algorithm for multi-machine power system," *Electric Power Systems Research,* 101 (2013) 36– 48.

Appendix 8–I

App. 8–I.1 Action of the Ramp Tracking Filter (RTF)

It is noted in Section 2.10.2.2 that, if there are two integrations in the forward path of the unity feedback system shown in Figure 8.25, the tracking error $[R(s) - C(s)]$ for a ramp input is zero.

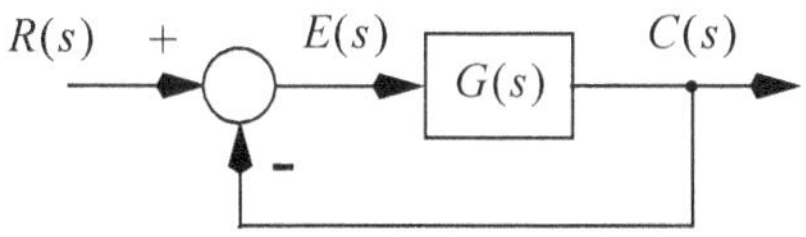

Closed-loop transfer function:

$$W(s) = \frac{C(s)}{R(s)} = \frac{G(s)}{1 + G(s)}$$

Figure 8.25 Structure of a closed-loop control system

Let us assume that the 5th order forward-loop transfer function of this unity feedback system is:

$$G(s) = (1 + sT_8)/[s^2 T_9^2(10 + 10sT_9 + 5s^2 T_9^2 + s^3 T_9^3)]. \tag{8.23}$$

(Note the double integration in $G(s)$.) The closed-loop transfer function is:

$$RTF(s) = W(s) = (1 + sT_8)/[1 + sT_8 + s^2 T_9^2(10 + 10sT_9 + 5s^2 T_9^2 + s^3 T_9^3)].$$

If $T_8 = 5T_9$ the closed-loop transfer function becomes:

$$RTF(s) = (1 + s5T_9)/(1 + sT_9)^5. \tag{8.24}$$

The transfer function of the ramp-tracking filter postulated in (8.14) is of the form:

$$RTF(s) = W(s) = [(1 + sT_8)/(1 + sT_9)^M]^N. \tag{8.25}$$

Comparing the last two equations, we note that they are identical if $N = 1$, $M = 5$ and $T_8 = MT_9 = 5T_9$.

Based on a formal method of analysis a general result for the open-loop transfer function of (8.23) with $N = 1$ and $M \geq 2$ is derived:

$$G(s) = \frac{1 + s(MT_9)}{(sT_9)^2 \times \sum_{k=2}^{M} \binom{M}{k} (sT_9)^{k-2}}. \tag{8.26}$$

Thus, provided $T_8 = MT_9$, the associated RTF has two integrations in the forward path and consequently the RTF will track a ramp input with zero following error in the steady-state.

For the case when $N > 1$ there are, in effect, N RTFs in cascade each satisfying the requirement $T_8 = MT_9$. An analysis similar to that for $N = 1$ confirms the validity of the latter result.

Note that the RTF will track, with a finite steady-state error, a signal which is the integral of a ramp, i.e. $\frac{1}{2Hs}(R_0/s^2)$. However, by extending the analysis of (8.26) it is a simple matter to derive the following transfer-function of a Parabolic Tracking Filter (PTF):

$$PTF(s) = \frac{1 + (MT_9)s + (M(M-1)/2 \cdot T_9^2)s^2}{(1 + sT_9)^M}, \tag{8.27}$$

which will track a *parabolic* input $P_0 t^2$, as well as a ramp input, with zero steady-state error.

App. 8–I.2 Steady-state conditions at the input and output of the RTF and associated tracking errors for mechanical power input $\Delta P_m(t) = a_n t^n$

App. 8–I.2.1 With and without an Ideal Integrator

The mechanical power changes of interest in the following analysis are: (i) a step of magnitude $a_0 = R_0$ (for $n = 0$); (ii) a ramp $R_0 t$, $a_1 = R_0$, ($n = 1$); (iii) a parabola $0.5R_0 t^2$, $a_2 = R_0/2$, ($n = 2$); and (iv) a cubic $(1/6)R_0 t^3$, $a_3 = R_0/6$, ($n = 3$). Note that each of the last three functions is an integral of the previous input function. Since the Laplace transform of t^n is $n!/s^{n+1}$, the Laplace transform of each of the input functions is simply R_0/s^{n+1}. Let us consider the alternatives of either an ideal integrator or of a pseudo-integrator being employed in the integration of the mechanical power signal.

Let us assume that for the path in the pre-filter shown in Figure 8.13 consists of k ideal integrators, $k = 0, 1$ and m washout filters, $m = 0, 1, 2$; assume for the mechanical power input $R_0 t^n$, $n = 0, 1, 2, 3$. The output of the ideal integrator, i.e. the input to the RTF, is then

$$U_{rtf}(s) = \left[\frac{sT_w}{(1 + sT_w)}\right]^m \cdot \left[\frac{1}{2Hs}\right]^k \cdot \frac{R_0}{s^{n+1}} = \frac{(T_w^m R_0)/(2H)^k}{s^{(n+k+1-m)}(1 + sT_w)^m}. \tag{8.28}$$

Applying the Final Value Theorem of (2.27) to (8.28), the general form of the expression for the steady-state input to the RTF, after the initial transients have decayed away, is found to be

$$U_{rtf}\big|_{ss} = \lim_{s \to 0} s(U_{rtf}(s)) = \frac{(T_w^m R_0)/(2H)^k}{s^{(n+k-m)}} \text{ as } s \to 0. \tag{8.29}$$

The associated output of the RTF, $V_{rtf}(s)$, is

$$V_{rtf}(s) = U_{rtf}(s) \cdot RTF(s), \text{ where } RTF(s) = \left[\frac{1 + sT_8}{(1 + sT_9)^M}\right]^N . \quad \text{((8.14) repeated)}$$

To ascertain how well the output of the RTF tracks its input, let $N = 1$ and let us calculate the tracking error $E_f(s)$ between the RTF's input and output:

$$\begin{aligned} E_f(s) &= U_{rtf}(s) - V_{rtf}(s) \\ &= U_{rtf}(s)[1 - RTF(s)] \\ &= U_{rtf}(s)\left[1 - \frac{G(s)}{1 + G(s)}\right], \\ &= U_{rtf}(s)\left[\frac{1}{1 + G(s)}\right] \end{aligned}$$

where $G(s)$ is given by (8.26). As $t \to \infty$, The tracking error in the steady state becomes

$$\begin{aligned} e_{fss} = \lim_{s \to 0} sE_f(s) &= \lim_{s \to 0} s\left(\left\{U_{rtf}(s)\left[1 - \frac{G(s)}{1 + G(s)}\right]\right\}\right) \\ &= \lim_{s \to 0} s\left\{U_{rtf}(s)\left[\frac{1}{1 + G(s)}\right]\right\} \end{aligned} . \quad (8.30)$$

However, from (8.26) we can deduce

$$\lim_{s \to 0} G(s) = \lim_{s \to 0} \left[\frac{(1 + sT_8)}{s^2 T_9^2 (10 + 10sT_9 + 5s^2 T_9^2 + s^3 T_9^3)}\right] = \left.\frac{1}{10 T_9^2 s^2}\right|_{s \to 0} . \quad (8.31)$$

Following substitution of (8.29) and (8.31) in (8.30), the latter reduces to a general expression for the tracking error between RTF input and output.

$$e_{fss} = \lim_{s \to 0} \left\{\left[\frac{(T_w^m R_0)/(2H)^k}{s^{(n+k-m-2)}}\right] \cdot 10T_9^2\right\} . \quad (8.32)$$

For example, for the case of an ideal integrator, two washout filters (column 8 of Table 8.6 on page 423), and a cubic mechanical power input, i.e. $n = 3$, $k = 1$ and $m = 2$, (8.32) becomes:

$$e_{fss} = \left[\frac{(T_w^m R_0)/(2H)^k}{s^0}\right] \cdot 10T_9^2 = K_{2I} K_G R_0 \text{, where } K_{2I} = T_W^2/(2H)$$

$$\text{and } K_G = 10T_9^2 .$$

Note, for the other inputs in column 8 when $n < 3$, $e_{fss} = 0$.

App. 8–I.2.2 With a Pseudo Integrator

The same analysis as in the previous section is conducted with the transfer function of a single pseudo integrator replacing that of the ideal integrator. It can be shown that

$$U_{rtf}\big|_{ss} = \lim_{s \to 0} s(U_{rtf}(s)) = \frac{(T_w^m T_H R_0)/(2H)}{s^{(n-m)}} \text{ as } s \to 0\text{, and} \tag{8.33}$$

$$e_{fss} = \lim_{s \to 0} \left\{ \left[\frac{(T_w^m T_H R_0)/(2H)}{s^{(n-m-2)}} \right] \cdot 10T_9^2 \right\}. \tag{8.34}$$

App. 8–I.3 Multi-Band PSS transfer function

Consider a differential filter of form shown in Figure 3 of [16]; it is also illustrated in Figure 8.26 in which R is a constant ratio.

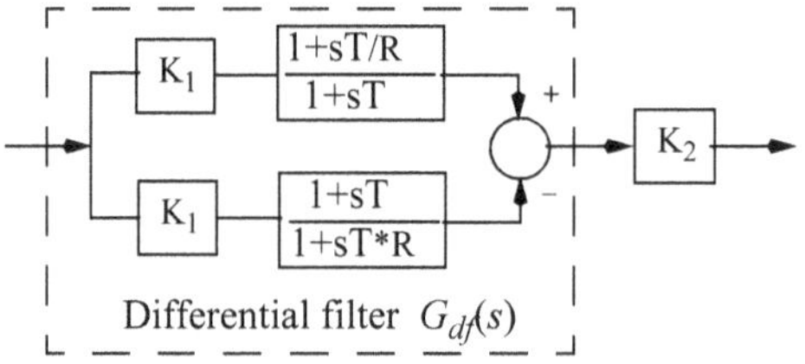

Figure 8.26 Differential filter G_{df}

The input-output transfer function is

$$\begin{aligned} G(s) = K_2 G_{df}(s) &= K_1 K_2 \left(\frac{1 + s(T/R)}{1 + sT} - \frac{1 + sT}{1 + s(TR)} \right) \\ &= K_1 K_2 (T/R) \cdot \frac{s(1-R)^2}{(1+sT)(1+sTR)} \end{aligned} \tag{8.35}$$

Equation (8.35) reveals that $G(s)$ is a band-pass filter. Let the frequency $\omega_m = 1/(T\sqrt{R})$ and $s = j\omega_m$; (8.35) then becomes:

$$G(j\omega_m) = K_1 K_2 \cdot \frac{(1-R)^2}{R(1+R)}. \tag{8.36}$$

However, at frequency ω_m rad/s it is required that the differential filter have unit gain; i.e. $|G_{df}(j\omega_m)| = 1$. Hence, from (8.36):

$$G_{df}(j\omega_m) = K_1 \cdot \frac{(1-R)^2}{R(1+R)} = 1\text{; thus}$$

$$K_1 = R(1+R)/(1-R)^2 \text{, or} \tag{8.37}$$

$$(K_1 - 1)R^2 - (2K_1 + 1)R + K_1 = 0. \tag{8.38}$$

Solving (8.38) for R we find:

$$R = [(2K_1 + 1) \pm \sqrt{8K_1 + 1}]/(2K_1 - 1), \text{ with } K_1 > 1.$$

Following substitution of (8.37) in (8.35) the transfer function $G(s)$ becomes:

$$G(s) = \frac{K_2(2\xi/\omega_m)s}{1 + (2\xi/\omega_m)s + (s/\omega_m)^2}, \tag{8.39}$$

where the damping ratio is $\xi = (R+1)/(2\sqrt{R})$. Equation (8.39) is that of a Q-filter with maximum gain K_2 at the centre frequency ω_m rad/s. Its normalized frequency response is shown in Figure 2.21 for $K_2 = 1$ and a range of damping ratios from 0.1 to 10.

Chapter 9

Basic Concepts in the Tuning of PSSs in Multi-Machine Applications

9.1 Introduction

The objective of the application of stabilizers in multi-machine power systems is to stabilize the system by providing adequate damping for the critical rotor modes of oscillation. These modes typically involve several power stations and their machines. In the case of inter-area modes many power stations, geographically widely separated, may participate in both the local and inter-area modes. It is therefore necessary that the stabilizer which, when fitted to a generator, contributes with stabilizers on other machines to the damping of the relevant modes. Furthermore, because operating conditions on the system continuously change, the performance of a fixed-parameter stabilizer should be robust to any such changes.

By employing the P-Vr method in the tuning of the PSS, as demonstrated in Chapter 5, the inherent magnitude and phase characteristics of the generator and power system are being utilized; for practical purposes these characteristics consistently lie in a relatively narrow band. Not only can the method account for variations over a wide range of loading conditions on the system, line outages, etc., but the resulting PSS is most effective and beneficial at the higher generator real power outputs as revealed in Table 5.6, and discussed in the associated text.

Prior to considering the application of the P-Vr method to the tuning of PSSs in multi-machine power systems, the use and significance of two valuable tools in the small-signal analysis of the dynamic performance of such systems are discussed. These tools concern the so-called "Mode Shape" and "Participation Factor" analyses of the system for a selected operating condition. Such analyses reveal the nature and significance of the various modes (both rotor or other modes), the involvement - and extent of involvement - of generators in the modes, and other insights such as the nature of the dynamic behaviour of other devices in the system (e.g. FACTS devices and their controls).

The application of other PSS tuning methods, namely the GEP Method and the Method of Residues, is discussed in Chapter 6. While these approaches can be adapted to the multi-machine system, for the reasons explained in the latter chapter the P-Vr method is considered to possess some significant advantages.

9.1.1 Eigenvalues and Modes of the system

It has been pointed out in Section 3.5, that the hth eigenvalue of the real, $n \times n$ system matrix $\boldsymbol{A}$ of the state equations is the real or complex scalar quantity, λ_h; it is the non-trivial solution of the equation

$$\boldsymbol{A}\boldsymbol{v}_h = \lambda_h \boldsymbol{v}_h . \tag{9.1}$$

The n -element column vector, $\boldsymbol{v}_h$, is the *right* eigenvector of the matrix $\boldsymbol{A}$ corresponding to the eigenvalue λ_h.

For low-order dynamic systems, typically with less than 2500 states, the eigenvalues are calculated using an algorithm that employs QR factorisation [1]. As the number of states approach 2500 the computation tends to become much slower. However, if fast computation is required to determine only those eigenvalues in a selected region of the complex s-plane, or if the system order is greater than 2500, methods such as Modified Arnoldi [2], Subspace Iteration [3] and Multiple-Shift-Point Sparse-Eigenanalysis [3] are available. Such facilities are normally included in software packages for the analysis of the small-signal dynamic performance and control of large power systems [4].

As has been discussed earlier, eigen-analysis is an extremely valuable tool because the n eigenvalues of the system characterize the nature of its dynamic behaviour in the following ways:

1. The time-domain responses of the system states and outputs to a disturbance are weighted sums of terms of the forms $a_i e^{\sigma_i t}$ and $b_k e^{\sigma_k t} \sin(\beta_k t + \phi_k)$, where $\lambda_i = \sigma_i$ is a real eigenvalue and $\lambda_k = \sigma_k \pm j\beta_k$ is a complex-conjugate pair of eigenvalues. The real and/or the real plus imaginary parts of the eigenvalues therefore clearly define the form of its responses.
2. The system is stable if the real parts, σ, of all n eigenvalues are negative.

3. The monotonic and oscillatory terms $a_i e^{\sigma_i t}$ and $b_k e^{\sigma_k t} \sin(\beta_k t + \phi_k)$ are the ith and k^{th} modes [1] of the system, respectively.

In the case of the oscillatory modes, assuming that the mode is unique, the right and left eigenvectors of the complex conjugate eigenvalues are also complex conjugates. Therefore, the mode shape and participation factors (as described below) of the mode can be identified by considering only the eigenvectors of one of the complex conjugate pair of eigenvalues. In addition, in later sections we will present the concept of the response of the system to a complex frequency, namely the modal frequency of a decaying oscillatory mode. Again, it can be shown that it suffices to evaluate the transfer function at one of the two complex conjugate eigenvalues. For these reasons, throughout the text we have sometimes referred to an oscillatory mode as, say, mode h where h is the index or number of the first of two complex conjugate eigenvalues which together constitute the mode.

The above items are valuable pieces of information but they do not answer the following questions concerning the modes of the system:

- What type of mode it is? (For example, is it primarily associated with the controller of a FACTS device?)
- What states participate in this mode, in what manner and to what extent? (Do the rotor speed states of generators i and j both participate significantly in the oscillatory mode λ_h?)
- Can analysis reveal the behaviour of one group of generators with respect to other groups in the case of the electro-mechanical modes?
- Which generators participate significantly in the lightly-damped or potentially unstable modes? (In practice it may be necessary to identify some or all of the rotor modes, particularly those that fall into the categories of being unstable or lightly damped.)

We shall therefore, in the following sections, examine two methods which are used to identify the modes by resolving the above issues, namely, Mode Shape and Participation Factor Analyses [5].

9.2 Mode Shape Analysis

In the analysis of the dynamic performance of multi-machine power systems, the concept of 'mode shapes' provides a practical and meaningful tool. In essence, mode shapes assist one to identify the mode type such as 'inter-area', 'local-area', 'inter-machine' / 'intra-station'.

---

1. See Section 3.5.2 concerning the distinction between 'eigenvalues' and 'modes'.

The theoretical basis for mode shapes is outlined in Section 3.9. It was shown that if the state equations of the dynamic system are excited by the *right eigenvector* $\boldsymbol{v}_i$ of a selected mode of rotor oscillation, λ_i, *only* that mode appears in the time-domain responses of the states - the responses for all other modes are zero; this is succinctly summarised by (3.45), namely

$$\boldsymbol{x}_i(t) = \boldsymbol{v}_i\, e^{\lambda_i t}. \tag{9.2}$$

The electro-mechanical or rotor modes of oscillation are usually identified with the perturbations of rotor speed about synchronous speed. The mode shape is therefore identified mainly from the phase of the elements of the right speed-eigenvector of the selected mode.

Rather than considering a complex multi-machine system, the significance and application of mode shapes are illustrated more simply - and in some detail - initially using a two-mass spring system.

9.2.1 Example 1: Two-mass spring system

A two-mass spring system which is constrained to move freely in the positive x-direction from a reference position is shown in Figure 9.1(a). The instantaneous position and speed of the centre of mass j is $x_j(t)$ (m) and $v_j(t)$ (m/s), respectively, are highlighted in Figure 9.1(b). M_j is the mass (kg), B_j is the viscous damping coefficient (N/m/s) between the mass and the ground plane, K_{jk} is the spring stiffness coefficient (N/m), and $f_j(t)$ is an externally applied force (N).

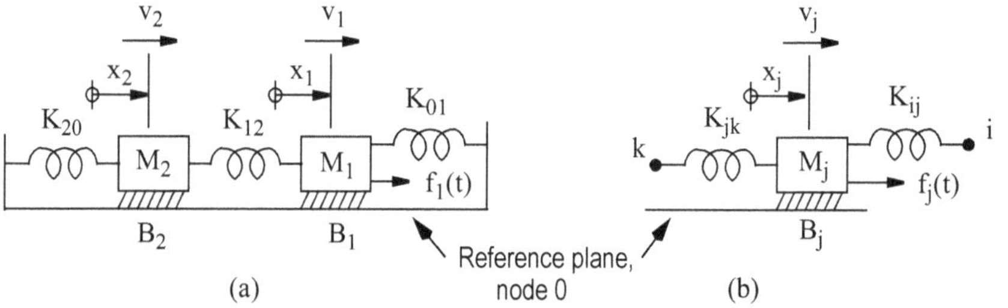

Figure 9.1 (a) A two-mass system free to move in the x-direction on a flat surface, (b) the general form of the parameters and variables for the jth mass.

Based on Figure 9.1(b), a general form of the equation of motion for mass M_j can be expressed as [6], [7], [8]:

$$f_j = -K_{ij}x_i + M_j\frac{dv_j}{dt} + B_j v_j + (K_{ij} + K_{jk})x_j - K_{jk}x_k. \tag{9.3}$$

This equation can be rewritten in state equation form as follows:

$$\dot{v}_j = -\frac{B_j}{M_j}v_j + \frac{K_{ij}}{M_j}x_i - \frac{K_{ij}+K_{jk}}{M_j}x_j + \frac{K_{jk}}{M_j}x_k + \frac{1}{M_j}f_j \,, \quad \text{and}$$

$$\dot{x}_j = v_j \,. \tag{9.4}$$

Applying the above relationships to the two masses in turn, a fourth-order set of state equations is formed in the state variables $[v_1, v_2, x_1, x_2]$; the derivation of the set of equations is left as an exercise to the reader.

Consider the following parameters for the four-mass spring system:

$M_1 = 2,\ M_2 = 4,\ B_1 = 1,\ B_2 = 0.5$, and $K_{01} = 10,\ K_{12} = 8,\ K_{20} = 10$.

For these values of the system parameters the eigenvalues of the system are given in Table 9.1.

Table 9.1 Eigenvalues of the two-mass spring system

Eigenvalue number and value			
1	2	3	4
-0.214+j3.21	-0.214-j3.21	-0.098+j1.77	-0.098-j1.77

We note that there are two stable oscillatory modes having damping ratios of 0.067 for mode A (which is associated with the complex conjugate eigenvalue pair 1,2) and 0.055 for mode B (eigenvalue pair 3,4). However, there is no information that reveals the nature of the system performance; for example, what is the relative characteristic behaviour of the masses for mode A?

The right speed-eigenvectors for the two oscillatory modes are shown in Table 9.2. It is noted for mode A, when it alone is excited, that the speed states v_1 and v_2 of masses 1 and 2 are essentially in anti-phase. The mass M_1 is said to 'swing against' mass M_2. The displacement states x_1 and x_2, which are almost in anti-phase, lag their respective speed states by nearly $90°$. When the right eigenvectors are normalised to $1\angle 0°$ for the state with the largest magnitude (the speed state v_1 for mode A, v_2 for mode B), the modal behaviour of the states is interpreted more easily using the polar plots for the relevant modes as shown in Figure 9.2.

Often in mode-shape analysis only the speed elements in the right eigenvector are plotted. In this event the plot is the same as that in Figure 9.2 except all other states are omitted.

Table 9.2 Right eigenvectors for the oscillatory modes

State	Right eigenvectors			
	Mode A: $-0.214 \pm j3.21$		Mode B: $-0.098 \pm j1.77$	
	Magnitude	Angle °	Magnitude	Angle °
v_1	0.904	180	0.492	-5.3
v_2	0.309	-9.6	0.719	0
x_1	0.281	86.2	0.277	-98.4
x_2	0.096	-103.4	0.405	-93.2

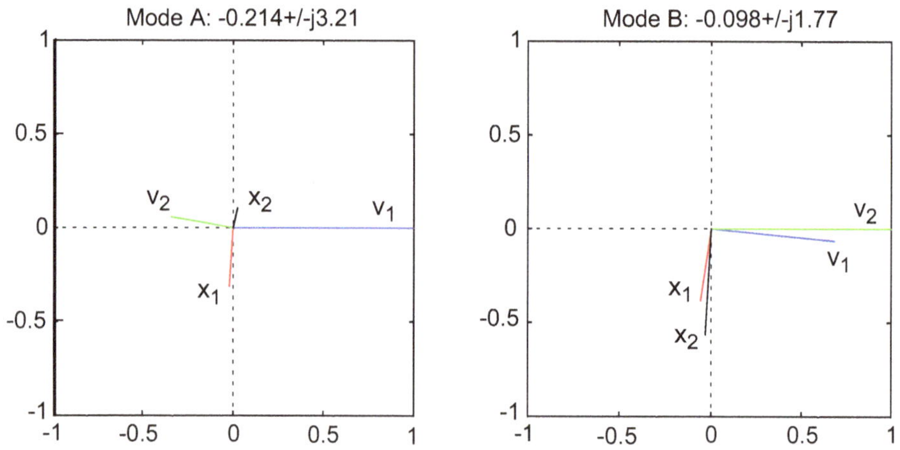

Figure 9.2 Normalised right eigenvectors of speed (v) and displacement (x) for the oscillatory modes

Let us now consider the time-domain responses of the states when the mass-spring system is excited by the right eigenvector consisting of the *real* parts of its elements for each of the modes in Table 9.2, e.g. by the initial condition $[-0.904\ 0.305\ 0.0188\ -0.0223]^T$ for mode A. The transient response to this initial condition is shown in Figure 9.3.

Note in Figure 9.3 the instantaneous phase relationship between the states is consistent with Figure 9.2 and/or Table 9.2. From the figure it is seen that

- the time constant and the period of the response are consistent with the single mode, $-0.214 \pm j3.21$;

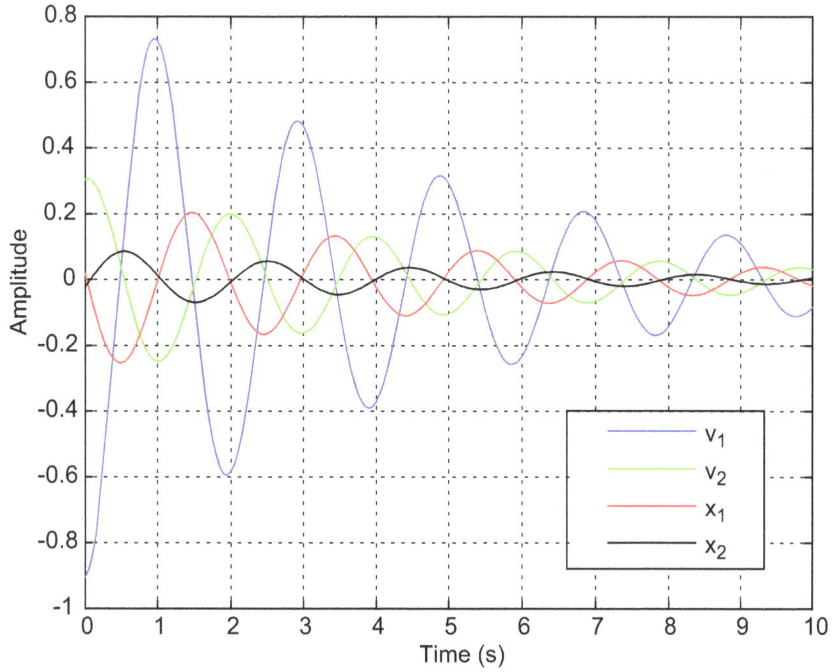

Figure 9.3 First ten seconds of the transient response speed (v) and displacement states (x) for the two-mass spring system to an initial condition which excites only mode A, $-0.214 \pm j3.21$.

- the speed states v_1 and v_2 as well as the displacement states x_1 and x_2 are, respectively, nearly in anti-phase;

- x_1 and x_2, respectively, lag v_1 and v_2 by nearly 90°;

- as might be expected for mode A, $-0.214 \pm j3.21$, in which the masses swing against each other as shown in Figure 9.3, the amplitude of the oscillation of the smaller mass is larger.

Likewise, as seen in Figure 9.4 if the system is excited by an initial condition $[0.490\ 0.719\ {-0.0407}\ {-0.0224}]^T$ on the four states in Table 9.2, only mode B is excited. In this case the speed states v_1 and v_2 as well as the displacement states x_1 and x_2 are, respectively, nearly in-phase, i.e. the two masses 'swing together' with respect to the reference frame. Again, the form of the responses is consistent with the results in Figure 9.2 and/or Table 9.2.

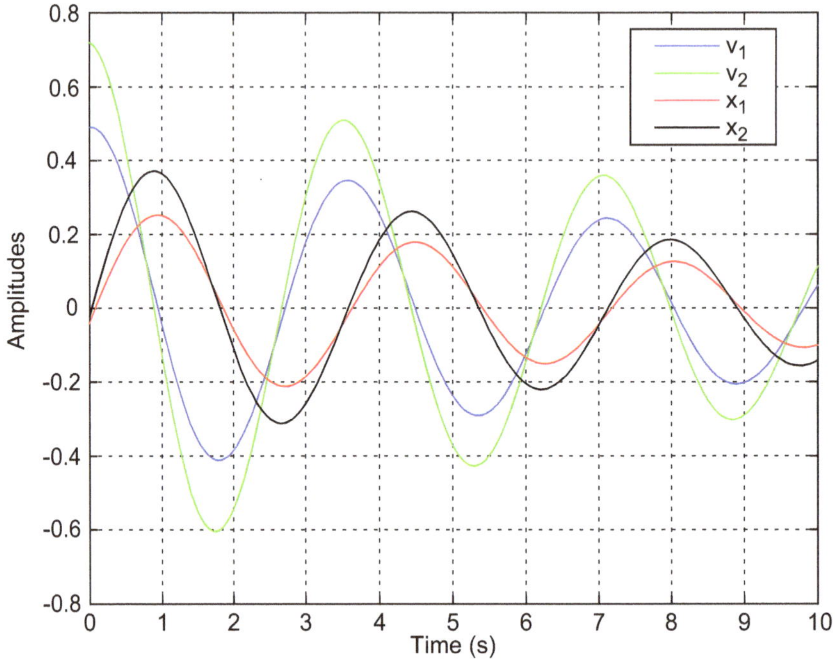

Figure 9.4 First ten seconds of the transient response of the two-mass spring system to an initial condition which excites only mode B, $-0.098 \pm j1.77$.

It is noted above for mode A, in which the masses swing in anti-phase, the amplitude of the oscillation of the smaller mass is larger. This suggests that the nature of the oscillations observed in the responses of Figure 9.3 and Figure 9.4 are associated with the interchange of energy between the energy storage elements. Let us calculate the instantaneous stored energies in the masses and the spring. The instantaneous stored energy in a mass is $M_j v_j^2/2$ and that in a spring is $K_{jk}(x_j - x_k)^2/2$. For mode A the time responses of the stored energy in each of the five elements for the relevant initial conditions are plotted in Figure 9.5.

As is to be expected, the envelope of the decay of the stored energies decays with a time constant of one-half of that of mode A [1]. Further we note:

- The stored energy in each of the two masses peak more-or-less simultaneously; at that time the stored energy in each of the three springs is zero;

1. Assume that the response of a speed state of a mass is $v(t) = V_0 e^{-\alpha t}$. The stored energy will decay as $v^2(t) = V_0^2 e^{-2\alpha t}$.

- A quarter cycle later of the modal frequency (3.21 rad/s, period approximately 2 s), the latter condition is reversed, i.e. the stored energies in the springs peak more-or-less simultaneously; at that time the stored energy in each of the masses is zero.

- If the losses during the interchange were zero (i.e. no viscous damping, $B = 0$), the system would oscillate indefinitely with constant amplitude and the peaks and troughs in the responses would coincide exactly.

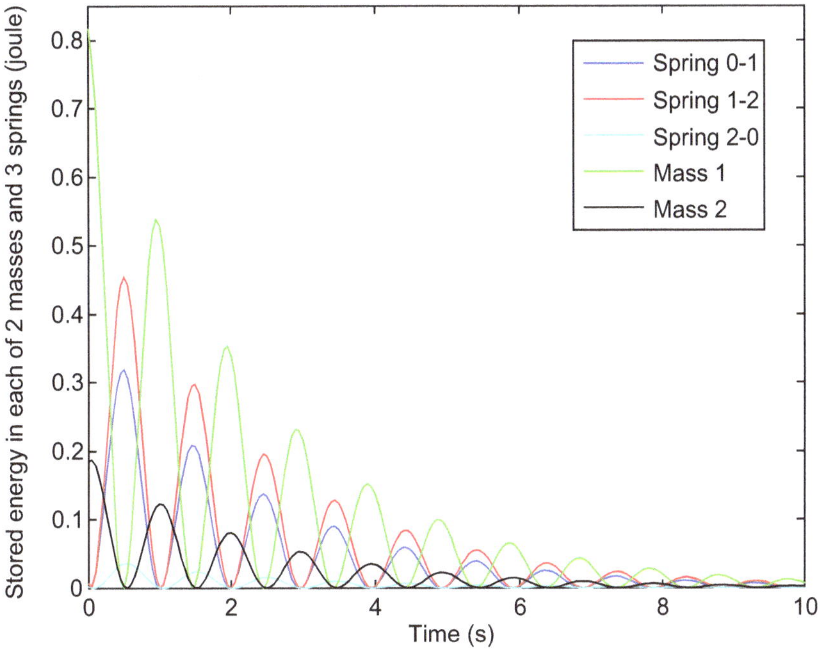

Figure 9.5 Stored energy response in each of the masses and springs for an initial condition which excites only mode A, $-0.214 \pm j3.21$.

A plot of the stored energy responses, similar to Figure 9.5, for mode B ($-0.098 \pm j1.77$) can be predicted from the mode shape shown in Figure 9.2 or the amplitude responses of Figure 9.4. This is left as an exercise to the reader.

The interchange of energy between energy storage elements every quarter of a cycle of the oscillatory behaviour is explained in any text book on the fundamentals in physics or engineering. The significance of mode shapes in the analysis of dynamic performance is that it reveals the nature of the behaviour of the masses (or inertias) in selected modes - normally the electro-mechanical modes in power system dynamic performance.

This example illustrates that, for the two-mass-spring system, there is one oscillatory mode representing the relative dynamic behaviour between the two masses. The second mode portrays the behaviour of the masses with respect to the reference frame.

9.2.2 Example 2: Four-mass spring system

To highlight some further relevant issues a somewhat more complex mass-spring system than that in Example 1 (Section 9.2.1) is analysed; the system is shown in Figure 9.6.

Unlike the previous example there are no springs restraining movement between the masses and the reference plane. The previous example of the two-mass-spring system is simple enough to demonstrate not only the concepts of mode shapes, but also the associated transient responses and the responses of the stored energy. However, the purpose of this example is to demonstrate for higher-order systems the types of interactions between elements that are revealed through mode-shape analysis. Moreover, the more complex system provides additional insight into the use of participation factor analysis described in Section 9.3.

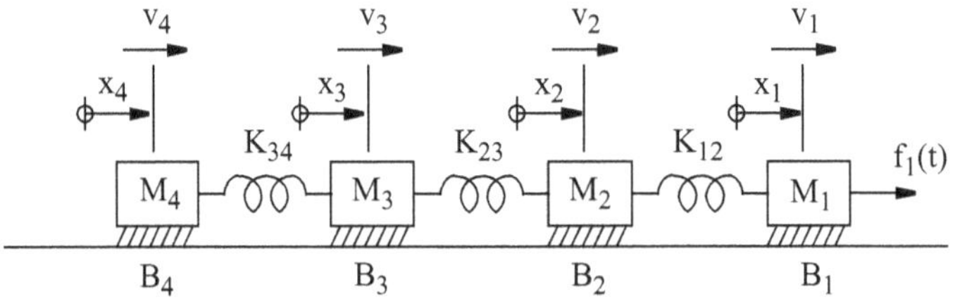

Figure 9.6 A four mass, three-spring system

The following are the parameters for the four-mass spring system:
$M_1 = 1$, $M_2 = 2$, $M_3 = 4$, $M_4 = 2$; $B_1 = 0.3$, $B_2 = 0.8$, $B_3 = 2.4$, $B_4 = 0.6$, and $K_{12} = 10$, $K_{23} = 8$, $K_{34} = 15$. The units of these parameters are supplied in Section 9.2.1.

The eight eigenvalues and five modes of system are given in Table 9.3.

Table 9.3 Eigenvalues and modes, A to E, of the four-mass system

Eigenvalue number and value							
Mode A (Oscillatory)		Mode B (Mono.)	Mode C (Mono.)	Mode D (Oscillatory)		Mode E (Oscillatory)	
1	2	3	4	5	6	7	8
-0.18+j4.12	-0.18-j4.12	-0.46	0	-0.19+j1.82	-0.19-j1.82	-0.21+j3.43	-0.21-j3.43
Mono.: a monotonically increasing or decaying mode							

We note that there is a pole at the origin in the case of mode C (eigenvalue 4), the remaining eigenvalues constitute modes that are stable, and that the oscillatory modes have a damping ratio between 0.04 for mode A (eigen-pair 1,2) and 0.10 for mode D (eigen-pair 5,6).

In the analysis of multi-machine power system dynamics it is common practice to employ the normalised right speed-eigenvectors in assessing mode shapes for the electro-mechanical modes. These eigenvectors of the oscillatory modes for the four-mass system are shown in both Table 9.4 and the plots of Figure 9.7. It is observed for oscillatory mode A, the only mode excited, that

- the speed states of masses M_1 and M_3 move together essentially in anti-phase with those of masses M_2 and M_4;
- the lighter masses M_1 and M_2 have the larger amplitudes;
- the frequency of oscillation of this mode is the highest of all the modes.

Table 9.4 Four mass system: Normalised right speed-eigenvectors for the three oscillatory modes

State	Normalised right speed-eigenvectors for oscillatory modes					
	Mode A: $-0.18 \pm j4.12$		Mode D: $-0.19 \pm j1.82$		Mode E: $-0.21 \pm j3.43$	
	Magnitude	Angle °	Magnitude	Angle °	Magnitude	Angle °
v_1	1.00	0	1.00	0	0.516	6.5
v_2	0.703	-177.9	0.668	-1.1	0.096	-161.6
v_3	0.167	8.2	0.306	-178.3	0.579	-174.9
v_4	0.131	-173.3	0.549	-176.6	1.00	0

The speed-eigenvector plot for the monotonically decaying real mode (B in Table 9.3) reveals that all masses move in-phase with respect to the reference when this mode is excited [1]. In the case of the oscillatory mode D (eigen pair 5,6), the lighter masses M_1 and M_2 swing together against M_4 and the heaviest mass M_3; the frequency of oscillation is the lowest of all the modes. Similarly, for mode E (eigen pair 7,8) the lighter masses M_4 and M_1

1. In Table 9.3 the elements of the speed eigenvector of mode C (eigenvalue 4), which represents a pole at the origin of the s-plane, are all zero. If mass 4 were attached to the reference plane through a spring with non-zero coefficient K_{40}, one is likely to find that modes B and C represent a fourth complex conjugate pair which would constitute a common oscillatory mode in which all four masses oscillate in-phase against the reference.

swing predominantly against the heaviest mass M_3. It is these types of phenomena that mode shapes are particularly useful in revealing when this analysis is applied to the electro-mechanical modes of a multi-machine power system. For example, when two large groups of generators swing against each other, the frequency of oscillation is typically low (e.g. 2 to 5 rad/s), but if a single generator swings against the rest of the machines the frequency tends to be relatively much higher (e.g. 7-10 rad/s).

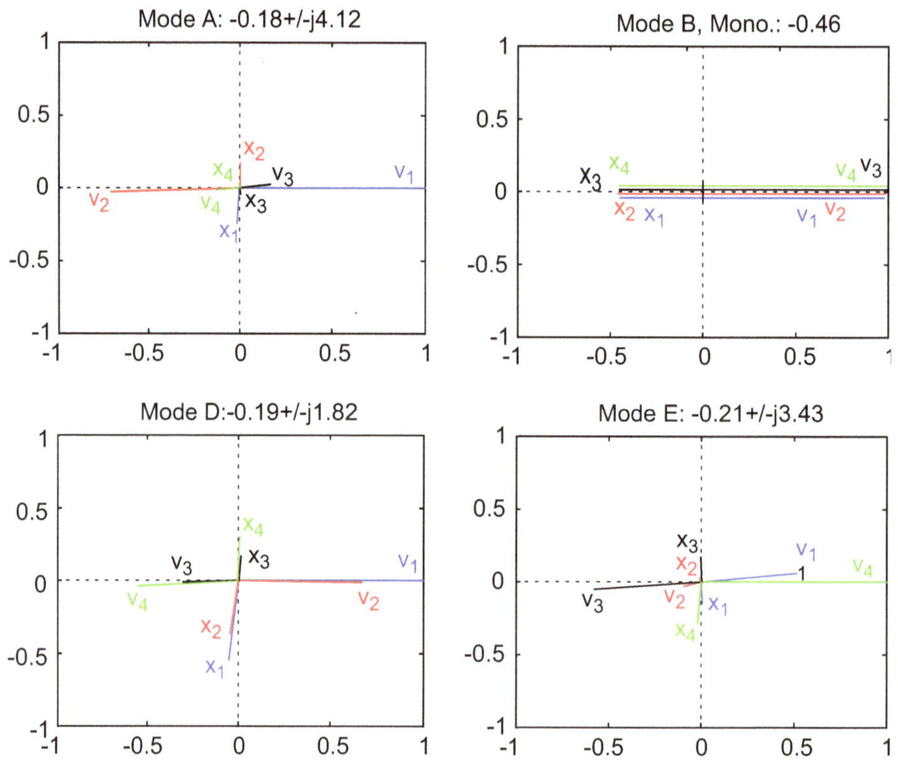

Figure 9.7 Normalised right speed-eigenvectors for the three oscillatory modes (A, D, E) and the monotonically decaying real mode B.

In this example of a four mass-spring system, there are three oscillatory modes representing the relative dynamic behaviour between the four masses. A fourth real or complex mode typically portrays the behaviour of all four masses with respect to a reference. Typically, if there are N masses, there are N-1 modes representing the dynamic characteristics of interactions between the masses. Instead of an analysis of masses which are subject to linear displacement an analysis of rotating masses can be conducted using a similar approach. Thus for N generators connected to a multi-machine system, there are N-1 modes of rotor - or electro-mechanical - oscillation representing the relative dynamic *interactions / behaviour between the N rotating masses*. The N modes are associated with N pairs of complex conjugate eigenvalues - that is, a total of $2N$ eigenvalues.

However, a note of warning is appropriate here. The mode shapes do not reveal the relative extent in which the set of states participate in a selected mode, or the relative extent for which a selected state participates in the set of system modes. As seen in Figures 9.3 and 9.4 the elements of the right eigenvector $\boldsymbol{v}_i$ in (9.2) represent the relative *amplitude* of the states at time zero and thereafter. These elements are not dimensionless; the first four elements have the dimensions of speed in *m/s*, the second four, displacement in *m*. The set of states in a model of a multi-machine power system is comprised of states of very different types, e.g. fluxes, control and excitation system variables, as well as rotor speed and angle. Again, the elements of the right eigenvector are not dimensionless and would alter if the per-unit system employed were changed. It is therefore not possible to measure the relative participation, say, of each of the system states in a selected mode based on the right eigenvector, unless the measure is expressed in a dimensionless form. This is achieved using the concept of participation factors.

9.3 Participation Factors

In the case of a multi-machine power system we may suspect from the mode shapes that the speed state or a field voltage state of a generator is involved in a particular rotor mode of oscillation, but how extensively are they involved, and how much relative to the involvement of other generators? Although an analysis of modes shapes reveals the phase relation between speed states in a given rotor mode of oscillation, the participation in that mode by some other state may be greater than that of the rotor speed state, e.g. field voltage (if the PSS damping gain is set to a high value).

Firstly, if $\boldsymbol{v}_h$ and $\boldsymbol{w}_k$ are the right and left eigenvectors, respectively, it is shown in Section 3.10, based on [5], that the dimensionless participation factor $p_{hk} = v_{kh} w_{hk}$ provides a measure of the relative extent to which each of the n states participates in the hth eigenvalue at time $t = 0$. Recall also that the sum of the participation factors for eigenvalue h is unity.

Secondly, it is also shown in Section 3.10.1 that the participation factor also provides a measure of the relative extent to which eigenvalue h participates in state k at time $t = 0$. Bear in mind, however, that at some later time $t = t_1$ some modes will have decayed away and only the more dominant modes prevail; consequently, the participation of the hth eigenvalue in state k will have changed from the values at $t = 0$. The participation factor is thus a meaningful measure only at $t = 0$ - which is the basis of its definition.

The first of the alternative forms of applying the participation factor concept is employed in the following chapters.

9.3.1 Example 4.3

In Example 4.2 the mode shapes for a four-mass spring system are derived. Due to the few number of states involved, let us now consider the associated participation factors for the states in the relevant modes for this simple system.

The participation matrix can be calculated as the element by element product of the left eigenvector matrix and the transpose of the right eigenvector matrix. The hth row then represents the participation factors of the states in the hth eigenvalue, as shown in Tables 9.5 and 9.6. Likewise, the kth column represents the participation factors of the modes in the kth state.

Table 9.5 Participation factors for the speed states v_i in the system modes.

Mode No.	Eigen-value No	Eigenvalue	Participation factors ($\times 10^{-2}$) for the eight states			
			v_1	v_2	v_3	v_4
A. (Osc)	1	-0.18+j4.12	23.5-j0.2	23.2+j1.5	2.5+j0.7	0.8+j0.2
B. (Mon)	3	-0.46	10.8	21.9	45.2	22.2
C. (Mon)	4	0	0	0	0	0
D. (Osc)	5	-0.19+j1.82	17.5+j1.4	15.6+j0.7	6.5+j0.9	10.4+j2.1
E. (Osc)	7	-0.21+j3.43	3.6+j0.7	0.2+j0.1	18.4+j2.8	27.8-j0.7
Osc.: Oscillatory mode. Mon.: Monotonic mode						

Table 9.6 Participation factors for the displacement states x_i in the system modes.

Mode No.	Eigen-value No	Eigenvalue	Participation factors ($\times 10^{-2}$) for the eight states			
			x_1	x_2	x_3	x_4
A. (Osc)	1	-0.18+j4.12	23.4-j1.9	23.2-j0.8	2.6+j0.3	0.8+j0.1
B. (Mon)	3	-0.46	3.7	2.7	-14.2	7.6
C. (Mon)	4	0	7.3	19.5	58.5	14.6
D. (Osc)	5	-0.19+j1.82	17.4-j1.5	15.4-j2.8	6.6-j1.2	10.5+j0.4
E. (Osc)	7	-0.21+j3.43	3.7+j0.4	0.2+j0.1	18.7-j0.4	27.5-j3.1

It can be confirmed that the sum of the participation factors covering the eight eigenvalues is $1+j0$. Note that the participation factors are complex but, as is often the case for the larger factors, they are almost real. Therefore, for ease of interpretation the *magnitudes* of the participation factors are plotted in the bar-chart form illustrated in Figure 9.8. The bar chart for mode C, $\lambda_4 = 0$, is not shown as the participation factors for the speed states are all zero.

For this simple case the participation of the set of modes in a selected state can be read from Figure 9.8, e.g. the magnitude of the participation factors for the eight eigenvalues, shown in Table 9.3, in the speed state v_2 are 0.232, 0.232, 0.219, 0, 0.156, 0.156, 0.002, 0.002 .

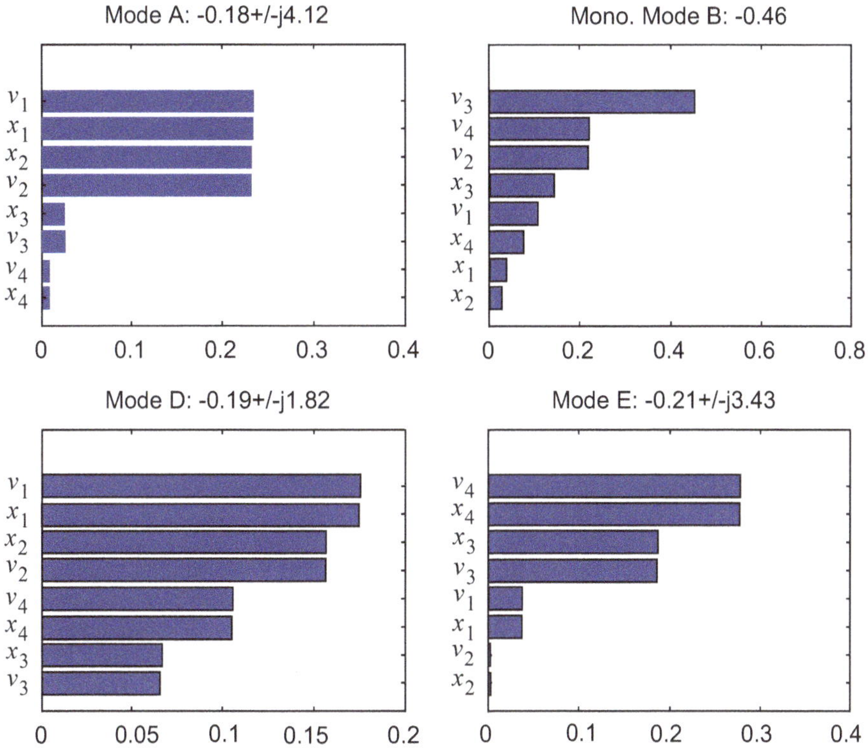

Figure 9.8 Participation factors of the set of states $[v_1, \ldots, v_4, x_1, \ldots, x_4]$ in selected modes.

A comparison between the plots of the participation factors and the modes shapes of Figures 9.8 and 9.7 is instructive. For oscillatory modes A, D and E the participation factors of the pairs of states v_i, x_i are almost identical but the magnitudes of the mode shapes for the same pairs differ significantly, typically by a factor of two or more to one in this example.

As demonstrated through the two examples, very pertinent information on the dynamic performance characteristics of the system is provided by the combination of modes shapes and the participation factors. It is the *relative phase information* provided by modes shapes that is particularly useful; the relative amplitude of the states depends the units of the states. As mentioned earlier if the system of units were changed, as in the selection of the per unit system, the relative amplitudes of the states could change. On the other hand, the participation factors indicate the relative degree of involvement not only of *all* the states in a mode on a dimensionless basis but also of the modes in a state; the participation factors are therefore a *characteristic* of the system, invariant to change in units. In the two examples the amplitudes of the right speed (note, *only* speed) eigenvectors for a given mode shape appear to correlate fairly well with the participation factors for the same mode; this may lead to the misconception that the amplitudes in the mode shape represent the 'participation' of the speed states in the selected mode.

The application of these tools will be demonstrated in analysing the dynamic behaviour of a multi-machine power system in Chapter 10.

9.4 Determination of the PSS parameters based on the P-Vr approach with speed perturbations as the stabilizing signal

9.4.1 The P-Vr transfer function in the multi-machine environment

Earlier, in Section 5.8, the tuning of a speed-input PSS for a generator in a single-machine infinite-bus system is outlined in detail. The P-Vr transfer function is introduced in Section 5.8.2, and its application to the tuning of the PSS compensating transfer function $G_c(s)$ is described in the subsequent subsections of Chapter 5. It is pointed out in those sections that the PSS tuning and use of the P-Vr transfer function are applicable to multi-machine systems; this is the case, but issues such as interactions between PSSs outlined in a later chapter need to be accounted for.

The comment in Section 5.8.2 for SMIB systems also applies to the tuning of PSSs in multi-machine systems, namely: *The PSS must be tuned to be robust to a full range of N and N-1 operating conditions. For this purpose it is necessary to select a set of appropriate operating conditions which encompass, and therefore include, the full range of conditions.* By examining the bordering conditions this approach reduces the number of cases for which the P-Vr characteristics must be evaluated.

In the case of the multi-machine system the P-Vr transfer function is defined as follows: *The P-Vr transfer function for generator i is the transfer function from the voltage reference* $\Delta V_{ri}(s)$ *of the AVR of generator i to the electrical torque output of the generator,* $\Delta P_{ei}(s)$, *with the shaft dynamics of all generators in the power system disabled.* Note that shaft dynamics on *all* generator are disabled for the purposes of the analysis.

Diagrammatically, this shown by a comparison of Figure 9.9(a), for the intact generators, and Figure 9.9(b) in which the shaft dynamics of all generators are disabled. An examination of Figure 9.9(b) reveals that there are a number of signal-flow paths associated with the P-Vr transfer function. For example, not only is there the direct path from the reference voltage input $\Delta V_{ri}(s)$ on generator *i* to its electrical torque output $\Delta P_{ei}(s)$ - but there are also paths through both the power system and the other generators to the electrical torque output on *i*. A question is: Compared to the SMIB system, do these additional paths diminish the effectiveness of the P-Vr approach to PSS tuning?

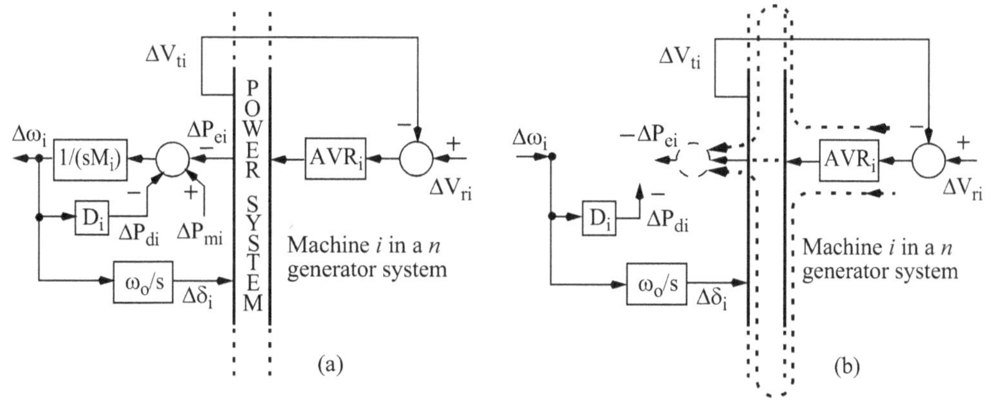

Figure 9.9 (a) Model of a generator in a multi-machine power system; (b) conceptually, with shaft dynamics on all machines disabled.

The terminology used here, i.e. `P-Vr transfer function', as defined above for the multi-machine context is that introduced in [9]. However, this same transfer function has been determined by different techniques elsewhere. For example, for the tuning of the PSS of a generator in a multi-machine power system, phase information on the P-Vr transfer function has been determined by field tests [10] or is based on SMIB models with the machine inertia constant set to a very large value on the generator of interest [11], [12]. In references [10], [11]and [12] no attempt is made to employ the P-Vr transfer function for the formal tuning of PSSs in a multi-machine system - including the concept and setting of the PSS damping gain, or as a basis for the coordination of PSSs.

The method adopted here for calculating the P-Vr transfer function is that presented in [9]. The significance of this approach is that a simple direct method is provided for determining both the *magnitude and phase* response of the P-Vr transfer function for each generator.

The theoretical basis for the P-Vr characteristic of generator *i*, $\Delta P_{ei}(s) = \alpha_i(s)\Delta V_{ri}(s)$, in a multi-machine system of N generators is considered in [14], [15]. With the shaft dynamics of all generators disabled it is shown that the electrical power or torque of the N generators is given by

$$\boldsymbol{\Delta P_e}(s) = \boldsymbol{A_v}(s)\boldsymbol{\Delta V_r}(s) + \boldsymbol{B_\omega}(s)\boldsymbol{\Delta\omega}(s). \tag{9.5}$$

Furthermore, it is shown that matrices $\boldsymbol{A_v}$ and $\boldsymbol{B_\omega}$ are essentially diagonal or block diagonal due to the diagonal dominance property of the reduced network admittance matrix into which the generator dynamic admittances are embedded as network elements. For brevity, let us now consider only the first term in (9.5) associated with the P-Vr-like matrix of generators in the multi-machine system, i.e.

$$\Delta \boldsymbol{P}_e(s) = \boldsymbol{A}_v(s)\Delta \boldsymbol{V}_r(s), \text{ where from [15],} \tag{9.6}$$

$$\begin{aligned} \boldsymbol{A}_v(s) &= \boldsymbol{G}_3(s)[\boldsymbol{Y}-\boldsymbol{G}_1(s)]^{-1}\boldsymbol{G}_5(s)+\boldsymbol{G}_6(s) \\ &= \boldsymbol{G}_3(s)\boldsymbol{Z}(s)\boldsymbol{G}_5(s)+\boldsymbol{G}_6(s) \end{aligned}, \tag{9.7}$$

$$\text{and } \boldsymbol{Z}(s) = [\boldsymbol{Y}-\boldsymbol{G}_1(s)]^{-1}, \tag{9.8}$$

and where $\boldsymbol{Y}$ is the reduced network admittance matrix. In the first term of the summation in (9.7), $\boldsymbol{G}_3(s)$ - and in (9.8) $\boldsymbol{G}_1(s)$ - are essentially functions of the steady-state conditions and are modified by the generator operational reactances $x_d(s)$ and $x_q(s)$. Moreover, in (9.7) the first term is determined mainly, and diagonally dominated, by the network admittance matrix $\boldsymbol{Y}$. The Thévenin equivalent of the network as seen from the terminals of generator i is not much affected by the dynamics of the other generators in the system. The second term $G_{6i}(s)$ in (9.7) depends only on the parameters of the generator i, its excitation system and a scalar multiplier v_{do_i}, the d-axis steady-state terminal voltage, i.e.:

$$G_{6i}(s) = v_{do\_i}G_{gen\_i}(s)G_{avr\_i}(s)/x_{di}(s), \tag{9.9}$$

where G_{gen_i}, G_{avr_i} and x_{di} are respectively the operational transfer functions of generator i, its AVR / exciter, and its direct-axis synchronous reactance; these functions are independent of the external system.

The *phase characteristic* of $G_{6i}(s)$ is independent of operating conditions in the external system, however, the *magnitude* of the *low-frequency* response varies only with the scalar gain v_{do}. The *magnitude characteristic* thus retains its shape over the range of operating conditions. Consider firstly the variation of v_{d0} with generator reactive power output at constant real power (P) and 1 pu terminal voltage as shown in Table 9.7. The low frequency gain of the P-Vr characteristics decreases with increasing lagging reactive power (Q); this observation is reflected in the P-Vr characteristics of Figure 5.16 for the SMIB system.

Likewise, as illustrated in Table 9.8 and manifested in the P-Vr characteristics of Figure 5.22, v_{d0} decreases with decrease in real power output at unity power factor. At rated power output v_{do} is relatively large, but tends to zero as the real power output is reduced.

A consideration of Table 9.7 suggests that it is prudent to include a range of reactive power outputs in the set of encompassing operating conditions.

Table 9.7 Variation of v_{d0} with increasingly lagging Q at constant real power outputs

	Corresponding values of Q and v_{d0}			
P=0.9, Q pu	-0.2	0	0.2	0.4
v_{d0} pu	0.930	0.851	0.766	0.686
P=0.7, Q pu	-0.2	0	0.2	0.4
v_{d0} pu	0.892	0.783	0.680	0.591

Table 9.8 Variation of v_{d0} with reduction in P at unity power factor

	Corresponding values of P and v_{d0}				
Q=0, P pu	0.9	0.7	0.5	0.3	0.1
v_{d0} pu	0.851	0.783	0.669	0.475	0.177

Because $G_{6i}(s)$ is the more significant term of the two in (9.7), it determines the consistently narrow bands of the frequency responses of the P-Vr characteristics of generator *i* given by $A_{v\_ii}(s) = \Delta P_{ei}(s)/\Delta V_{ri}(s)$. For example, for the SMIB system the relatively minor variations in the phase of the P-Vr characteristic with steady-state operating conditions observed in Figure 5.16 are caused by the contribution of first term in (9.7), $[\boldsymbol{G}_3(s)\boldsymbol{Z}(s)\boldsymbol{G}_5(s)]$, over the modal frequency range.

In [15] the authors imply that the P-Vr characteristic of generator *i* can be calculated if the network is represented by a SMIB system connected at the generator terminals. However, in multi-machine cases, it may not be clear *what value should be attributed to the impedance* of the Thévenin equivalent, particularly as it will change with line outages, whether electrically close-by machines are on/off line, the effect of close-by loads, etc. It is then simpler and more efficient to calculate the P-Vr characteristics of generator *i*[1] for each operating condition using the complete model of the multi-machine system. Moreover, each generator may participate in a range of local- and inter-area modes as well as intra-station modes, not in a single mode as is the case in the SMIB system.

These results in [15] provide a theoretical basis for the observation in [13] that the P-Vr transfer function is relatively robust to changes in the system operating conditions in multi-machine systems. That is, for *higher values* of generator real power outputs both the gain, phase and the shapes of the frequency response of the P-Vr transfer functions do not vary appreciably, for practical purposes, over a wide range of operating conditions and system

1. Or the characteristics of *generating station i* if there are a number of identical units in the station.

configurations. *Consequently, in multi-machine systems, individual PSS designs that are based on the synthesized P-Vr transfer function using the methodology adopted in Section 5.10 are also robust over a wide range of operating conditions.* Typically, this applies for generator real power outputs exceeding 0.5 pu. An examination of Figures 5.21, 5.22 and Tables 5.5 and 5.6 reveals that the mode shifts are essentially real, an observation which supports the above statement.

The robustness and application of the P-Vr characteristic has been demonstrated and verified for generators on very large systems [20].

For the multi-machine system the P-Vr characteristics of each generator, $H_{PVri}(j\omega_f) = \Delta P_{ei}(j\omega_f)/\Delta V_{ri}(j\omega_f)$, are calculated in a similar fashion to that for the single-machine infinite-bus system, except that the characteristics are calculated for the entire network with the shaft dynamics of *all* machines disabled. The calculation is similar to that described in Section 5.10.3 in which rows and columns of the ***A***, ***B*** and ***C*** matrices associated with the speed states in the states equations (3.9) are eliminated; the ***D*** matrix is usually a null matrix. The relationship between perturbations in electric power (or torque) as the output quantity and voltage reference as the input quantity can then be formed, and the frequency response evaluated for the set of encompassing operating conditions and over the range of modal frequencies.

The derivation of the *synthesized* transfer function,

$$H_{PVrS_i}(s) = \Delta P_{ei}(s)/\Delta V_{ri}(s)\big|_{\text{Synth}}, \tag{9.10}$$

which is selected from the family of P-Vr frequency response characteristics as the most suitable basis for the tuning of the PSS, has been covered in Section 5.10.6.

9.4.2 Transfer function of the PSS of generator *i* in a multi-machine system

The basic concepts for the determination of the parameters of a PSS in a single machine system have been outlined in Chapter 5. The approach in the case of a generator in a multi-machine systems follows along similar lines in Section 5.8.1 and therefore can be summarized fairly briefly.

Consider the model of generator fitted with a speed-PSS in multi-machine system as shown in Figure 9.10. Note that the transfer functions from ΔV_{ri} or ΔV_{si} to ΔP_{ei} are identical.

It has been established in Section 5.9.1 that the PSS transfer function for a generator - say the ith - takes the form

$$H_{PSS_i}(s) = k_i G_i(s) = k_i \cdot G_{Wi}(s) \cdot G_{ci}(s) \cdot G_{LPi}(s), \tag{9.11}$$

where k_i is the damping gain; $G_{ci}(s)$ is the PSS compensation block; $G_{Wi}(s)$ and $G_{LPi}(s)$ are the transfer functions of the washout and low-pass filters, respectively.

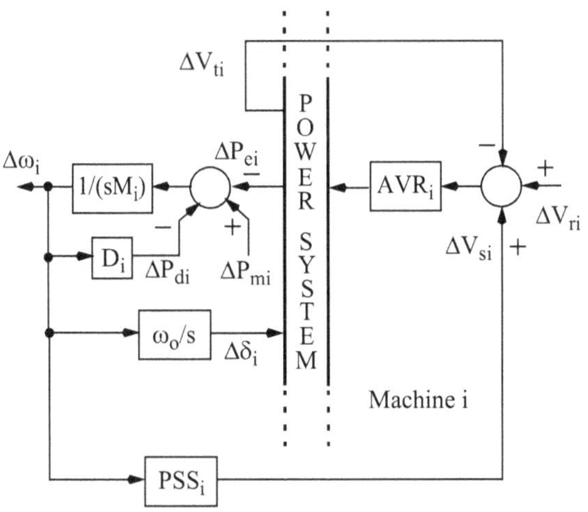

Figure 9.10 Model of generator *i* fitted with a PSS in a multi-machine power system

As has been outlined earlier in Section 5.14:

1. The aim of the tuning procedure is to introduce on the generator shaft a damping torque (a torque proportional to machine speed); this causes the modes of rotor oscillation to be shifted directly to the left [1] in the complex *s*-plane.

2. The compensation transfer function $G_{ci}(s)$ is tuned to achieve the desired *left-shift* in the complex *s*-plane of the relevant modes of rotor oscillation.

3. The damping gain k_i (on machine MVA rating) of the PSS determines the *extent* of the left-shift.

Based on item 1, the ideal transfer function between speed $\Delta\omega_i$ and the electrical damping torque perturbations ΔP_{ei} due to the action of the PSS *i* over the range of modal frequencies should ideally be:

$$\Delta P_{ei}(s)|_{PSSi} = D_{ei}\Delta\omega_i(s), \tag{9.12}$$

where D_{ei} is a damping torque coefficient and is a real number (p.u. on generator MVA rating). The transfer function $G_{ci}(s)$ compensates in *magnitude as well as phase* for the synthe-

1. By 'direct left-shift' is implied that the mode shift is $-\alpha \pm j0$, $\alpha \geq 0$. As explained in Chapter 13, deviations from the 'direct left-shift' of modes are mainly due to interactions between multi-machine PSSs and non-real generator participation factors.

sized P-Vr transfer function of machine i, $H_{PVrS_i}(s)$, defined in (9.10). With rotor speed being used as the input signal to the PSS, whose output is $\Delta V_{si}(s)$, the expression (9.12) for D_{ei} can also be expressed in terms of the P-Vr and PSS transfer functions as:

$$D_{ei} = \frac{\Delta P_{ei}(s)}{\Delta V_{si}(s)} \cdot \frac{\Delta V_{si}(s)}{\Delta \omega_i(s)} = H_{PVrS_i}(s)[k_i G_{ci}(s)]; \qquad (9.13)$$

hence, rearranging (9.13), we find

$$[k_i G_{ci}(s)] = D_{ei}/(H_{PVrS_i}(s)). \qquad (9.14)$$

It follows from an examination of (9.14) that

$$k_i = D_{ei} \text{ and } G_{ci}(s) = 1/(H_{PVrS_i}(s)), \qquad (9.15)$$

Note from (9.15) that k_i is a damping torque coefficient. Assuming that the synthesized transfer function $H_{PVrS_i}(s)$ is of the *general* form

$$H_{PVrS_i}(s) = k_{ci}\frac{(1+sT_{b1_i})\ldots}{(1+c_{1i}s+c_{2i}s^2)(1+sT_{a1_i})\ldots}, \qquad (9.16)$$

then, from (9.15), the compensation block transfer function is

$$G_{ci}(s) = \frac{1}{k_{ci}} \cdot \frac{(1+c_{1i}s+c_{2i}s^2)(1+sT_{a1_i})\ldots}{(1+sT_{b1_i})\ldots}, \qquad (9.17)$$

where k_{ci}, c_{1i}, c_{2i}, T_{a1_i}, ..., T_{b1_i}, ... are parameters determined from the synthesized P-Vr characteristic in the tuning procedure. Note that in the form of (9.16) (i) real and complex zeros can be accommodated in the synthesized P-Vr transfer function; (ii) the coefficient of s^0 is unity.

Substitution of (9.17) in (9.11), and incorporating the washout and low pass filters, yields the PSS transfer function:

$$H_{PSSi}(s) = k_i\left[\frac{sT_{Wi}}{1+sT_{Wi}} \cdot \frac{1}{k_{ci}} \cdot \frac{(1+c_{1i}s+c_{2i}s^2)(1+sT_{a1_i})\ldots}{(1+sT_{b1_i})\ldots} \cdot \frac{1}{(1+sT_{1i})(1+sT_{2i})\ldots}\right]. \qquad (9.18)$$

For generator i, T_{Wi} is the time constant of the washout filter; T_{1i}, T_{2i}, ... are the time constants of the low-pass filter which may be added (i) to ensure $H_{PSSi}(s)$ is proper, (ii) to mitigate against excitation of the torsional modes of the rotating turbine/generator/exciter shaft system.

Earlier the gain k_i has been referred to as the '*damping gain*' of the PSS. The gain k_{ci} is the DC gain of P-Vr characteristic of generator *i*. If the washout filter is ignored the DC gain of the PSS transfer function is k_i/k_c; *conventionally this is referred to as the 'PSS Gain'*. (However, the PSS gain k_i/k_{ci} has been attributed little meaning because the significance of the gain k_{ci} has not been recognized.)

Note, assuming that the synthesized P-Vr characteristic $H_{PSSi}(s)$ for generator *i* closely matches that for the selected operating condition, $H_{PVr_i}(s)$, it follows that

$$H_{PVr_i}(s)[k_i G_{ci}(s)] \approx k_i + j0\,, \tag{9.19}$$

(i.e. equal to the damping gain) over the modal frequency range [1]. In the next chapter, by examining the damping torque coefficient, this result will be used to confirm that the designed damping gain k_i of PSS *i* is, in fact, achieved (see Section 10.6).

In the multi-machine PSS tuning methodology the PSS is designed not only to swamp any negative (destabilizing) inherent damping torque coefficients on that machine over the range of frequencies of the rotor modes, but also to provide sufficient damping so that the associated damping criteria of the multi-machine system are satisfied [9], [13]. These issues, together with the contribution to damping by stabilizers installed on FACTS devices, are considered in a later chapters.

Examples of the application of the P-Vr approach to the tuning of PSSs in multi-machine power systems have been presented in several publications [9], [13], [16], [17], [18] and [19]. In Chapter 10 the tuning the PSSs of generators in an inherently unstable 14-generator, multi-machine power system is described. Based on this system, the features of the PSS tuning technique discussed above are illustrated through an example for which the complete system data is provided.

9.5 Synchronising and damping torque coefficients induced by PSS *i* on generator *i*

The concepts of synchronising and damping torques over a range of frequencies ($s = j\omega_f$) are explained in the context of the single-machine system in Sections 5.3 and 5.5. The same concepts are employed in the multi-machine application to assess the synchronising and damping torque coefficients developed by PSS *i* on generator *i*. However, let us first assess the significance of Figure 9.11 which is derived from Figure 9.10.

---

1. Note $G_{ci}(s)$ is the compensation which applies over the range of modal frequencies. The washout and low-pass filter time constants lie outside the latter range and are not included in $G_{ci}(s)$, but are included in the PSS transfer function, $H_{PSS}(s)$.

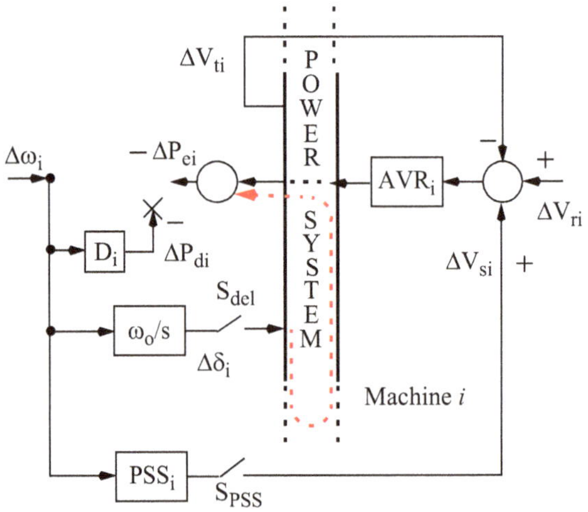

Figure 9.11 Model of generator *i* in a multi-machine system with (i) shaft dynamics on all machines disabled and (ii) switches S_{del} and S_{PSS} in rotor angle and PSS paths, respectively.

Consider generator *i*. When the shaft dynamics on all generators are disabled with switch S_{del} in the rotor angle path closed and the PSS out of service, the signal flow paths can be deduced from Figure 9.11. As in the case of the SMIB system there are signal flow paths directly from $\Delta\omega_i$ through $\Delta\delta_i$ to ΔP_{ei} (or to ΔP_2 in Figure 5.2). However, in the multi-machine case there are paths from $\Delta\omega_i$ through the network to perturbations in rotor angles and terminal voltages on other generators, then to the *inherent* torque output ΔP_{ei}. The principle of superposition in linear systems analysis says that these paths remain when the shaft dynamics are enabled and the full system is reinstated.

The object of the following analysis is to determine in the multi-machine cases if the PSS performance is consistent with its design basis, i.e. if $H_{PVr_i}(j\omega_f)[k_i G_{ci}(j\omega_f)] \approx k_i + j0$, *over the modal frequency range* for an encompassing range of operating conditions. In other words, is the per unit damping gain k_i of the PSS the realised? This objective is illustrated in Figure 9.11 when the PSS is in service with switch S_{PSS} closed and the rotor angle path is open by means of switch S_{del}.

The synchronising and damping torque coefficients for generator *i* are defined in Section 5.3 and apply to generator *i* in Figure 9.11:

$$k_{si} = -\Im\left\{\frac{\omega_f}{w_0} \cdot \frac{\Delta P_{ei}(j\omega_f)}{\Delta\omega_i(j\omega_f)}\right\} \quad \text{and} \quad k_{di} = \Re\left\{\frac{\Delta P_{ei}(j\omega_f)}{\Delta\omega_i(j\omega_f)}\right\}, \tag{9.20}$$

i.e. components on generator *i* of torques in quadrature and in phase with rotor speed on unit *i*.

The theoretical basis for concept of the *inherent* synchronising and damping torques for generators in a multi-machine is derived from [14], [15] and the associated equation (9.5). Recall that the shaft dynamics of all generators are disabled (i.e. switch S_{del} is closed and S_{PSS} is open in Figure 9.11. The torque-like relationship of interest for generator *i* is extracted from (9.5), i.e.

$$\boldsymbol{\Delta P_e}(s) = \boldsymbol{B_\omega}(s)\boldsymbol{\Delta\omega}(s)\text{, where} \tag{9.21}$$

$$B_{\omega\_ii}(s) \approx G_{3i}(s)Z_{ii}(s)G_{2i}(s) + G_{4i}(s)\,. \tag{9.22}$$

In (9.22) $G_{2i}(s)$, $G_{3i}(s)$ and $G_{4i}(s)$are essentially functions of the steady-state conditions and are modified by the generator operational reactances $x_d(s)$ and $x_q(s)$. Notice that the first term, through (9.8), is determined mainly by the reduced network admittance matrix ***Y***.

According to [14] both terms in (9.22) are essentially functions of the steady-state conditions at the generator terminals. However, the first term in (9.22) also involves the Thévenin equivalent of the network as seen from the terminals of generator *i*, and the generator parameters. Unlike the second term $G_{6i}(s)$ in (9.7), $G_{4i}(s)$ does not display phase invariance in its frequency response over a wide range of operating conditions. This implies that, unlike the P-Vr characteristics, the *inherent* synchronising and damping torques coefficients for generators in a multi-machine system do vary with changes in operating conditions. Moreover, both $G_{2i}(s)$ and $G_{4i}(s)$ incorporate a multiplying factor ω_0/s with the result that at high and low frequencies the frequency response rolls off at 20 dB/decade and exhibits a constant phase of –90°. This observation is illustrated in the SMIB cases by a comparison of the inherent torque coefficients in Figures 5.5(a) and 5.18.

As implied in Figure 9.9(a), in reality with the shaft dynamics of all generators enabled, there are also paths through the power system, the AVRs, and the other generators. A disturbance in the speed of generator *j* will therefore induce a torque on the generator *i*, the unit of interest, at the inter-area and other modal frequencies. The components of this torque in phase, or lagging by 90°, with the speed of generator *i* are damping or synchronizing torques induced on unit *i* by *j*. The associated torques coefficients, without and with PSSs in operation and *with the shaft dynamics of all generators enabled*, are analysed in Chapter 12 and are referred to as Modal Induced Torque Coefficients.

9.6 References

[1] G. Golub and C. F. Van Loan, *Matrix Computations*, Third Edition, John Hopkins University Press, 1996.

[2] W. E. Arnoldi, "*The principle of minimized iterations in the solution of the matrix eigenvalue problem*," Quarterly of Applied Mathematics, vol. 9, pp. 17–29, 1951.

[3] K. J. Bathe, "The subspace iteration method – Revisited", Comput Struct (2012), http://dx.doi.org/10.1016/j.compstruc.2012.06.002.

[4] D. J. Vowles and M. J. Gibbard, *Mudpack User Manual: Version 10S-03*, School of Electrical and Electronic Engineering, The University of Adelaide, July 2014.

[5] F. L. Pagola, I. J. Perez-Arriaga and G. Verghese, "On sensitivities, residues and participations: Applications to oscillatory stability analysis and control", *Power Systems, IEEE Transactions on*, vol. 4, pp. 278-285, 1989.

[6] J.D'Azzo and C.H. Houpis, *Linear Control System Analysis and Design, Conventional and Modern*, McGraw-Hill International Editions, 3rd Edition, 1988.

[7] Gene Franklin, J.D. Powell and Abbas Emami-Naeini, *Feedback Control of Dynamic Systems*, 5th edition, Prentice Hall, October 2005.

[8] R.C. Dorf and R.H. Bishop, *Modern Control Systems*, 10th edition, Prentice Hall, April 2004.

[9] M. J. Gibbard, "Coordinated design of multimachine power system stabilisers based on damping torque concepts", *IEE Proc. Part-C*, pages 276-284, July 1988.

[10] E. V. Larsen and D.A. Swann, "Applying power system stabilizers: Part I-III", *IEEE Trans. PAS*, pp. 3017-3046, 1981.

[11] P. Kundur, M. Klein, G. J. Rogers, and M. S. Zywno, "Application of power system stabilizers for enhancement of overall system stability," *Power Systems, IEEE Transactions on*, vol. 4, pp. 614-626, 1989.

[12] F.P. De Mello, J. S. Czuba, P. A. Rusche, and J. Willis, "Developments in application of stabilizing measures through excitation control", *International Conference on Large High Voltage Electric Systems,* CIGRE, paper 38-05, 1986.

[13] M. J. Gibbard, "Robust design of fixed-parameter power system stabilizers over a wide range of operating conditions", *Power Systems, IEEE Transactions on*, vol. 6, pp 794-800, 1991.

[14] D. M. Lam, *Eigenvalue analysis and stabilizer design for electrical power systems,* PhD Thesis, University of Sydney, 1995.

[15] D. M. Lam and H. Yee, "A study of frequency responses of generator electrical torques for power system stabilizer design," *Power Systems, IEEE Transactions on*, vol. 13, pp. 1136-1142, 1998.

[16] M.J. Gibbard, D.J. Vowles and P. Pourbeik, "Interactions between, and effectiveness of, power system stabilizers and FACTS stabilizers in multimachine systems", *Power Systems, IEEE Transactions on*, vol. 15,pp. 748-755, 2000.

[17] M.J. Gibbard, D.J. Vowles, "Reconciliation of methods of compensation for PSSs in multimachine systems", *Power Systems, IEEE Transactions on,* vol. 19, pp. 463-472, 2004.

[18] CIGRE Technical Brochure no. 166 prepared by Task Force 38.02.16, "*Impact of Interactions among Power System Controls*", published by CIGRE in August 2000.

[19] M.J. Gibbard, N. Martins, J.J. Sanchez-Gasca, N. Uchida, V. Vittal, and L. Wang, "Recent Applications in Linear Analysis Techniques", *Power Systems, IEEE Transactions on*, vol. 16, pp. 154-162 2001.

[20] P. Pourbeik, T. Cain, and R. Bottoms, "Application of small-signal stability tools and techniques to a large power system," in *The Proceedings of the IEEE Power & Energy Society General Meeting, 2009.* IEEE PES '09, pp. 1-11.

Chapter 10

Application of the PSS Tuning Concepts to a Multi-Machine Power System

10.1 Introduction

The previous chapter introduced some important concepts in the tuning of PSSs in multi-machine power systems. The purpose of this chapter is to demonstrate the application of the associated techniques for the analysis and tuning of PSSs in a fourteen-generator power system which, without continuously acting PSSs, is inherently unstable. Each 'generator' in this system, in fact, represents a power station which accommodates between one and twelve units; the number of units in-service (n_u) depends on the particular operating condition. The units in a power station are assumed to be identical, therefore the rating of the equivalent generator for a station is n_u times the rating of a single unit. It is assumed that the individual generators in each power station are fitted with identical excitation systems and PSSs.

In a later chapter a class of stabilizers known as Power Oscillation Dampers (PODs) are discussed; these are stabilizers that can be fitted to power-electronic based transmission devices such as FACTS (e.g. Static Var Compensators) and HVDC transmission. The analysis and tuning of POD stabilizers are demonstrated by means of examples in Chapter 11. In the fourteen-generator power system described in this chapter the Static Var Compensators (SVCs) are fitted with continuously acting voltage regulators controlling bus voltage, but are not fitted with stabilizers.

The steps in the tuning of PSSs of machines in a multi-machine system are explored, commencing with (i) the eigen-analysis of the system with all PSSs out of service, and (ii) the associated analysis based on Mode Shapes and Participation Factors. The PSSs are then tuned using the P-Vr approach discussed in Section 9.4. Having completed the determination of the PSS parameters, the effect on the shifts of eigenvalues associated with the rotor modes are assessed as the damping gains of the PSSs are increased; ideally over the range of operating conditions such shifts are directly to the left in the complex s-plane.

In practice a new power station is built to supply energy to an existing power system in which many of the existing generators may already be fitted with PSSs. The latter PSSs would have been tuned and their parameters set to fixed values. The PSSs in a new power station have to be tuned to satisfy the damping and other performance criteria of the system operators over the range of system operating conditions and contingencies. However, in the following example the PSSs fitted to all generators are tuned at the same time, and the effect on damping established as the PSS damping gains are increased from zero to 30 pu on machine base. This analysis reveals a number of issues that are not found in the analysis associated with the new power station in an existing system. Nevertheless, the approach adopted in the example is applicable to the tuning of PSSs for additional generation.

Earlier work has investigated the tuning of PSSs to adequately damp both local- and inter-area modes [1] [2]. It will be demonstrated that the design of PSSs based upon the P-Vr concept inherently damps both types of modes.

Although each power station is represented by a single composite generator formed from the n_u units in service, it is often necessary to represent the individual machines in the station. Because the PSS tuning techniques do not directly determine the nature of the intra-station modes (i.e. modes of oscillation between machines in a single power plant), the effects of the PSS tuning on these modes are examined in Section 10.8 to assess their characteristics. If the damping of the intra-station modes is poor, it will be necessary to determine what action needs to be taken to remedy the problem.

Normally the main emphasis is placed on the dynamic performance of the multi-machine power system following large-signal disturbances. Such disturbances are major faults on the system, switching of heavily-loaded transmission lines, the tripping of a generator, the loss of a significant load, etc. Notwithstanding the non-linear nature of the limiting action of controllers immediately following the fault, the dynamic performance is determined by the non-linear nature of differential-algebraic equations. The question arises: what is the relevance and significance of small-signal analysis to the analysis and understanding of large-signal dynamic behaviour? These issues are investigated in Section 10.9 by examining the transient response of the 14-generator system following the incidence of a major fault in a critical location. Furthermore, interesting recent developments are establishing a "bridge" between small- and large-signal analysis; this has been achieved by including the second-order terms in the Taylor series expansion about the steady-state operating condition as discussed in Section 10.9.2. To provide further understanding of the nature of the system behaviour, concepts of "modal interactions" and their significance with respect to both large- and small-signal dynamic performance are briefly discussed.

10.2 A fourteen-generator model of a longitudinal power system

The simplified system of 14 power stations [1] is shown in Figure 10.1. It represents a long, linear system as opposed to the more tightly meshed networks found in Europe and the USA. For convenience, the system has been divided into 5 areas in which areas 1 and 2 are more closely coupled electrically. There are in essence 4 main areas and hence 3 inter-area modes, as well as 10 local-area modes. Without PSSs installed on generators in this system, many of these modes are unstable.

For the purpose of designing generator PSSs in practice a wide range of both normal operating conditions and contingencies [2] are considered. However, to simplify the procedures for illustrative purposes in this and the following chapters, a limited number of cases encompassing a range of fairly diverse, *normal* conditions is employed. The encompassing range of operating conditions [3], system loads and major inter-area flows are listed in Table 10.1.

Table 10.1 Six normal steady-state operating conditions

	Case 1	Case 2	Case 3	Case 4	Case 5	Case 6
Load Condition	Heavy	Medium-heavy	Peak	Light	Medium	Light
Total generation (MW)	23030	21590	25430	15050	19060	14840
Total load (MW)	22300	21000	24800	14810	18600	14630
Inter-area flows	(North to south)	(South to north)	(Area 1 to N & S)	(Area 2 to N & S)	(N & S to Area 1)	(~Zero transfers)
Area 4 to Area 2 (MW)	500	500	-500	-200	300	0
Area 2 to Area 1 (MW)	1134	1120	-1525	470	740	270
Area 1 to Area 3 (MW)	1000	1000	1000	200	-200	0
Area 3 to Area 5 (MW)	500	500	250	200	250	0

1. In the analysis a power station with *n* units on-line is represented as a single generator. Consequently the station is often referred to as a 'generator'
2. These are referred to as N and N-1 conditions respectively.
3. The term "encompassing range of operating conditions" in defined Section 5.1. It is assumed that, for the subsequent analysis, a reduced set of steady-state conditions are selected which encompass those conditions for which the stabilizers are to be tuned.

For the six cases the ratings of generators, the number of units on line, and their real and reactive power outputs are listed in Table 10.2. The power stations are designated *PS_<area number>, e.g. HPS_1 refers a power station (PS) called 'H' in area 1. Note that the number of units on-line in certain stations can vary considerably over the range of operating conditions. A number of the units in the hydro station HPS_1 can operate as synchronous compensators, or as synchronous motors driving pumps in a pump-storage mode of operation.

Table 10.2 Generation conditions for six power flow cases.

Power Station/Bus # Rating Rated power factor	Case 1 No. units MW Mvar	Case 2 No. units MW Mvar	Case 3 No. units MW Mvar	Case 4: No. units MW Mvar	Case 5: No. units MW Mvar	Case 6: No. units MW Mvar
HPS_1 / 101 12 x 333.3 MVA 0.9 power factor lag	4 75.2 77.9	3 159.6 54.4	12 248.3 21.8	2 0 -97.4 Syn.Cond	3 -200.0 -26.0 Pumping	2 0 -102.2 Syn.Cond
BPS_2 / 201 6 x 666.7 MVA 0.9 power factor lag	6 600.0 95.6	5 560.0 38.9	6 550.0 109.1	4 540.0 -30.8	5 560.0 38.7	3 560.0 -53.5
EPS_2 / 202 5 x 555.6 MVA 0.9 power factor lag	5 500.0 132.7	4 480.0 60.5	5 470.0 127.6	3 460.0 -2.5	4 480.0 67.2	3 490.0 -7.3
VPS_2 / 203 4 x 555.6 MVA 0.9 power factor lag	4 375.0 132.8	3 450.0 82.4	2 225.0 157.0	3 470.0 9.4	2 460.0 83.1	3 490.0 3.7
MPS_2 / 204 6 x 666.7 MVA 0.9 power factor lag	6 491.7 122.4	4 396.0 17.8	6 536.0 96.5	4 399.3 -43.6	4 534.4 55.2	3 488.6 -61.2
LPS_3 / 301 8 x 666.7 MVA 0.9 power factor lag	7 600.0 142.3	8 585.0 141.1	8 580.0 157.6	6 555.0 16.6	8 550.0 88.1	6 550.0 9.4
YPS_3 / 302 4 x 444.4 MVA 0.9 power factor lag	3 313.3 51.5	4 383.0 63.3	4 318.0 49.6	2 380.0 -9.3	3 342.0 43.8	2 393.0 -6.9
TPS_4 / 401 4 x 444.4 MVA 0.9 power factor lag	4 350.0 128.7	4 350.0 116.5	4 350.0 123.2	3 320.0 -21.9	4 346.0 84.9	3 350.0 -32.6

Power Station/Bus # Rating Rated power factor	Case 1 No. units MW Mvar	Case 2 No. units MW Mvar	Case 3 No. units MW Mvar	Case 4: No. units MW Mvar	Case 5: No. units MW Mvar	Case 6: No. units MW Mvar
CPS_4 / 402 3 x 333.3 MVA 0.9 power factor lag	3 279.0 59.3	3 290.0 31.4	3 290.0 32.0	2 290.0 -2.4	3 280.0 45.4	3 270.0 4.7
SPS_4 / 403 4 x 444.4 MVA 0.9 power factor lag	4 350.0 52.3	4 350.0 47.2	4 350.0 47.3	3 320.0 14.2	4 340.0 46.3	2 380.0 25.2
GPS_4 / 404 6 x 333.3 MVA 0.9 power factor lag	6 258.3 54.5	6 244.0 39.8	6 244.0 40.0	3 217.0 -3.5	5 272.0 50.4	3 245.0 3.9
NPS_5 / 501 2 x 333.3 MVA 0.9 power factor lag	2 300.0 25.3	2 300.0 -8.8	2 300.0 6.5	2 280.0 -52.5	2 280.0 -35.2	1 270.0 -42.2
TPS_5 / 502 4 x 250 MVA 0.8 power factor lag	4 200.0 40.1	4 200.0 53.0	4 180.0 48.8	3 180.0 -1.8	4 190.0 0.1	4 200.0 -9.7
PPS_5 / 503 6 x 166.7 MVA 0.9 power factor lag	4 109.0 25.2	5 138.0 36.9	6 125.0 32.6	1 150.0 2.2	2 87.0 3.5	2 120.0 -11.2

10.2.1 Power flow analysis

Data for the power flow analysis of the six normal operating conditions given in Table 10.1 are supplied in Appendix 10–I.2. Included in Appendix 10–I.2 are relevant results of the analysis such as reactive outputs of generators and SVCs, together with tap positions on generator and network transformers. This information permits the power flows to be set up on any power-flow platform and the results checked against those provided in this document.

10.2.2 Dynamic performance criterion

The dynamic performance criterion requires

- that all modes are stable;
- for all normal and N-1 system conditions the damping of the electro-mechanical modes is to be such that the associated halving times are 5 s or less.

The 'halving time' is defined as the time for the mode or its envelope to decay to half its initial amplitude. The real parts of the electro-mechanical modes must therefore be less than $\sigma = -0.139$ (since $\exp(\sigma 5) = 0.5$) to satisfy the latter requirement.

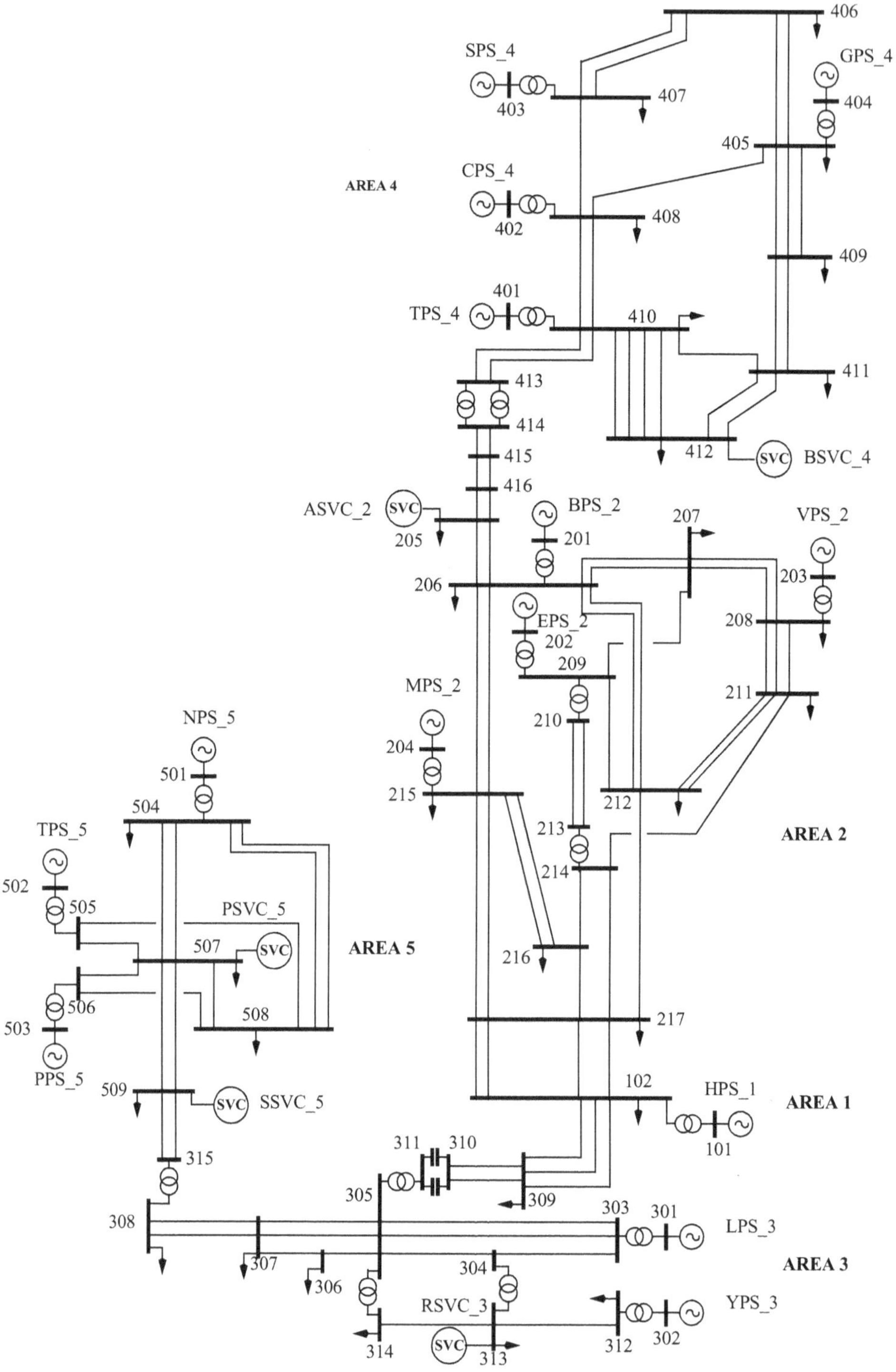

Figure 10.1 Simplified fourteen-generator system.

10.3 Eigen-analysis, mode shapes and participation factors of the 14-generator system, no PSSs in service

In order to gain some insight into the dynamic performance and characteristics of the system, a series of analyses is conducted without - and later with - PSSs in service on all generators.

10.3.1 Eigenvalues of the system with no PSSs in service

The preliminary objective of the eigen-analysis is to identify the nature of the unstable and lightly-damped modes.

Let us consider the eigen-analysis of Case 1, a heavy load condition, in the 14-generator power system with no PSSs or SVC stabilizers in service. In this case there are 125 states and consequently 125 modes. Because there are N_g=14 generators in service there are N_g-1=13 rotor modes that reflect the modal interplay between generation. A fourteenth real or complex mode typically portrays the behaviour common to all fourteen rotating masses with respect to a reference (see Section 9.2.2).

For this simple system the eigenvalues are calculated using an algorithm that employs QR factorisation. The unstable modes are displayed either as a listing of the eigenvalues, or on a plot in the complex s-plane. For Case 1 such a plot, with eigenvalue designations, is shown in Figure 10.2 for a limited region about the positive imaginary axis.

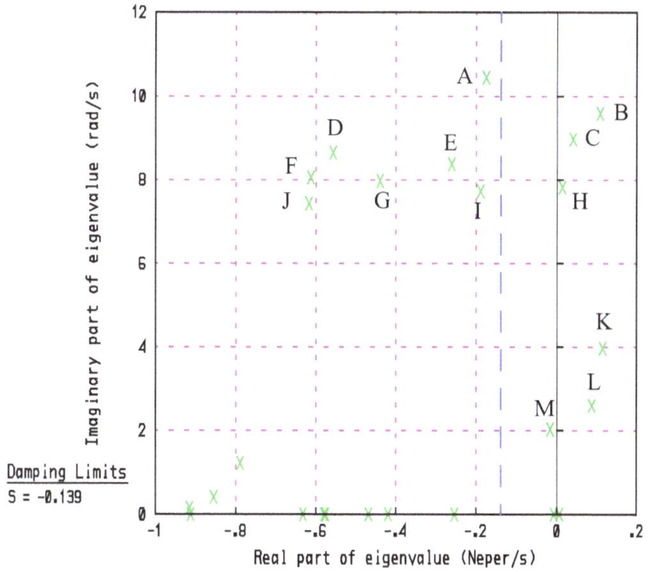

Figure 10.2 Plot of eigenvalues for Case 1, no PSSs in service

Figure 10.2 reveals that there are five unstable oscillatory modes, one stable oscillatory mode that does not satisfy the dynamic performance criterion and seven other lightly-damped oscillatory modes with damping ratios less than 0.1.

Valuable information on the stability of the modes is provided by the eigenvalue plot but it does not reveal the type or nature of modes. In this case it is desirable to identify all thirteen electro-mechanical modes, particularly those which are unstable or are lightly damped. Participation Factor and Mode Shape Analyses are employed for this purpose.

10.3.2 Application of Participation Factor and Mode Shape Analyses to Case 1

Consider in Figure 10.2 the unstable, oscillatory mode $0.088 + j2.60$ (designated 'Mode L'). Let us view the plots not only of the magnitudes of its participation factors (PFs) but also of its mode shape (MS); the plots are shown in Figure 10.3.

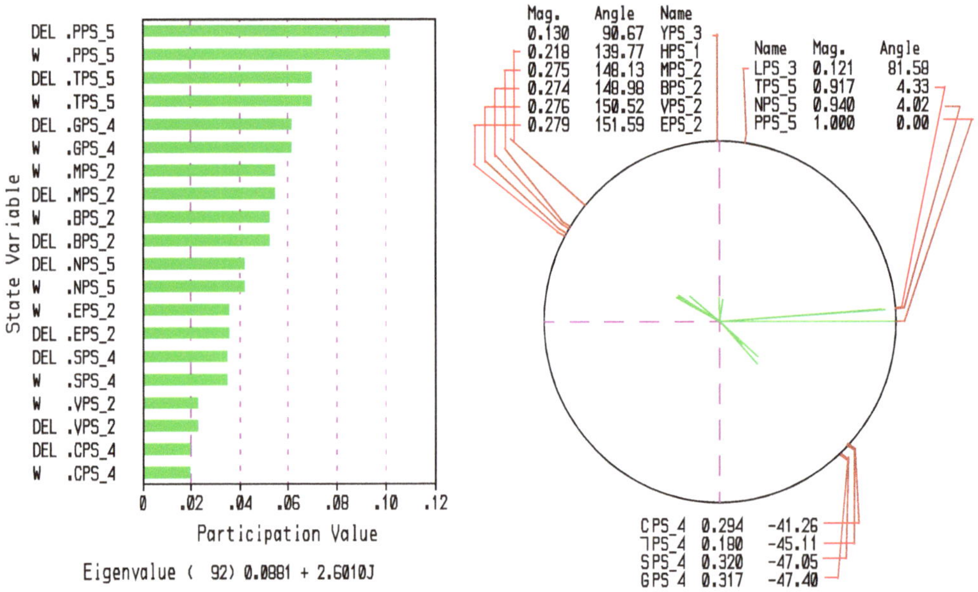

Figure 10.3 Magnitude of the participation factors (left) and the mode shape (right) for the unstable mode L, $0.088 + j2.60$. No PSSs are in service.

(In the plot of the participation factors 'W' and 'DEL' are the rotor speed and angle perturbations, respectively.)

Recall that the concepts of participation factors (PFs) and mode shapes (MSs) were discussed in Chapters 3 and 9. In this case the participation factor is the participation of the states in the selected mode arranged in decreasing values of the magnitude of the PFs. For the selected mode the mode shape is the plot of the normalised magnitude and phase of the right speed eigenvectors and reveals, for example, that a group of generators swing with - or against - another group of machines.

According to the plot of the PFs in Figure 10.3 the two states, rotor speed and angle, of a number of generators dominate the involvement of the states in mode L; this mode is therefore an electro-mechanical mode. There are a total of 125 states in this system model. The MS reveals that the generators in Areas 5 and 4 swing against those in Areas 1 and 2; the participation factors of those in Area 3 are small. Mode L is therefore classified as an 'inter-area' mode. However, note that:

- although the magnitude of the MS phasor of generator NPS_5 is the second largest, the PF of its speed state is the twelfth largest;
- as highlighted in Section 9.2.2, some care should therefore be attached to interpreting the lengths of the MS phasors. The length of the MS phasors for some generators is shorter than for others because their inertias on system MVA base may be significantly greater. In Figure 10.3, for example, the relative lengths of the MS phasors for BPS_2 and PPS_5 are in the ratio 0.274:1, the ratio of their inertias is 2.56:1. The most useful feature of the MS plot is therefore the relative phase information that it provides.
- the PFs for mode L are nearly real (e.g. PF is $0.101 + j0.013$ for both the speed and rotor angle states of PPS_5). When in a later Chapter 13 we analyse the mode shift contributed by the PSS of a given generator we shall find that the complex value of its PF plays a major role [4].

For a second unstable mode, $0.115 + j3.97$, the PF plot in Figure 10.4 reveals that this mode is also an electro-mechanical mode; the MS shows that generators in Areas 3 and 1 swing against machines in Areas 2 and 5. This mode, called 'K', is also an inter-area mode.

For reference in later studies the PFs and MS for the third inter-area mode ('M') are shown in Figure 10.5. In this case Areas 5, 3 and 2 swing against Area 4.

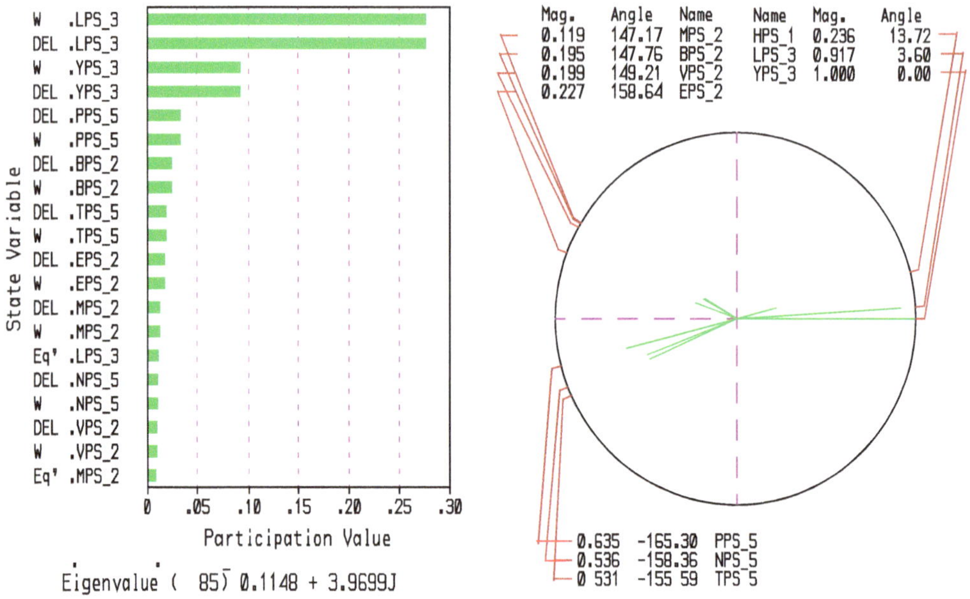

Figure 10.4 Participation factors and mode shape for the unstable mode K ($0.115 + j3.97$), no PSSs are in service.

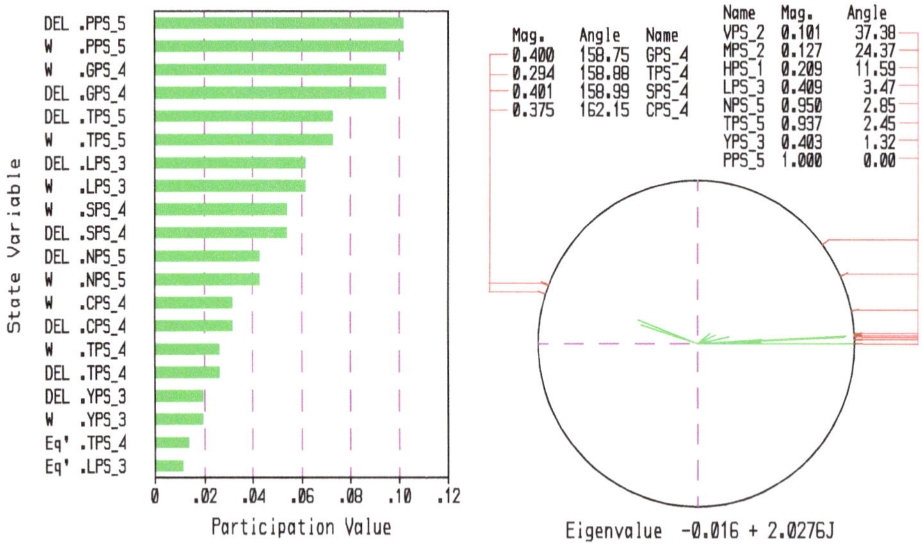

(a) (b)

Figure 10.5 Case 1. Participation factors and mode shape for the inter-area mode M ($-0.016 + j2.03$), no PSSs are in service.

To determine the nature of other lightly-damped or unstable oscillatory modes with frequencies between 7 and 11 rad/s, shown in Figure 10.2, the plots of their MSs and PFs are examined. Such plots are displayed in Figure 10.6; each plot reveals a rotor mode of oscillation. All three are found to be local-area modes:

- in mode $-0.17+j10.4$ VPS_2 swings against EPS_2;
- in the unstable mode $0.11+j9.58$ SPS_4 swings mainly against CPS_4 and GPS_4;
- in the unstable mode $0.04+j8.96$ BPS_2 swings mainly against EPS_2, VPS_2 and TPS_4.

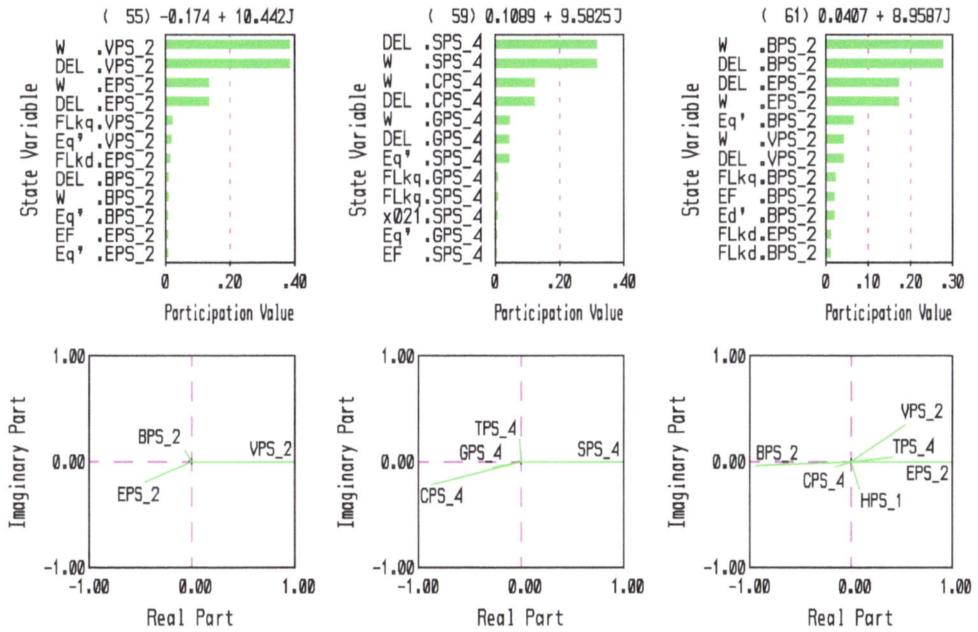

Figure 10.6 Case 1. Participation factors and mode shapes for three lightly-damped modes, A, B & C, respectively $[-0.17+j10.4, 0.11+j9.58, 0.04+j8.96]$; the latter two modes are unstable.

The behaviour and type of the thirteen electro-mechanical modes in the fourteen machine system are summarised in Table 10.3.

Though not shown in the eigen-plot of Figure 10.2 there is an oscillatory mode at about $-1.4 \pm j2.8$ which could be of interest since it lies in the frequency range of the rotor modes. In the PF plot, shown in Figure 10.7, it is observed that the states mainly participating in the mode are associated with the direct axis of the generators at NPS_5, i.e. the field voltage and the AVR. Thus, this mode is likely to be a controller mode associated with the AVR and generator dynamics of NPS_5. Because such an examination of the PFs of a selected mode quickly establishes the nature of the mode, participation factor analysis proves to be a very useful tool.

Table 10.3 Behaviour and type of the electromechanical modes, Case 1; no PSSs in service

Mode				Mode Behaviour	Mode Type
No.	Real	Imag	ξ		
A	-0.17	10.44	0.02	VPS_2<-->EPS_2, BPS_2	Local Area
B	0.11	9.58	-0.01	SPS_4<-->CPS_4, GPS_4	“
C	0.04	8.96	-0.01	EPS_2, VPS_2<-->BPS_2	“
D	-0.56	8.63	0.06	NPS_5<-->TPS_5	“
E	-0.26	8.37	0.03	CPS_4, SPS_4<-->GPS_4, TPS_4,	“
F	-0.61	8.05	0.08	HPS_1, MPS_2<-->EPS_2, VPS_2, LPS_3	“
G	-0.44	7.96	0.06	MPS_2, HPS_1<-->EPS_2, BPS_2, VPS_2	“
H	0.01	7.81	-0.00	TPS_4<-->GPS_4, SPS_4, MPS_2	“
I	-0.19	7.72	0.02	YPS_3, MPS_2<-->LPS_3, EPS_2	“
J	-0.62	7.43	0.08	PPS_5<-->TPS_5, NPS_5	Local Area
K	0.12	3.97	-0.03	Area 3 <--> Area 5, Area 2	Inter-area
L	0.09	2.60	-0.03	Area 5, Area 4 <--> Area 2	“
M	-0.02	2.03	0.01	Area 5, Area 3 <--> Area 4	Inter-area

<--> means ‘... swings against ...’. ξ - damping ratio.
In ‘Mode Behaviour’, generators or areas are listed in descending order of their participation factors.

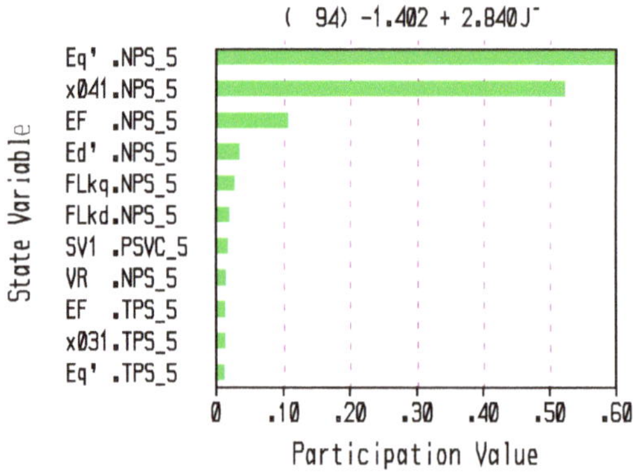

Figure 10.7 Participation factor plot for an oscillatory mode that participates mainly in states associated with the direct axis of NPS_5.

The analysis of the behaviour of the electro-mechanical modes demonstrated above for Case 1, Table 10.3, is repeated for the other cases 2 to 6 for all PSSs out of service. In Tables 10.4, 10.15 and 10.16 the modes for each case are sorted such that each row contains the modes of the same behaviour and type. For example,. in Table 10.4 the modes 'J' in row 10 for Cases 3 and 4, $-0.58 \pm j7.62$ and $-0.19 \pm j7.20$, respectively, are modes in which the same generators are the main participants and both are local-area modes. This type of information will prove useful in a later chapter.

Table 10.4 Rotor modes of oscillation and damping ratios, Cases 3 and 4, peak and light loads. No PSSs in service

Mode No.	Case 3. Peak load			Case 4. Light load		
	Real	Imag	ξ	Real	Imag	ξ
A	-0.38	11.11	0.03	0.20	10.48	-0.02
B	0.10	9.56	-0.01	0.03	9.67	-0.00
C	-0.30	9.02	0.03	-0.17	9.37	0.02
D	-0.58	8.66	0.07	-0.51	8.52	0.06
E	-0.18	8.48	0.02	-0.18	8.78	0.02
F	-0.13	6.31	0.02	-1.54	8.28	0.18
G	-0.14	8.26	0.02	-0.56	8.58	0.07
H	-0.19	7.91	0.02	-0.43	8.21	0.05
I	-0.08	7.38	0.01	-0.21	8.28	0.03
J	-0.58	7.62	0.07	-0.19	7.20	0.03
K	0.01	4.08	-0.00	0.17	4.74	-0.03
L	0.02	2.67	-0.01	0.02	3.57	-0.01
M	-0.03	2.05	0.01	-0.01	2.68	0.00

ξ is the damping ratio

10.4 The P-Vr characteristics of the generators and the associated synthesized characteristics

The basis for the P-Vr characteristics is outlined and illustrated in Chapter 5 for single-machine infinite-bus systems. The extension for their application to multi-machine systems is explained and illustrated in Section 9.4. This section highlights the calculation of the P-Vr characteristics, an examination of their forms, and the synthesis of a P-Vr characteristic representing a set of P-Vrs of a particular generator.

For each of the 14 generators the P-Vr characteristics are determined (with all shaft dynamics disabled) and are shown in each of Figures 10.8 to 10.21 for the six operating conditions; each characteristic is in per unit on the generator rating given in Table 10.23. These characteristics, determined with all shaft dynamics disabled, are calculated using a software package for the analysis of the small-signal dynamic performance and control of large power systems [5].

To avoid unnecessary complexity it should be noted in this analysis that the limited number of encompassing operating conditions on which the power flows - and thus the P-Vr characteristics - are based are normal operating conditions. In practice, the P-Vr characteristics for a relevant encompassing set of contingency conditions must be included in the determining the synthesized characteristic.

Examination of Figures 10.8 to 10.21 reveals that, over the modal frequency range of 1 to 15 rad/s, the bands of P-Vr characteristics [1] for any generator under normal operating conditions may possess the following features:

- Magnitude plots: The width of the bands is typically less than 6 dB; the variation about a characteristic lying in the centre of the band is therefore ±3 dB or less.
- Phase plots: The maximum width of the bands at the relevant frequency is typically less than 15° ; the variation about a central characteristic is thus 7.5° or less.

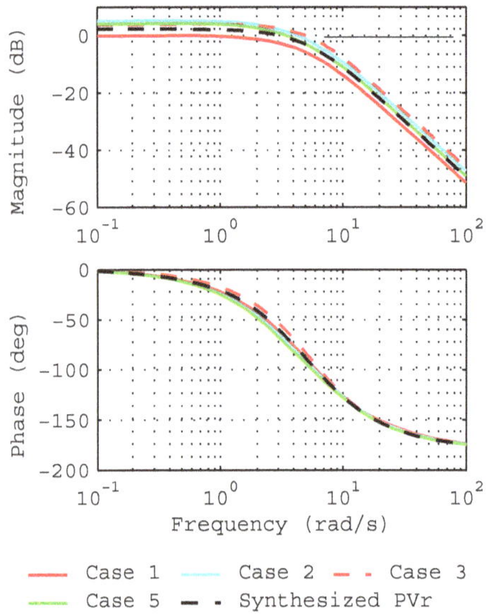

Figure 10.8 P-Vr Xstics, HPS_1

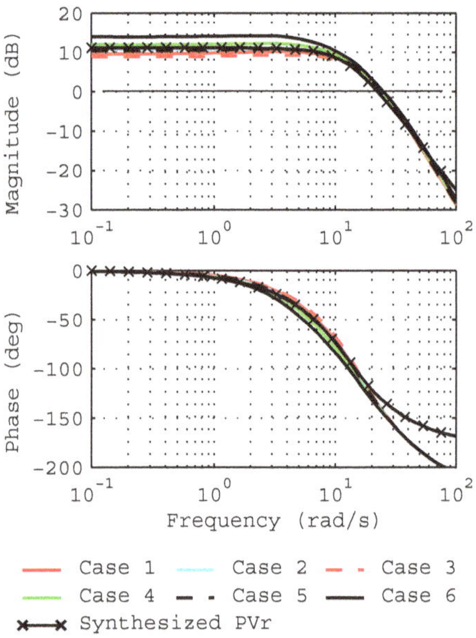

Figure 10.9 P-Vr Xstics, BPS_2

1. The word "characteristics" is shortened to "Xstics" in the following figure captions.

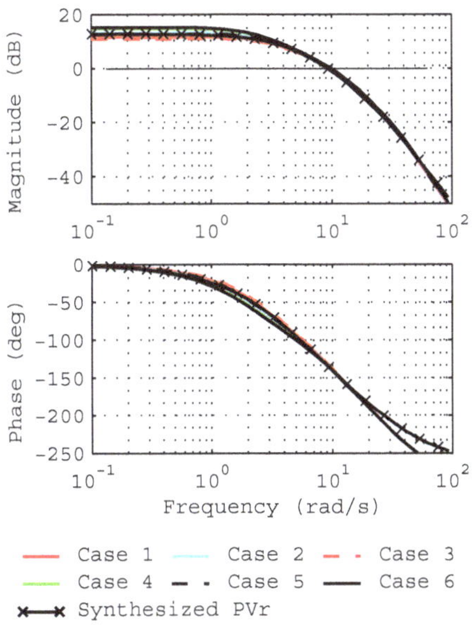

Figure 10.10 P-Vr Xtics, EPS_2

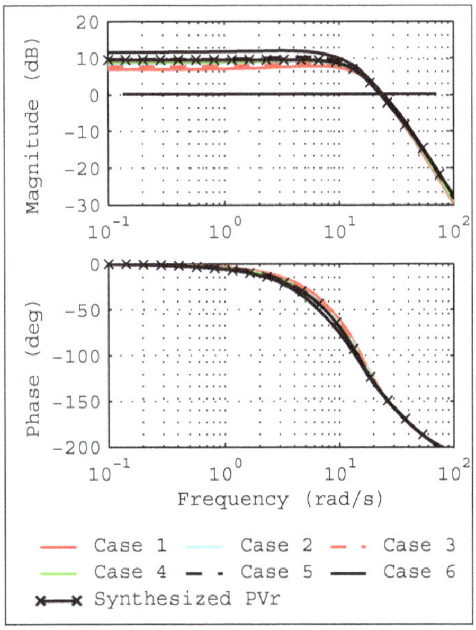

Figure 10.11 P-Vr Xtics, MPS_2

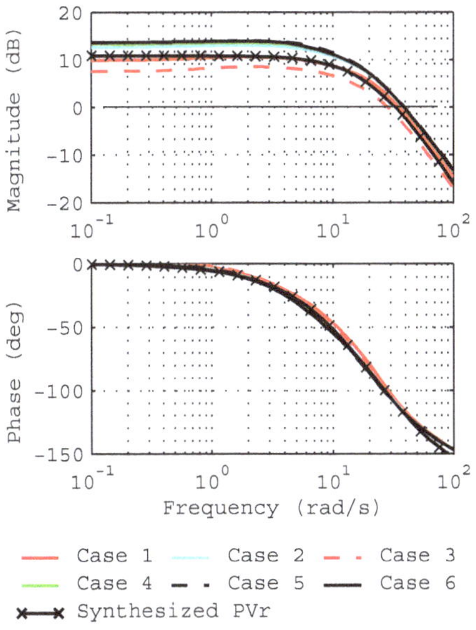

Figure 10.12 P-Vr Xtics,VPS_2

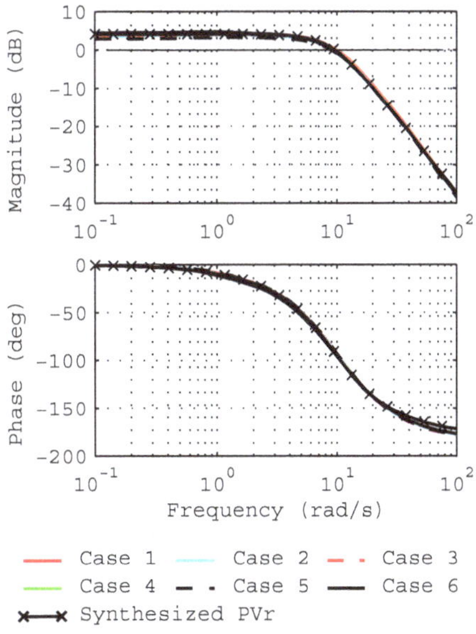

Figure 10.13 P-Vr Xtics, LPS_3

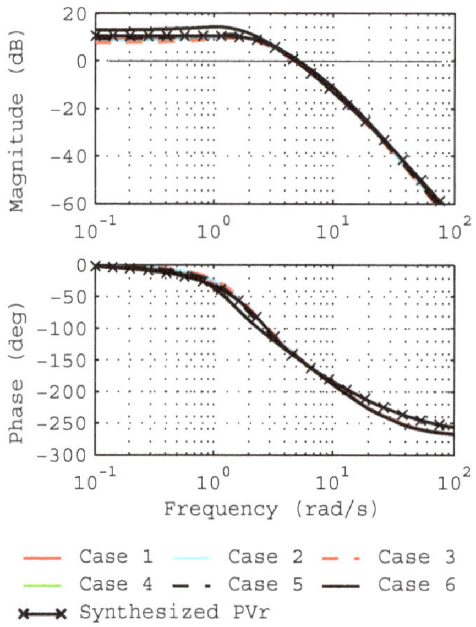

Figure 10.14 P-Vr Xtics, YPS_3

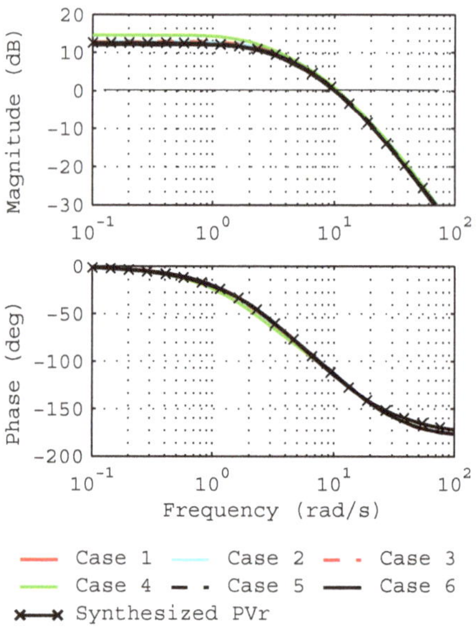

Figure 10.15 P-Vr Xtics, CPS_4

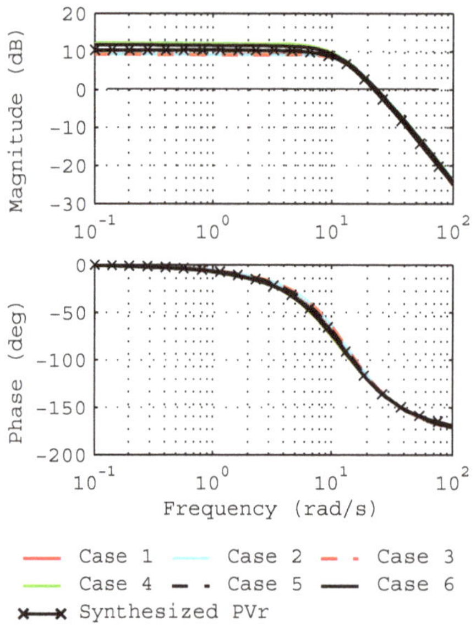

Figure 10.16 P-Vr Xtics, GPS_4

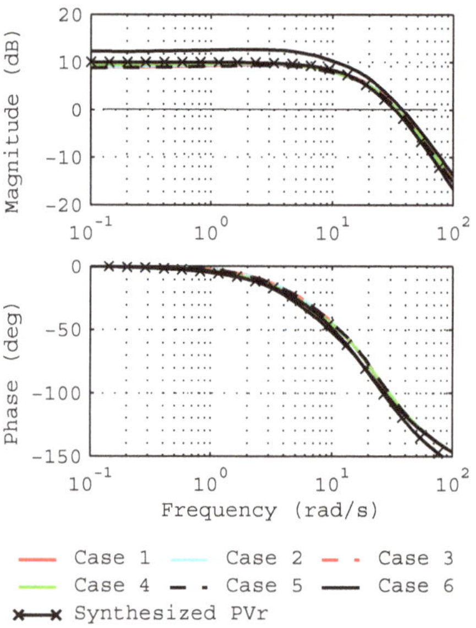

Figure 10.17 P-Vr Xtics, SPS_4

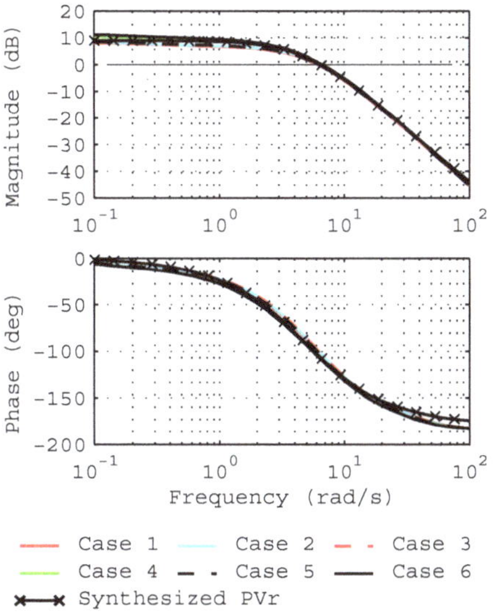

Figure 10.18 P-Vr Xtics, TPS_4

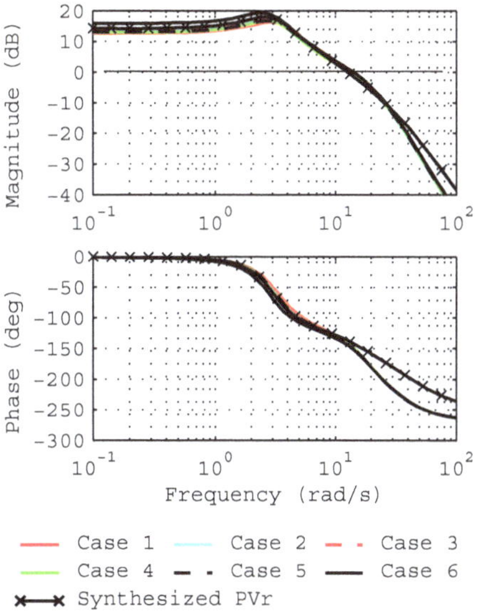

Figure 10.19 P-Vr Xtics, NPS_5

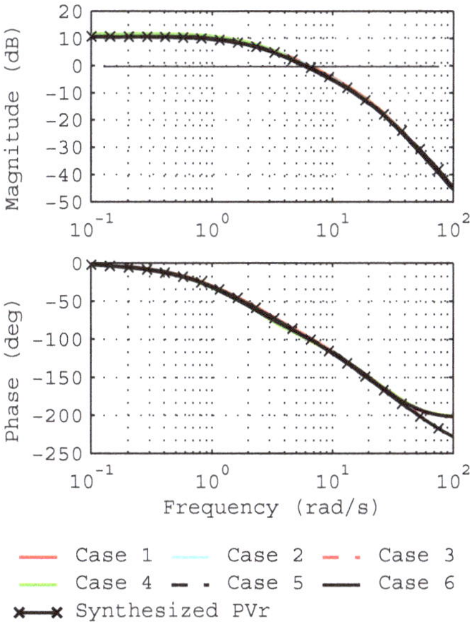

Figure 10.20 P-Vr Xtics,TPS_5

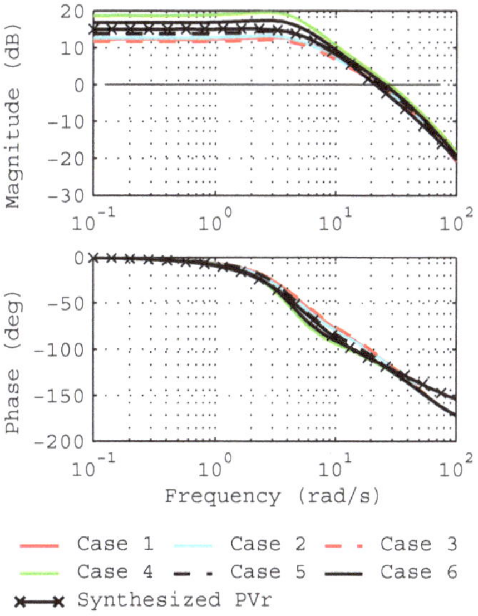

Figure 10.21 P-Vr Xtics, PPS_5

For each set of generator P-Vr characteristics a synthesized P-Vr characteristic is derived based on the following:

- The synthesized characteristic is a best fit of a generator's P-Vr characteristics for the range of cases examined over the modal frequency range of interest, 1.5 to 15 rad/s. As outlined in Section 5.10.6.1 the 'best fit' characteristic for these studies is considered to be that lying in the centre of the magnitude and phase bands formed by the P-Vr characteristics [1].

- If particular P-Vr characteristics tend to lie outside the bands formed by the majority of the characteristics, the synthesized P-Vr may be offset towards the band formed by the majority (e.g. see Figures 10.17 and 10.19). However, weighting of P-Vrs depends on knowledge of the system, the contingencies and engineering judgement.

The transfer function of the synthesized P-Vr characteristic, *PVR*(*s*), for each of the 14 generators is given in Table 10.5.

In several figures, e.g. Figures 10.12 and 10.21 for generators VPS_2 and PPS_5 respectively, the bands of the low-frequency responses for the *magnitude plots* are much wider than those in other figures, e.g. Figure 10.20 for TPS_5. An examination of the generation conditions for the six power flow cases in Table 10.2 reveals that the generator real power outputs vary from 45% to 98% of rated real power for VPS_2, and 60% to 100% for PPS_5; on the other hand, the variation for TPS_5 is much smaller, 90-100%. These observations are consistent with those in Section 5.11 and Figure 5.16, namely, that the low-frequency magnitude response (the gain) of the P-Vr characteristic decreases as the real power output of the generator is reduced. This phenomenon is explained in Section 9.4.1. It is shown that the gain of the P-Vr characteristic varies only with the scalar gain v_{do}, the steady-state d-axis component of the terminal voltage, but retains its shape over the range of operating conditions. At rated power output v_{do} is relatively large, but tends to zero as the real power output is reduced. However, from 70% to 100% of real power output the magnitude characteristic is, for practical purposes, lie within a band of less than ±3 dB from the Design Case [2]. (Similarly, at constant real power output the magnitude of the gain decreases as the reactive power output is varied from maximum leading to maximum lagging power factor. See Tables 9.7 and 9.8).

The more-or-less invariant nature of the *phase responses* of the P-Vr characteristics is also explained in Section 9.4.1.

---

1. A least squares estimation procedure, or the MATLAB® Signal Processing Toolbox routine 'invfreqs.m', can be employed to determine.the parameters for the synthesized transfer function.
2. 'Design Case' is defined in Section 5.10.6.1.

Table 10.5 Transfer functions of synthesized P-Vr characteristics, $PVR(s)$

Generation	$PVR(s)$
HPS_1	$PVR(s) = 1.3/(1 + s0.373 + s^2 0.0385)$
BPS_2	$PVR(s) = 3.6/(1 + s0.128 + s^2 0.0064)$
EPS_2	$PVR(s) = 4.3/[(1 + s0.286)(1 + s0.111)(1 + s0.040)]$
MPS_2	$PVR(s) = 3.0/[(1 + s0.01)(1 + s0.1 + s^2 0.0051)]$
VPS_2	$PVR(s) = 3.5/[(1 + s0.0292)(1 + s0.0708)]$
LPS_3	$PVR(s) = 1.6/(1 + s0.168 + s^2 0.0118)$
YPS_3	$PVR(s) = 3.35/[(1 + s0.05)(1 + s0.509 + s^2 0.132)]$
CPS_4	$PVR(s) = 4.25/[(1 + s0.278)(1 + s0.100)]$
GPS_4	$PVR(s) = 3.3/(1 + s0.115 + s^2 0.00592)$
SPS_4	$PVR(s) = 3.16/(1 + s0.0909 + s^2 0.00207)$
TPS_4	$PVR(s) = 2.8/[(1 + s0.208)(1 + s0.208)]$
NPS_5	$PVR(s) = 5.13(1 + s0.3)/[(1 + s0.033)^2(1 + s0.3 + s^2 0.111)]$
TPS_5	$PVR(s) = 3.4/[(1 + s0.500)(1 + s0.0588)(1 + s0.0167)]$
PPS_5	$PVR(s) = \dfrac{5.62(1 + s0.350)(1 + s0.0667)}{(1 + s0.02)(1 + s0.167)(1 + s0.187)(1 + s0.2)}$

10.5 The synthesized P-Vr and PSS transfer functions

Because the forms of the transfer functions of the synthesized P-Vr transfer functions - and consequently those of the PSS compensation and low-pass filters - vary significantly between generators, it is instructive to list the parameters that have been evaluated for all PSSs.

The parameters of the compensation transfer function of the PSS are based on those of the synthesized P-Vr transfer function given by (5.45). However, a more general form of the synthesized function, which includes (say) additional poles and zeros as required by the form of the design-case P-Vr, is

$$H_{PVrS_i}(s) = k_{ci}\frac{(1 + sT_{b1_i})\ldots}{(1 + c_{1i}s + c_{2i}s^2)(1 + sT_{a1_i})\ldots}.$$

The transfer function of the associated speed PSS, the structure of which is shown in Figure 10.22, incorporates the compensation transfer function and the other elements as described by (10.1), i.e.

$$H_{PSSi}(s) = k_i\left[\frac{sT_{Wi}}{1+sT_{Wi}}\cdot K_{ci}\cdot\frac{(1+c_{1i}s+c_{2i}s^2)(1+sT_{a1i})\ldots}{(1+sT_{b1i})\ldots}\cdot\frac{1}{(1+sT_{1i})\ldots}\right] \tag{10.1}$$

where $K_{ci} = 1/k_{ci}$.

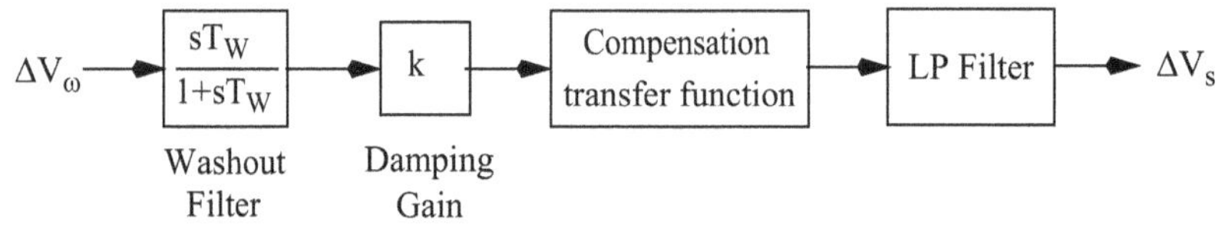

Figure 10.22 Structure of the PSS for analysis and design purposes

The damping gain, the compensation transfer function and the low-pass filter can be included in the one structure if the relevant number and type of blocks are provided in the PSS. Blocks which can accommodate complex poles and zeros are desirable in the PSS structure, as will be seen from the form of PSS transfer functions in (10.4) and (10.6) below.

The damping gain for all PSSs is assumed to be $k_i = 20$ pu on generator MVA rating, a value that is assumed to be a medium value of gain. Similarly, in all PSSs the washout time constant T_W is set at 7.5 s. Its corner frequency of 0.133 rad/s is more than a decade below the lowest inter-area modal frequency of about 2 rad/s; the phase lead it introduces at the latter frequency is therefore small, less than $5°$. Omitting the damping gain k_i and the washout filter, the compensation transfer function and the low-pass filter are combined in the following transfer function (TF), i.e.

$$H_{ci}(s) = K_{ci}\cdot\frac{(1+c_{1i}s+c_{2i}s^2)(1+sT_{a1_i})\ldots}{(1+sT_{b1i})\ldots(1+sT_{1i})(1+sT_{2i})\ldots}. \tag{10.2}$$

However, for simplicity and illustrative purposes in this example the time constants of the low-pass filters are all selected to be 0.00667 s; the associated corner frequency is 150 rad/s [1]. The reason for the selection of the value of the corner frequency is to reduce phase lags introduced by the filter on the phase lead provided by the compensation at the frequencies of the local-area modes, e.g. for three low-pass filter poles the filter contributes a phase lag of $11°$ at 10 rad/s. These issues have been discussed in more detail for the single machine case in Sections 5.8.5 and 5.8.6.

Form of the third-/fourth-order compensation TF having real zeros, and a low-pass filter:

Based on (10.2) the form of this TF follows in (10.3); its parameters are listed in Table 10.6.

1. For illustrative purposes the very short time constants (6.7 ms) of the low-pass filter are used here to minimise its influence in the range of modal frequencies. However, such time constants should typically be 3 or more times the cycle time of the PSS processor to reduce phase errors at higher frequencies.

$$H_c(s) = K_c \cdot \frac{1+sT_a}{1+sT_e} \cdot \frac{1+sT_b}{1+sT_f} \cdot \frac{1+sT_c}{1+sT_g} \cdot \frac{1+sT_d}{1+sT_h}. \tag{10.3}$$

Table 10.6 Compensation and LP Filter Parameters for PSS based on (10.3).

Generator	K_c	T_a	T_b	T_c	T_d	T_e	T_f	T_g	T_h
EPS_2	0.233	0.286	0.111	0.040	0	0.00667*	0.00667*	0.00667*	0
PPS_5	0.178	0.200	0.187	0.167	0.020	0.350	0.0667	0.00667*	0.00667*
TPS_5	0.294	0.500	0.0588	0.0167	0	0.00667*	0.00667*	0.00667*	0
* Low-pass filter parameters									

Form of the fourth-order compensation TF having real and complex zeros, and a low-pass filter: Based on (10.2) the form of this TF is:

$$H_c(s) = K_c \cdot \frac{1+sT_a}{1+sT_d} \cdot \frac{1+sT_b}{1+sT_e} \cdot \frac{1+as+bs^2}{(1+sT_f)(1+sT_g)}. \tag{10.4}$$

The associated parameters are given in Table 10.7:

Table 10.7 Compensation and LP Filter Parameters for PSS based on (10.4)

Generator	K_c	T_a	T_b	a	b	T_d	T_e	T_f	T_g
MPS_2	0.333	0.010	0	0.10	0.0051	0.00667*	0	0.00667*	0.00667*
YPS_3	0.298	0.050	0	0.5091	0.1322	0.00667*	0	0.00667*	0.00667*
NPS_5	0.195	0.033	0.033	0.30	0.1111	0.300	0.00667*	0.00667*	0.00667*
* Low-pass filter parameters									

Form of the second-order compensation TF having real zeros, and a low-pass filter. The form of this low-order TF is:

$$H_c(s) = K_c \cdot \frac{1+sT_a}{1+sT_e} \cdot \frac{1+sT_b}{1+sT_f}. \tag{10.5}$$

Its parameters are provided in Table 10.8.

Table 10.8 Compensation and LP Filter Parameters for PSS based on (10.5)

Generator	K_c	T_a	T_b	T_e*	T_f*
TPS_4	0.357	0.2083	0.2083	0.00667	0.00667
CPS_4	0.235	0.2777	0.1000	0.00667	0.00667
VPS_2	0.286	0.0708	0.0292	0.00667	0.00667
* T_e and T_f are low-pass filter parameters					

Form of the second-order compensation TF having complex zeros, and a low-pass filter
The TF is:

$$H_c(s) = K_c \cdot \frac{1 + as + bs^2}{(1 + sT_e)(1 + sT_f)} . \tag{10.6}$$

Table 10.9 Compensation and LP Filter Parameters for PSS based on (10.6)

Generator	K_c	a	b	T_e*	T_f*
HPS_1	0.769	0.3725	0.03845	0.00667	0.00667
BPS_2	0.278	0.1280	0.00640	0.00667	0.00667
LPS_3	0.625	0.1684	0.01180	0.00667	0.00667
GPS_4	0.303	0.1154	0.005917	0.00667	0.00667
SPS_4	0.316	0.0909	0.002067	0.00667	0.00667
* T_e and T_f are low-pass filter parameters					

10.6 Synchronising and damping torque coefficients induced by PSS *i* on generator *i*

The concepts of synchronising and damping torques coefficients are explained in the context of the single-machine system in Sections 5.3 and 5.10.6.3. The basis for the application of the same concepts in the multi-machine case is explained in Section 9.5 to assess the synchronizing and damping torque coefficients developed by PSS i on generator i. The object of the following analysis is to determine if the PSS transfer function $k_i G_{ci}(j\omega_f)$ is consistent with its design basis, i.e. $H_{PVr_i}(j\omega_f)[k_i G_{ci}(j\omega_f)] \approx k_i + j0$ (eqn. (9.19)), over the modal frequency range for the selected operating condition. In other words, recalling that k_i is also a damping torque coefficient, is the desired per unit damping gain k_i of PSS *i* realized?

The relevant part of Figure 9.11 is shown in Figure 10.23 in which the rotor dynamics on all generators are disabled.

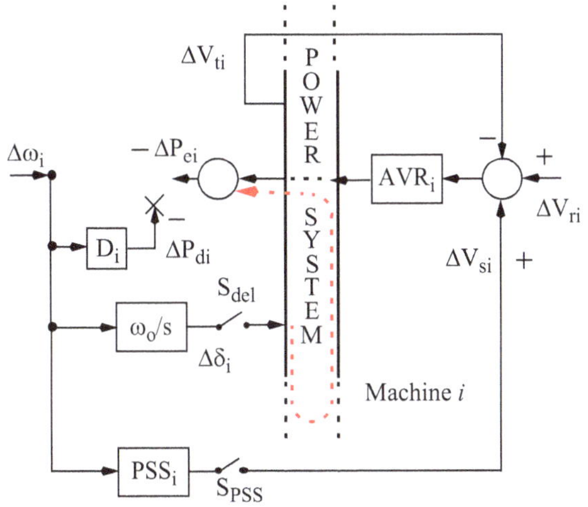

Figure 10.23 Model of generator *i* fitted with a PSS in a multi-machine power system; shaft dynamics on all machines are disabled.

The damping torque coefficient is defined in Section 5.3 and applies to generator i in Figure 10.23:

$$k_{di} = \Re\left\{\frac{\Delta P_{ei}(j\omega_f)}{\Delta\omega_i(j\omega_f)}\right\}. \tag{10.7}$$

Firstly, with the path through S_{del} in Figure 10.23 closed, and S_{PSS} open, let us examine the inherent frequency responses of the torque coefficients $\Delta P_{ei}/\Delta\omega_i$ for several generators in the fourteen-generator system when the individual machines are either heavily or lightly loaded (see Tables 10.2 and 10.10). The responses are shown in Figure 10.24 in per unit on generator base.

As anticipated in Section 9.5, at high and low frequencies the frequency response characteristically rolls off at 20 dB/decade and exhibits a constant phase of –90°. Over the range of modal frequencies, however, the phase varies about –90° implying that the inherent damping torque coefficient is negative when the phase is less than –90°, and positive when greater than –90°. Unlike the P-Vr characteristics, it appears that it is not possible to characterize the variation of the damping torque coefficients; as foreshadowed in Section 9.5 the torque coefficients depends mainly on the steady-state conditions of the generator.

Table 10.10 Generator loading and study case-numbers

Units	Heavily loaded		Lightly loaded	
	Case	Number & output*	Case	
BPS_2	1	6 @ 100%	4	4 @ 90%
VPS_2	6	3 @ 98%	3	2 @ 45%
PPS_5	4	1 @ 100%	5	2 @ 58%
* Number of equally loaded units on-line and percentage of rated real power output				

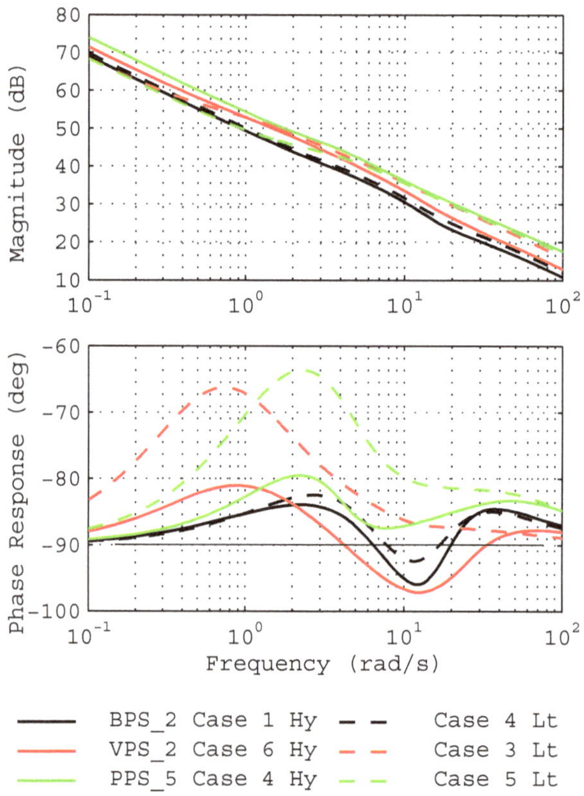

Figure 10.24 Frequency responses of the inherent torque coefficients $\Delta P_{ei}/\Delta\omega_i$ when the unit is either heavily (Hy) or lightly (Lt) loaded (in pu on unit rating)

Secondly, let us examine separately the frequency responses of the *inherent* and the *PSS-induced* synchronizing and damping torque coefficients for generators whose inherent torque coefficients are negative at low frequencies. For the calculations the shaft dynamics are disabled and, as above, in the case of the inherent torque coefficients the PSS path is open. For the PSS-induced synchronizing and damping torque coefficients the rotor-angle path is open and the PSS path closed (see Figure 10.23). The responses are shown in Figure 10.25 for two generators for the operating condition Case 1; the coefficients are in per unit on generator rating.

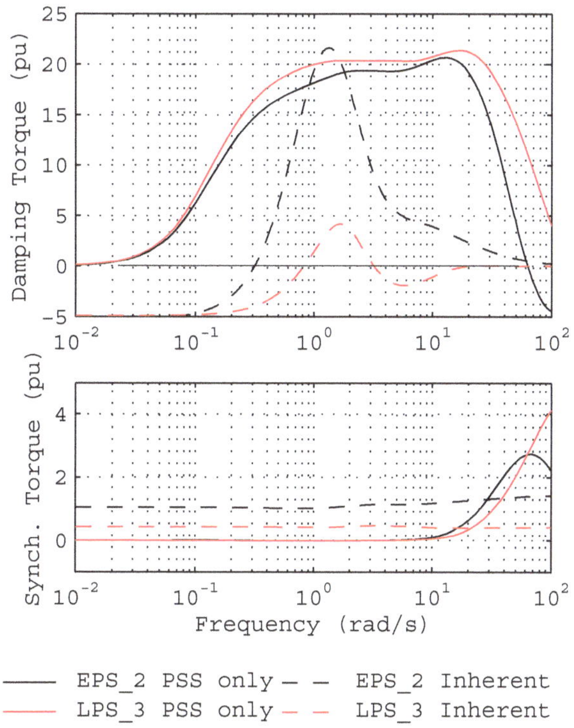

Figure 10.25 Case 1. Synchronizing and damping torque coefficients for EPS_2 (5x555.6 MVA) and LPS_3 (7x666.7 MVA); the coefficients are in per unit on generator base.

Over the range of modal frequencies, 1.5 to 15 rad/s, the following are observed from Figure 10.25.

(i) The PSS gain is more-or-less flat at the desired damping gain setting of 20 pu on generator MVA rating; the deviations from 20 pu are typically accounted for by the factors listed in Section 5.10.6.3. (Thus the question raised at the beginning of this section (is the damping gain k_i realized?) - is successfully answered.)

(ii) For each machine the positive damping torque coefficient induced by the PSS swamps the inherent negative damping torque coefficient.

(iii) It is desirable to attenuate the PSS output signal at higher frequencies to avoid exciting torsional modes at 50 rad/s or greater. This is achieved by means of an integral-of-accelerating-power PSS (see Section 8.5); other types of PSSs may require the use of torsional notch-filters as mentioned in Chapter 8.

The significance of the above analysis and observations is that they confirm - or provide a check - that the PSS transfer function designed for each machine is being realized.

It will be shown that more meaningful information on synchronizing and damping torque coefficients can be derived through 'Modal Induced Torque Coefficients' - which are the subject of Chapter 12.

10.7 Dynamic performance of the system with PSSs in service

10.7.1 Assessment of dynamic performance based on eigen-analysis

PSSs are assumed installed on all generators in each of the power stations. Two units only are on-line at HPS_1 in Cases 4 and 6 and are operating as synchronous compensators. In Case 5, however, three units at HPS_1 are operating as pumps with their PSSs in service. Due to the motoring action in the latter case the sign on the PSS output for HPS_1 units is negated.

The damping gain on each PSS is set to 20 pu on generator MVA rating. In Table 10.11 is shown the values of the modes of rotor oscillation for a heavy and a light load condition, Cases 1 and 4, with the PSSs out of and in service. Also shown are mode shifts and damping ratios for each of the modes. The corresponding set of results are provided for remaining cases in Tables 10.15 and 10.16.

It is instructive to track the shifts in the rotor modes as the PSS damping gains are jointly increased from zero to 150% of the damping gain of 20 pu on machine MVA rating. For Cases 1 to 6 the modes shifts for each of modes A to M are tracked in Figure 10.26 as the damping gain on all unit is increased in 25% (5 pu) steps; the value of 100% corresponds to 20 pu on machine MVA rating. The plots in the figure are
(a) Case 1 (heavy load),
(b) Case 2 (medium-heavy load),
(c) Case 3 (peak load),
(d) Case 4 (light load),
(e) Case 5 (medium load), and
(f) Case 6 (light load).

See Table 10.3 for the details of the nature and types of modes A to M.

Table 10.11 Rotor modes and modes shifts for heavy and light loads, Cases 1 and 4. All PSS damping gains are 20 pu on generator MVA rating..

No. *	Case 1. Heavy load								Case 4. Light load †							
	No PSSs			All PSSs in service			Mode Shift		No PSSs			All PSSs in service			Mode Shift	
	Real	Imag	ξ	Real	Imag	ξ	Real	Imag	Real	Imag	ξ	Real	Imag	ξ	Real	Imag
A	-0.17	10.44	0.02	-2.19	10.39	0.21	-2.02	-0.06	0.20	10.48	-0.02	-2.37	10.77	0.22	-2.57	0.29
B	0.11	9.58	-0.01	-1.98	9.74	0.20	-2.09	0.16	0.03	9.67	-0.00	-2.16	9.95	0.21	-2.19	0.29
C	0.04	8.96	-0.01	-1.93	9.29	0.20	-1.97	0.33	-0.17	9.37	0.02	-2.27	9.81	0.23	-2.10	0.44
D	-0.56	8.63	0.06	-2.51	8.86	0.27	-1.95	0.22	-0.51	8.52	0.06	-2.49	8.83	0.27	-1.98	0.30
E	-0.26	8.37	0.03	-1.95	8.26	0.23	-1.69	-0.11	-0.18	8.78	0.02	-2.27	8.79	0.25	-2.09	0.01
F	-0.61	8.05	0.08	-1.97	8.49	0.23	-1.36	0.44	-1.54	8.28	0.18	-1.69	8.17	0.20	-0.15	-0.11
G	-0.44	7.96	0.06	-1.87	7.76	0.23	-1.44	-0.21	-0.56	8.58	0.07	-2.50	9.06	0.27	-1.94	0.48
H	0.01	7.81	-0.00	-1.78	7.64	0.23	-1.79	-0.17	-0.43	8.21	0.05	-2.28	8.28	0.27	-1.85	0.07
I	-0.19	7.72	0.02	-2.06	7.87	0.25	-1.87	0.15	-0.21	8.28	0.03	-2.55	8.44	0.29	-2.34	0.17
J	-0.62	7.43	0.08	-1.89	7.59	0.24	-1.26	0.16	-0.19	7.20	0.03	-1.32	7.49	0.17	-1.13	0.29
K	0.12	3.97	-0.03	-1.04	3.64	0.28	-1.16	-0.33	0.17	4.74	-0.03	-1.08	4.58	0.23	-1.25	-0.16
L	0.09	2.60	-0.03	-0.39	2.40	0.16	-0.47	-0.20	0.02	3.57	-0.01	-0.56	3.32	0.17	-0.59	-0.25
M	-0.02	2.03	0.01	-0.52	1.80	0.28	-0.51	-0.23	-0.01	2.68	0.00	-0.59	2.51	0.23	-0.58	-0.17

* Mode Number. ξ is the damping ratio..

† In Case 4 the PSS of HPS_1 is OFF as the machine operates as a synchronous compensator.

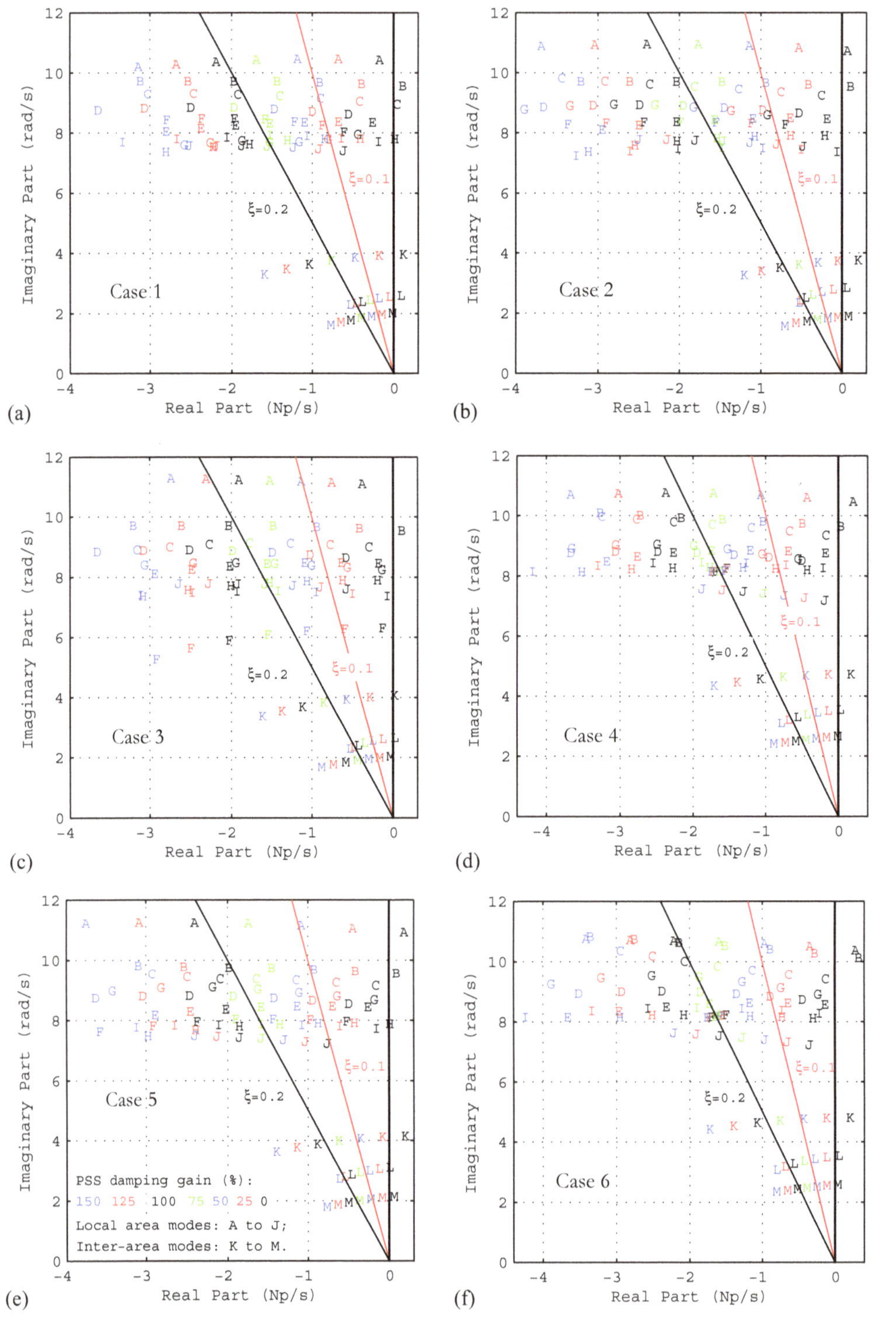

Figure 10.26 Tracking of rotor modes for values of PSS damping gain 0 to 150% (30 pu)

It is noted:

1. the modes, particularly the local-area modes, shift more-or-less horizontally to the left in the complex *s*-plane;

2. the extent of the left shift is least for the peak and heavily-loaded conditions, and most for the light-load cases;

3. for a given Case and a selected mode the extent of the left shift for each 25% increment in damping gain is fairly uniform; the amount of the left shift for the selected mode varies from Case to Case;

4. items 1 to 3 above satisfy the definition for *robustness* in item 3 of Section 1.2;

5. a variation to items 1 to 3 above applies to the inter-area mode L. The increment in the left shift progressively reduces at the higher values of the damping gains; this phenomenon will be discussed in Chapter 14 (there is inadequate damping support for mode L). Employing the P-Vr concept to determine the parameters for decentralized PSSs is shown in the above studies to improve the damping performance of both the local- and inter-area modes. It will be demonstrated in Chapter 13 that the smaller increments in the mode shift of the inter-area modes with PSS gain are due (i) the smaller values of the participation factor of generators participating in the mode, and (ii) the affect of interactions between their PSSs [4].

10.7.2 Assessment of dynamic performance based on participation and mode-shape analysis

It is interesting to establish if the nature of the rotor modes for this system have changed between the case when all PSSs are out of service to that when all are in service with the damping gain set to 20 pu on machine MVA rating.

The participation factor and mode shape plots for representative modes are therefore re-examined for Case 1, the medium-heavy load condition for the system. The plots for an inter-area mode and a set of three local-area modes are shown in Figures 10.27 and 10.28, respectively.

The mode shapes in both figures reveal that the nature of the four modes has not changed from the case when the PSSs are out of service (see Figures 10.4 and 10.6). For example, for the inter-area mode the Area 3 generators continue to 'swing against' generators in Areas 5 and 2. However, a comparison of the participation plots with PSSs out and in service in the four figures demonstrates that, with the PSSs in service, the following states associated with the action of the PSSs participate noticeably:

- internal states in the PSSs (e.g. x081);
- field voltage (EF);
- voltage behind d-axis transient reactance (Eq').

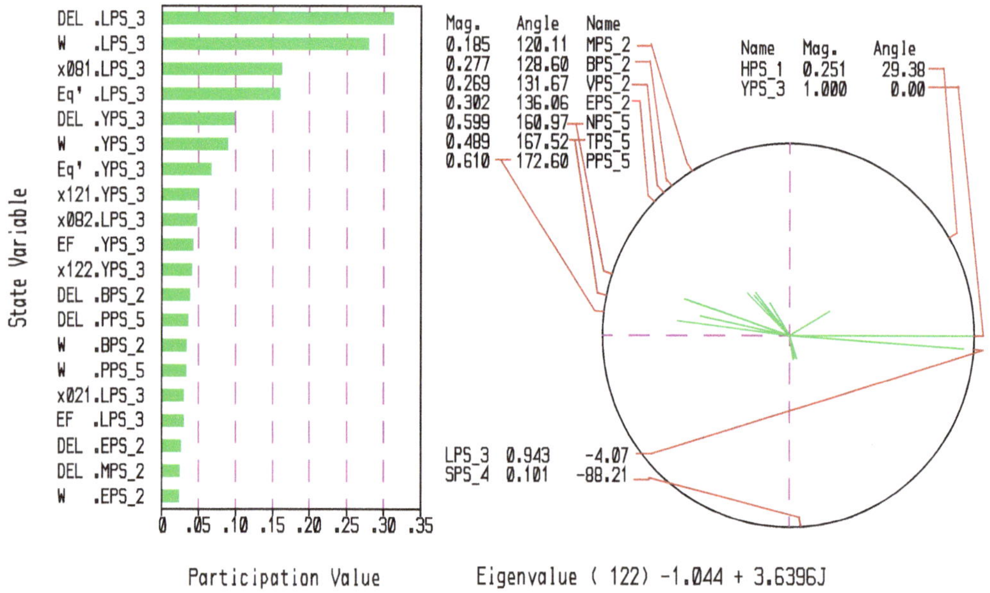

Figure 10.27 Case 1. Participation factors and mode shape for the inter-area mode K ($-1.04+j3.64$) when all PSS damping gains are set to 20 pu. (Compare with Figure 10.4)

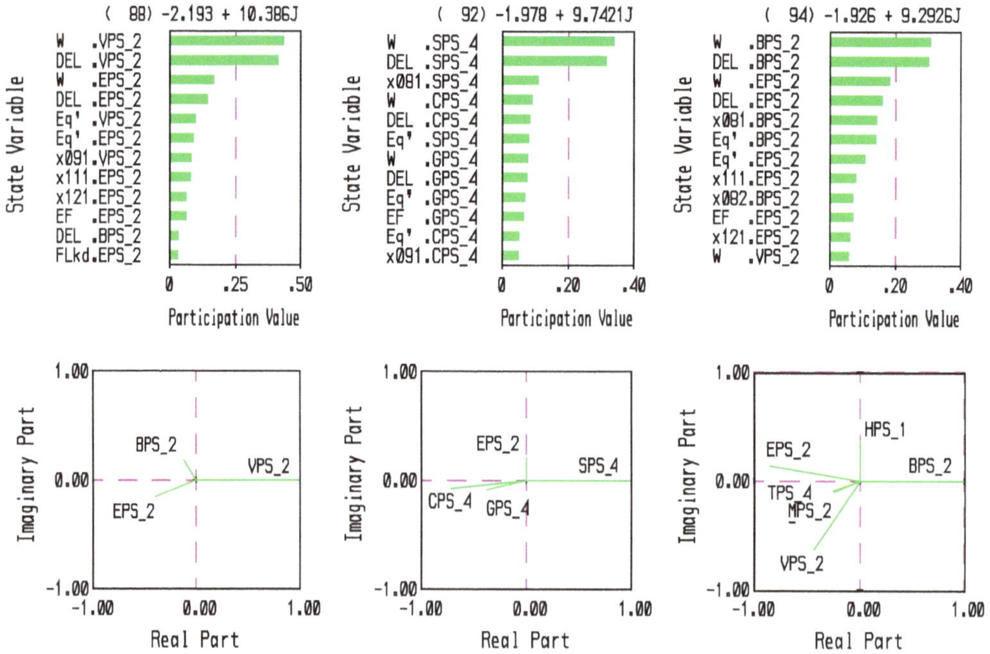

Figure 10.28 Case 1. Participation factors and mode shapes for the local-area modes A ($-2.19+j10.4$), B ($-1.98+j9.74$), and C ($-1.93+j9.29$); PSSs in service, damping gains are 20 pu. (Compare with Figure 10.6)

10.7.3 Assessment of dynamic performance based on time responses

The transient responses of a two-mass spring system to an initial condition which only excites a single mode are demonstrated in Figures 9.3 and 9.4. As an example of the same concept applied to the dynamics of a multi-machine power system let us consider the case of a stable, lightly-damped system when the PSS damping gains for Case 1 are each set to 5 pu on machine MVA rating. This low gain setting is chosen because the oscillatory nature of the responses will be more pronounced than at higher gains.

The plot of the rotor modes for Case 1 with increasing gain is shown in Figure 10.26(a). Let us consider the inter-area mode labelled 'M' associated with a PSS damping gain 25% (5 pu); the value of this mode is $-0.144 \pm j1.98$. The initial conditions for the transient response are the real parts of all elements of the right speed eigenvector; none of the control inputs is excited. The mode shape for this scenario and the transient response of representative machines which are the most responsive in this mode are shown in Figure 10.29.

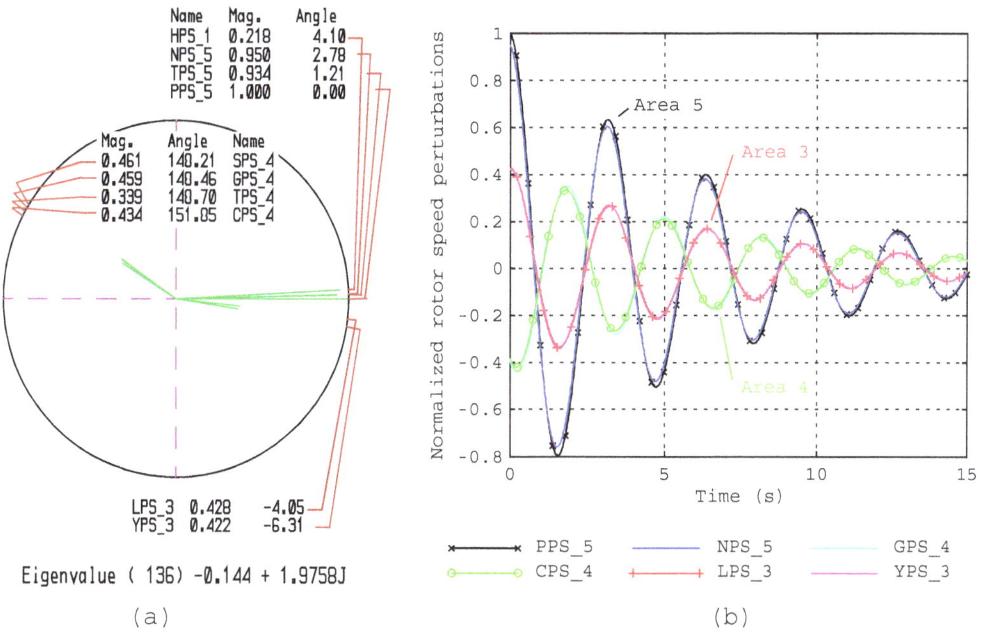

Figure 10.29 Case 1. PSS damping gains 5 pu on machine MVA rating.
(a) Mode shape for the inter-area mode M ($-0.144 \pm j1.98$).(Compare with Figure 10.5)
(b) Time responses of rotor speed perturbations for initial conditions comprising the real parts of all elements of the right speed eigenvector.

The nature of the transient response reflects the relative phase and magnitude information provided not only by the mode shape but also confirms the values of the real and imaginary parts of the complex eigenvalue (by rate of decay and frequency of oscillations). The relative magnitude and phase relationships apply for all $t \geq 0$. Thus, as emphasized earlier, these tools provide valuable aids for the rapid assessment of the characteristics of the dynamic behaviour of the system.

An alternative method of exciting mainly this mode in an analysis of the transient response is to apply small step changes in mechanical power to appropriate generating units. In this case, guided by the mode shape, step increases in power are applied to units which swing together in phase and step decreases in power to those that swing together in anti-phase. The magnitudes of the steps must be adjusted to accentuate the mode of interest and to reduce the influence of other modes which might also be excited, such as some local area modes. The sum of the positive and negative changes in mechanical power should amount to zero.

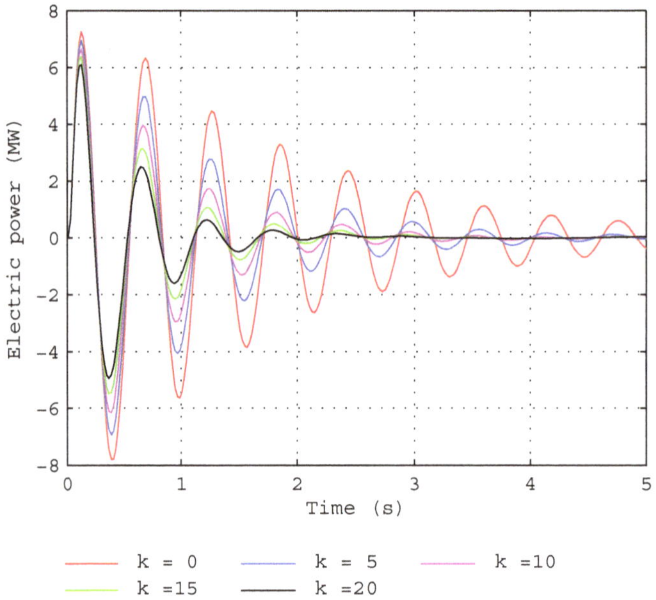

Figure 10.30 Case 6, light load operating condition. Simulated time responses for a step change in reference voltage on the single generator on-line and under test at SPS_4. The PSS damping gain (k) is varied from 0 to 20 pu on machine MVA rating; all other PSS gains are set to 20 pu.

In commissioning a PSS the recording of the time responses of generator outputs to small step changes in the generator's reference voltage is often used to verify that the parameters of the PSS have been correctly set. Such verification is conducted by comparing the measured response with those of the time responses predicted by simulation. For illustrative purposes, the nature of the time responses for a set of step changes are displayed in Figure 10.30 for a range of gain settings on the PSS of a generator at SPS_4 when only the unit on test at the generating station is on-line under a light load condition. The damping gains on the latter PSS are varied from zero to 20 pu (100%) on machine MVA rating; the damping gains of all other PSSs are set to 20 pu. For a given setting of the PSS damping gain the superimposed step responses from the commissioning test and that derived from simulation should agree closely.

10.8 Intra-station modes of rotor oscillation [6], [7]

In the previous studies and the associated eigen-analysis the generators in a power station have been treated as a single generator which is assumed to represent the number of identical, equally-loaded units on-line.

In the studies all units within a power station could have been individually represented. However, for our purposes this would added complexity to both the analysis and assessment of results. In practice, representation of individual units may be necessary, (i) if the loadings on individual unit differs markedly for different operating conditions, (ii) if there are machines of different rating and parameters in the station, and (iii) in order to understand the nature of the intra-station modes and how the PSS tuning affects these modes. If there are m machines in a station, there are $m-1$ modes of rotor oscillations; we will refer to these as the intra-station or inter-machine modes.

In Table 10.12 are shown the three intra-station modes when the four unequally loaded units at SPS_4 and PPS_5 are represented individually for the heavy load condition, Case 1. The three intra-station modes in each station are well damped when all machine PSSs are set to 20 pu on machine MVA rating; the values of the other 13 modes, both local-area and inter-area, are close to those given in Table 10.11 for Case 1. The frequencies of the intra-station modes for the SPS_4 machines are significant higher than all other rotor modes, primarily because the inertia constant of each unit is relatively low at 2.6 MWs/MVA. On the other hand, the frequencies of the intra-station modes for the PPS_5 machines are relatively lower, the inertia constant of each unit being greater at 7.5 MWs/MVA.

By means of the participation factors and mode shapes the nature of the intra-station modes is demonstrated in Figure 10.31 for SPS_4. Unit #1 in SPS_4 predominantly swings against the other three machines in the case of mode 105. For mode 107 SPS_4 unit #2 swings mainly against machine #3 whilst for mode 109 unit #4 swings mainly against machine #3. Because of the level of the damping gain of the PSSs, the PSS and d-axis states participate more markedly in these modes.

For a light-load condition, Case 4, three units at SPS_4 are in service and one at PPS_5 (see Table 10.2). From a comparison of light and heavy load conditions in Tables 10.13 and 10.12, respectively, it is noted that the two intra-station modes for SPS_4 are comparable.

Table 10.12 Case 1: Heavy load. Intra-station modes for four units at SPA_4 and PPS_5.

Generator	Generator output		Inter-machine modes		
	MW	Mvar	PSSs off	PSSs on †	Mode shift
SPS_4 no. 1	400	58.3	$0.08 \pm j12.2$	$-2.75 \pm j13.6$	$-2.83 \pm j1.42$
SPS_4 no. 2	367	54.4	$0.21 \pm j12.2$	$-2.78 \pm j13.6$	$-2.99 \pm j1.41$
SPS_4 no. 3	333	50.7	$0.33 \pm j12.2$	$-2.79 \pm j13.5$	$-3.12 \pm j1.38$
SPS_4 no. 4	300	47.4			
PPS_5 no. 1	149	30.2	$-0.43 \pm j7.25$	$-1.34 \pm j7.43$	$-0.91 \pm j0.18$
PPS_5 no. 2	122	26.7	$-0.81 \pm j7.17$	$-1.57 \pm j7.40$	$-0.76 \pm j0.22$
PPS_5 no. 3	96	24.0	$-1.10 \pm j7.05$	$-1.81 \pm j7.22$	$-0.71 \pm j0.17$
PPS_5 no. 4	69	21.9			
† All PSS damping gains set to 20 pu on machine MVA rating					

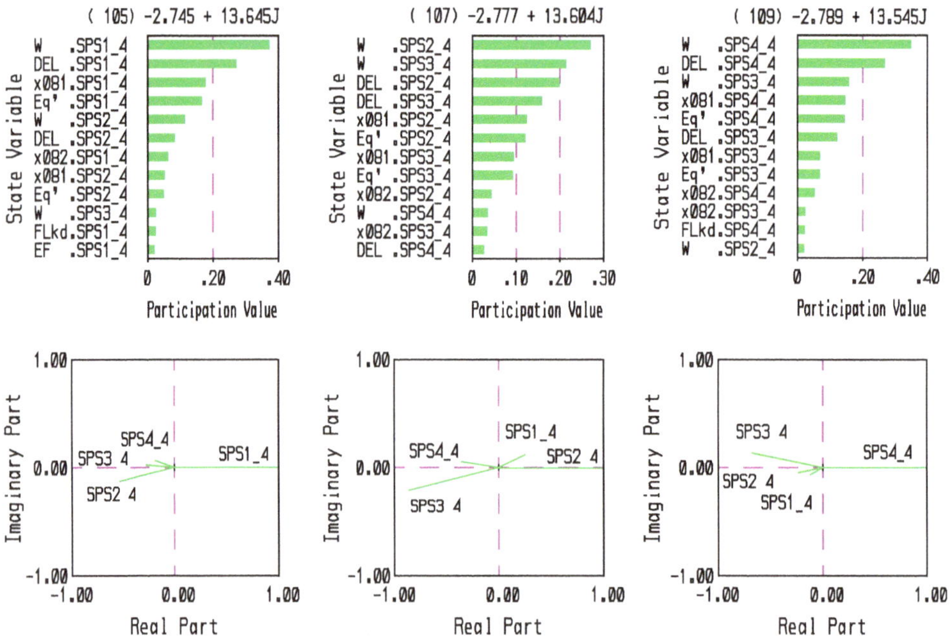

Figure 10.31 Case 1. Participation factors and mode shapes for intra-station modes $-2.75 + j13.6$, $-2.78 + j13.6$ and $-2.79 + j13.5$ for 4 units on-line at SPS_4. All PSS damping gains are all set to 20 pu.

Table 10.13 Case 4: Light load. Intra-station modes for the three units at SPS_4

Generator	Generator output		Intra-station modes		
	MW	Mvar	PSSs off	PSSs on †	Mode shift
SPS_4 no. 1	330	4.5	$0.20 \pm j12.0$	$-2.79 \pm j13.3$	$-2.99 \pm j1.36$
SPS_4 no. 2	320	3.4	$0.24 \pm j12.0$	$-2.80 \pm j13.3$	$-3.04 \pm j1.35$
SPS_4 no. 3	310	2.4			
† All PSS damping gains set to 20 pu on machine MVA rating					

Note that the PSS design procedure based on the P-Vr characteristic does not explicitly attempt to shift the intra-station modes directly to the left in the complex s-plane. For the SPS_4 units, from the condition in Table 10.12 when all PSSs are off to that when all PSSs are in service and damping gains set to 20 pu, there is a marked increase in modal frequency in the intra-station mode shifts (i.e. by ~ 1.4 rad/s); however, such a mode shift does not apply to the intra-station modes for the four PPS_5 units (~ 0.20 rad/s). As a matter of course in the design process the effects of PSS tuning on the intra-station modes should be assessed to ensure they are adequately damped, and that there are no unexpected interactions between controllers.

The design of an ancillary controller specifically to damp the intra-station modes is proposed in [8].

10.9 Correlation between small-signal dynamic performance and that following a major disturbance

In Section 1.10 the question: "how small is small" in small-signal analysis is discussed. In practice, of particular concern is the stability and dynamic performance of the power system following a major disturbance, i.e. a "large-signal" disturbance. To name a few examples, such disturbances are faults, tripping of a large generators, the opening of transmission lines, the loss of significant loads. It is important to realize that the nature of the responses following a major disturbance correlates with the performance predicted from small-signal analysis, particularly after limiting action by controllers has ceased. In other words, small-signal analysis provides significant insights into, and understanding of, the nature of the large-signal dynamic performance - or the transient stability - of the system.

10.9.1 A transient stability study based on the fourteen-generator system

We shall examine the dynamic behaviour of the simplified fourteen-generator system of Figure 10.1 to a three-phase fault at a major busbar on the high-voltage side of a large power station, i.e. busbar #206 at BPS_2. Because there is no line switching or other system changes associated with this busbar fault, which is cleared in 0.120 s, the system configuration and steady-state operating conditions in the post- and pre-fault periods are the same. The system modes are therefore unchanged.

In order to reveal features of the dynamic responses following the clearance of the fault, the low value of the damping gain of 5 pu on machine MVA rating is adopted for all the PSSs. As is seen in Table 10.14 or Figure 10.26(a) for Case 1, a heavy load condition, the system is stable and the real parts of the rotor modes lie between -0.05 and -1.00. The mode behaviour shown in the table does not differ significantly from that of Table 10.3 when all PSSs are out of service.

Table 10.14 Behaviour and type of the rotor modes, Case 1; damping gain of PSSs is 5 pu.

Mode				Mode Behaviour	Mode Type
No.	Real	Imag	ξ		
A	-0.68	10.47	0.065	VPS_2<-->EPS_2	Local Area
B	-0.39	9.65	0.041	SPS_4<-->CPS_4, GPS_4	"
C	-0.42	9.06	0.046	BPS_2<-->EPS_2, VPS_2	"
D	-1.00	8.73	0.114	NPS_5<-->TPS_5	"
E	-0.68	8.38	0.081	CPS_4, SPS_4<-->TPS_4, GPS_4,	"
F	-0.88	8.27	0.106	HPS_1, EPS_2<-->MPS_2, LPS_3	"
G	-0.81	7.80	0.103	HPS_1, MPS_2<-->EPS_2, BPS_2	"
H	-0.40	7.82	0.052	TPS_4<-->GPS_4, SPS_4, MPS_2	"
I	-0.64	7.83	0.082	YPS_3, MPS_2, HPS_1<-->LPS_3, EPS_2	"
J	-0.92	7.48	0.123	PPS_5<-->TPS_5, NPS_5	Local Area
K	-0.18	3.93	0.046	Area 3 <--> Area 5, Area 2	Inter-area
L	-0.05	2.57	0.021	Area 4, Area 5 <--> Area 2	"
M	-0.14	1.98	0.073	Area 5, Area 3 <--> Area 4	Inter-area

<--> means '... swings against ...'. Generators or areas are listed under 'Mode Behaviour' are in descending order of their participation factors.

The responses of speed perturbations about synchronous speed following the incidence of the three-phase fault are shown in Figure 10.32 for selected generators. As stated, the fault occurs at the 330 kV bus at BPS_2 (bus 206) and is cleared in 0.120 s. The responses are divided into three time intervals so that the various features of the modal behaviour in each interval can be examined; the time intervals are (a) 0 to 7 s, (b) 7 to 16 s, (c) 16 to 30 s. (Note the changes of scales on both axes.)

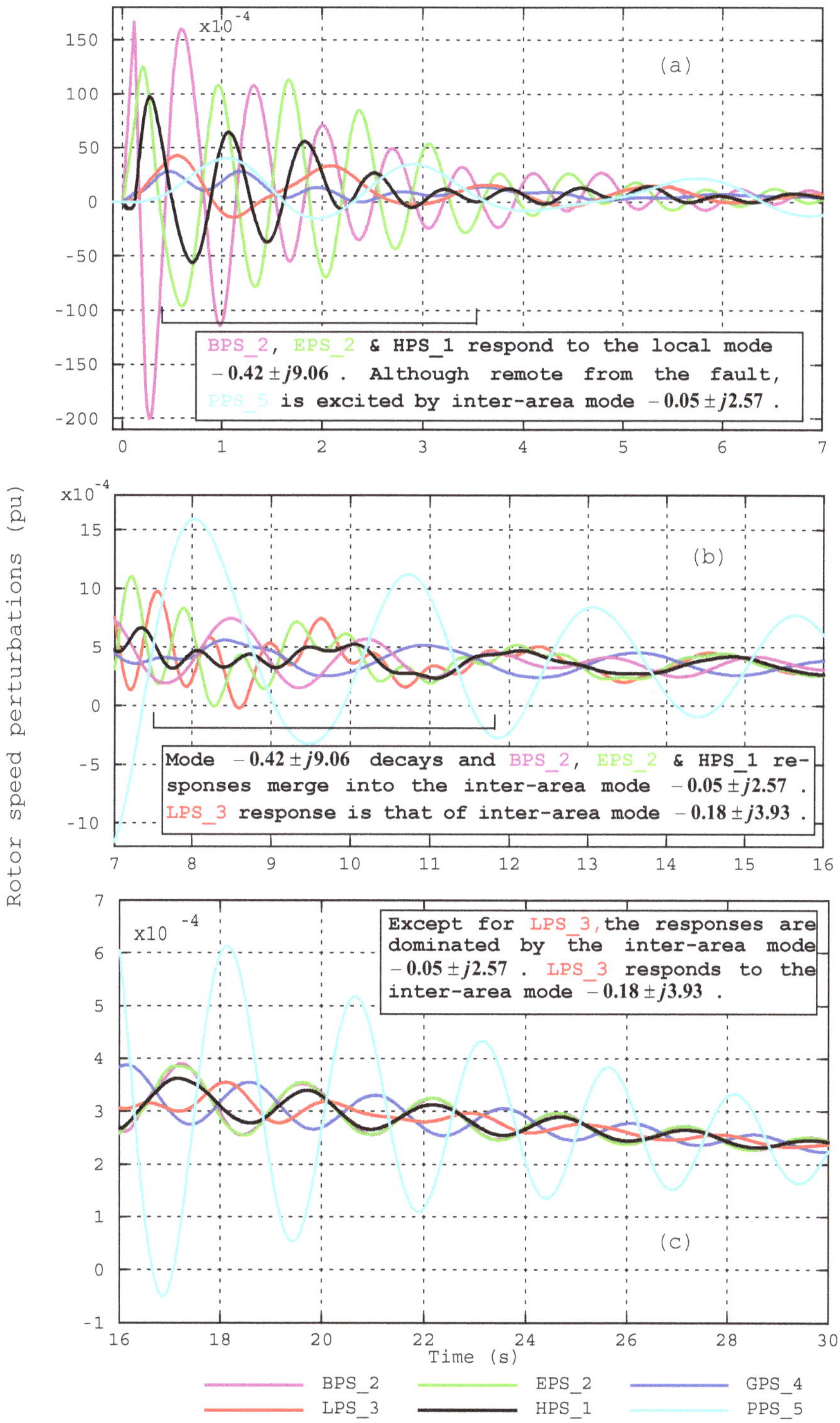

Figure 10.32 Rotor speed perturbations of selected generators following a 3-phase fault

During the interval 0 to 4 s the responses shown in Figure 10.32(a) for selected rotor speed perturbations is dominated by the mode C, $-0.42 \pm j9.06$, subject to the caveat discussed later. The phase relationship between the principal participants in the response appears close to that predicted by the mode shape in Figure 10.33. Although remote from the faulted bus, PPS_5 is excited by the inter-area mode L, $-0.05 \pm j2.57$, in which machines in Area #2 also participate, as revealed in Figure 10.33. The same comment applies to LPS_3 with respect to the inter-area mode K, $-0.18 \pm j3.93$.

During the interval 7 to 14 s shown in Figure 10.32(b) the responses principally associated with mode C, $-0.42 \pm j9.06$, decay away and merge into the modal behaviour revealed in the mode shape in Figure 10.33 for the inter-area mode L, $-0.05 \pm j2.57$. After 16 s, except for LPS_3, the machines participate in the slowly decaying mode L, with a 5% settling time of ~56 s. LPS_3 continues to participate in the more rapidly-decaying mode K, $-0.18 \pm j3.93$, (see Figure 10.33(c)).

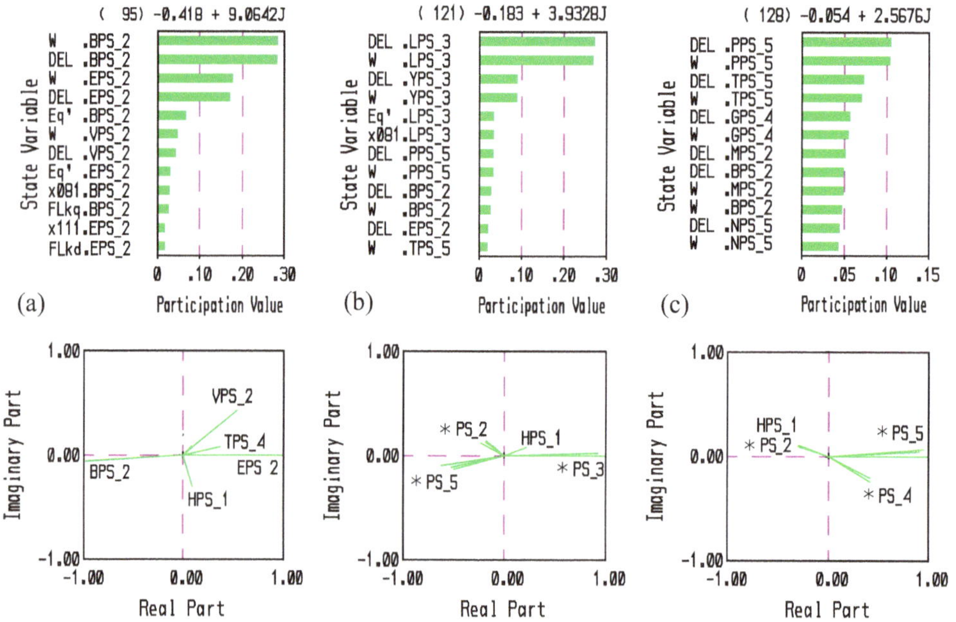

Figure 10.33 Participation factors and mode shapes for the principal modes in the response, the local-area mode C ($-0.42 + j9.06$) and the inter-area modes K ($-0.18 + j3.93$) and L ($-0.05 + j2.57$); all PSS damping gains in Case 1 are all set to 5 pu on machine MVA rating. (<*PS_area> refers to all generators in the numbered area.)

10.9.1.1 Benefits of small-signal analysis of large power systems

This example demonstrates how small-signal analysis complements that based on transient stability studies.

The above example reveals the important features of small-signal analysis, that is, it furnishes not only an understanding of the underlying modal structure of the power system and but also provides insights into a system's dynamic characteristics that cannot easily be derived from time-domain simulations for large magnitude disturbances. It is the case in Figure 10.33 that only a few of the thirteen modes appear to be excited; the nature and location of the fault does not significantly excite the local-area modes outside the faulted area at all. Understanding the nature of the small-signal modal behaviour therefore yields a synoptic view of the system characteristics which would require many large-signal studies of faults in different locations to gain similar, but not exact, information [9].

Knowledge of the behaviour of certain local and inter-area modes has revealed the nature of the responses of the speed states following a major disturbance on the system. However, as stated earlier, the behaviour of the system is highly non-linear during the initial phase of the response. During the first 0.6 s certain exciters reach their ceiling voltages and some PSSs, together with most SVCs, hit limits on their outputs. In the context of the magnitude of rotor speed oscillations, the question is asked in Section 1.10, “how small is small?”. The peak amplitudes of the speed perturbations in Figure 10.32(a) are 1.5 to 2% which are not small. The functional non-linearities come into play and therefore the small-signal analysis is based is not strictly accurate. In the following section the applicability and validity of the small-signal analysis that has been conducted in this section is reviewed.

10.9.2 The analysis of modal interactions [10], [11], [12]

As has been discussed earlier, small-signal analysis is based on the first-order approximation of the non-linear power system equations, both differential and algebraic, about a steady-state operating condition. Strictly speaking, such analysis is valid as the perturbations in variables become vanishingly small. Consequently, once limiting by controllers has ceased following a large-magnitude disturbance, techniques based on linear analysis are unlikely to provide accurate information on the dynamic behaviour of the system when the variations in system variables is large. This is likely to be valid particularly for so-called stressed conditions in which the system is heavily loaded and/or the system performance is bordering on instability in the period immediately following the disturbance.

In [10] the significance and application of extending first-order (linear) system analysis to include the second-order terms is reviewed. The Taylor series expansion about the steady-state operating condition now includes both the first- and second-order terms, but no third- or higher-order terms. Based on the second-order form of the expansion for a state equation, and employing Normal Form analysis, it is shown in [10] that the ith state equation can be expressed as:

$$x_i(t) = \sum_{j=1}^{n} v_{ij} z_{j0} e^{\lambda_j t} + \sum_{j=1}^{n} v_{ij} \left[\sum_{k=1}^{n} \sum_{l=1}^{n} h2^j_{kl} z_{k0} z_{l0} e^{(\lambda_k + \lambda_l)t} \right], \tag{10.8}$$

where λ_j, λ_k and λ_l are 'conventional' eigenvalues of the state matrix $\boldsymbol{A}$; v_{ij} is an element of the ***right*** eigenvector corresponding to the eigenvalue or mode λ_j ; z_{p0} is a function of the initial conditions; $h2^j_{kl}$ is a function of $1/(\lambda_k + \lambda_l - \lambda_j)$, $\lambda_k + \lambda_l - \lambda_j \neq 0$.

Equation (10.8) reveals the relation between the state variables $x_i(t)$, the first-order system modes $\lambda_1 \ldots \lambda_n$, and the second-order modes,

$$(\lambda_1 + \lambda_1), (\lambda_1 + \lambda_2), \ldots, (\lambda_{n-1} + \lambda_n), (\lambda_n + \lambda_n).$$

Note the following:

- The terms associated with the mode pairs $\lambda_k + \lambda_l$ represent "modal interactions" that arise due to the inclusion of the second-order terms.
- The second-order terms supplement information provided from the first-order linear approximation of the power system equations.
- If the system is stable, the second-order mode $\lambda_k + \lambda_l$ lies to the left of either of its constituent modes, λ_k or λ_l, in the complex *s*-plane; it therefore decays more rapidly than either of the individual modes.
- The "interaction coefficients", $h2^j_{kl} z_{k0} z_{l0}$, of the exponential terms $\lambda_k + \lambda_l$ in (10.8) provide a measure of the participation of any of the mode pairs in the state variable.

Firstly, for the first- and second-order modes discussed in the following, let us assume the linear coefficient term, $v_{ij} z_{j0}$, and the interaction coefficients in (10.8) are not negligible. Secondly, we will assume $\lambda_j = -\alpha_j + j\omega_j$ and $\lambda_{j+1} = -\alpha_j - j\omega_j$ are the complex conjugate pair of the dominant first-order mode, normally an inter-area mode. When $k = j$ and $l = j$ in (10.8), the mode pair $\lambda_k + \lambda_l = -2\alpha_j + j2\omega_j$; likewise when $k = j + 1$ and $l = j + 1$, the mode pair $\lambda_k + \lambda_l = -2\alpha_j - j2\omega_j$. Thus, due to modal interactions, a second-order mode of double the frequency and double the damping constant of the first order mode is introduced into the response; significantly, however, it decays in *half* the settling time of the linear mode. Thirdly, let us assume there is some other, more heavily damped first-order mode present, $\lambda_{r,r+1} = -\alpha_r \pm j\omega_r$. When $k = j$ and $l = r$ in (10.8), the second-order mode pair $\lambda_k + \lambda_l = -(\alpha_j + \alpha_r) + j(\omega_j + \omega_r)$ is introduced. Thus the resulting complex second-order mode will be of higher frequency than the dominant first-order mode and, because $\alpha_r > \alpha_j$, it will decay with a settling time of *less than half* the settling time of the linear mode.

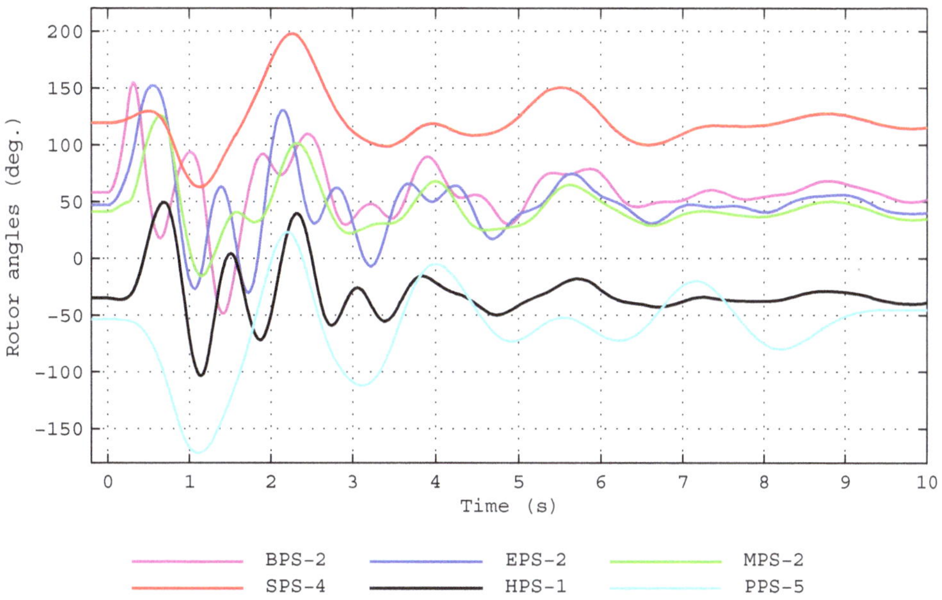

Figure 10.34 Case 1. Responses of rotor angles of selected generators to a three-phase fault at the hv bus at the terminals of the transformers at BPS_2 cleared in 0.250 s. Angles are relative to that of LPS_3. All PSS gains set to 20 pu on machine MVA rating.

The transient response of the rotor angles of generators is shown in Figure 10.34 for a three-phase fault at bus #206, the high-voltage bus at the terminals of the generator transformers at BPS_2. The system conditions are the same as those in Figure 10.32 except the fault clearing time has been increased from 120 ms to 250 ms. The system, which is marginally stable with rotor angle differences across the system reaching 260° at about 1 s, is heavily stressed in the immediate post-fault period. The *essential point of the analysis in this section* is that in this period, during or after which no limiting by controllers occurs, one should be aware that second-order modes of some significance may arise. However, because such modes are better damped than their constituent first-order modes, they tend to decay more rapidly.

The analysis of modal interactions based on Normal Forms for a large system is compute-intensive and complex, mainly because of the size of the system and the number of combinations of both the second-order modes $\lambda_k + \lambda_l$ and the associated interaction coefficients. Moreover, given the identical system conditions and type of disturbance, the latter coefficients will vary depending on the instant in the transient response at which initial conditions are selected.

In the particular cases of Figures 10.32 and 10.34 it is clear from (10.8) that the first-order modes exist in the responses. However, without a detailed analysis based on Normal Forms it is unclear what modal interactions are present, and their magnitude at any instant - at least up to one half of the settling time of the dominant mode when the responses of second-order modes $\lambda_k + \lambda_l$ have effectively decayed away. From studies in the literature it appears that for less stressed systems the effects of modal interactions dissipate well within the latter time. For the Study Case 1, shown in Figure 10.32, this may well be the situation. In [14], [15] interesting comparisons are made between the transient response of a stressed system to major disturbances and the first- and second-order responses based on the results of Normal Form analysis. For the scenarios considered the second-order responses agree closely with those derived from the transient responses based on the step-by-step simulation.

The above summary of modal interactions and their significance is necessarily very brief. More extensive details are provided in other papers referenced in [10], [14], [15], [16].

10.10 Summary: Tuning of PSSs based on the P-Vr approach

The case study illustrates the basis and benefits of the P-Vr method but also helps to identify some of the limitations of the basic design technique. (These limitations motivate certain developments which follow in later chapters.)

- The P-Vr method provides a systematic approach and a formal basis for the design of PSSs. The *phase* of P-Vr characteristics are, for practical purposes, more-or-less invariant over the prudently-selected set of encompassing operating conditions (see Section 9.4.1). At higher real power outputs, typically 0.5 to 1 pu of rated power, the *magnitude* response of the P-Vr characteristic retains its shape and consistently lies in a band of ± 3 dB from the Design Characteristic.

- These encompassing conditions should not only cover normal operation but also include various contingencies, line outages, and perhaps some potentially extreme conditions in order to ensure that the PSS is adequately tuned.

- The calculation of P-Vrs for normal operation and for contingencies is easily automated, resulting in the display of a full set of P-Vr characteristics and providing a basis for the synthesizing of the PSS transfer function.

In tuning fixed-parameter PSSs using the P-Vr approach the concept of robustness is based on the following considerations:

- There are two important components of a fixed-parameter PSS transfer function $kG_c(s)$ which are essentially decoupled for practical purposes;

    (a) the rotor modes are more-or-less directly left-shifted by the PSS compensating transfer function $G_c(s)$ with increase in the PSS damping gain, k;

    (b) the extent of the left-shift of the rotor modes is determined by the damping gain, k;

(c) the value of the damping gain should be such that the damping torque contributions induced by the PSS swamp the negative inherent contributions by the generator.

- Ideally, the incremental left-shifts of the rotor modes are linearly related to increments in PSS gain for changes about selected nominal values.

- Such considerations should apply over the set of encompassing operating conditions and an appropriate range of rotor modes.

In Section 10.7.1, and from an examination of the modes shifts induced by the PSSs as shown in Tables 10.11, 10.15 and 10.16, we observe that the above considerations for robustness are - in essence - satisfied. However, there are two factors which cause the a deviation from a direct left shift of the modes, an increase or decrease on modal frequency with increase in damping gain.

Firstly, in Figure 10.26 it is noted that mode shifts for the selected gain increment vary with the type of mode (e.g. a local mode) and the machines participating in the mode. In Section 5.9.3 it is foreshadowed that the shift in the complex rotor mode λ_h is given by

$$\Delta\lambda_h = -\frac{\rho(\lambda_h)}{2H} \cdot H_{PVr}(\lambda_h) \cdot G_c(\lambda_h) \cdot \Delta k, \quad \text{((5.36) repeated)} \qquad (10.9)$$

where Δk is an increment in the damping gain of the PSS and $\rho(\lambda_h)$ is the complex participation factor of the generator's speed state in the mode λ_h, evaluated with the PSS in service with a damping-gain setting, k_0. It is shown in Section 13.3 that $\rho(\lambda_h)$ is essentially real for generators participating strongly in the mode, but for those participating with a relatively small participation factor it ($\rho(\lambda_h)$) may acquire a not insignificant positive or negative imaginary component. However, being small the contribution by the generator to the mode shift $\Delta\lambda_h$ may be minor.

Secondly, in Chapter 13 it is shown that in the multi-machine environment the modes shift in (10.9) can either be enhanced or degraded by the action of PSSs installed on other generators. This is caused by the production of a positive or negative damping torque being induced on generator i by the action of the PSS fitted to machine j [4]. Furthermore, we observe in Figure 10.26 that the mode shifts associated with the inter-area modes are smaller than those of the local modes. This feature is also considered in Chapter 13.

We have examined the intra-station modes and emphasized that their damping should be examined because information on how the design methods (including the GEP and Residue Methods) influence these modes is not readily available. Though exciter modes, which can become lightly damped or unstable, have not been examined, the same comments are relevant [6], [7].

10.11 References

[1] P. Kundur, M. Klein, G.J. Rogers and M.S. Zywno, "Application of power system stabilizers for enhancement of overall system stability", *Power Systems, IEEE Transactions on*, vol. 4, no. 2, May 1989, pp. 614-626.

[2] M. J. Gibbard, "Co-ordinated design of multimachine power system stabilisers based on damping torque concepts," *IEE Proceedings C Generation, Transmission and Distribution*, vol. 135, pp. 276-284, 1988.

[3] P. Pourbeik, *Design and Coordination of Stabilisers for Generators and FACTS devices in Multimachine Power Systems*, PhD Thesis, The University of Adelaide, Australia, 1997.

[4] M.J. Gibbard, D.J. Vowles and P. Pourbeik, "Interactions between, and effectiveness of, power system stabilizers and FACTS stabilizers in multimachine systems", *Power Systems, IEEE Transactions on*, vol 15, pp. 748-755, 2000.

[5] D.J. Vowles and M.J. Gibbard, *Mudpack User Manual: Version 10S-03*, School of Electrical and Electronic Engineering, The University of Adelaide, July 2014.

[6] G.J. Rogers, "The application of power system stabilizers to a multigenerator plant", *Power Systems, IEEE Transactions on*, vol 15, pp. 350-355, 2000.

[7] M.J. Gibbard, D.J. Vowles, G.J. Rogers, Discussion of "The application of power system stabilizers to a multigenerator plant," [and reply], *Power Systems, IEEE Transactions on*, vol 15, pp. 1462 - 1464, 2000.

[8] N. Martins and T. H. S. Bossa, "A Modal Stabilizer for the Independent Damping Control of Aggregate Generator and Intraplant Modes in Multigenerator Power Plants," *Power Systems, IEEE Transactions on*, vol. 29, pp. 2646-2661, 2014.

[9] P. Pourbeik, T. Cain, and R. Bottoms, "Application of small-signal stability tools and techniques to a large power system," in *Power & Energy Society General Meeting, 2009. PES '09. IEEE*, pp. 1-11.

[10] J.J. Sanchez-Gasca, V.Vittal, M.J. Gibbard, A.R. Messina, D.J. Vowles, B.S. Liu and U.D. Annakkage, "Inclusion of higher-order terms for small signal (modal) analysis: Committee Report - Task Force on assessing the need to include higher-order terms for small-signal (modal) analysis", *Power Systems, IEEE Transactions on*, vol. 20, pp. 1886-1904, 2005.

[11] *ibid*, "Analysis of Higher Order Terms for Small Signal Stability Analysis", *Conference Proceedings, IEEE Power Engineering Society General Meeting*, San Francisco, California, USA, 12-16 June 2005.

[12] V. Vittal, N. Bhatia, and A. A. Fouad, "Analysis of the Inter Area Mode Phenomenon in Power Systems Following Large Disturbances," *Power Systems, IEEE Transactions on*, vol. 6, pp. 1515-1521, 1991.

[13] IEEE Recommended Practice for Excitation System Models for Power System Stability Studies, *IEEE Standard No:421.5-2005*, ISBN:0-7381-4787-7.

[14] E. Barocio and A. R. Messina, "Application of perturbation methods to the analysis of low frequency inter-area oscillations," in *Power Engineering Society Summer Meeting, 2000.* IEEE, pp. 1845-1850 vol. 3.

[15] A. R. Messina, J. Arroyo, and E. Barocio, "Analysis of modal interaction in power systems with FACTS controllers using normal forms," in *Power Engineering Society General Meeting, 2003*, IEEE, p. 2117 Vol. 4.

[16] L. Shu, A. R. Messina, and V. Vittal, "A Normal Form Analysis Approach to Siting Power System Stabilizers (PSSs) and Assessing Power System Nonlinear Behavior," *Power Systems, IEEE Transactions on*, vol. 21, pp. 1755-1762, 2006.

Appendix 10–I

App. 10–I.1 Modes of rotor oscillation for Cases 2, 3, 5 and 6

The following Tables 10.15 and 10.16, together with Table 10.11, show the values of the modes of rotor oscillation for Cases 1 to 6 with the PSSs out and in service. When the PSSs are in service, the damping gains are all set to 20 pu on machine MVA rating.

Table 10.15 Rotor modes and modes shifts for medium-heavy and peak loads, Cases 2 & 3. All PSS damping gains are 20 pu on generator MVA rating.

No. *	Case 2. Medium-heavy load								Case 3. Peak Load							
	No PSSs			All PSSs in service			Mode Shift		No PSSs			All PSSs in service			Mode Shift	
	Real	Imag	ζ	Real	Imag	ζ	Real	Imag	Real	Imag	ζ	Real	Imag	ζ	Real	Imag
A	0.07	10.74	-0.01	-2.40	10.96	0.21	-2.47	0.22	-0.38	11.11	0.03	-1.91	11.24	0.17	-1.53	0.14
B	0.10	9.56	-0.01	-2.04	9.72	0.21	-2.14	0.16	0.10	9.56	-0.01	-2.04	9.72	0.21	-2.14	0.16
C	-0.25	9.26	0.03	-2.37	9.64	0.24	-2.12	0.38	-0.30	9.02	0.03	-2.28	9.10	0.24	-1.98	0.08
D	-0.53	8.67	0.06	-2.49	8.94	0.27	-1.96	0.27	-0.58	8.66	0.07	-2.52	8.91	0.27	-1.94	0.25
E	-0.18	8.48	0.02	-2.04	8.38	0.24	-1.86	-0.10	-0.18	8.48	0.02	-2.03	8.38	0.24	-1.84	-0.10
F	-0.70	8.29	0.08	-2.44	8.37	0.28	-1.74	0.08	-0.13	6.31	0.02	-2.03	5.91	0.32	-1.90	-0.41
G	-0.92	8.61	0.11	-2.81	8.96	0.30	-1.88	0.35	-0.14	8.26	0.02	-1.95	8.49	0.22	-1.81	0.27
H	-0.21	7.93	0.03	-2.03	7.74	0.25	-1.82	-0.19	-0.19	7.91	0.02	-2.01	7.73	0.25	-1.82	-0.18
I	-0.06	7.39	0.01	-2.02	7.49	0.26	-1.96	0.11	-0.08	7.38	0.01	-1.93	7.53	0.25	-1.86	0.15
J	-0.49	7.57	0.06	-1.81	7.77	0.23	-1.33	0.20	-0.58	7.62	0.07	-1.93	7.80	0.24	-1.36	0.18
K	0.19	3.77	-0.05	-0.77	3.54	0.21	-0.96	-0.24	0.01	4.08	-0.00	-1.12	3.71	0.29	-1.13	-0.37
L	0.05	2.86	-0.02	-0.45	2.54	0.17	-0.50	-0.32	0.02	2.67	-0.01	-0.43	2.42	0.17	-0.45	-0.25
M	0.08	1.92	-0.04	-0.43	1.76	0.24	-0.51	-0.16	-0.03	2.05	0.01	-0.58	1.86	0.30	-0.55	-0.19

* Mode Number ζ - damping ratio.

Table 10.16 Rotor modes and modes shifts for medium and light loads, Cases 5 & 6.
All PSS damping gains are 20 pu on generator MVA rating.

No. *	Case 5. Medium load								Case 6. Light load							
	No PSSs			All PSS in service			Mode Shift		No PSSs			All PSS in service †			Mode Shift	
	Real	Imag	ζ	Real	Imag	ζ	Real	Imag	Real	Imag	ζ	Real	Imag	ζ	Real	Imag
A	0.18	10.94	-0.02	-2.41	11.26	0.21	-2.59	0.32	0.28	10.39	-0.03	-2.22	10.71	0.20	-2.49	0.32
B	0.09	9.57	-0.01	-1.99	9.76	0.20	-2.07	0.19	0.32	10.14	-0.03	-2.14	10.65	0.20	-2.50	0.51
C	-0.16	9.17	0.02	-2.09	9.39	0.22	-1.90	0.22	-0.13	9.42	0.01	-2.07	10.02	0.20	-1.94	0.59
D	-0.50	8.55	0.06	-2.47	8.83	0.27	-1.98	0.27	-0.46	8.74	0.05	-2.38	9.02	0.26	-1.93	0.29
E	-0.26	8.45	0.03	-2.02	8.38	0.24	-1.76	-0.07	-0.14	8.58	0.02	-2.32	8.51	0.26	-2.18	-0.07
F	-0.52	7.98	0.07	-2.38	7.97	0.29	-1.86	-0.01	-1.51	8.24	0.18	-1.70	8.17	0.20	-0.19	-0.07
G	-0.18	8.70	0.02	-2.19	9.12	0.23	-2.01	0.42	-0.23	8.92	0.03	-2.52	9.54	0.26	-2.29	0.62
H	0.01	7.90	0	-1.85	7.81	0.23	-1.86	-0.09	-0.30	8.13	0.04	-2.08	8.24	0.24	-1.78	0.11
I	-0.16	7.74	0.02	-2.12	7.87	0.26	-1.96	0.13	-0.21	8.29	0.03	-2.58	8.45	0.29	-2.37	0.17
J	-0.77	7.24	0.11	-1.86	7.45	0.24	-1.09	0.21	-0.36	7.25	0.05	-1.60	7.55	0.21	-1.24	0.30
K	0.19	4.15	-0.05	-0.88	3.90	0.22	-1.08	-0.25	0.20	4.81	-0.04	-1.08	4.64	0.23	-1.28	-0.17
L	0.01	3.12	-0	-0.46	2.89	0.16	-0.46	-0.23	0.05	3.55	-0.02	-0.57	3.30	0.17	-0.62	-0.25
M	0.06	2.15	-0.03	-0.50	1.96	0.25	-0.56	-0.20	0.04	2.60	-0.01	-0.52	2.45	0.21	-0.56	-0.15

* Mode Number ζ - damping ratio. † PSS of HPS_1 is OFF as it operates as a synchronous compensator in this case.

App. 10–I.2 Data for steady-state power flow analysis

Table 10.17 SVC bus numbers, ratings and operating conditions for Cases 1 to 6.

SVC name / Bus No.	Reactive Power Range (Mbase)	Qmax	Qmin	Case 1 Voltage Mvar	Case 2 Voltage Mvar	Case 3 Voltage Mvar	Case 4 Voltage Mvar	Case 5 Voltage Mvar	Case 6 Voltage Mvar
	Mvar @ 1.0 pu voltage								
ASVC_2 / 205	650.0	430.0	-220.0	1.055 -68.3	1.055 41.8	1.02 -5.2	1.045 -39.3	1.045 -118.3	1.045 -29.4
RSVC_3 / 313	800.0	600.0	-200.0	1.015 71.4	1.015 129.4	1.015 158.8	1.015 86.7	1.015 54.9	1.015 54.2
BSVC_4 / 412	1430.0	1100.0	-330.0	1.000 58.2	1.000 63.9	1.000 83.8	1.000 -52.2	1.000 22.8	1.000 -0.2
PSVC_5 / 507	500.0	320.0	-180.0	1.015 22.6	1.040 36.8	1.043 18.0	1.010 -4.0	1.015 13.8	1.000 -3.7
SSVC_5 / 509	550.0	400.0	-150.0	1.030 10.6	1.027 50.2	1.050 -63.4	1.030 -109.3	1.030 -123.8	1.030 -109.3
Note: System frequency is 50 Hz.									

Table 10.18 Switched Shunt Capacitor / Reactor banks (C/R) in service, Cases 1-6 (Mvar)

Bus Number	Case 1	Case 2	Case 3	Case 4	Case 5	Case 6
211	-	-	100 C	-	-	-
212	400 C	150 C	150 C	400 C	400 C	400 C
216	300 C	150 C	150 C	300 C	300 C	300 C
409	60 C	60 C	60 C	60 C	60 C	60 C
411	30 C	30 C	30 C	30 C	30 C	30 C
414	30 R	30 R	30 R	30 R	30 R	30 R
415	60 R	60 R	60 R	60 R	60 R	60 R
416	60 R	60 R	60 R	60 R	60 R	90 R
504	-	90 R	90 R	-	-	-

Table 10.19 Transmission Line Parameters; Values per circuit

From bus / to bus	Line No.	Line: r+jx, b (pu on 100MVA)		
102 217	1,2	0.0084	0.0667	0.817
102 217	3,4	0.0078	0.0620	0.760
102 309	1,2	0.0045	0.0356	0.437
102 309	3	0.0109	0.0868	0.760
205 206	1,2	0.0096	0.0760	0.931
205 416	1,2	0.0037	0.0460	0.730
206 207	1,2	0.0045	0.0356	0.437
206 212	1,2	0.0066	0.0527	0.646
206 215	1,2	0.0066	0.0527	0.646
207 208	1,2	0.0018	0.0140	0.171
207 209	1	0.0008	0.0062	0.076
208 211	1,2,3	0.0031	0.0248	0.304
209 212	1	0.0045	0.0356	0.437
210 213	1,2	0.0010	0.0145	1.540
211 212	1,2	0.0014	0.0108	0.133
211 214	1	0.0019	0.0155	0.190
212 217	1	0.0070	0.0558	0.684
214 216	1	0.0010	0.0077	0.095
214 217	1	0.0049	0.0388	0.475
215 216	1,2	0.0051	0.0403	0.494
215 217	1,2	0.0072	0.0574	0.703
216 217	1	0.0051	0.0403	0.494
303 304	1	0.0010	0.0140	1.480
303 305	1,2	0.0011	0.0160	1.700
304 305	1	0.0003	0.0040	0.424
305 306	1	0.0002	0.0030	0.320
305 307	1,2	0.0003	0.0045	0.447
306 307	1	0.0001	0.0012	0.127
307 308	1,2	0.0023	0.0325	3.445
		continued ...		
... cont'd				
309 310	1,2	0.0090	0.0713	0.874
310 311	1,2	0.0000	-0.0337	0.000
312 313	1	0.0020	0.0150	0.900
313 314	1	0.0005	0.0050	0.520
315 509	1,2	0.0070	0.0500	0.190
405 406	1,2	0.0039	0.0475	0.381
405 408	1	0.0054	0.0500	0.189
405 409	1,2,3	0.0180	0.1220	0.790
406 407	1,2	0.0006	0.0076	0.062
407 408	1	0.0042	0.0513	0.412
408 410	1,2	0.0110	0.1280	1.010
409 411	1,2	0.0103	0.0709	0.460
410 411	1	0.0043	0.0532	0.427
410 412	1 to 4	0.0043	0.0532	0.427
410 413	1,2	0.0040	0.0494	0.400
411 412	1,2	0.0012	0.0152	0.122
414 415	1,2	0.0020	0.0250	0.390
415 416	1,2	0.0037	0.0460	0.730
504 507	1,2	0.0230	0.1500	0.560
504 508	1,2	0.0260	0.0190	0.870
505 507	1	0.0008	0.0085	0.060
505 508	1	0.0025	0.0280	0.170
506 507	1	0.0008	0.0085	0.060
506 508	1	0.0030	0.0280	0.140
507 508	1	0.0020	0.0190	0.090
507 509	1,2	0.0300	0.2200	0.900

Note: System frequency is 50 Hz.

Table 10.20 Transformer Ratings and Reactances.

Buses		Number	Rating, each Unit (MVA)	Reactance per transformer	
From	To			% on Rating	per unit on 100MVA
101	102	g	333.3	12.0	0.0360
201	206	g	666.7	16.0	0.0240
202	209	g	555.6	16.0	0.0288
203	208	g	555.6	17.0	0.0306
204	215	g	666.7	16.0	0.0240
209	210	4	625.0	17.0	0.0272
213	214	4	625.0	17.0	0.0272
301	303	g	666.7	16.0	0.0240
302	312	g	444.4	15.0	0.0338
304	313	2	500.0	16.0	0.0320
305	311	2	500.0	12.0	0.0240
305	314	2	700.0	17.0	0.0243
308	315	2	370.0	10.0	0.0270
401	410	g	444.4	15.0	0.0338
402	408	g	333.3	17.0	0.0510
403	407	g	444.4	15.0	0.0338
404	405	g	333.3	17.0	0.0510
413	414	3	750.0	6.0	0.0080
501	504	g	333.3	17.0	0.0510
502	505	g	250.0	16.0	0.0640
503	506	g	166.7	16.7	0.1000

g - Generator/transformer unit; in service if associated generator is online.
Note: System frequency is 50 Hz.

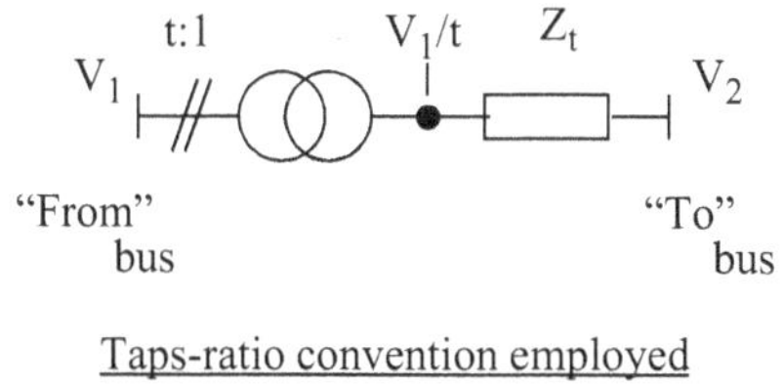

Figure 10.35 Transformer Taps Convention

The transformer tap ratios listed in Table 10.21 are based upon the convention shown in Figure 10.35.

Table 10.21 Transformer Tap Ratios for power flow Cases 1 to 6

Buses		Case 1	Case 2	Case 3	Case 4	Case 5	Case 6
From	To						
101	102	0.939	0.948	0.948	1.000	1.000	1.000
201	206	0.943	0.948	0.939	1.000	0.971	1.010
202	209	0.939	0.948	0.939	1.000	0.971	1.010
203	208	0.939	0.948	0.939	1.000	0.971	1.010
204	215	0.939	0.948	0.939	1.000	0.971	1.010
209	210	0.976	0.990	0.976	0.976	0.976	0.976
213	214	1.000	1.000	1.000	1.000	1.000	1.000
301	303	0.939	0.935	0.930	1.000	0.961	1.000
302	312	0.952	0.952	0.952	1.000	0.961	1.000
304	313	0.961	0.961	0.948	0.961	0.961	0.961
305	311	1.000	1.000	1.000	1.000	1.000	1.000
305	314	1.000	1.000	1.000	1.000	1.000	1.000
308	315	1.000	0.960	1.000	1.000	1.000	1.000
401	410	0.939	0.939	0.939	1.000	0.952	1.010
402	408	0.952	0.952	0.952	1.000	0.952	1.000
403	407	0.952	0.952	0.952	1.000	0.952	1.000
404	405	0.952	0.952	0.952	1.000	0.952	1.000
413	414	1.000	1.000	1.000	1.000	1.015	1.000
501	504	0.952	0.952	0.952	1.000	0.985	1.015
502	505	0.962	0.930	0.930	1.000	0.995	1.020
503	506	0.962	0.930	0.930	1.000	0.985	1.020

For simplicity, loads are assumed to behave as constant impedances in the small-signal analysis.

Table 10.22 Busbar Loads (P MW, Q Mvar) for Cases 1 to 6

	Case 1		Case 2		Case 3		Case 4		Case 5		Case 6	
Bus No.	P	Q	P	Q	P	Q	P	Q	P	Q	P	Q
102	450	45	380	38	475	50	270	30	340	35	270	30
205	390	39	330	33	410	40	235	25	290	30	235	25
206	130	13	110	11	140	15	80	10	100	10	80	10
207	1880	188	1600	160	1975	200	1130	120	1410	145	1110	120
208	210	21	180	18	220	25	125	15	160	20	125	15
211	1700	170	1445	145	1785	180	1060	110	1275	130	1035	110
212	1660	166	1410	140	1740	180	1000	110	1245	125	1000	110
215	480	48	410	40	505	50	290	30	360	40	290	30
216	1840	184	1565	155	1930	200	1105	120	1380	140	1105	120
217	1260	126	1070	110	1320	140	750	80	940	95	750	80
306	1230	123	1230	123	1450	150	900	90	1085	110	900	90
307	650	65	650	65	770	80	470	50	580	60	470	50
308	655	66	655	66	770	80	620	100	580	60	620	100
309	195	20	195	20	230	25	140	15	170	20	140	15
312	115	12	115	12	140	15	92	10	105	15	92	10
313	2405	240	2405	240	2840	290	1625	165	2130	220	1625	165
314	250	25	250	25	300	30	180	20	222	25	180	20
405	990	99	1215	120	1215	120	730	75	990	100	730	75
406	740	74	905	90	905	90	540	55	740	75	540	55
407	0	0	0	0	0	0	0	0	0	0	0	0
408	150	15	185	20	185	20	110	10	150	15	110	10
409	260	26	310	30	310	30	190	20	260	30	190	20
410	530	53	650	65	650	65	390	40	530	55	390	40
411	575	58	700	70	700	70	420	45	575	60	420	45
412	1255	126	1535	155	1535	155	922	100	1255	130	922	100
504	300	60	200	40	300	60	180	20	225	25	170	20
507	1000	200	710	140	1100	220	640	65	750	75	565	65
508	800	160	520	105	800	160	490	50	600	60	450	50
509	200	40	70	15	100	20	122	15	150	15	117	15

Load Characteristics: Constant Impedance

App. 10–I.3 Data for dynamic performance analysis

The parameters of the fourteen generators are listed in Table 10.23.

Table 10.23 Generator Parameters ##

Generator	Bus	Order	Rating MVA	No. of Units	H MWs / MVA	Xa pu	Xd pu	Xq pu	Xd' pu	Tdo' s	Xd" pu	Tdo" s	Xq' pu	Tqo' s	Xq" pu	Tqo" s
HPS_1	101	5	333.3	12	3.60	0.14	1.10	0.65	0.25	8.50	0.25	0.050	-	-	0.25	0.200
BPS_2	201	6	666.7	6	3.20	0.20	1.80	1.75	0.30	8.50	0.21	0.040	0.70	0.30	0.21	0.080
EPS_2	202	6	555.6	5	2.80	0.17	2.20	2.10	0.30	4.50	0.20	0.040	0.50	1.50	0.21	0.060
MPS_2	204	6	666.7	6	3.20	0.20	1.80	1.75	0.30	8.50	0.21	0.040	0.70	0.30	0.21	0.080
VPS_2	203	6	555.6	4	2.60	0.20	2.30	1.70	0.30	5.00	0.25	0.030	0.40	2.00	0.25	0.250
LPS_3	301	6	666.7	8	2.80	0.20	2.70	1.50	0.30	7.50	0.25	0.040	0.85	0.85	0.25	0.120
YPS_3	302	5	444.4	4	3.50	0.15	2.00	1.80	0.25	7.50	0.20	0.040	-	-	0.20	0.250
CPS_4	402	6	333.3	3	3.00	0.20	1.90	1.80	0.30	6.50	0.26	0.035	0.55	1.40	0.26	0.040
GPS_4	404	6	333.3	6	4.00	0.18	2.20	1.40	0.32	9.00	0.24	0.040	0.75	1.40	0.24	0.130
SPS_4	403	6	444.4	4	2.60	0.20	2.30	1.70	0.30	5.00	0.25	0.030	0.40	2.00	0.25	0.250
TPS_4	401	6	444.4	4	2.60	0.20	2.30	1.70	0.30	5.00	0.25	0.030	0.40	2.00	0.25	0.250
NPS_5	501	6	333.3	2	3.50	0.15	2.20	1.70	0.30	7.50	0.24	0.025	0.80	1.50	0.24	0.100
TPS_5	502	6	250.0	4	4.00	0.20	2.00	1.50	0.30	7.50	0.22	0.040	0.80	3.00	0.22	0.200
PPS_5	503	6	166.7	6	7.50	0.15	2.30	2.00	0.25	5.00	0.17	0.022	0.35	1.00	0.17	0.035

Classically-defined operational parameters (see Section 4.2.12.2 and Section 4.2.13).
Generator reactances in per unit on machine rating as base. System frequency is 50 Hz.
For all generators the stator winding resistance (Ra) and damping torque coefficient (D) are both assumed to be zero.

App. 10–I.3.1 Excitation System Parameters

Two basic types of excitation systems are employed, AC4A and AC1A [13]. The parameters of the AVR have been tuned to ensure that the open-circuit generator under closed-loop voltage control is stable and satisfies the performance specifications.

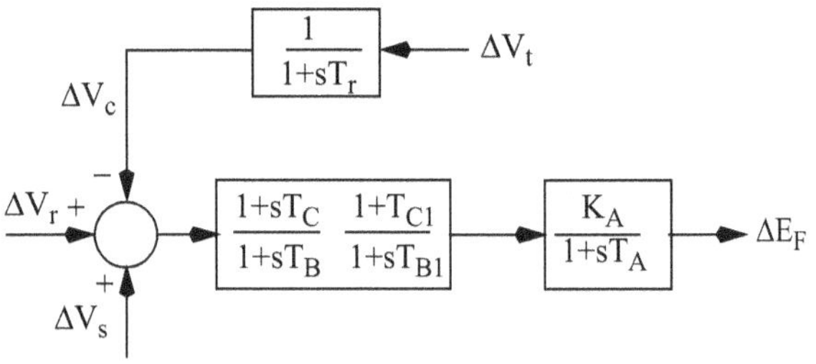

Figure 10.36 Small-signal model of a type AC4A Excitation System

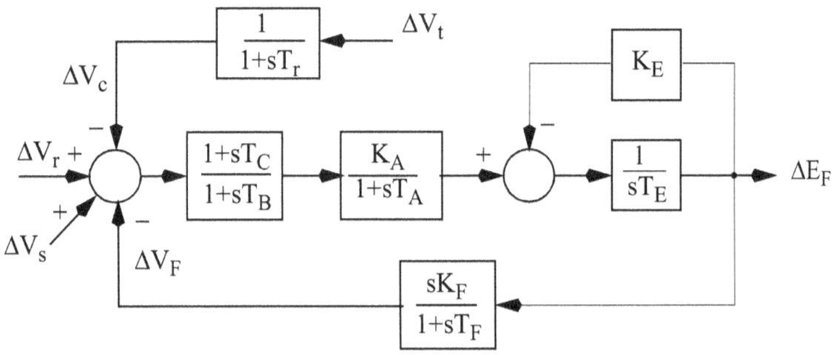

Figure 10.37 Small-signal model of a type AC1A Excitation System; demagnetizing effect of field current neglected

Table 10.24 Excitation System Parameters: 14-generator system

	HPS_1	BPS_2	EPS_2	MPS_2	VPS_2	LPS_3	YPS_3
Type	AC4A	AC4A	AC1A	AC4A	AC4A	AC4A	AC1A
T_r (s)	0	0	0	0	0	0	0
K_A (s)	200	400	400	400	300	400	200
T_A (s)	0.10	0.02	0.02	0.02	0.01	0.05	0.05
T_B (s)	13.25	1.12	0	1.12	0.70	6.42	0
T_C (s)	2.50	0.50	0	0.50	0.35	1.14	0
K_E	-	-	1.0	-	-	-	1.0
T_E (s)	-	-	1.0	-	-	-	1.333
K_F	-	-	0.029	-	-	-	0.020
T_F (s)	-	-	1.0	-	-	-	0.8

Table 10.25 Excitation System Parameters: 14-generator system (continued)

	CPS_4	GPS_4	SPS_4	TPS_4	NPS_5	TPS_5	PPS_5
Type	AC4A	AC4A	AC4A	AC4A	AC1A	ST5B	AC4A
T_r (s)	0.02	0	0	0	0	0	0
K_A	300	250	300	300	1000	400	300
T_A (s)	0.05	0.20	0.01	0.10	0.04	0.50	0.01
T_B (s)	9.80	0.0232	0.70	40.0	0	16.0	0.8
T_C (s)	1.52	0.1360	0.35	4.00	0	1.40	0.2
T_{B1} (s)	0	0	0	0	0	0.05	0
T_{C1} (s)	0	0	0	0	0	0.60	0
K_E	-	-	-	-	1.00	-	-
T_E (s)	-	-	-	-	0.87	-	-
K_F	-	-	-	-	0.004	-	-
T_F (s)	-	-	-	-	0.27	-	-

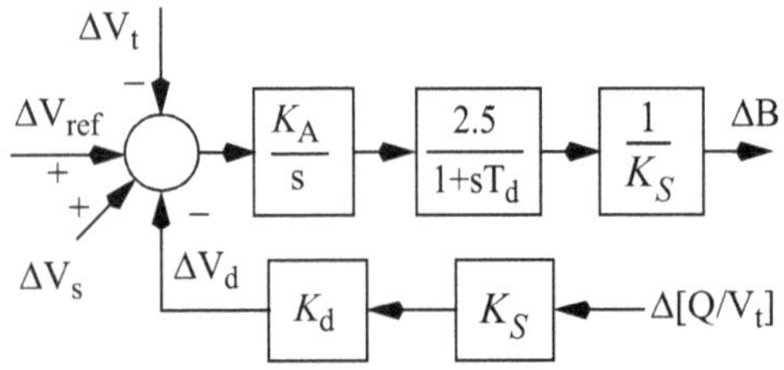

SVC name / bus number	Mbase (Mvar)	K_A	K_S
ASVC_2 / 205	650	500	6.5
RSVC_3 / 313	800	500	8.0
BSVC_4 / 412	1430	500	14.3
PSVC_5 / 507	500	250	5.0
SSVC_5 / 509	550	250	5.5

ΔB and $\Delta[Q/V_t]$ are in per-unit on MBASE.

$K_d = 0.01$ pu on SBASE and $T_d = 0.005$ s

Figure 10.38 Small-signal model of the controller for the SVCs.

Chapter 11

Tuning of FACTS Device Stabilizers

11.1 Introduction

In the 1990s the development of high power semiconductor devices found application in power electronic equipment in power systems. Such transmission systems and associated devices are generally known as Flexible AC Transmission Systems (FACTS); a comprehensive description of the technology, the devices and references to the literature are given in [1] (published in 2000).

In this chapter the tuning of stabilizers is outlined for FACTS devices such as Static Var Compensators (SVCs), the converters at the ends of High Voltage Direct Current (HVDC) transmission lines, Thyristor-Controlled Series Capacitor (TCSC), and other similar FACTS devices. Such stabilizers are generally known as Power Oscillation Dampers (PODs), however, the role of PSSs is also to act as power oscillation dampers - hence we will refer to PODs as FACTS Device Stabilizers (FDSs) to emphasize the application to FACTS devices.

Consider the studies for Cases 1 to 6 presented in the previous chapter. Referring to Tables 10.11, 10.15 and 10.16 it is noted that, for all PSSs in service with the damping gain set to 20 pu on machine MVA rating, the real parts of the mode shifts for the local-area modes typically vary from -1.3 to -2.5 Np/s over the encompassing range of operating conditions covered by the six cases. However, the real parts of the mode shifts for the inter-area modes, modes K, L and M, roughly vary over a much smaller range, from -0.4 to -1.1 Np/s for the same operating conditions. The damping of all modes in these cases is good, the lowest damping ratio being about 15%. However, because the damping of some modes may be poor, stabilizers installed on FACTS devices can provide a significant improvement in the damping of targeted modes. By reducing PSS damping gains to 5 and/or 10 pu on machine MVA ratings, cases of poorer damping are also examined in which the damping ratios of the inter-area modes are in the range 2 to 8%.

The common configuration of the FACTS device and controllers is shown in Figure 11.1. In the case of a Static Var Compensator (SVC), for example, the controller regulates the voltage at its terminals or at an electrically close, high-voltage busbar where voltage support is required [1], [2]. The location of the SVC in the network may be such that a stabilizer installed on the SVC is effective in improving the damping of certain inter-area modes. An effective stabilizing signal may be the perturbations in frequency at its terminals, an appropriate power flow, etc. [3].

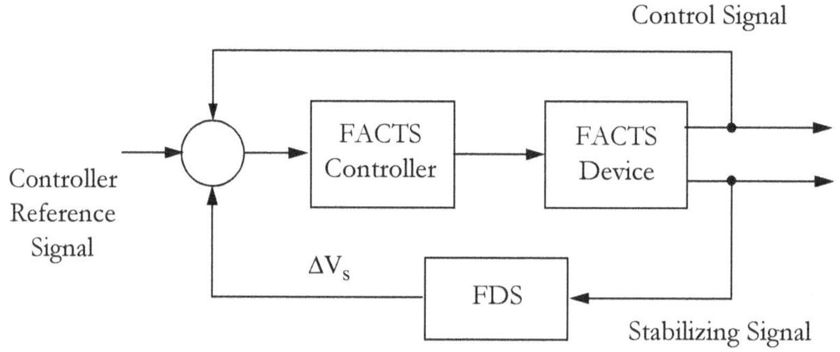

Figure 11.1 Configuration of the FACTS device, its controller and stabilizer (FDS)

The objective of FDS tuning is to improve the damping of lightly damped modes, ideally without degrading the damping of other modes, or compromising the performance of the primary control function of the device. As foreshadowed above, an inter-area mode is typically - but not necessarily - the mode which is targeted for enhanced damping.

As background, a 'simplistic' tuning procedure for a SVC is considered to illustrate the intent of the FDS tuning methods. The theoretical basis of the Method of Residues, already analysed in Chapter 6, is briefly summarized and will provide the basis for the tuning of FDSs [4], [5], [8]. However, in the multi-machine case there is a major difference with respect to the SMIB case of Section 6.3; the FDS may be tuned to provide damping over a range of modal frequencies.

A variety of other methods for tuning the stabilizers of a range of FACTS devices are described in the literature, [9] to [19]. Reference [20] provides a more detailed account of modelling shunt FACTS devices such as SVCs and Static Compensator. The tuning methods and approaches investigated in this chapter are presented using simple models for the FACTS devices; however, the methods are equally applicable for more sophisticated systems. The models presented in [20] provide detailed descriptions of modern control features such as coordinated control of nearby switched capacitor banks by the SVC. Such functions are not considered any further in this book in which only small deviations about a steady-state operating condition are relevant to small-signal analysis.

11.2 A 'simplistic' tuning procedure for a SVC

The application of a 'simplistic' procedure to the tuning of a SVC, BSVC_4, at bus number 410 in the fourteen-generator network in Figure 10.1 is now examined. As stated in Table 10.17, the maximum and minimum reactive power generation for BSVC_4 is 1100 and –330 Mvar, respectively, giving a reactive range (Mbase) of 1430 Mvar. The operating condition selected is the heavy load condition of Case 1 (see Table 10.2). For all PSSs in service, with their damping gains set to 20 pu on machine MVA rating, the local and inter-area modes are shown in Table 10.11. For illustrative purposes we will consider the more lightly damped of the complex inter-area modes, M $(-0.52 \pm j1.80)$.

The terminal voltage bus frequency, ω_{freq} (pu of system frequency), which is the rate of change of the terminal-voltage angle, α rad, is employed as the stabilizing signal. As shown in (8.9) the transfer function of the bus-frequency pre-filter is:

$$\frac{\Delta\omega_{freq}(s)}{\Delta\alpha(s)} = (1/\omega_0)\frac{s}{1+sT_F}, \tag{11.1}$$

where $\omega_0 = 2\pi f_0$ (rad/s) and f_0 is the system frequency (Hz). At $f_0 = 50$ Hz, $(1/\omega_0) = 0.003183$; T_F is normally set so that high frequency noise above the selected corner frequency $(1/T_F)$ is attenuated, say, $T_F = 0.005$ s [1].

The block diagram of the FACTS device controller and stabilizer is shown in Figure 11.2. Let us assume that the transfer function of the FDS is $\Delta V_s(s)/\Delta F_{rq}(s) = k_{fds}$, a real gain (i.e. omitting compensation, washout and low-pass filters). Let's calculate the values of the mode M for a range of gain values (not knowing as yet what constitute high gain values). As noted in Figure 10.38, the value of Mbase is 1430 Mvar, Sbase = 100 Mvar.

1. The time constant T_F (5 ms) is very short. Such time constants should typically be 3 or more times the cycle time of the PSS processor to reduce phase errors at higher frequencies.

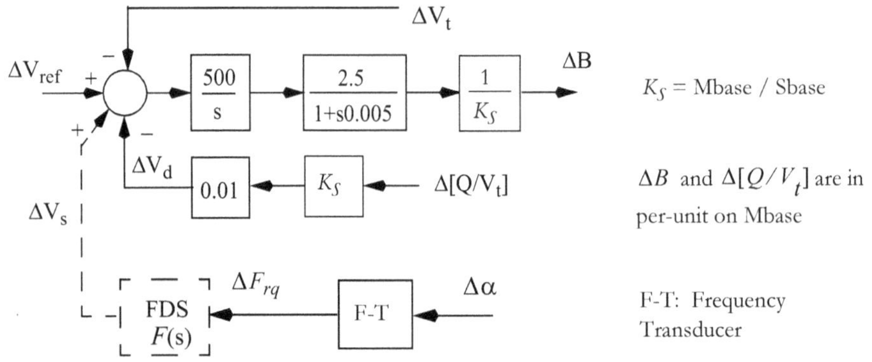

Figure 11.2 The controller and stabilizer, $F(s)$, for SVC BSVC_4 showing terminal voltage control, the provision of droop, and the frequency stabilizing signal ΔF_{rq}.

In Table 11.1 the mode shifts in mode M for Case 1 are shown as the stabilizer gain k_{fds} is increased from zero with the stabilizer in service. The mode shift for a gain of 30 pu is shown in Figure 11.3. Ideally, to introduce pure damping to the mode, the mode shift should lie at 180°. Phase lag compensation must therefore be provided for the multi-machine system in this example noting that the required lag compensation angle increases with increasing gain. Although the lag compensation which the stabilizer transfer function should provide is as much as 16° for the selected gain range, let us derive the transfer function of the lag compensation with a lag angle of 11° at $s = 0 + j1.8$ (1.8 rad/s) for the stabilizer gain of 30 pu.

Table 11.1 Case 1. Shifts in inter-area mode M with increasing FDS transfer function gain k_{fds} [a]

k_{fds} (pu)	Mode M	Mode Shift	Angle $\phi°$ [b]
0	$-0.522 \pm j1.797$	-	-
10	$-0.649 \pm j1.786$	$-0.127 \mp j0.011$	5.0
20	$-0.778 \pm j1.763$	$-0.256 \mp j0.034$	7.6
30	$-0.905 \pm j1.722$	$-0.383 \mp j0.075$	11.1
40	$-1.024 \pm j1.656$	$-0.502 \mp j0.141$	15.7
Note: (a) FDS is a pure gain transfer function. (b) Required lag compensation angle			

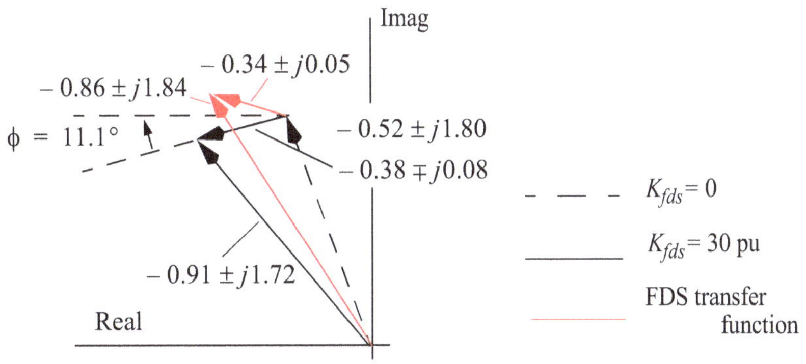

Figure 11.3 Shift in mode M both for K_{fds}= 30 pu and for FDS transfer function $F(s)$ (11.2).

The calculation of the transfer function of the lag compensator is similar to that for lead compensation in the example in Section 2.12.1.4 and is based on frequency response analysis with $s = 0 + j\omega$. The simple compensator transfer function for the lag angle of $11°$ at 1.8 rad/s is $(1 + s0.458)/(1 + s0.674)$. When washout and low pass filters, with corner frequencies 0.17 and 30 rad/s respectively, are included the transfer function of the FDS is:

$$F(s) = \frac{\Delta V_s}{\Delta F_{rq}} = k_{fds} \cdot \frac{6s}{1+6s} \cdot \frac{1+s0.458}{1+s0.674} \cdot \frac{1}{1+s0.033}, \text{ with } k_{fds} = 30 \text{ pu.} \qquad (11.2)$$

With the FDS of BSVC_4 in service with the above transfer function the resulting value of mode M is $-0.86 \pm j1.84$ for $k_{fds} = 30$ pu compared to the value of $-0.91 \pm j1.72$ for the scalar transfer function $k_{fds} = 30$ in Table 11.1. While the FDS enhances the damping of mode M relative to the case when the FDS is out of service, the mode shift $-0.34 \pm j0.05$ is not quite that desired; moreover, its modal frequency is increased from that with the stabilizer off-line. There are therefore a number of observations that can be found in this 'simplistic' procedure.

- The agreement between the value of the targeted mode using the 'simplistic' procedure to evaluate the stabilizer transfer function is not as close as desirable. (Further iterations of the procedure could improve the result.)
- The lag compensation of the stabilizer transfer function is based on the frequency response calculation using $s = j\omega_f$ rather than the complex value in the vicinity of the targeted mode, $s = \alpha + j\omega_f$. This problem is compounded when the washout and low-pass filters are added. A more rigorous, iterative process is required to converge on a lag transfer function for the stabilizer - with the specified filters - in the vicinity of the targeted mode. (With the FDS out of service, mode M varies between $-0.43 \pm j1.76$ and $-0.59 \pm j2.51$ over the six operating conditions, see Tables 10.11, 10.15 and 10.16.)

- No cognizance has been given to the suitability of the stabilizer transfer function (11.2) over an encompassing range of operating conditions (including outages, etc.) in enhancing the damping of the targeted mode.
- Although the damping of the targeted mode may be enhanced over the range of operating conditions, the damping of other modes may be degraded.
- Under some operating conditions the presence of zeros or modes (other than rotor modes), in the vicinity of the targeted mode may significantly affect the trajectory of the mode as the stabilizer gain is increased.
- From Figure 10.26 it is observed that the frequency of the inter-area mode M decreases with increasing PSS damping gains. To improve synchronizing torques it may be desirable to tune the FDS to enhance not only the damping of the targeted mode but also to increase its oscillatory frequency.

It is clear that a method for tuning the stabilizers is desirable that better takes account of the range of operating conditions, the filters and the complex value of the targeted mode.

11.3 Theoretical basis for the tuning of FACTS Device Stabilizers

Some of the relevant theoretical material, based on the 'Method of Residues', is described in Section 6.2.1 and is summarized here for ease of reference.

Let the stabilizer transfer function be:

$$F(s) = k_{fds}H(s) = k_{fds}G_c(s) \cdot G_W(s) \cdot G_{LP}(s), \tag{11.3}$$

where the transfer function $G_c(s)$ of the stabilizer in this application is tuned to provide the appropriate phase compensation and is assumed to consist of m lead or lag blocks of the form:

$$G_c(s) = \left[\frac{1 + T_n s}{1 + T_d s}\right]^m, \quad s = \alpha + j\omega. \tag{11.4}$$

The FDS gain setting in (11.3) is k_{fds} (note, this is not the 'damping gain' value). The washout and low-pass filter transfer functions, $G_W(s)$ and $G_{LP}(s)$, are given by (5.29) and (5.30), respectively. It is assumed that the values of the time constants in the latter two transfer functions have been appropriately selected (see Section 5.8.6). The objective of the tuning procedure for the ith stabilizer is to determine the values of the parameters k_{fdsi}, T_{ni} and T_{di}, $i = 1 \ldots m$ in (11.4) that satisfy the relevant requirements on damping.

The following analysis (which repeats part of that in Section 6.2.1) assumes that (i) initially the FDS is out of service, and then it is in service with the FDS gain set to k_{fds}, (ii) the FDS feedback is positive (see Figure 6.1 and 11.2). It is shown in (6.8) that the mode shift in the targeted mode λ_h is:

$$\Delta\lambda_h = \frac{k_{fds} r_h H(\lambda_h)}{1 - r_h k_{fds} \frac{\partial}{\partial \lambda_h} H(\lambda_h)}, \tag{11.5}$$

where r_h is the residue of the transfer function $\Delta V_{stab}(\lambda_h)/V_{ref}(\lambda_h)$ (no FDS); ΔV_{stab} is the stabilizing signal selected to be the input to the FDS.

If in (11.5) the gain k_{fds} is chosen such that $\left| r_h k_{fds} \frac{\partial}{\partial \lambda_h} H(\lambda_h) \right| \ll 1$ then (11.5) reduces to yield the approximate value of the mode shift, i.e.:

$$\Delta\lambda_h \approx k_{fds} r_h H(\lambda_h). \tag{11.6}$$

Let the value of k_{fds} for which $\left| r_h k_{fds} \frac{\partial}{\partial \lambda_h} H(\lambda_h) \right| = 1$ be k_{Rm}, a quantity which provides a *nominal* measure of an upper value of the gain. According to (11.6) for values of $k_{fds} \ll k_{Rm}$ *the mode shift increases linearly with stabilizer gain.* Thus it follows from the definition of k_{Rm} that

$$k_{Rm} = 1 / \left| r_h \frac{\partial}{\partial \lambda_h} H(\lambda_h) \right|. \tag{11.7}$$

In order for the mode shift $\Delta\lambda_h$ in (11.6) to be $\pm 180°$, i.e. a direct left-shift of λ_h in the complex *s*-plane,

$$\arg\{r_h H(\lambda_h)\} = \pm 180°. \tag{11.8}$$

Therefore the compensation angle ϕ provided by the FDS is

$$\phi = \arg\{H(\lambda_h)\} = \pm 180 - \arg\{r_h\} \quad (°). \tag{11.9}$$

Typically k_{fds} is selected to be less than $0.1 k_{Rm}$. However, in multi-machine cases the effect on the actual modal trajectories of other system poles and zeros, as the FDS gain is increased from $k_{fds} = 0$, may result in mode shifts estimated from the above analysis differing substantially from actual shifts, even at gains much less than $0.1 k_{Rm}$.

Other comments in Section 6.2.1 are also applicable to FDS tuning.

Consider now the application of the above results to the tuning of a FDS in a multi-machine system. It may be necessary to tune the FDS to improve the damping of several rotor modes and to accommodate the associated variation in magnitude and phase of the associated residues.

The application of the Method of Residues is now illustrated by a number of studies; two studies illustrate the tuning of a FDS for a SVC using bus frequency or real power flow as stabilizing signals. A study on a different FACTS device concerns the tuning of a stabilizer for a thyristor-controlled series capacitor (TCSC). In the latter case the stabilizer transfer function is required to accommodate power flows in both directions through the TCSC. In all studies the trajectories of selected inter-area modes are tracked as the stabilizer gain is increased from zero to an appropriate value. The aims of mode-tracking studies are (i) to determine the stability of the system, (ii) to investigate the characteristics of the mode shifts with increasing stabilizer gain, (iii) to compare the estimated mode shifts calculated using (11.5) or (11.6) with those calculated by eigen-analysis, and in some cases (iv), to account for the nature of the deviation between estimated and calculated values.

11.4 Tuning SVC stabilizers using bus frequency as a stabilizing signal

As mentioned earlier, a SVC is primarily installed for voltage support and control, typically in areas more remote from generation - such in the vicinity of loads or at intermediate substation buses on higher voltage transmission lines.

In this application of FDS tuning it is assumed that there are inter-area modes whose damping may be improved by a FDS installed on a SVC close to a major load centre. Conceptually, when close to a major load centre the FDS should modulate the load-area voltage such that load real power is reduced concomitant with a fall in system frequency - thereby enhancing the damping of the mode. This suggests that frequency may be a suitable stabilizing signal.

Because it has been the basis of a number of studies the 14-generator power system employed in Chapter 10 is used as the study system. From Tables 10.11, 10.15 and 10.16 it is observed that the inter-area modes L and M typically have values in the vicinity of $-0.5 \pm j2.8$ and $-0.5 \pm j1.9$, respectively, over the range of the normal cases 1 to 6 with all PSS damping gains set to 20 pu. Inter-area mode K is generally well damped, but may be enhanced by the FDSs.

With reference to the system diagram in Figure 10.1, the SVCs 'BSVC_4' in Area 4 and 'PSVC_5' in Area 5 will be used to establish what improvements in damping of the inter-area modes can be achieved using perturbations in local frequency as a stabilizing signal. It will also be found that it is desirable to install a SVC in Area 2; this is considered in Chapter 14.

Based on the results in Chapter 10 when all PSS damping gains are set to 20 pu it may be considered that it is not necessary to install stabilizers on any FACTS device. On the other hand, say, can the PSS damping gain settings be reduced with the installation of FDSs? Let us therefore establish a whether a FDS transfer function tuned for 20 pu PSS damping gains adequately covers a lower range of PSS damping gain settings, say 10 to 20 pu.

For illustrative purposes and to avoid complexity, line outages and other conditions which cause a degradation in the damping of the inter-area modes have not been included in the limited analysis which follows. The tuning of a frequency-stabilized FDS for BSVC_4 is now investigated.

11.4.1 Use of bus frequency as a stabilizing signal for the SVC, BSVC_4

The perturbations in local bus frequency is synthesized from angular perturbations $\Delta\alpha$ (rad) in the terminal voltage of the SVC at bus 412 in Figure 10.1. The basis for employing bus-frequency perturbations $\Delta F_{frq} = \Delta\omega_{freq}$ (pu of system frequency) as a stabilizing signal is outlined in Section 11.2; the transfer function of the frequency transducer is given by (11.1).

Initially it is of interest to learn which of the inter-area modes over the encompassing range of normal operating conditions, Cases 1 to 6, are best damped by means of the FDS on BSVC_4. Mbase for BSVC_4 is 1430 Mvar (see Figure 10.38).

11.4.1.1 Determination of the stabilizer transfer function for BSVC_4

Referring to (11.3) and (11.4), the aim of the analysis is to determine $k_{fds}G_c(s)$, the transfer function of the compensation, as well as the parameters of the washout and low-pass filters such that the damping of the mode(s) satisfies the relevant performance criteria. Furthermore, the improvement in damping of any inter-area mode should not lead to an unacceptable degradation in the damping of other inter-area modes or of local modes in the vicinity of the SVC.

Let us assume (i) all SVCs are in service and controlling the voltage on their respective buses, (ii) the FDS path in Figure 11.2 is open, and (iii) all PSSs are in service and their damping gains are set to 20 pu on machine MVA rating. The residues of the transfer function $\Delta F_{rq}(s)/\Delta V_{ref}(s)$ for the inter-area modes K, L and M are then calculated for the operating conditions 1 to 6. Depending on the characteristics of the residues as revealed by their polar plots, it is of interest to ascertain if the compensation should in fact target any one of the three inter-area modes. It is also possible, for example, that compensation based on the residues for mode L may enhance or degrade the damping on mode M, or vice-versa.

Using the Mudpack small-signal, power system dynamic performance package [21] the polar plot of the residues for modes L and M is shown in Figure 11.4. The residues are of the SVC transfer function $\Delta F_{rq}(s)/\Delta V_{ref}(s)$ for the range of operating conditions, Cases 1 to 6. The residues of the inter-area mode K are negligible and are omitted from the plot. The values of modes L and M are listed in Tables 10.11, 10.15 and 10.16.

As foreshadowed in the Section 11.4, it is desirable to establish whether the same FDS transfer function adequately covers the 10 and 20 pu sets of PSS damping gains. The polar plot of the residues for the lower set of PSS damping gains is shown in Figure 11.5.

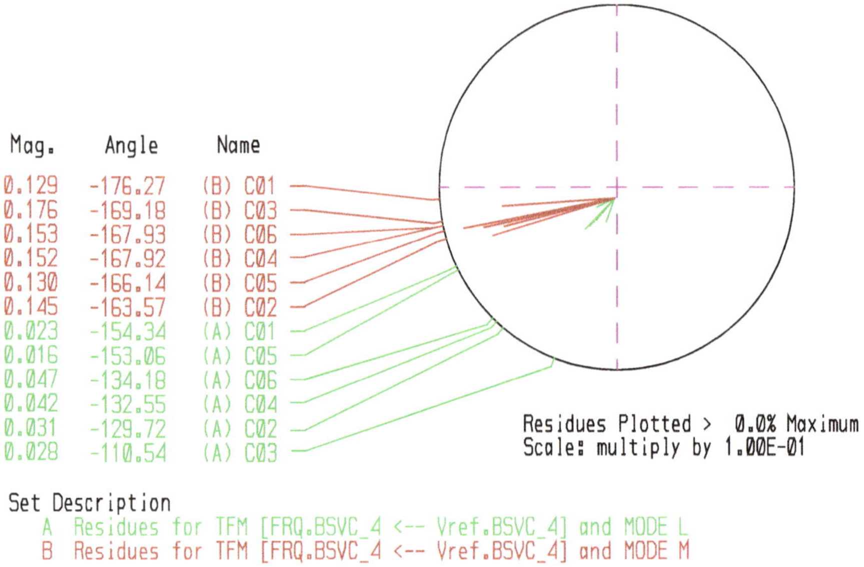

Figure 11.4 Polar plot of the residues for the transfer function, $\Delta F_{rq}/\Delta V_{ref}$, for modes L and M and six operating conditions. All PSS damping gains set to 20 pu on machine MVA rating. The values of modes L and M are in the vicinity of $-0.47 \pm j2.8$ and $-0.5 \pm j1.9$, respectively. Note: the magnitude scale is to be multiplied by 0.1.

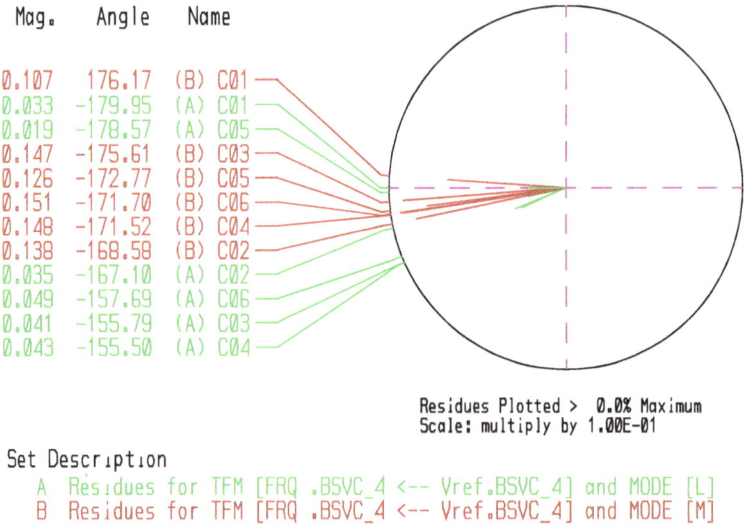

Figure 11.5 Polar plot of the residues as for Figure 11.4 with all PSS damping gains set to 10 pu on machine MVA rating. Note scaling.

A comparison of the magnitudes of the residues in Figures 11.4 and 11.5 reveals (i) the residues for mode M are about four times greater than those of mode L, and (ii) the band of phase angles of the residues is narrower than that for mode L. The compensation will therefore target mode M for which, as revealed in Figure 11.4, the residues lie in a relatively narrow phase-band of approximately 13° with a spread in magnitudes between 0.013 and 0.018 units.

For the range of operating conditions it is now necessary to select representative values of (i) the compensation angle for the calculation of the compensator transfer function, (ii) the magnitude of the residues for determining the nominal upper gain value, and (iii) a single mode value considered to cover the modes of interest or concern.

From Figure 11.4 for PSS damping gains set to 20 pu a representative angle for the residues of mode M is selected to be $-168°$ which lies in the mid-range of values. The required compensation angle is therefore $-12°$ (or $12°$ lagging). For mode M the maximum value of the residues of 0.0176 pu (on SVC base) is selected from Figure 11.4. (These decisions may depend on the application, e.g. whether to weight certain operating conditions more heavily, or whether to abide by the system criteria which specify the minimum level for damping, say, for the outage of a critical circuit.) For mode M, and for the range of modal values over the encompassing operating conditions, a targeted value of complex frequency is selected to be $-0.5 \pm j1.9$, a value which tends to favour the heavier load conditions.

An associated set of representative values can be deduced from Figure 11.5 when all PSS damping gains are set to 10 pu; similarly a set for 5 pu is derived. The values are summarized in Table 11.2.

Table 11.2 Representative values for evaluation of compensation transfer function, mode M

PSS gain (pu)	Phase spread (deg)	Representative phase angle (deg)	Compensation angle (deg)	Maximum residue	Representative modal frequency*
20	-164 --> -176	-168	-12	0.0176	$-0.5 \pm j1.9$
10	-168 --> -176	-172	-8	0.0151	$-0.25 \pm j2.2$
5	-171 --> -184	-178	-2	0.0149	$-0.12 \pm j2.2$

* Representative modal frequencies over the range of operating conditions

In Mudpack [21] there are facilities to calculate iteratively the compensation transfer function of the stabilizer given the desired compensation angle, a representative or target complex tuning frequency, the order of the lag or lead compensator, and the required number of washout and low-pass filters and their parameters (see Appendix 6–I.1). Based on (11.3) and (11.4) the form of stabilizer transfer function is given in (11.10).

$$F(s) = k_{fds} \cdot H(s) = k_{fds} \cdot \left[\frac{T_W s}{1 + T_W s}\right]^w \cdot \left[\frac{1 + T_n s}{1 + T_d s}\right]^m \cdot \left[\frac{1}{1 + T_{LP} s}\right]^z . \qquad (11.10)$$

There are also facilities in the software to *estimate*, for a selected stabilizer gain, the mode shifts for the target and other selected modes, according to (11.5), for each of the operating conditions. The provision of estimates of local as well as inter-area modes can reveal if a local mode, say, is unduly degraded by the stabilizer and therefore may be of concern. However, it is also necessary to establish through Bode- or eigen-analysis the range of FDS gains k_{fds} for which the closed-loop system is stable. It may happen that a mode other than a rotor mode becomes unstable.

Two cases of FDS tuning could now be considered. In the first case the washout and low-pass filters are selected to cover a wide band of modal frequencies; in the second the filters provide a narrow band which specifically targets mode M. Wide-band compensation only is now considered; a practical example of narrow band compensation is analysed in [6] using the Method of Residues. The objectives of the former are to improve the damping of the inter-area modes as well as local-area modes, if possible. To cover the range of operating conditions and rotor modes in Tables 10.11, 10.15 and 10.16, a first-order compensator is specified, together with first-order washout and low-pass filters with parameters $T_W = 6$ s and $T_{LP} = 0.033$ s. The phase shifts of the filters, which lie a decade above and below the corner frequencies of 0.17 and 30 rad/s, respectively, are less than 5°. The FDS transfer function thus takes the form:

$$F(s) = k_{fds} H(s) = k_{fds}\left[\frac{sT_W}{1 + sT_W}\right]\left[\frac{1 + T_n s}{1 + T_d s}\right]\left[\frac{1}{1 + T_{LP} s}\right] . \qquad (11.11)$$

The representative values for 20 pu PSS damping gain settings in Table 11.2 is used to calculate the compensation transfer function because

1. In practice the PSS damping gain settings may tend towards the higher value of 20 pu because normal, outage and N-1 operating conditions must all satisfy the system damping performance specifications.

2. The range of residue angles in Table 11.2 for 10 and 5 pu PSS damping gains are essentially covered by that for the 20 pu gain settings.

3. The representative residue angles differ by 10° at most, and the associated range of compensation angles lie between 2° to 12° lagging. 'Over compensation' in this study is likely to increase the frequency of the inter-area modes at the lower gain settings and thereby improve synchronizing torques. (This may help to offset the decrease in the inter-area frequencies, observed in Figure 10.26, with increase in PSS damping gains.)

Based on the representative values in Table 11.2 for the evaluation of the compensation transfer function, the iterative procedure described in Appendix 6–I.1 is used to calculate the parameters of the lag transfer function for PSS damping gains of 20 pu. The PSS transfer function (11.11) is found to be

$$F(s) = k_{fds}H(s) = k_{fds}\left[\frac{s6}{1+s6}\right]\left[\frac{1+s0.346}{1+s0.498}\right]\left[\frac{1}{1+s0.033}\right], \tag{11.12}$$

the nominal upper gain value being $k_{Rm} = 398$ pu.

11.4.1.2 Range of stabilizer gains for stability

Before the trajectories of the selected modes are calculated by eigen-analysis it is desirable to ascertain for what range of gains the system with the stabilizer transfer function calculated above is stable. It may not be clear if some other mode (e.g. a controller mode) becomes unstable as the stabilizer gain is increased - or if instability occurs, say, for some value of gain less than $0.1k_{Rm}$ where k_{Rm} is the nominal upper value determined by the Residues Method.

From Tables 10.11, 10.15 and 10.16 for the six operating conditions with all PSS damping gains set to 20 pu on machine MVA rating, and for no FDSs in service, it is known (i) that the system is stable, and (ii) that the stabilizer transfer function of (11.12) possesses left-half plane poles. Therefore, with no right-half plane poles, we can use the open-loop Bodes plots to determine closed-loop stability as well as the gain and phase margins for a selected stabilizer gain.

Lest us insert the stabilizer transfer function (11.12) in the feedback path in Figure 11.2. The feedback path at the summing junction is left open in order to calculate the open-loop transfer functions $\Delta V_s(j\omega_f)/\Delta V_{ref}(j\omega_f)$. For the Case 1, a heavy load condition, the associated Bode plot is shown in Figure 11.6 remembering that, for stability analysis using the Bode plot, negative feedback is assumed (positive feedback of the stabilizer output is specified in Figure 11.2).

Note in Case 1 that the gain margin for stability (673 pu) is greater than 10% (i.e. ~40 pu) of the upper gain value ($k_{Rm} = 398$ pu) necessary to satisfy the nominal upper value of gain as determined by the Method of Residues. The stability limits for cases 1 to 6 are confirmed by eigen-analysis. Within the gain range of 0 to 40 pu the selection of the gain setting k_{fds} is dependent on a number of factors: for example: (i) Can the desired damping of the target mode be achieved with lower gain settings such that the reactive power output of the SVC is not continually hitting limits for acceptable variations in frequency? (ii) Is the damping of other modes unduly degraded? (iii) Can we be confident about the accuracy of the models of the devices and the system? (iv) What are the effects of high controller gains on unmodelled dynamics, etc.

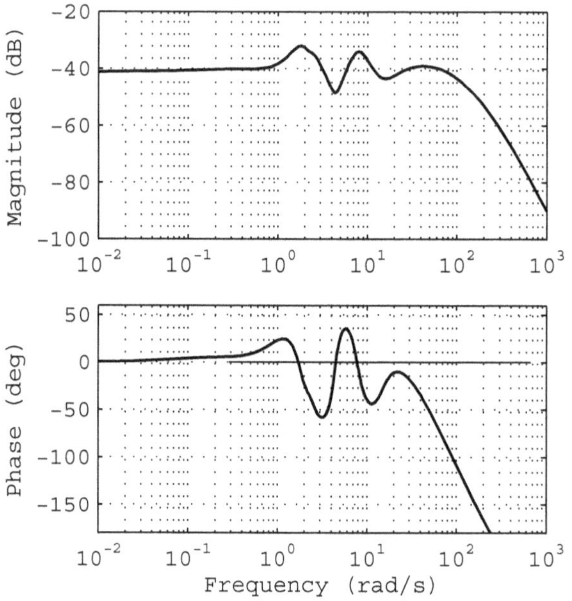

Figure 11.6 Case 1: FDS plus SVC. Open-loop frequency response $\Delta V_s/\Delta V_{ref}$ for k_{fds} = -1 pu on device base. The gain margin is 56.6 dB (673 pu on device base) at 241 rad/s. All PSS damping gains are set to 20 pu.

11.4.1.3 Inter-area modal trajectories as the stabilizer gain is increased

To ascertain the effectiveness of the FDS tuning, which assumes all PSS damping gains are set to 20 pu on machine MVA ratings, the eigenvalue trajectories are calculated as the FDS gain is increased from zero to 100 pu on the SVC base. For the same FDS parameters the trajectories are also evaluated for the case when all PSS damping gains are set to 10 pu. Based respectively on Cases 1 and 4 both heavy and light load conditions are considered. The trajectories of modes L and M are shown in Figure 11.7.

From the modal trajectories, it is observed that:

- For increases of stabilizer gain up to 40 pu the shift in the inter-area mode M is more-or-less directly to the left in the *s*-plane with small changes in modal frequency at higher gains.

- The shift in mode L is negligible. It may be necessary to investigate whether stabilizers on other SVCs in the system enhance the damping of mode L.

- The use of the FDS transfer function, whose tuning is based on a damping gain setting of 20 pu on all PSSs, is satisfactory for (i) both the heavy and light load cases investigated, (ii) both PSS damping gain settings of 10 and 20 pu.

- As predicted from Figure 11.6 no evidence of closed-loop instability is found over the FDS gain range 0 - 100 pu for the six cases investigated.

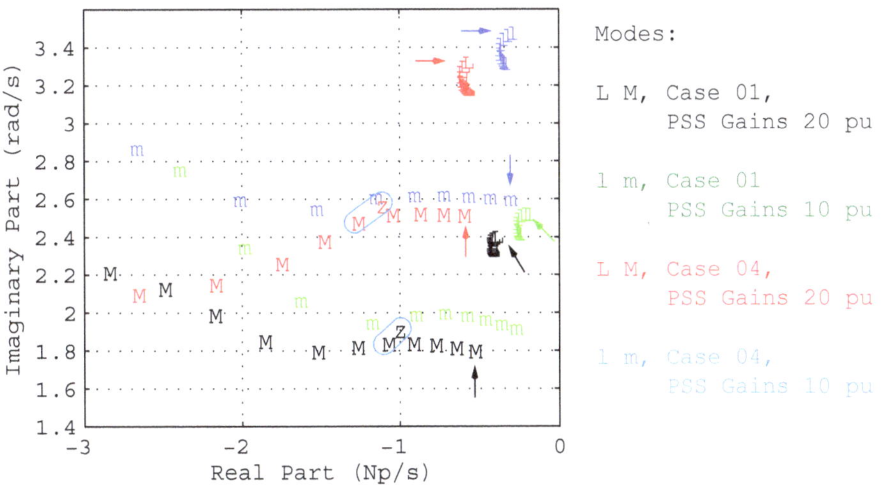

Figure 11.7 Cases 1 and 4. Trajectories of the inter-area modes L and M as the stabilizer gain k_{fds} is increased in 10 pu steps from zero (shown by an arrow) to 100 pu on the SVC base. All PSS damping gains are set to 10 or 20 pu on machine base.
Z: Estimated mode values from (11.5) for stabilizer gain k_{fds} = 40 pu.

11.5 Use of line real-power flow as a stabilizing signal for a SVC

Consider the case when a major load is connected to two separate areas of generation through high voltage transmission lines. Associated with these areas there is an inter-area mode which is assumed to be lightly damped. If a SVC is located close to the load centre the FDS may be able to improve the damping of the oscillatory power flow between the areas by modulating the load-area voltage - and hence the real power flow into the load.

Referring to the system diagram in Figure 10.1, it is observed that a major load and the SVC BSVC_4 are both connected to bus 412. It is also noted that the power flow on transmission lines between buses 410 and 412 and between buses 411 and 412 supply the net real power to the load at 412. The total net power flow perturbations, ΔP_{tot}, at bus 412 into the load bus will be considered to be a potential stabilizing signal.

For the purposes of calculating the residues of the transfer function $\Delta P_{tot}(\lambda_h)/\Delta V_{ref}(\lambda_h)$, the SVC is placed under closed-loop voltage control with the FDS path in Figure 11.8 open. All PSSs are in service with their damping gains set to 20 pu on machine base. A polar plot of the residues is shown in Figure 11.9 from which it is noted that, by comparison with that of Figure 11.4, the spread of amplitude and phase over the range of operating conditions is greater. Note: we cannot compare the residues derived from power signals with those derived from bus frequency (see Section 3.7).

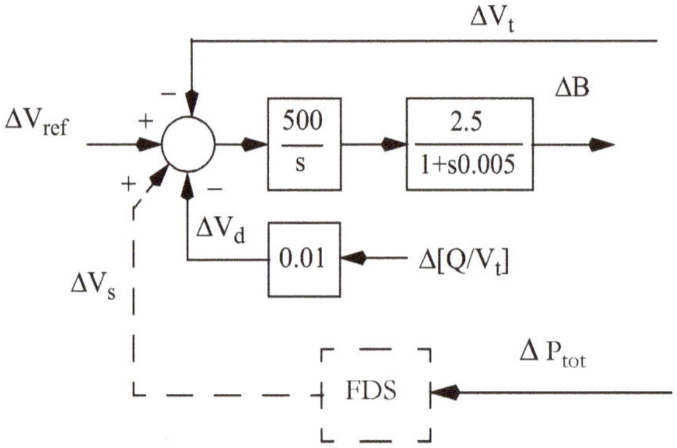

Figure 11.8 The controller and stabilizer for SVC BSVC_4 showing terminal voltage and droop controls, as well as the real power stabilizing signal ΔP_{tot}.

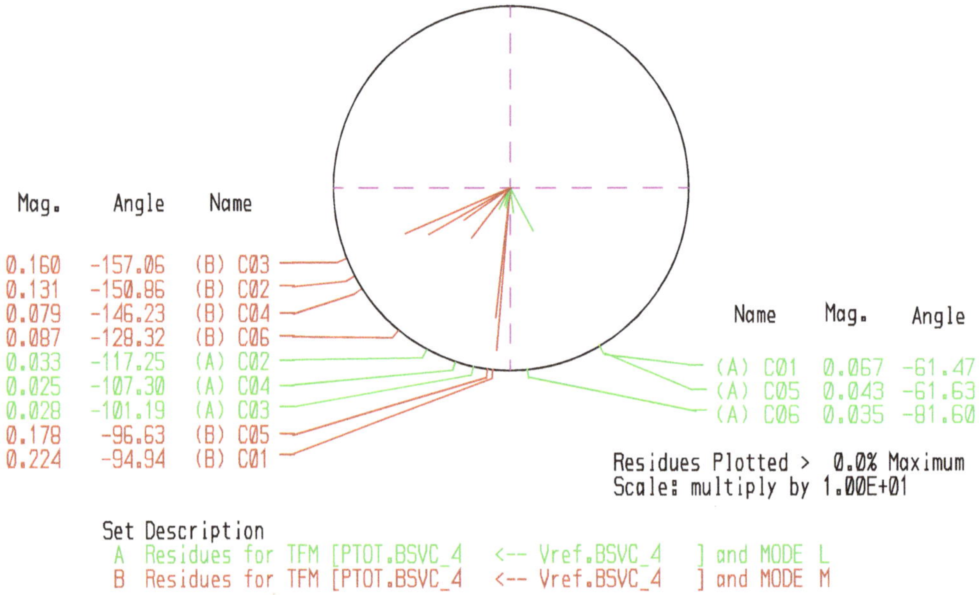

Figure 11.9 BSVC_4. Polar plot of the residues of the transfer function $\Delta P_{tot}/\Delta V_{ref}$ for modes L and M and six operating conditions. PSS damping gains 20 pu. Note: the magnitude scale is to be multiplied by 10.

As previously discussed, the objectives of the compensation is to improve the damping of the inter-area mode M and, if feasible, mode L and the local-area modes as well. To cover the range of rotor modes (see Tables 10.11, 10.15 and 10.16), a compensation angle of –45°

is selected together with a representative value for the residue of 2.0 pu on SVC base. A first-order compensator is specified, together with parameters of the first-order washout and low-pass filters of $T_W = 7$ s and $T_{LP} = 0.02$ s, respectively. The complex tuning frequency is again $-0.5 \pm j1.9$, targeting mode M.

Using an iterative procedure in Mudpack the parameters of the compensator are calculated based on (11.11). The resulting FDS transfer function is

$$F(s) = k_{fds}H(s) = k_{fds}\left[\frac{s7}{1+s7}\right]\left[\frac{1+s0.188}{1+s0.799}\right]\left[\frac{1}{1+s0.02}\right];$$

the nominal value of the upper gain is $k_{Rm} = 2.02$ pu.

With the FDS in service, the trajectories of the inter-area modes L and M, together with that of a controller mode X associated with the FDS, are plotted in Figure 11.10 for Case 1. It is seen that mode X becomes unstable when the stabilizer gain $k_{fds} = 3.47\%$ or 0.070 pu on SVC base. This result is confirmed from the Bode plot of the open-loop transfer function $\Delta V_s(j\omega_f)/\Delta V_{ref}(j\omega_f)$.

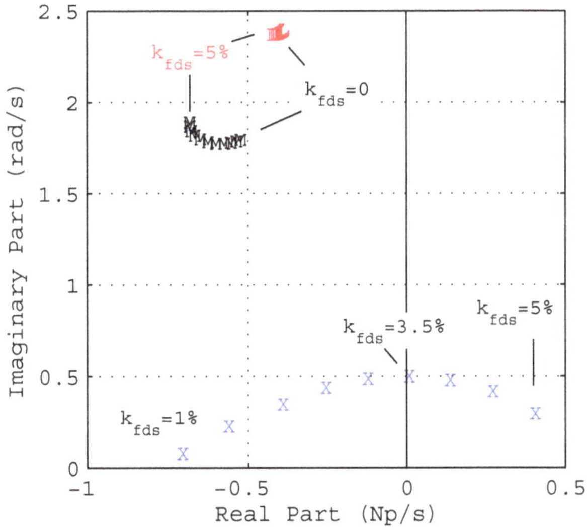

Figure 11.10 Case 1 with power flow FDS. Trajectories of the inter-area modes L and M and stabilizer mode X as the stabilizer gain k_{fds} is increased in 0.5% steps from zero to 5% (0.101 pu on SVC base). All PSS damping gains set to 20 pu.

In Table 11.3 the estimated rotor mode shifts for the inter-area mode M are compared with the eigen-analysis-based ('actual') values for a range of stabilizer gains. The estimated mode shifts are based on (11.5) and the 'actual' shifts are calculated by the software package for the full system.

Table 11.3 Estimated [a] and actual[b] mode shifts in mode M for Case 1, k_{fds} = 1 to 4%

Rotor mode shift	Stabilizer gain, k_{fds}			
	1% (0.020 pu)	2% (0.040 pu)	3% (0.061 pu)	4% (0.081 pu)
Estimated shift	$-0.02 \mp j0.02$	$-0.05 \mp j0.04$	$-0.07 \mp j0.06$	$-0.09 \mp j0.07$
Estimated mode	$-0.54 \pm j1.78$	$-0.57 \pm j1.76$	$-0.59 \pm j1.74$	$-0.61 \pm j1.72$
Actual mode shift	$-0.03 \mp j0.02$	$-0.08 \mp j0.02$	$-0.12 \pm j0.00$	$-0.15 \pm j0.05$
Actual mode value	$-0.55 \pm j1.78$	$-0.60 \pm j1.78$	$-0.65 \pm j1.80$	$-0.68 \pm j1.85$

a: Based on (11.5). b: Calculated from eigen-analysis
Value of mode M at $k_{fds} = 0$ is $-0.52 \pm j1.80$.

From the table it is observed: (i) With increasing gain the left shift in the real part of the mode calculated from eigen-analysis is 30 to 70% greater than the corresponding estimated shifts. (ii) The system is unstable at k_{fds} = 3.47 % (0.070 pu), however, the Method of Residues does not indicate that the instability of a controller mode occurs.

Allowing for a 10 dB (3.1 times) margin the gain setting k_{fds} should be 0.070/3.1=0.022 pu (1.1%). The associated mode shift is small and therefore the operation of a power-stabilized FDS may not be justified for this system based on this study alone.

11.6 Use of bus frequency as a stabilizing signal for the SVC, PSVC_5

It was noted in Section 11.4.1.1 that the frequency-stabilized FDS installed on the SVC, BSVC_4, did not usefully contribute to the damping of inter-area modes K and L. For the purposes of coordination of stabilizers in Chapter 14 it is of interest to ascertain if the SVC, PSVC_5 at bus 507, contributes to the damping of any of the inter-area modes. The maximum and minimum reactive power generation for PSVC_5 is 320 and −180 Mvar, respectively, giving a reactive range (Mbase) of 500 Mvar. The relevant details are provided in Table 10.17 and in the block diagram of Figure 11.2; Sbase = 100 Mvar.

A similar procedure to that outlined for BSVC_4 is followed: (i) to determine the parameters of the FDS, and (ii) to evaluate the damping performance of PSVC_5 by means of the modal trajectories for increasing FDS gain. For this purpose the FDS at BSVC_4 is out of service.

The local bus frequency F_{rq} is used as the stabilizing signal and is synthesized from angular perturbations $\Delta\alpha$ (rad) in the terminal voltage of the SVC, bus 507 in Figure 10.1. For Cases

1 to 6 the polar plot of the residues of the inter-area modes K, L and M are shown in Figure 11.11; the damping gain of the PSSs is set to 20 pu.

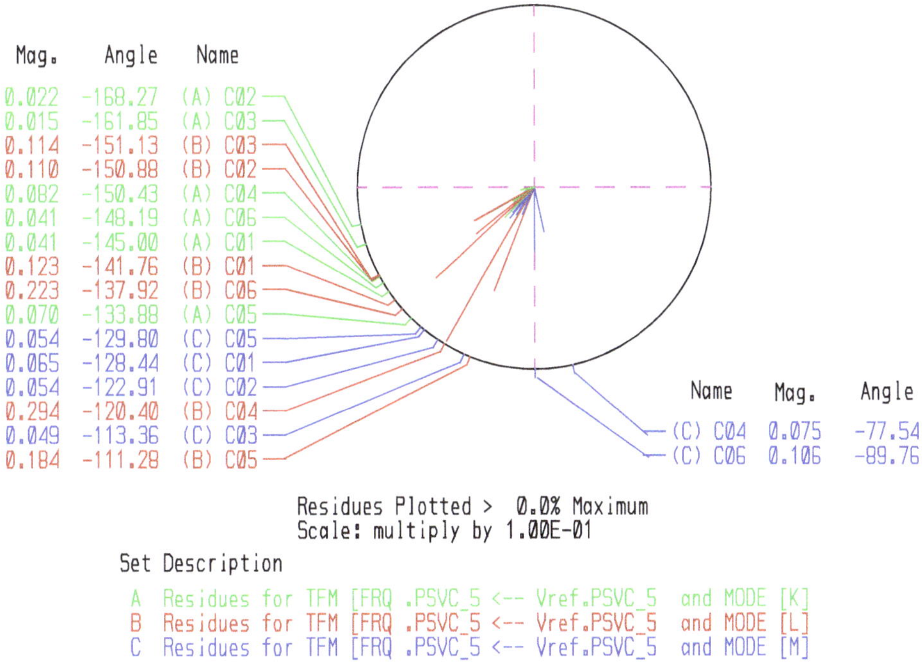

Figure 11.11 PSVC_5. Polar plot of the residues for the transfer function $\Delta F_{rq}/\Delta Vref$ for modes K, L and M and six operating conditions. All PSS damping gains are set to 20 pu on machine MVA rating. The values of modes K, L and M are in the vicinity of $-1.0 \pm j4.0$, $-0.45 \pm j2.6$ and $-0.5 \pm j1.9$, respectively.
Note: the magnitude scale is to be multiplied by 0.1.

Because the magnitudes of the residues for mode L in Figure 11.11 are two to three times greater than those for modes M and K, the FDS tuning targets mode L. It is evident from Table 11.4 that the representative values for PSS damping gains of 20 pu are likely to lead to a satisfactory FDS design for the case when all PSSs are set to the lower value of 10 pu.

Table 11.4 Representative values for evaluation of compensation transfer function, mode L

PSS gain (pu)	Phase spread (deg)	Representative phase angle (deg)	Compensation angle (deg)	Maximum residue	Representative modal frequency
20	-111 --> -151	-131	-49	0.0294	$-0.45 \pm j2.6$
10	-123 --> -148	-136	-44	0.0251	$-0.29 \pm j2.7$

The FDS transfer function, based on the PSS damping gains of 20 pu is found to be:

$$F(s) = k_{fds}\left[\frac{6s}{1+6s}\right]\left[\frac{1+0.140s}{1+0.685s}\right]\left[\frac{1}{1+0.033s}\right], \quad k_{Rm} = 312 \text{ pu}; \tag{11.13}$$

the same washout and low-pass filter time constants as for the FDS of BSVC_4 are employed to cover the ranges of the inter-area modal frequencies.

To evaluate the effectiveness in the damping introduced by the FDS on PSVC_5 a similar set of modal trajectories to those in Figure 11.7 are plotted in Figure 11.12.

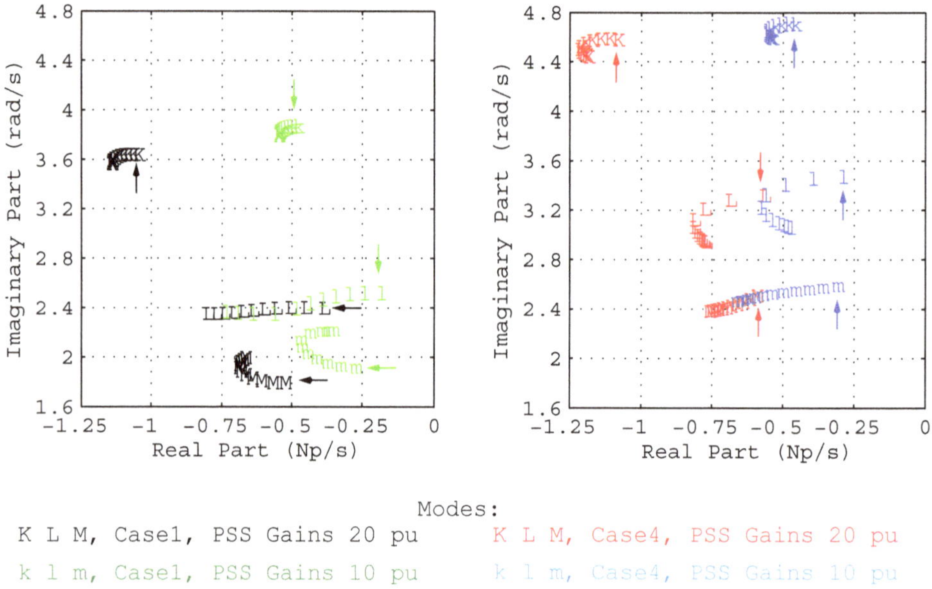

Figure 11.12 Mode trajectories for Cases 1 and 4 as the FDS gain on PSVC_5 varies from zero (shown by an arrow) to 100 pu on device base in 10 pu steps. In each case all PSS damping gains are set to 10 or 20 pu on generator MVA rating.

For the range of FDS gains 0 to 100 pu this system is stable.

From the trajectories of the inter-area modes the following are observed.

- Mode L is left-shifted in the s-plane with a slight decrease in frequency when all PSSs are set to 20 pu. However, its improvement is limited in the light-load condition, Case 4, when the FDS gain exceeds 30 pu.

- Improvement in the damping of mode M is also limited for Case 1, the heavy load condition, when the stabilizer gain exceeds 30 to 40 pu.

- There are marginal improvements in the damping of mode K, but are limited for FDS gains exceeding 30 pu.

Because the FDSs on BSVC_4 and PSVC_5 induce marked shifts in modes M and L respectively, coordination between the FDSs is desirable to achieve the best improvement in the damping of these modes within the limitations which have been observed. The coordination of stabilizers, PSSs and FDSs, will be considered in Chapter 14.

11.7 Tuning a FDS for a TCSC using a power flow stabilizing signal

A series capacitor is primarily installed to reduce the series inductive reactance of transmission lines thereby improving both the voltage and rotor-angle stability of the interconnected system [2], [22] and [23]. It also reduces the voltage drop between buses straddling the line and series capacitor - as well as reducing the I^2X losses in the circuit. The proportion of the line's series inductive reactance which the series capacitor cancels out depends on a number of factors which are determined by the characteristics of the system, [23]; such factors are beyond the scope of this discussion.

For present purposes it is assumed that in a Thyristor-Controlled Series Capacitor (TCSC) the series reactance is effectively perturbed by an amount ΔX through the action of the stabilizer [24]. Conceptually, for perturbations in real power flow in the line the action of the FDS is to reduce the effective series reactance of the line when the power flow tends to increase, and vice-versa. Damping of both the relevant modes and the line flow perturbations is thereby improved. Due to the action of its washout filter the FDS does not respond to relatively slow changes in the line's real power flow associated with changes in load or in generation dispatch.

The Method of Residues is again employed for the tuning of the stabilizer for the TCSC; this technique is applied in [4], [5], [25], [26], and Appendix A of [27]. Other techniques are covered in [28], [30] and [31].

It is assumed that equivalent single series capacitance, located between buses 310 and 311 in the simplified 14-generator system (see Figure 10.1), is thyristor controlled [29]. The MVA base (Mbase) for the TCSC is selected to be 300 Mvar, The relevant section of the network and the format of the stabilizing controls are shown in Figure 11.13 (a) and (b), respectively.

In Figure 11.13(b), for the purposes of analysis, (i) a dummy reference is inserted in the controller, and (ii) the gain in the forward path is $K_A = 1$ pu on Mbase. Based on the transmission line data in Table 10.19, the effective series reactance of the two capacitors in parallel is $-j0.01685$ pu on system base (Sbase = 100 MVA).

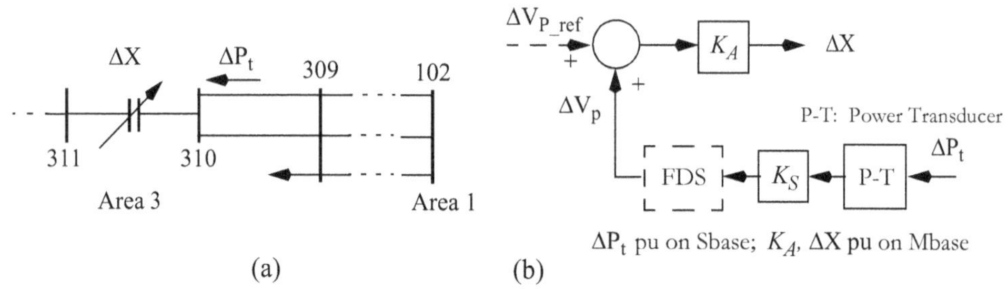

Figure 11.13

(a) TCSC in the simplified 14-generator system (see Figure 10.1 for bus numbering).
(b) Model of FDS using the perturbation in the real power flow ΔP_t (pu) through the TCSC as the stabilizing signal. (K_S = Sbase/Mbase=100/300)
ΔX (pu) is the perturbation in the series reactance of the equivalent single capacitor.

For the purposes of testing the variation of the residues with PSS damping gains it is assumed that the inter-area modes may be heavily or lightly damped for all the operating conditions. Such damping is implemented by setting all PSS damping gains to 20 pu or by reducing all such gains to 5 pu on generator MVA rating. The effect of the gain reduction on all the rotor modes in Cases 1 to 6 can be seen in Figure 10.26; for the inter-area modes the associated eigenvalues are listed in Table 11.5. It is of interest to learn if an improvement in the damping of the inter-area modes such that their damping ratios exceed 0.1 is achievable with the FDS installed on the TCSC.

In order to derive a transfer function for the FDS we follow the procedure outlined in the previous studies.

With no stabilizers in service on the SVCs and at the TCSC, the values of the lightly- and heavily damped inter-area modes are listed in Table 11.5 for PSS damping gains of 5 and 20 pu; the residues are calculated for these PSS damping gains and modes.

According to Table 11.5 for the operating conditions in Cases 1, 3 and 4 the power flow through the TCSC is from Area 1 to Area 3, and from Area 3 to Area 1 for Cases 2, 5 and 6. It is therefore proposed to use the modulus of the total power $\Delta Pmod = |\Delta P_t|$ through the TCSC as the stabilizing signal with the object of deriving a single FDS transfer function covering flows in both directions. For the purpose of calculating the residues of the transfer function $\Delta Pmod(\lambda_h)/\Delta V_{ref}(\lambda_h)$ the FDS path in Figure 11.13(b) is open. Polar plots of the residues are shown in Figures 11.14 and 11.15 for PSS damping gain settings of 20 and 5 pu, respectively.

Table 11.5 Modes K, L, M. No FDSs on SVCs and at TCSC.
PSS gains 5 and 20 pu on generator MVA rating.

Case	TCSC/ Line* (MW)	PSS gains (pu)	Mode K		Mode L		Mode M	
			Value	ξ	Value	ξ	Value	ξ
1	763/ 984	5	$-0.18 \pm j3.93$	0.05	$-0.05 \pm j2.57$	0.02	$-0.14 \pm j1.98$	0.07
		20	$-1.04 \pm j3.64$	0.28	$-0.39 \pm j2.40$	0.16	$-0.52 \pm j1.80$	0.28
2	-1291/ -1023	5	$-0.05 \pm j3.75$	0.04	$-0.11 \pm j2.81$	0.03	$-0.050 \pm j1.88$	0.03
		20	$-0.77 \pm j3.54$	0.21	$-0.45 \pm j2.54$	0.17	$-0.43 \pm j1.76$	0.24
3	730/ 984	5	$-0.28 \pm j4.02$	0.07	$-0.12 \pm j2.63$	0.05	$-0.16 \pm j2.01$	0.08
		20	$-1.12 \pm j3.71$	0.29	$-0.43 \pm j2.42$	0.17	$-0.58 \pm j1.86$	0.30
4	58 / 199	5	$-0.14 \pm j4.74$	0.03	$-0.13 \pm j3.53$	0.04	$-0.16 \pm j2.64$	0.06
		20	$-1.08 \pm j4.58$	0.23	$-0.56 \pm j3.32$	0.17	$-0.59 \pm j2.51$	0.23
5	-379 / -201	5	$-0.08 \pm j4.13$	0.02	$-0.12 \pm j3.08$	0.04	$-0.09 \pm j2.12$	0.04
		20	$-0.88 \pm j3.90$	0.22	$-0.46 \pm j2.89$	0.16	-0.50 ± 1.96	0.25
6	-141 / 0	5	$-0.12 \pm j4.80$	0.03	$-0.11 \pm j3.51$	0.03	$-0.11 \pm j2.57$	0.04
		20	$-1.08 \pm j4.64$	0.23	$-0.57 \pm j3.30$	0.17	$-0.52 \pm j2.45$	0.21

* Total power flow through (i) TCSC at and from bus 310 (upper value); (ii) Line, from bus 102 to 309 at 309 (lower value) ξ - Damping ratio

Figure 11.14 reveals that a lag compensation angle exceeding 60° would cause a degradation in the damping of Mode L for Case 2. Four assumptions are therefore made in determining the representative values for calculating the compensation transfer function. (i) The degradation in mode L is ignored unless it becomes excessive (i.e. other stabilizers are capable of providing additional damping for this mode). (ii) The target mode for improvement in damping is mode K. (iii) The FDS on the TCSC is switched off-line only when the steady-state power flow in the TCSC is less than 200 MW, i.e. in Cases 4 and 6. (iv) It is anticipated that PSS damping gains are normally in the vicinity of 20 pu on machine base.

For the FDSs designed for SVCs it is noted in mode trajectories, such in Figure 11.12, that the imaginary parts of the modes tend to decrease with increasing gain when the compensation shifts the residue such that the imaginary part of the residue is negative. For example, in Figure 11.14 the residue for mode K, Case 5, is $6.1\angle{-70.7°}$; if the compensation angle were 90° lagging, say, the residue is shifted to $-161°$. It is therefore decided to provide over-compensation with a compensation angle of $-110°$ in order to increase the frequency of oscillation of the inter-area modes; this applies to Cases 1, 2, 3, and 5.

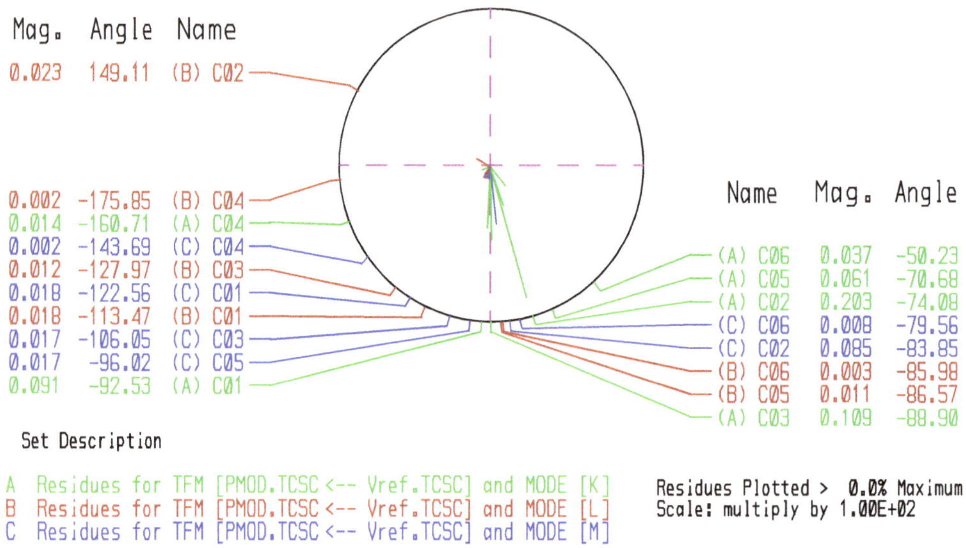

Figure 11.14 TCSC: Polar plot of the residues for the transfer function $\Delta Pmod(\lambda_h)/\Delta V_{ref}(\lambda_h)$ for modes K, L and M and six operating conditions. All PSS damping gains set to 20 pu. Note: magnitude scale is to be multiplied by 100.

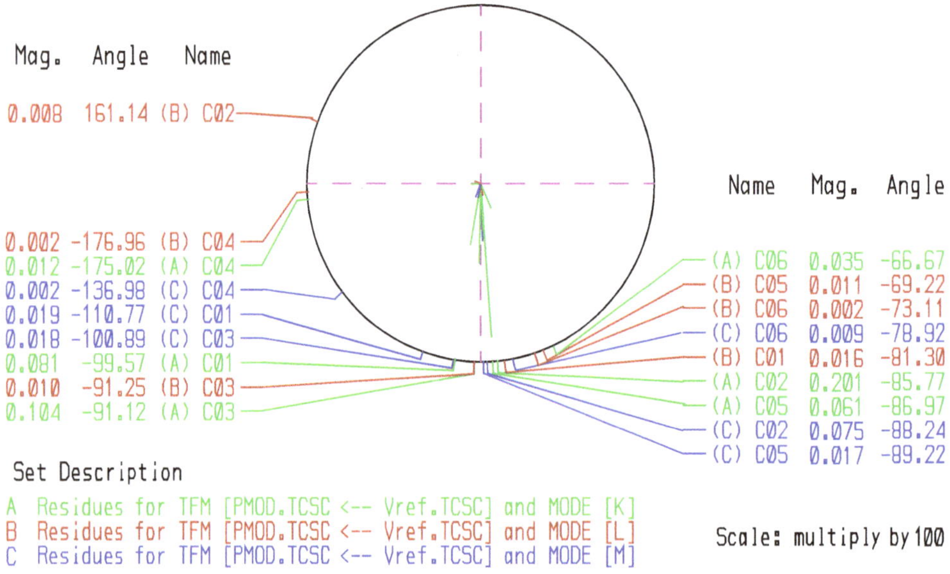

Figure 11.15 TCSC: Polar plot of the residues for the transfer function $\Delta Pmod(\lambda_h)/\Delta V_{ref}(\lambda_h)$ for modes K, L and M and six operating conditions. All PSS damping gains set to 5 pu. Note: magnitude scale is to be multiplied by 100.

For the FDSs designed for SVCs it is noted in mode trajectories, such in Figure 11.12, that the imaginary parts of the modes tend to decrease with increasing gain when the compensation shifts the residue such that its imaginary part is negative. For example, in Figure 11.14 the residue for mode K, Case 5, is $6.1\angle{-70.7°}$; if the compensation angle were 90° lagging, say, the residue is shifted to $-161°$. It is therefore decided to provide over-compensation with a compensation angle of $-110°$ in order to increase the frequency of oscillation of the inter-area modes; this applies to Cases 1, 2, 3, and 5.

It is decided to base the calculation of the compensation transfer function on PSS damping gains of 20 pu because:

- the nature of the residues in Figures 11.14 and 11.15 are comparable for PSS damping gains set to 20 and 5 pu, respectively;
- the representative values in Table 11.6 for 5 pu PSS damping gains are essentially covered by those for the 20 pu gain settings.

Table 11.6 Representative values for evaluation of the compensation transfer function, mode K (excluding Cases 4 and 6)

PSS damping gain (pu)	Phase spread for K (deg)	Representative phase angle (deg)	Compensation angle (deg)	Revised Comp. angle (deg)	Representative residue	Representative modal value
20	-71 --> -93	-82	-98	-110	20.3	$-1 \pm j4$
5	-86 --> -100	-93	-87	-110	20.1	$-0.14 \pm j4.2$

The transfer function is therefore based on the representative values for PSS gains of 20 pu. Because the required phase lag is greater than 60° and less than 120° a second-order transfer function is selected (see Appendix 6–I.1). The transfer function is therefore:

$$\frac{\Delta Vref}{\Delta Pmod} = k_{fds} \cdot H(s) = k_{fds} \cdot \frac{6s}{1+6s} \cdot \left[\frac{1+0.078s}{1+0.429s}\right]^2 \cdot \frac{1}{1+0.025s} \quad k_{Rm} = 0.404 \text{ pu.} \quad (11.14)$$

11.7.1 Gain range for the stability of TCSC with the FDS in service

Maximum power flow through the TCSC from buses 310 to311 occurs in Cases 1 and 2 (763 and -1291 MW, respectively). Analysis to establish the range of gains for which the system is stable is based on the Bode plot of the open-loop transfer function $\Delta V_p(j\omega_f)/\Delta V_{ref}(j\omega_f)$ and assumes (i) negative feedback at the open-loop summing junction [1], (ii) there are no open-loop poles in the right-half of the s-plane. For Cases 1 and 2 and with all PSS damping gains set to 20 pu the Bode plot of the open-loop transfer function $\Delta V_p(j\omega_f)/\Delta V_{ref}(j\omega_f)$ is shown in Figure 11.16. With the FDS in closed-loop operation the plot establishes for

1. Positive feedback of the FDS transfer function is assumed in Figure 11.13.

Case 1 that the system is stable - theoretically - over the gain range zero to infinity. In practice due to unmodelled dynamics the gain range for system stability may be limited. However, for Case 2 the Bode Plot reveals that the system is unstable for stabilizer gains exceeding 0.21 pu on the TCSC base (this limit corresponds to 51.5% of the nominal upper gain of 0.404 pu). Eigen-analysis of the closed-loop system reveals that a mode associated with a stabilizer state migrates into the right-half of the *s*-plane at the limiting value of gain.

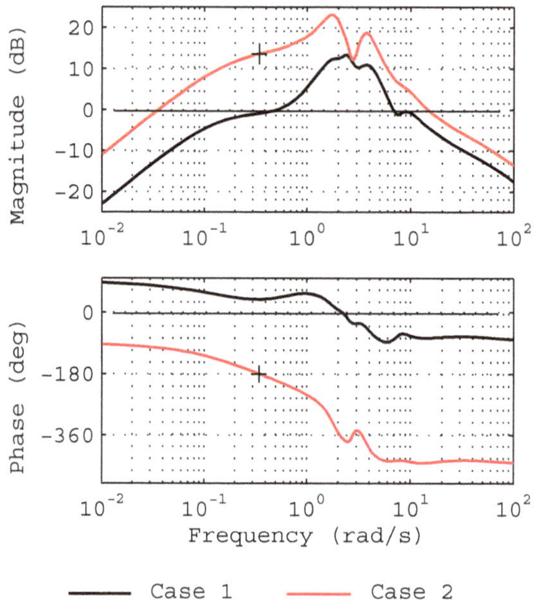

Figure 11.16 TCSC. Cases 1 and 2: Open-loop frequency responses of the FDS and SVC, $\Delta V_p / \Delta V_{ref}$, k_{fds} = -1. Case 1 is stable over the gain range; for Case 2 the gain margin is −13.6 dB at 0.35 rad/s. All PSS damping gains 20 pu on generator MVA ratings.

11.7.2 Inter-area mode trajectories with increasing stabilizer gain

The stabilizer transfer function is given by (11.14). With the stabilizer loop closed, it is desirable (i) to assess the nature of the variation of the inter-area modes K, L and M as the value of the gain k_{fds} is increased, and (ii) to compare modal values with those estimated based on (11.5) of the Residue Method. For the heavier load Cases 1 and 2, and for stabilizer gain settings between zero and 0.135 pu, the trajectories of the inter-area modes are plotted in Figure 11.17. The modal trajectories are the eigen-value plots calculated by the Mudpack software package.

In Figure 11.17 for Case 1 the estimated values based on (11.5) of the Residue Method agree closely with those from eigen-analysis. However, for Case 2 in Figure 11.17 the modal values calculated from eigen-analysis diverge from those estimated for FDS gains greater than

0.045 pu. As predicted from the plot of residues for Case 2, the damping of mode L degrades slightly with increasing FDS gain.

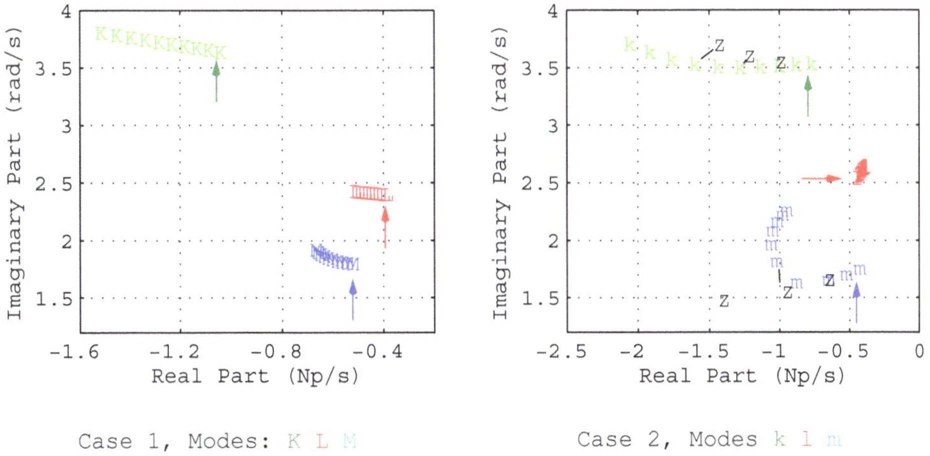

Figure 11.17 TCSC, Cases 1 and 2. Trajectories of the inter-area modes K, L and M as the stabilizer gain is increased from zero (shown by an arrow) to 0.135 pu in 0.015 pu steps. All PSS damping gains are set to 20 pu on machine MVA rating.
Z: Estimated mode values from (11.5) for stabilizer gain k_{fds} of 0.03, 0.06 and 0.09 pu on TCSC base.

For this study there are only a few feasible operating conditions on which to base the tuning of the FDS with confidence. In addition to line and other outages conditions it would be desirable to include operating conditions in which the flow through the TCSC from Area 1 to Area 3 is of a comparable magnitude to that in Case 2, i.e. about 1300 MW. Clearly the range of encompassing conditions needs to be widened - and the benefits established - in order to justify an expensive FACTS device such as a TCSC with stabilizing controls.

11.8 Concluding comments

11.8.1 Improving the damping of inter-area modes using FACTS devices

A number of papers on the tuning of a stabilizer for a FACTS device consider a SMIB system or simple four machine system with a single inter-area mode. However, the examples in this chapter illustrate the potential difficulties of developing such a stabilizer at a particular location in a multi-modal, multi-machine system, particularly as the location of the device is chosen primarily for reasons other than damping rotor modes. Consequently, the stabilizer may not be able to provide the desired damping for some or all the lightly-damped modes of concern. The examples do suggest that a number of locations at which FACTS devices are situated could be examined to determine their suitability for improving the damping particular modes. While the frequency-stabilized FDS might be feasible for BSVC_4 for improving the damping of the inter-area mode M (in Section 11.4.1), it has a much smaller effect on mode L. A FDS in some other part of the network may, because of its location, provide superior damping for mode L but may be less effective for mode M. It might therefore be useful to screen the appropriate FACTS devices to ascertain if a stabilizer installed on the device might be effective in damping particular modes.

In considering the FACTS devices in the 14-generator system analysed in Chapter 10 there are a number of sites at which SVCs are located. However, an examination - for example - of the residues of the SVC in Area 3, RSVC_3, reveals that over the range of operating conditions a FDS installed on this SVC produces relatively much greater mode shifts for the inter-area mode K than for modes L and M. Furthermore, other analyses show that incrementing the PSS damping gain of the generators at LPS_3 in Area 3 is very effective in enhancing the damping of mode K. A stabilizer installed on this SVC would have been of more interest if it were relatively more effective in damping mode L, say.

11.8.2 Robustness of FDSs

In tuning fixed-parameter FDSs the concept of robustness is based on the following considerations:

- there are two important components of a fixed-parameter FDS transfer function $k_{fds}G_c(s)$ which should be decoupled for practical purposes;
  - (a) the rotor modes are more-or-less directly left-shifted by the FDS compensating transfer function $G_c(s)$ with increase in the FDS gain, k_{fds} [1];
  - (b) the extent of the left-shift of the rotor modes is determined by the gain, k_{fds};
- ideally, the incremental left-shifts of the rotor modes are linearly related to increments in FDS gain for changes about selected nominal values.
- such considerations should apply over the set of encompassing operating conditions and an appropriate range of rotor modes.

The securing of a predominately left-shift of the relevant modes with increasing stabilizer gain is a requirement for the simultaneous coordination of PSSs and FDSs in Chapter 14, [32].

From the various studies presented in this chapter it is clear that the task of ensuring robustness is complex and time-consuming. In particular the FDS typically enables the damping of certain modes only, such damping being found to be dependent upon the location of the FACTS device in the system and the type of stabilizing signal employed.

---

1. This is not a 'damping' gain which is associated with P-Vr based PSS tuning.

Unlike the P-Vr characteristics which contributes robustness in the analysis of generator PSS parameters, there appears to be no corresponding system characteristics for the determination of FDS parameters. The issue of robustness of FDSs is a deficiency associated with the Method of Residues.

11.8.3 Estimated versus calculated mode shifts

From the mode trajectories for increasing stabilizer gains in Figures 11.7 and 11.17 it is observed that the value of the estimated mode, derived from the mode shift calculated using (11.5), diverge significantly at higher gains from those calculated using eigen-analysis.

The estimated mode shifts are based on the simple relations in (11.5) or (11.6). These do not account of the characteristics of a multi-machine system. In particular, as is well-known in the root-locus analysis of transfer functions, with increasing gain the modes migrate from the open-loop poles to the finite system zeros or those zeros at infinity. As the gain of the FDS is increased from zero, the influence of system zeros arises and the mode trajectories deviate from the ideal direct left-shift, or approach a close-by zero. Moreover, these modes may diverge from the estimated left-shift at relatively low or high gains. If there are open-loop zeros in the right-half s-plane, poles may migrate towards them resulting in instability, possibly at a relatively low value of stabilizer gain.

11.8.4 The notion of a 'nominal upper gain' for FDSs

The value of the stabilizer gain can be expressed as a fraction or a percentage of the so-called 'nominal upper gain', k_{Rm} (per unit). Note that (11.7) shows that k_{Rm} is inversely proportional to the representative value selected for the magnitude of the residue. If the largest magnitude is chosen then the value of $k_{fdsR} = 0.1 \times k_{Rm}$ may provide a notional warning that the mode shift may no longer increase linearly with increase in stabilizer gain for certain modes or operating conditions. As stated above, it has been shown that in multi-machine systems a number of other factors may result in linear changes in mode shift ceasing at lower values of k_{fdsR}. Depending on the application, the latter values may provide a convenient or meaningful gain limit - or a warning for the user.

11.9 References

[1] Narian G. Hingorani and Laszlo Gyugyi, *Understanding FACTS: concepts and technology of flexible AC transmission systems*, IEEE Press, Piscataway, NJ, 2000.

[2] P. Kundur, *Power System Stability and Control,* McGraw-Hill, Inc., 1994.

[3] E. V. Larsen, J. J. Sanchez-Gasca, and J. H. Chow, "Concepts for design of FACTS controllers to damp power swings," *Power Systems, IEEE Transactions on,* vol. 10, pp. 948-956, 1995.

[4] P. Pourbeik, *Design and Coordination of Stabilisers for Generators and FACTS devices in Multimachine Power Systems*, PhD Thesis, The University of Adelaide, Australia, 1997.

[5] P. Pourbeik and M.J. Gibbard, "Tuning of SVC stabilisers for the damping of inter-area modes of rotor oscillation", in *Proceedings of Australasian Universities Engineering Conference*, Perth, Sept. 1995, pp. 265-270.

[6] S. Hiley, *Redesign of the Blackwell Static Var Compensator Power Oscillation Damper Controller.* Undergraduate Thesis Report, School of Information Technology and Electrical Engineering, The University of Queensland, Australia, 2007.

[7] T. Smed and G. Andersson, "Utilizing HVDC to damp power oscillations," *Power Systems, IEEE Transactions on*, vol. 8, pp. 620-627, 1993.

[8] P. Pourbeik and M.J. Gibbard, "Damping and synchronising torques induced on generators by FACTS stabilizers in multimachine power systems", *Power Systems, IEEE Transactions on*, vol. 11, pp. 1920-1925, 1996.

[9] H. F. Wang and F. J. Swift, "Capability of the static VAr compensator in damping power system oscillations," *Generation, Transmission and Distribution, IEE Proceedings-*, vol. 143, pp. 353-358, 1996.

[10] H. F. Wang and F. J. Swift, "FACTS-based stabilizer designed by the phase compensation method. Part II: Multi-machine power systems", *Proc. 4th International Conference on Advances in Power System Operation and Management, APSCOM-97*, pp.644-649, Hong Kong, 1997.

[11] H. F. Wang, "Phillips-Heffron model of power systems installed with STATCOM and applications," *Generation, Transmission and Distribution, IEE Proceedings-*, vol. 146, pp. 521-527, 1999.

[12] H. F. Wang, "Selection of operating conditions for the coordinated setting of robust fixed-parameter stabilisers," *Generation, Transmission and Distribution, IEE Proceedings-*, vol. 145, pp. 111-116, 1998.

[13] J. J. Sanchez-Gasca, "Coordinated control of two FACTS devices for damping inter-area oscillations," *Power Systems, IEEE Transactions on*, vol. 13, pp. 428-434, 1998.

[14] H. F. Wang and F. J. Swift, "A unified model for the analysis of FACTS devices in damping power system oscillations. I. Single-machine infinite-bus power systems," *Power Delivery, IEEE Transactions on*, vol. 12, pp. 941-946, 1997.

[15] H. F. Wang, F. J. Swift, and M. Li, "A unified model for the analysis of FACTS devices in damping power system oscillations. II. Multi-machine power systems," *Power Delivery, IEEE Transactions on*, vol. 13, pp. 1355-1362, 1998.

[16] H. F. Wang, "Selection of robust installing locations and feedback signals of FACTS-based stabilizers in multi-machine power systems," *Power Systems, IEEE Transactions on*, vol. 14, pp. 569-574, 1999.

[17] N. Mithulananthan, C. A. Canizares, J. Reeve, and G. J. Rogers, "Comparison of PSS, SVC, and STATCOM controllers for damping power system oscillations," *Power Systems, IEEE Transactions on*, vol. 18, pp. 786-792, 2003.

[18] N. Tambey and M. L. Kothari, "Damping of power system oscillations with unified power flow controller (UPFC)," *Generation, Transmission and Distribution, IEE Proceedings-*, vol. 150, pp. 129-140, 2003.

[19] J. Guo, M. L. Crow, and J. Sarangapani, "An Improved UPFC Control for Oscillation Damping," *Power Systems, IEEE Transactions on*, vol. 24, pp. 288-296, 2009.

[20] P. Pourbeik, D. J. Sullivan, A. Bostrom, J. Sanchez-Gasca, Y. Kazachkov, J. Kowalski, A. Salazar, A. Meyer, R. Lau, D. Davies, and E. Allen, "Generic Model Structures for Simulating Static Var Systems in Power System Studies - A WECC Task Force Effort," *Power Systems, IEEE Transactions on*, vol. 27, pp. 1618-1627, 2012.

[21] D.J. Vowles and M.J. Gibbard, *Mudpack User Manual: Version 10S-03*, School of Electrical and Electronic Engineering, The University of Adelaide, July 2014.

[22] G. D. Breuer, H. M. Rustebakke, R. A. Gibley, and H. O. Simmons, "The Use of Series Capacitors to Obtain Maximum EHV Transmission Capability," *Power Systems, IEEE Transactions on*, vol. 83, pp. 1090-1102, 1964.

[23] P. M. Anderson and R. G. Farmer, *Series Compensation of Power Systems*, Encinitas, Calif.: PBLSH Inc.,1996.

[24] Yong Hua Song and A. T. Johns (Editors), *Flexible ac transmission systems (FACTS)*, The Institution of Electrical Engineers, UK, 1999.

[25] N. Yang, Q. Liu and J.D. McCalley, "TCSC controller design for damping interarea oscillations", *Power Systems, IEEE Transactions on*, Vol. 13, no. 4, 1998, pp. 1304 - 1310.

[26] L. Rouco and F. L. Pagola, "An eigenvalue sensitivity approach to location and controller design of controllable series capacitors for damping power system oscillations," *Power Systems, IEEE Transactions on*, vol. 12, pp. 1660-1666, 1997.

[27] "Impact of the Interactions Among Power System Controls", *CIGRE TF 38.02.16*, CIGRE, Paris, France, Technical Report 166, 2000.

[28] J. J. Paserba, N. W. Miller, E. V. Larsen, and R. J. Piwko, "A thyristor controlled series compensation model for power system stability analysis," *Power Delivery, IEEE Transactions on*, vol. 10, pp. 1471-1478, 1995.

[29] E. V. Larsen, K. Clark, S. A. Miske, and J. Urbanek, "Characteristics and rating considerations of thyristor controlled series compensation," *Power Delivery, IEEE Transactions on*, vol. 9, pp. 992-1000, 1994.

[30] A. M. Simoes, D. C. Savelli, P. C. Pellanda, N. Martins, and P. Apkarian, "Robust Design of a TCSC Oscillation Damping Controller in a Weak 500-kV Interconnection Considering Multiple Power Flow Scenarios and External Disturbances," *Power Systems, IEEE Transactions on*, vol. 24, pp. 226-236, 2009.

[31] A. B. Leirbukt, J. H. Chow, J. J. Sanchez-Gasca, and E. V. Larsen, "Damping control design based on time-domain identified models," *Power Systems, IEEE Transactions on*, vol. 14, pp. 172-178, 1999.

[32] P. Pourbeik and M. J. Gibbard, "Simultaneous coordination of power system stabilizers and FACTS device stabilizers in a multimachine power system for enhancing dynamic performance," *Power Systems, IEEE Transactions on*, vol. 13, pp. 473-479, 1998.

Chapter 12

The Concept, Theory, and Calculation of Modal Induced Torque Coefficients

12.1 Introduction

In this chapter the concept, the theory, and calculation of modal induced torque coefficients (MITCs) in multi-machine power systems are introduced. The concept of a modal induced torque coefficient is new [1], [2]. It forms the basis for calculation of the shifts in rotor modes when the stabilizer gains of one or more PSSs and/or FDSs are incremented by Δk (pu) on device base. Based on the concept of MITCs, the background theory of the rotor modes shifts, together with analysis of the effectiveness of, and interactions between, PSSs and FDSs in multi-machine systems are described in Chapter 13.

The theoretical development of MITCs in this chapter is fairly detailed and can be omitted if the practical applications of the analysis of rotor modes shifts are of primary interest. Where relevant, references are made in Chapter 13 to the results and equations that are developed in this chapter. A case study in the latter chapter demonstrates the significance of the MITCs and the insights that they provide into the dynamic performance of a multi-machine power system.

In essence, the concept of a modal induced torque coefficient is a further development of the concepts of damping and synchronising torque coefficients based on frequency response analysis (i.e. $s = j\omega_f$) [1]. In this chapter the torque coefficients are evaluated at the complex rotor modes ($\lambda_h = \alpha \pm j\omega$). The frequency-analysis-based torque coefficients are introduced in Section 5.3 for a SMIB system and in Sections 9.5 and 10.6 for the multi-machine case. *With shaft dynamics enabled*, the modal induced torque coefficient, T_{ij}^{h}, for the complex rotor mode λ_h is defined as a complex torque coefficient which is induced on generator *i* due to a perturbation in the stabilizing signal of stabilizer *j*. The stabilizer in question may be a power system stabilizer (PSS) or a FACT device stabilizer (FDS) installed on a FACTS device. Such FACTS devices are static var compensators (SVCs), high voltage DC links, thyristor controlled series capacitors (TCSCs) among others. It will be shown that the effect of any such stabilizer on the damping of any mode of rotor oscillation can be quantified. The concept can be extended to other devices such as wind turbine generators, photovoltaics and any other power-converter based transmission or generation equipment. The calculation of the MITC is from any controller to the effective induced torque coefficient on a specific synchronous generator. The controller may be installed on another synchronous generator on or any other dynamic device.

The analysis of the torque coefficients for generator *i* in the earlier chapters is based on the frequency response of the transfer function $\Delta P_{0i}(s)/\Delta\omega_i(s)|\ s = j\omega_f$, *all machine dynamics being disabled.* However, it is possible that due to perturbations in the speed of machine *j* a torque coefficient is induced on generator *i*. However, because we cannot relate speed perturbations on machine *j* to those on generator *i* when the shaft dynamics are disabled, the component of electro-magnetic torque induced by perturbations on generator *i*, $\Delta P_{0ij}(j\omega_f)$. in phase with speed perturbations on generator *i*, $\Delta\omega_i(j\omega_f)$, cannot be calculated, i.e. the damping torques induced by other machines on generator *i* are not available. However, the concept of *modal* induced torque coefficients (MITCs) overcomes this problem and facilitates, among other outcomes, the calculation of synchronizing and damping torque coefficients at modal frequencies. Essentially the analysis is divided in parts, (i) analysis *with rotor dynamics enabled*; (ii) analysis based on part (i) to derive the MITCs; (iii) analysis based on the MITCs to derive mode shifts due to stabilizer gain increments (this analysis is conducted in Chapter 13). *The advantage of this approach is that it facilitates the study of (i) the effects of controls on individual rotor modes, and (ii) the relative effects of controls on a set of selected modes.* It should be emphasized that the analysis of a large system *with shaft dynamics enabled* is complex, and the effects of controls on selected modes may be difficult to separate out.

For the purposes of generality in the initial analysis of MITCs, it is assumed that the controller to which all n speed stabilizing signals are fed is a "centralized" PSS, a full $n \times n$ matrix transfer function. Following the derivation of a set of general results, "decentralized" PSSs are employed in which the PSS matrix transfer function is diagonal. The decentralized stabilizer is, of course, the practical form of the PSS. Likewise for FACTS devices, the z local stabilizing signals are transmitted to a centralized FDS which is represented by a full $z \times z$ matrix transfer function. Each output of the centralized FDS is then fed to the summing junction of the controller on each FACTS device. In the practical form of the decentralized FDS the matrix transfer function is diagonal.

This chapter is structured as follows. In Sections 12.2.2 to 12.5 the concept of the modal induced torque coefficient (MITC) is introduced and its physical significance is explained. In order to apply the concepts to a multi-machine system a transfer function model of the system and its controllers is derived. in Section 12.3. Furthermore, a method is outlined for calculating MITCs when either a centralized PSS or FDS is in service. The application to decentralized controllers follows in Section 12.6. Using parameter-perturbation analysis in the remaining parts of the chapter, the relationship between MITCs and stabilizer gains is established. On this relationship is based the calculations in Chapter 13 of the shifts in the rotor modes of oscillation caused by an increment in the gain of any or all stabilizers.

12.2 The Concept of Modal Induced Torque Coefficients (MITCs)

12.2.1 Conventional frequency response techniques versus modal analysis

In previous chapters the analysis of decentralised, fixed-parameter PSSs and FDSs has been based mainly on frequency response techniques, i.e. with $s = j\omega_f$. However, in the analysis that follows we are interested in the components of torques of electromagnetic origin induced on the shaft of a generator at a *modal* frequency of rotor oscillation, $s = \lambda = \alpha \pm j\omega$, through the action of a generator or FACTS device stabilizer. Because certain relationships [(12.67) and (12.69)] apply only at a modal frequency of rotor oscillation $s = \lambda$, modal analysis must be employed in the associated analysis. [It is easy to show that for a complex modal frequency $s = \lambda = \alpha \pm j\omega$, which is a damped-sinusoid, the response of the transfer function $G(s)$ at that modal frequency (and only at that modal frequency), is $G(\lambda)$. The derivation of this result is similar to that in Section 2.11 for $G(s)$]. For the sake of generality, however, in parts of Sections 12.2.4 to 12.6 which are applicable in both the domains $s = j\omega_f$ and $\lambda = \alpha \pm j\omega$ the relevant expressions are expressed as functions of the form $F(s)$.

12.2.2 Modal torque coefficients induced by the action of a power system stabilizer

Two generators in a linearized representation of a multi-machine power system are shown in Figure 12.1(a). The speed-input PSS on machine *j* is assumed to be in service; machine *i* is not fitted with a stabilizer. A small system disturbance is assumed to occur which results in (i) only the hth complex mode of rotor oscillation, λ_h, being excited; (ii) the rotors of the generators being perturbed from synchronous speed. For machines *i* and *j*, the relative magnitude and phase of perturbations in their speeds, $\Delta\omega_i$ and $\Delta\omega_j$, respectively, are related by (12.67) in Appendix 12–I.1 (repeated here):

$$\frac{\Delta\omega_j(\lambda_h)}{\Delta\omega_i(\lambda_h)} = \frac{v_{jh}}{v_{ih}}, \tag{12.1}$$

where, for the eigenvalue λ_h, v_{ih} and v_{jh} are the ith and j^{th} elements of the right-eigenvector corresponding to the respective speed states of the two generators.

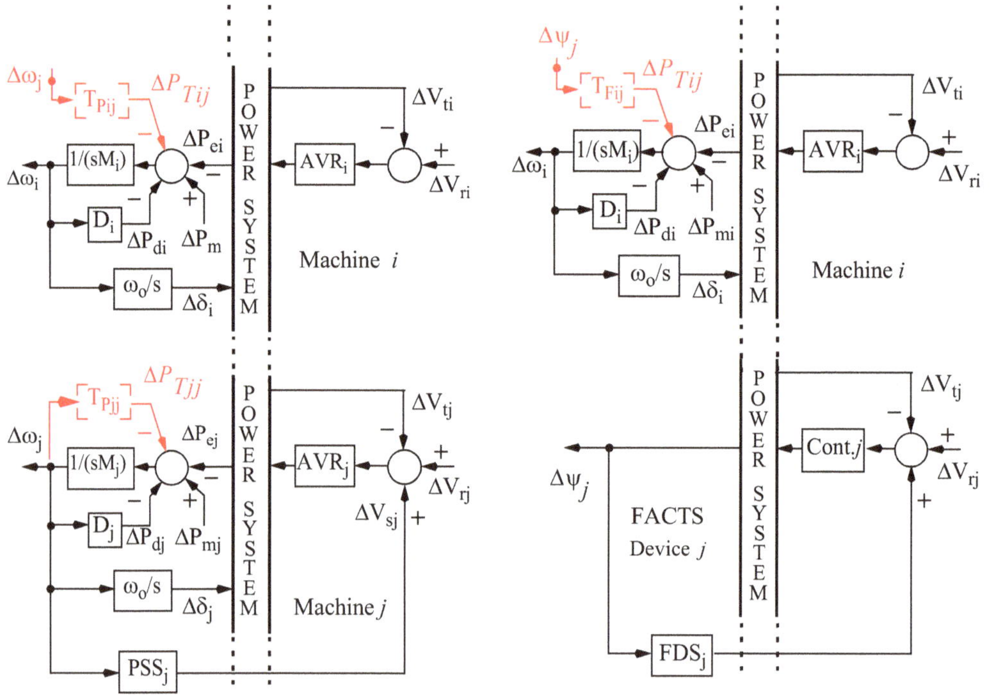

(a) Action of PSS j

(b) Action of FDS j

Figure 12.1 (a) Conceptually, at complex modal frequency $s = \lambda_h$, due to speed perturbation on machine j acting through its PSS, the MITCs T^h_{Pij} and T^h_{Pjj} for mode λ_h are induced on machines i and j respectively.
(b) Likewise, the electrical torque coefficient T^h_{Fij} is induced on machine i due to the perturbation in the local stabilizing signal $\Delta\psi_j$ acting through the FDS on FACTS device j.

Consider in Figure 12.1(a) the signal path from the speed perturbation $\Delta\omega_j$ through PSS j to the torque of electro-magnetic origin acting on the rotor of generator j. As has been discussed in Section 10.6, there will be in the latter torque a component ΔP_{Tjj} induced on the shaft of generator j by its own PSS. However, there is also a signal path from $\Delta\omega_j$ through PSS j to the electrical torque component ΔP_{Tij} on the rotor of generator i. *The component of* ΔP_{Tij} *on generator i in phase with the speed perturbation* $\Delta\omega_i$ *on generator i is a damping torque induced on that generator; this torque arises from the perturbation in* $\Delta\omega_j$ *and the action of PSS j* [3]. Consequently, for the complex modal frequency λ_h, a complex modal induced torque coefficient, T^h_{Pij}, can be defined which relates the change in a component of the torque of electro-magnetic origin ΔP_{Tij} on generator *i* to the change in speed $\Delta\omega_i$ on the same machine, the electrical torque being a result of the perturbation, say, of the input signal on stabilizer *j*, i.e.

$$\Delta P_{Tij}(\lambda_h) = T^h_{Pij}\Delta\omega_i(\lambda_h). \tag{12.2}$$

Using (12.1) the torque coefficient, T^h_{Pij}, may be expressed as follows:

$$T^h_{Pij} = \frac{\Delta P_{Tij}}{\Delta\omega_i}(\lambda_h) = \frac{\Delta P_{Tij}}{\Delta\omega_j}(s)\Bigg|^{pss_j}_{s=\lambda_h}\frac{\Delta\omega_j}{\Delta\omega_i}(\lambda_h) = \frac{\Delta P_{Tij}}{\Delta\omega_j}(s)\Bigg|^{pss_j}_{s=\lambda_h}\frac{v_{jh}}{v_{ih}}. \tag{12.3}$$

The term in (12.3), $\left[\Delta P_{Tij}/\Delta\omega_j\Big|^{pss_j}_{s=\lambda_h}\right]$, is the transfer function from the speed of machine *j*, $\Delta\omega_j$, through PSS *j* to the electrical torque component ΔP_{Tij}. The effect of the term, v_{jh}/v_{ih}, is to relate the perturbation in ΔP_{Tij} to a perturbation in speed $\Delta\omega_i$ on machine *i*, rather than machine *j*. Because, in (12.3), there are components of ΔP_{Tij} on machine *i* in phase and quadrature with $\Delta\omega_i$, the concept of damping and synchronising torques can be employed [4]. Conceptually, the complex modal induced torque coefficient, T^h_{Pij}, can be considered to be embedded in the linearized model of machine *i*, as shown in the dashed blocks of Figure 12.1(a).

An examination of the linearized model of Figure 12.1(a) reveals that a modal torque coefficient is also induced on machine *i* due to a speed perturbation on machine *j* acting *through the feedback path of the rotor angle* $\Delta\delta_j$ of machine *j*; this path is parallel to that of PSS *j*. Though not shown in Figure 12.1(a), this is an *inherent* torque coefficient. Since the inherent torque coefficient is associated with a path which is independent of the PSS feedback path, it is induced on the generators both in the absence and presence of the PSS. However, the emphasis in this and Chapter 13 is on the role of stabilizers in enhancing the damping performance of the multi-machine power system. Nevertheless, it is important to ensure that negative inherent modal torque coefficients are not significant enough to swamp out the positive torque coefficients induced by the PSS.

12.2.3 Modal torque coefficients induced by the action of a FACTS device stabilizer

Consider a FACTS device in service in a multi-machine power system. In Figure 12.1(b) are shown representative elements of a linearized model of the power system consisting of *n* generators and z FACTS devices and z FDSs. An examination of the figure reveals there is a signal path from the local stabilizing signal, $\Delta\psi_j$, through FDS j to the electric torque component ΔP_{Tij} on generator i.

In the analysis of the matrix transfer function representation of the multi-machine system in Section 12.3 a system output vector is defined in (12.10) as:

$$\Delta Y = [\Delta P_{e1}, \ldots, \Delta P_{en}, \Delta\psi_1, \ldots, \Delta\psi_z]^T = [\Delta y_1, \ldots, \Delta y_n, \Delta y_{n+1}, \ldots, \Delta y_{n+z}]^T, \quad (12.4)$$

where ΔP_{ei} is the electrical power output of generator i; the vector of the z FDS stabilizing signals is $\mathbf{\Delta\Psi} = [\Delta\psi_1, \ldots, \Delta\psi_z]^T$. As a result of a perturbation in the stabilizing signal $\Delta\psi_j$ a complex modal torque of electromagnetic origin is induced on the shaft of the generator i. The electric torque component ΔP_{Tij} induced on the shaft of machine i can be expressed as:

$$\Delta P_{Tij}(\lambda_h) = \left.\frac{\Delta P_{Tij}}{\Delta\psi_j}(\lambda_h)\right|_{fds_j} \Delta\psi_j(\lambda_h), \quad j = 1, z. \quad (12.5)$$

Again, of particular interest for generator i is the component of ΔP_{Tij} in phase with the speed on machine i. To determine the induced damping and synchronising torques, (12.5) must be modified such that the induced electrical torque component on machine i is related to its own speed perturbation, $\Delta\omega_i$. If a single mode of rotor oscillation, λ_h, is excited then the *relative magnitude and phase* of the two signals, $\Delta\omega_i$ and the FDS stabilizing signal, $\Delta\psi_j$, can be calculated using (12.69) in Appendix 12–I.1,

$$\frac{\Delta\psi_j}{\Delta\omega_i}(\lambda_h) = \frac{\Delta y_q}{\Delta\omega_i}(\lambda_h) = \frac{c_{q*}v_{*h}}{v_{ih}}, \text{ where} \quad (12.6)$$

- $q = (n+j)$ since $\Delta\psi_j$ is both the $(n+j)^{th}$ element of the output-vector $\mathbf{\Delta Y}$ and the jth of $\Delta\mathbf{\Psi}$;
- c_q is the qth row vector of the output state matrix $\mathbf{C}$;
- $\mathbf{v}_h$ is the right-eigenvector of the eigenvalue λ_h; and
- v_{ih} is the ith element of $\mathbf{v}_h$ corresponding to the speed-state of generator i.

For the purposes of analysis, however, it is again more convenient to define a complex induced torque *coefficient* rather than consider the induced torque itself. The modal torque coefficient induced on machine i due to the action of FDS j is T^h_{Fij} therefore defined as

$$\Delta P_{Tij}(\lambda_h) = T^h_{Fij}\Delta\omega_i(\lambda_h). \quad (12.7)$$

Using (12.6) an expression for the torque coefficient T^h_{Fij} is derived:

$$T^h_{Fij} = \frac{\Delta P_{Tij}}{\Delta\omega_i}(\lambda_h) = \frac{\Delta P_{Tij}}{\Delta\psi_j}(\lambda_h)\bigg|_{fds_j} \frac{\Delta\psi_j}{\Delta\omega_i}(\lambda_h) = \frac{\Delta P_{Tij}}{\Delta\psi_j}(\lambda_h)\bigg|_{fds_j} \frac{c_{q*}v_{*h}}{v_{ih}}. \tag{12.8}$$

The term, $\Delta P_{Tij}(\lambda_h)/\Delta\psi_j(\lambda_h)$, in (12.8) is the transfer function from the stabilizing signal $\Delta\psi_j$ through FDS *j* to the torque component ΔP_{Tij} on machine *i*, evaluated at the mode of interest. This gives the component of electrical torque induced on machine *i* due to perturbation in $\Delta\psi_j$. The second term, $c_{q*}v_{*h}/v_{ih}$, relates the perturbation in torque ΔP_{Tij} to the perturbation in speed $\Delta\omega_i$ on machine *i*. Because in (12.7) there are components of ΔP_{Tij} on machine *i* in phase and quadrature with $\Delta\omega_i$, the concept of damping and synchronizing torques can again be applied. As was the case in Figure 12.1(a), the complex MITC T^h_{ij} can also be considered, conceptually, to be embedded in the linearized model of machine *i*, as shown in Figure 12.1(b).

12.2.4 Modal torque coefficients induced by centralized stabilizers

In the following analysis it is assumed that a single centralized PSS accepts *n* speed input signals and supplies *n* output signals to the *n* generators. Similarly, a centralized FDS accepts *z* stabilizing signals and delivers *z* output signals at the *z* summing junctions of the FDSs. The assumption that the stabilizing signals of the centralized PSS are shaft speeds is made for convenience, but is not a necessary condition for the analysis; the theory of modal induced torque coefficients can be developed for an appropriate stabilizing signal.

In order to analyse the contributions to the MITCs on generators by the PSSs and FDSs in a multi-machine system a transfer function matrix model of a power system is first derived.

12.3 Transfer function matrix representation of a linearized multi-machine power system and its controllers

To calculate the modal induced torque coefficients it is necessary to identify the signal paths which cause torques of electromagnetic origin to be developed on the rotors of generators, and to be able to evaluate the associated components of torque. Consequently, in this section a transfer function matrix (TFM) model of a power system is derived which facilitates the calculation of the shaft torques that result from the action of PSSs and FACTS Device Stabilizers (FDSs) [1], [2]. Moreover, the associated state equations not only form the basis for the calculation of the MITCs, but also the design of stabilizers for FACTS devices, the coordination of FDSs with the PSSs, the interactions between stabilizers, and the development of the so-called 'stabilizer damping contribution diagrams'.

Let the number of states be *N*, the number of generators be *n* and the number of FACTS devices be *z*. The input vectors $\Delta \boldsymbol{V}_r$ and $\Delta \boldsymbol{U}_r$ represent voltage reference inputs to the generator AVR and the FACTS controllers, respectively. Similarly the vectors $\Delta \boldsymbol{P}_e$ and $\Delta \boldsymbol{\Psi}$ represent, respectively, the generator electrical power outputs and the local signals which act as inputs to FACTS device stabilizers. The vector of system states can be divided into three groups, $\Delta \boldsymbol{\omega}$, $\Delta \boldsymbol{\delta}$ and $\Delta \boldsymbol{q}$ which represent the generator speeds, the generator rotor-angles and the set of all other system states, respectively.

Consider the state-space model:

$$\Delta \dot{x} = \boldsymbol{A}\Delta \boldsymbol{x} + \boldsymbol{B}\Delta \boldsymbol{u}, \quad \Delta y = \boldsymbol{C}\Delta \boldsymbol{x} + \boldsymbol{D}\Delta \boldsymbol{u}, \tag{12.9}$$

where

$$\Delta \boldsymbol{x} = \begin{bmatrix}\Delta\omega \;\Delta\delta\; \Delta q\end{bmatrix}^T \quad (N \times 1);\; \Delta y = \begin{bmatrix}\Delta \boldsymbol{P}_e \;\Delta\Psi\end{bmatrix}^T \quad (n+z)\times 1,$$

$$\Delta \boldsymbol{u} = \begin{bmatrix}\Delta \boldsymbol{V}_r \;\Delta \boldsymbol{U}_r\end{bmatrix}^T \quad (n+z)\times 1; \tag{12.10}$$

$$\boldsymbol{A} = \begin{bmatrix} \boldsymbol{A}_{\omega\omega} & \boldsymbol{A}_{\omega\delta} & \boldsymbol{A}_{\omega q} \\ \omega_o \boldsymbol{I} & \boldsymbol{0} & \boldsymbol{0} \\ \boldsymbol{0} & \boldsymbol{A}_{q\delta} & \boldsymbol{A}_{qq} \end{bmatrix} \quad (N\times N);\; \boldsymbol{A}_{\omega\omega} = diag\{-K_{di}/M_i\} \quad (n\times n); \tag{12.11}$$

$$\boldsymbol{B} = \begin{bmatrix} \boldsymbol{0} & \boldsymbol{0} & \boldsymbol{B}_{qVr} \\ \boldsymbol{0} & \boldsymbol{0} & \boldsymbol{B}_{qUr} \end{bmatrix}^T \quad N\times(n+z); \tag{12.12}$$

$$\boldsymbol{C} = \begin{bmatrix} \boldsymbol{0} & \boldsymbol{C}_{P\delta} & \boldsymbol{C}_{Pq} \\ \boldsymbol{0} & \boldsymbol{C}_{\Psi\delta} & \boldsymbol{C}_{\Psi q} \end{bmatrix} \quad (n+z)\times N; \tag{12.13}$$

$$\boldsymbol{D} = \boldsymbol{0} \quad (n+z)\times(n+z). \tag{12.14}$$

Note that it is assumed that the system matrix, ***D***, is a zero matrix; the physical interpretation of which is that there is no instantaneous relationship between system outputs and system inputs. This is a valid assumption in the context of power system models. Furthermore, assume that in the above state-space model:

(i) there are no PSSs fitted to the generators,
(ii) governors and turbines are not modelled, and
(iii) there are no stabilizers fitted to the FACTS devices in the system.

In order to disable the dynamics of all generator shafts, let us (i) temporarily remove the state equations describing the shaft dynamics from (12.9), and (ii) treat $\Delta\boldsymbol{\delta}$ as an input vector. As explained in Sections 5.10.2 and 5.10.3 this is similarly achieved by eliminating the rows associated with $\Delta\boldsymbol{\omega}$ in matrices ***A***, ***B*** and ***C***, and transferring the columns associated with the speed states in these matrices to expanded ***B*** and ***D*** matrices to form a new set of state equations. Therefore, (12.9) reduces to

$$\Delta\dot{q} = A_{qq}\Delta q + \begin{bmatrix} B_{qVr} & B_{qUr} & A_{q\delta} \end{bmatrix} \begin{bmatrix} \Delta V_r \\ \Delta U_r \\ \Delta\delta \end{bmatrix}, \text{ and} \tag{12.15}$$

$$\Delta y = \begin{bmatrix} C_{Pq} \\ C_{\Psi q} \end{bmatrix} \Delta q + \begin{bmatrix} 0 & 0 & C_{P\delta} \\ 0 & 0 & C_{\Psi\delta} \end{bmatrix} \begin{bmatrix} \Delta V_r \\ \Delta U_r \\ \Delta\delta \end{bmatrix}. \tag{12.16}$$

Using (12.15) and (12.16) we can write a TFM equation which relates the vector of system outputs, Δy in (12.10), to the vector of system inputs, $\begin{bmatrix} \Delta V_r & \Delta U_r & \Delta\delta \end{bmatrix}^T$, with shaft dynamics disabled. That is,

$$\begin{bmatrix} \Delta P_e(s) \\ \Delta\Psi(s) \end{bmatrix} = \begin{bmatrix} H_{PVr}(s) & H_{PUr}(s) & H_{P\delta}(s) \\ H_{\Psi Vr}(s) & H_{\Psi Ur}(s) & H_{\Psi\delta}(s) \end{bmatrix} \begin{bmatrix} \Delta V_r \\ \Delta U_r \\ \Delta\delta \end{bmatrix}, \tag{12.17}$$

where

$$H_{PVr}(s) = C_{Pq}(sI - A_{qq})^{-1} B_{qVr}; \tag{12.18}$$

$$H_{PUr}(s) = C_{Pq}(sI - A_{qq})^{-1} B_{qUr}; \tag{12.19}$$

$$H_{P\delta}(s) = C_{Pq}(sI - A_{qq})^{-1} A_{q\delta} + C_{P\delta}; \tag{12.20}$$

$$H_{\Psi Vr}(s) = C_{\Psi q}(sI - A_{qq})^{-1} B_{qVr}; \tag{12.21}$$

$$H_{\Psi Ur}(s) = C_{\Psi q}(sI - A_{qq})^{-1} B_{qUr}; \tag{12.22}$$

$$H_{\Psi\delta}(s) = C_{\Psi q}(sI - A_{qq})^{-1} A_{q\delta} + C_{\Psi\delta}. \tag{12.23}$$

The above equations represent the TFMs *from* perturbations (i) in generator reference voltages ΔV_r, (ii) in the FACTS device reference inputs ΔU_r, and (iii) in generator rotor-angles $\Delta\delta$, *to* perturbations both in generator electric power outputs ΔP_e and in stabilizing signals $\Delta\Psi$ of the local FACTS devices. Details of the calculation of the matrices $C_{P\delta}$ and C_{Pq} are supplied in Appendix 12–I.3.

The shaft dynamics equations, which were temporarily eliminated from (12.9) in forming (12.15), can be expressed as

$$\Delta\omega(s) = J(s)[\Delta P_m(s) - \Delta P_e(s)], \qquad \Delta\delta(s) = N(s)\Delta\omega(s), \tag{12.24}$$

where

$$\left.\begin{aligned} \boldsymbol{J}(s) &= diag\{1/(M_i s + K_{di})\} \quad \forall i \in [1, n] \\ \boldsymbol{N}(s) &= diag\{\omega_0/s\} \quad (n \times n) \end{aligned}\right\}. \tag{12.25}$$

The generator shaft dynamics represented by (12.25), and the rest of the system dynamics represented by equations (12.18) to (12.23), are combined with the TFM representing the PSSs, FDSs and governors to form a TFM model of the power system. This model is shown in Figure 12.2.

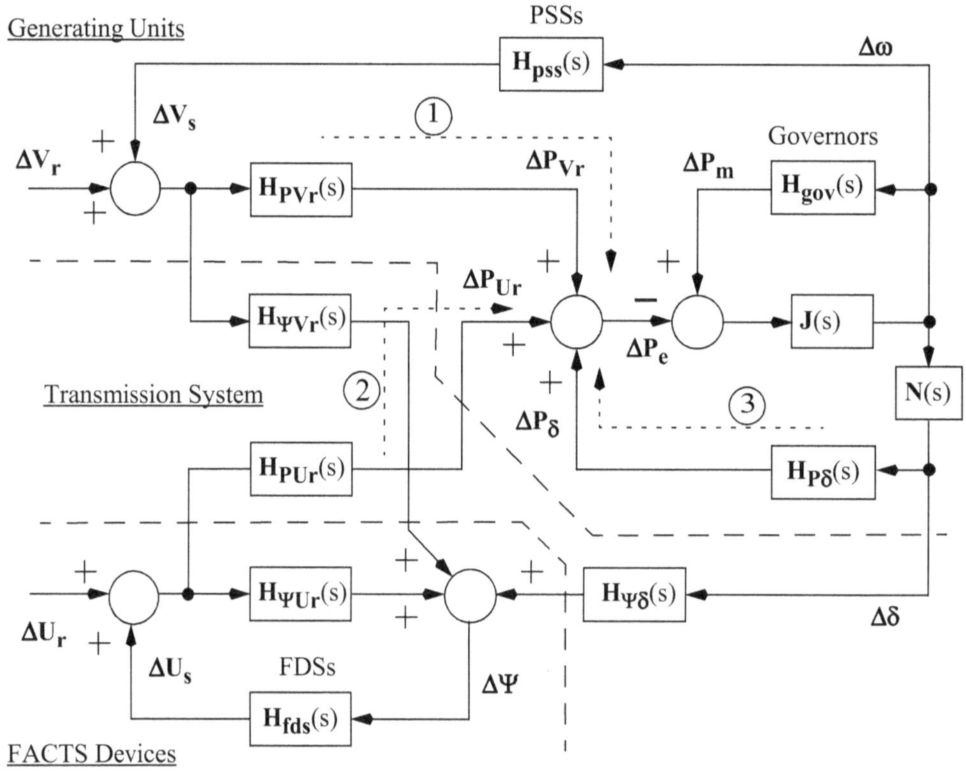

Figure 12.2 Transfer function matrix model showing three paths through which components of electrical modal torques are induced on the generator rotors.

The significance of each TFM block in the figure is examined below.

$\boldsymbol{H}_{pss}(s)$: is initially a TFM of a centralized speed PSS and is a full $n \times n$ matrix. In later sections, because in practice each generator is fitted with a single, decentralized PSS, $\boldsymbol{H}_{pss}(s)$ becomes a diagonal matrix.

$\boldsymbol{H}_{fds}(s)$: is initially a TFM of a centralized FACTS device stabilizer and is assumed to be a full $z \times z$ matrix. The inputs are the local stabilizing signals given by the vector $\Delta\boldsymbol{\Psi}$, $z \times 1$. The outputs of the centralized FDS, $\Delta\boldsymbol{U}_s$, are inputs to the summing junctions of the FACTS device controllers. Again in later sections, $\boldsymbol{H}_{fds}(s)$ becomes a diagonal matrix representing the transfer functions of decentralized FDSs.

$\boldsymbol{H}_{gov}(s)$ is a TFM of the governors; it is a $n \times 1$ matrix.

$\boldsymbol{H}_{PVr}(s)$: is a full $n \times n$ TFM which relates the perturbations in torque contributions of electromagnetic origin on generators, $\Delta\boldsymbol{P}_{Vr}(s)$, $n \times 1$, due to perturbations in the reference inputs, $\Delta\boldsymbol{V}_r$, $n \times 1$, on the AVRs when $\Delta\boldsymbol{V}_s = 0$. Of particular significance is that the diagonal element, $H_{PVr\_ii}(s)$, is the P-Vr transfer function of machine *i*.

$\boldsymbol{H}_{PUr}(s)$: is a full $n \times z$ TFM which relates the perturbations in torque contributions of electromagnetic origin on generators, $\Delta\boldsymbol{P}_{Ur}$ $n \times 1$, due to perturbations in the reference signals on the FACTS device controllers, $\Delta\boldsymbol{U}_r$, $z \times 1$, when $\Delta\boldsymbol{U}_s = 0$.

$\boldsymbol{J}(s)$: is a $n \times n$ diagonal TFM which represents the shaft dynamics of the units.

$\boldsymbol{N}(s)$: is a $n \times n$ TFM of diagonal elements ω_0/s; it relates the perturbations in rotor angle due to speed to perturbations on the units.

$\boldsymbol{H}_{P\delta}(s)$: is a full $n \times n$ TFM which relates the perturbations in torque contributions of electromagnetic origin, $\Delta\boldsymbol{P}_{\delta}$, due to perturbations in the rotor angles of machines. It is through this TFM that the inherent torques are produced.

$\boldsymbol{H}_{\psi\delta}(s)$, $\boldsymbol{H}_{\psi Vr}(s)$ and $\boldsymbol{H}_{\psi Ur}(s)$: are full TFMs which relate the perturbations in the stabilizing signals, $\Delta\boldsymbol{\psi}$, due to perturbations in the rotor angle ($\Delta\boldsymbol{\delta}$), the machine reference voltage, ($\Delta\boldsymbol{V}_r$), and the FACTS controller reference signals, ($\Delta\boldsymbol{U}_r$), respectively, when $\Delta\boldsymbol{V}_s$ and $\Delta\boldsymbol{U}_s = 0$.

In Figure 12.2 three distinct paths are shown through which components of electrical torque are induced on the shafts of generators. The first path, #1, is through the speed-PSS feedback path, the second, #2, is through the FDS feedback path and the third, #3, is via the rotor-angle feedback path. From Figure 12.2 it is revealed that the torque induced by the third path is given by

$$\Delta\boldsymbol{P}_{\delta}(s) = \boldsymbol{H}_{P\delta}(s)\boldsymbol{N}(s)\Delta\boldsymbol{\omega}(s). \tag{12.26}$$

Substitution of (12.20) and (12.25) for $H_{P\delta}(s)$ and $N(s)$, respectively, into (12.26) yields

$$\begin{aligned}\Delta P_{\delta}(s) &= (C_{P\omega}(sI-A_{\omega\omega})^{-1}A_{\omega\delta}+C_{P\delta})\cdot \mathrm{diag}(\omega_0/s)\cdot \Delta\omega(s)\\ &= (C_{P\omega}(sI-A_{\omega\omega})^{-1}A_{\omega\delta}+C_{P\delta})\Delta\delta(s)\end{aligned} \qquad (12.27)$$

The term $C_{P\delta}$ is a matrix of real, constant, inherent synchronising torque coefficients (compare this with the coefficient K_1 in the SMIB case of Figure 5.1). The term $C_{P\omega}(sI-A_{\omega\omega})^{-1}A_{\omega\delta}+C_{P\delta}$ represents the inherent feedback paths from $\Delta\delta$ through the generator electromagnetic circuits, the network and the AVR/exciter to the electrical torque induced on the rotors of the generators. Because the term is complex at any frequency, its real and imaginary components represent the inherent damping and synchronising torque coefficients on each machine.

Following a large-magnitude disturbance, high gain AVRs tend to increase synchronising power flows and thus enhance first swing stability [6]; however, high-gain AVRs have a tendency to reduce damping torques [7]. As has been discussed in earlier chapters, the objective of PSS and FDS design in small-signal analysis is to induce positive damping torques on generator rotors for all modal frequencies of rotor oscillation. In order to achieve a constant damping torque coefficient over the range of rotor modes of oscillation, the frequency response (for $s = j\omega_f$) of the coefficient induced by the PSSs should ideally be flat with negligible phase shift and must swamp the negative inherent damping torque coefficients (see Section 10.6). Furthermore, the PSS or FDS should not significantly reduce the inherent synchronising torque coefficients. Paths #1 and #2 in Figure 12.2 will therefore be analysed in more detail to determine quantitatively the modal torque coefficients induced on each generator by the PSSs and FDSs.

12.4 Modal torque coefficients induced by a centralized speed PSS

The analysis in the previous section is formulated on a centralized stabilizers. The modal induced torque coefficients for the centralized PSS are derived below. The results for the decentralized PSS are then developed in Section 12.6.

Based on Figure 12.2 it is seen that the total electrical torques induced on the rotors of generators are the sum of the torque components associated with paths #1 to #3 and are expressed by the vector

$$\Delta P_e(s) = \Delta P_{Vr}(s)+\Delta P_{Ur}(s)+\Delta P_{\delta}(s), \quad \text{where} \qquad (12.28)$$

$\Delta P_{Vr}(s)$ - is the vector of contributions to the electrical torques resulting from rotor speed perturbations being fed back through the centralized PSS, path #1;
$\Delta P_{Ur}(s)$ - is the vector of contributions resulting from the perturbations in the local stabilizing signals for FACTS devices being fed back through the centralized FDS, path #2;
$\Delta P_{\delta}(s)$ - is the vector of contributions to the inherent electrical torques, path #3.

Note the effects of speed-governors are ignored here, but they can be included in the TFM $\boldsymbol{G}_m(s)$, as shown in Appendix 12–I.2, without affecting the following analysis.

Let us consider the contribution made by the TFM of the centralized speed-PSS to electrical torques on the generators. It is shown in Appendix 12–I.2 that the block diagram in Figure 12.2 can be reduced to that illustrated in Figure 12.3; the latter shows only the loop associated with the centralized PSS feedback path. The rest of the system dynamics, including that of the centralized FDSs, etc., has been incorporated in the TFM $\boldsymbol{G}_{\boldsymbol{p}}(s)$.

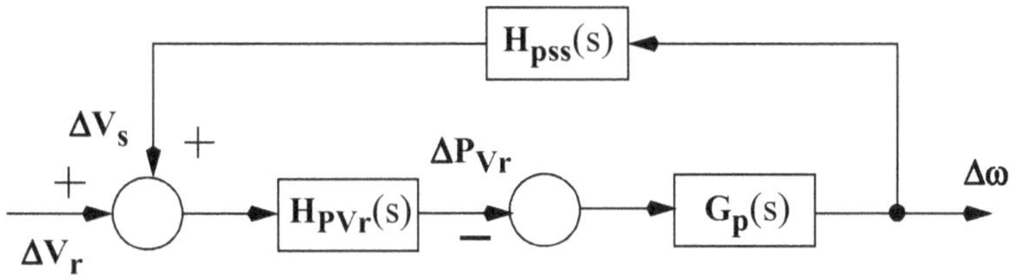

Figure 12.3 TFM representation of a linearized multi-machine system showing the loop associated with PSS feedback paths. All other dynamics are included in $\mathbf{G_p}$(s).

From Figure 12.3, the vector of contributions of electrical torque, $\Delta\boldsymbol{P}_{Vr}$, resulting from generator speed perturbations, $\Delta\boldsymbol{\omega}$, fed to the centralized speed PSS $\boldsymbol{H}_{pss}(s)$ is

$$\Delta\boldsymbol{P}_{Vr}(s) = \boldsymbol{H}_{PVr}(s)\boldsymbol{H}_{pss}(s)\Delta\boldsymbol{\omega}(s) = \boldsymbol{\tau}_{\boldsymbol{P}}(s)\Delta\boldsymbol{\omega}(s), \tag{12.29}$$

where $\boldsymbol{\tau}_{\boldsymbol{P}}(s) = \boldsymbol{H}_{PVr}(s)\boldsymbol{H}_{pss}(s)$ is a $n \times n$ TFM which relates speed perturbations on any machine to the PSS-induced torque perturbations on any machine. Based on (12.29), the total PSS-induced torque on the rotor of generator i due to the speed perturbations on all n generators is given by

$$\Delta P_{Vr\_i}(s) = \sum_{j=1}^{n} \Delta P_{Vr\_ij}(s) = \sum_{j=1}^{n} \{\tau_{Pij}(s)\Delta\omega_j(s)\}, \tag{12.30}$$

where the elements of the speed-torque TFM $\boldsymbol{\tau}_{\boldsymbol{P}}(s)$ are given by

$$\tau_{Pij}(s) = \sum_{l=1}^{n} \{H_{PVr\_il}(s)H_{pss\_lj}(s)\}. \tag{12.31}$$

The jth element of the summation in (12.30), ΔP_{Vr_ij}, which is the component of the PSS-induced torque on machine i due to speed perturbation on machine j, is

$$\Delta P_{Vr\_ij}(s) = \tau_{Pij}(s)\Delta\omega_j(s). \tag{12.32}$$

The above equation relates to torques induced on generator i by speed perturbations on i and all other machines. However, to calculate the synchronizing and damping torques induced on generator i we need to explicitly determine torques in-quadrature and in-phase with the speed perturbations on generator i, caused by speed perturbations on all generators.

Bearing in mind the comments in Section 12.2.1, consider now a *complex mode of rotor oscillation,* $s = \lambda_h$. The torque ΔP_{Vr_ij} can therefore be defined in terms of the modal induced torque coefficient T^h_{Pij} in (12.2), i.e.

$$\Delta P_{Vr\_ij}(\lambda_h) = T^h_{Pij}\,\Delta\omega_i(\lambda_h)\,, \tag{12.33}$$

where the MITC is given in (12.3) by

$$T^h_{Pij} = \frac{\Delta P_{Vr\_ij}}{\Delta\omega_i}(\lambda_h) = \frac{\Delta P_{Vr\_ij}}{\Delta\omega_j}(s)\Bigg|^{pss_j}_{s\,=\,\lambda_h}\frac{v_{jh}}{v_{ih}}\,. \tag{12.34}$$

Substitution in (12.34) of $\tau_{Pij}(s)$ for $\Delta P_{Vr\_ij}(s)/\Delta\omega_j(s)$ from (12.32) thus yields

$$T^h_{Pij} = \tau_{Pij}(\lambda_h)\frac{v_{jh}}{v_{ih}}\,. \tag{12.35}$$

This equation provides an expression for calculating the MITC on machine *i* due to speed perturbations on machine *j*, at the single modal frequency λ_h. Equation (12.35) is substituted into (12.30) to give,

$$\Delta P_{Vr\_i}(\lambda_h) = \sum_{j=1}^{n}\left\{\tau_{Pij}(\lambda_h)\frac{v_{jh}}{v_{ih}}\Delta\omega_i(\lambda_h)\right\} = \left(\sum_{j=1}^{n}T^h_{Pij}\right)\Delta\omega_i(\lambda_h)\,. \tag{12.36}$$

The term $\sum_{j=1}^{n} T^h_{Pij}$ is the total MITC on machine *i* due to perturbations in the speed of all *n* generators being fed back through the PSS TFM, $\boldsymbol{H}_{pss}(s)$. The total MITC on machine *i* is defined as

$$\Gamma^h_{Pi} = \sum_{j=1}^{n} T^h_{Pij}\,. \tag{12.37}$$

Using this definition, (12.36) becomes

$$\Delta P_{Vr\_i}(\lambda_h) = \Gamma^h_{Pi}\Delta\omega_i(\lambda_h)\,. \tag{12.38}$$

Hence, considering all *n* machines in the system, (12.29) can be written TFM form as

$$\Delta\boldsymbol{P}_{Vr}(\lambda_h) = \boldsymbol{\Gamma_P}^h\Delta\boldsymbol{\omega}(\lambda_h) \tag{12.39}$$

at the modal frequency λ_h, where the matrix $\boldsymbol{\Gamma_P}^h = \text{diag}[\Gamma^h_{Pi}],\ i \in [1,n]$.

At this point it is important to reiterate the physical significance of the MITCs. Because the real part of T^h_{Pij} represents a damping torque coefficient, the *total* damping torque coefficient induced on generator *i* by the action of *all* stabilizers is $\Re(\Gamma^h_i)$.

In summary, for a centralized speed-PSS whose elements $H_{pss\_lj}(s)$ are known, the total induced torque coefficient Γ^h_{Pi} for generator i and mode λ_h can be calculated by successively evaluating (12.31), (12.35) and (12.37).

12.5 Modal torque coefficients induced by a centralized FDS

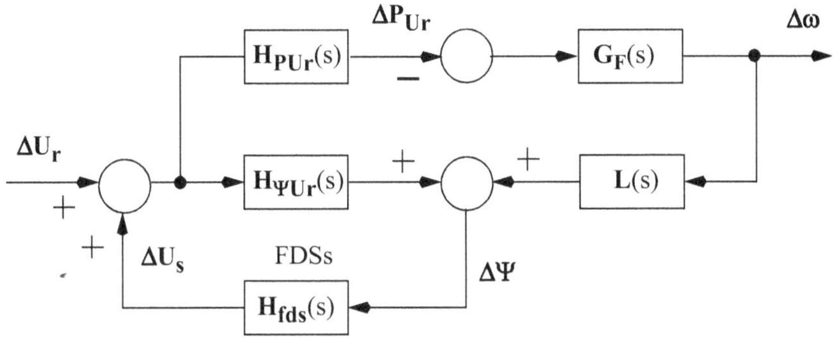

Figure 12.4 TFM representation of a linearized multi-machine system showing the loop associated with FDS feedback paths; other dynamics are included in TFMs $\mathbf{G_F}$(s) and **L**(s).

Let us consider now the contribution made by the TFM of a centralized FDS to the modal electrical torque induced on the generators, $\Delta \boldsymbol{P}_{Ur}$. It is shown in Appendix 12–I.2 that the transfer function block diagram in Figure 12.2 can be simplified to that illustrated in Figure 12.4; the latter shows only the loops associated with the centralized FDS feedback path. All other system dynamics and feedback paths, such as the PSSs, have been absorbed into the TFMs $\boldsymbol{G}_F(s)$ and $\boldsymbol{L}(s)$. This figure shows that there is a closed path from $\Delta\boldsymbol{\omega}$, the vector of generator speeds, to $\Delta \boldsymbol{P}_{Ur}$, through the TFM of the FDSs. Therefore, associated with speed perturbations there will be components of torque in phase and quadrature with speed induced on the shaft of each machine through the feedback path of the FDSs. Based on Figure 12.4 the stabilizing signals supplied to the FDSs are given by the vector

$$\Delta\boldsymbol{\Psi}(s) = \boldsymbol{L}(s)\Delta\boldsymbol{\omega}(s) + \boldsymbol{H}_{\Psi Ur}(s)[\Delta\boldsymbol{U}_r(s) + \boldsymbol{H}_{fds}(s)\Delta\boldsymbol{\Psi}(s)], \qquad (12.40)$$

where $\boldsymbol{H}_{fds}(s)$ is the centralized FDS TFM defined in Section 12.2.4. The components of torque induced through the feedback path of the FDSs are

$$\Delta\boldsymbol{P}_{Ur}(s) = \boldsymbol{H}_{PUr}(s)\boldsymbol{H}_{fds}(s)\Delta\boldsymbol{\Psi}(s) + \boldsymbol{H}_{PUr}(s)\Delta\boldsymbol{U}_r(s). \qquad (12.41)$$

The vector $\Delta\boldsymbol{U}_r$ is the vector of reference set points at the summing junction of the FACTS device controllers (see Figure 12.2). Let us assume that there is no perturbation in these signals, i.e. $\Delta\boldsymbol{U}_r = 0$. Therefore, (12.40) and (12.41) become

$$\Delta\boldsymbol{\Psi}(s) = \boldsymbol{L}(s)\Delta\boldsymbol{\omega}(s) + \boldsymbol{H}_{\Psi Ur}(s)\boldsymbol{H}_{fds}(s)\Delta\boldsymbol{\Psi}(s), \qquad (12.42)$$

$$\begin{aligned}\Delta \boldsymbol{P}_{Ur}(s) &= \boldsymbol{H}_{PUr}(s)\boldsymbol{H}_{fds}(s)\Delta\boldsymbol{\Psi}(s)\\ &= \boldsymbol{\tau}_{fds}\Delta\boldsymbol{\Psi}(s)\end{aligned} \tag{12.43}$$

Solving (12.42) for $\Delta\boldsymbol{\Psi}(s)$ and substituting this into (12.43) we find

$$\begin{aligned}\Delta \boldsymbol{P}_{Ur}(s) &= \Big(\underbrace{\boldsymbol{H}_{PUr}(s)\boldsymbol{H}_{fds}(s)[\boldsymbol{I} - \boldsymbol{H}_{\Psi Ur}(s)\boldsymbol{H}_{fds}(s)]^{-1}\boldsymbol{L}(s)}_{(n\times n)}\Big)\Delta\boldsymbol{\omega}(s)\\ &= \boldsymbol{\tau}_{F}(s)\Delta\boldsymbol{\omega}(s)\end{aligned} . \tag{12.44}$$

AN ASIDE: As in the case of the centralized speed-PSS (12.30), the effective action of the transfer function matrix of the centralized FDS, $\boldsymbol{H}_{fds}(s)$ in Figure 12.4, can be replaced by the equivalent $n \times n$ speed-torque TFM $\boldsymbol{\tau}_F(s)$ of (12.44).

If the power system model used here represented a SMIB system, with a single FACTS device in the system, evaluating $\tau_F(s)$ would be relatively easy; $\tau_F(s)$ would represent the transfer function between the generator's speed and electrical torque, due to the action of the FDS. Therefore, for this case, a FDS could be designed based on damping torque concepts as for PSSs. That is, based on (12.44), if the FDS transfer function were designed such that

$$\frac{\Delta P_{Ur}}{\Delta\omega} = H_{PUr}(s)H_{fds}(s)[1 - H_{\psi Ur}(s)H_{fds}(s)]^{-1}L(s) = k_{fds}, \tag{12.45}$$

then ideally a real damping torque coefficient, equal to k_{fds}, would be induced on the shaft of the generator. By solving (12.45) for $H_{fds}(s)$ we find the FDS transfer function to be

$$H_{fds}(s) = \frac{k_{fds}}{H_{PUr}(s)L(s) + k_{fds}H_{\Psi Ur}(s)} . \tag{12.46}$$

This is equivalent to the design procedure, based on a frequency response approach, derived in [8].

In the case a multi-machine system evaluating the TFM $\boldsymbol{\tau}_F(s)$ in (12.44) is not a trivial task, especially since $\boldsymbol{H}_{\Psi Ur}(s)$ is not a diagonal matrix [9]. Therefore, no attempt will be made here to evaluate $\boldsymbol{\tau}_F(s)$; moreover (12.46) cannot be extended to the multi-machine case in an approach analogous to that for PSSs in Section 12.4.

In contrast to the tuning of PSSs, which is described in Chapters 5 and 10 and which is based on frequency response methods, a different approach to the tuning of FDSs in a multi-machine system is adopted here. First an expression will be derived which relates the perturbation in the electrical torque induced on the shaft of generator i to the perturbations in the signal $\Delta\psi_j$. Then, as in section Section 12.4, an expression will be derived for this component of torque at a selected mode of rotor oscillation, λ_h.

Based on (12.43), the torque induced on generator i due to the action of the z FDSs is

$$\Delta P_{Ur\_i}(s) = \sum_{j=1}^{z} \Delta P_{Ur\_ij}(s) = \sum_{j=1}^{z} \{\tau_{fds\_ij}(s)\Delta\psi_j(s)\}, \tag{12.47}$$

where, from (12.43), the elements of the speed-torque TFM $\boldsymbol{\tau}_{fds}(s)$ are given by

$$\tau_{fds\_ij} = \sum_{l=1}^{z} \{H_{PUr\_il}(s)H_{fds\_lj}(s)\}. \tag{12.48}$$

Let us consider the jth element of the summation in (12.47), i.e.

$$\Delta P_{Ur\_ij}(s) = \tau_{fds\_ij}(s)\Delta\psi_j(s). \tag{12.49}$$

Equation (12.49) relates to torques induced on generator i by stabilizing signal perturbations on FDS j. However, to calculate the synchronizing and damping torques induced on generator i we need to explicitly determine torques on generator i in quadrature and in-phase with the speed perturbations on generator i, caused by stabilizing signal perturbations on all FDSs.

At a selected rotor mode of oscillation, $s = \lambda_h$, the torque of electromagnetic origin ΔP_{Ur_ij} can also be defined in terms of the MITC T^h_{Fij}, as in (12.33),

$$\Delta P_{Ur\_ij}(\lambda_h) = T^h_{Fij}\,\Delta\omega_i(\lambda_h). \tag{12.50}$$

Referring to (12.69) in Appendix 12–I.1, the MITC is given by (12.8) becomes

$$T^h_{Fij} = \frac{\Delta P_{Ur\_ij}}{\Delta\omega_i}(s)\Bigg|_{s=\lambda_h}^{fds_j} = \frac{\Delta P_{Ur\_ij}}{\Delta\psi_j}(s)\Bigg|_{s=\lambda_h}^{fds_j}\frac{\Delta\psi_j}{\Delta\omega_i}(\lambda_h) = \frac{\Delta P_{Ur\_ij}}{\Delta\psi_j}(s)\Bigg|_{s=\lambda_h}^{fds_j}\frac{\boldsymbol{c}_{q*}\boldsymbol{u}_{*h}}{u_{ih}} \tag{12.51}$$

where $q = (n+j)$ and $\Delta\psi_j$ is the jth element of the vector $\Delta\boldsymbol{\Psi}$ and the $(n+j)^{th}$ element of the system output vector $\Delta\boldsymbol{Y}$ (see (12.4)).

Following substitution of the transfer function $\tau_{fds\_ij}(\lambda_h) = \Delta P_{Ur\_ij}(\lambda_h)/\Delta\psi_j(\lambda_h)$ from (12.49) into (12.51) the MITC on machine i associated with the signal path through FDS j becomes

$$T^h_{Fij} = \tau_{fds\_ij}(\lambda_h)(\boldsymbol{c}_{q*}\boldsymbol{u}_{*h}/u_{ih}). \tag{12.52}$$

We can also define the total MITC on machine i, T^h_{Fij}, associated with signal paths through all z FDSs, i.e.

$$\Gamma_{Fi}^{h} = \sum_{j=1}^{z} T_{Fij}^{h}. \qquad (12.53)$$

Consequently, at the single modal frequency λ_h, (12.49) can be written as

$$\Delta P_{Ur\_i}(\lambda_h) = \Gamma_{Fi}^{h} \Delta\omega_i(\lambda_h). \qquad (12.54)$$

Once again, an examination of (12.50) and (12.54) reveals that the real parts of both T_{Fij}^{h} and Γ_{Fi}^{h} represent damping torque coefficients on the ith generator.

In summary, for a centralized FDS whose elements $H_{fds\_lj}(\lambda_h)$ are known, the total MITC for generator Γ_{Fi}^{h} can be calculated by successively evaluating (12.48), (12.52) and (12.53) for the mode λ_h.

12.6 General expressions for the torque coefficients induced by conventional, decentralized PSSs & FDSs

In Sections 12.4 and 12.5 centralized speed-PSSs and FDSs are assumed. However, for a decentralized PSS, the stabilizing signal on machine i is derived only from the speed signal of machine i. Consequently, the TFM $\boldsymbol{H}_{pss}(s)$ is a diagonal matrix, i.e. $H_{pss\_lj} = 0,\ l \neq j$, and thus (12.31) becomes

$$\tau_{Pij}(s) = \sum_{l=1}^{n} \{H_{PVr\_il}(s) H_{pss\_lj}(s)\} = H_{PVr\_ij}(s) H_{pss\_jj}(s). \qquad (12.55)$$

An expression for the modal torque coefficient induced on machine i due to speed perturbation, $\Delta\omega_j$, on machine j acting through PSS j is derived by substituting (12.55) into (12.35), i.e.

$$T_{Pij}^{h} = H_{PVr\_ij}(\lambda_h) H_{pss\_j}(\lambda_h) \frac{v_{jh}}{v_{ih}}, \qquad (12.56)$$

where $H_{pss\_jj}(\lambda_h)$ is simply written as $H_{pss\_j}(\lambda_h)$. The total MITC for generator i, Γ_{Pi}^{h}, is given by (12.37).

Similarly, decentralized, practical FDSs are designed for the various FACTS devices. That is, for each FACTS device a local stabilizing signal is selected as an input to its FDS and thus the TFM $\boldsymbol{H}_{fds}(s)$ becomes a diagonal matrix. Consequently, (12.48) becomes

$$\tau_{fds\_ij}(s) = \sum_{l=1}^{z} \{H_{PUr\_il}(s) H_{fds\_lj}(s)\} = H_{PUr\_ij}(s) H_{fds\_jj}(s). \qquad (12.57)$$

Substituting (12.57) into (12.52) we find the modal torque coefficient induced on machine i due to perturbations in the local stabilizing signal of FDS j is

$$T_{Fij}^{h} = H_{PUr\_ij}(\lambda_h)H_{fds\_j}(\lambda_h)\cdot(c_{q*}v_{*h}/v_{ih}), \tag{12.58}$$

where $H_{fds\_j}(\lambda_h)$ is the jth element of the diagonal TFM $\boldsymbol{H}_{fds}(\lambda_h)$.

Let us assume that the transfer function of any decentralized stabilizer is $k_jG_j(s)$. This is the transfer function between the qth system output (which represents a local stabilizing signal) and the jth system input (the summing junction of either a generator or FACTS device). Based on (12.56) and (12.58), the following generalised form which applies to both PSSs and FDSs is proposed for the MITC:

$$T_{ij}^{h} = H_{ij}(\lambda_h)G_j(\lambda_h)\cdot(c_{q*}v_{*h}/v_{ih})\cdot k_j. \tag{12.59}$$

Note that for PSSs $H_{ij}(\lambda_h) = H_{PVr\_ij}(\lambda_h)$; for FACTS devices $H_{ij}(\lambda_h) = H_{PUr\_ij}(\lambda_h)$. The transfer function $H_{ij}(\lambda_h)$ is evaluated at the modal frequency λ_h *with all machine shaft dynamics disabled.* To calculate the TFMs $\boldsymbol{H}_{PVr}(\lambda_h)$ and $\boldsymbol{H}_{PUr}(\lambda_h)$ at a number of modal frequencies, (12.18) and (12.19) are employed, e.g.

$$\boldsymbol{H}_{PVr}(s) = \boldsymbol{C}_{Pq}(s\boldsymbol{I}-\boldsymbol{A}_{qq})^{-1}\boldsymbol{B}_{qVr}, \qquad \text{(12.18) repeated.}$$

Following the calculation of the eigenvalues and eigenvectors of $\boldsymbol{A}_{qq}$ the TFM is decomposed into a form which allows it to be evaluated easily at the required modal frequencies.

12.6.0.1 *Relationship between: MITC T_{ii}^{h} and PSS damping gain k_i*

Any appropriate stabilizing signal can be used for a PSS. Note that a speed-input PSS is a special case because the speed of a machine is both a state and an output of the system. Therefore if a speed-PSS is installed on the lth generator then the lth row of the output matrix of the state equations (12.9) is $c_{l\circ} = [0\ 0 \ldots 1 \ldots 0\ 0]$; hence the term $\dfrac{c_{l\circ}v_{\circ h}}{v_{ih}}$ in (12.59) reduces to $\dfrac{v_{ih}}{v_{ih}} = 1$. Furthermore, it is of interest to note from (12.59) that the MITC induced on generator i due to its own PSS is

$$T_{i}^{h} = H_{PVr\_ii}(\lambda_h)G_i(\lambda_h)\cdot\frac{v_{ih}}{v_{ih}}\cdot k_i \approx k_i, \tag{12.60}$$

since for a well-tuned PSS $G_i(s) \approx 1/H_{PVr\_ii}(s)$ [1]. That the MITC is equal to the nominal damping gain of PSS i is consistent with the design objectives for PSSs discussed in Chapter 5 and summarized in Section 5.14.

---

1. As in Chapter 5 this assumes that for the mode of interest $s = \lambda_h \approx j\omega_f$.

12.6.1 The total modal induced torque coefficients for systems with both PSSs and FDSs

The total induced torque coefficient on generator *i* due to PSSs and FDSs is given in (12.37) and (12.53), respectively. Because in both these equations the MITC T_{ij}^{h} can be replaced by the generalised expression of (12.59), the total induced modal torque coefficient for mode λ_h can be evaluated from the contribution of all *n* PSSs and *z* FDSs, i.e.

$$\Gamma_i^h = \sum_{j=1}^{n+z} T_{ij}^h . \tag{12.61}$$

12.6.2 A relationship between modal induced torque coefficients and incremental stabilizer gains

Let the transfer function of the jth stabilizer (PSS or FDS) be $H_{stab\_j}(s) = k_j G_j(s)$; the torque coefficient induced by this stabilizer on machine *i* is then given by (12.59). For a selected operating condition and stabilizer gain setting k_j, the calculation of the eigenvalues λ_h and eigenvectors in this equation is based on the model of the system dynamics described by (12.9). However, the relationship between T_{ij}^h and the stabilizer transfer function $H_{stab\_j}(s)$, given by (12.59), is based on this system *with its shaft dynamics and rotor angle feedback paths disabled.* Into this disabled and therefore different system are injected the mode λ_h and right-eigenvector components v_{ih} and v_{jh} selected from eigenanalysis of the original system. Therefore, in differentiating (12.59) with respect to k_j neither λ_h nor its eigenvector components are functions of k_j in the disabled system. Hence, for small changes in the stabilizer transfer functions,

$$\Delta T_{ij}^h = \frac{dT_{ij}^h}{dH_{stab\_j}(\lambda_h)} \cdot \Delta H_{stab\_j}(\lambda_h) = H_{ij}(\lambda_h)\left(\frac{c_{q*}v_{*h}}{v_{ih}}\right)\Delta H_{stab\_j}(\lambda_h) . \tag{12.62}$$

The increment $\Delta H_{stab\_j}(\lambda_h)$ is given by

$$\Delta H_{stab\_j}(\lambda_h) = \frac{\partial H_{stab\_j}(\lambda_h)}{\partial \lambda_h} \cdot \Delta\lambda_h + \frac{\partial H_{stab\_j}(\lambda_h)}{\partial k_j} \cdot \Delta k_j . \tag{12.63}$$

As is shown in [1], the term $\frac{\partial H_{stab\_j}(\lambda_h)}{\partial \lambda_h} \cdot \Delta\lambda_h \ll \frac{\partial H_{stab\_j}(\lambda_h)}{\partial k_j} \cdot \Delta k_j$ for PSSs. For FDSs the stabilizer gain is chosen to satisfy a gain criterion which ensures that the term $\frac{\partial H_{stab\_j}(\lambda_h)}{\partial \lambda_h} \cdot \Delta\lambda_h$ is negligible. Therefore ignoring the latter term, equation (12.63) reduces to:

$$\Delta H_{stab\_j}(\lambda_h) = \frac{\partial H_{stab\_j}(\lambda_h)}{\partial k_j} \cdot \Delta k_j ,$$

and, by substituting it into (12.62), we find

$$\Delta T_{ij}^{h} = H_{ij}(\lambda_h)\left(\frac{c_{q*}v_{*h}}{v_{ih}}\right)\left(\frac{\partial H_{stab\_j}(\lambda_h)}{\partial k_j}\cdot\Delta k_j\right) = H_{ij}(\lambda_h)\left(\frac{c_{q*}v_{*h}}{v_{ih}}\right)G_j(\lambda_h)\cdot\Delta k_j\,. \qquad (12.64)$$

This is a first-order Taylor series approximation, and is based on parameter perturbation analysis, i.e. ΔT_{ij}^{h} and Δk_j represent increments in the parameters T_{ij}^{h} and k_j, respectively.

An array of incremental MITCs, $\Delta \boldsymbol{T}^{h} = [\Delta T_{ij}^{h}]$, $n \times (n+z)$, is now defined. Its elements are complex numbers. The n rows of the array represent the n generators on which the incremental MITCs are induced, while the $n+z$ columns represent the stabilizers (the n PSSs and z FDSs) on which the gain increments are made. The significance of the array can be explained as follows:

- The element, ΔT_{ii}^{h}, is the incremental MITC on machine i due to the increment in the nominal gain of its own PSS i, Δk_i.
- The jth element of row i, ΔT_{ij}^{h}, is the incremental MITC on machine i due to the increment Δk_j on stabilizer j.
- All elements ΔT_{ij}^{h} for which $j \leq n$ are due to PSSs and all those for which $j > n$ are due to the FDSs.

For mode number h the summation of all elements in row i of the array yields the total incremental MITC, $\Delta\Gamma_i^{h}$, on machine i due to the gain increments on all PSSs and FDSs in the system It is derived from (12.61), i.e.

$$\Delta\Gamma_i^{h} = \sum_{j=1}^{n+z} \Delta T_{ij}^{h}\,. \qquad (12.65)$$

The practical application and significance of the arrays of incremental and total MITCs, which are given by (12.64) and (12.65), are addressed in Chapter 13 and are illustrated with several case studies.

12.7 References

[1] P. Pourbeik, *Design and Coordination of Stabilisers for Generators and FACTS devices in Multimachine Power Systems*, PhD Thesis, The University of Adelaide, Australia, 1997.

[2] P. Pourbeik and M. J. Gibbard, "Damping and synchronizing torques induced on generators by FACTS stabilizers in multimachine power systems," *Power Systems, IEEE Transactions on*, vol. 11, pp. 1920-1925, 1996.

[3] M. J. Gibbard, "Coordinated design of multimachine power system stabilisers based on damping torque concepts," *IEE Proceedings C Generation, Transmission and Distribution*, vol. 135, pp. 276-284, 1988.

[4] F. P. Demello and C. Concordia, "Concepts of Synchronous Machine Stability as Affected by Excitation Control," *Power Apparatus and Systems, IEEE Transactions on*, vol. PAS-88, pp. 316-329, 1969.

[5] H. F. Wang, "Selection of operating conditions for the coordinated setting of robust fixed-parameter stabilisers," *Generation, Transmission and Distribution, IEE Proceedings-*, vol. 145, pp. 111-116, 1998.

[6] P. M. Anderson and A. A. Fouad, *Power System Control and Stability*, IEEE Press, 1993.

[7] R. L. Bolden, P. J. Wallace, and A. W. Grainger, "Considerations in the improvement of damping in the South East Australian interconnected system," *CIGRE, paper 31-05,* 1982.

[8] K. R. Padiyar and R. K. Varma, "Damping torque analysis of static VAR system controllers," *Power Systems, IEEE Transactions on*, vol. 6, pp. 458-465, 1991.

[9] J. M. Maciejowski, *Multivariable feedback design*, Electronic systems engineering series. Addison Wesley, 1989.

[10] P. Pourbeik, M. J. Gibbard, and D. J. Vowles, "Proof of the Equivalence of Residues and Induced Torque Coefficients for Use in the Calculation of Eigenvalue Shifts," *Power Engineering Review, IEEE*, vol. 22, pp. 58-60, 2002.

Appendix 12–I

App. 12–I.1 Appendix: System response at a single modal frequency

The natural response of a linearized system is given by the differential equation $\Delta\dot{\boldsymbol{x}}(t) = \boldsymbol{A}\Delta\boldsymbol{x}(t)$. For a given vector of initial conditions $\Delta\boldsymbol{x}(0)$ the solution of this equation for the vector of states is

$$\Delta\boldsymbol{x}(t) = \boldsymbol{V}\, diag\left\{ e^{\lambda_1 t}, e^{\lambda_2 t}, \ldots, e^{\lambda_N t} \right\} \boldsymbol{W}\,\Delta\boldsymbol{x}(0), \tag{12.66}$$

where $\boldsymbol{V}$, $\boldsymbol{W}$ are the right and left modal matrices of $\boldsymbol{A}$, respectively. The condition for exciting only one mode, e.g. mode $\lambda_h = \alpha_h + j\omega_h$, is $\Delta\boldsymbol{x}(0) = \xi\boldsymbol{v}_{*h}$, where $\boldsymbol{v}_{*h}$ is the hth column vector of $\boldsymbol{V}$, and ξ is an arbitrary real constant, e.g. unity. When only mode λ_h is excited, (12.66) reduces to

$$\Delta\boldsymbol{x}(t) = \boldsymbol{v}_{*h} e^{\lambda_h t}\xi .$$

Then the ratio of the qth and k^{th} responses is $\dfrac{\Delta x_q(t)}{\Delta x_k(t)} = \dfrac{v_{qh} e^{\lambda_h t}\xi}{v_{kh} e^{\lambda_h t}\xi} = \dfrac{v_{qh}}{v_{kh}}$.

Thus in the case of the linearized power system model (12.9), the speed perturbation of machine *i* is related to the speed perturbation of machine *j*, when only one mode is excited, by

$$\frac{\Delta\omega_j(\lambda_h)}{\Delta\omega_i(\lambda_h)} = \frac{v_{jh}}{v_{ih}}. \tag{12.67}$$

Furthermore, since the vector of system outputs is given by $\Delta\boldsymbol{y}(t) = \boldsymbol{C}\Delta\boldsymbol{x}(t)$, then the ratio of the qth to the kth outputs is

$$\frac{\Delta y_q(\lambda_h)}{\Delta y_k(\lambda_h)} = \frac{\boldsymbol{c}_{q*}\boldsymbol{v}_{*h}}{\boldsymbol{c}_{k*}\boldsymbol{v}_{*h}}, \quad \text{and} \tag{12.68}$$

$$\frac{\Delta y_q(\lambda_h)}{\Delta\omega_i(\lambda_h)} = \frac{\boldsymbol{c}_{q*}\boldsymbol{v}_{*h}}{v_{ih}}, \tag{12.69}$$

where $\boldsymbol{c}_{q*}$ is the qth row vector of the $\boldsymbol{C}$ matrix. Thus, if only the hth mode, λ_h, is excited by setting $\Delta\boldsymbol{x}(0) = \xi\boldsymbol{v}_{*h}$, at any instant of time the ratios of any two system output responses is related by constant magnitude and phase. By mathematical induction it may be shown that if $\Delta\boldsymbol{x}(0)$ is set to some linear combination of right-eigenvectors then all of the corresponding modes will be excited.

App. 12–I.2 Reducing the TFM model in Figure 12.2 to those in 12.3 and 12.4

The multi-machine TFM power system model in Figure 12.2 can be reduced to that in Figure 12.3 by closing the mechanical torque loop, i.e.

$$G_m(s) = J(s)(I - H_{gov}(s)J(s))^{-1} \qquad (n \times n),$$

where $H_{gov}(s)$ is a diagonal TFM representation of the governor/turbine dynamics and $J(s)$ is the diagonal TFM of machine shaft dynamics given by (12.25), i.e.

$$\left.\begin{aligned} J(s) &= diag\{1/(M_i s + K_{di})\} \\ N(s) &= diag\{\omega_0/s\} \end{aligned}\right\} \qquad (n \times n).$$

Based on Figure 12.2, the following expressions may be written (Note: for convenience the Laplace operator s is omitted):

$$\Delta P_\delta = H_{P\delta} N \Delta\omega, \qquad \Delta P_{Ur} = H_{PUr}[\Delta U_r + \Delta U_s], \qquad \Delta U_s = H_{fds}\Delta\Psi, \text{ and}$$

$$\Delta\Psi = H_{\Psi\delta} N \Delta\omega + H_{\Psi Vr}[H_{pss}\Delta\omega + \Delta V_r] + H_{\Psi Ur}[\Delta U_r + \Delta U_s].$$

The expressions for the TFM $H_{P\delta}$, $H_{\Psi\delta}$, etc. are given by (12.18) to (12.23). Therefore

$$\begin{aligned} \Delta P_\delta + \Delta P_{Ur} = {} & H_{P\delta} N \Delta\omega + H_{PUr}\Delta U_r + (I - H_{\Psi Ur} H_{fds})^{-1} \times \\ & H_{fds}[H_{\Psi\delta} N \Delta\omega + H_{\Psi Vr}(H_{pss}\Delta\omega + \Delta V_r) + H_{\Psi Ur}\Delta U_r] \end{aligned} . \tag{12.70}$$

Let the perturbation in the reference-set-point inputs at the summing junctions on generators and FACTS devices, ΔV_r and ΔU_r, respectively, be zero-vectors. Hence, (12.70) becomes

$$\begin{aligned} \Delta P_\delta + \Delta P_{Ur} &= \left\{ H_{P\delta} N + H_{PUr}(I - H_{\Psi Ur} H_{fds})^{-1} H_{fds}[H_{\Psi\delta} N + H_{\Psi Vr} H_{pss}] \right\} \Delta\omega \\ &= Z\Delta\omega \end{aligned} .$$

Close loops #2 and #3 in Figure 12.2, around the shaft and governor /turbine dynamics, $G_m(s)$, by letting $G_P = G_m(I + ZG_m)^{-1}$. Figure 12.2 thus reduces to Figure 12.3.

A similar analysis can be adopted to reduce Figure 12.2 to Figure 12.4, i.e.

$$\begin{aligned} \Delta P_\delta + \Delta P_{Vr} &= [H_{P\delta} N + H_{PVr} H_{pss}]\Delta\omega \\ &= Z_F \Delta\omega \end{aligned} .$$

and by letting $\quad G_F = G_m(I + Z_F G_m)^{-1}$, and $L = H_{\Psi\delta} N + H_{\Psi Vr} H_{pss}$.

App. 12–I.3 Elements of the output matrix, *C*

The following is an important aside. In the above formulation of the TFM power system model, the matrix partitions $\boldsymbol{A}_{\omega\delta}$ and $\boldsymbol{A}_{\omega q}$ may appear to have been ignored, because neither of the two matrix partitions appear in equations (12.18) to (12.23). It is important to show that these terms are not ignored because by doing so it will become apparent that certain partitions of the system matrices ***A*** and ***C*** are related. These relationships are then used in a proof in [10]. To show that they have not been ignored, consider the following argument. From (12.20) and (12.21):

$$\Delta\dot{\omega} = \boldsymbol{A}_{\omega\omega} + \boldsymbol{A}_{\omega\delta}\Delta\boldsymbol{\delta} + \boldsymbol{A}_{\omega q}\Delta\boldsymbol{q} = \begin{bmatrix} (-K_{d1}/M_1)\Delta\omega_1 \\ \cdots \\ (-K_{dn}/M_n)\Delta\omega_{1n} \end{bmatrix} + \boldsymbol{A}_{\omega\delta}\Delta\boldsymbol{\delta} + \boldsymbol{A}_{\omega q}\Delta\boldsymbol{q}\,. \qquad (12.71)$$

Comparing this with (4.64) on page 115, and noting that governor/turbines have not been modelled in (12.9) (i.e. $\Delta\boldsymbol{P}_m = \boldsymbol{0}$), it is clear that

$$-diag\{1/(M_i)\}\Delta\boldsymbol{P}_a = \boldsymbol{A}_{\omega\delta}\Delta\boldsymbol{\delta} + \boldsymbol{A}_{\omega q}\Delta\boldsymbol{q} \qquad (12.72)$$

where $\Delta\boldsymbol{P}_a$ is the accelerating power acting on the generator shaft, which is equal to the machine electric power output, $\Delta\boldsymbol{P}_e$, plus the stator copper losses. Neglecting copper losses and since, based on (12.13) and (12.9), $\Delta\boldsymbol{P}_e = \boldsymbol{C}_{P\delta}\Delta\boldsymbol{\delta} + \boldsymbol{C}_{P\delta q}\Delta\boldsymbol{q}$ it can be shown that

$$\boldsymbol{C}_{P\delta} = -diag\{M_i\}\boldsymbol{A}_{\omega\delta}\text{, and} \qquad (12.73)$$

$$\boldsymbol{C}_{Pq} = -diag\{M_i\}\boldsymbol{A}_{\omega q}\,; \qquad (12.74)$$

the latter two equations can be substituted in (12.18) to (12.20) as required.

Chapter 13

Interactions between, and effectiveness of, PSSs and FDSs in a multi-machine power system

13.1 Introduction

In this chapter the theoretical basis and a case study are used to illustrate the concepts of interactions between, and effectiveness of, PSSs and FDSs in a multi-machine power system. The theoretical relationships between the incremental modal induced torque coefficients (MITCs), the associated mode shifts, and increments in stabilizer gains are outlined. The case study will illustrate how the method developed for estimating rotor mode (eigenvalue) shifts can be used to assess the relative effectiveness of stabilizers and, thereby, gain some important insights which form a basis for the coordination of stabilizers [1], [2], [3], [4].

Techniques have been described in the literature for determining shifts in the modes of rotor oscillation due to changes in stabilizer parameters [5], [6], [7], [8]. These techniques have been used not only for determining optimal locations for PSSs and FACTS devices [6], [7], [8] but also for tuning PSS parameters [9], [10].

In this chapter the theory and analysis is used to:

- develop a new method, based on incremental MITCs, for estimating the mode or eigenvalue shifts;

- develop, for a given rotor mode, a method for estimating the contributions to damping of selected stabilizers for a selected increment in stabilizer gain, be they PSSs or FDSs;
- deduce the relative effectiveness of selected stabilizers in contributing to damping, say, of an inter-area mode;
- assess the effect of interactions [1] between PSSs, particularly for inter-area and local modes;
- provide a basis for the systematic coordination of both PSSs and FDSs in multi-machine systems [4].

13.2 Relationship between rotor mode shifts and stabilizer gain increments

Recall from Section 12.6 that the MTIC T_{ij}^h is the modal (complex) torque coefficient for the hth mode, λ_h, this torque coefficient being induced on the shaft of the ith generator by the jth stabilizer. It is established in (12.59) for both PSSs and FDSs that T_{ij}^h is dependent on the gain setting k_j (in pu on device base) of stabilizer j. In Section 12.6.2 a relationship is developed between the incremental MITCs and increments in stabilizer gains; this is then employed in the following analysis to determine the eigenvalue shifts due to increments in any or all of the gains of the n PSSs and z FDSs. This allows us to calculate for generator i the change in T_{ij}^h, ΔT_{ij}^h, caused by a change in stabilizer gain Δk_j. An expression will now be derived relating the shift in the h^{th}mode of rotor oscillation, $\Delta\lambda_{ij}^h$, due to the change ΔT_{ij}^h. The gain increments Δk_j may differ in magnitude and sign, i.e. they may be positive, negative or zero.

The proposed technique provides a direct relationship between the eigenvalue shift, $\Delta\lambda_{ij}^h$, and an increment in gain Δk_j on any stabilizer, be it a PSS or FDS. Note that $\Delta\lambda_{ii}^h$ is the contribution to the mode shift by generator i due to an increment in gain on its own PSS; this is, in essence, the objective of the PSS design methods discussed in Chapters 5 and 9. The shift $\Delta\lambda_{ij}^h$ also represents the contribution to the shift of the mode λ_h by generator i due to a gain increment, Δk_j, on some other stabilizer j. Depending on its sign, the shift $\Delta\lambda_{ij}^h$ may enhance or degrade the damping of the mode λ_h; the mode shift $\Delta\lambda_{ij}^h$ therefore

---

1. Such interactions have been observed earlier [11]

represents the contribution of the interaction between stabilizer *j* and generator *i* to the shift in the hth mode. The proposed method provides a basis for quantifying the effects of such interactions for selected modes of rotor oscillation; this analysis is discussed in Section 13.2.2. Moreover, by appropriately summing the mode shifts, $\Delta\lambda_{ij}^{h}$, the contribution made either by each generator or by each stabilizer to the total shift in a rotor mode can be calculated for a set of stabilizer gain increments. This provides a basis for the coordination of all stabilizer gains, both PSSs and FDSs [4].

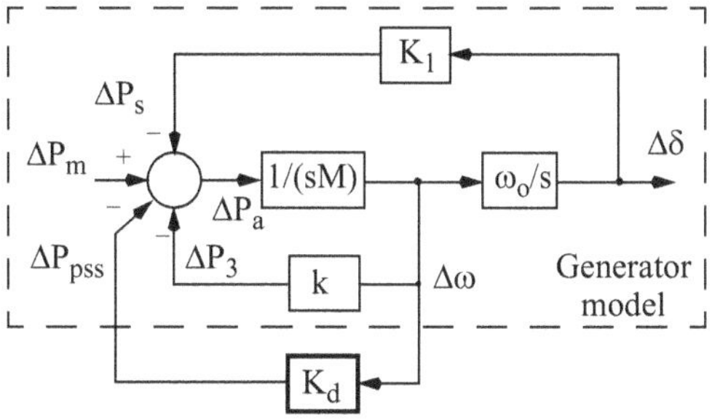

Figure 13.1 A simple model of the generator with an ideal speed-PSS

Consider the simple generator model with an ideal PSS in Figure 13.1. The PSS damping gain K_d shown in this figure is a real damping torque coefficient because it induces on the shaft of machine *i* a component of torque, $\Delta P_{pss} = K_d \Delta\omega_i$, which is in phase with the machine's speed perturbation. Using a first-order Taylor series approximation, the change in the eigenvalue for an incremental change in the torque coefficient K_d, is

$$\Delta\lambda^h = \frac{\partial\lambda_h}{\partial K_d} \cdot \Delta K_d. \tag{13.1}$$

Assume that *n* generators and *z* FACTS devices are fitted with stabilizers in a multi-machine system; consider an ideal PSS on generator *i*. Though the torque coefficient, K_{di}, by definition is real and the modal induced torque coefficient ΔT_{ij}^{h} on generator *i* by stabilizer *j* is complex, conceptually they both have the similar effect on machine *i*. Thus, by using the artifice of replacing ΔK_{di} in (13.1) by the total incremental MITC $\Delta\Gamma_i^h = \sum_{j=1}^{n+z} \Delta T_{ij}^{h}$ due to all $n+z$ stabilizers, the variation in λ_h due to the increments in gains of all stabilizer gains is given by the total differential,

$$\Delta\lambda_h = \sum_{j=1}^{n+z}\left\{\frac{\partial\lambda_h}{\partial K_{d1}}\Delta T_{1j}^h + \ldots + \frac{\partial\lambda_h}{\partial K_{dn}}\Delta T_{nj}^h\right\}$$
$$= \frac{\partial\lambda_h}{\partial K_{d1}}\Delta\Gamma_1^h + \ldots + \frac{\partial\lambda_h}{\partial K_{dn}}\Delta\Gamma_n^h \qquad (13.2)$$

Alternatively $\Delta\lambda_h = \sum_{j=1}^{n+z}\sum_{i=1}^{n}\Delta\lambda_{ij}^h$, where $\Delta\lambda_{ij}^h = \frac{\partial\lambda_h}{\partial K_{di}}\cdot\Delta T_{ij}^h$. (13.3)

If λ_h is an eigenvalue of the system matrix A, and if $\boldsymbol{w}_{h\circ}$ and $\boldsymbol{v}_{\circ h}$ are respectively the associated left and right eigenvectors, it is shown in Section 3.11 that

$$\frac{\partial\lambda_h}{\partial K_{di}} = \boldsymbol{w}_{h\circ}\frac{\partial \boldsymbol{A}}{\partial K_{di}}\boldsymbol{v}_{\circ h}. \qquad \text{((3.51) repeated)} \qquad (13.4)$$

The system matrix $\boldsymbol{A}$ is given by (12.11). Differentiation of (12.11) with respect to K_{di} yields

$$\frac{\partial A}{\partial K_{di}} = \begin{bmatrix} 0 & \circ & \circ & \circ & 0 \\ \circ & \circ & \circ & \circ & \circ \\ \circ & \circ & (-1/M_i) & \circ & \circ \\ \circ & \circ & \circ & \circ & \circ \\ 0 & \circ & \circ & \circ & 0 \end{bmatrix},$$

and thus (13.4) becomes

$$\frac{\partial\lambda_h}{\partial K_{di}} = -\frac{w_{hi}\,v_{ih}}{M_i} = -\frac{\rho_{ih}}{M_i}, \qquad (13.5)$$

where $\rho_{ih} = w_{hi}\,v_{ih}$ is the complex participation factor of the ith system state - namely the rotor speed perturbation of the ith generator $\Delta\omega_i$ - in the rotor mode λ_h; M_i is twice the inertia constant (H) of generator i. Substitution of (13.5) into (13.3) yields the expression for estimating the contribution to the eigenvalue or mode shift by generator i due to an incremental change in the torque coefficient, ΔT_{ij}^h, on machine i, i.e.

$$\Delta\lambda_{ij}^h = -(\rho_{ih}/M_i)\,\Delta T_{ij}^h. \qquad (13.6)$$

Recall for mode λ_h, that the incremental MITCs for generator i are related to the incremental gains on the n PSSs and the z FDSs by

$$\Delta T_{ij}^h = H_{ij}(\lambda_h)(\boldsymbol{c}_{j*}\boldsymbol{v}_{*h}/v_{ih})G_j(\lambda_h)\cdot\Delta k_j, \qquad \text{((12.64) repeated)).} \qquad (13.7)$$

where $j = 1, \ldots, n, n+1, \ldots, z$, the gains being in pu on the base of the device.

Finally, by substitution of (13.7) in (13.6), a general expression is derived for the contribution to the mode shift by generator *i* due to an increment in gain on stabilizer *j*, namely

$$\Delta\lambda_{ij}^{h} = -(\rho_{ih}/M_i)H_{ij}(\lambda_h)(\mathbf{c}_{j*}\mathbf{v}_{*h}/v_{ih})[G_j(\lambda_h)\Delta k_j], \tag{13.8}$$

where stabilizer *j* may be a PSS or a FDS. Note that for PSSs $H_{ij}(s) = H_{PVr\_ij}(s)$, $j = 1, \ldots, n$; for FACTS device stabilizers, $H_{ij}(s) = H_{PUr\_ij}(s)$, $j = 1+n, \ldots, z$. The transfer function $H_{ij}(\lambda_h)$ is evaluated at the modal frequency λ_h *with all machine shaft dynamics disabled.*

In addition to (13.8) three additional expressions will be employed in later sections. The first is the contribution to the mode shift by all *n* generators as a result of an increment in gain on any PSS *j*,

$$\Delta\lambda^{h}\Big|_{\text{stab\_j}} = -\left(\sum_{i=1}^{n} (\rho_{ih}/M_i)H_{ij}(\lambda_h)(\mathbf{c}_{j*}\mathbf{v}_{*h}/v_{ih})\right)[G_j(\lambda_h)\Delta k_j]. \tag{13.9}$$

The second expression is the contribution by generator *i* to the mode shift caused by increments in the gains in some or all of the *n* PSSs and *z* FDSs:

$$\Delta\lambda^{h}\Big|_{\text{gen\_i}} = -(\rho_{ih}/M_i)\cdot\sum_{j=1}^{n+z} H_{ij}(\lambda_h)(\mathbf{c}_{j*}\mathbf{v}_{*h}/v_{ih})[G_j(\lambda_h)\Delta k_j]. \tag{13.10}$$

Thirdly, the total contribution to the mode shift by all *n* generators as a result of increments in the gain on some or all *n*+*z* stabilizers is:

$$\Delta\lambda^{h} = -\sum_{j=1}^{n+z}\sum_{i=1}^{n} (\rho_{ih}/M_i)H_{ij}(\lambda_h)(\mathbf{c}_{j*}\mathbf{v}_{*h}/v_{ih})[G_j(\lambda_h)\Delta k_j]. \tag{13.11}$$

13.2.1 Relationship between residues and MITCs in calculation of mode shifts

It is shown in Chapters 6 and 11 that, for mode *h* and generator *i*, the contribution to the mode shift by generator *i* is related to an increment in gain on stabilizer *j* by:

$$\Delta\lambda_{ij}^{h} = r_{ij}^{h}G_j(\lambda_h)\Delta k_j, \tag{13.12}$$

where r_{ij}^{h} is the residue of the transfer function from the input to the summing junction of generator *i* (e.g. V_{ref_i}) to the output used as the stabilizing signal (e.g. speed for a PSS, bus frequency for a FDS).

In comparing (13.8), i.e. $\Delta\lambda_{ij}^{h} = -(\rho_{ih}/M_i)H_{ij}(\lambda_h)(\mathbf{c}_{j*}\mathbf{v}_{*h}/v_{ih})[G_j(\lambda_h)\Delta k_j]$ with (13.12), it is observed in [1], [12] that the residue is

$$r_{ij}^{h} = -(\rho_{ih}/M_i)H_{ij}(\lambda_h)(\mathbf{c}_{j*}\mathbf{v}_{*h}/v_{ih}). \tag{13.13}$$

The expression in (13.13) contains much more information than the form of the expression in (13.12) which is the basis for the analysis in Chapters 6 and 11. The roles of the participation factor ρ_{ih} and the inertia constant H_i (= $M_i/2$) in (13.6), the MITCs (through (13.7)), and the P-Vr or P-Ur characteristics in the residue of (13.13), are isolated. These components will help to explain certain characteristics, including 'interactions', derived in this chapter.

13.2.2 Concept of 'interactions'

Based on (13.10) the concept and implications of 'interactions' between stabilizers and of 'interactions' between generators will now be defined. Let us assume for simplicity that a gain increment is made on the PSS fitted to generator i and then on other stabilizers, $j = 1, \ldots, n, n+1, \ldots, z \;\; j \neq i$. The contribution to the mode shift by generator i is

$$\Delta\lambda^h\big|_{\text{gen\_i}} = -(\rho_{ih}/M_i) \cdot H_{ii}(\lambda_h)\,[G_i(\lambda_h)\Delta k_i] \\ -(\rho_{ih}/M_i) \cdot \sum_{j=1 \neq i}^{n+z} H_{ij}(\lambda_h)(c_{j*}v_{*h}/v_{ih})[G_j(\lambda_h)\Delta k_j] \tag{13.14}$$

The first term in (13.14), an alternative form of (13.10), is the contribution to the mode shift by an increment Δk_i in the gain of the PSS fitted to generator i. However, if both the stabilizer gains Δk_i and Δk_j are increased, it is apparent from the second term in (13.14) that the gain increment Δk_j can be considered to modify the effect of the gain increment Δk_i on the mode shift. The net effect on the mode shift depends on the resulting sign of the real part of second term. If the net effect of the increment Δk_j is to enhance the damping of the mode λ_h then there is a *positive interaction of stabilizer j* with PSS i. It is important to note that the jth stabilizer can be either a PSS or FDS. Clearly, if the increment Δk_j degrades the damping of the mode λ_h then the interaction between stabilizers is a negative. Let us call all such interactions *stabilizer interactions* [3].

It is also insightful to consider a scenario in which the gain of the PSS fitted to the ith generator is unchanged (i.e. $\Delta k_i = 0$). Then, the contribution to the mode shift by generator i is due only to its interactions with the other stabilizers in the system, i.e.

$$\Delta\lambda^h\big|_{\text{gen\_i}} = -(\rho_{ih}/M_i) \cdot \sum_{j=1 \neq i}^{n+z} H_{ij}(\lambda_h)(c_{j*}v_{*h}/v_{ih})[G_j(\lambda_h)\Delta k_j]\,. \tag{13.15}$$

In the event that gain increments are restricted only to FDSs the summation in (13.15) is restricted to $j = n+1$ to $n+z$. This reveals that FDSs contribute indirectly to the shift in mode h by their interactions with the generators.

It is at times more informative to assess the contributions to the mode shifts by the n generators due a gain increment on stabilizer *i only*. If the stabilizer is a PSS, these contributions can be expressed by an alternative form of (13.9):

$$\Delta\lambda^h\Big|_{\text{pss\_i}} = -(\rho_{ih}/M_i)\cdot H_{ii}(\lambda_h)\,[G_i(\lambda_h)\Delta k_i] - \sum_{j=1\neq i}^{n} (\rho_{jh}/M_j)H_{ji}(\lambda_h)(c_{j*}v_{*h}/v_{ih})\,[G_i(\lambda_h)\Delta k_i]\cdot \tag{13.16}$$

The above equation reveals that the contribution to damping of generator i due to its own PSS (the first term in (13.16)) may be enhanced or degraded due to contributions from, or *interactions with, the remaining generators* (the second term) through the network. Let us call such interactions *generator interactions* [3].

Consider now the case in which a gain change is restricted to stabilizer FDS j only. The resulting contribution to the mode shift by an increment Δk_j in the gain of FDS j is given by (13.9).This result reveals that FDS j acts to shift mode h only by means of its interactions through the network with each of the n generators.

In addition to the cases associated with the three equations, (13.9) to (13.11), a case of special interest is the contribution by generator i to the mode shift due to an increment in the damping gain Δk_i of its PSS (assumed ideal). By substitution of (12.64) in (13.6), this self-contribution is found to be:

$$\Delta\lambda_{ii}^h \approx -(\rho_{ih}/M_i)\Delta k_i\,. \tag{13.17}$$

This result [1] provides a type of benchmark for the contribution of an *ideal* PSS to damping. Typically if a machine participates significantly in a mode (usually a local-area mode), the speed-state participation is about 0.5 or less. The mode shift is then directly to the left in the s-plane and is equal to $\Delta k_i/(2M_i)$ or $\Delta k_i/(4H_i)$. (The latter result is consistent with that which was derived based on an analysis of the block diagram for a SMIB system in Section 5.4.) Clearly from (13.17), with low participation in the speed state, the extent of the mode shift is reduced. However, a reduced contribution to the mode shift may also be attributed in part to the effect of interactions as explained above. This will be illustrated in Section 13.3.

Note from (13.17) that the extent of the mode shift for low-inertia generating units is greater than that of high inertia units of the same rating, all else being equal. Fitting PSSs to the former units are likely to more effective than to the latter.

13.2.3 Relationships between mode shifts, MITCs, participation factors and stabilizer gains

Eventually, our aims are: (i) to determine the mode shift $\Delta\lambda_{ij}^h$ in a selected mode λ_h due to gain increments on a single or on a number of PSSs and/or FDSs, and (ii) to assess the rel-

1. This result is also employed in Section 5.9.3.

ative effectiveness and contribution of individual stabilizers to the enhancement or degradation of modal damping. However, let us firstly review some of the background equations.

Consider generator *i* and its PSS. From (13.6) it has been shown that incremental mode shift $\Delta\lambda_{ii}^{h}$ is related to the incremental MITC by:

$$\Delta\lambda_{ii}^{h} = -(\rho_{ih}/M_i)\ \Delta T_{ii}^{h}, \tag{13.18}$$

that is, through a complex factor which is the *inertia weighted, speed participation factor*, $\rho_{ih}/(2H_i)$. For mode λ_h, it has been shown using (13.7) that the incremental MITC for generator *i* is in turn related to the incremental gain on PSS *i*, i.e.

$$\Delta T_{ii}^{h} = H_{PVr\_ii}(\lambda_h) \cdot G_i(\lambda_h)\Delta k_i \quad \text{or,} \tag{13.19}$$

$$\text{for a ideally-tuned PSS,} \quad \Delta T_{ii}^{h} = \Delta k_i\,. \tag{13.20}$$

The three equations, (13.6), (13.18) and (13.20) are of particular interest in the following discussions.

A case study is now used to illustrate some of the physical insights provided by the theoretical analysis. In particular, based on (13.8) to (13.17), the concept and effects of interactions between stabilizers will be discussed. Furthermore, it will be demonstrated how the method developed above for estimating eigenvalue shifts can be used to assess the relative effectiveness of stabilizers and thereby gain some important insights which form a basis for the coordination of PSSs and FDSs.

13.3 Case Study: Contributions to MITCs/Mode Shifts by PSSs and generators

The purpose of this study is to demonstrate the insights that the incremental MITCs and the associated mode shifts provide in the dynamic performance of a multi-machine power system. This study illustrates a basis for the tuning and coordination of PSSs, and of PSSs and FDSs. Recall that the basic approaches to methods for the tuning of PSSs has been discussed in the earlier Chapters 5 and 9, and for FDSs in Chapter 11.

The single-line diagram of the fourteen generator power system employed in the case study is shown in Figure 10.1. Only Case 1, which is a heavy-load operating condition - and which has been the subject of studies for both PSSs and FDSs in earlier chapters - is now examined. Two modes, an inter-area mode M and the local-area mode B will be the initial focus of the studies.

For illustrative purposes all PSS damping gains are set to 5 pu on machine base; the PSS-related parameters are listed in Tables 10.5 to 10.9. The FACTS device is the SVC, BSVC_4, located at bus 412 in Figure 10.1; the parameters for its bus-frequency FDS are provided in

(11.12) of Section 11.4.1.1. Based on Figure 11.7 the gain of the FDS is set to 30 pu on device base, a value at the lower end of its potential gain range. The characteristics of the associated rotor modes are given in Table 13.1.

Table 13.1 Characteristics of the electromechanical modes, Case 1; PSSs and FDS of BSVC_4, in service.

Mode				Mode Behaviour	Mode Type
No.	Real	Imag	ξ		
A	-0.68	10.47	0.065	VPS_2<-->EPS_2, BPS_2	Local-area
B	-0.40	9.66	0.041	SPS_4<-->CPS_4, GPS_4, TPS_4	"
C	-0.42	9.06	0.047	BPS_2<-->EPS_2, VPS_2, TPS_4	"
D	-1.00	8.73	0.114	NPS_5<-->TPS_5	"
E	-0.65	8.32	0.078	CPS_4, SPS_4<-->GPS_4, TPS_4,	"
F	-0.88	8.27	0.105	MPS_2, LPS_3<-->HPS_1, EPS_2, VPS_2	"
G	-0.84	7.80	0.107	HPS_1,MPS_2,<-->YPS_3, EPS_2, VPS_2	"
H	-1.24	8.09	0.151	TPS_4<-->GPS_4, SPS_4, EPS_2	"
I	-0.64	7.83	0.081	YPS_3, MPS_2<-->LPS_3, EPS_2,	"
J	-0.92	7.48	0.123	PPS_5<-->TPS_5, NPS_5	Local-area
K	-0.18	3.93	0.046	Area 3 <--> Area 5, Area 2	Inter-area
L	-0.14	2.56	0.056	Area 5, Area 4 <--> Area 2	"
M	-0.42	2.04	0.201	Area 5, Area 3 <--> Area 4	Inter-area

Nominal gain settings. All PSSs: 5 pu damping gain on machine MVA rating
FDS of BSVC_4: 30 pu on the device base.
<--> means '... swings against ...'. ξ - damping ratio.
In 'Mode Behaviour', generators or areas are listed in descending order of their participation factors.

In order to assess the effects of the incremental changes in the MITCs and associated mode shifts due to changes in stabilizer gains, let us increase the damping gain of all PSSs by 1 pu (20% of 5 pu) and the gain of the FDS by 0.9 pu (3% of 30 pu), such gains being in per unit on the device bases. Although the increases in PSS gains are 1 pu, an examination of Figure 10.26 reveals that the left shifts in the associated modes are close to being linearly related to the PSS gain increments. Similarly, for the FDS the left shift in the mode is likewise related to FDS gain increments in the vicinity of the nominal gain setting of 30 pu (Figure 11.7). As foreshadowed, with the PSSs and the FDS in operation, it is of interest to analyse for Case 1 the MITCs and mode shifts for the inter-area mode M, $\lambda_M = -0.42 \pm j2.04$, and for the local mode B, $\lambda_B = -0.40 \pm j9.66$; the generators in Area 4 participate fairly significantly in both modes. (It is shown in Table 10.14 that without a FDS in operation on BSVC_4 these modes are respectively $\lambda_M = -0.14 \pm j1.98$ and $\lambda_B = -0.39 \pm j9.65$.)

Let us examine the state participation factors, shown in Figure 13.2, for modes B and M.

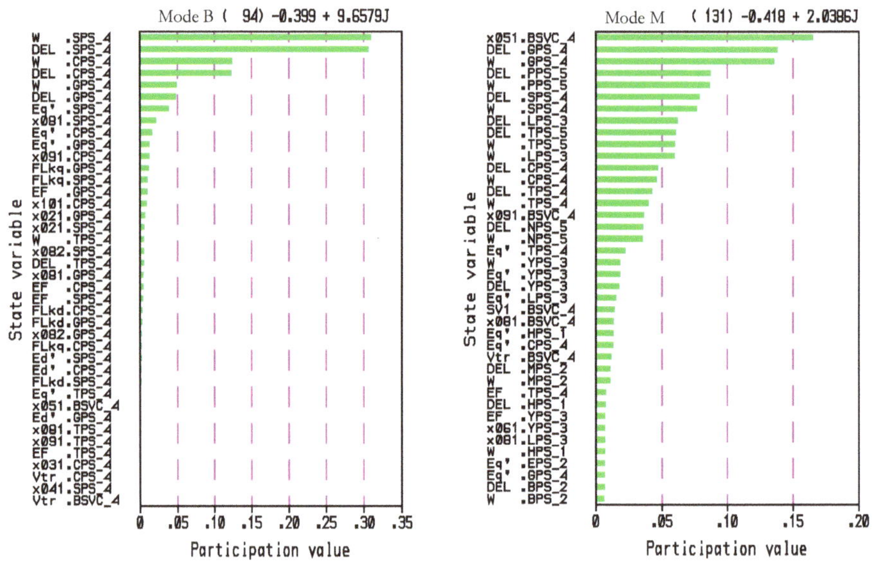

(a) Local-area mode B (b) Inter-area mode M

Figure 13.2 Case 1. Magnitudes of the state participation factors for local-area mode B and inter-area mode M. The PSSs and the FDS on BSVC_4 are in service.

For the local mode B the speed and rotor angle states of generators in Area 4 dominate the participation of the state in the mode. In the case of mode M, the inter-area mode, a controller state in the FDS of the SVC is dominant, followed by speed and rotor angle states of generators in Areas 3, 4 and 5. The mode shape for mode M is similar to that shown in Figure 10.29(a) when the FDS is out of service. It reveals that machines in Areas 5 and 3 swing against those in Area 4. However, it is the complex *inertia-weighted*, speed participation factors that are of interest in (13.6); these participation factors are shown in Figure 13.3(a) and (b). In this figure it is noteworthy that:

- because fewer machines participate in the local mode, the magnitudes of the participation factors in Figure 13.3(a) are much greater than those for the inter-area mode in Figure 13.3(b). (Recall that the sum of the complex participation factors for mode λ_h is unity.) Larger values of the speed participation factors of local-area modes are typically a characteristic which differentiates them from inter-area modes.

- in Figure 13.3(b) for the inter-area mode M, the dominant speed participation factors of generators in Areas 3, 4 and 5 are - for most purposes - real (or nearly real), and thus so is the factor $\rho_{ih}/(2H_i)$ in (13.6).

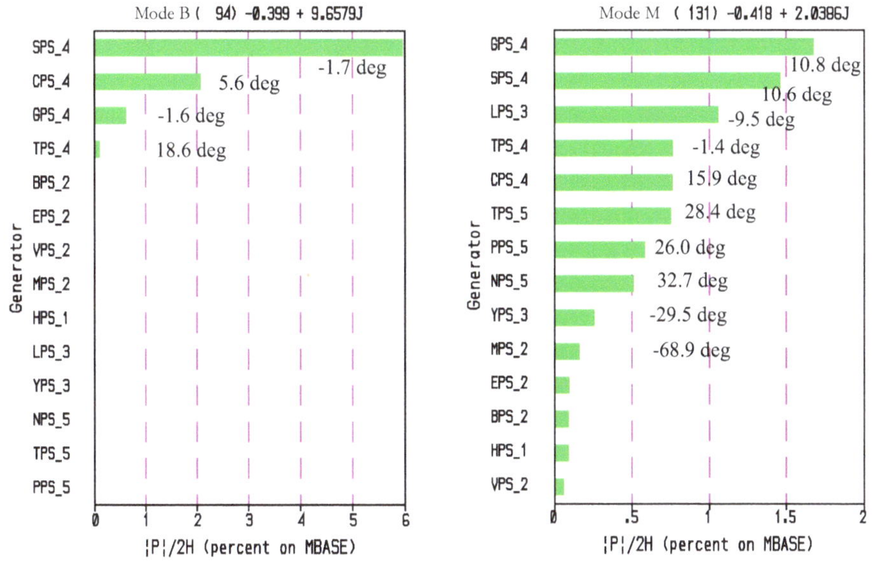

(a) Local-area mode B (b) Inter-area mode M

Figure 13.3 Case 1. Magnitude and angle of inertia-weighted, speed participation factors, $|\rho|/(2H)$, $\angle\rho$, of generators for local mode B and inter-area mode M.

13.3.1 Contributions to the MITC of each generator, local mode B

Based on (13.7) the full $n \times (n+z)$ array of incremental MITCs for the local mode B ($\lambda_B = -0.40 \pm j9.66$) is shown in Table 13.2 on page 601. The full array contains a number of features. (i) For the increments in gains of all PSSs (1 pu) and of the single FDS (0.9 pu) the array of *incremental torque coefficients* have a direct bearing on the associated *mode shifts* in Table 13.3 on page 603. (ii) Interesting aspects of the MITCs for the local mode B will be compared with similar aspects for the MITCs of the inter-area mode M, $\lambda_M = -0.42 \pm j2.04$. Figures 13.2 and 13.3 reveal that the generators in Area 4 participate fairly significantly in both modes B and M.

The rows of Table 13.2, which are on the MVA ratings of the generator nominated in the left-hand column, show the contributions to the total incremental MITC of each generator resulting from gain increments: (i) on individual stabilizers (a PSS or the FDS) or (ii) on *all* PSSs and the FDS (that is, the row sum,

$$\Delta\Gamma_i^h = \sum_{j=1}^{n+z} \Delta T_{ij}^h). \tag{13.21}$$

Alternatively, a column in the table reveals the components of the complex incremental MITCs induced on each generator due to a gain increment on a selected stabilizer. Note: if the MITCs in the array were expressed on *system* base MVA then the sum of MITCs in each column would yield the total MITC induced by an individual stabilizer.

A diagonal element, which is the incremental torque coefficient induced on generator *i* by the PSS on generator *i*, is essentially real [1], i.e. it is a pure damping coefficient; the diagonal element is referred to as the 'self' contribution. It is also observed that for such elements the incremental damping torque coefficients are - in most cases - close to unity; ideally, this result is predicted by the incremental form of (13.20),

$$\Delta T^h_{ii} \approx \Delta k_i = 1 \text{ pu.} \tag{13.22}$$

Strictly speaking, the above result applies to the operating condition for which the PSS of the particular generator is tuned, usually at or close to rated power output (typically 0.7 to 0.9 pu on generator MVA rating, see Chapters 5 or 10). Assume for the encompassing range of operating conditions the P-Vr Design Case may lie at 0.7 pu on generator MVA rating. Typically, over the range of power outputs the band of P-Vr characteristics lie within ±3 dB and ±10° of the desired P-Vr design characteristic. At lower levels of power output, less than 0.7 pu, the low frequency gain of the associated P-Vr characteristic is less than that of the P-Vr Design Case. Consequently, in Case 1 for generating stations on part load, such as HPS_1, MPS_2 and PPS_5, the value of ΔT^h_{ii} in the table is less than $1\angle 0°$ pu [2]. Conversely, for generators operating at rated power output such as LPS_3 and NPS_5, the gain of the associated P-Vr characteristic typically exceeds that of the P-Vr Design Case. As a result, the value of ΔT^h_{ii} in the table for such machines exceeds $1\angle 0°$ pu.

An examination of the Table 13.2 on page 601 provides an insight, for example, why a PSS is not contributing to damping of mode(s) to the extent expected, or by what mechanism are FDSs contributing to damping.

Consider generator SPS_4.

1. From the row we note that if all stabilizers are incremented by 1 pu the incremental damping component of the total incremental MITC for SPS_4 is enhanced from that induced by its own PSS, 0.93, by 0.21 to 1.14 pu. According to (13.6) the damping of mode B would also be enhanced, however, the extent of improvement is determined by the inertia-weighted, speed participation factor of SPS_4.

2. A gain increment *only* on the PSS of CPS_4 (col. 9) increases the damping component of the MITC of SPS_4 by 0.13 pu on the latter's MVA rating - for which there will be an associated improvement in the damping of mode B as implied by (13.6).

---

1. This is the objective of the PSS design procedures discussed in Chapters 5 or 10.
2. See Section 9.4.1. It is shown that with decreasing real power output (P) at constant reactive power (Q) the scalar voltage v_{d0} decreases. Correspondingly the low frequency gain of the P-Vr characteristic reduces. However, v_{d0} also decreases at constant P as Q becomes more lagging (see Table 9.7). Both these effects influence the P-Vr characteristics.

Table 13.2 Case 1, Mode B. MITCs on each unit due to gain increments on PSSs/FDS

	Contributions to incremental modal induced torque coefficients ΔT^B_{ij} of each generator i listed in the left-hand column due to an increment in the gain of the stabilizers j listed in the column headings (pu on generator MVA rating)														
	Local-area mode B, $\lambda_B = -0.40 \pm j9.66$														
Generator	HPS_1 PSS	BPS_2 PSS	EPS_2 PSS	VPS_2 PSS	MPS_2 PSS	LPS_3 PSS	YPS_3 PSS	TPS_4 PSS	CPS_4 PSS	SPS_4 PSS	GPS_4 PSS	NPS_5 PSS	TPS_5 PSS	PPS_5 PSS	BSVC 4 FDS	Row Sum
Col.	**1**	**2**	**3**	**4**	**5**	**6**	**7**	**8**	**9**	**10**	**11**	**12**	**13**	**14**	**15**	**16**
HPS_1	0.72	0.27	0.01	0.23	0.04	0.01	0	-0.44	0.01	-0.04	0.04	0	0	0	0.03	0.87
	-j0.07	j0.82	j0.31	j0.06	-j0.18	j0.01	j0	j0.07	-j0.42	j0.46	-j0.18	j0	j0	j0	-j0.05	j0.20
BPS_2	-0.01	0.99	0.04	0.07	0.02	0	0	0.35	0.18	-0.17	0.07	0	0	0	0.01	1.52
	j0.01	-j0.06	-j0.07	-j0.03	-j0.02	j0	j0	j0.39	j0.36	j0.43	j0.19	j0	j0	j0	j0	-j0.43
EPS_2	0.02	0.23	1.03	0.04	0.01	0	0	-0.13	-0.18	0.19	-0.07	0	0	0	0	1.14
	-j0.04	-j0.05	-j0.13	j0.07	j0	j0	j0	j0.25	-j0.19	j0.21	-j0.09	j0	j0	j0	-j0.02	j0.02
VPS_2	-0.02	0.02	-0.01	0.99	-0.02	0	0	0.19	-0.26	0.29	-0.13	0	0	0	-0.02	1.03
	-j0.02	-j0.71	j0.33	-j0.03	j0.02	j0	j0	j0.29	j0.10	-j0.08	j0.02	j0	j0	j0	j0.01	-j0.08
MPS_2	0.14	1.36	-0.17	-0.17	0.87	0	0	-0.44	-0.86	0.91	-0.38	0	0	0	-0.03	1.22
	-j0.18	-j1.75	j0.35	j0.11	-j0.03	j0	j0	j1.27	-j0.62	j0.78	-j0.36	j0	j0	j0	-j0.03	-j0.46
LPS_3	-1.28	-1.90	0.58	-0.05	0.35	1.04	-0.04	0.40	0.81	-0.85	0.32	0	0	0	0.06	-0.56
	j0.45	j0.29	j0.02	j0.24	j0.09	-j0.06	j0.03	-j1.03	j0.53	-j0.65	j0.31	j0	j0	j0	j0.08	j0.30
YPS_3	0.56	-2.41	1.21	0.30	0.74	-0.50	1.03	-0.93	1.19	-1.34	0.56	0	0	0	0.21	0.61
	-j0.16	j2.08	j0.45	j1.05	-j0.14	j0.90	-j0.26	-j1.07	-j0.72	j0.71	-j0.21	j0	j0	j0	-j0.01	j1.74
TPS_4	0	0.01	0	0	0	0	0	1.03	-0.06	0.08	-0.03	0	0	0	0.17	1.18
	j0	j0	j0	j0	j0	j0	j0	j0.05	-j0.59	j0.65	-j0.21	j0	j0	j0	-j0.06	-j0.16
CPS_4	0	0	0	0	0	0	0	0.02	0.99	0.15	0	0	0	0	0	1.16
	j0	j0	j0	j0	j0	j0	j0	-j0.03	-j0.04	-j0.08	j0.01	j0	j0	j0	j0.01	-j0.15
SPS_4	0	0	0	0	0	0	0	-0.01	0.13	0.93	0.08	0	0	0	0	1.14
	j0	j0	j0	j0	j0	j0	j0	j0	-j0.04	-j0.03	-j0.04	j0	j0	j0	0	-j0.10
GPS_4	0	0	0	0	0	0	0	0.09	-0.31	0.86	0.89	0	0	0	0	1.54
	j0	j0	j0	j0	j0	j0	j0	-j0.01	j0.20	-j0.71	j0.05	j0	j0	j0	-j0.02	-j0.49
NPS_5	0.92	-3.20	1.64	0.33	0.92	-1.10	-0.16	-0.47	0.75	-0.82	0.33	1.08	-0.18	-0.02	0.21	0.21
	-j2.33	j1.03	-j0.11	j1.51	j0.10	j1.02	-j0.03	-j0.59	-j0.49	j0.50	-j0.12	-j0.19	j0.15	j0.04	j0	j0.50
TPS_5	1.18	-4.38	2.37	0.74	1.35	-1.41	-0.23	-0.95	1.03	-1.14	0.48	-0.19	1.06	-0.04	0.31	0.18
	-j3.46	j2.21	-j0.48	j2.18	-j0.04	j1.66	-j0.01	-j0.77	-j0.99	j1.02	-j0.30	j0.14	-j0.15	j0.04	-j0.04	j1.02
PPS_5	10.22	-18.00	9.02	-0.38	4.76	-7.66	-0.81	-2.29	4.24	-4.63	1.76	-1.16	-1.81	0.87	1.12	-4.73
	j12.35	j0.88	j1.81	j8.97	j1.84	j3.80	-j0.42	-j3.09	-j2.30	j2.36	-j0.56	j0.51	j0.52	j0.05	j0.24	j2.27

Col. 16, the Row Sum, is given by (13.21). All PSS damping gains=5 pu, increment 1 pu on machine ratings.
Gain of FDS=30 pu, increment 0.9 pu on SVC base.
The incremental MITC induced by a generator's PSS is shaded in yellow.

Consider now generators PPS_5 and LPS_3.

3. An increment in the PSS gain on BPS_2 (col. 2) or LPS_3 (col. 6) significantly degrades the incremental MITCs on PPS_5, while increments in the gain of other PSSs enhance or degrade the incremental MITCs relatively less. The net effect is a significant degradation in the damping component of MITC of PPS_5 (in col. 16). However, a reference to Figure 13.3 on page 599 reveals that the inertia-weighted, speed participation factor of PPS_5 for mode B is exceedingly small so that the effects of any stabilizer, including its own, on the associated mode shifts is negligible.

   A similar comment applies to an increment in PSS gain on generator LPS_3.

Consider SVC BSVC_4.

4. For an increment in the gain of the FDS on BSVC_4 the induced damping coefficients are small or negligible on those generators in Area 4 which have significant inertia-weighted, speed participation factors. Therefore their contribution to damping of mode B by the FDS will be small; for other generators - for which the latter participation factor is negligible - the contribution will also be negligible.

13.3.2 Contributions of the mode shifts of each generator to mode B damping

Let us now consider the full array of *mode shift* contributions for the *local mode* B, $\lambda_B = -0.40 \pm j9.66$, and compare its features with those highlighted for its associated array of MITCs. The mode shift array is shown in Table 13.3 on page 603.

Further to the observations made for the MITCs of mode B, the following comments are offered on the associated mode shifts.

5. The contribution to the incremental mode shift of generator *i* is related to the MITC generated by its own PSS by (13.6), $\Delta\lambda_{ii}^h = -(\rho_{ih}/M_i)\,\Delta T_{ii}^h$. Bearing that in mind, the earlier observations on the MITCs (numbered 1 to 4 concerning the implications for the associated mode shifts) are confirmed by an examination of Table 13.3.

6. Ideally, the diagonal elements of the MITC array are $\Delta T_{ii}^h = 1$ pu. According to (13.6) above, the incremental mode shift for the mode is ideally the value for generator *i* given by the inertia-weighted, speed participation factor of Figure 13.3. Accordingly, this mode shift for unity MITC for generator SPS_4 is ideally $-\rho/(2H) = -59.57\angle{-1.69°}$ units (1 unit $= 1 \times 10^{-3}$). As shown in Table 13.2 on page 601 the value of the incremental MITC is actually $\Delta T_{ii}^h = 0.9328 - j0.0293$ thus the mode shift according to (13.6) is $\Delta\lambda_{ii}^h = -55.49 + j3.38$ or $-55.59\angle{-3.49°}$ units; the value in Table 13.3 is $\Delta\lambda_{ii}^h = -55.5 + j3.40$.

Table 13.3 Case 1. Mode B: Mode Shifts on each unit due to gain increments on PSSs & FDS.

	Contributions to Mode Shift $\Delta\lambda_{ij}^{B}$ by each generator (row) and by each PSS or FDS (col.) (x 10^{-3})																
	Local-area mode $\lambda_B = -0.40 \pm j9.66$																
Generator	HPS_ 1 PSS	BPS_ 2 PSS	EPS_ 2 PSS	VPS_ 2 PSS	MPS2 PSS	LPS_ 3 PSS	YPS_ 3 PSS	TPS_ 4 PSS	CPS_ 4 PSS	SPS_4 PSS	GPS_ 4 PSS	NPS_ 5 PSS	TPS_ 5 PSS	PPS_ 5 PSS	Sum PSSs	SVC4 FDS	Row Sum
	1	2	3	4	5	6	7	8	9	10	11	12	13	14	15	16	17
HPS_ 1	0	0	0	0	0	0	0	0	0	0	0	0	0	0	0	0	0
	j0	j0	j0	j0	j0	j0	j0	j0	j0	j0	j0	j0	j0	j0	j0	j0	j0
BPS_ 2	0	0.01	0	0	0	0	0	0	0.01	-0.01	0	0	0	0	0	0	0
	j0	-j0.02	j0	j0	j0	j0	j0	-j0.01	j0	j0	j0	j0	j0	j0	-j0.03	j0	-j0.03
EPS_ 2	0	0	0	0	0	0	0	0	0	0	0	0	0	0	0	0	0
	j0	j0	-j0.01	j0	j0	j0	j0	j0	j0	j0	j0	j0	j0	j0	-j0.01	j0	-j0.01
VPS_ 2	0	0	0	0	0	0	0	0	0	0	0	0	0	0	0	0	0
	j0	j0	j0	j0	j0	j0	j0	j0	j0	j0	j0	j0	j0	j0	j0	j0	j0
MPS 2	0	0	0	0	0	0	0	0	0	0	0	0	0	0	0	0	0
	j0	j0	j0	j0	j0	j0	j0	j0	j0	j0	j0	j0	j0	j0	j0	j0	j0
LPS_ 3	0	0	0	0	0	0	0	0	0	0	0	0	0	0	0	0	0
	j0	j0	j0	j0	j0	j0	j0	j0	j0	j0	j0	j0	j0	j0	j0	j0	j0
YPS_ 3	0	0	0	0	0	0	0	0	0	0	0	0	0	0	0	0	0
	j0	j0	j0	j0	j0	j0	j0	j0	j0	j0	j0	j0	j0	j0	j0	j0	j0
TPS_ 4	0	-0.01	0	0	0	0	0	-1.04	-0.14	0.15	-0.04	0	0	0	-1.09	-0.18	-1.27
	j0	j0	j0	j0	j0	j0	j0	-j0.41	j0.62	-j0.70	j0.23	j0	j0	j0	-j0.26	j0.01	-j0.25
CPS_ 4	0	0.01	0	0	0	0	0	-0.43	-20.6	-3.33	0.03	0	0	0	-24.3	0.08	-24.3
	j0	j0	j0	j0	j0	j0	j0	j0.53	-j1.10	j1.37	-j0.29	j0	j0	j0	j0.52	j0.21	j0.73
SPS_	0	0	0	0	0	0	0	0.44	-7.61	-55.5	-4.57	0	0	0	-67.2	-0.19	-67.4
	j0	j0	j0	j0	j0	j0	j0	-j0.01	j2.36	j3.40	j2.38	j0	j0	j0	j8.13	j0.25	j7.88
GPS_ 4	0	0	0	0	0	0	0	-0.57	1.85	-5.21	-5.50	0	0	0	-9.43	0.03	-9.39
	j0	j0	j0	j0	j0	j0	j0	j0.06	-j1.31	j4.54	-j0.14	j0	j0	j0	j3.15	j0.14	j3.30
NPS_ 5	0	0	0	0	0	0	0	0	0	0	0	0	0	0	0	0	0
	j0	j0	j0	j0	j0	j0	j0	j0	j0	j0	j0	j0	j0	j0	j0	j0	j0
TPS_ 5	0	0	0	0	0	0	0	0	0	0	0	0	0	0	0	0	0
	j0	j0	j0	j0	j0	j0	j0	j0	j0	j0	j0	j0	j0	j0	j0	j0	j0
PPS_ 5	0	0	0	0	0	0	0	0	0	0	0	0	0	0	0	0	0
	j0	j0	j0	j0	j0	j0	j0	j0	j0	j0	j0	j0	j0	j0	j0	j0	j0
Col. Sum	0	0.01	0	0	0	0	0	-1.61	-26.5	-63.9	-10.1	0	0	0	-102.	-0.26	-102.
	0	-j0.02	-j0.01	j0	j0	j0	j0	j0.16	j0.58	j8.62	j2.18	j0	j0	j0	j11.5	j0.11	j11.6

All PSS damping gains=5 pu, increment 1 pu on machine MVA rating.
Gain of FDS=30 pu, increment 0.9 pu on device base.
The incremental mode shift induced by the generator's own PSS is shaded in yellow.
The box highlights the PSSs and generators which are the main contributors to the damping of mode B.

7. From the columns in Table 13.3 we can assess the contribution to the shift of mode B due to a gain increment on any PSS or the FDS, or a selected group of stabilizers. For instance, a 1 pu increment in the gain of the SPS_4 PSS only (col. 10) causes a real mode shift of -55.5 units to be induced on SPS_4. However, due to an 'interaction' between the PSS of SPS_4 and the other PSSs in Area 4, a real mode shift contribution is also induced on each of the other generators, i.e. -5.2, -3.3 and 0.15 units on GPS_4, CPS_4 and TPS_4, respectively. Therefore, due to the increment in the SPS_4 PSS gain, the units in Area 4 contribute a real mode shift of -64 units.

8. The total mode shift comprising the sum of the real components of all the diagonal (self) terms in the table is -82.6 units. Including the mode shifts induced by all PSS interactions, the real part of total mode shift due to PSSs is -102 units (col. 15). Interactions have thus enhanced the damping of the local mode B.

9. From the inertia-weighted, speed participation factors of Figure 13.3 it is observed that the generators SPS_4, CPS_4 and GPS_4, in that order, are the dominant participants in mode B; all other generators participate in small (e.g. TPS_4) or negligible amounts. It is therefore not surprising that, for a 1 pu increment in all PSS gains, the three dominant participants contribute a real component of -101 units out of a total real contribution of -102 units (col. 15) to the enhancement of the damping of mode B.

10. In assessing the effectiveness of PSSs on the damping of mode B note that, according to the 'column sum', a 1 pu increment in the gain of the PSS on SPS_4 is about 2.5 times and 6 times more effective than a similar change on the PSSs of CPS_4 and GPS_4, respectively.

13.3.3 Contributions to the MITCs of each generator, inter-area mode M

Let us now assess the full array of the contributions to the MITC of each generator for the *inter-area* mode M, $\lambda_M = -0.42 \pm j2.04$. The array for 1 pu and 0.9 pu gain increments on the PSSs and the FDS, respectively, is shown in Table 13.4 on page 605.

From Table 13.3 on page 599 for the inertia-weighted, speed participation factors it is noted that generators in Areas 4, 5 and 3 are the dominant participants in the inter-area mode M. Let us examine the MITCs for units within the latter Areas, initially ignoring the contributions from the FDS, and remembering that for each row the MVA rating is the base quantity of the associated generator.

Table 13.4 Case 1, Mode M. MITCs on each unit due to gain increments on PSSs/FDS

Contributions to incremental modal induced torque coefficients ΔT_{ij}^{M} of each generator i due to an increment in the gain of the stabilizers listed in the column headings listed in the left-hand column (pu on generator MVA rating)

Inter-area mode M, $\lambda_M = -0.42 \pm j2.04$

Generator	HPS_1 PSS	BPS_2 PSS	EPS_2 PSS	VPS_2 PSS	MPS_2 PSS	LPS_3 PSS	YPS_3 PSS	TPS_4 PSS	CPS_4 PSS	SPS_4 PSS	GPS_4 PSS	NPS_5 PSS	TPS_5 PSS	PPS_5 PSS	BSVC 4 FDS	Row Sum
cols	1	2	3	4	5	6	7	8	9	10	11	12	13	14	15	16
HPS_1	0.74	-0.04	-0.02	0	-0.08	-0.33	-0.02	0.02	0	0	0	0	0	0	0.01	0.26
	j0	j0.01	j0.01	-j0.01	j0.01	-j0.03	j0	j0.00	j0	j0	j0	j0	j0	j0	j0.01	-j0.01
BPS_2	0.21	0.90	-0.20	-0.11	-0.30	0.01	0	0.07	0.01	0.01	0.01	0	0	0	0.04	0.56
	j0.07	j0.10	j0.03	-j0.03	-j0.08	j0.04	j0.01	j0.10	j0.01	j0.01	j0.02	j0	j0	j0	j0.04	j0.32
EPS_2	0.13	-0.20	1.00	-0.10	-0.08	-0.08	-0.01	0.03	0.01	0	0	0	0	0	0.05	0.65
	j0.21	j0.07	j0.07	j0.19	j0.12	j0.09	j0.01	j0.04	j0.01	j0.01	j0.01	j0	j0	j0	j0.01	j0.84
VPS_2	0.31	-0.31	-0.31	1.00	-0.02	0.01	0	0.06	0.01	0.01	0.01	0	0	0	0.07	0.68
	j0.17	-j0.06	j0.01	j0.08	-j0.06	j0.09	j0.01	j0.05	j0.01	j0.01	j0.01	j0	j0	j0	j0.04	j0.37
MPS_2	0.05	-0.18	-0.06	-0.01	0.77	-0.08	0	0.03	0	0	0	0	0	0	0.02	0.50
	j0.07	j0.02	j0.02	-j0.01	j0.09	-j0.02	j0	j0.02	j0	j0	j0	j0	j0	j0	j0.02	j0.22
LPS_3	-0.01	0	0	0	-0.01	1.04	-0.12	0	0	0	0	-0.01	-0.02	-0.01	0	0.85
	j0.02	j0	j0	j0	j0	j0.04	j0.06	j0	j0	j0	j0	j0	j0	j0	j0	j0.12
YPS_3	0.03	0	0	0	-0.01	-0.59	1.04	0	0	0	0	-0.01	-0.01	-0.01	0	0.46
	j0.03	j0	j0	j0	j0	j0.21	-j0.07	j0	j0	j0	j0	j0	j0	j0	j0	j0.18
TPS_4	-0.01	0.01	0	0	0	-0.06	0	0.71	-0.14	-0.11	-0.13	0	0	0	0.61	0.93
	j0.01	-j0.01	j0	j0	j0	j0	j0	j0.14	j0.02	-j0.04	j0.04	j0	j0	j0	j0.07	j0.11
CPS_4	0	0	0	0	0	0	0	-0.11	0.92	-0.30	-0.19	0	0	0	0.10	0.41
	j0	j0	j0	j0	j0	j0	j0	j0.16	j0.08	j0.12	j0.16	j0	j0	j0	j0.08	j0.45
SPS_4	0	0	0	0	0	0	0	0.01	-0.14	0.90	-0.24	0	0	0	0.14	0.67
	j0	j0	j0	j0	j0	j0	j0	-j0.01	-j0.01	j0.10	-j0.01	j0	j0	j0	-j0.09	-j0.02
GPS_4	0	0	0	0	0	0	0	-0.05	-0.12	-0.33	0.77	0	0	0	0.11	0.37
	j0	j0	j0	j0	j0	j0	j0	j0	-j0.01	-j0.03	j0.07	j0	j0	j0	-j0.09	-j0.07
NPS_5	0	0	0	0	0	-0.05	-0.01	0	0	0	0	0.76	-0.61	-0.32	0	-0.24
	j0	j0	j0	j0	j0	j0.03	j0	j0	j0	j0	j0	j0.06	j0.09	-j0.03	j0	j0.15
TPS_5	0	0	0	0	0	-0.07	-0.01	0	0	0	0	-0.15	1.06	-0.17	0	0.65
	j0	j0	j0	j0	j0	j0.03	j0	j0	j0	j0	j0	j0.02	j0.06	j0.02	j0	j0.12
PPS_5	0	0	0	0	0	-0.07	-0.01	0	0	0	0	-0.20	-0.47	0.74	0	0.00
	j0	j0	j0	j0	j0	j0.01	j0	j0	j0	j0	j0	-j0.05	-j0.05	j0.12	j0	j0.02

All PSS damping gains=5 pu, increment 1 pu on machine MVA rating..
Gain of FDS=30 pu, increment 0.9 pu on device base.
The incremental MITC induced by the generator's PSS is shaded in yellow.

11. In the case of mode M, let us assume that only the PSSs gains on each machine in Area 4 are raised 1 pu on generator MVA rating; the remaining PSS gains remain unchanged. The off-diagonal terms for Area 4 generation in the box in Table 13.4

are negligible or negative - and therefore the associated interactions between the Area 4 units are destabilizing. As is evident in col. 16, the net effect for each Area 4 generator is that the resulting real contribution to the MITC of each is less than that contributed by its own PSS. In the case of local mode B on the other hand, and with the exception for units in Area 5, it was found that the off-diagonal terms tend to enhance the contribution to the damping (see Table 13.2 on page 601.)

12. The same comments apply to generators in Areas 3 and 5 when only the PSS gains in the same Area are incremented.

13. An increment in gain on the FDS mainly increases the MITCs on Area 4 generators and consequently improves the damping of mode M.

14. The relative MITCs of both the local-area mode B and the inter-area mode M are demonstrated in the bar chart shown in Figure 13.4 on page 607. Note (i) the charts reflects the 'Row Sum', $\Delta\Gamma_i^h = \sum_{j=1}^{n+z} \Delta T_{ij}^h$, in column 16 for each generator in Tables 13.2 and 13.4; (ii) the Row Sum includes the FDS contribution; (iii) due to interactions the value of $\Delta\Gamma_i^h$ hovers about 1 pu for mode B, but is significantly less for than 1 pu for mode M.

13.3.4 Contributions of the mode shifts of each generator to the Mode M damping

As was examined for mode B in Section 13.3.2, let us now consider the full array of *mode shift* contributions for the *inter-area* mode M bearing in mind that mode shifts are directly related to the MITCs through (13.6). The components of contributions to the mode shifts are listed in Table 13.5 on page 608. From the latter table the following are noted.

15. If, for exploratory purposes, it is desirable to increase the gains on all PSSs by 1 pu and the FDS gain by 0.9 pu, the total shift in the real part of mode M is found from the table to be -40.8 units (col. 17), where 1 unit = 10^{-3}. Under the same conditions, the total shift in the real part of mode B is -102 units, a factor of 2.5 times that of mode M. This result emphasizes the relatively poorer damping characteristics of the inter-area mode compared to the local mode for the same increments in stabilizer gains.

16. Gain increments of 1 pu on the PSSs of generators in Area 2 provide a relatively small net improvement in damping of mode M (see cols. 2 to 5, 15).

17. If damping of Mode M is to be improved by increasing the gains on PSSs by varying amounts, the PSSs which ought to be selected are revealed by an examination of the columns of Table 13.5. It is evident that 1 pu increment in the gain of the LPS_3 PSS followed by GPS_4, SPS_4 and TPS_4 PSSs provide a net greater boost to the damping of mode M (-8.4, -6.5, -4.1 and -4.0 units, respectively) (see the 'Col. Sum'). In comparison, for the local mode B, the same increment in gain only on the SPS_4, CPS_4, and GPS_4 PSSs boosts the damping of mode B (-63.9, -26.5 and -10.1

units, respectively). To boost the damping of mode M, it may be desirable to coordinate the increase the gain, say, of LPS_3 by 3 pu, GPS_4 by 3pu and SPS_4 by 2 pu.

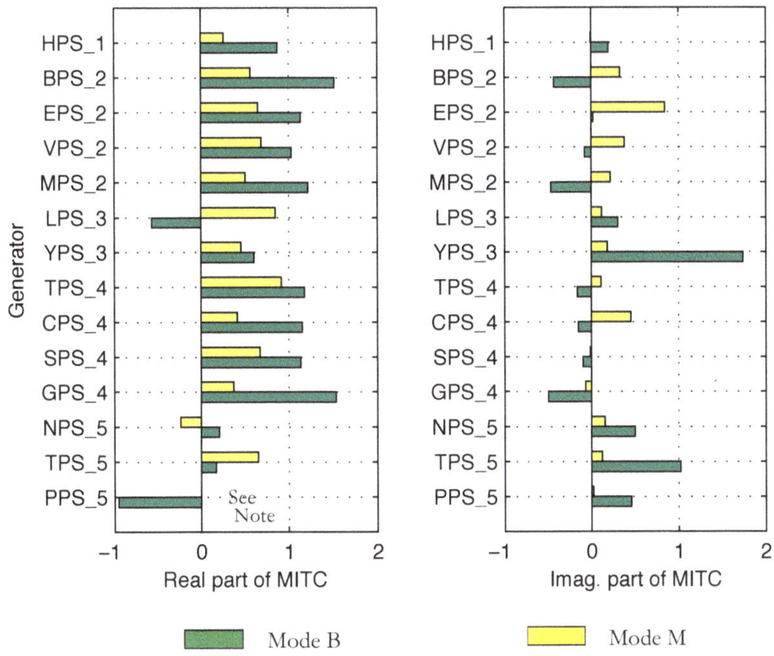

Figure 13.4 Total incremental MITCs, $\Delta\Gamma_i^h$, on generators produced by increments of 1 pu on all PSSs and 0.9 pu on the FDS from gain settings of 5 pu on PSSs and 30 pu on the FDS, inter-area & local modes M & B. Note: 20% of MITC for PPS_5 for mode B is shown (actual is -4.73+j2.27 pu on machine MVA rating).

18. The FDS provides a relatively significant boost to the total incremental damping of mode M. In Table 13.5 the PSSs gain increments contribute shifts of -30.7 units and the single FDS -10.1 units, i.e about 25% of the total mode shift of -40.8 units.

19. In Section 13.3.1 it was noted that with the FDS on BSVC_4 out of and in operation (gain: 0 and 30 pu) the real part of mode M was enhanced from -0.14 to -0.42 Np/s, a change of -274 units. In Table 13.5, for gain increment of 1 pu on the FDS, the total boost of the damping of mode M is -10.1 units. *Decreasing* its gain by 30 pu to zero yields an estimate of the change in the real part of mode M of $-30 \times (-10.1) = 303$ units. For exploratory purposes, the magnitude of these changes are close enough to suggest that the relationship between gain increments in the FDS and shifts in mode M is reasonable linear in the vicinity of 30 pu. Moreover, if the effect of an increase in the FDS gain on the damping of mode M is being investigated, with some confidence it can be assumed that the boost in the real part of mode M, say a 10 pu gain change, is likely to be from -0.42 to about -0.52 Np/s.

Table 13.5 Case1, Mode M. Mode Shift contributions for gain increments on PSSs/FDS

	Contributions to Mode Shift $\Delta\lambda_{ij}^{M}$ by each generator (row) and by each PSS or FDS (col.) (x 10^{-3})																
	Inter-area mode M, $\lambda_M = -0.42 \pm j2.04$																
Generator	HPS_1 PSS	BPS_2 PSS	EPS_2 PSS	VPS_2 PSS	MPS2 PSS	LPS_3 PSS	YPS_3 PSS	TPS_4 PSS	CPS_4 PSS	SPS_4 PSS	GPS_4 PSS	NPS_5 PSS	TPS_5 PSS	PPS_5 PSS	Sum PSSs	SVC4 FDS	Row Sum
	1	2	3	4	5	6	7	8	9	10	11	12	13	14	15	16	17
HPS_1	-0.52 j0.46	0.02 -j0.03	0.01 -j0.02	0 j0.01	0.06 -j0.06	0.26 -j0.19	0.01 -j0.01	-0.01 j0.01	0 j0	0 j0	0 j0	0 j0	0 j0	0 j0	-0.17 j0.19	0 -j0.01	-0.18 j0.17
BPS_2	-0.07 j0.20	-0.11 j0.90	-0.03 -j0.20	0.03 j0.11	0.08 -j0.30	-0.04 j0.01	-0.01 j0	-0.10 j0.07	-0.01 j0.01	-0.01 j0.01	-0.02 j0	0 j0	0 j0	0 j0	-0.28 j0.60	-0.04 -j0.04	-0.33 j0.55
EPS_2	-0.23 j0.12	-0.05 -j0.21	-0.14 j1.02	-0.19 -j0.12	-0.11 -j0.09	-0.09 -j0.09	-0.01 -j0.01	-0.04 j0.03	-0.01 j0.01	-0.01 j0	-0.01 j0	0 j0	0 j0	0 j0	-0.89 j0.65	0.01 -j0.05	-0.90 j0.60
VPS_2	-0.13 j0.18	0.06 -j0.19	0.01 -j0.18	-0.13 j0.62	0.04 -j0.01	-0.06 j0	-0.01 j0	-0.04 j0.03	-0.01 j0.01	-0.01 j0	-0.01 j0	0 j0	0 j0	0 j0	-0.26 j0.45	-0.02 -j0.05	-0.28 j0.40
MPS_2	-0.15 j0.04	0.08 -j0.29	0.01 -j0.11	0.03 j0.02	-0.61 j1.2	0.08 -j0.12	0 -j0.01	-0.04 j0.03	-0.01 j0.01	-0.01 j0	-0.01 j0	0 j0	0 j0	0 j0	-0.62 j0.70	-0.02 -j0.04	-0.64 j0.66
LPS_3	0.05 -j0.18	0.02 -j0.03	0.02 -j0.02	0.01 j0.01	0.08 -j0.06	-11.0 j1.37	1.20 -j0.82	-0.01 j0.01	0 j0	0 j0	0 j0	0.11 j0.01	0.23 -j0.03	0.11 j0.01	-9.12 j0.28	0.01 -j0.02	-9.13 j0.26
YPS_3	-0.12 -j0.03	0.01 -j0.01	0 j0	0 j0	0.01 -j0.01	1.05 -j1.24	-2.27 j1.50	0 j0	0 j0	0 j0	0 j0	0.01 -j0.01	0.02 -j0.02	0.01 -j0.01	-1.27 j0.18	0 j0	-1.27 j0.18
TPS_4	0.09 -j0.05	-0.10 j0.10	0.01 -j0.02	0.03 -j0.04	0.03 -j0.04	0.04 j0.01	0 j0	-5.54 -j0.92	1.11 j0.14	0.84 j0.28	0.99 j0.29	0 j0	0 j0	0 j0	-2.50 -j0.23	-4.73 -j0.44	-7.23 -j0.68
CPS_4	0.02 -j0.02	-0.02 j0.03	0 j0	0 -j0.01	0 -j0.01	0.01 j0	0 j0	1.15 -j0.97	-6.68 -j2.48	2.50 -j0.29	1.74 -j0.80	0 j0	0 j0	0 j0	-1.28 -j4.55	-0.90 j0.37	-2.18 -j4.19
SPS_4	0 j0	0 j0	0 j0	0 j0	0 j0	0 j0	0 j0	0.22 j0.04	2.04 j0.53	-12.8 -j4.04	3.41 j0.87	0 j0	0 j0	0 j0	-7.59 -j2.61	-2.26 j0.95	-9.85 -j1.65
GPS_4	0.02 j0	-0.02 j0.01	0 j0	0.01 -j0.01	0.01 -j0.01	0.01 j0	0 j0	0.85 j0.24	1.93 j0.56	5.46 j1.62	-12.5 -j3.61	0 j0	0 j0	0 j0	-4.29 -j1.19	-2.15 j1.18	-6.44 j0.01
NPS_5	0 j0.01	0 j0	0 j0	0 j0	0 j0	0.32 j0	0.03 j0	0 j0	0 j0	0 j0	0 j0	-3.09 -j2.35	2.87 j1.31	1.30 j1.02	1.42 j0	0 j0	1.42 -j0
TPS_5	-0.01 j0.02	0 j0	0 j0	0 j0	0 j0	0.58 j0.06	0.05 j0.01	0 j0	0 j0	0 j0	0 j0	1.09 j0.43	-6.79 -j4.17	1.20 j0.51	-3.87 -j3.14	0 j0	-3.87 -j3.14
PPS_5	0 j0.01	0 j0	0 j0	0 j0	0 j0	0.37 j0.13	0.03 j0.02	0 j0	0 j0	0 j0	0 j0	0.89 j0.78	2.30 j1.47	-3.54 -j2.53	0.01 -j0.12	0 j0	0.06 -j0.11
Col. Sum	-1.03 j0.79	-0.10 j0.27	-0.10 j0.45	-0.18 j0.34	-0.40 j0.57	-8.42 -j0.05	-0.95 j0.68	-4.01 -j1.42	-1.65 j1.21	-4.09 -j2.41	-6.46 -j3.24	-0.99 -j1.14	-1.37 -j1.44	-0.91 -j1.00	-30.7 -j8.79	-10.1 j1.84	-40.8 j6.96

All PSS damping gains=5 pu, increment 1 pu on machine MVA rating.
Gain of FDS=30 pu, increment 0.9 pu on device base.
The incremental mode shift induced by the generator's own PSS is shaded in yellow

20. Let us compare, for a significant participant in each of the modes B and M, the real parts of the *diagonal element* with that of the *column sum* for a selected PSS. For example, for a 1 pu gain increment in the PSS of SPS_4, Table 13.3 on page 603

(col. 10) and Table 13.5 (col. 10), reveal that interactions not only increase the left-shift of the local mode B from -55.5 to -63.9 units, but also reduce significantly the left-shift of the inter-area mode M from -12.8 to -4.1 units.

13.4 Stabilizer damping contribution diagrams

In order to assess the relative effectiveness of the stabilizers concerned, the contributions to the mode shifts of many PSSs and FDSs may need to be examined jointly and compared. Much of the information inherent in the arrays of the contributions to mode shifts for the local- and inter-area modes can be meaningfully displayed graphically in *stabilizer damping contribution diagrams* (SDCDs). From a SDCD it is possible to assess the joint effects of positive, zero or negative gain increments of differing magnitudes on individual stabilizers. Such diagrams relate either to the column or row sums of mode shift contributions in the arrays of Tables 13.3 or 13.5 and therefore take the effect of interactions into account. Earlier, in Section 13.2.2, the concepts of *stabilizer interactions* and *generator interactions* are introduced based on the associated forms of the equations (13.14) and (13.16) which relate the complex mode shifts to incremental stabilizer gains.

The SDCDs are now examined for the two types of interactions for three inter-area modes and one local mode; the variety of information provided by each SDCD is illuminating.

Consider firstly the mode shift on generator *i* resulting from an increment in its PSS gain; the shift is derived from (13.16):

$$\Delta\lambda^{n}\big|_{\text{pss\_i}} = -(\rho_{ih}/M_i)\cdot H_{ii}(\lambda_h)\,[G_i(\lambda_h)\Delta k_i] - \sum_{j=1\neq i}^{n} (\rho_{jh}/M_j)H_{ji}(\lambda_h)(c_{j*}v_{*h}/v_{ih})\,[G_i(\lambda_h)\Delta k_i]$$

As mentioned in Section 13.2.2 the first term in the above equation shows the mode shift on generator *i* resulting from a gain increment Δk_i on its own PSS. The second term reveals that, for the same PSS gain increment Δk_i, there are also contributions to the mode shift of generator *i* by the other $n-1$ generators. The relevant information is provided in the columns for each PSS in Tables 13.3 and 13.5. For example, consider col. 6 in Table 13.5 for the mode M, $\lambda_M = -0.42 \pm j2.04$. Listed in col. 6 are the component mode shifts resulting from a gain increment of 1 pu on the PSS of generator LPS_3. The mode shift $\Delta\lambda_{ii}^{h}$ which is displayed in row 6 of col. 6 is $-11.0 \pm j1.37$ units (1 unit = 10^{-3}).The mode shifts $\Delta\lambda_{ij}^{h}$ which are associated with interactions with the other generators are listed in the remaining $n-1$ elements in col. 6. The net mode shift, $\Delta\lambda^{h}\big|_{\text{pss\_i}}$, is the sum of all the contributions in col. 6, i.e. $-8.42 - j0.05$ units; in this case the effect of interactions is to degrade the damping contribution of the PSS on generator LPS_3 by 2.6 units.

Secondly, consider the SDCD associated with a gain increment on each stabilizer as displayed in Figure 13.5 for the inter-area modes M ($\lambda_M = -0.42 \pm j2.04$) and L ($\lambda_L = -0.14 \pm j2.56$). As noted above for the inter-area mode M the net mode shift, $\Delta\lambda^h|_{\text{pss\_i}}$, for the increment in the damping gain of the LPS_3 PSS is $-8.42 - j0.05$ units; this result is also shown in the figure. The main contributors (found from the column sums in the last row of Table 13.5) to the damping of the inter-area mode M are not only the PSSs in Areas 3 & 4 but also the FDS on BSVC_4. Note that in col. 16 of Table 13.5 for mode M the contribution from the FDS to the mode shift is $-10.1 + j1.84$ units. This is the dominant contribution which results from the increment in the FDS gain and is clearly observed in Figure 13.5.

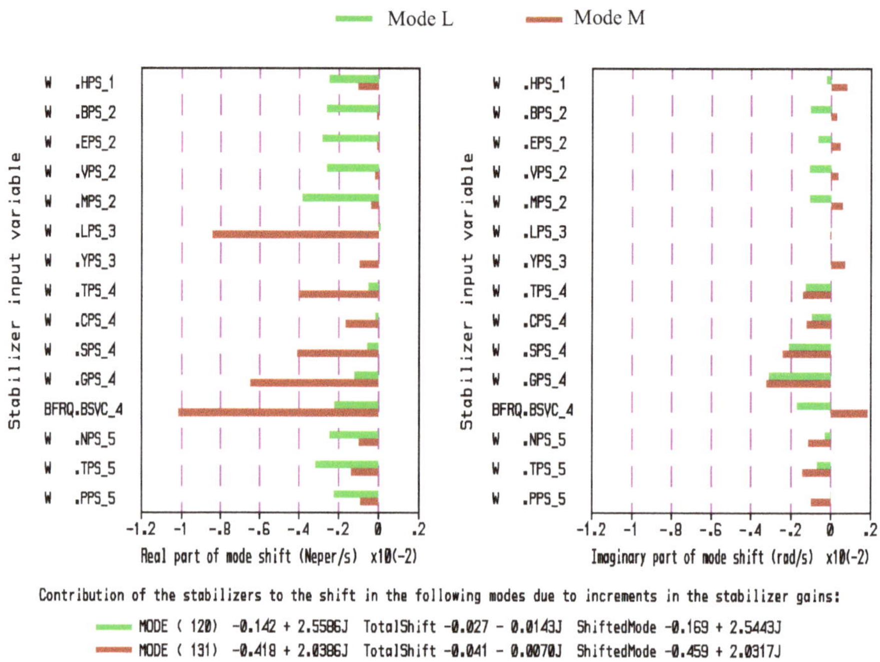

Figure 13.5 Case 1. Stabilizer damping contribution diagram for inter-area modes L & M for a damping gain increment of 1 pu on each PSS and 0.9 pu on the FDS. Nominal gain settings: All PSSs 5pu, FDS 30 pu. Note scale.

As an aside, consider the contribution of the FDS derived from (13.8) in which the gain of FDS j is incremented by Δk_j. The resulting contribution of FDS j to the mode shift is:

$$\Delta\lambda^h|_{\text{fds\_j}} = -\left(\sum_{i=1}^{n} (\rho_{ih}/M_i) H_{ij}(\lambda_h) (\boldsymbol{c}_{j*}\boldsymbol{v}_{*h}/v_{ih})\right)[G_j(\lambda_h)\Delta k_j],$$

and demonstrates that the FDS interacts with all of n generators to produce a shift in the mode.

The SDCD in Figure 13.5 further reveals that the damping of mode L is effectively improved by increasing the gains of PSSs in Areas 2 and 5 as well as the gain of the FDS.

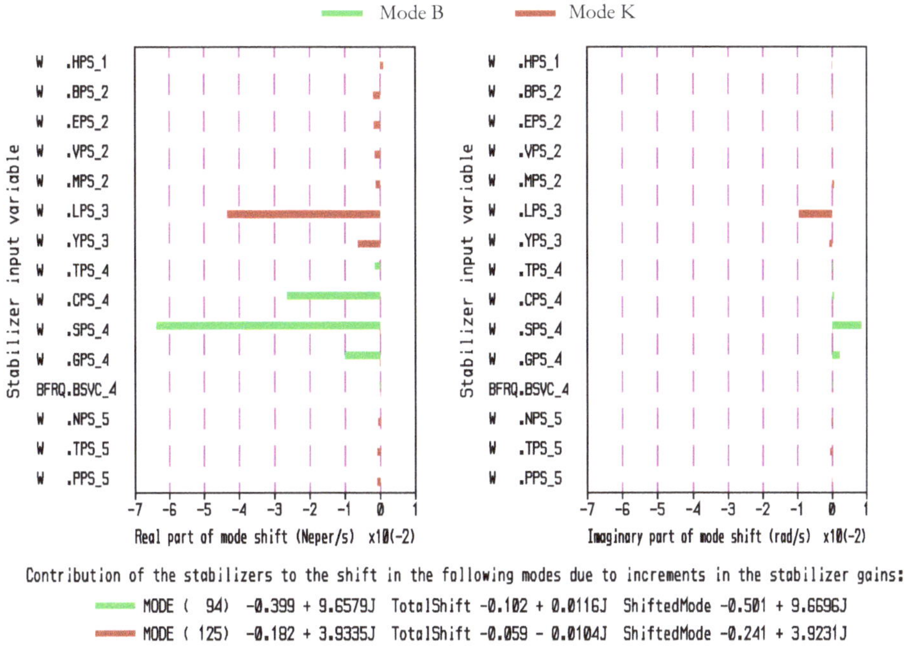

Figure 13.6 Case 1. Stabilizer damping contribution diagram for local- and inter-area modes B & K for damping a gain increment of 1 pu on each PSS and of 0.9 pu on the FDS. Nominal gain settings: All PSSs 5pu, FDS 30 pu. Note scale.

In Figure 13.6 is shown the SDCD associated with a gain increment on each stabilizer for the local-area mode B, $\lambda_B = -0.40 \pm j9.66$, and the inter-area mode K, $\lambda_K = -0.18 \pm j3.93$. For mode B the main contributors to the enhanced damping of this local-area mode (found from the column sums in the final row of Table 13.3 on page 603) are the PSSs in Area 4. Useful information revealed by the SDCD for the inter-area mode K is that increasing the PSS gain on unit LPS_3 is an effective way of boosting the damping of that mode. Incrementing the gain of the FDS, however, is ineffective in improving the damping of both modes. Note that the SDCDs do not show the extent of the interactions between PSSs which may enhance or degrade the damping of the mode; such effects are revealed in the off-diagonal elements in the columns of Tables 13.3 and 13.5.

Note that the relative amounts of the increments in individual stabilizer gains, whether they be positive, zero or negative, depend on a number of factors. In particular, before gain increments are decided upon, the effects of gain changes on the pertinent local- and inter-area modes need to be examined for an encompassing range of operating conditions, normal and outage.

An alternative set of SDCDs, shown in Figures 13.7 and 13.8, is based on (13.14), namely:

$$\Delta\lambda^h\big|_{\text{gen\_i}} = -(\rho_{ih}/M_i)\cdot H_{ii}(\lambda_h)\,[G_i(\lambda_h)\Delta k_i] \\ -(\rho_{ih}/M_i)\cdot \sum_{j=1\neq i}^{n+z} H_{ij}(\lambda_h)(\boldsymbol{c}_{j*}\boldsymbol{u}_{*h}/u_{ih})[G_j(\lambda_h)\Delta k_j]\cdot$$

The contributions to the mode shift by generator *i* resulting from a gain increment on each of the $n+z$ stabilizers are shown in the rows for that generator in Tables 13.3 and 13.5. For each generator in these tables the net mode shift, $\Delta\lambda^h\big|_{\text{gen\_i}}$ (the 'Row Sum' in col. 17), is displayed in the SDCD.

Figure 13.7 shows the contributions to damping of modes L and M by each generator when all stabilizer gains are increased.

In Figure 13.7 it is of interest to note that generator PPS_5 makes a negligible contribution to mode M although it is a dominant participant in the state participation factors in Figure 13.2. This is seen to apply also to mode L in Figure 10.33 when all PSS gains are 5 pu on machine base but the FDS is out of operation. From these observations it is concluded that strong participation in a mode does not necessarily mean that the unit contributes to the damping of that mode - as is evident in Figure 10.32. Examination of the row for the generator PPS_5 in Table 13.4 on page 605 for mode M reveals that the real component of the incremental MITC of 0.74 pu due to the PSS installed on that generator is cancelled by negative interactions with the other stabilizers, notably TPS_5, NPS_5 and LPS_3. Consequently, even though the inertia-weighted participation factor of about 0.06 pu for PSS_5 in Figure 13.3 is not insignificant the mode shift due to PPS_5 is negligible because the net incremental MITCs for the generator are negligible.

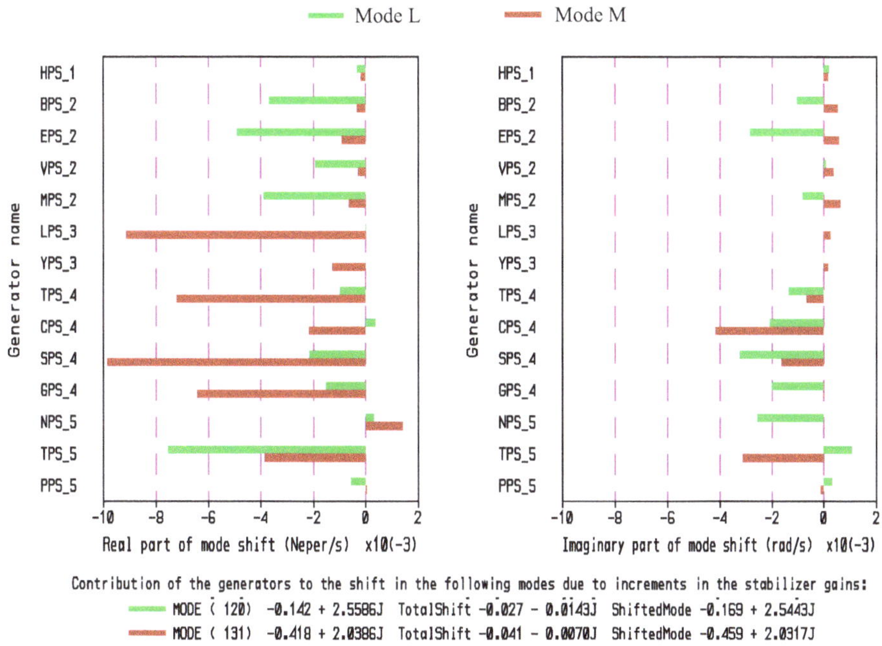

Figure 13.7 Case 1. Contributions to mode shifts of modes L & M by *each generator* for a damping gain increment of 1 pu on all PSSs and 0.9 pu on the FDS. Nominal gains: All PSSs 5pu, FDS 30 pu. Note scale.

The contributions to damping of modes B and K by each *generator* in Figure 13.8 on page 614 appear to be similar to those in Figure 13.6 for the contributions to damping of modes B and K by each *stabilizer*. This not the case in Figures 13.5 and 13.7 for modes M and L. The observation on modes B and K is valid for those units for which the interactions with other stabilizers are small compared to the mode shifts induced on generators by their own PSSs. Furthermore, note that the contribution by the FDS to the damping of modes B and K is negligible; it is significant for modes M and L.

While Tables 13.3 and 13.5 and the SDCDs provide - for tuning or exploratory purposes - similar useful information on the effectiveness of stabilizers for small increments in gain, the question arises 'for what size gain increments is that information valid?'. The following illustrates an approach for answering the question.

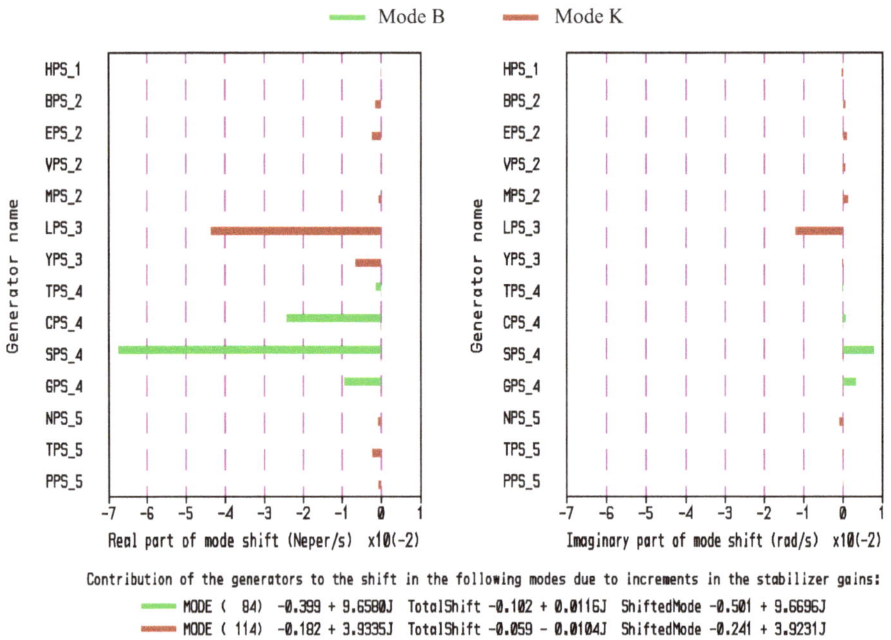

Figure 13.8 Case 1. Contributions to mode shifts of modes B and K by *each generator* for a damping gain increment of 1 pu on all PSSs and a gain of 0.9 pu on the FDS. Gains: All PSSs 5pu, FDS 30 pu. Note scale.

13.5 Comparison of the estimated and actual mode shifts for increments in stabilizer gain settings

In the studies of Section 13.3 the increments in PSS and FDS gains are set to 1 pu and 0.9 pu, respectively. A comparison of the estimated modal trajectories from the SDCDs and the actual trajectories is therefore made (i) for the local-area mode B, and (ii) for the inter-area mode M, for gain increments Δk of 1, 5 and 10 pu for all PSSs, and 0.9, 4.5 and 9.0 pu for the FDS based on the nominal gain settings of 5 and 30 pu for the PSSs and the FDS, respectively. The estimated mode shifts, $\Delta\lambda_{est}$, the associated estimated mode value, λ_{est}, and the actual mode values, λ_{act}, are shown in Table 13.6. The actual values are determined by recalculating the eigenvalues with the nominal plus the incremented values of the stabilizer gain settings.

Table 13.6 Estimated changes from the SDCDs and actual changes in modes B & M for increasing increments in PSS and FDS gains

Gain increments		Local-area mode B, $\lambda_B = -0.40 + j9.66$			
		$\Delta\lambda_{est}$	λ_{est}	λ_{act}	$\lambda_{act} - \lambda_{est}$
PSSs	FDS	col. 1	col. 2	col. 3	col. 4
20%, 1pu *	3%, 0.9 pu	$-0.102 + j0.012$	$-0.501 + j9.670$	$-0.502 + j9.669$	$-0.001 - j0.001$
100%, 5pu	15%, 4.5 pu	$-0.512 + j0.058$	$-0.911 + j9.716$	$-0.918 + j9.705$	$-0.007 - j0.011$
200%, 10pu	30%, 9.0 pu	$-1.023 + j0.116$	$-1.422 + j9.774$	$-1.451 + j9.731$	$-0.029 - j0.403$

* Increments in stabilizer gain settings used for Table 13.3 and for the SDCDs
Nominal gain settings for PSSs and the FDS are 5 and 30 pu, respectively.

$$\lambda_{est} = \Delta\lambda_{est} + (-0.40 + j9.66)$$

Gain increments		Inter-area mode M, $\lambda_M = -0.42 + j2.04$			
		$\Delta\lambda_{est}$	λ_{est}	λ_{act}	$\lambda_{act} - \lambda_{est}$
PSSs	FDS	col. 1	col. 2	col. 3	col. 4
20%, 1pu *	3%, 0.9pu	$-0.041 - j0.007$	$-0.459 + j2.032$	$-0.459 + j2.031$	$0 - j0.001$
100%, 5pu	15%, 4.5pu	$-0.204 - j0.035$	$-0.622 + j2.004$	$-0.635 + j1.997$	$-0.013 - j0.007$
200%, 10pu	30%, 9.0pu	$-0.408 - j0.070$	$-0.826 + j1.969$	$-0.887 + j1.929$	$-0.061 - j0.040$

* Increments in stabilizer gain settings used for Table 13.5 and for the SDCDs
Nominal gain settings for PSSs and the FDS are 5 and 30 pu, respectively.

$$\lambda_{est} = \Delta\lambda_{est} + (-0.42 + j2.04)$$

From Table 13.6 it is observed that both the real and imaginary parts of the actual and estimated mode values agree within 5% for gain increments up to five times the increments of 1 and 0.9 pu for the PSS and FDS gains. As the gain increments increase towards ten times the latter gain increments, the actual and estimated mode values agree within 7.5% for mode M; the agreement for mode B remains within 5%. The nature of the actual trajectory for mode M observed in Table 13.6 is consistent with that observed in Figure 11.7, i.e. from values of the FDS gain above 35 pu the change in the mode value does not increase linearly with change in gain.

As seen in the above tables and in figures such as Figure 10.26, the predominately left shift of the modes with increasing stabilizer gain is the objective of the tuning procedures outlined in Chapters 5, 6 and 11. Given that the stabilizer transfer functions are of the form $k_i G_i(s)$, where k_i is a real gain (in pu on device base) and $G_i(s)$ is the compensation transfer function, then ideally, (i) $G_i(s)$ ensures the left shift of all modes over a range of modal frequencies, and (ii) the gain k_i determines the extent of the mode shift.

The basis for the tuning of PSS transfer functions $k_iG_i(s)$ using the P-Vr method is explained and applied in Chapters 5, 9 and 10. Adopting this approach, and employing the appropriate SDCDs, suggest that the coordination of stabilizers can be achieved through the coordination of their gains, k_i. This is the basis of the analysis in Chapter 14.

The acceptable extent to which the actual and estimated modal trajectory diverge as the stabilizer gains are incremented depends on the user's application and objectives. For example, in a formal procedure for the coordination of gain settings for PSSs and FDSs it is necessary to confirm that the gain increments selected do not result in the difference between the estimated and actual mode shifts exceeding acceptable limits over a range of operating conditions [4].

13.6 Summary

13.6.1 Interactions [3] [1]

The analysis of interactions is based on PSS and FDS transfer functions being of the form $k_iG_i(s)$. As stated, the transfer function $G_i(s)$ is tuned to effect a left-shift of the rotor modes and the gain k_i determines the extent of the shift. It is shown that, for an increment in PSS damping gain Δk_i the self-induced modal torque coefficient on generator *i* and the associated self-contribution to the shift in mode λ_h are $\Delta T_{ii}^h \approx \Delta k_i$ and $\Delta\lambda_{ii}^h = -(\rho_{ih}/M_i)\,\Delta T_{ii}^h$, respectively. This is consistent with tuning techniques based on the P-Vr method, however, due to interactions from other generators, ΔT_{ii}^h and $\Delta\lambda_{ii}^h$ may be enhanced or degraded. Hence:

- The machines with higher inertia-weighted participation factors, ρ/M, are the more effective contributors to damping.
- For local modes, which typically have only a few machines participating, the magnitudes of the factors ρ/M for the dominant machines are significantly larger than those for the inter-area modes, which may have numerous machines participating. Thus the self-contributions to damping $\Delta\lambda_{ii}^h$ by dominant machines are likely to be less for inter-area modes than for local modes.

---

1. © 2000 IEEE. Reprinted with permission [3].

- For inter-area modes, the effect of interactions is to degrade further the already lower self-damping contribution $\Delta\lambda_{ii}^{h}$ of generators. As is shown in the study, and observed in practice, the damping of inter-area modes is generally poorer than local modes and is more difficult to improve using PSSs. The damping of local modes maybe enhanced by interactions (as observed in [13]).

- The studies show that the damping of several inter-area modes can be enhanced significantly by fitting a FACTS device placed at a suitable location with a tuned FDS [14]. The FDSs may have little effect on the damping of local-area modes, however, this is location, system and operating condition dependent.

- For the inter-area mode M the FDS in the study induces positive damping torques on the generators and thus contribute to damping by each generator. In this case, the interactions between the FDS and the PSSs are positive, i.e. an increment in FDS gain enhances the self-damping resulting from an increment in PSS gain on generator *i*.

- It can be shown that the term $r_{ij}^{h} = -(\rho_{ih}/M_i)H_{ij}(\lambda_h)(c_{j*}v_{*h}/v_{ih})$ in (13.13) is the residue from the voltage reference to the speed output on machine *j* [12]. However, associated information on interactions provided by (13.8), which incorporates the residue r_{ij}^{h}, is not available through the Method of Residues in Chapters 6 and 11.

13.6.2 Relative Effectiveness of Stabilizers [3]

- The stabilizer damping contribution diagram is a simple, productive tool for displaying simultaneously the contributions to damping by some or all of the PSSs and FDSs. Hence those stabilizers which make the most significant contributions to the damping of rotor modes can be identified rapidly.

- Such diagrams provide the engineering insight and basis for the simultaneous coordination of PSSs with PSSs, and PSSs with FDSs [4]. These aspects have been found to be particularly valuable in practical applications and will be employed in Chapter 14.

- For practical applications the increments in stabilizer gains can be applied to one, some or all stabilizers simultaneously. This facility permits the coordination of all stabilizers in Area 2 only, say, to investigate the improvement in the damping of the relevant local- and inter-area modes.

- The disadvantage of the diagram is that it applies to small increments in stabilizer gains. However it has been found, for example, that mode shifts due to PSS gain increments of ±5 pu on machine base are accurate typically within 5%; such information needs to be confirmed for the power system under study.

13.7 References

[1] P. Pourbeik, *Design and Coordination of Stabilisers for Generators and FACTS devices in Multimachine Power Systems*, PhD Thesis, The University of Adelaide, Australia, 1997.

[2] P. Pourbeik and M. J. Gibbard, "Damping and synchronizing torques induced on generators by FACTS stabilizers in multimachine power systems," *Power Systems, IEEE Transactions on*, vol. 11, pp. 1920-1925, 1996.

[3] M. J. Gibbard, D. J. Vowles, and P. Pourbeik, "Interactions between, and effectiveness of, power system stabilizers and FACTS device stabilizers in multimachine systems," *Power Systems, IEEE Transactions on*, vol. 15, pp. 748-755, 2000.

[4] P. Pourbeik and M. J. Gibbard, "Simultaneous coordination of power system stabilizers and FACTS device stabilizers in a multimachine power system for enhancing dynamic performance," *Power Systems, IEEE Transactions on,* vol. 13, pp. 473-479, 1998.

[5] E. Z. Zhou, "Functional sensitivity concept and its application to power system damping analysis," *Power Systems, IEEE Transactions on*, vol. 9, pp. 518-524, 1994.

[6] N. Martins and L. T. G. Lima, "Determination of suitable locations for power system stabilizers and static var compensators for damping electromechanical oscillations in large scale power systems", *Power Systems, IEEE Transactions on*, vol. 5, pp. 1455-1469, 1990.

[7] F. L. Pagola, I. J. Perez-Arriaga, and G. C. Verghese, "On sensitivities, residues and participations: applications to oscillatory stability analysis and control," *Power Systems, IEEE Transactions on*, vol. 4, pp. 278-285, 1989.

[8] E. Z. Zhou, O. P. Malik, and G. S. Hope, "Theory and Method for Selection of Power System Stabilizer Location," *Power Engineering Review,* IEEE, vol. 11, p. 45, 1991.

[9] E. Zhou, O. P. Malik, and G. S. Hope, "Design of stabilizer for a multimachine power system based on the sensitivity of PSS effect," *Energy Conversion, IEEE Transactions on*, vol. 7, pp. 606-613, 1992.

[10] D. R. Ostojic, "Stabilization of multimodal electromechanical oscillations by coordinated application of power system stabilizers," *Power Systems, IEEE Transactions on,* vol. 6, pp. 1439-1445, 1991.

[11] H. B. Gooi, E. F. Hill, M. A. Mobarak, D. H. Thorne, and T. H. Lee, "Coordinated Multi-Machine Stabilizer Settings Without Eigenvalue Drift," *Power Apparatus and Systems, IEEE Transactions on*, vol. PAS-100, pp. 3879-3887, 1981.

[12] P. Pourbeik, M. J. Gibbard, and D. J. Vowles, "Proof of the Equivalence of Residues and Induced Torque Coefficients for Use in the Calculation of Eigenvalue Shifts," *Power Engineering Review,* IEEE, vol. 22, pp. 58-60, 2002.

[13] M. J. Gibbard, "Coordinated design of multimachine power system stabilisers based on damping torque concepts," *IEE Proceedings C Generation, Transmission and Distribution*, vol. 135, pp. 276-284, 1988.

[14] S. Hiley, *Redesign of the Blackwell Static Var Compensator Power Oscillation Damper Controller.* Undergraduate Thesis Report, School of Information Technology and Electrical Engineering, The University of Queensland, Australia, 2007.

Chapter 14

Coordination of PSSs and FDSs using Heuristic and Linear Programming Approaches

14.1 Introduction

Various techniques have been reported in the literature for the coordination of PSSs in multi-machine power systems [1], [2], [3], [4]. Some of these techniques have used linear programming solutions for coordinating PSS gains [5], [6]. However, little attention has been given to the simultaneous coordination of PSSs and FDSs [7], [8], [9] [10]; this aspect is the subject of this chapter. It must be emphasized that in the current context the term 'coordination' is used to mean coordinating the gains of stabilizers installed on generators and FACTS devices, say, in an area of interest for the purpose of improving the damping of rotor modes. This is as opposed to coordination in the context of coordinating controllers, e.g. AVR-PSS coordination, within a single generating unit [11]. In the following text the damping gains of PSSs and the gains of FDSs are collectively referred to as *stabilizer gains.*

It has been emphasized that the predominately left shift of the modes with increasing stabilizer gain is the objective of the design procedures outlined in Chapters 5 and 10 for PSSs and Chapter 11 for FDSs. In essence, because the stabilizer transfer functions are of the form $kG(s)$, where k is a real gain and the transfer function $G(s)$ provides the phase compensation, then ideally, (i) $G(s)$ ensures the left shift of all modes over the selected range of modal frequencies, and (ii) the gain k determines the extent of the left-shift of the mode. This basic approach to the tuning of stabilizers provides the following rationale for the methods of heuristic and automated coordination.

- In both the heuristic and automated coordination procedures the stabilizer gain and the phase compensation are the two important components which are *essentially decoupled for practical purposes.* Therefore, in the coordination procedures that follow, the stabilizer gains are the adjustable quantities and the parameters of the compensation transfer functions $G(s)$ remain unchanged.

- For the process of stabilizers coordination the PSSs and FDSs should be robust over an encompassing range of operating conditions, normal and outage (see Section 1.2 item 3 and Section 11.8.2, respectively).

- Ideally, the incremental left-shifts of the rotor modes should be more-or-less linearly related to increments in stabilizer gain for small changes about the nominal values. (See 10.26 for PSSs.) For FDSs certain rotor modes may be insensitive to changes in stabilizer gain (see Figure 11.7).

- The nature of the trajectories for the inter-area modes K, L and M are shown in Figure 11.7 for BSVC_4 (for mode M), Figure 11.12 for PSVC_5 (modes L, M), and in Figure 14.2 for SVC2 (modes K, L). At the lower values of gain the incremental left-shift of the nominated modes increases linearly with the increments in gain.

In Chapter 13 a basis is developed for the heuristic coordination of PSSs and FDSs employing stabilizer damping contribution diagrams (SDCDs). Nevertheless, it is apparent from the example in Section 13.4 that the approach presented in that chapter would require a series of calculations to determine the appropriate stabilizer gain settings to enhance the damping of each rotor mode; the approach can therefore be tedious for larger systems where certain damping criteria are to be met for a large number of rotor modes. Moreover, in the approach presented in Chapter 13 it is not clear how the following constraints can be satisfied: (i) limiting any right-shift in exciter/controller modes that may occur as a result of increasing stabilizer gains; (ii) constraining the shift in the frequency of rotor modes (i.e. the imaginary part of their eigenvalues) which may result from increases in stabilizer gains. Since it has been established that the coordination of the stabilizers can be achieved by coordinating their gains, then an appropriate method for automating the gain selection procedure subject to a series of constraints is to use linear programming [7], [8], or genetic algorithms [4], for example. Nevertheless, it is informative to reveal the insights provided through heuristic coordination.

Based on the SDCDs an example of the analysis and process of heuristic coordination of PSSs and FDSs over six operating conditions is presented in Section 14.3.

The method for heuristic coordination is extended in Section 14.4 to the application of linear programming (LP) for stabilizer coordination. Employing this approach the calculation of the gain settings becomes automated, and thus less tedious. Furthermore, the gain settings given by the solution of the LP problem is an optimal set; however, it is not guaranteed that the solution is unique [12].

For both the heuristic and the automated approaches the two-stage coordination procedure is adopted to determine the parameters of the PSS and FDS transfer functions $k_j G_j(s)$. In stage one, the transfer functions $G_j(s)$ are designed to left-shift the relevant modes by providing the appropriate phase compensation (as explained in Chapters 5, 10 and 11). In the heuristic approach in stage two, information on the sensitivity of the real part of a selected mode to an increment in any stabilizer gain is derived from the SDCDs. The necessary increments in stabilizer gains can then be calculated to achieve a desired left-shift in the mode. For the automated analysis in stage two, the stabilizer gains k_j are determined by solving a LP problem. The objective function of the LP problem is selected such that the weighted sum of the stabilizer gains is minimised, subject to (i) satisfying a desired level of damping for selected modes of rotor oscillation, (ii) constraining the right shift of the exciter or other controller modes, and (iii) limiting the allowable change in the frequency of oscillation of the rotor modes. The objective function is chosen because, for small system disturbances, low stabilizer gains reduce not only the effect of limiting action on the output of the stabilizer, the AVR and excitation systems, but also the reactive power swings on generators [1].

To illustrate the two methods of stabilizer coordination the studies are based on the multi-machine power system used in the studies in Chapters 10, 11, and 13.

14.2 The 14-generator power system

The 14-generation system described in Section 10.2 again serves as an example to illustrate the procedures for heuristic and automated coordination. The parameters of the PSSs are provided in Tables 10.5 to 10.9; the transfer functions for the bus-frequency FDSs on BSVC_4 and PSVC_5 on buses 412 and 507, respectively, are given in equations (11.12) and (11.13).

Because separate studies have shown the need for a SVC and stabilizer to provide voltage control and to assist in providing damping for mode L (frequency ~2.6 rad/s), a SVC (base MVA is 200 Mvar) is installed at bus 212, a major load bus, located in the vicinity of other load buses in Area 2. It is therefore necessary (i) to install in the power flow analysis a SVC on the selected bus, (ii) to provide voltage regulation at bus 212, and (iii) to include voltage droop. The procedures developed in Chapter 11 will then be employed to the evaluate the parameters of its bus-frequency stabilizer. In the studies in Chapter 10 the SVCs, BSVC_4 and PSVC_5, are on-line under closed-loop voltage control but with their FDSs out of service. For Cases 1 to 6 the associated voltages, real and reactive power flows, and other variables in the steady-state power flows are provided in Table 10.2, together with the rotor modes in Tables 10.11, 10.15 and 10.16. Note that these quantities will be slightly modified with the addition of the SVC, called SVC_2. For Case 1 with SVC_2 in service under closed-

1. Reactive power swings on generators can occur with certain types of PSS (see Section 8.3.1)

loop voltage control the inter-area modes are K $-1.05 \pm j3.68$, L $-0.37 \pm j2.45$, and M $-0.51 \pm j1.80$ (all FDSs out of service).

The small-signal model of the controller for SVC_2 is that shown in Figure 10.38. Its parameters are: 200 MVAr (Mbase), $K_S = 2.0$, $K_A = 1000$, $K_d = 0.005$ pu on Sbase (100 MVA) and $T_d = 0.005$ s The parameters of its bus-frequency FDS for SVC_2 are to be determined in the following section.

14.2.1 Evaluation of the transfer function for the SVC at bus 212.

The polar plot of the residues for modes K, L and M is shown in Figure 14.1 for the SVC transfer function $\Delta Frq(\lambda_h)/\Delta V_{ref}(\lambda_h)$ for the range of operating conditions, Cases 1 to 6. The damping gains of PSSs are all set to 20 pu or all to 10 pu; no other FDSs are in operation. The selection of representative tuning parameters is weighted towards mode L; the residues for mode M are negligible.

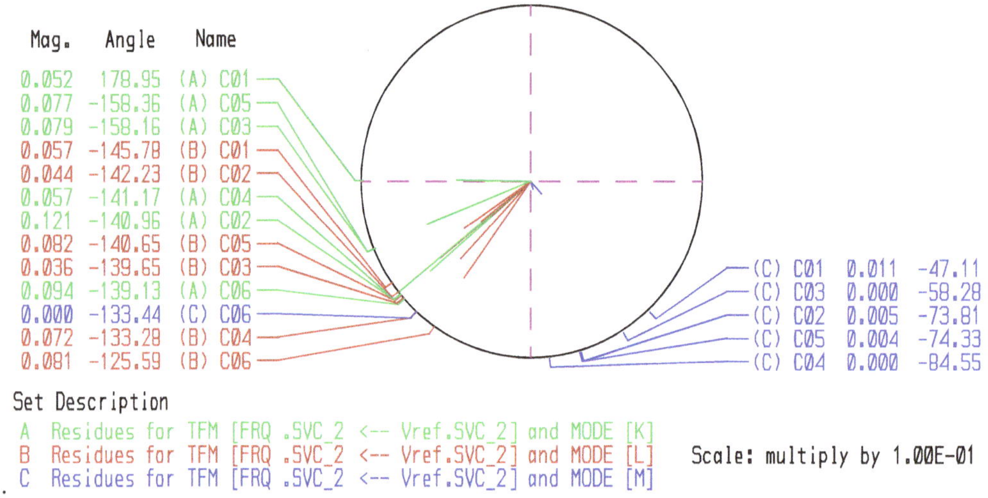

Figure 14.1 Polar plot of the residues for the transfer function $\Delta Frq/\Delta V_{ref}$ for SVC_2, modes K, L and M and six operating conditions. All PSS damping gains set to 20 pu on machine base. For L the values of the modes are in the vicinity of $-0.45 \pm j2.6$.
Note: the magnitude scale is to be multiplied by 0.1.

The calculation of the FDS transfer function is based on the representative values, shown in Table 14.1, with the PSS gains of 20 pu rather than 10 pu, because (i) after coordination of PSSs and FDSs the PSS damping gains are expected to be greater than 10 pu and in the vicinity of 20 pu; (ii) based on previous studies, the damping contributions of the FDS at the lower PSS gains are unlikely to be reduced markedly.

Table 14.1 Representative values for evaluation of compensation transfer function, mode L

PSS Gains (pu)	Mode & Phase spread (deg)	Represent-ative phase angle (deg)	Compensa-tion angle (deg)	Maximum residue	Represent-ative mode
20	L: 20 *	-136	-44	0.008	$-0.45 \pm j2.6$
10	L: 17 *	-152	-28	0.008	
* Note: The residues for mode M are small.					

Based on the algorithm in Appendix 6–I.1, the FDS transfer function for SVC_2 when all PSS damping gains are set to 20 pu is:

$$F(s) = k_{fds}\left[\frac{6s}{1+6s}\right]\left[\frac{1+0.157s}{1+0.628s}\right]\left[\frac{1}{1+0.033s}\right], \quad k_{Rm} = 781 \text{ pu}; \qquad (14.1)$$

the same washout and low-pass filter time constants are employed as for the FDSs of BSVC_4 and PSVC_5 to cover the range of the rotor modal frequencies.

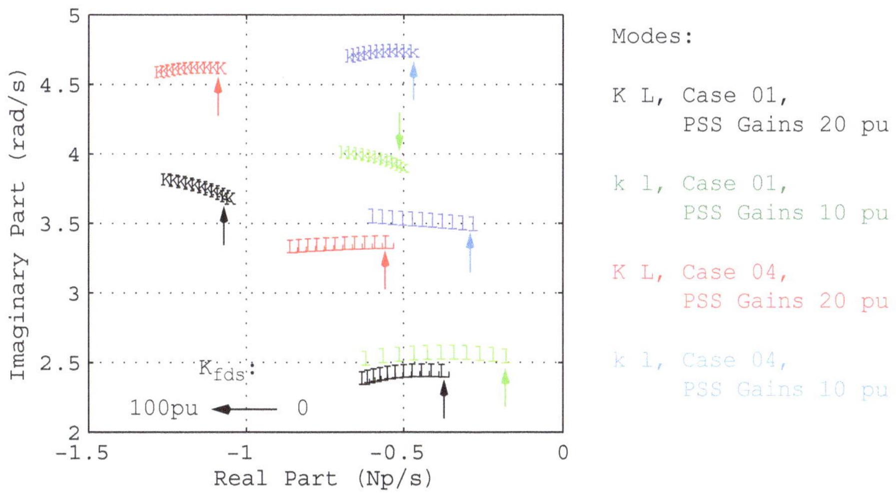

Figure 14.2 SVC_2. Cases 1 and 4. Trajectories of modes K and L as the stabilizer gain, k_{fds}, is increased from zero (shown by an arrow) to 100 pu in steps of 10 pu on the SVC base; changes in Mode M are negligible. PSS damping gains are set to 20 and 10 pu.

To establish the effectiveness of the FDS on SVC_2 in improving the damping of the inter-area modes K and L, the modes are tracked in Figure 14.2 as the FDS gain is increased from zero to 100pu. The trajectories of the modes are illustrated for both Cases 1 and 4 (heavy and light-load conditions), and for all PSS gains set to 20 or 10pu. Over the gain range the modes shift more-or-less linearly with increments in FDS gain; all other modes are stable.

14.3 A Heuristic Coordination Approach

14.3.1 Coordination of stabilizers for damping the inter-area modes

The object of the coordination procedure is to satisfy the criteria for the damping and other small-signal requirements of the system. Such criteria should include both normal and the relevant outage conditions; for simplicity, the latter conditions are excluded from the analysis but are notionally accounted for in the formulation of performance criteria.

The dynamic performance criteria in the following procedure are (i) the damping ratio of the inter-area modes is to be greater than 0.2 (20%), (ii) the PSS and FDS gains are be held at 'low' values for reasons explained earlier; e.g. initially set all PSS damping gains to 5 or 10 pu. In the following study the damping criterion has been chosen to be somewhat high so that the system is likely to be small-signal stable for the outage of a major transmission element.

As has been emphasized, there is no unique method for the heuristic coordination of the gain settings of those PSSs and FDSs units which are selected for the purpose. However, to reveal the insights that can be derived, the following procedure is adopted.

For the FDSs that participate in the coordination of stabilizers the characteristics of their eigen-trajectories should be noted, namely;

- In Figure 11.7 for BSVC_4 the shift in mode M is more-or-less to the left in a gain range of 0 to 40 pu on device base; the left-shifts in modes K & L are negligible.

- For PSVC_5 it is observed in Figure 11.12 that the shift in mode L is to the left over a gain range 0 to 100 pu on device base. There is a less extensive left shift in mode K over a gain range 0 to 50 pu on device base.

- The left shifts in modes K and L for SVC_2 are seen in Figure 14.2 to increase linearly with gain increments from 0 to 100 pu on device base. The shifts in mode M are negligible.

Consider the operating conditions, Cases 1 to 6. Initially, using Case 1, the stabilizer gains will be adjusted to satisfy the damping criterion for the inter-area modes. Proceeding to Case 2 the inter-area modes will be evaluated using the same gains to establish if the criterion is infringed; if so, the gains are appropriately adjusted. This process, covered by a set of steps, is continued on in Cases 3 to 6 until an acceptable set of gains is found that cover all cases - possibly after several iterations. The steps, and the inputs and outcomes of each step in the process, are listed in Tables 14.2 and 14.3.

In Step 1 for Case 1 the effects on the inter-area modes of selecting an initial or nominal set of stabilizer gains are examined. These studies will form the basis for the subsequent analysis. Details of the analysis in Steps 1 to 5 for Case 1 are shown in Table 14.2.

Table 14.2 Steps in the coordination procedure, Case 1

Step	Gains: Nominal, PSSs & FDSs / Incremental (pu)						Inter-area mode / damping ratio ζ		
	PSSs	Gain/ Inc.	BSVC 4	PSVC 5	SVC 2	###	K	L	M
1a	All	10 0	0 0	0 0	0 0	Act.	$-0.50\pm j3.91$ 0.126	$-0.18\pm j2.55$ 0.069	$-0.25\pm j1.92$ 0.131
1b		10 0	20 0	20 0	20 0	Act.	$-0.55\pm j3.94$ 0.14	$-0.41\pm j2.59$ 0.16	$-0.55\pm j1.98$ 0.27
2	All	10 / 0	20 / 2	20 / 2	20 / 2	Est.	$-0.55\pm j3.94$	$-0.44\pm j2.59$	$-0.58\pm j1.99$
3	All	10 / 1	0 / 0	0 /0	0 / 0	Est.	$-0.56\pm j3.89$	$-0.20\pm j2.54$	$-0.28\pm j1.91$
4a	All	10 / 0	20 / 2	20/12	20/14	Est.	$-0.58\pm j3.96$	$-0.53\pm j2.61$	$-0.63\pm j2.00$
4b		10	22	32	34	Act.	$-0.58\pm j3.95$ 0.15	$-0.53\pm j2.61$ 0.20	$-0.63\pm j2.00$ 0.30
5a	LPS_3, YPS_3	10 / 5	22	32	34	Est	$-0.82\pm j3.89$	$-0.53\pm j2.61$	$-0.67\pm j1.99$
5b	Other PSSs Nom. 10pu	15	22	32	34	Act.	$-0.81\pm j3.88$ 0.21	$-0.53\pm j2.60$ 0.20	$-0.67\pm j1.99$ 0.32

Notes. ## Est: Estimated values of modes from SDCD.
Act: Actual values of modes from eigen-analysis

Step 1a, Case 1. It is assumed that the required PSS damping gains will exceed 10 pu. With the FDSs on the SVCs off, the inter-area modes and their damping ratios are calculated; the latter are less than 0.2

Step 1b, Case 1. It is assumed that the gains of the PSSs and the FDS will exceed, respectively, the nominal values of 10 pu and 20 pu on device bases; the gains are set initially to these values. However, the performance criterion ($\zeta > 0.2$) is not satisfied by the nominal gains although there is a significant left-shift in modes L and M due to the action of the FDSs.

Step 2, Case 1. In order to determine the effectiveness of the FDSs their gains are incremented by 10% (2 pu). The stabilizer damping contribution diagram (SDCD) is shown in

Figure 14.3; it is clear from the figure that the FDSs have little influence on mode K. However, BSVC_4 contributes significantly to the damping of mode M.

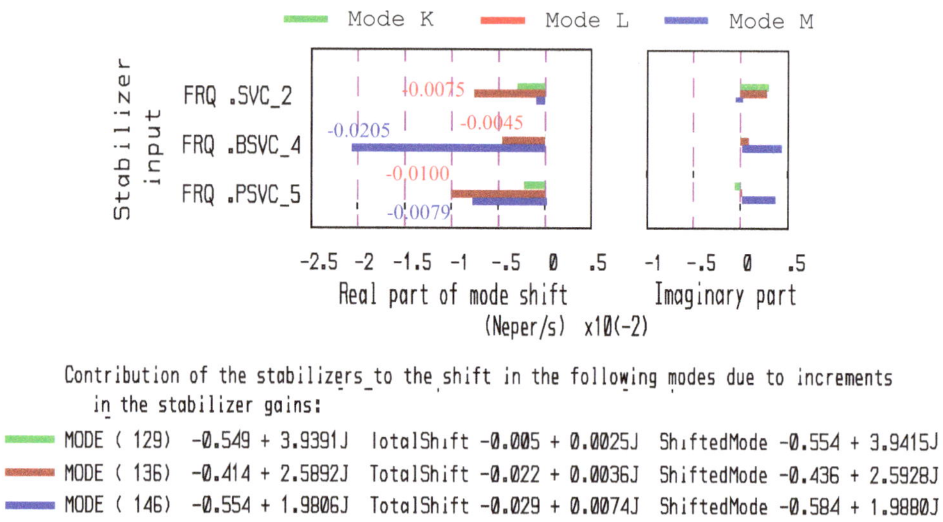

Figure 14.3 Step 2, Case 1. Contributions to the mode shifts of inter-area modes K, L & M by each SVC for a gain increment of 2 pu on the bus-frequency FDSs. Nominal gains: All PSSs 10 pu, FDSs 20 pu on device bases. Note scale.

Step 3, Case 1. Based on Step 1b the aim is to establish, with all FDSs out of service, (i) the relative effectiveness of the PSSs in improving damping, (ii) which PSSs influence mode K in particular when the gains of all PSSs are incremented by 10% (1 pu). The resulting SDCD is given in Figure 14.4.

From the figure it is observed that an increase in the gain on the PSS of LPS_3 is most effective in improving the damping of mode K. PSSs in Area 2 also contribute marginally to the damping of modes K and L. Likewise, PSSs in Areas 3 and 4 are of some minor benefit to mode M. In comparison, in Figure 14.3 it is demonstrated that all three SVC FDSs improve the damping of mode L.

Step 4a, Case 1. In Step 1b of Table 14.2 it is noted that the damping of mode L is improved significantly when the FDSs are on-line with all gains set to a nominal value of 20 pu. For the damping ratio of mode L to increase to 0.2, stabilizers must shift the real part of mode from $-0.41 \pm j2.59$ by -0.12 Np/s to $-0.53 \pm j2.6$. Based on the mode shifts for a gain increment of 2 pu produced by FDSs in Figure 14.3, gain increments in PSVC_5, SVC_2 and BSVC_4 of 12 pu, 14 pu and 2 pu, respectively, would yield a shift in mode L of -0.12 Np/s (i.e. $(12/2)(pu) \times (-0.01) + (14/2)(pu) \times (-0.0075) + (2/2)(pu) \times (-0.0045)$). Likewise modes K and M should benefit by real shifts of -0.034 and -0.075 Np/s, respectively. The estimated real shifts are confirmed in Figure 14.5; there are small positive shifts in the imaginary components.

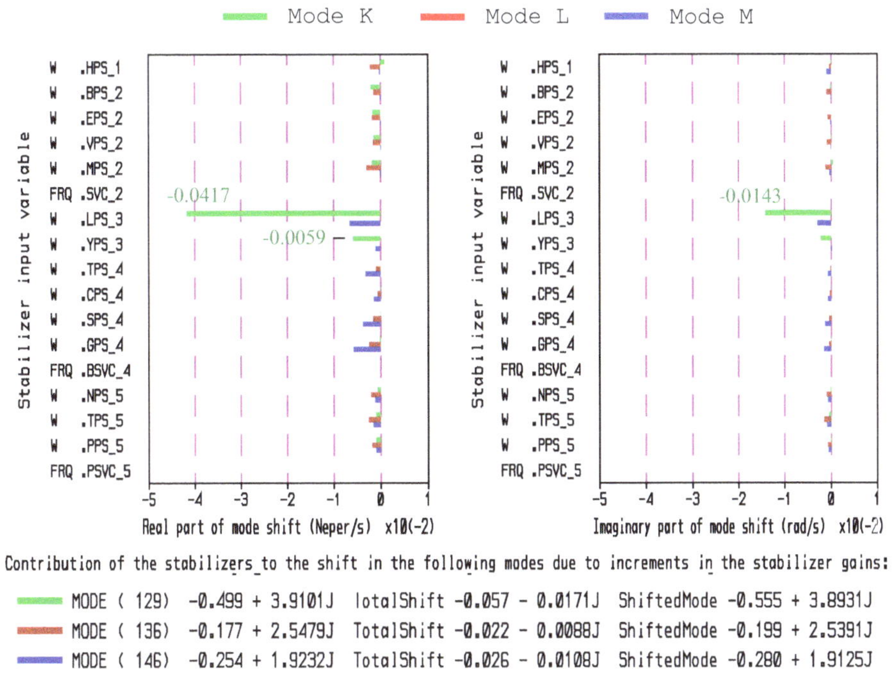

Figure 14.4 Step 3, Case 1. Contributions to the mode shifts of inter-area modes K, L & M by each PSS for a gain increment of 1 pu.
Nominal gains: All PSSs 10 pu, FDSs 0 pu. Note scale.

Step 4b, Case 1. Eigen-analysis reveals (i) that there is close agreement with the estimated mode values from Step 4a, and (ii) that the damping ratios of modes L and M are equal to or better than 0.2; however, the damping ratio of K is 0.15 which does not satisfy the performance criterion.

Step 5a, Case 1. The aim now is to increase the damping ratio of mode K at Step 4b from 0.15 to 0.2. In Figure 14.4 it is noted that the PSS on LPS_3 is most effective in increasing the shift of the latter mode. An increase of 5 pu in the damping gain of the PSSs on LPS_3 and YPS_3 could shift the mode from $-0.58 \pm j3.95$ by -0.24 to $-0.82 \pm j3.95$, say, to yield a damping ratio of 0.2 (i.e $5(pu) \times (-0.0417) + 5(pu) \times (-0.0059) = -0.24$). It is shown that the estimated inter-area modes in Step 5a agree closely with the actual values in Step 5b in Table 14.2 and also satisfy the criterion on the damping ratios.

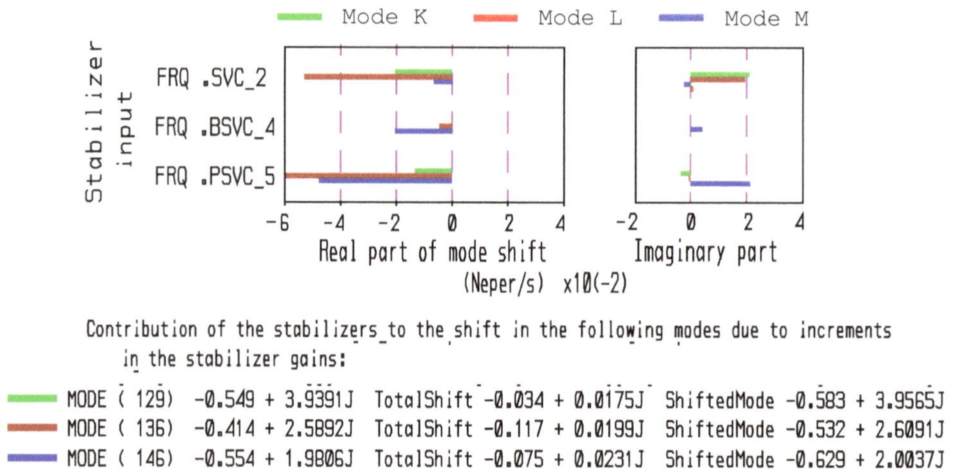

Figure 14.5 Step 4a, Case 1. Contributions to the mode shifts of inter-area modes K, L & M by SVC_2, BSVC_4 and PSVC_5 for gain increments of 14 pu, 2 pu and 12 pu on the respective FDSs. Nominal gains: All PSSs 10 pu, FDSs 20 pu on device base. Note scale.

It is now necessary to check if the stabilizer gain settings at Step 5b satisfy the performance criterion for Case 2. The coordination of the stabilizer gains for Cases 2 to 6 continues in Table 14.3.

Table 14.3 Steps in the coordination procedure, Cases 2 to 6

Step/ Case	Gains: Nominal / Incremental (pu)					Inter-area mode damping ratio ζ			
	PSS	Gain	BSVC 4	PSVC 5	SVC 2	##	K	L	M
S6/ Case 2	Gens_3** Other PSSs	15 10	22	32	34	Act.	$-0.58\pm j3.75$ 0.15	$-0.55\pm j2.75$ 0.19	$-0.62\pm j1.84$ 0.32
S7/ Case 2	Gens_3 Other PSSs	15/1.5 10/1.0	22/2.2	32/3.2	34/3.4	Est	$-0.66\pm j3.74$	$-0.60\pm j2.73$	$-0.70\pm j1.83$
S8/ Case 2	Gens_3 Gens_2 Other PSSs	15 / 4.5 10 / 5 10 / 0	22 / 0	32 / 0	34 / 6	Est.	$-0.80\pm j3.71$	$-0.60\pm j2.70$	$-0.68\pm j1.84$
S9/ Case 2	As above	19.5 15, 10	22	32	40	Act.	$-0.80\pm j3.70$ 0.21	$-0.59\pm j2.70$ 0.21	$-0.68\pm j1.85$ 0.34
S10/ Case 3-6	As above	19.5, 15, 10	22	32	40	Act.	All damping ratios exceed 0.2		

Step/ Case	Gains: Nominal Incremental (pu)						Inter-area mode damping ratio ζ		
	PSS	Gain	BSVC 4	PSVC 5	SVC 2	##	K	L	M
Check S11/ Case1	As above	As above					$-1.06\pm j3.79$ 0.27	$-0.58\pm j2.58$ 0.22	$-0.71\pm j1.96$ 0.34

Notes. ## Est: Estimated values of modes from SDCD. Act: Actual values of modes from eigen-analysis.
** Gens_A implies all generators in the area number A,

Step 6, Case 2. Eigen-analysis based on the stabilizer gains in Step 5b reveals that the damping ratios of mode K and L are less than 0.2.

Step 7, Case 2. In order to ascertain the relative effectiveness of the stabilizers in this case, all PSS and FDS gains are increased by 10%. The resulting contributions to the mode shifts of the inter-area modes are revealed in Figure 14.6.

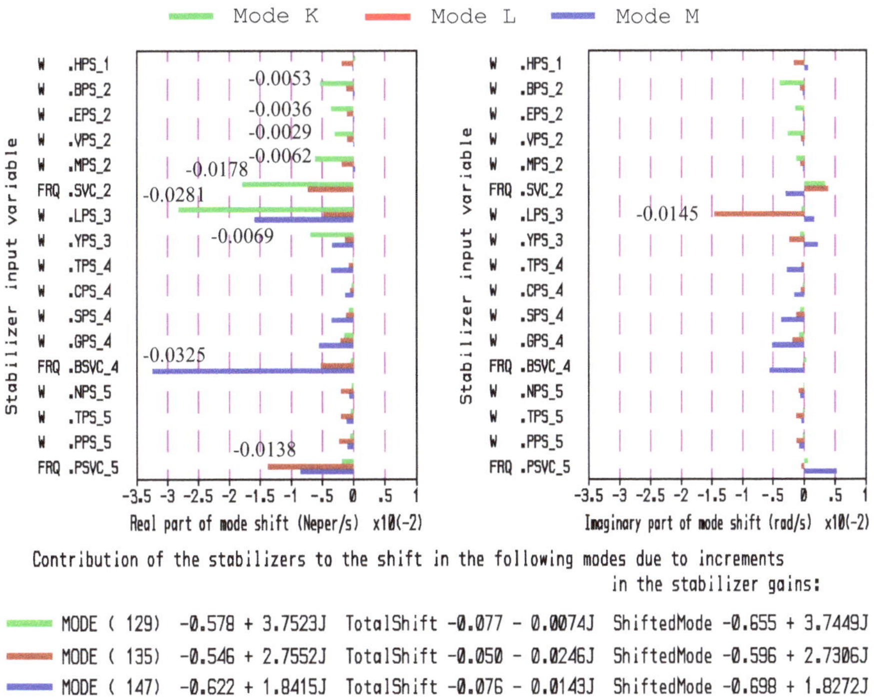

Figure 14.6 Step 7, Case 2. Contributions to the mode shifts of inter-area modes K, L & M by all stabilizers for gain increments of 10% in nominal gains. Note scale.

From Step 6, Case 2 it is noted that is necessary to shift mode K from $-0.58 \pm j3.75$ by -0.21 Np/s to $-0.79 \pm j3.75$ to yield a damping ratio of 0.2. It is observed from Figure 14.6 for Step 7, Case 2, that the combined real shift in mode K is -0.0350 Np/s for a 10% gain increment (1.5 pu) on the PSSs of LPS_3 and YPS_3. Likewise, for a 10% increment (1.0 pu) on the PSSs of Gens_2 [1] the total real shift is -0.0180 Np/s. Furthermore, the gain increment of 3.4 pu on SVC_2 (nominal gain 34 pu) produces a real shift in mode K of -0.0178 Np/s. To achieve the desired mode shift in mode K let us increase the gain on the stabilizers from the nominal values as follows:

Gens_3 by 3x1.5=4.5 pu from 15 pu.	Real shift = 3x(-0.0350) =	-0.1050 Np/s
Gens_2 by 5x1.0=5.0 pu from 10 pu.	Real shift = 5x(-0.0180) =	-0.0900
SVC_2 by 6.0 pu from 34 pu.	Real shift = (6/3.4)x(-0.0178) =	-0.0314
	Total real shift =	-0.2264

The new stabilizer gain settings are:
Gens_3: 15+4.5 = 19.5 pu; Gens_2: 10+5 = 15 pu; SVC_2: 34+6 = 40 pu.
Unchanged: Other PSSs: 10 pu; BSVC_4: 22 pu; PSVC_5: 32 pu.

Although the resulting FDS gain increases on SVC_2 and PSVC_5 are comparable with the original nominal gains of 20 pu for each FDS, the left shift of the inter-area modes are closely linearly related to the increments in gain over the gain ranges (see the modal trajectories for these modes in Figures 11.7, 11.12 and 14.2).

Step 8, Case 2. Using the SDCD shown in Figure 14.7 the mode shifts and modes are estimated with the gain increments proposed in Step 7; note that the required mode shift of -0.2264 Np/s for mode K is achieved. The damping ratios of all inter-area modes in Case 2 now satisfy the dynamic performance criterion.

Step 9, Case 2. Eigen-analysis confirms the validity of the results of Step 8.

Step 10, Cases 3 to 6. Based on the stabilizer gains confirmed in Step 9 the damping performance of the inter-area modes is validated using eigen-analysis, i.e. in these Cases the criterion ($\zeta > 0.2$) is satisfied.

Step 11, Case 1. With the same gain settings of Step 9 the values of the modes and damping ratios for Case 1 are recalculated and compared with those in Step 5b of Table 14.2. The damping of the inter-area modes calculated in this Step is an improvement over that in Step 5b.

---

1. Gens_*m* covers all generators in Area *m*.

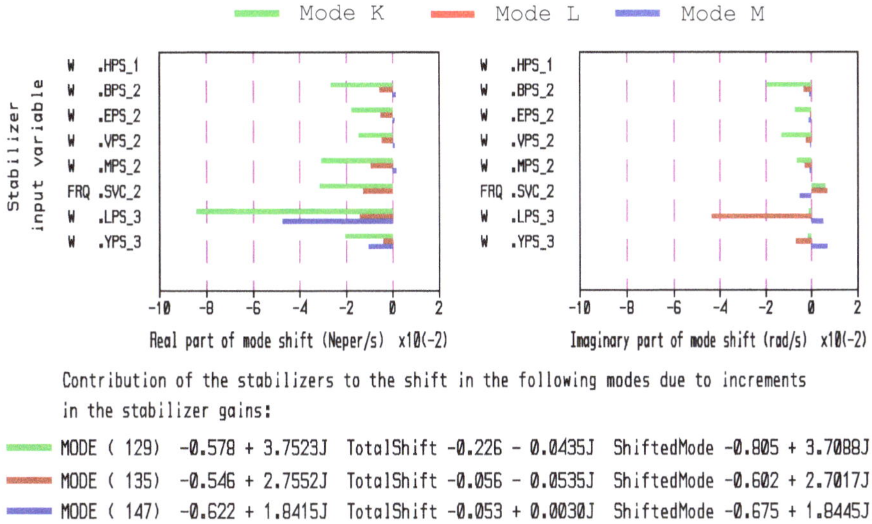

Figure 14.7 Step 8, Case 2. Contributions to the mode shifts of inter-area modes K, L & M for the stabilizer gains increments given in Table 14.3. Note scale.

In comparing the eigenvalues K, L, M in Step 1a, Case 1, at the start of the procedure of coordination, with those for the final stabilizer settings in Step 11, Case 1, it is observed:

(i) the real parts of eigenvalues have left-shifted by some 100 to 200% [1];

(ii) the imaginary parts have remained within a band of about 3%.

These observations are consistent with the rationale of the P-Vr approach to the tuning PSSs and of the Method of Residues for tuning FDSs.

14.3.2 Coordination of local-area modes.

The effect of the increases in stabilizer gains on the local-area modes, which are well damped, has so far been ignored. However, it may be the case that certain modes have been degraded during the above procedure designed to ensure the damping criterion is satisfied for the inter-area modes. Also of interest is which local-area modes are affected by the FDSs, and the extent of the resulting mode shifts.

The SCDC in Figure 14.8 for local modes I, G and H is based on the nominal gain settings for Step 9. Increments of 1 pu on all PSSs and 4 pu on all FDSs are assumed. The SDCD demonstrates that increases in stabilizer gains enhance the damping of the three local modes. In particular, a 4 pu increment in the FDS of BSVC_4 causes a significant shift in mode H, a mode in which TPS_4 (a generating unit electrically relatively close to the SVC) swings against the other units in Area 4. The FDSs have little effect on the seven remaining local modes apart from BSVC_4 which slightly degrades mode E (not shown - it is the second of

1. With the higher gain settings on the FDSs, a possible extension of the Mvar range and ratings of the SVCs may be required.

the three local-area modes in Area 4). If only the gain on FDS of BSVC_4 were increased, it may occur that the damping on mode E is more severely degraded; this should be checked if the mode is relatively poorly damped.

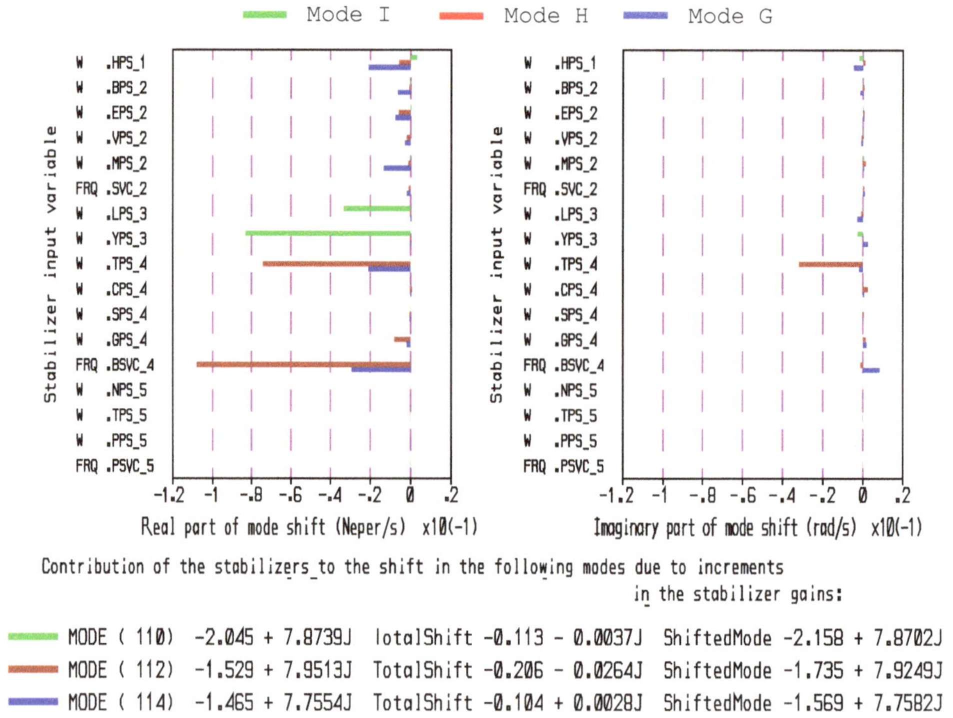

Figure 14.8 Case 1. Contributions to the mode shifts of local-area modes I, H & G for increments of 1 pu on all PSSs and 4 pu on all FDSs. Nominal gains: as for Step 9. Note scale.

14.4 Simultaneous Coordination of PSSs and FDSs using Linear Programming

14.4.1 Introduction

In Section 14.3 a heuristic approach is outlined for the coordination of PSSs and FDSs. Nevertheless, it is apparent from the case studies in Section 14.3 that the approach requires a series of calculations to determine the appropriate stabilizer gain settings to enhance the damping of the selected rotor modes. This approach may be tedious for larger systems where the criterion for damping is to be satisfied for a large number of rotor modes. Moreover, in the heuristic approach it is not clear how the following constraints can be met:

(i) limiting any right-shift in exciter/controller modes that may occur as a result of increasing stabilizer gains;

(ii) constraining the shift in the frequency of rotor modes (i.e. the imaginary part of their eigenvalues) which may result from increases in stabilizer gains.

Since it has been established that the coordination of the stabilizers can be achieved by coordinating their gains, then an appropriate method for automating the gain selection procedure subject to a series of constraints is to use linear programming. Thus, the heuristic method for coordination is extended in the following by the use of linear programming (LP). Employing this approach the calculation of the gain settings becomes automated, and thus less tedious. Furthermore the gain settings given by the solution of the LP problem is an optimal set; however, it is not guaranteed that it is a unique solution [12]. Moreover, a useful feature in the LP solution is a set of limitations or constraints that can be applied to the solution, namely:

- (i) a minimal level of damping for selected rotor modes;
- (ii) a right-shift in the eigenvalues associated with generator exciters or other controllers, due to increasing stabilizer gains;
- (iii) changes in the left-shift of selected rotor modes;
- (iv) changes in the frequency of selected rotor modes.

Through a LP solution some unexpected insights into the support provided by certain stabilizers to the damping of nominated modes may be experienced.

A two-stage coordination algorithm is developed. Stage one is the same as for heuristic coordination. That is, for the transfer functions of the form $k_j G_j(s)$ for the PSSs and FDSs, the transfer functions $G_j(s)$ are designed to provide appropriate phase compensation as explained in Chapters 5, 10 and 11. However, in stage two the stabilizer gains k_j are now determined by solving a LP problem. The objective function of the LP problem is selected such that the weighted sum of the stabilizer gains is minimised, subject to any of the above set of constraints. For practical reasons this function is chosen because, for small system disturbances, low stabilizer gains reduce not only the effects of limiting in the stabilizer and the AVR and excitation systems, but also reduce swings on the reactive output of generators.

14.4.2 Comment on the LP solution: optimality versus uniqueness

The algorithm used to solve the LP problem is the *revised two-phase simplex algorithm* [12], [13]. The process of solving a LP problem can be summarised as follows. The search begins at an extreme point of the solution space called a *basic feasible solution*. The procedure then determines if a shift to an adjacent point in the solution space can improve the objective function. If so, the algorithm moves the solution to the point which offers the greatest improvement. This procedure continues until an optimal solution is found or it is determined that the problem is unbounded or infeasible. Typically it is difficult to determine a basic feasible solution by observation. Therefore to start the procedure, a set of artificial variables are introduced into the problem. This allows us to manipulate the problem such that zero becomes a basic feasible solution. However, if another basic feasible solution were found, and the LP algorithm started at that point, the algorithm may converge to a different optimal solution. The practical implication of this is that other optimal solutions may exist. Moreover, it is possible to establish whether more than one optimal solution exists. Nevertheless, having determined an optimal solution it is not generally possible to ascertain what the other optimal solutions are.

14.4.3 Coordination of PSSs and FDSs

As stated, the aim of both the PSS and FDS design methods is to achieve a left-shift in the modes of rotor oscillation by inducing pure damping torques on the shafts of generators. The aim of the coordination procedure is to determine the minimum required stabilizer gains to achieve desired damping criteria for selected rotor modes and to prevent undesirable right shifts in other modes.

Because the stabilizer transfer functions cannot provide ideal compensation, and due to the effect of PSS interactions, the shift in rotor modes will not be a pure left-shift. Consequently there will be changes in the frequency of oscillation of the rotor modes. Any excessive such changes will be constrained by the LP problem in order to limit undesirable changes in synchronising torques.

Assume that the transfer functions, $G_j(s)$, have been tuned for the relevant set of the n generators and z FACTS devices using the methods described in Chapters 10 and 11. Let each stabilizer be a fixed-parameter device of the form $k_jG_j(s)$. For a given vector of small gain increments, $\mathbf{\Delta k}$ $(n+z)\times 1$, the shift in selected modes of the system can be estimated based on (13.9), or:

$$\Delta\boldsymbol{\lambda}_h = [\phi_{h1}\ \phi_{h2}\ \cdots\ \phi_{h(n+z)}]\mathbf{\Delta k}\text{, where} \tag{14.2}$$

$$\phi_{hj} = -\sum_{i=1}^{n}\left\{\frac{\rho_{ih}}{M_i}\left(H_{ij}(\lambda_h)G_j(\lambda_h)\frac{\boldsymbol{c}_{j*}\boldsymbol{u}_{*h}}{u_{ih}}\right)\right\}. \tag{14.3}$$

Based on this equation a linear programming problem is formulated for the simultaneous coordination of the $(n+z)$ stabilizer gains.

Let the set of m modes of interest be

$$\boldsymbol{\lambda} = [\lambda_1, \lambda_2, \ldots, \lambda_m],$$

which may be a combination of rotor modes of oscillation and exciter/controller modes. This set is a subset of the N system eigenvalues (the system being represented by a $N \times N$ state-matrix for which there are N eigenvalues). Without loss of generality let the desired damping criteria be achieved by left-shifting the m modes of interest by

$$\Delta\boldsymbol{\sigma} = [\Delta\sigma_1, \Delta\sigma_2, \ldots, \Delta\sigma_m].$$

Also, let the changes in the modal frequencies be bounded by

$$\Delta\boldsymbol{\Omega} = [\Delta\Omega_1, \Delta\Omega_2, \ldots, \Delta\Omega_m].$$

Assuming each stabilizer gain is weighted by a coefficient w_j the LP problem can be stated:

$$\text{minimise} \quad \sum_{j=1}^{n+z} w_j \Delta k_j, \quad \text{subject to:} \tag{14.4}$$

$$\Re\{\boldsymbol{\Phi}\}\Delta\boldsymbol{k} \leq -\Delta\boldsymbol{\sigma} \ \ \text{Np/s}, \tag{14.5}$$

$$\Delta\boldsymbol{\Omega} \geq \Im\{\boldsymbol{\Phi}\}\Delta\boldsymbol{k} \geq -\Delta\boldsymbol{\Omega} \ \ \text{rad/s}, \tag{14.6}$$

$$\Delta\boldsymbol{K}_{max} \geq \Delta\boldsymbol{k} \geq \Delta\boldsymbol{K}_{min} \geq \boldsymbol{0}, \tag{14.7}$$

where $(\boldsymbol{\Phi} = [\phi_{hj}])\ (m \times (n+z))$. This problem is solved by using the two-phase simplex algorithm. Note that the LP problem becomes infeasible if:

- The set of gain limits $\Delta\boldsymbol{K}_{max}$ are too small for the required left-shifts $\Delta\boldsymbol{\sigma}$ Np/s.
- The allowable modal frequency deviations $\Delta\boldsymbol{\Omega}$ rad/s are too small. This may imply that the tuning of some stabilizers may be poor and hence result in excessive frequency shifts in the modes of interest.

Also note that the coefficients, w_j, in (14.4) can be chosen to weight all stabilizer gain increments equally (with unity values) - or they may be chosen to bias the solution in favour of the most effective stabilizers. Furthermore, it is important to note that the accuracy of the estimated shift in an eigenvalue, given by (14.2), diminishes as the gain Δk_j becomes larger. This is because (14.2) is a linear approximation to the non-linear eigenvalue trajectory. To reduce the error due to this linear approximation, the coordination procedure may be carried out in a number of steps.

Let the total required left shift in the modes of interest be $(\Delta\boldsymbol{\sigma}_{total})$ Np/s. Then the LP problem can be split into n_s steps, with a shift of $(\Delta\boldsymbol{\sigma}_{total})/n_s$ required per step. The following algorithm is proposed for the coordination of PSSs and FDSs having transfer functions of the form $k_j G_j(s)$.

Two-stage Coordination Algorithm Using a LP Solution

Stage 1: Determining stabiliser transfer functions, $G_j(s)$

Follow the tuning procedures that have been described in Chapters 10 and 11 for PSSs and FDSs, respectively.

Stage 2: Determining stabilizer gains, k_j:

1. Set the required vectors of the total shift in the real parts and the allowable total shift in the imaginary parts of the modes of interest to $-\Delta\boldsymbol{\sigma}_{total}$ Np/s and $\Delta\boldsymbol{\Omega}_{total}$ rad/s, respectively, in order to achieve a desired damping ratio (say) for the modes. The vector $-\Delta\boldsymbol{\sigma}_{total}$ also contains the constraints on right-shifts in exciter/controller modes. Set $\Delta\boldsymbol{K}_{max}$ and $\Delta\boldsymbol{K}_{min}$.

2. Set the number of steps n_s to two or more; the gain weighting vector $\boldsymbol{w}$ to unity (say); the step counter i_s to zero; the initial stabilizer gains to a vector of nominal values (e.g. 5 pu) or zero.

3. Calculate $\boldsymbol{\Phi}$ based on (14.3).

4. Form the LP problem given by (14.4)-(14.7) with

$$-\Delta\boldsymbol{\sigma} = -\Delta\boldsymbol{\sigma}_{total}/n_s\,,\ \Delta\boldsymbol{\Omega} = \Delta\boldsymbol{\Phi}_{total}/n_s\,;$$

$$\Delta\boldsymbol{K}_{max/step} = \Delta\boldsymbol{K}_{max}/n_s \text{ and } \Delta\boldsymbol{K}_{min/step} = \Delta\boldsymbol{K}_{min}/n_s\,.$$

Initialise the vector of gains $\Delta\boldsymbol{k}$ to zero and solve the LP problem.

5. If the LP problem is infeasible then:
(a) choose a larger value of $\Delta\boldsymbol{K}_{max/step}$, and/or
(b) allow a slightly greater value of $\Delta\Omega_h$ for the mode λ_h which is most tightly constrained. (If the increase in modal frequency $\Delta\Omega_h$ is too large then certain stabilizers may need retuning, i.e. return to Stage 1).

6. Set the step counter $i_s = i_s + 1$ and increment the vector of gains $\boldsymbol{k} = \boldsymbol{k} + \Delta\boldsymbol{k}$.

7. If $i_s < n_s$ then recalculate system eigenvalues/vectors with the new gain settings and go to 3.

There is an opportunity to view the intermediate results after line 7, following the eigen-analysis. According to (14.2) the contribution to the shift in the hth mode by the jth stabilizer (a PSS or FDS) is

$$\Delta\lambda_j^h = \phi_{hj}\Delta k_j\,. \tag{14.8}$$

This result is based on (14.2). At the end of each of the n_s steps of the coordination algorithm (i.e. following item 7 above), the user may assess the effectiveness of the contributions of each stabilizer and each generator to the damping of a selected mode. This and other physical insights provided by this technique will be demonstrated in the case study in the next section.

14.5 Case study: Simultaneous coordination in a multi-machine power system of PSSs and FDSs using linear programming

The purpose of this study is to demonstrate:

1. the method of simultaneous coordination of selected PSSs and FDSs;
2. the insights - some unexpected - revealed by the action of the step-by-step LP procedure of determining the stabilizer gains;
3. the contributions of selected stabilizers to the damping of the rotor modes (and the inter-area modes, in particular);
4. the benefits and disadvantages of the automatic process of determining stabilizer gain settings.

The system under study in this section is the fourteen generator system employed in Section 14.2 in which three SVCs and their FDSs are in service, namely SVC_2 at bus 212, BSVC_4 at bus 412, and PSVC_5 at bus 507.

A study is conducted on Case 1 in which a number of scenarios are examined as the constraints on modal damping ratios, modal damping constants, and stabilizer gains are varied. The nominal gains of the PSSs and the FDSs are all set to 5 pu on device base. Reducing the gains from the higher values employed in Section 14.3.1[1] for heuristic coordination allows for more flexibility in the optimization of the gain settings. The rotor modes are listed in Table 14.4 for Case 1 with the stabilizers out of service and then in service with their gains set to the nominal values.

In order to understand the action of the LP algorithm in adjusting the stabilizer gains let us firstly consider the SDCD for Case 1 with the nominal gain settings. Because the emphasis in the following analysis concerns the inter-area modes K, L, and M, the SDCD for these modes is shown in Figure 14.9 to ascertain the effects on the modes of a 1 pu increment on all stabilizers.

1. See step 1b, Table 14.2; nominal PSS gains 10 pu, FDS gains 20 pu.

Table 14.4 Rotor modes and modes shifts for heavy load condition, Case 1 [1]. Nominal gains of all stabilizers 5 pu on device base.

No. *	Case 1. Heavy load							
	Stabilizers off			All PSSs & FDSs in service. Nominal stabilizer gains on device bases: all 5.0 pu			Mode Shift	
	Real	Imag	ξ	Real	Imag	ξ	Real	Imag
A	-0.16	10.45	0.01	-0.67	10.49	0.06	-0.51	0.04
B	0.11	9.58	-0.01	-0.39	9.65	0.04	-0.50	0.07
C	0.03	8.93	-0.00	-0.43	9.03	0.05	-0.46	0.10
D	-0.56	8.63	0.06	-1.00	8.74	0.11	-0.45	0.09
E	-0.26	8.37	0.03	-0.69	8.37	0.08	-0.43	0.00
F	-0.68	8.00	0.08	-0.85	8.25	0.10	-0.17	0.25
G	-0.40	8.05	0.05	-0.85	7.87	0.11	-0.45	-0.18
H	0.02	7.81	-0.00	-0.53	7.86	0.07	-0.54	0.05
I	-0.19	7.72	0.02	-0.65	7.82	0.08	-0.46	0.10
J	-0.62	7.42	0.08	-0.92	7.49	0.12	-0.31	0.07
K	0.08	4.02	-0.02	-0.22	3.99	0.06	-0.31	-0.03
L	0.06	2.61	-0.02	-0.12	2.60	0.05	-0.19	-0.01
M	0.01	2.03	-0.00	-0.20	1.99	0.10	-0.21	-0.03
* Mode Number. ξ is the damping ratio.								

For insight, based on Figure 14.9, consider the stabilizers which dominate - in descending order left to right - the contributions to the damping (i.e. left-shifts) for each of the inter-area modes.

:

Mode	Stabilizer				
K	LPS_3	YPS_3	EPS_2	MPS_2	
L	PSVC_5	SVC_2	MPS_2	GPS_4	HPS_1
M	BSVC_4	GPS_4	PSVC_5	SPS_4	TPS_4

1. The values of the modes in the table differ slightly from those in Table 10.14 due to: (i) the addition of a SVC on bus 212, and (ii) the FDSs on all three SVCs being in service at their nominal gains of 5 pu.

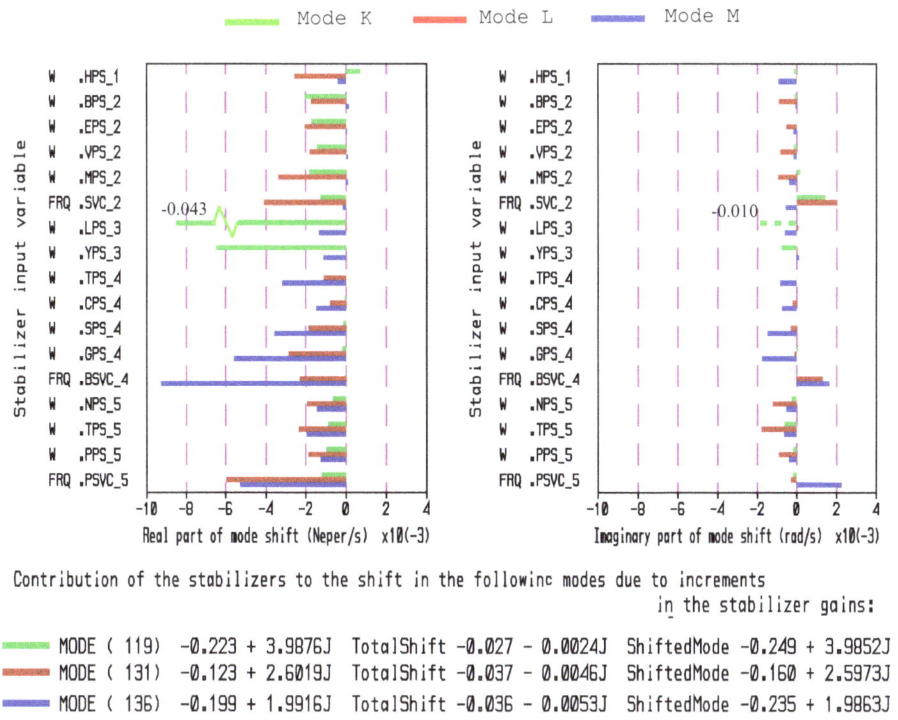

Figure 14.9 Case 1. Contributions to modes shifts of inter-area modes K, L & M by each stabilizer for a gain increment of 1 pu. Nominal gain settings for all stabilizers is 5 pu on device base. (Note: contribution of LPS_3 is off-scale; the value is

$$5 \times (8.57 \pm j1.9) \times 10^{-3} = -0.043 \pm j0.01.)$$

Notice there are several stabilizers that can affect the contributions to the damping of more than one mode. For example, if the gain on PSVC_5 needs to be increased by the LP algorithm to satisfy a requirement on mode L, it also produces a contribution to damping on mode M. Similar implications apply to MPS_2 and GPS_4; such observations may help to explain what may be unexpected results.

To examine the action and performance of the LP algorithm five scenarios are considered in which all stabilizer weightings $w_j = 1$ and the following constraints or limits are varied:

1. the maximum gain on selected stabilizers,
2. the type of mode (local- or inter-area),
3. the *real part* of the mode-shift is limited by
   - the modal damping constant $-\sigma$ which must be less than or equal to a specified value;

- the damping ratio ξ of the rotor mode which must be greater than or equal to a nominated limit;
- the *change* in the damping constant $-\sigma$ which must be less than or equal to a specified value.

In the following five scenarios shown in Table 14.5 for Case 1 the first three implement a real left-shift on the inter-area modes K, L and M with the constraint that their damping ratios are $\xi \geq 0.2$. The maximum stabilizer gains are reduced over Scenarios 1 to 3 to investigate the reallocation of stabilizer gains between stabilizers. In Scenario 4 the limits on modes K, L and M require that the real parts of the modes are $\sigma \leq -0.4$ with maximum stabilizer gains of 20 pu on device base. Finally, in Scenario 5 the stabilizer gains must satisfy (i) a limit $\sigma \leq -0.4$ on the real parts of modes K, L and M, (ii) the constraint $\xi \geq 0.1$ for the more lightly damped local-area modes A, B, C, I and H.

The optimum stabilizer gains (in pu on device base) derived from the LP analysis and the associated constraints for the five scenarios are summarized in Table 14.5

14.5.1 Scenario 1: Inter-area modes. Maximum PSS & FDS gain 40 pu.

In this scenario the damping ratios of the inter-area modes are to be equal to or greater than 0.2. From col. 2 of the table it is noted that the FDSs on two of the three SVCs are at their limiting gains, and that the gains of only three of the fourteen PSSs are increased from their nominal values. The condition $\xi \geq 0.2$ requires a significant left-shift on the higher frequency inter-area modes as revealed in the eigen-trajectories in Figure 14.10. Some local-area modes are only marginally left-shifted by the increases in stabilizer gains, some significantly.

It is evident from the SDCD of Figure 14.9 that, of the three modes, for 1 pu gain increment on any one stabilizer the contributions to the shifts in modes K and M are generally greater than or comparable to those for mode L. Because stabilizers PSVC_5, SVC_2, MPS_2 and GPS_4 contribute most to mode L, their gains are increased and, as a result, their contributions to modes K and M are also raised. As a consequence a lower contribution to mode K by LPS_3 (the largest potential contributor to the mode) is required.

Table 14.5 Case 1. Optimum stabilizer gains determined by LP for five scenarios.

	PSS or FDS gain				
	Scenario 1	Scenario 2	Scenario 3	Scenario 4	Scenario 5
Modes -->	Inter-area K, L, M				K L M & local area A B C H I
K_{max} PSSs (pu) -->	40	40	20	20	20
K_{max} FDSs (pu)-->	40 (All)	20 (FDS)	20 (All)	20(All)	20(All)
Mode constraints -->	Damping ratio, KLM: $\xi \geq 0.2$			KLM: $\sigma \leq -0.4$	KLM: $\sigma \leq -0.4$ ABCIH: $\xi \geq 0.1$
Generator / SVC |	Gain (pu on device base)				Gain (pu on device base)
col. 1	col. 2	col. 3	col. 4	col. 5	col. 6
HPS_1	5	14.0	20	11.7	5
BPS_2	5	5	5	5	7.7
EPS_2	5	5	16.9	5	15.0
VPS_2	5	5	5	5	6.1
MPS_2	18.8	40	20	20	19.4
SVC_2	40	5	20	20	20
LPS_3	15.7	15.7	15.9	7.4	6.6
YPS_3	5	5	5	5	6.9
TPS_4	5	5	5	5	6.0
CPS_4	5	5	5	5	5
SPS_4	5	5	6.8	5	11.5
GPS_4	8.7	40	20	20	19.9
BSVC_4	5	5	20	5.4	6.1
NPS_5	5	18.7	17.1	5	5
TPS_5	5	34.2	20	16.6	9.4
PPS_5	5	5	12.5	5	5
PSVC_5	40	20	20	20	20
Sum of device gains	183	233	234	166	174
The yellow shading indicates that the stabilizer gain is a maximum.					

The stabilizer gains are listed in col. 2. Gain limits of 40 pu occur only on the FDSs of PSVC_5 and SVC_2 which are the main contributors to mode L.

In Figure 14.10 the estimated and actual values of the three inter-area modes at each step in LP procedure are in close agreement. The associated initial and final values of the ten local-area modes, A to J, are also recorded.

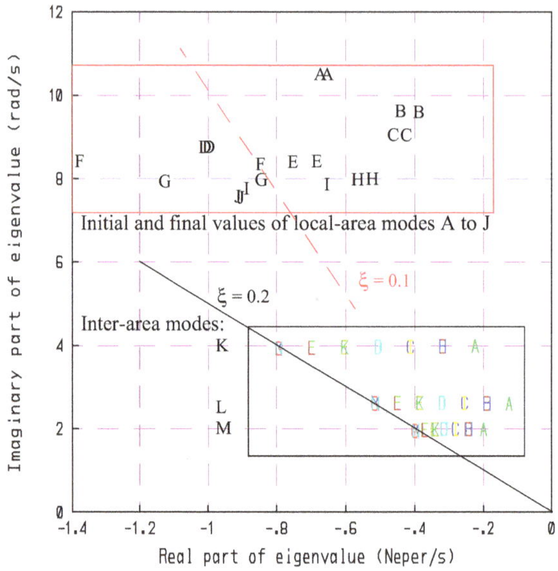

1

Figure 14.10 Case 1, Scenario 1. In the lower box the eigen-trajectories of inter-area modes K, L and M are plotted ($\xi \geq 0.2$). The six steps of the trajectory are shown; at each step the *estimated* values B to G of the mode are shown to agree closely with the *actual* values H to M. In the upper box the initial and final eigenvalues of the left-shifted local-area modes A to J are also marked.

14.5.2 Scenario 2: Inter-area modes. Limits PSS gains 40 pu, FDSs 20 pu

For comparison with Scenario 1 the effect of the reallocation of gains to other stabilizers resulting from the reduction in the maximum gain on all FDSs from 40 to 20 pu is of interest. (See col. 3 of Table 14.5.) The gain limit of 40 pu now occurs only on the PSSs of MPS_2 and GPS_4, and limit of 20 pu on the FDS of PSVC_5. It is insightful to examine the characteristics of the main contributions by the stabilizers to the real shifts in modes K, L and M at steps one to six of the LP calculation. Such contributions are shown in Table 14.6 in which it is noted:

- (i) the relative real shifts contributed by a stabilizer to the inter-area modes correspond closely to those shown in the SDCD in Figure 14.9 (e.g. for MPS_2 in the SDCD the ratio of the contributions to L and K are about 2:1; this is reflected in the table);

- (ii) the resulting stabilizer gain is

$$k = \left(1 + \sum_{j=1}^{6} f_j\right) k_{nom},$$

where f_j is the incremental gain factor and k_{nom} is the nominal gain setting of the stabilizer. The resulting gains are listed in col. 3 of Table 14.5.

- (iii) Also, from Table 14.6, the associated contribution to the real shift in inter-area mode h by stabilizer j is given by

$$\Delta\sigma^{h}{}_{total} = \sum_{j=1}^{6} \Delta\sigma_{h_j}.$$

- (iv) Stabilizers MPS_2, PSVC_5, GPS_4 and TPS_5 play a major role in contributing to the real left-shift in the mode L. In so doing, GPS_4 and PSVC_5 are also major contributors to real shift in mode M. Similar observations also apply to LPS_3 and MPS_2 for mode K.

14.5.3 Scenario 3: Inter-area modes. Limits: all stabilizer gains 20 pu

As to be expected in comparison with Scenario 2, additional stabilizers - namely the FDSs on the three SVCs and PSSs on four generators - operate at the maximum gain of 20 pu. The gain of the PSS of LPS_3 is more or less constant (15.7 - 15.9 pu) over scenarios 1 to 3 for reasons discussed in Section 14.5.1.

14.5.4 Scenario 4: Inter-area modes. Limits $\sigma \le -0.4$; all stabilizer gains 20 pu

The limit on the inter-area modes is much less onerous on modal damping than in Scenarios 1 to 3 for which the limit is $\xi \ge 0.2$. Note that in latter studies the real part of the eigenvalue for mode K is -0.8 at the limit. Consequently in this scenario the PSS gain of LPS_3 is approximately halved, but the FDS gains of SVC_2, PSVC_5 and PSSs on MPS_2 & GPS_4 are at the limit of 20 pu.

14.5.5 Scenario 5: Inter-area modes; local-area modes A, B, C, H, I

The limits for this scenario are: inter-area modes $\sigma \le -0.4$; the selected local-area modes $\xi \ge 0.1$; limit on all stabilizer gains 20 pu.

The object of this scenario is to assess, in comparison with Scenario 4, the influence of the constraint on the *selected local modes* on the distribution of the gains between stabilizers. The local-area modes are selected on the basis that they are the more lightly damped. The eigen-trajectories for both inter- and local-area modes are plotted in Figure 14.11.

An expanded and more detailed plot of the eigen-trajectories of the local-area modes are displayed in Figure 14.12. Note that in both this and previous figure the shifts in the trajectories are more-or-less directly to the left [1]. For practical purposes in the LP procedure, the extent of left-shift associated with increases in stabilizer gain remains decoupled from the stabilizer phase compensation over the range of frequencies of the rotor modes.

---

1. By 'directly to the left' is implied that the mode shift is $-\alpha \pm j0$, $\alpha \ge 0$.

Table 14.6 Case 1, Scenario 2. Contributions by stabilizer to the real shifts on Modes K, L and M Real left-shifts greater than 10 units (0.01 Np/s) are high-lighted.

Step	HPS_1				MPS_2				LPS_3				GPS_4			
	f	$\Delta\sigma_K$	$\Delta\sigma_L$	$\Delta\sigma_M$	f	$\Delta\sigma_K$	$\Delta\sigma_L$	$\Delta\sigma_M$	f	$\Delta\sigma_K$	$\Delta\sigma_L$	$\Delta\sigma_M$	f	$\Delta\sigma_K$	$\Delta\sigma_L$	$\Delta\sigma_M$
1	1.093	+4.0	-14.0	-2.1	1.167	-10.6	-19.7	+0.6	0.393	-84.2	0	-13.1	1.167	-1.1	-16.7	-32.6
2	0.219	+0.8	-2.5	-0.5	1.167	-10.7	-18.4	+0.5	0.359	-75.7	0	-12.0	1.167	-1.0	-15.1	-34.2
3	0.405	+1.6	-4.5	-2.1	1.167	-10.6	-17.9	+0.5	0.364	-76.7	0	-12.2	1.167	-1.1	-14.4	-35.1
4	0	0	0	0	1.167	-10.5	-17.2	+0.6	0.347	-73.2	0	-11.8	1.167	-1.1	-13.6	-36.3
5	0	0	0	0	1.167	-10.3	-16.8	+0.8	0.343	-72.8	0	-11.8	1.167	-1.2	-13.2	-37.0
6	0.088	+0.4	-0.9	-0.1	1.167	-9.9	-16.5	+1.0	0.339	-72.7	0	-11.9	1.167	-1.2	-12.8	-38.1

f - Stabilizer incremental gain factor. σ_m - Modal real shift (units), 1 unit = 10^{-3} Np/s.

Step	NPS_5				TPS_5				PSVC_5				SVC_2 & BSVC_4			
	f	$\Delta\sigma_K$	$\Delta\sigma_L$	$\Delta\sigma_M$	f	$\Delta\sigma_K$	$\Delta\sigma_L$	$\Delta\sigma_M$	f	$\Delta\sigma_K$	$\Delta\sigma_L$	$\Delta\sigma_M$	f	$\Delta\sigma_K$	$\Delta\sigma_L$	$\Delta\sigma_M$
1	0	0	0	0	0	0	0	0	0.5	-3.0	-14.9	-13.2	0	0	0	0
2	0	0	0	0	1.167	-5.3	-14.5	-10.0	0.5	-3.0	-15.2	-12.3	0	0	0	0
3	0	0	0	0	1.167	-5.2	-14.5	-9.4	0.5	-2.9	-14.7	-11.7	0	0	0	0
4	0.659	-2.1	-6.8	-3.4	1.167	-5.2	-14.5	-8.8	0.5	-2.9	-14.3	-11.7	0	0	0	0
5	0.918	-3.0	-9.3	-4.5	1.167	-5.1	-14.0	-8.5	0.5	-2.8	-13.5	-10.5	0	0	0	0
6	1.167	-3.9	-11.7	-5.5	1.167	-5.1	-13.3	-8.2	0.5	-2.7	-12.7	-10.0	0	0	0	0

f - Stabilizer incremental gain factor. σ_m - Modal real shift (units), 1 unit = 10^{-3} Np/s.

Clearly, in comparison to Scenario 4 and in satisfying the additional constraints imposed by the local-area modes, the gains of the stabilizers change somewhat (see Table 14.5, col. 6). The degradation in the figure of merit - i.e. the sum of the gains - increases slightly from 166 for Scenario 4 to 174.

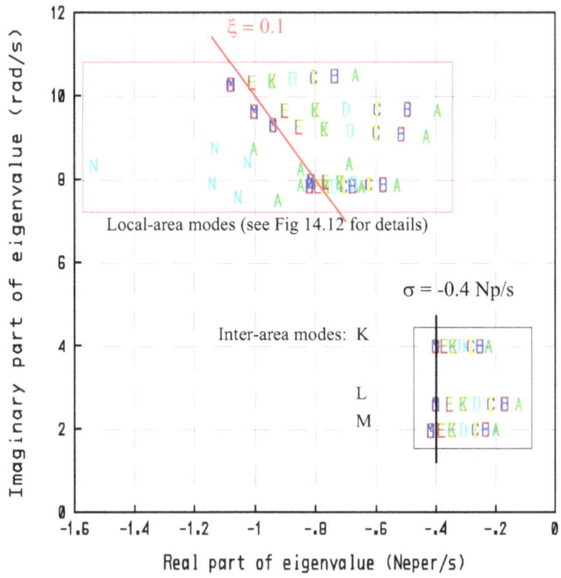

Figure 14.11 Case 1, Scenario 5. The eigen-trajectories are plotted of inter-area modes K, L and M as well as local-area modes A, B, C, H and I. The six steps of the trajectories are marked; at each step the *estimated* values B to G of the mode and the *actual* values H to M agree closely as shown. The initial (A) and resulting final (N) eigenvalues of the remaining local-area modes are also indicated.

14.6 Concluding remarks

The approaches for tuning PSSs in Chapters 5 and 10 and FDSs in Chapter 11 provide the means of the coordination of stabilizers by the coordination of their gains using either the SDCD- or LP-based procedures. The preliminary tuning of stabilizers constitutes Stage 1 in their coordination. The purpose of Stage 2 in the either of the two procedures of coordination is to satisfy certain criteria on modal damping and on stabilizer gain values.

Heuristic coordination

In the heuristic based approach the SDCDs provide information on the extent of the left-shift available on a selected rotor mode for gain increments on a range of stabilizers, PSSs and FDSs. Such information permits the user to estimate the gain increments required to produce an adequate left-shift in the mode that satisfies the modal damping criteria within a nominated range of gains.

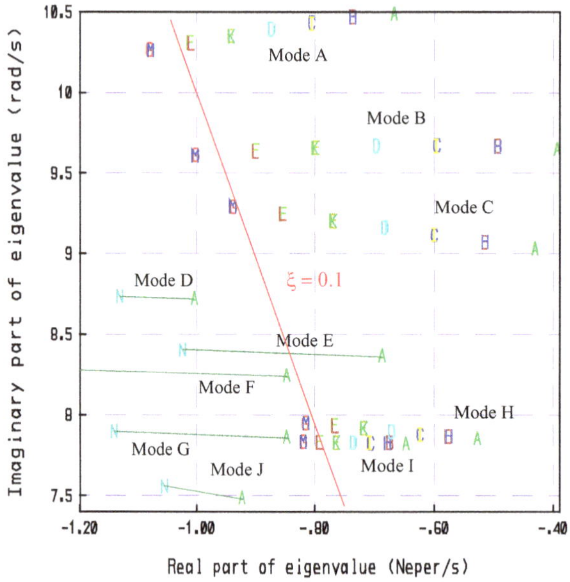

Figure 14.12 Case 1, Scenario 5. Trajectories of local mode A-J. The six steps are shown for the constrained modes A, B, C, I and H; at each step the estimated values B to G and the actual mode values H to M are plotted. For the modes D, E, F, G and J the initial values A and the final values N are displayed.

Having established a set of stabilizer gains required for a selected operating condition, this set can be tested on other operating conditions to ascertain if the criteria on the modes are satisfied or not. If not, the SDCDs for the new operating condition are invoked to establish the changes in the stabilizer gains required to satisfy the relevant criteria. By iterating through the range of encompassing operating conditions a set of stabilizer gains can be found which satisfies all conditions; this is illustrated in Section 14.3. Clearly the process may be tedious and does not lead to a unique solution as it depends on the user's methodology and experience.

It is assumed that those stabilizers that are required to be in service are initially set to certain minimal gains. In the scenarios demonstrating the LP procedure the nominal stabilizer gains are all set to 5 pu on device base. This allows for flexibility in the procedure, e.g. it may reveal that the gains of some stabilizers remain at the minimum value and therefore do not effectively contribute to any improvement in damping.

Automated coordination

The LP-based approach reveals aspects of the coordination process that are not obvious or accounted for in the SDCD-based procedure. It reveals that certain stabilizers are (unexpectedly) more influential in satisfying the modal criteria than those stabilizers which may appear to be the more obvious candidates. This is particularly the case in dealing with modes that are initially more poorly damped. Such information may fruitfully be incorporated into the SDCD-based analysis.

The example of applying the LP-based procedure to the 14-generator system suggests alternative approaches can be adopted to the setting of stabilizer gains to satisfy the rotor modal criteria. For example, if is intended to determine a set of gains for PSSs and FDSs for Area 2 only, the settings in other Areas remaining unchanged, a tailored approach may be required.

The LP-based procedure offers the following benefits.

- (i) The criteria for damping of some rotor modes may differ from those for others; this, together with a larger number of local- and inter-area modes, complicate the analysis in the SDCD-based approach.
- (ii) Criteria can be placed on exciter or controller modes to limit their right-shift.
- (iii) The roles and merits of certain stabilizers are high-lighted, e.g. MPS_2, SVC_2; these roles may not be readily apparent from the SDCD of Figure 14.9.
- (iv) The stabilizer gains are no higher than necessary.
- (v) Stabilizers that are the more critical to the support and improvement in the damping of a poorly-damped rotor mode are revealed.
- (vi) The number of trial-and-error studies required for coordination are reduced.
- (vii) The comparison between scenarios of the sum of stabilizer gains (the quantity minimized in the objective function) is an indicator of the 'gain loading' on the stabilizers; the larger the sum the more likely it is that further maximum stabilizers gains are imposed.
- (viii) Information gleaned from LP-based procedure, such as in (iii) above, together with the knowledge of the practical implications, can be incorporated into the analysis based on the SDCD-based approach. The two approaches can complement each other.

The scenarios demonstrating the performance of the LP-based procedure highlight the effectiveness of the bus-frequency stabilized FDSs. In the studies their contributions to the damping of certain inter-area modes is more extensive and effective than PSSs, and no deleterious effects are observed on the local-area modes. The potential degradation in damping on rotor modes when SVCs or the FDSs are out of service requires investigation. Likewise, a study would be required into the effect of the outage of generation on a difficult-to-damp mode, e.g. the loss on MPS_2 on mode L.

The disadvantage of the LP-based procedure is that it provides a set of stabilizer gains, subject to certain constraints, that are optimum for the selected operating condition; it does not have the facility to optimise the gains over a range of operating conditions (See [4], [6]). However, at other encompassing operating conditions for which the stabilizers are tuned, the rotor modes are likely to be left-shifted with increases in stabilizer gains as long as the gains are in the acceptable range of values.

Using the automated approach, a case study such as that in Section 14.5 can used to examine stabilizer coordination in a selected area of a larger power system, e.g. Area 2 in Figure 10.1. This would provide some guidance on how better to improve damping on the system and where additional PSSs and/or FDSs may be located to achieve better damping.

The SDCD- and the LP-based approaches are together a useful set of tools because they provide information and insight into the power system's dynamic characteristics. Such information and the guidance allows the user to make judicious, practical decisions on the parameter settings of stabilizers. Of note, the automated approach to stabilizer coordination has been extended to cover the set of encompassing operating scenarios in [6] and [16].

Studies on the coordination of the controllers for other FACTS installations such as multiple HVDC links are reported in the literature, e.g. [14], [15].

14.7 References

[1] F. P. de Mello, P. J. Nolan, T. F. Laskowski, and J. M. Undrill, "Coordinated Application of Stabilizers in Multimachine Power Systems," *Power Apparatus and Systems, IEEE Transactions on*, vol. PAS-99, pp. 892-901, 1980.

[2] M. J. Gibbard, "Co-ordinated design of multimachine power system stabilizers based on damping torque concepts," *IEE Proceedings C Generation, Transmission and Distribution*, vol. 135, pp. 276-284, 1988.

[3] Z. Ao, T. S. Sidhu, and R. J. Fleming, "Stability investigation of a longitudinal power system and its stabilization by a coordinated application of power system stabilizers," *Energy Conversion, IEEE Transactions on*, vol. 9, pp. 466-474, 1994.

[4] J. N. Taranto, A. L. B. do Bomfim, D. M. Falcao, and N. Martins, "Automated design of multiple damping controllers using genetic algorithms," in *Power Engineering Society 1999 Winter Meeting, IEEE*, 1999, pp. 539-544 vol.1.

[5] A. Doi and S. Abe, "Coordinated Synthesis of Power System Stabilizers in Multimachine Power Systems," *Power Apparatus and Systems, IEEE Transactions on*, vol. PAS-103, pp. 1473-1479, 1984.

[6] R. A. Jabr, B. C. Pal, and N. Martins, "A Sequential Conic Programming Approach for the Coordinated and Robust Design of Power System Stabilizers," *Power Systems, IEEE Transactions on*, vol. 25, pp. 1627-1637, 2010.

[7] P. Pourbeik, *Design and Coordination of Stabilisers for Generators and FACTS devices in Multimachine Power Systems,* PhD Thesis, The University of Adelaide, Australia, 1997.

[8] P. Pourbeik and M. J. Gibbard, "Simultaneous coordination of power system stabilizers and FACTS device stabilizers in a multimachine power system for enhancing dynamic performance," *Power Systems, IEEE Transactions on*, vol. 13, pp. 473-479, 1998.

[9] M. J. Gibbard, N. Martins, J. J. Sanchez-Cascara, N. Uchida, V. Vittal, and W. Lei, "Recent applications of linear analysis techniques," *Power Systems, IEEE Transactions on*, vol. 16, pp. 154-162, 2001.

[10] C. Li-Jun and I. Erlich, "Simultaneous coordinated tuning of PSS and FACTS damping controllers in large power systems," *Power Systems, IEEE Transactions on*, vol. 20, pp. 294-300, 2005.

[11] H. Bourles, S. Peres, T. Margotin, and M. P. Houry, "Analysis and design of a robust coordinated AVR/PSS," *Power Systems, IEEE Transactions on*, vol. 13, pp. 568-575, 1998.

[12] J. P. Ignizio, *Linear programming in single- and multiple-objective systems*, Prentice-Hall, 1982.

[13] G. B. Dantzig and M. N. Thapa, *Linear Programming,* Springer-Verlag, 1997.

[14] L. A. S. Pilotto, M. Szechtman, A. Wey, W. F. Long, and S. L. Nilsson, "Synchronizing and damping torque modulation controllers for multi-infeed HVDC systems," *Power Delivery, IEEE Transactions on*, v ol. 10, pp. 1505-1513, 1995.

[15] A. M. Ersdal, L. Imsland, and K. Uhlen, "Coordinated control of multiple HVDC links using backstepping," in *Control Applications (CCA), 2012 IEEE International Conference on*, 2012, pp. 1118-1123.

[16] J. C. R. Ferraz, N. Martins, and G. N. Taranto, "Coordinated Stabilizer Tuning in Large Power Systems Considering Multiple Operating Conditions," in *Power Engineering Society General Meeting, 2007.* IEEE, 2007, pp. 1-8.

Index

A

Automated coordination *634*
- Benefits *649*
- Constraints on solution *635*
- Eigen-trajectories *644*, *647*, *648*
- Optimum gains *642*
- PSSs and FDSs *Sec. 14.4.3*, *Sec. 14.5*
- Simplex algorithm *635*
- Two stage algorithm *637*
- using Linear Programming *635*

AVR - Analysis
- Conversion from PI to PID Compensation *352*
- Phase-matching method *392*
- PI Compensation *351*
- PID compensation *327–351*
- Rate feedback compensation *354–370*
- see also PID compensation *329*
- Transient gain reduction *320–327*
- Type 2B PID compensation *371–383*

AVR compensation types
- PID and PI *316*
- Rate feedback *316*
- Transient gain reduction (TGR) *316*

B

Benchmark for damping by ideal PSS *595*

Bode Plots *51–56*
- Gain and frequency margins *59*
- Gain cross-over frequency *59*
- Stability Analysis *59*

Brushless exciter
- Frequency response *373*
- Models *386*, *App. 7–I.2*

Bus-frequency stabilizers *407*
- Degradation in mode shifts *410*
- PSSs *407*
- Stability with PSS *410*

C

Characteristics of systems
- Characteristic equation *41*
- Damping constant *41*
- Damping ratio *41*
- First- and second-order systems *39*
- Second-order response *42*
- Time constant *39*
- Undamped natural frequency *40*

Classical Parameter Model *89*
- Block diagram form *146*
- Equation summary *147*
- Formulation *89*, *145*
- Linearization em equations *155*
- Relation to cc parameters *141*, *143*
- Steady-state conditions *Sec. 4.2.13.3*
- Transient flux linkages *150*

Coordination of stabilizers *621*
- - using Linear Programming *Sec. 14.4*
- Heuristic approach *Sec. 14.3*

Coupled-Circuit Model *89*
- Relation to classical parameters *144*

D

Definition
- Automatic voltage regulator (AVR) *315*
- Excitation control system (ECS) *314*
- Excitation system (ES) *314*
- Frequency stability *10*
- Halving time *479*
- Per unit definitions for ECS *315*, *App. 7–I.1*
- Power system stability *7*
- Rotor angle stability *8*
- Transient stability *8*
- Voltage stability *9*

E

Eigen-analysis
14 generator system *481*
Calculation of Eigenvalues *448*
Eigenvalue sensitivities *85*
Eigenvalues of state matrix *74*
Eigenvalues, modes and stability *Sec. 3.5.2*
Modified Arnoldi *448*
Multiple-Shift-Point Eigenanalysis *448*
QR factorisation *448*
Errors in the steady-state
Alignment *46*
Following error *48*

F

FACTS device models *157*
Base quantities at AC terminals *160*
Linearized equations, AC quantities *159–163*
Modelling assumptions *158*
Representation of interface with AC network *158*
SVC model *163–165*
VSC model *165–169*
FACTS Device Stabilizer
- called FDS *4*
Configuration of controller and stabilizer *532*
Estimate FDS mode shifts *542*
Form of FDS transfer function *542*
Improving damping *Sec. 11.8.1*
Method of Residues *537*
Modal trajectories *544*, *547*, *550*, *625*
Mode shift in targeted mode *536*
Polar plot of residues *539*, *545*, *549*, *552*, *624*
Robustness of FDSs *Sec. 11.8.2*
See also FDS *4*
Stability *543*, *547*, *550*
SVC - tuning *538*, *545*
TCSC - Modal trajectories *556*
TCSC - Polar plot of residues *554*
TCSC - Stability *555*
Theoretical basis for tuning *536*
Tuning *531*
Tuning of a SVC FDS *538*
Upper value of gain *537*
FDS
FACTS Device Stabilizer *4*
Filter characteristics
Washout and low-pass filters *246*
Frequency response diagram
see Bode Plot
Frequency stabilizing signal
for a PSS *407*
for a SVC *539*, *548*

G

Generator Models
Airgap flux linkages *121*
Alternative dq-axis rotor structures *111*
Base values for generator quantities *98*
Canay inductances *95*
d-q equivalent circuits diagrams *97*
d-q-axis equivalent circuits. *97*
Definitions of reciprocal field quantities, etc *115*
Demagnetizing effects of saturation *105–110*
Flux linkage distribution *95*
Functions for saturation characteristics *120*
Initial conditions *90*
Interface between generator and network frames *130*
Lad-base reciprocal pu field quantities *116*
Linearization of the coupled-circuit equations *107*
Linearized coupled-circuit equations *133*
Model code *91*
Modelling assumptions *93*
Modelling generator saturation *118–124*
Modelling magnetic saturation *122*
Non-reciprocal pu field quantities *116*
Open-circuit characteristic *117*
Parameter and variable definitions *99*
Park-Blondel reference frame *130*
Park-Blondel transform *93*
Per-unit linearized model *90*
Per-unit model structure *92*
Procedure to compute coefficients in linearized c-c equations *134*
Rankin base values *97*
Rrotor equations of motion in pu *114*
Saturation Methods *123–124*
See also Classical Parameters
See also Operational Parameters
Standard parameters *93*
Steady-state operating conditions *125–129*
Summary of the per-unit system *96*
Unequal mutual coupling, rotor windings *94*
GEP method

Algorithm - calculation of stabilizer parameters *309*, *543*
Basis of tuning *302*
Comparison - GEP and P-Vr characteristics *303*
Conclusions *306*
PSS-induced damping torque coefficient *305*

H

Heuristic coordination *626*
Features - Features *647*
for inter-area modes *Sec. 14.3.1*
for local-area modes *Sec. 14.3.2*
Use of SDCDs *627*, *634*
HVDC link modelling *177–180*
AC current and apparent power *Sec. 4.3.8.5*
Commutating reactance in TCCX *182*
Commutating voltage *182*
Converter model for inverter operation *Sec. 4.3.8.6*
HVDC models, internal - external interface values *Sec. 4.3.8.9*
HVDC models, Power factor *186*
HVDC models, Power factor angle *186*
with VCSs *180*
with Voltage commutated converters *181–185*
Hydro-Québec
Multi-path, multi-band PSS *397*

I

Integral-of-accelerating-power PSS *413*
Action of pre-filter, no washouts *416*
Analysis of tracking errors *442*
Block diagram of prefilter *419*
Corrupted speed signal *414*
Degradation in performance of the pre-filter *429*
Dynamic performance of the pre-filter *424*
Effect of the washouts and integrators *419*
Lack of fidelity in Prefilter *398*
Ramp Tacking Filter *415*
Torsional modes *414*
Integrator Wind-up Limiting *391*
Interactions *594*
between PSSs & FDSs *589*
Case Study *Sec. 13.3*
Concept of *Sec. 13.2.2*
Definition - Generator interactions *595*
Definition - Stabilizer interactions *594*
Summary *Sec. 13.6.1*

L

Laplace Transform *27*
Linearized power system model
Form of linearized DAEs *Sec. 4.4.1*
Formof nodal current equations *Sec. 4.4.2*
Modular and sparse structure *Sec. 4.23*
Perturbation in load bus frequency *203*
Structure of linearized DAEs *Sec. 4.4.4*
Load Models
Alternative modelling *204*
DAEs *203*, *204*
Frequency dependence *201*
Linearized equations *202*
Network connection *202*
Types of static loads *201*
ZIP types *201*

M

Method of Residues
Conclusions *300*
FDS tuning *536*
Form of compensation transfer function *294*
Form of PSS transfer function *298*
Theoretical basis *294*
Tuning a speed-PSS *297*
MITCs *4*
Concepts *563*, *Sec. 12.2*
Contibutions - local mode *Sec. 13.3.1*
Contibutions to inter area mode *Sec. 13.3.3*
Definition, MITC *576*
Features & significance *564*
Modal torque coefficients induced by FDS *Sec. 12.2.3*
Modal torque coefficients induced by PSS *Sec. 12.2.2*
Reductions of system TFM model *App. 12–I.2*
Relationship, MITCs and stabiliser gains *Sec. 12.6.2*
System response for single modal frequency *App. 12–I.1*
Torque coefficients induced by centralized FDS *Sec. 12.5*
Torque coefficients induced by centralized PSS *Sec. 12.4*

Torque coefficients induced by centralized stabilizers *Sec. 12.2.4*
Torque coefficients induced by decentralized stabilizers *Sec. 12.6*
Mode Shape Analysis *449*
Four-mass spring system *456*
Normalised right speed-eigenvectors *452*
Note of warning *459*
Stored energy in masses and springs *454*
Theoretical basis *450*
Two-mass spring system *450*
Mode shapes *80*
14 generator system *482*
Theoretical basis *80*
Mode Shifts
- with stabilizer gain increments *590*, *593*, *Sec. 13.2.3*
Analysis of variation *251*
Modes
Types of modes *10*
Units of Modal Frequency *18*
Multi-band PSS *397*
Block diagram of PSS *433*
Frequency response *436*
Objective of PSS structure *433*
Q-filter *435*
Source Hydro-Québec *397*
Transfer function analysis *App. 8–I.3*
Multi-machine PSSs *462*
Definition P-Vr transfer function *462*
Derivation of PSS transfer functions *468*
Inherent torques *471*
P-Vr transfer function *462*
Multi-machine system *475*
14 generator model system *477*
Behaviour and types of rotor modes *486*, *510*
Composite generator *476*
Correlation, small- & large-signal dynamic performance *509*
Dynamic performance criterion *479*
Dynamic performance from eigen-analysis *500*
Dynamic performance from participation, mode-shape analysis *503*
Dynamic performance from time responses *505*
Encompassing range of operating conditions *477*
Inherent torque coefficients *498*
Intra-station modes of oscillation *507*
Line diagram 14-generator system *480*
P-Vr characteristics of generators *487*
Parameters of PSS transfer functions *493*
Parameters of synthesized P-Vr transfer functions *493*
Participation factors and mode shape , inter-area mode *484*
Participation factors and mode shape, local-area mode *485*
Participation factors of a d-axis mode *486*
PSS-induced torque coefficients *498*
Synchronising and damping torques *498*
Synchronous compensators/motors *478*
Synthesized characteristic of P-Vrs *492*
Tables of data for dynamic performance analysis *527–530*
Tables of data for power-flow analysis *522–526*
Tables of PSS parameters *494–496*
Tracking of rotor modes. PSSs in service *502*

N
Normal Form analysis *513*

O
Operational Parameter Model *89*
Exactly-defined parameters *89*

P
Participation Factors *83*
14 generator system *482*
Bar charts *461*
Inertia weighted *596*, *599*
of a four-mass spring system *460*
Theoretical analysis *83*
Theoretical basis *459*
Performance criteria
Control systems *14*
Damping performance *13*
ECS *316–319*
Multi-machine power system *479*
PI compensation *351*
Conversion to PID *352*
Using positive feedback *388*
PID 2B compensation
Example Three-generator power system *371–383*
Frequency response characteristics *349*
Phase-matching method *392*
Purpose of example *371*
Tuning of Type 2B *348*
PID compensation

Characteristics Type 1 *330*
Characteristics Type 2 *333*
Characteristics Type 3 *335*
Purpose *327*
Theoretical background *329–335*
Tuning of Type 1 *335*
Tuning of Type 2A *337*
Tuning of Type 2B *348*
Types 1 to 3 *329*
Types 2A and 2B *334*
POD *2, 4*
Poincaré *2, 9*
Power System Stabilizer
Bus-frequency PSS *407*
Comparison speed & bus frequency PSSs *409*
Compensating transfer function *241, 243, 278*
Concepts for speed-PSS *237*
Damping gain *231, 245*
Damping torque coefficient *231*
Electrical power stabilizing signal *404*
Ideal PSS *232, 236*
Low-pass filter *246*
Simple concepts *230*
Synthesis of PSS compensating transfer function *266*
Transfer function of PSS *248*
Washout filter *246*
PSS & FDS tuning using
Method of Residues *225, 294*
PSS tuning using
GEP(s) transfer function *225*
P-Vr approach *225*
P-Vr approach to tuning (Summary) *284*
P-Vr characteristics
6th order generator model *264*
Basis in multi-machine systems *463*
Nature of *242*
of individual generators *488–491*
Relative invariance of phase and magnitude *464*
Selection of Design Case P-Vr characteristic *265*
Synthesis of P-Vr transfer function *266, 466*

Q

Q-filter *61*

R

Ramp tracking filter *415*
Analysis of following errors *App. 8–I.2*
Frequency responses *415*
Steady-state tracking - and errors *421*
Theoretical basis *App. 8–I.1*
Rate feedback compensation
Application of Root Locus Method *368*
Applications *358–368*
Desired form of transfer function *356*
Method of analysis *354*
Tuning of Excitation System *354*
Relative effectiveness of stabilizers *Sec. 13.6.2*
Residues
Calculation of Residues *Sec. 2.5*
Method of Residues *Sec. 6.2*
Relation - residues and MITCs *593*
Robustness
Multi-machine PSSs *516*
Multi-machines robust PSSs *462*
Robust Controller *3*
Robust Controllers *15*
Robustness of FDSs *Sec. 11.8.2*

S

SDCDs *609*
Damping contributions by FDSs *610*
Damping contributions by generators *612*
Damping contributions by PSSs *609*
Stabilizer damping contribution diagrams *Sec. 13.4*
Shaft dynamics
Disabled *229*
Disabled for GEP method (GEPSDD) *303*
SMIB System
Heffron and Phillips' Model *226*
Inherent synchronizing and damping torques *232*
K-coefficients, Heffron and Phillips' Model *288*
Synchronizing and damping torques *227*
Stability of linear systems *44*
STATCOM model
Assumptions *172*
Simplified model *Sec. 4.3.4*
State equations
Concept of state *67*
Determination of residues *77*
Forced response *74*
Modes & eigenvalues *Sec. 3.5.2*
Natural response *71*
Note on stability *76*

Solution of *71*
Stability of operating point *69*
SVCs
Simple model *Sec. 4.3.2*
Simplistic tuning *532*

T
TCSC
Modulus of real power FDS *552*
Tuning a FDS *Sec. 11.7*
Thyristor Controlled Series Capacitor (TCSC)
Network nodal equations *193*
Representation of TCSC in steady state *193*
Transfer function
Poles and zeros *30*
Proper / strictly proper *27*
Transfer function form *93*
Transfer function form
Exact definition *93*
Relation to cc form of em equations *139*
Transfer function matrix (TFM)
Model - centralized stabilizers *572*
Transient stability studies *68*, *509*, *515*

V
Voltage Commutated Converters
Converter control input *Sec. 4.3.8.7*
Extinction-angle advance *188*
Extinction-delay angle *186*
Firing-angle advance *188*
Firing-delay angle *186*
Linearized algebraic equations *Sec. 4.3.8.8*
Overlap angle *187*
VSC model
Converter, AC and DC quantities *182*
Summary linearized equations *170–172*

W
Washout and low-pass filters *Sec. 5.8.6*
Washout filters (1 or 2) *399*
Characteristics of 1 & 2 filters *403*
Frequency-domain responses *401*
Time-domain responses *399*